2017 IEEE Asian Solid-State Circuits Conference (A-SSCC 2017)

Seoul, South Korea
6-8 November 2017

IEEE Catalog Number: CFP17SSC-POD
ISBN: 978-1-5386-3179-9

**Copyright © 2017 by the Institute of Electrical and Electronics Engineers, Inc.
All Rights Reserved**

Copyright and Reprint Permissions: Abstracting is permitted with credit to the source. Libraries are permitted to photocopy beyond the limit of U.S. copyright law for private use of patrons those articles in this volume that carry a code at the bottom of the first page, provided the per-copy fee indicated in the code is paid through Copyright Clearance Center, 222 Rosewood Drive, Danvers, MA 01923.

For other copying, reprint or republication permission, write to IEEE Copyrights Manager, IEEE Service Center, 445 Hoes Lane, Piscataway, NJ 08854. All rights reserved.

*** This is a print representation of what appears in the IEEE Digital Library. Some format issues inherent in the e-media version may also appear in this print version.

IEEE Catalog Number: CFP17SSC-POD
ISBN (Print-On-Demand): 978-1-5386-3179-9
ISBN (Online):   978-1-5386-3178-2

Additional Copies of This Publication Are Available From:

Curran Associates, Inc
57 Morehouse Lane
Red Hook, NY 12571 USA
Phone: (845) 758-0400
Fax: (845) 758-2633
E-mail: curran@proceedings.com
Web: www.proceedings.com

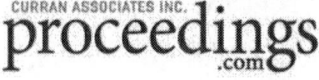

2017 IEEE Asian Solid-State Circuits Conference (A-SSCC 2017)

Seoul, South Korea
6-8 November 2017

IEEE Catalog Number: CFP17SSC-POD
ISBN: 978-1-5386-3179-9

Table of Contents

Session 1

PLENARY SPEECH

P1-1   **Technology Trends and Challenges in the Development of Future Automobiles**
Dr. Joseph Yoon
Senior Vice President, Head of Vehicle Component Tech. Center, LG Electronics, Korea -------- N/A

P1-2   **The Development of China's IC Industry - Its influence on global semiconductor community**
Dr. Prof. Shaojun Wei
Dean of the Dept. of Micro- and Nano-Electronics, Tsinghua University, China
Vice president, China Semiconductor Industry Association, China -------- N/A

Session 2

Low-Power Programmable SoCs and Embedded Memories

S2-1 (2027)   **A Programmable RFSoC in 16nm FinFET Technology for Wideband Communications**
Brendan Farley, Christophe Erdmann, Bruno Vaz, John McGrath, Edward Cullen, Bob Verbruggen, Roberto Pelliconi, Daire Breathnach, Peng Lim, Ali Boumaalif, Patrick Lynch, Conrado Mesadri, David Melinn, Kwee Peng Yap, and Liam Madden
Xilinx Ireland, Ireland -------- 1

S2-2 (2166)   **A Reconfigurable Analog Baseband Transformer for Multistandard Applications in 14nm FinFET CMOS**
Jongmi Lee, Jongwoo Lee, Chilun Lo, Jaehoon Lee, In-Young Lee, Byungki Han, Seunghyun Oh, and Thomas Cho
Samsung Electronics, Korea -------- 5

S2-3 (2136)   **A 1.4Mb 40-nm embedded ReRAM macro with 0.07um^2 bit cell, 2.7mA/100MHz low-power read and hybrid write verify for high endurance application**
Chia-Fu Lee, Hon-Jarn Lin, Chiu-Wang Lien, Yu-Der Chih, and Jonathan Chang
Taiwan Semiconductor Manufacturing Company, Taiwan -------- 9

S2-4 (2045)   **A Dynamic Power Reduction in Synchronous 2RW 8T Dual-Port SRAM by Adjusting Wordline Pulse Timing with Same/Different Row Access Mode**
Yoshisato Yokoyama, Yuichiro Ishii, Haruyuki Okuda, and Koji Nii
Renesas Electronics Corporation, Japan -------- 13

S2-5 (2067) **14nm Broadwell Xeon® Processor family: Design methodologies and optimizations**
Mahesh K Kumashikar, Shridhar G Bendi, Srikanth Nimmagadda, Anup J Deka, and Anil Agarwal
Intel Corporation, India 17

Session 3

Circuits and Systems for Sensing and Security

S3-1 (2107) **A Dual-Axis MEMS Vibratory Gyroscope ASIC with 0.0061°/s/√Hz Noise Floor over 480 Hz Bandwidth**
Zhichao Tan, Khiem Nguyen, Jeff Yan, Howard Samuels, Shane Keating, Paul Crocker, and Bill Clark
Analog Devices, Inc., USA 21

S3-2 (2141) **Chaos, Deterministic Non-Periodic Flow, for Chip-Package-Board Interactive PUF**
Noriyuki Miura, Masanori Takahashi, Kazuki Nagatomo, and Makoto Nagata
Kobe University, Japan 25

S3-3 (2089) **A 93μW 11Mbps Wireless Vital Signs Monitoring SoC with 3-Lead ECG, Bio-Impedance, and Body Temperature**
Yuxuan Luo[1], Kok-Hin Teng[1], Yongfu Li[1], Wei Mao[1], Yong Lian[2], and Chun-Huat Heng[1]
[1]National University of Singapore, Singapore
[2]York University, Canada 29

S3-4 (2090) **A 16-Channel TDM Analog Front-end with Enhanced System CMRR for Wearable Dry EEG Recording**
Tao Tang[1,2], Wang Ling Goh[1], Lei Yao[2], and Yuan Gao[2]
[1]Nanyang Technological University, Singapore
*[2]A*STAR, Singapore* 33

S3-5 (2169) **An Area-Efficient Amplifier-Less Digitally-Controlled Li-Ion Battery Charger in 0.35-μm CMOS**
Sheng-Ying Lin and Tsung-Hsien Lin
National Taiwan University, Taiwan 37

Session 4

Sensor Interface

S4-1 (2123) **A 0.5V BJT-Based CMOS Thermal Sensor in 10-nm FinFET Technology**
Da Shin Lin[1] and Hao Ping Hong[2]
[1]MediaTek,Taiwan,
[2]MediaTek USA, USA 41

S4-2 (2138) **An Ultra-low Power 169-nA 32.768-kHz Fractional-N PLL**
Chun-Yu Lin, Tun-Ju Wang, Tzu-Hsuan Liu, and Tsung-Hsien Lin
National Taiwan University, Taiwan 45

S4-3 (2062) A 10kHz-BW 93.7dB-SNR Chopped ΔΣ ADC with 30V Input CM Range and 115dB CMRR at 10kHz

Long Xu, Johan H. Huijsing, and Kofi A.A. Makinwa

Delft University of Technology, The Netherlands --- 49

S4-4 (2143) An Energy-Efficient Self-Charged Crystal Oscillator with a Quadrature-Phase Shifter Technique

Wei-Sung Chang, Dai-En Jhou, Yu-Hong Yang, and Tai-Cheng Lee

National Taiwan University, Taiwan --- 53

S4-5 (2209) An Area-Efficient Capacitively-Coupled Sensor Readout Circuit with Current-Splitting OTA and FIR-DAC

Chih-Chan Tu[1], Feng-Wen Lee[1,2], Han-Chun Chen[1], Yu-Kai Wang[1,2], and Tsung-Hsien Lin[1]

[1]*National Taiwan University, Taiwan*
[2]*Mediatek, Taiwan* --- 57

Session 5

Digital Building Blocks

S5-1 (2049) 25 fJ/bit, 5Mb/s, 0.3V True Random Number Generator With Capacitively-Coupled Chaos System and Dual-Edge Sampling Scheme

Anh Tuan Do and Xin Liu

Institute of Microelectronics, Agency for Science, Technology and Research (A*STAR), Singapore
--- 61

S5-2 (2142) A 1.25pJ/bit 0.048mm^2 AES Core with DPA Resistance for IoT Devices

Shengshuo Lu[1], Zhengya Zhang[1], and Marios Papaefthymiou[1,2]

[1]*University of Michigan, USA*
[2]*University of California, USA* --- 65

S5-3 (2218) A 0.40 pJ/cycle 981 μm^2 Voltage Scalable Digital Frequency Generator for SoC Clocking

Martin Cochet[1,2], Sylvain Clerc[2], Guénolé Lallement[1,2], Fady Abouzeid[2], Philippe Roche[2], and Jean-Luc Autran[1]

[1]*Aix-Marseille University & CNRS, France*
[2]*STMicroelectronics, France* --- 69

S5-4 (2188) A 10-GHz Multi-purpose Reconfigurable Built-in Self-Test Circuit for High-Speed Links

Myungguk Lee, Seungho Han, Jae-Yoon Sim, Hong-June Park, and Byungsub Kim

Pohang University of Science and Technology, Korea --- 73

Session 6

PAM-4 Receiver Techniques

S6-1 (2103) **A 56Gbps PAM-4 Optical Receiver Front-end**
Kuan-Lin Fu and Shen-Iuan Liu
National Taiwan University, Taiwan 77

S6-2 (2129) **A Low-Power PAM4 Receiver Using 1/4-Rate Sampling Decoder with Adaptive Variable-Gain Rectification**
Guang Zhu[1], Quan Pan[2], John Zhuang[3], Charlie Zhi[3], and C. Patrick Yue[1]
[1]The Hong Kong University of Science and Technology, Hong Kong
[2]Etopus, USA
[3]Brite Semiconductor, China 81

S6-3 (2050) **A 82 mW 28 Gb/s PAM-4 Digital Sequence Decoder with built-in Error correction in 28nm FDSOI**
Masum Hossain[1], Aurangozeb[1], AKM Delwar Hossain[1], and Maruf Mohammad[2]
[1]University of Alberta, Canada
[2]Qualcomm Atheros, USA 85

S6-4 (2226) **A 51Gb/s, 320mW, PAM4 CDR with Baud-Rate Sampling for High-Speed Optical Interconnects**
Nan Qi[1,2], Yuhang Kang[3], Qipeng Lin[3], Jianxu Ma[4], Jingbo Shi[2], Bozhi Yin[2], Chang Liu[2], Rui Bai[4], Shang Hu[2], Juncheng Wang[2], Jiangbing Du[5], Lin Ma[5], Zuyuan He[5], Ming Liu[3], Feng Zhang[3], and Patrick Yin Chiang[2,6]
[1]Institute of Semiconductors, Chinese Academy of Sciences, Beijing, China
[2]Fudan University, China
[3]Institute of Microelectronics, Chinese Academy of Sciences, Beijing, China
[4]PhotonIC Technologies, China
[5]Shanghai Jiao Tong University, China
[6]Oregon State University, USA 89

Session 7

Building Blocks for Frequency Synthesizers

S7-1 (2052) **A 15-μW, 103-fs step, 5-bit Capacitor-DAC-based Constant-Slope Digital-to-Time Converter in 28nm CMOS**
Peng Chen[1], Feifei Zhang[1], Zhirui Zong[2], Hao Zheng[1], Teerachot Siriburanon[1], and Robert Bogdan Staszewski[1]
[1]University College Dublin, Ireland
[2]Delft University of Technology, the Netherlands 93

S7-2 (2047) **A 173–200 GHz Quadrature Voltage-Controlled Oscillator in 130 nm SiGe BiCMOS**
Paul Stärke, Vincent Rieß, Corrado Carta, and Frank Ellinger
Technischen Universität Dresden, Germany 97

S11-4 (2207) A 2.79-mW 0.5%-THD CMOS Current Driver IC for Portable Electrical Impedance Tomography System

Jaeeun Jang[1], Minseo Kim[1], Joonsung Bae[2], and Hoi-Jun Yoo[1]
[1]Korea Advanced Institute of Science and Technology (KAIST), Korea
[2]Kangwon National University, Korea ------- 145

S11-5 (2186) 0.5 and 1.5 THz Monolithic Imagers in a 65 nm CMOS Adopting a VCO-Based Signal Processing

Suna Kim, Kyoung-Yong Choi, Dae-Woong Park, Joo-Myoung Kim, Seok-Kyun Han, and Sang-Gug Lee

Korea Advanced Institute of Science and Technology (KAIST), Korea ------- 149

Session 12

Memory System

S12-1 (2009) Dual-Loop 2-step ZQ Calibration for Dedicated Power Supply Voltage in LPDDR4 SDRAM

Chang-Kyo Lee, Junha Lee, Ki-Ho Kim, Jin-Seok Heo, Gil-Hoon Cha, Jin-Hyeok Baek, Dae-Sik Moon, Yoon-Joo Eom, Tae-Sung Kim, Hyunyoon Cho, Younghoon Son, Seonghwan Kim, Jong-Wook Park, Sewon Eom, Si-Hyeong Cho, Young-Ryeol Choi, Seungseob Lee, Kyoung-Soo Ha, Youngseok Kim, Bo-Tak Lim, Dae-Hee Jung, Eungsung Seo, Kyoung-Ho Kim, Yoon-Gyu Song, Youn-Sik Park, Tae-Young Oh, Seung-Jun Bae, In-Dal Song, Seok-Hun Hyun, Joon-Young Park, Hyuck-Joon Kwon,Young-Soo Sohn, Jung-Hwan Choi, Kwang-Il Park, and Seong-Jin Jang

Samsung Electronics, Korea ------- 153

S12-2 (2224) MLC/3LC NAND Flash SSD Cache with Asymmetric Error Reduction Huffman Coding for Tiered Hierarchical Storage

Hikaru Watanabe, Yoshiaki Deguchi, and Ken Takeuchi

Chuo University, Japan ------- 157

S12-3 (2175) Word-line Batch V_{TH} Modulation of TLC NAND Flash Memories for Both Write-Hot and Cold Data

Yoshiaki Deguchi and Ken Takeuchi

Chuo University, Japan ------- 161

S12-4 (2065) A 16kb Column-based Split Cell-VSS, Data-Aware Write-Assisted 9T Ultra-Low Voltage SRAM with Enhanced Read Sensing Margin in 28nm FDSOI

M. Sultan M. Siddiqui, Zhao Chuan Lee, and Tony Tae-Hyoung Kim

Nanyang Technological University, Singapore ------- 165

S12-5 (2204) An Energy-optimized (37840, 34320) Symmetric BC-BCH Decoder for Healthy Mobile Storages

Seokha Hwang[1], Jaehwan Jung[2], Daesung Kim[3], Jeongseok Ha[2], In-Cheol Park[2], and Youngjoo Lee[4]
[1]Kwangwoon University, Korea
[2]Korea Advanced Institute of Science and Technology (KAIST), Korea
[3]SK Hynix, Korea
[4]Pohang University of Science and Technology (POSTECH), Korea ------- 169

S12-6 (2178)   **A 130nm 1Mb HfO$_X$ Embedded RRAM Macro Using Self-Adaptive Peripheral Circuit System Techniques for 1.6X Work Temperature Range**

Feng Zhang[1], Dongyu Fan[1,2], Yuan Duan[1], Jin Li[1], Cong Fang[1], Yun Li[1], Xiaowei Han[3], Lan Dai[2], Chengying Chen[1], Jinshun Bi[1], Ming Liu[1], and Meng-Fan Chang[4]

[1]*Institute of Microelectronics Chinese Academy of Sciences, China*
[2]*North China University of Technology, China*
[3]*Xi'an UniIC Semiconductors Co., Ltd., China*
[4]*National Tsing Hua University, Taiwan* -- 173

Session 13

Wireless Receivers and Transmitters

S13-1 (2076)   **A Reconfigurable Dual-Band WiFi/BT Combo Transceiver with Integrated 2G/BT SP3T, LNA/PA Achieving Concurrent Receiving and Wide Dynamic Range Transmitting in 40nm CMOS**

Meng-Hsiung Hung, Yi-Shing Shih, Chin-Fu Li, Wei-Kai Hong, Ming-Yeh Hsu, Chih-Hao Chen, Yu-Lun Chen, Chun-Wei Lin, and Yuan-Hung Chung

MediaTek Inc, Taiwan --- 177

S13-2 (2195)   **A High-Speed DDFS MMIC with Frequency, Phase and Amplitude Modulations in 65nm CMOS**

Abdel Martinez Alonso, Masaya Miyahara, and Akira Matsuzawa

Tokyo Institute of Technology, Japan --- 181

S13-3 (2160)   **A −121dBm Sensitivity, 2.8µJ/bit Rx, 8.8µJ/bit Tx, Narrowband transceiver for ARIB STD and IoT**

M. Kumarasamy Raja, Zhao Bin, Yan Dan Lei, Zhang Hongbao, Lim Wei Yi, and Chemmanda John Leo

A*STAR (Agency for Science, Technology and Research), Singapore -------------------------- 185

S13-4 (2121)   **Detection of 3.0 THz wave with a detector in 65 nm standard CMOS process**

Tong Fang, Zhao-yang Liu, Li-yuan Liu, Yuan-yuan Li, Jun-qi Liu, Jian Liu, and Nan-jian Wu

University of Chinese Academy of Sciences, China -- 189

S13-5 (2217)   **A 0.6-V 200-kbps 429-MHz Ultra-low-power FSK Transceiver in 90-nm CMOS**

Chun-Yuan Chiu, Zhen-Cheng Zhang, and Tsung-Hsien Lin

National Taiwan University, Taiwan -- 193

Session 14

Energy-efficient & Variation resilient Digital Circuits

S14-1 (2028)   **An 82% Energy-Saving Change-Sensing Flip-Flop in 40nm CMOS for Ultra-Low Power Applications**

Van Loi Le[1,2], Juhui Li[2], Alan Chang[2], and Tony T. Kim[1]

[1]*Nanyang Technological University, Singapore*
[2]*NXP Semiconductors, Singapore* --- 197

S14-2 (2150) NBTI/PBTI separated BTI monitor with 4.2x Sensitivity by Standard Cell Based Unbalanced Ring Oscillator

Mitsuhiko Igarashi, Yoshio Takazawa, Yasumasa Tsukamoto, Kan Takeuchi, and Koji Shibutani

Renesas Electronics Corporation, Japan 201

S14-3 (2034) A 0.44V-1.1V 9-Transistor Transition-Detector and Half-Path Error Detection Technique for Low Power Applications

Xinchao Shang, Weiwei Shan, Longxing Shi, Xing Wan, and Jun Yang

Southeast University, China 205

S14-4 (2155) HTD: A Light-Weight Holosymmetrical Transition Detector Based In-situ Timing Monitoring Technique for Wide-Voltage-Range in 40nm CMOS

Wentao Dai, Weiwei Shan, Xinning Liu, and Jun Yang

Southeast University, China 209

Session 15

Nyquist-rate ADCs

S15-1 (2133) A 0.5V 12-bit SAR ADC using Adaptive Time-Domain Comparator with Noise Optimization

Chen-Che Kao, Sung-En Hsieh, and Chih-Cheng Hsieh

National Tsing Hua University, Taiwan 213

S15-2 (2203) Range Pre-selection Sampling technique to reduce input drive energy for SAR ADCs

Harijot Singh Bindra[1], Joeri Lechevallier[1], Anne-Johan Annema[1], Simon Louwsma[2], Ed van Tuijl[1,2], and Bram Nauta[1]

[1]*University of Twente, The Netherlands*
[2]*Teledyne DALSA, The Netherlands* 217

S15-3 (2064) A 5-bit 2 GS/s Binary-Search ADC with Charge-Steering Comparators

U-Fat Chio, Sai-Weng Sin[1], Seng-Pan U[1,2], Franco Maloberti[3], and R. P. Martins[1,4]

[1]*University of Macau, Macao, China*
[2]*Synopsys Macau Ltd., Macao, China*
[3]*University of Pavia, Italy*
[4]*Instituto Superior Técnico/Universidade de Lisboa, Portugal* 221

S15-4 (2083) A 13-bit 160MS/s Pipelined Subranging-SAR ADC with Low-Offset Dynamic Comparator

Weitao Li, Fule Li, Jia Liu, Hongyu Li, and Zhihua Wang

Tsinghua University, China 225

S15-5 (2114) A 1.5fJ/Conv-step 10b 100kS/s SAR ADC with Gain-Boosted Dynamic Comparator

Xiyuan Tang, Long Chen, Jeonggoo Song, and Nan Sun

The University of Texas at Austin, USA 229

Session 16

Machine Learning and Recognition SoCs

S16-1 (2137) A 2.56mm^2 718GOPS Configurable Spiking Convolutional Sparse Coding Processor in 40nm CMOS

Chester Liu, Sung-Gun Cho, and Zhengya Zhang

University of Michigan, USA ———————————————————————— 233

S16-2 (2206) A 21mW Low-power Recurrent Neural Network Accelerator with Quantization Tables for Embedded Deep Learning Applications

Jinmook Lee, Dongjoo Shin, and Hoi-Jun Yoo

Korea Advanced Institute of Science and Technology (KAIST), Korea ———————— 237

S16-3 (2015) EQSCALE: Energy-Quality Scalable Feature Extraction Engine for Sub-mW Real-time Video Processing with 0.55 mm2 Area in 40nm CMOS

Anastacia B. Alvarez[2], Gopalakrishnan Ponnusamy[1], and Massimo Alioto[1]

[1]*National University of Singapore, Singapore*
[2]*University of the Philippines, Philippines* ——————————————————— 241

S16-4 (2227) A Self-Powered Always-On Vision-based Wake-up Detector for Wearable Gesture User Interfaces

Suhwan Cho, Seongrim Choi, Junsik Woo, Ara Kim, and Byeong-Gyu Nam

Chungnam National University, Korea ———————————————————————— 245

Session 17

Advanced Wireline Clock Generators and Transmitters

S17-1 (2094) A 18-to-23 GHz -253.5dB-FoM Sub-Harmonically Injection-Locked ADPLL with ILFD Aided Adaptive Injection Timing Alignment Technique

Zhao Zhang, Jincheng Yang, Liyuan Liu, Peng Feng, Jian Liu, and Nanjian Wu

University of Chinese Academy of Sciences, China ——————————————————— 249

S17-2 (2115) A 1.5-GHz Sub-Sampling Fractional-N PLL for Spread-Spectrum Clock Generator in 0.18-µm CMOS

Chun-Yu Lin, Tun-Ju Wang, and Tsung-Hsien Lin

National Taiwan University, Taiwan ———————————————————————— 253

S17-3 (2191) A 2.1Gbps 12-Channel Transmitter with Phase Emphasis Embedded Serializer for UHD Intra-panel Interface

Jihwan Park, Joo-Hyung Chae, Yong-Un Jeong, Jae-Whan Lee, and Suhwan Kim

Seoul National University, Korea ———————————————————————————— 257

S17-4 (2099) A Low-Power Dual-Mode 20-Gb/s NRZ and 28-Gb/s PAM-4 Voltage-Mode Transmitter

Hae-Woong Yang[1], Ashkan Roshan-Zamir[1], Young-Hoon Song[2], and Samuel Palermo[1]

[1]*Texas A&M University, USA*
[2]*NXP Semiconductor, USA* ——————————————————————————————— 261

Session 18

Analog Techniques

S18-1 (2117) Subthreshold Voltage Reference With Nwell/Psub Diode Leakage Compensation for Low-Power High-Temperature Systems

Inhee Lee, Dennis Sylvester, and David Blaauw

University of Michigan, USA 265

S18-2 (2192) A Smart-Offset Analog LDO with 0.3V Minimum Input Voltage and 99.1% Current Efficiency

Saurabh Chaubey and Ramesh Harjani

University of Minnesota, USA 269

S18-3 (2151) A 762-µW 16.3-ps Resolution Digital Pulse Width Modulator Using Zooming Phase-Interpolator

Masanobu Tsuji

ROHM Co., Ltd., Japan 273

S18-4 (2157) Fully-Integrated AMLED Micro Display System With a Hybrid Voltage Regulator

Junmin Jiang, Liusheng Sun, Xu Zhang, Shing Hin Yuen, Xianbo Li, Wing-Hung Ki, C. Patrick Yue, and Kei May Lau

The Hong Kong University of Science and Technology, Hong Kong 277

S18-5 (2197) A Low-Voltage Low-Offset Dual Strong-Arm Latch Comparator

Aikaterini Papadopoulou[1], Vladimir Milovanović[2], and Borivoje Nikolić[1]

[1]*University of California at Berkeley, USA*
[2]*University of Kragujevac, Serbia* 281

Session 19

High-resolution ADCs

S19-1 (2032) A 5.35 mW 10 MHz Bandwidth CT Third-Order ΔΣ Modulator with Single Opamp Achieving 79.6/84.5 dB SNDR/DR in 65 nm CMOS

Wei Wang[1], Yan Zhu[1], Chi-Hang Chan[1], Seng-Pan U[1,2], and Rui Paulo Martins[1,3]

[1]*University of Macau, China*
[2]*Synopsys Macau Ltd., China*
[3]*Instituto Superior Técnico/Universidade de Lisboa, Portugal* 285

S19-2 (2120) A 72.9-dB SNDR 20-MHz BW 2-2 Discrete-Time Sturdy MASH Delta-Sigma Modulator Using Source-Follower-Based Integrators

Yong-Sik Kwak, Kang-Il Cho, Ho-Jin Kim, Seung-Hoon Lee, and Gil-Cho Ahn

Sogang University, Korea 289

S19-3 (2042) A Compact 87.1-dB DR Bandwidth-Scalable Delta-Sigma Modulator Based on Dynamic Gain-Bandwidth-Boosting Inverter for Audio Applications

Young-Ha Hwang, Jun-Eun Park, and Deog-Kyoon Jeong

Seoul National University, Korea -- 293

S19-4 (2176) A 172dB-FoM Pipelined SAR ADC Using a Regenerative Amplifier with Self-Timed Gain Control and Mixed-Signal Background Calibration

Miguel Gandara[1], Paridhi Gulati[2], and Nan Sun[1]

[1]The University of Texas at Austin, USA

[2]Analog Devices, Inc., USA --- 297

Session 20

IPs for Emerging Applications

S20-1 (2016) A Fully-Synthesizable C-Element Based PUF Featuring Temperature Variation Compensation with Native 2.8% BER, 1.02fJ/b at 0.8-1.0V in 40nm

Sachin Taneja, Anastacia Alvarez, Gopalakrishnan Sadagopan, and Massimo Alioto

National University of Singapore, Singapore --- 301

S20-2 (2202) A 0.37mm^2 LTE/Wi-Fi Compatible, Memory-Based, Runtime-Reconfigurable $2^n3^m5^k$ FFT Accelerator Integrated with a RISC-V Core in 16nm FinFET

Angie Wang, Brian Richards, Palmer Dabbelt, Howard Mao, Stevo Bailey, Jaeduk Han, Eric Chang, James Dunn, Elad Alon, and Borivoje Nikolić

University of California, USA --- 305

S20-3 (2200) A 65nm 376nA 0.4V Linear Classifier Using Time-Based Matrix-Multiplying ADC with Non-Linearity Aware Training

Anvesha A and Arijit Raychowdhury

Georgia Institute of Technology, USA --- 309

S20-4 (2167) A 1GHz Fault Tolerant Processor with Dynamic Lockstep and Self-recovering Cache for ADAS SoC Complying with ISO26262 in Automotive Electronics

Jinho Han[1,2], Youngsu Kwon[1], Yong Cheol Peter Cho[1], and Hoi-Jun Yoo[2]

[1]Electronics and Telecommunications Research Institute(ETRI), Korea

[2]Korea Advanced Institute of Science and Technology(KAIST), Korea ----------------------- 313

Session 21

High-performance RF Frequency-generation Techniques

S21-1 (2194) A 77-GHz Mixed-Mode FMCW Signal Generator Based on Bang-Bang Phase Detector

Jianfu Lin[1], Zheng Song[1], Nan Qi[2], Woogeun Rhee[1], and Baoyong Chi[1]

[1]Tsinghua University, China

[2]Chinese Academy of Sciences, China --- 317

S21-2 (2125) **A 7GHz-Bandwidth 31.5 GHz FMCW-PLL with Novel Twin-VCOs Structure in 65nm CMOS**

Shunli Ma, Jili Sheng, Ning Li, and Junyan Ren

Fudan University, China 321

S21-3 (2071) **A -245 dB FOM 48 fs rms jitter semi-digital PLL with intrinsic temperature compensation in 130 nm CMOS**

J. Anders[1], S. Bader[1], M. Dietl[2], P. Sareen[2], G. Rombach[2], S. Tambouris[2], and M. Ortmanns[1]

[1]University of Ulm, Germany

[2]Texas Instruments Germany, Germany 325

S21-4 (2080) **An Ultra-Low Phase Noise All-Digital Multi-Frequency Generator Using Injection-Locked DCOs and Time-Interleaved Calibration**

Suneui Park, Heein Yoon, and Jaehyouk Choi

Ulsan national Institute of Science and Technology(UNIST), Korea 329

A Programmable RFSoC in 16nm FinFET Technology for Wideband Communications

Brendan Farley, Christophe Erdmann, Bruno Vaz, John McGrath, Edward Cullen, Bob Verbruggen, Roberto Pelliconi, Daire Breathnach, Peng Lim, Ali Boumaalif, Patrick Lynch, Conrado Mesadri, David Melinn, Kwee Peng Yap, Liam Madden

Analog and Digital-RF Systems
Xilinx Ireland, Dublin, Ireland

Email: brendan.farley@xilinx.com

Abstract—In this paper, we present a Programmable SoC device with monolithically integrated RF-ADCs and RF-DACs in a 16nm FinFET process. The device includes quad ARM Cortex-53 and dual ARM Cortex-R5 processing subsystem, 750K programmable logic cells, 4000 DSP slices and 4 32Gb/s serial transceivers. Each 14-bit RF-DAC operates at a sample rate of up to 6.4GS/s and can directly synthesize RF carriers up to 4GHz. Each 12-bit RF-ADC operates at a sample rate of up to 4GS/s and can directly digitize RF carriers up to 4GHz. Other RF functions generally associated with analog / RF such as complex mixing and filtering have also been integrated and implemented in the digital domain (digital-RF), enabling lower system cost and power and increased flexibility.

Keywords—Digital-RF; Data Converter; RFSoC; FinFET

I. INTRODUCTION

In future applications from 5G base stations to next generation cable and imaging, the RF bandwidth and channel count will expand rapidly. Any commercial solution must be delivered with constant or lower power and footprint and in a cost-effective manner. The emergence of discrete RF data converters has enabled custom analog-RF radios to be replaced with a more efficient and flexible digital-RF implementation through direct digitisation and synthesis of wideband RF signals. This creates the need to move large amounts of digital data across the system in a very efficient manner. Indeed as RF data converters have become more efficient their digital interface to the digital front-end (DFE) has become the power bottleneck in the system.

This paper will describe an RFSoC device which results in dramatic reduction in system power and footprint and increased flexibility for wide bandwidth communications applications. This is achieved through monolithic integration of the RF data converters with a programmable FPGA-SoC in 16nm FinFET as illustrated in fig. 1.

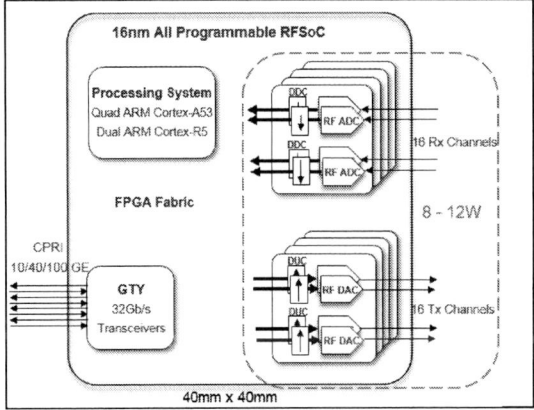

Fig. 1: RFSoC System Level Diagram

The format for the rest of the paper is as follows: in section II we discuss technology factors driving integration; in section III and IV the FPGA / SoC and digital-RF capability; sections V and VI describe the RF-ADC and RF-DAC architectures respectively; in section VII we show RFSoC applications and measurement results including chip level isolation; and we conclude in section VIII.

II. TECHNOLOGY FACTORS DRIVING AND ENABLING MONOLITHIC INTEGRATION

Data converters exhibit continuous power efficiency improvements [1] as illustrated in fig. 2. Analog bandwidths and sampling rates now extend to multi-GHz with increasing efficiency, allowing viable digital-RF implementation in advanced nodes.

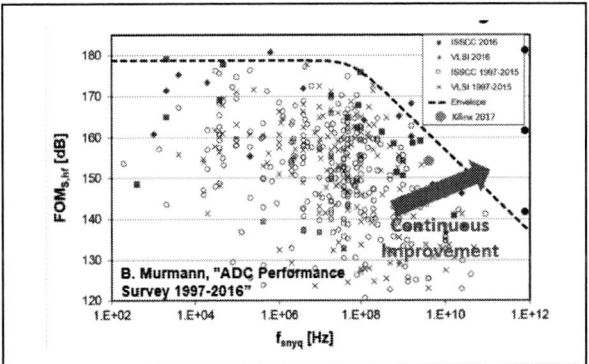

Figure 2: ADC Figure of Merit versus Nyquist bandwidth

FinFET technology enables the traditional, custom, analog-RF radio architecture to be replaced by a digital-RF implementation as illustrated in fig. 3. Digital complex mixing and filtering can now be implemented in a highly power efficient and cost effective manner. The design can be highly flexible and support very wide bandwidths. Efficient support for multi-band radio can be achieved entirely within the digital domain, provided there is sufficient data converter dynamic range. And the solution can be implemented with lower size, weight and system cost. The system performance bottleneck is in the RF data converters which require much more aggressive sampling rate and analog bandwidth relative to the traditional radio architecture.

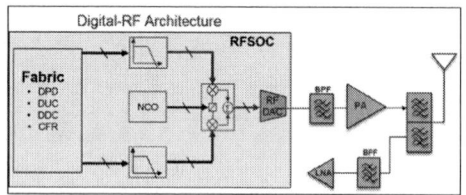

Figure 3: Digital-RF architecture enabled with RFSoC

For analog circuits, low FinFET switch resistance (R_{ON}) allows linear wideband, direct-RF sampling. A diminishing body effect also contributes to linearity improvements. A much higher inherent R_{OUT} of FinFET devices, allows the use of smaller, lower capacitance devices for maximum linear RF bandwidth. Technology scaling / shrinking of basic analog functions such as comparators and amplifiers enables faster analog signal processing and high density logic allows non-idealities in these circuits to be digitally calibrated in a cost and power efficient manner. And improved transistor and capacitor matching allows higher inherent linearity of array based structures at baseband frequencies.

III. INTEGRATED FPGA AND SoC CAPABILITY

The UltraScale+ RFSoC devices offer up to 2.4x performance per Watt, or up to 1.6x fabric Fmax, when compared with previous generations [1], as illustrated in table I. The processing capabilities have increased from dual-core ARM-A9s at 1GHz in 28nm to quad-core ARM-A53s at 1.5GHz in 16nm FinFET and the SoC capability includes dual-core R5 'Real-Time' Processing Units (RPUs. As a result of this high performance integration, the ADC and DAC blocks can interface directly with the FPGA fabric at the full data-converter bandwidth. This translates to 410 Gb/s for a DAC tile, and 256 Gb/s for an ADC tile, or over 2 Tera-bits/s of raw data-converter data in a single device.

Device Family	7-series (28nm)	UltraScale (20nm)	UltraScale+ (16nm)	
Operating Voltage	1V	0.95V	0.85V	0.72V
Fabric Performance	1.0x	1.2x	1.6x	1.2x
Total Power	1.0x	0.7x	0.8x	0.5x
Performance / Watt	1.0x	1.7x	2.0x	2.4x

Table I: Comparison of FPGA Technologies

IV. INTEGRATED DIGITAL-RF CAPABILITY

The density and power-efficiency of the 16nm FinFET process allows for hardening of flexible digital-RF data path functions, co-located with the data-converters [3]. DACs and ADCs are grouped into tiles, with up to 4 channels per tile. Each DAC channel provides selectable interpolation from 1x to 8x with >89dB image rejection, 48-bit NCOs, full complex-to-complex mixing, inverse-sinc compensation and QMC correction. Each ADC channel provides selectable decimation from 1x to 8x with >89dB alias rejection, 48-bit NCOs, full complex to complex mixing, threshold detection and QMC correction. Both ADC and DAC tiles offer integrated multi-band support, which allows flexible interconnection of the data-converters and digital DUC/DDC data paths. All of these functions are hardened within the ADC and DAC tiles, and in total provide up to 10 Tera-MACs/s of effective DSP processing in a single device. This is without consuming any of the >4,000 FPGA DSP resources in an RFSoC device, which allows them to be fully dedicated to the customer's application.

V. RF-ADC ARCHITECTURE

In [4] we describe the integrated RF-ADC architecture, illustrated in Fig. 4, which can be programmed to support IQ (2x2GS/s) and direct (4GS/s) RF communications architectures. Each 2GS/s ADC unit consists of four interleaved 500MS/s sub-ADC slices and a sampling network composed of a single front-end switch, to avoid time-skew calibration requirements, and four channel-switches. Each 500MS/s sub-ADC slice is supported by foreground (FG) and background (BG) calibration loops. In 4GS/s mode, a digitally controllable delay cell is used to minimize the sampling time-skew between 2GS/s units.

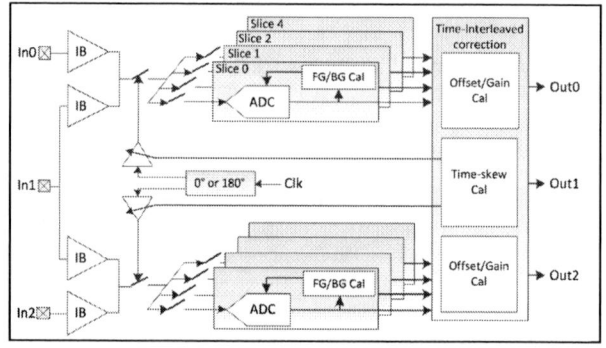

Figure 4: RF-ADC interleaved architecture

VI. RF-DAC ARCHITECTURE

In [5] we describe the integrated RF-DAC architecture, which is illustrated in Fig. 5. The output current is carried to the 50Ω on-die termination through a 1.7nH shunt inductor to extend the analog bandwidth to 4GHz.

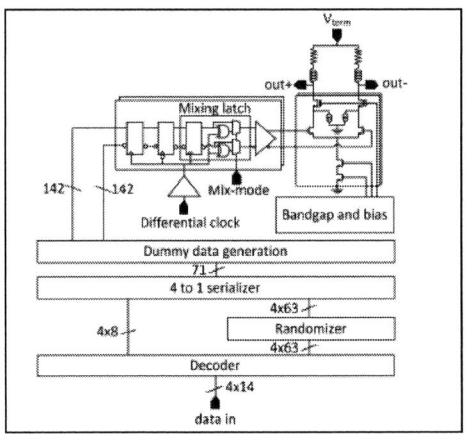

Figure 5: Current steering DAC with 6-8 unary-binary segmentation

Fig. 6 illustrates the desired high-level transistor characteristics for each stage of the DAC.

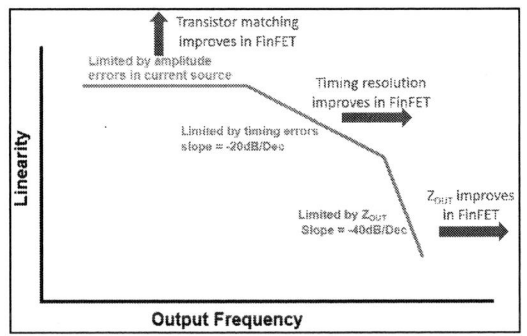

Figure 6: Desired transistor characteristics for each DAC stage

Improved FinFET speed and matching affects all DAC stages, improving low output frequency (F_{OUT}) linearity and mitigating timing mismatch limitations at high F_{OUT}. A significant improvement in FinFET output impedance (Z_{OUT}) results in improved linearity at high F_{OUT}. Fig. 7 illustrates how FinFET technology benefits DAC performance across the output spectrum.

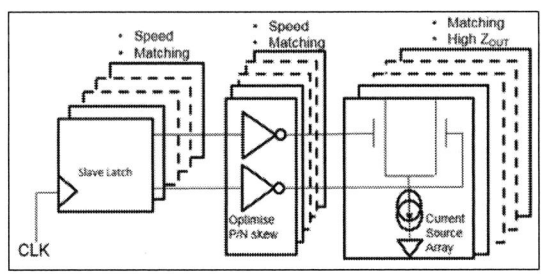

Figure 7: Advantages of FinFET for the current steering DAC

VII. RFSoC Applications & Measurement Results

A. Remote Radio Head (RRH)

The Direct RF-sampling base station architecture is an emerging use case for RFSoC. Features include RF data converters, digital up-conversion and down-conversion with programmable complex mixing, digital pre-distortion of the power amplifier (PA) and 25Gbps serial interfacing to the baseband. A high-level implementation is illustrated in fig. 8.

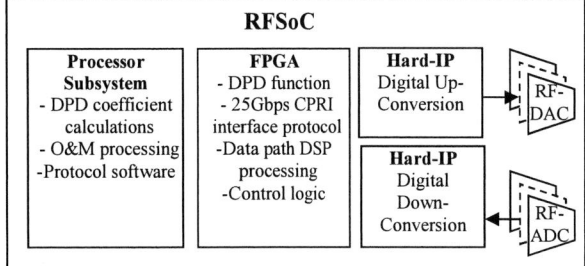

Figure 8: RFSoC implementation of RRH incorporating DPD

The power benefit of integration of the RF data converters with the FPGA for a number of RRH use cases is illustrated in table II. The key contributors to the power reduction are the removal of the serdes interfaces and the relatively low power implementation of the RF data converters in 16nm FinFET as compared to discrete data converters which are typically implemented in larger geometry technologies.

RRH Use Cases			
	4x4 100MHz	4x4 200MHz	8x8 100MHz
Discrete Implementation			
FPGA	15W	23W	23W
Discrete ADC/DAC	16W	16W	32W
Total Power	31W	39W	55W
Fully Integrated RFSoC Implementation			
FPGA + RF Subsystem	18W	25W	27W
Total Power Saving	41%	37%	51%

TABLE II: Power dissipation of discrete versus integrated RRH [6]

For sub-6GHz 5G applications, Band 42 (3.4-3.6GHz) provides 200MHz of contiguous TDD bandwidth. RX sensitivity is key performance requirement for RRH to enable wide coverage. Fig.9 illustrates excellent IM3 performance at 3.5GHz for very low signal amplitude, with all interleaved components highly suppressed relative to the noise floor.

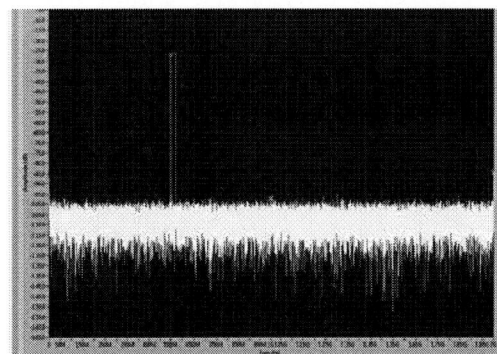

Figure 9: 2-tone F_{IN}=3.5GHz @-26dBFS, FS=4GS/s, IM3=-84dBc

B. DOCSIS Remote PHY Cable Modem

Cable infrastructure is moving from a central office scheme servicing 1000's of end-points to a distributed remote PHY architecture, where each remote node services perhaps 64 end-points. The remote PHY is an emerging RFSoC use case and the high-level implementation is illustrated in Fig. 10.

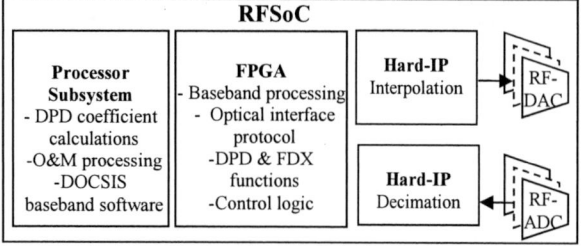

Figure 10: RFSoC implementation of DOCSIS remote PHY application

A key requirement for cable modem applications is to synthesize up to 160x 5.4MHz-wide QAM modulated carriers with minimal leakage between channels. Fig. 11 illustrates 59dBc noise power ratio (NPR) in the center of the band.

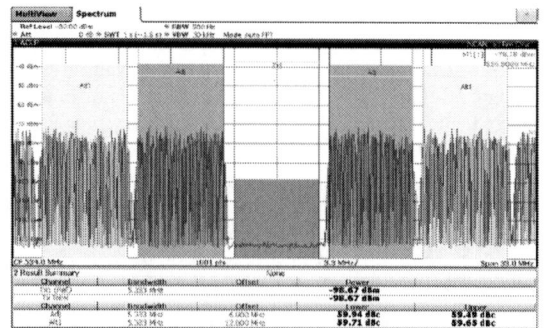

Figure 11: Cable noise power ratio (NPR) measurement for 160x6MHz channels (54-1008MHz) with empty channel @534MHz

C. Isolation Measurements

The isolation of the data converters from FPGA digital noise was evaluated by measuring interference power in the DAC spectrum in the presence of up to 30W of digital power clocking at up to 550MHz, as illustrated in fig. 12. Direct interference products are typically better than -105dBm with the largest at the resonant frequency of the package.

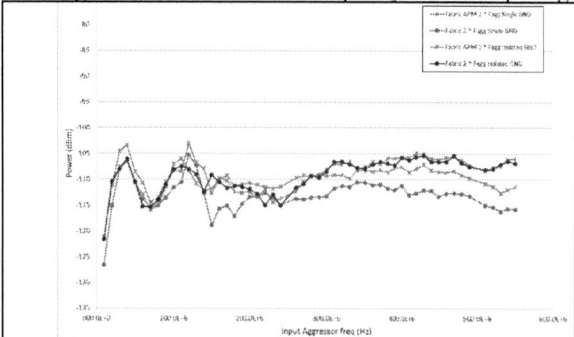

Figure 12: DAC isolation from digital fabric

VIII. CONCLUSIONS

A programmable FPGA-SOC with integrated RF data converters in 16nm FinFET has been described. The measurements demonstrate that state-of-the-art data converter performance can be achieved on a FinFET process. The excellent isolation results show that these circuits can co-exist when monolithically integrated with large scale digital systems including digital-RF, FPGA fabric and integrated processor subsystem.

The RFSoC enables the implementation of cost effective, power efficient, wideband communications systems with the most demanding RF performance requirements. Applications include multi-band and wide-band wireless base stations with flexible digital pre-distortion algorithms, wireless backhaul and DOCSIS 3.1 remote-PHY cable modems. The device was integrated in a 1517 pin 40x40mm BGA package and the chip photograph is illustrated in fig. 13.

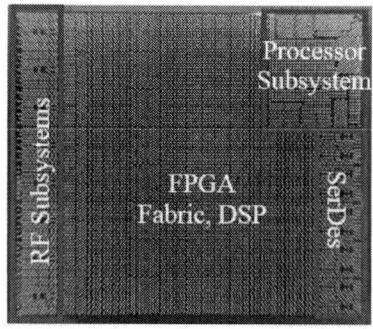

Figure 13: Die photo of RFSoC

REFERENCES

[1] B. Murmann, "ADC Performance Survey 1997-2016"

[2] https://www.xilinx.com/products/technology/power.html

[3] https://www.xilinx.com/support/documentation/white_papers/wp489-rfsampling-solutions.pdf

[4] B. Vaz et al., "A 13b 4GS/s digitally assisted dynamic 3-stage asynchronous pipelined-SAR ADC," *ISSCC Dig. Tech. Papers*, pp. 276-277, Feb. 2017

[5] C. Erdmann et al., "16.3 A 330mW 14b 6.8GS/s dual-mode RF DAC in 16nm FinFET achieving −70.8dBc ACPR in a 20MHz channel at 5.2GHz," *ISSCC Dig. Tech. Papers*, pp. 280-281, Feb. 2017.

[6] A. Collins et al., "RFSoC Integrates RF Sampling Data Converters for 5G New Radio," *Microwave Journal*, June, 2017.

A Reconfigurable Analog Baseband Transformer for Multi-standard Applications in 14nm FinFET CMOS

Jongmi Lee[1], Jongwoo Lee, Chilun Lo, Jaehoon Lee, In-Young Lee, Byungki Han,
Seunghyun Oh, and Thomas Cho
Samsung Electronics, Hwasung, Korea
[1]jongmi06.lee@samsung.com

Abstract—This paper proposes the reconfigurable RX analog baseband transformer that supports multi-standard applications. The proposed ABB can transform its structure between a delta-sigma modulated ADC for narrow band and a baseband LPF for wide band with a simple switch configuration without extra cost. Thus, the ABB obtains efficiency in both size and power aspects. It occupies only 0.11mm² of active area by avoiding large capacitor for narrow band filtering and consumes 1.33mW and 12.1mW in GSM and LTE40M modes, respectively.

Index Terms—Active filters, delta-sigma modulation, low-pass filters, analog-digital conversion, data engineering

I. INTRODUCTION

With the evolution of cellular standards from GSM to LTE, the system requirements of RF transceiver has become intricate to support multi-standard operation. Particularly, for multi-mode signal conditioning of the down-converted data, the analog-baseband (ABB) of the direct conversion cellular receiver (RX) requires large area and power consumption to meet a very wide range of performance requirement depending on the mode.

Fig.1 shows two representative ABB structures for baseband receivers. One is an active low pass filter (LPF) with ADC, and the other is delta-sigma modulated (DSM) ADC with decimator. These two structures have their own pros and cons depending on the input signal bandwidth and blocker requirements.

The first structure is composed of the LPFs which attenuate out-of-band undesired signal and ADCs which digitize filter's output. In wide signal-bandwidth (WB) mode, such as LTE which has higher signal bandwidth and lower blocker rejection requirement, LPF based architecture is the preferred structure in terms of power consumption since it does not require higher oversampling clock with low jitter [1-2]. However, in narrow signal-bandwidth (NB) mode, active RF LPFs typically require high order and large capacitance to support its higher blocker rejection requirement and low signal bandwidth with low thermal noise.

To reduce the expensive capacitor area in NB mode, DSM ADC can replace the large-sized LPFs and ADCs by noise-shaping technique which can get high SNR, since the

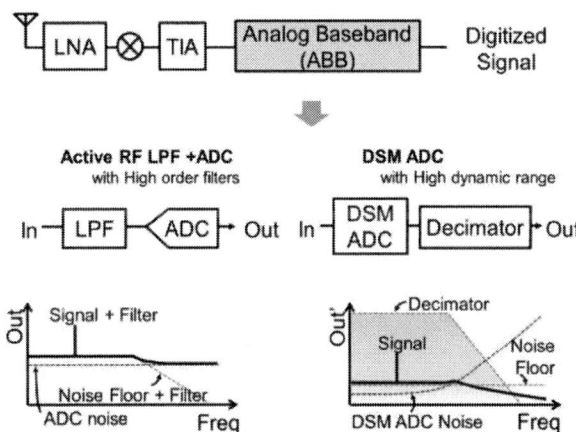

Fig.1. Conventional analog-baseband topologies in RX chain

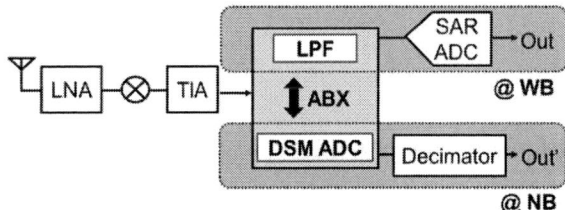

Fig.2. Proposed analog baseband transformer

blocker filtering can be pushed to digital domain with decimation filters [3-5]. On the other hand, for WB mode, a DSM ADC is not preferred due to its high power consumption because of the required oversampling clock frequency that is much higher than Nyquist ADCs such as SAR or Pipelined. Some papers have been tried to reduce power consumption by smaller oversampling ratio with increasing the number of integrator [3] or using multi-level quantizers [4-5]. However, they bring up undesired instability or area increase.

This paper proposes a reconfigurable analog baseband circuit, named analog baseband transformer (ABX), which can operate in an area-and-power-efficient DSM ADC for NB mode and in an active RF LPF for WB mode for low power consumption as shown in Fig.2. Most of the required analog building blocks such as operational amplifiers (opamp) and passive component can be shared for different

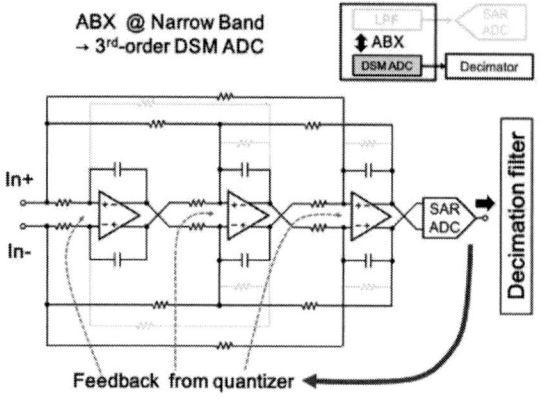

Fig.3.The 3rd-order CT-DSM ADC mode in NB mode

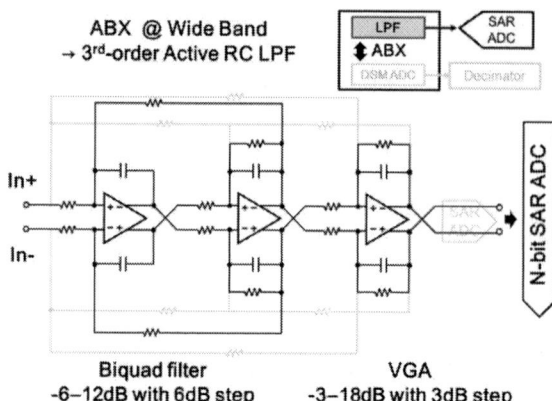

Fig.4.The 3rd-order active RC filter mode in WB mode

ABX modes, requiring no additional power, complexity and area increase.

II. CIRCUIT IMPLEMENTATION

A. Overall scheme of ABX

For the ABX, which transforms its structure according to the input frequency, in order for the ABX to save power and area more effectively the threshold frequency of the input signal for switching between NB and WB modes is determined as 2.5 MHz. The absence of NB-LPF in ABX results in a dramatic size reduction, since the size of the RC LPF ABB has been mostly occupied by the capacitors in NB mode. Additionally, the power of ABX in WB mode can be low due to its transformation from DSM ADC to RC LPF.

Fig.3 and 4 describe the two different modes of ABX, 3rd-order DSM ADC and 3rd-order active RC LPF, respectively. The simplified path switch configurations in each transformer are shown in dark (on) or light shading (off). As shown in Fig. 3 and 4, the ABX can be easily reconfigured to a different mode of operation by simply switching the path connections without additional complex circuits.

The LPF's capacitors for bandwidth control in WB are reused as the integrating capacitors of the DSM ADC in NB while the resistors for LPF gain control in WB are disconnected in DSM ADC mode. Besides, the two-stage opamp can efficiently operate for both low bandwidth in NB mode and high bandwidth in WB mode with its programmable bias current as seen in Fig.5. The opamp has a PMOS differential input pair with a degeneration source load to achieve high SNR performance in NB mode. In WB mode, the resistor for degeneration becomes inactive.

B. CTDSM ADC mode of ABX for NB (NB-ABX)

In NB mode, the DSM ADC adopts a 3rd-order feedback topology with input feed-forward path to reduce the

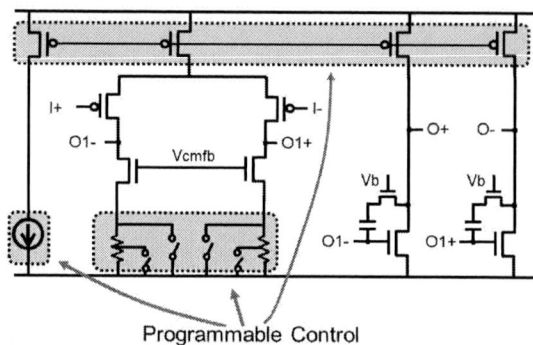

Fig.5.Programmable amplifier for supporting wide bandwidth and low noise

integrator internal swing (Fig.3). To avoid anti-aliasing issue, continuous type (CT) DSM ADC is implemented. The 4bit SAR ADC is used for quantizer, and the feedback current is injected by 4bit DAC with data weighted averaging (DWA) technique to improve linearity. With a 52 MHz of the sampling frequency the state scaling and feedback coefficients of the DSM ADC are decided by MATLAB simulation to have the best SNDR performance and guarantee the stability of the circuit at the same time.

C. Filter mode of ABX for WB (WB-ABX)

In WB mode, the 3rd-order active RC filter consists of a biquad (BQ) filter and a variable gain amplifier (VGA) (Fig.4). Since all switches used for mode conversion should have low leakage and low resistance at the same time, so they are connected to the input node of opamps which is virtual ground and implemented only with NMOS. The switch size is decided to be optimal in the max gain setup and low bandwidth mode to avoid the gain degradation and the unexpected zero of the filter. In addition, the bias of dac is tied to ground and power in WB mode to ensure that leakage from the NB is eliminated. The RC filter is designed to have a programmable gain from −9dB to 30dB

978-1-5386-3179-9/17 $31.00 © 2017 IEEE

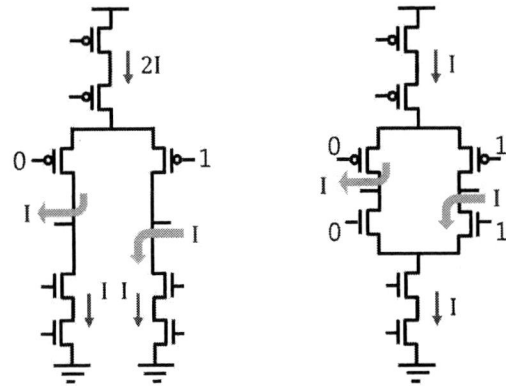

Fig.6 Unipolar and bipolar DAC unit for example

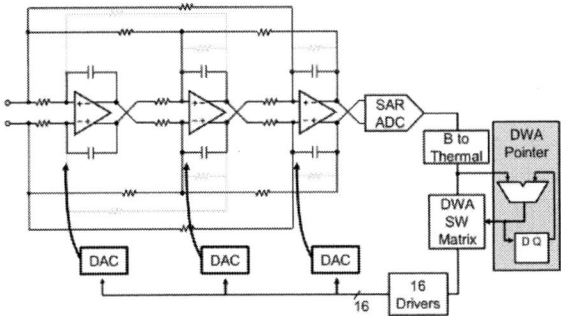

Fig.7 Feedback DAC topology with DWA technique

in a 3dB-step size where the BQ filter gain and VGA gain range from -6 to 12dB with 6dB resolution and from -3 to 18dB with 3dB resolution respectively. In order to implement the 3dB gain step the logarithmic-tuning technique [1] is adopted for the resistance control, which also have benefits in the aspect of area. Meanwhile, the filter bandwidth is designed to cover from 5MHz to 20MHz without pass-band ripple to support all channel bandwidths (LTE 10/ 15/ 20/ 40M).

D. Feedback DAC with DWA technique

The DAC unit cell design is based on bipolar topology instead of the unipolar one. By using bipolar switching in DAC unit cell, the DAC current can be the half of that in the unipolar version as shown in Fig.6. The cascode transistors of the current source in Fig. 6 are also implemented in both NMOS and PMOS to increase output impedance of DAC unit cell.

In order to achieve high SNR, linearity of DAC should be carefully taken into account. In this chip, data weighted averaging (DWA) technique [6] is adopted to alleviate DAC linearity problem that allows avoiding the huge area of DAC caused by the intrinsic device size matching (Fig.7).

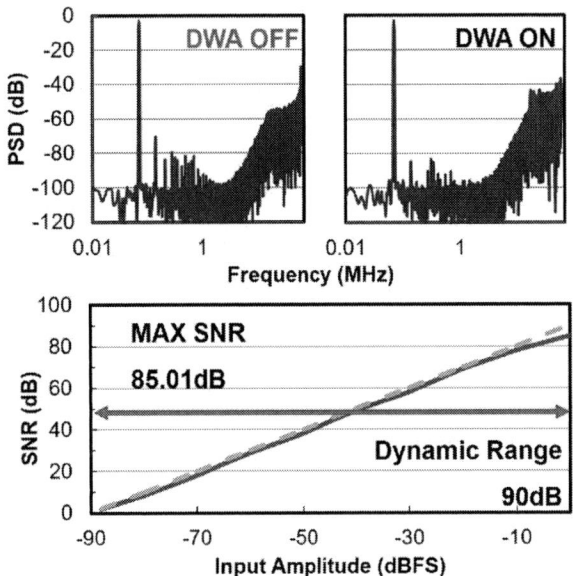

Fig.8 Measured output spectra and dynamic range of the DSM ADC in NB mode

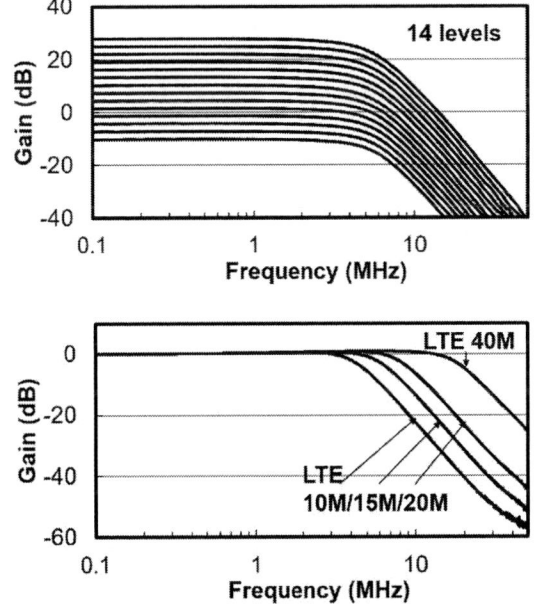

Fig.9. Measured gain and bandwidth of the 3rd-order filter in WB mode

III. MEASUREMENT RESULTS

Fig.8 depicts the measured output spectrum and SNR of the CT-DSM ADC in NB-ABX. The input full scale of NB-ABX is 1.2Vpp, and 0dBR is -2.55dBFS. With DWA technique, the harmonic components from mismatch are remarkably reduced. In 100kHz BW set-up with 52MHz

TABLE I COMPARISON TABLE

	NB-ABX I&Q	Active RC Filter	
		[1] 2014 ASSCC	[2] 2015 TCAS2
Tech (nm)	14	65	65
Application	NB	NB / WB	NB / WB
BW(MHz)	0.1 - 2.5	0.1 - 14.2	0.2 - 20
Size(mm²)	0.11	2.03	1.6
Power(mW) @ GSM	1.33	7.3	3.44
	WB-ABX +ADC* I&Q	DSM ADC		
		[4] 2012 VLSI	[3] 2016 ISSCC	[5] 2015 ISSCC
Tech (nm)	14	65	65	28
Application	WB	WB		
Max BW(MHz)	5-20	20	25	50
Size(mm²)	0.29 (=0.11+0.18*)	0.15	0.5	0.68
Power(mW)	16.1 (12.1+4*)	52	82.8	156
DR+Filter's Max Gain (dB)	89 (=59+30)	76	77	85

The reported DSM ADC's power and size are doubled to compare with ABX which has I&Q path.
* Additional ADC power & area.

	NB-ABX	[4]	[3]	[5]
FOM(dB)	171.77	160	164.8	173.1

*FOM=DR+10log(BW/P)

Fig.10.Layout of the 12bit SAR ADC (I&Q) in 14nm CMOS process

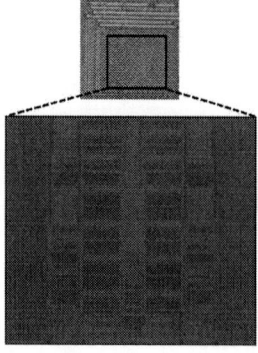

Fig.11.Die photograph of ABX

sampling frequency, the measured SNR and dynamic range is 85.01dB and 90dB. Fig.9 shows the measured gain and BW of the 3rd-order filter in WB- ABX. The measured gain and BW range from -9 to 30dB in 3dB step and from 5MHz to 20MHz, respectively.

Table I compares the ABX's performance with previously reported state-of-the-art ABBs. Since the ABX can support narrow band, where a large area is typically required, the area comparison of ABX should be done with ABBs that support narrow bands. The areas of the referred NB-ABBs[1-2] are about 14-18 times larger than the proposed ABB, which is too large even considering the process difference. In addition, the active RC filters also require ADCs to generate additional digitized outputs. In terms of power consumption, the NB-ABX with a feed-forward path consumes 1.33mW, and the FOM of NB-ABX is 171.77dB in GSM mode.

For a fair comparison of WB-ABX which has I and Q paths with the reported DSM ADCs, the power and area of the reported ADCs are multiplied by 2. In addition, unlike DSM ADCs, WB-ABX requires additional ADCs, so the performance of the WB-ABX is compared to other DSM ADCs by adding ADC's power and area in the same process (Fig.10). The additional ADC used in calculation consumes 4mW and occupies 0.18mm². With the ADC, the WB-ABX in a LTE 40M mode consumes only 16.1mW, which is much lower than the power consumption of the previous DSM ADCs. The WB-ABX with ADC occupies 0.29mm², and acquires a high sum of filter's max gain and ADC's DR, 30dB and 59dB, respectively.

IV. CONCLUSION

The implemented ABX in 14nm FinFET CMOS consumes 1.33mW in GSM, and 12.1mW in LTE 40M with 1V supply voltage, and occupies 0.11mm² (Fig.11). The ABX, which maximizes the efficiency in area and power by transforming its structure depending on the signal bandwidth, is a novel architecture to support multi-standard.

REFERENCES

[1] Jongwoo Lee, et al., "A Reconfigurable Analog Baseband for Single-chip, Saw-less, 2G/3G/4G Cellular Transceivers with Carrier Aggregation," IEEE Asian Solid-State Circuits Conference,pp.9-12, Nov, 2014.

[2] Yixiao Wang, et al., "Highly Reconfigurable Analog Baseband for Multi-standard Wireless Receivers in 65-nm CMOS," IEEE Transactions on Circuits and Systems II, pp.296-300, Mar, 2015.

[3] Lucien Breems, et al., "A 2.2GHz Continuous-Time ΔΣ ADC with -102dBc THD and 25MHz BW," ISSCC Dig. Tech. Papers, pp. 272-273, Feb. 2016.

[4] Gerry Taylor, et al., "A Reconfigurable Mostly-Digital ΔΣ ADC with a Worst-Case FOM of 160dB," Dig. Symp. VLSI Circuits, pp. 66-167, June 2012

[5] Do-YeonYoong, et al., "An 85dB-DR 74.6dB-SNDR 50MHz-BW CT MASH ΔΣ ADC Modulator in 28nm CMOS," ISSCC Dig. Tech. Papers, pp. 272-273, Feb. 2015

[6] Sheng-Jui Huang, et al., "A 1.2V 2MHz BW 0.084mm² CT ADC with -97.7dBc and 80dB DR using low – latency DEM." ISSCC Dig. Tech. Papers, pp. 172-173, 2009

A 1.4Mb 40-nm embedded ReRAM macro with 0.07um² bit cell, 2.7mA/100MHz low-power read and hybrid write verify for high endurance application

Chia-Fu Lee, Hon-Jarn Lin, Chiu-Wang Lien, Yu-Der Chih, Jonathan Chang

Memory Solution Division (MSD)
Taiwan Semiconductor Manufacturing Company, Taiwan

Abstract—Resistive RAM (ReRAM) is an attractive candidate for next generation embedded nonvolatile memory [1][2], with several advantages compared to conventional flash technology. First, ReRAM is a CMOS-compatible low temperature back-end of line (BEOL) memory. There is almost no mutual impact between ReRAM element and front-end CMOS devices during the wafer processing. Second, it only needs 2~4 extra masks, resulting in lower chip cost compared to embedded flash. Third, its byte-alterability makes it capable of being used as a unified memory for storage of instruction code and real-time data.

In this work, a 1.4Mb HfOx-based embedded ReRAM macro with 0.07um² bit cell is fabricated in a 40nm CMOS process. A hybrid write-verify algorithm (HWVA) is used to tighten the resistance distribution and avoid over-write damage for achieving fast write and high endurance. Cycling test results show a write endurance of 10^6 cycles for the proposed HWVA. For low-power applications, a low power sensing scheme is presented for this 1.4Mb macro, achieving 100MHz read speed and 2.7mA/100MHz read current for 44-bits read operation. An area-efficient row-redundancy scheme and double-error-correction ECC are implemented for further improvement of yield and reliability.

Keywords—*ReRAM, HWVA, WCMS, MRWV*

I. INTRODUCTION

Fig. 1. shows the cross-section of ReRAM cell structure, schematic of 2-by-4 bit-cell array, and characteristics of Set/Reset/Forming of HfOx-based ReRAM in 40nm CMOS process. Two additional masks for top electrode (TE) etch and photo of bottom electrode (BE) VIA are needed. Forming voltage is ~ 3V, set and reset voltages are ~ +1.5V and -1.5V respectively. Typical resistance values of low-resistance state (LRS) and high-resistance state (HRS) are ~10^4 ohms and 10^5 ohms respectively. Dedicated source-line (DSL) array architecture is adopted for concurrent set/reset and parallel forming operation.

From our experimental results, it was found that a long write pulse for set or reset operation is more effective than multiple short pulses. However, due to variability of ReRAM bit cells, a long write pulse may cause over-write, which either impacts the speed of subsequent write operations or damages the ReRAM element. In order to prevent over-write, a write-current monitor scheme (WCMS) is proposed.

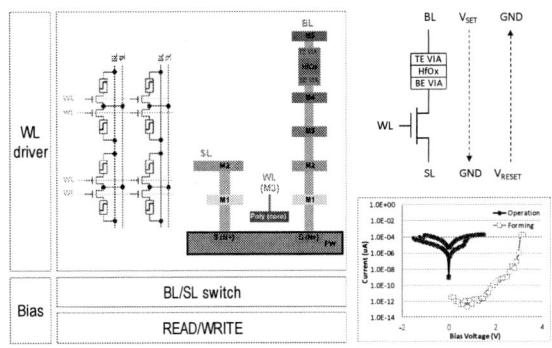

Fig. 1. Cross-section of cell structure, schematic of 2-by-4 bit-cell array and characteristics of Set/Reset/Forming of HfOx-based ReRAM

II. WRITE CURRENT MONITOR SCHEME

As shown in Fig. 2., it includes the feedback-control voltage-mode write buffer and write-current monitor circuit. Compared with conventional write buffer circuits [3][4], the proposed write buffer circuit can achieve lower operation voltage (<1.8V). During set/reset operation, the current mirrored from device M1 to M2 is used to monitor the switching of resistance. For example, after the cell state has been switched from HRS to LRS by set operation, the current of M2 will increase and the signal of WCM will change from "0" to "1" to stop the write operation. Likewise, the signal of WCM will change from "1" to "0" after cell state is switched from LRS to HRS by reset operation. Although a tight resistance distribution of LRS/HRS was expected using WCMS, a long distribution tail was observed in our test results.

To investigate the root cause, we measured the resistance (or cell current) of the normal bits and the tail bits under different bias voltages. The example measurement results in Fig. 3. shows that the resistance of the tail bit under "read" bias (0.3V) deviates from the resistance of the normal bits even though their resistance values converge to ~15K ohm under "write" bias (1.6V) by WCMS. As a result, the resistance of bit cells may have a wide distribution under read bias condition if only WCMS is used to control the write operation. To solve this problem, a hybrid write-verify algorithm (HWVA) is proposed.

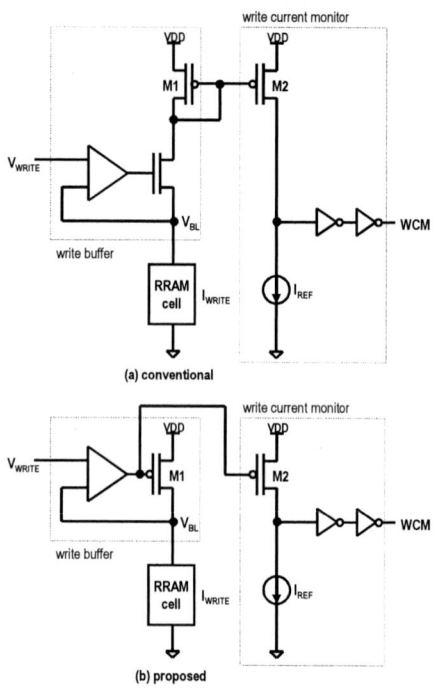

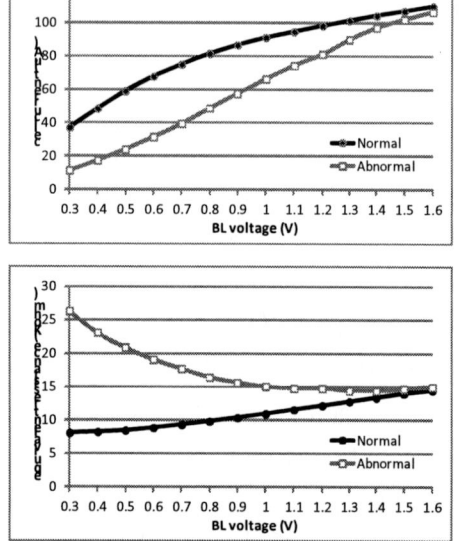

Fig. 2.   Conventional and proposed write buffer circuit

Fig. 3.   Measured resistance of LRS by set operation controlled by WCMS

III.   MARGIN-READ WRITE VERIFY

In this hybrid algorithm as Fig. 4., the WCMS is used only for modulating the write pulse width for prevention of over-write during a long write pulse but it is not used for terminating the whole write cycle.

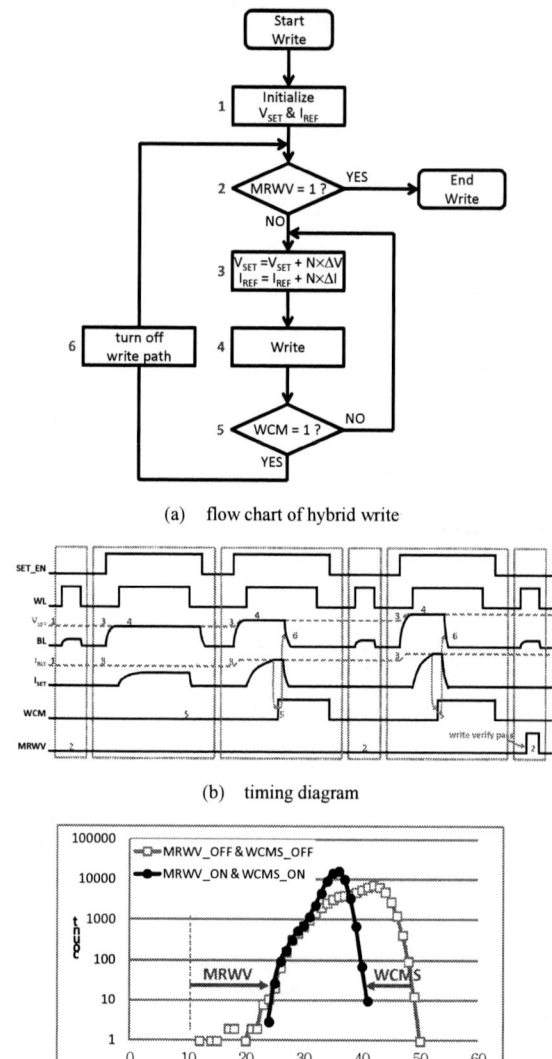

Fig. 4.   Flow chart and timing diagram of hybrid write scheme and cell current distribution with and without hybrid write scheme

A margin-read write verify (MRWV) is added to verify if the cell resistance is programmed to the targeted value. If margin-read verification passes, write operation will be stopped. But if margin-read verification fails, HWVA will repeat the write operation with a higher V_{SET} (V_{RESET}) and I_{REF}. Experimental results show that WCMS can effectively prevent the over-write for fast-switching bits and the additional

MRWV can effectively tighten the LRS/HRS cell resistance distribution for the slow-switching bits (Fig. 4.). Most of the bits just require one shot, therefore on the average the hybrid write does not impact write latency much.

IV. ENDURANCE CHARACTERISTICS

The endurance performance for cycling with two write schemes, cycling without write-current monitor and cycling with the hybrid write verify were measured for 2K-bits arrays. The resulting cell current distribution measurement in Fig.5. shows that the proposed hybrid write verify can effectively improve the cycling endurance, and also the current distribution is reduced from 25uA to 15uA.

Fig. 6. shows the endurance characteristics using MRWV combine with WCMS. 2K-bits endurance check to 100K cycling, and 256-bits endurance check up to 1M cycling. The current separation is larger than 17uA after 1M cycling.

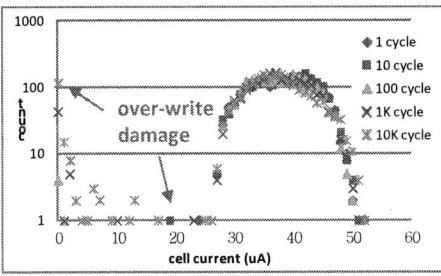

(a) MRWV only and without WCMS

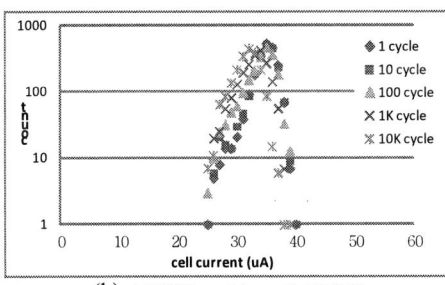

(b) MRWV combine with WCMS

Fig. 5. Cell current distirbution measurements of MRWV combine without and with WCMS

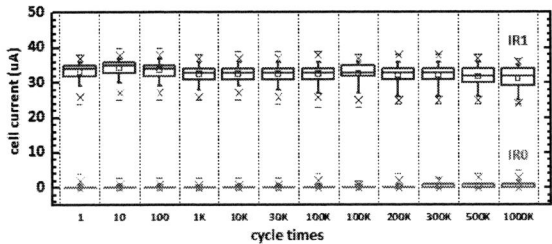

Fig. 6. Endurance characteristics of MRWV combine with WCMS

V. LOW POWER CURRENT-MODE SENSE AMPLIFIER

For battery-powered applications such as mobile phone, portable medical device and smart meters, low-power read operation and fast read speed are essential. Fig. 7.(a) shows the schematic of the low power current-mode sense amplifier in our design, which is constructed with one global reference/bias scheme and x44 current-mode sense amplifier.

The reference current generator and replica bias circuit for bit-line clamp in the left are shared by x44 sense amplifiers. Reference current for sensing is generated by a poly resistor string with adjustable resistance values, for setting different reference currents for different read modes. To avoid read disturb, bit-line voltage during read is clamped to ~0.3V by an NMOS source follower with a fixed gate bias. The fixed gate bias is generated with a close loop feedback circuit such that the bit-line voltage will not change with process, voltage and temperature.

Fast sensing is achieved by the two-stage current mode amplifier. The first-stage compares the cell current with the mirrored reference cell, and then feed the small voltage difference into next stage. And the second-stage amplifies the voltage difference to a digital signal to indicate the selected cell. For simplicity, the bit-line pre-charge circuit isn't shown here.

The word width for read and write is 44 bits, including 12 parity bits for double-error correction and double-error detection (DECDED) ECC, implemented by BCH code. The measured read current is less than 2.7mA at 100MHz operation. The shmoo plot in Fig.7.(b) shows that the access time can be within 10ns from -40'C to 125'C.

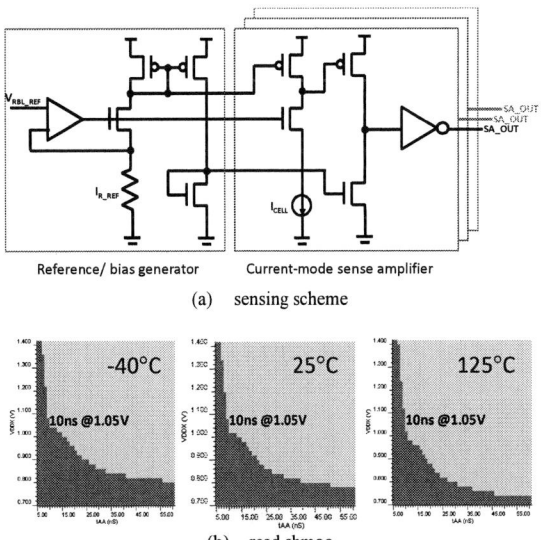

(a) sensing scheme

(b) read shmoo

Fig. 7. Low power sensing amplifier including reference current generator and Shmoo plot of access time

VI. REDUNDANCY

For yield improvement, row redundancy including repair control circuit has been implemented. Fig. 8. shows the block diagram of the repair scheme. The repair information is stored in a dedicated information row in the ReRAM array to save area compared to conventional e-Fuse based repair. In case this dedicated row contains a few defect bits, a repair-recall scheme based on triple modular redundancy (TMR) is used to achieve a reliable reading of repair information from ReRAM array during power up. Comparing to using eFuse to store repair information, the 3-cell-per-bit has less area overhead.

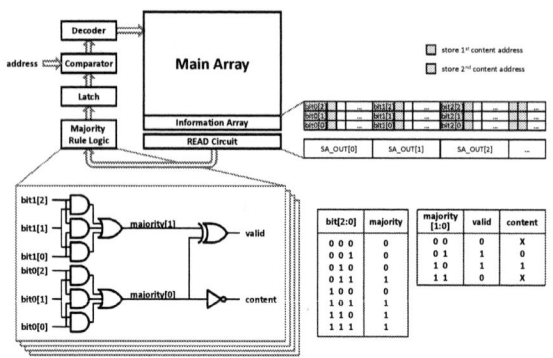

Fig. 8. Repair scheme and repair-recall scheme based on triple modular redundancy (TMR)

VII. CONCLUSION

This paper presents a 1.4Mb HfOx-based embedded ReRAM macro with 0.07um^2 bit cell is fabricated in a 40nm CMOS process. Using the proposed hybrid write verify scheme and a low-power read scheme, a cycling endurance more than 10^6 and a read current of 2.7mA @ 100MHz have been achieved. An area-efficient row-redundancy scheme and double-error-correction ECC are implemented. Fig. 9. shows the micrograph, and Fig. 10. shows the main features of this 1408 columns x 1024 rows ReRAM macro.

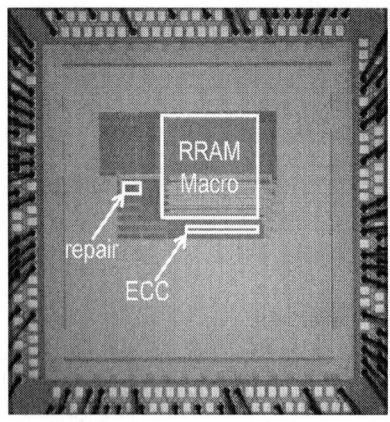

Fig. 9. Chip micrograph of ReRAM macro

Process	TSMC 40nm low power
Supply voltage	1.1V/2.5V
Memory cell	HfOx-based, TiN BE and Ti TE.
Memory size	0.07um^2
Memory capacity	1.4Mb
Macro size	0.28mm^2
I/O width	X44
Read speed	10ns
Read power	2.7mA @ 100MHz
Endurance	10^6 within 2Kb 10^7 within 256b
ECC	DECDED

Fig. 10. ReRAM macro main features

REFERENCES

[1] Akifumi Kawahara1, et al., "Filament Scaling Forming Technique and Level-Verify-Write Scheme with Endurance Over 10^7 Cycles in ReRAM," ISSCC, pp. 200–201, 2013

[2] Meng-Fan Chang1,, et al., "A 0.5V 4Mb Logic-Process Compatible Embedded Resistive RAM (ReRAM) in 65nm CMOS Using Low-Voltage Current-Mode Sensing Scheme with 45ns Random Read Time," ISSCC, pp. 434–435, 2012

[3] Xiaoyong Xue, et al, "A 0.13 μm 8 Mb Logic-Based Cu Si O ReRAM With Self-Adaptive Operation for Yield Enhancement and Power Reduction, IEEE JSSCC, Vol. 48, No. 5, May 2013

[4] Y. L. Song,et al, "Reliability Significant Improvement of Resistive Switching Memory by Dynamic Self-adaptive Write Method", VLSI Technology Symposium . pp. T102 - T103,2013

S2-4 (2045)

A Dynamic Power Reduction in Synchronous 2RW 8T Dual-Port SRAM by Adjusting Wordline Pulse Timing with Same/Different Row Access Mode

Yoshisato Yokoyama, Yuichiro Ishii, Haruyuki Okuda and Koji Nii

Renesas Electronics Corporation, Tokyo, Japan

{yoshisato.yokoyama.jx, koji.nii.uj}@renesas.com

Abstract—**An effective method is proposed to reduce dynamic power for synchronous 2-read/write (2RW) 8T dual-port (DP) SRAM. Adjusting the wordline (WL) pulse timing control circuit is newly introduced for both reading and writing operations. Row addresses of port-A and port-B are compared. The same row access is detected or not in each cycle, which is an inherent access mode of 2RW 8T DP SRAM. In different row access, the WL pulse width is shortened to reduce excessive bitline (BL) discharging power. A well balanced 8T DP SRAM bitcell layout is demonstrated using 40-nm technology. A test chip including 512 w × 73 b 36 kbit and 2 kw × 19 b (38 kbit) 2RW DP SRAM macros was designed and fabricated using 40-nm technology. The measured data show that reading and writing powers are reduced, respectively, by ~7% and ~18% using the proposed scheme.**

Keywords—Dual-port SRAM, Disturbance, Dynamic power, 8T, 40-nm

I. INTRODUCTION

Along with the device scaling down, the multi-processing is more required than higher clock frequency f to achieve energy-efficient SoCs with high performance. Embedded SRAMs are key devices for reducing power consumption, especially being essential not only single-port but also multi-port access for parallel processing. As a data cache or shared cache memory, 2-read/write (2RW) dual-port (DP) SRAMs are used frequently for these multi-CPU and many-core architectures [1–3]. Furthermore, DP SRAMs also function in several applications such as FPGAs [4] or dynamically reconfigurable processors [5,6]. Many reports of the relevant literature describe studies related to the DP SRAM designs [7–16]. Enhancement of write/read margins for lower voltage operation, improvement of the density, and prevention or detection of disturbance issues between port accesses have been discussed mainly in earlier reports of the literature.

In this paper, we propose a dynamic power reduction scheme for both reading and writing operation by detecting the same and different row access, which is the inherent mode in an 8T DP SRAM.

II. 8T SRAM BITCELL OPTIMIZATION

Fig.1 presents an illustration of a simple block diagram of 2RW DP SRAM macro. AA (AB) and DA/QA (DB/QB) respectively show address inputs and data inputs/outputs for port-A (port-B). The 1-write/1-read (1W1R) 2-port (2P) 8T

bitcell (BC) with decoupled read bitline (BL) is used widely for buffer memory of imaging processing or cache of multi-core CPUs [15]. However, this type of 2P 8T BC is limited for read-only access in port-B. The differential type of DP 8T BC is required to support both read and write operations for each port-A and port-B. A related schematic is presented in Fig. 2(d). Conventional layouts of two types have been reported [7–10, 14]: a high-density area, but not good symmetry between each true/bar BL pair, or one between port-A/port-B. To improve that imbalance, we propose a new BC layout [7], as presented in Fig. 2(c). The proposed BC has a shorter wordline (WL) than the conventional layout (a), thereby reducing the WL metal resistance. It helps to reduce the BL delay differences between the far column and the near one from the WL driver. If marked differences of delay exist between near and far columns, the excessive discharge power of BL is consumed because of the faster BL swing of steeping WL rise/fall change in the near column.

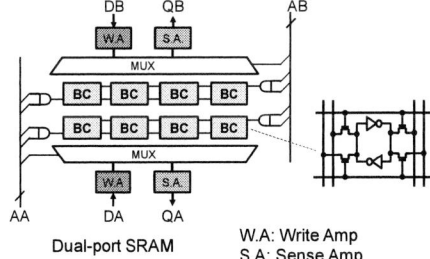

Fig. 1 Block diagrams for DP SRAM.

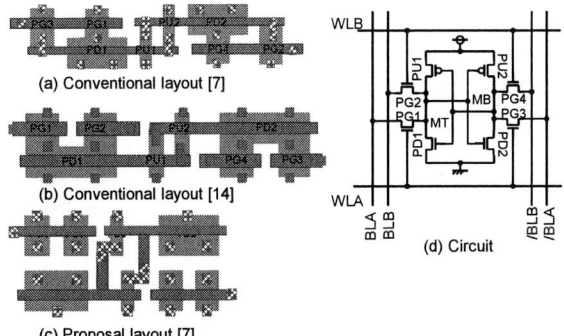

(a) Conventional layout [7]

(b) Conventional layout [14]

(c) Proposal layout [7]

(d) Circuit

Fig. 2 40nm 8T SRAM bitcell (BC).

978-1-5386-3179-9/17 $31.00 © 2017 IEEE   13

To validate of the symmetry of the proposed DP 8T BC layout, we implemented an isolated test structure for direct measurement of each cell current (I_{read}). Fig. 3 presents the measured distributions of I_{read} for A-true, A-bar, B-true and B-bar BLs implemented using 40-nm technology. In the graph, all values of I_{read} are normalized by median and sigma. The proposed layout was observed to have small differences among all I_{read}, whereas the two conventional types of layouts have offsets between true/bar or port-A/-B BLs. From the measured I_{read} distributions, the write-margin (WM), read-margin (RM), and static-noise margin (SNM) of the proposed BC are expected to be improved by virtue of the small offsets, resulting in a good minimum operating voltage (V_{min}). However, the offsets in the conventional layouts should occur extra design margins, inducing the overhead of area, power and timing, and deterioration of V_{min}. In the following chapters, we simulated or measured SRAM macros using the proposed BC.

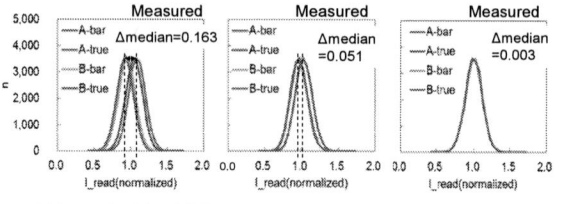

(a) Conventional bitcell [14]   (b) Conventional bitcell [14]   (c) Proposed bitcell

Fig. 3 Measured I_{read} distributions of A/B-true/bar BL (n=20320).

III. DISTURBANCE ISSUES

The DP 8T SRAM has two disturbance modes: read disturbance and write disturbance. Fig. 4 presents an illustration of the read disturbance issue by activating the same-row address. The different row access mode works as a single port (SP) 6T SRAM: disturbance does not occur. Meanwhile the undesirable cell current flows through the other BL into the PD NMOS in the accessed BC in the same row access mode. In that case, the I_{read} of target BL through the PG and PD NMOSs decreases by additional cell current from the other BL. Using the same isolated test structure as that presented in the section II, we measured the I_{read} differences between the different row access mode and the same row access mode.

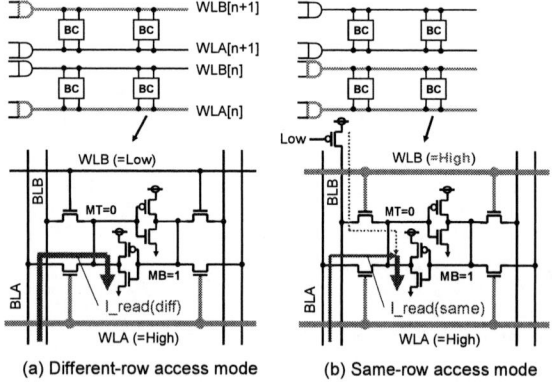

(a) Different-row access mode   (b) Same-row access mode

Fig. 4 Read disturbance issue.

Fig. 5 presents the measured distributions of I_{read} (n=20320). Both distributions in the same and different row access modes are observed on the normal distributions. The delta of I_{read} depends on the sigma value: it is 18% at 0 sigma (mean value) and 30% at -5 sigma as presented in Fig. 5(b). Fig. 6 shows the other disturbance issue in the write operation. It is also disturbed by another BL current as well as a read-disturbance issue [12]. In a different row access mode, the write operation will be performed if the pulling down current (IPD) is larger than pulling up current (IPU) (Fig. 6(a)). In the same row access mode, IPU is increased by the pre-charge circuit of BLB, disturbing the write ability.

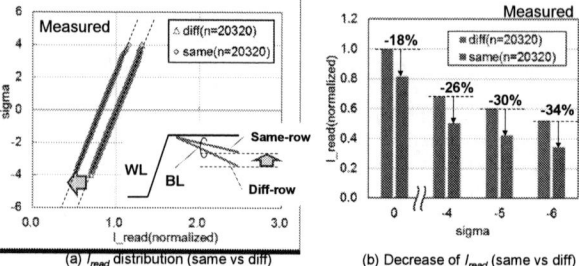

(a) I_{read} distribution (same vs diff)   (b) Decrease of I_{read} (same vs diff)

Fig. 5 Measured I_{read} distribution (different-row *vs* same-row access).

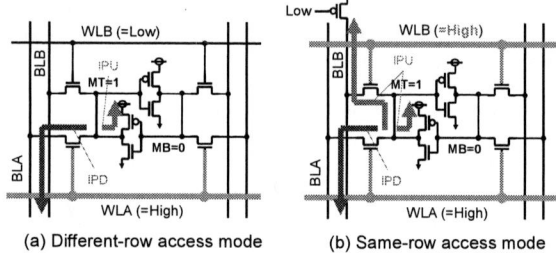

(a) Different-row access mode   (b) Same-row access mode

Fig. 6 Write disturbance issue.

IV. DETECTING SAME-ROW AND ADJUSTING MARGIN CIRCUIT

The probability of occurrence of the same row access is 1/row if the row address is accessed randomly. For example, it is 0.4% in the 256-row macro configuration. Although there is a small probability of the same row access mode, it should be considered for maintaining the design margin. In other words, 99.6% of operation have excessive timing margins because the different row access has better timing margins for both reading and writing. This over-timing of margins consumes undesirable dynamic power by an extra discharge of BL capacitances. To eliminate this extra dynamic power consumption, we propose an adjusting WL pulse timing control circuit. Fig. 7 presents a block diagram of the proposed circuit, where ARA is the row address of the A-port, and ARB is the row address of the B-port. If the input row addresses of port-A and port-B are the same, then the output signal SR of the row address comparator (XNOR) becomes high. For that reason, the parasitic capacitance of the replica BL increases. As a result, the WL pulse width will be longer by increased replica delay. The triggered timing of sense enable for sense amplifier is also shifted in the read operation. In the write operation, the WL pulse width and enabling write driver period increase if the same row address is applied.

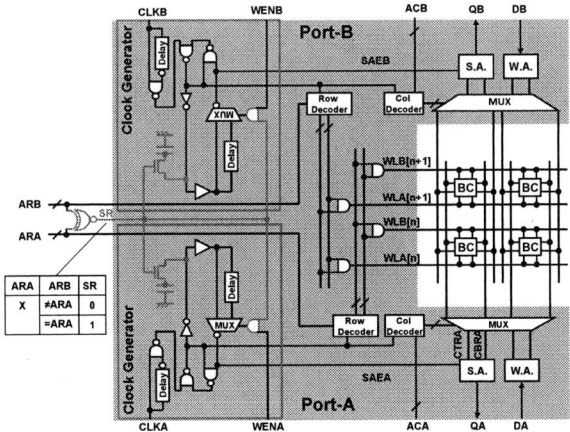

Fig. 7 Block diagram of proposed DP SRAM.

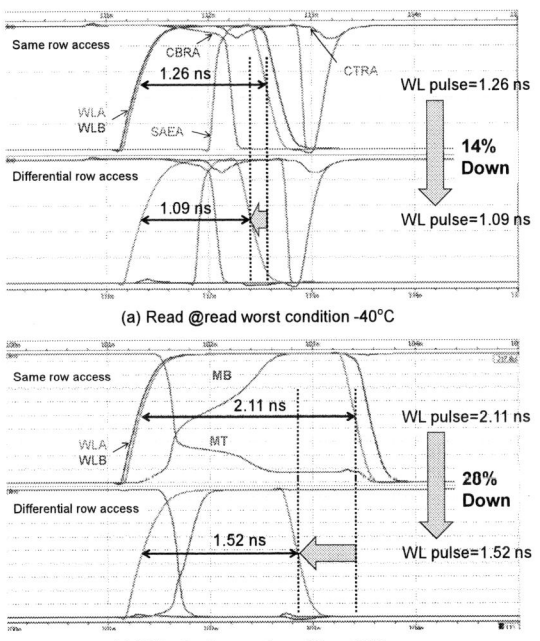

(a) Read @read worst condition -40°C

(b) Write @write worst condition -40°C

Fig. 8 Simulation waveforms of 38-kbit MUX8 (2048-word x 19-bit) with adjusting WL pulse timing control.

Figs. 8(a) and 8(b) respectively depict SPICE simulation waveforms at the worst PVT conditions in the read operation and write operation. The simulated WL pulse width for the read operation, as presented in Fig. 8(a), can be shorten by 14% in the different row access mode compared to the same row access mode. Here, we assume that the same BL swing: delta voltage between CBRA and CTRA in Fig. 8(a) is almost identical to the different row access mode. However, the simulated WL pulse width for the write operation, as presented in Fig. 8(b), can be shorten further by 28% in the different row access mode compared to the same row access mode. This simulation demonstrated that the write disturbance strongly affects the

differences of WL width between the same and different row access modes. From the simulation results presented in Figs. 8(a) and 8(b), we estimated a dynamic power comparison in the two types of MUX configurations under the worst PVT conditions. Fig. 9 portrays the estimated results of expecting dynamic power reduction by introducing proposed circuits. In read and write operations, respectively, the dynamic power is reduced by ~7% and ~18%.

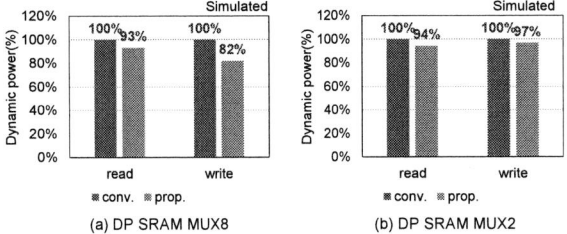

(a) DP SRAM MUX8   (b) DP SRAM MUX2

Fig. 9 Simulation result of dynamic power consumption at typical condition.

V. DESIGN AND EVALUATION OF TEST CHIP

Fig. 10 is a die photograph of a test chip using 40 nm CMOS technology. DP SRAM macros of two kinds, MUX2 and MUX8 are implemented in the test chip. Table 1 presents a summary of the test chip features. We observed full read/write functions at temperatures of -40°C to 125°C. Fig. 11(a) and 11(b) show cumulative distribution functions (CDFs) of V_{min} at -40°C for two macros. The total number of measured dies is 70. The median of V_{min} for TT process is 0.71 V for each macro. Fig. 12 shows the measured dynamic power consumptions of write and read operations vs. the supply voltage. The power consumption measurements obtained using the proposed circuitry are, respectively, 19.5 µW/MHz (read operation) and 19.6 µW/MHz (write operation), under the random address accessing (probability of same row access is less than 4%), at the typical condition (TT/1.1 V/25°C). The measured data show that the dynamic powers of proposed DP SRAM macros are reduced by 7% and 18%, respectively, for reading and writing operations compared to the conventional macros. The proposed circuit shown in Fig. 7 has no disadvantage for access time or address setup time. Furthermore, the area overhead is less than 1%.

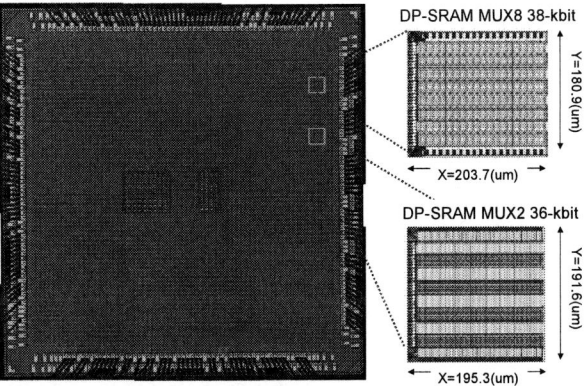

Fig. 10 Microphotograph of the Test chip and layout plots of proposed two DP SRAM macros using 40-nm technology.

978-1-5386-3179-9/17 $31.00 © 2017 IEEE   15

Table 1 Features of the test chip.

		Features
Technology		40-nm CMOS process
Macro configuration	MUX2	36-kbit (512 word x 73 bit)
	MUX8	38-kbit (2048 word x 19 bit)
Macro size	MUX2	195.3 μm x 191.6 μm 37419 μm²@36-kbit
	MUX8	180.9 μm x 203.7 μm 36849 μm²@38-kbit
Bit density	MUX2	0.95 Mbit/mm²
	MUX8	1.01 Mbit/mm²
Access time @typical	MUX2	1.4 ns
	MUX8	1.6 ns
Cycle time @typical	MUX2	2.0 ns
	MUX8	2.3 ns
Dynamic power @typical	MUX8	Read: 19.5 μW/MHz Write: 19.6 μW/MHz

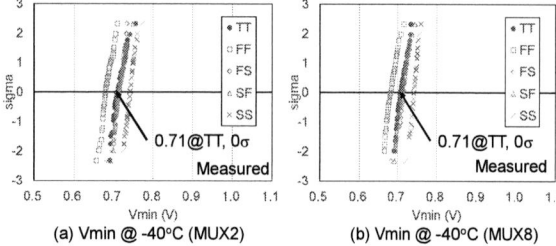

(a) Vmin @ -40°C (MUX2) (b) Vmin @ -40°C (MUX8)

Fig. 11 Distribution of measured V_{min} at -40°C worst temperature.

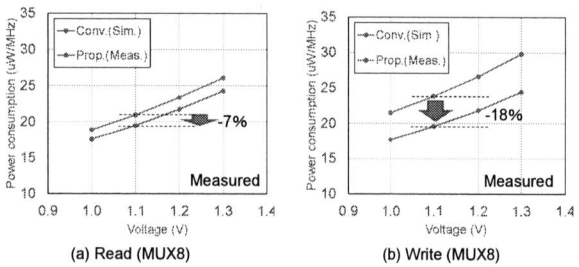

(a) Read (MUX8) (b) Write (MUX8)

Fig. 12 Measured dynamic power vs supply voltage at 25°C, TT-process.

VI. CONCLUSION

We proposed an adjusting WL pulse timing control circuit to reduce read and write dynamic power for a 2RW DP SRAM. Row addresses for port-A and port-B were compared and the same row access was detected, or not, which is an inherent access mode of 2RW 8T DP SRAM design. Well-balanced 8T DP SRAM bitcell was demonstrated using 40-nm technology. Test chips including macros of two kind were designed and fabricated using 40-nm technology. The measured data show that read and write powers were reduced, respectively, by ~7% and ~18%.

REFERENCES

[1] T. Shiota et al., ISSCC Digest, pp. 194-195, 593, Feb. 2005.

[2] M. Nakajima, T. Yamamoto, M.Yamasaki, K. Kaneko, T. Hosoki, "Homogeneous Dual-Processor core with Shared L1 Cache for Mobile Multimedia SoC," in VLSI Circuits, 2007. Digest of Technical Papers, pp. 216-217, June 2007.

[3] Igor Loi, Luca Benini, "A multi banked - Multi ported - Non blocking shared L2 cache for MPSoC platforms," 2014 Design, Automation & Test in Europe Conference & Exhibition (DATE), pp.1-6, Mar. 2014.

[4] https://www.xilinx.com/support/documentation/data_sheets/ds099.pdf

[5] T.Toi, N.Nakamura, T.Fujii, T.Kitaoka, K.Togawa, K.Furuta and T.Awashima, "Optimizing Time and Space Multiplexed Computation in a Dynamically Reconfigurable Processor", ICFPT 2013, pp.106-111, Dec. 2013.

[6] T.Toi, N.Nakamura, Y.Kato, T.Awashima, K.Wakabayashi, "High-level Synthesis Challenges for Mapping a Complete Program on a Dynamically Reconfigurable Processor", IPSJ Trans. on System LSI Design Methodology, Vol. 3, pp.91-104 , Feb. 2010.

[7] Patent US6529401, **2003.**

[8] K. Nii, Y. Tsukamoto, S. Imaoka, and H. Makino, "A 90 nm dual-port SRAM with 2.04 μm² 8T-thin cell using dynamically-controlled column bias scheme," *ISSCC Dig. Tech. Papers*, pp. 508-509, 543, 2004.

[9] K. Nii, Y. Masuda, M. Yabuuchi, Y. Tsukamoto, S. Ohbayashi, S. Imaoka, M. Igarashi, K. Tomita, N. Tsuboi, H. Makino, K. Ishibashi, and H. Shinohara, "A 65 nm Ultra-High-Density Dual-Port SRAM with 0.71μm² 8T-Cell for SoC," *VLSI Cir. Symp. Dig.*, pp. 130-131, 2006.

[10] K. Nii, M. Yabuuchi, Y. Tsukamoto, S. Ohbayashi, Y. Oda, K. Usui, T. Kawamura, N. Tsuboi, T. Iwasaki, K. Hashimoto, H. Makino, and H. Shinohara, "A 45-nm Single-port and Dual-port SRAM family with Robust Read/Write Stabilizing Circuitry under DVFS Environment," *VLSI Cir. Symp. Dig.*, pp. 212-213, 2008.

[11] Y. Ishii, H. Fujiwara, S. Tanaka, T. Doguchi, O. Kuromiya, H. Chigasaki,Y. Tsukamoto, K. Nii, Y. Kihara and K. Yanagisawa, "A 28 nm dual-port SRAM macro with screening circuitry against write–read disturb failure issues," *in Proc. ASSCC*, pp. 1-4, 2010.

[12] Y. Ishii, H. Fujiwara, S. Tanaka, Y. Tsukamoto, K. Nii, Y. Kihara and K. Yanagisawa "A 28 nm dual-port SRAM macro with screening circuitry against write-read disturb failure issues" JSSCC, pp. 2535-2544, vol.46, issue 11, Nov. 2011.

[13] S.Tanaka, Y.Ishii, M.Yabuuchi, T.Sano, K.Tanaka, Y.Tsukamoto, K.Nii and H.Sato, "A 512-kb 1-GHz 28-nm Partially Write-Assisted Dual-Port SRAM with Self-Adjustable Negative Bias Bitline," *VLSI Cir. Symp. Dig.*, pp. 113-114, 2014.

[14] Y.Yokoyama, Y.Ishii, K.Tanaka, T.Fukuda, Y.Tsujihashi, A.Miyanishi, S.Asayama, K.Maekawa, K.Shiba and K.Nii, "40 nm Dual-port and two-port SRAMs for automotive MCU applications under the wide temperature range of -40 to 170°C with test screening against write disturb issues," in Proc. A-SSCC, pp25-28, Nov. 2014.

[15] Yen-Huei Chen, Kao-Cheng Lin, Ching-Wei Wu, Wei-Min Chan, Jhon-Jhy Liaw, Hung-Jen Liao and Jonathan Chang, "A 16nm Dual-Port SRAM with Partial Suppressed Word-line, Dummy Read Recovery and Negative Bit-line Circuitries for Low VMIN Applications," *VLSI Cir. Symp. Dig.*, 2016.

[16] Y.Ishii, M.Yabuuchi, Y.Sawada, M.Morimoto, Y.Tsukamoto, Y.Yoshida, K.Shibata, T.Sano and K.Nii, "A 5.92-Mb/mm2 28-nm Pseudo 2-Read/Write Dualport SRAM using Double Pumping Circuitry," in Proc. A-SSCC, pp. 17-20, Nov. 2016.

14nm Broadwell Xeon® Processor family: Design methodologies and optimizations

Mahesh K Kumashikar, Shridhar G Bendi, Srikanth Nimmagadda, Anup J Deka, Anil Agarwal

Intel Corporation, Bangalore, India

mahesh.k.kumashikar@intel.com, shridhar.g.bendi@intel.com, srikanth.nimmagadda@intel.com, anup.j.deka@intel.com, anil.agarwal@intel.com

Abstract

This paper provides details on the new Broadwell server product family designed on 14nm Intel process. This was the first Xeon® server product on this process node. Low power, density optimized product, Broadwell-DE, has upto 8 dual-threaded 64b Broadwell cores [1], 12MB L3 cache, 2 10GT/s Ethernet KR lanes, 2 DDR4-2400MHz memory channels and 24 8GT/s PCIe lanes. Broadwell-SP, a high performance server CPU, has upto 24 dual-threaded cores, 60MB L3 cache, 4 DDR4 channels and 48 PCIe lanes. While introducing servers with varying segment types on a lead process node provides business advantages, it entails designs ability to operate across larger window of operation especially on evolving process collaterals. This paper shows a case study of server chips on 14nm process node [2], and methodologies that were used to debug and productize. The paper will show silicon results which support the strength of the methodology.

Keywords: 14nm server, process changes, design methodology, circuit configuration

I. Introduction

Broadwell server is the 1st server processor to be productized on Intel 14nm process. Market requirements mandated a very wide process targeting: low leakage process for Xeon-D® and high performance process for Xeon®. Design methodology for circuits had to address the requirements for both process versions. In addition, critical process collateral (transistor models, passives etc.) was still maturing towards expected end-of-life goals. Design methodology responded to late changes and combined with circuit knobs to meet product requirements.

The Density Efficient (DE) family targeted high density through integration of 10G Ethernet on SoC and Platform Control Hub (PCH) on a Ball Grid Array (BGA) muti-chip package (MCP) and has 20-45W TDP. Figure 1 shows a block diagram of this MCP.

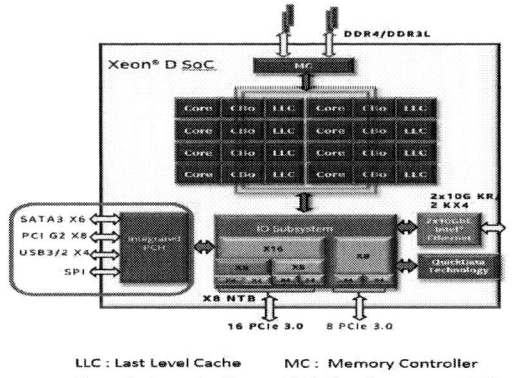

LLC : Last Level Cache   MC : Memory Controller
Cbo : Cache controller   NTB : Non Transparent Bridge

Figure 1 Broadwell-DE MCP Architecture

The Scalable Performance (SP) family targeted high performance with higher core count, DDR4 channels and services upto 150W TDP. Key IPs are shared between two families and designed to encompass wide performance and process range. Both products are introduced on advanced process node to get intrinsic power/area benefits. The design methodology comprehended wide dynamic range of shared IPs and provided adaptation to evolving process collaterals. Various circuit configuration knobs were provided to fix unanticipated process/design interaction issues. These methods helped to successfully productize IPs in server SoCs ranging from 20W-150W power envelopes.

The rest of the paper is organized as follows. Section II provides details of DE & SP family design & systems and show key optimization points. Section III provides design methodologies of analog & digital IPs and Section IV shows silicon results that showcase the strength of methodologies. Conclusions are presented in Section V.

II. Design optimization

Figure 2 shows die photograph of Broadwell-DE SoC die photograph. Low power and density optimization were achieved through integration of high power platform components like Ethernet, design optimizations and with low leakage variant of process.

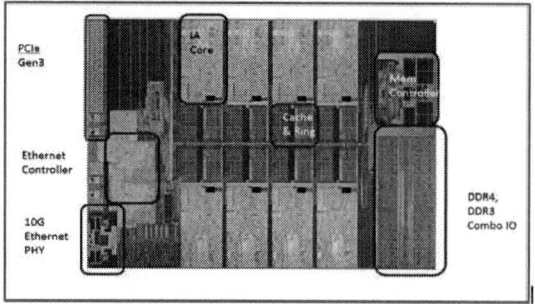

Figure 2 Broadwell-DE die photograph & key IPs

Design optimizations looked at SoC and system level aspects to balance power and performance. Typically SoC power is dominated by dynamic power, which is given by CV^2F. Here C is effective capacitance for a given application. V & F are the operating voltage and frequency. There are two major operating regions that depend SoC frequency. First region is linear power scaling region where operating voltage is limited by circuit operations (Vmin) and not frequency. In this region, power is linearly proportional to frequency. Second region is where increasing frequency requires increased voltage. In this region power is a cubic function of frequency. SoC operating point was set to optimal performance/watt and cubic power region was avoided. This resulted in low power SoC and enabled more SoCs at system level.

The product integrates 10GT/s Ethernet (KR) physical layer (PHY) for supporting backplane applications, with up to 3 connectors, exceeding the IEEE 802.3 spec to accommodate signal integrity challenges relating to package and server blade PCB and unique noise profiles. The PHY (Figure4) uses FFE for the front-end equalization and signal conditioning, adaptive gain stages (with inductive peaking for bandwidth optimization) and an extensive DFE for enhanced ISI tackling capability. Timing extraction is done using a Muller-Muller scheme and a fully digital voltage controlled oscillator based (no phase interpolator) clock data recovery. PHY is microprocessor controlled, employing modern DSP techniques for equalization control while being highly flexible to tackle extreme channels and complex interoperability issues.

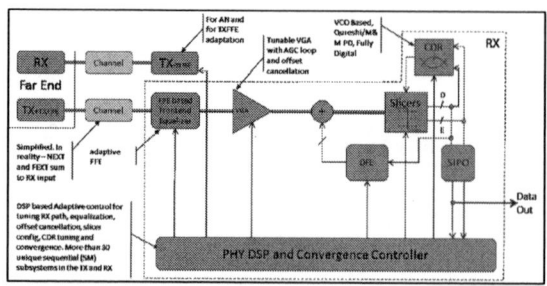

Figure 4 10G (KR) PHY architecture

Figure 5 shows block diagram of Broadwell-SP SoC with 24 cores. There are 4 DDR channels on south and increased high speed IO (PCI-E and coherent links) on the North. The cores are connected over a dual ring configuration, where each ring has 12 cores.

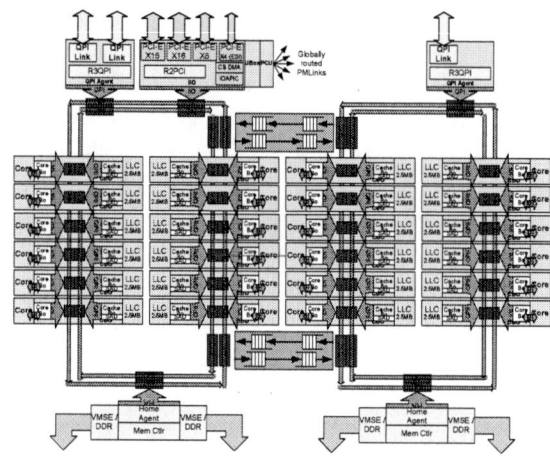

Figure 5 24-Core Broadwell-SP Architecture

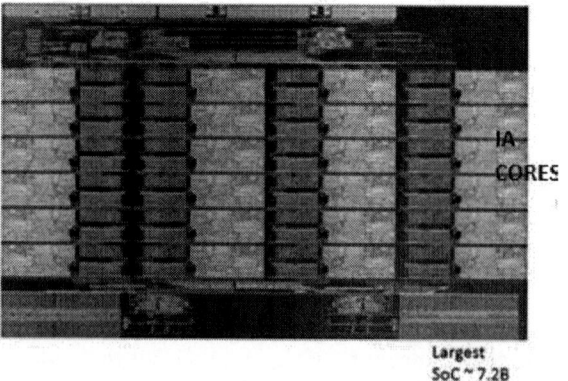

Figure 6 Broadwell-SP die photograph

This server processor is targeted to provide high performance with higher core count, higher core frequency, and higher TDP (upto 150W). The operating point for this SoC will be in the cubic power scaling region. Faster process is targeted to reduce operating voltage at the expense of leakage. The IPs used in the both products are designed once and have to cover full range of process and voltage & frequency operating conditions. Figure 6 show die photograph of Broadwell-SP SoC.

III. Design methodology

Analog IPs: Key IPs (DDR, PLL, PCIe IO etc.) were designed to be meet performance across full range of process using early collateral. In the anticipation of maturing process, design knobs were incorporated on gain, BW, & frequency. Examples include increased resiliency to passive electrical variation compensation in PLLs, 40% increase in PCI front-end gain to deal with >6-sigma low Rout corner. Digital domain PLLs are equipped with duty-cycle adjusters with a range of <=25% of phase. Duty Cycle (DC) corrector circuits are implemented on most clock domains of CPU die. The main intent of the circuit is to ensure 50% duty cycle for the clocks by correcting for any offset. A Duty Cycle Corrector (DCC) circuit consists of two parts: DC Sensing Circuit and DC Adjusting Circuit. The adjusting circuit is implemented by a current starved inverter as shown in Fig 7. The PMOS/NMOS sizes can be varied independently to either increase or decrease the Pull Up or Pull Down Path.

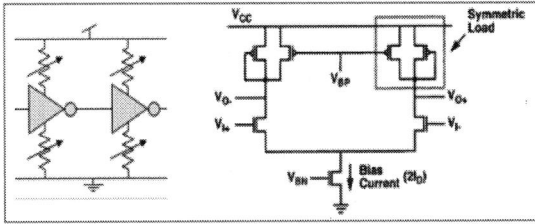

Figure 7 Duty cycle corrector and differential buffer

The DC Sensing Circuit's FSM based on the voltage levels sensed helps infer whether DC is >50% or <50%.This Duty Cycle Corrector supports three modes of operation: (a) Normal Mode: It auto-corrects DC to 50% (b) Bypass Mode: It adjusts the DC by altering the P/N strength based on programmed fuses (c) Offset Mode: It first auto-corrects the DC to 50% and then adds an offset to either increase/decrease the high phase to a desired value.

Digital IPs: Design/signoff methodology was loosely dependent on precision of library collateral. Design convergence target was based on nominal corner in the

low leakage process, while clock/cycle time derates were used to model other corners. R/RC derates were used to model interconnect corners. As the process matures, generating library collateral has such a long lead time (Si→Collateral→Library→ design), we relied on heuristic methods to model silicon observations on design and fixed the outliers. Heuristics were based on analysis of miscorrelation between silicon/simulation and covered relative/absolute transistor speed changes and RC variation over time. Goal of heuristic based methodology is to fix hold timing violations and obtain product yield consistent with expectations. Thus heuristics were used to model changing silicon instead of generating updated library collateral. Table1 shows the list of heuristics and their usage.

Silicon/Simulation Miscorrelation	Heuristic Introduced	Design fixes made
Transistor speed (Slow down, Speed Up from model)	Cell delay scaling - clock, data Assessment of timing profile Update cycle time target for yield	Improve margins for vulnerable paths
Interconnect changes (R, RC, C change from model)	RC delay scaling - clock, data Assessment of timing profile	Improve margins on flagged paths Re-center of interconnect dependent 'LC' PLLs
Frequency limiters (String of slow transition devices)	Added delay penalty on high transition cases - Update in library	Improve path margins for cells Fix transition times
P Vs. N drive changes (Asymmetric delay from model)	Model based on library Restrict skewed gates	Change Vt type Improve noise margins Use balanced P/N ratio devices
Analog templates (Delta in median values and variation)	NA	Increased range to handle variations based on early testchip data

Table1: List of heuristics developed.

Two design steppings were needed to co-optimize design/silicon as shown in Figure 8. First stepping accounted for heuristics based on simulation and second stepping accounted for silicon based observations. Fully Integrated Voltage Regulator [3] provided per part voltage control for handling large transistor speed variations.

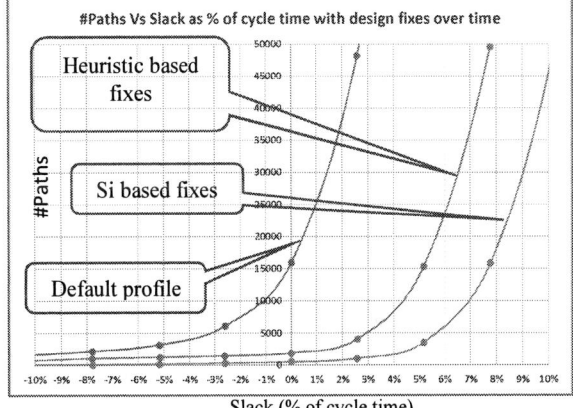

Figure 8 Timing wall improvement of high frequency blocks

IV. Silicon results

We observed several design/process interplay issues. Critical issues gating productization were handled via design knobs. Examples below:

Example1: Figure 9 shows Shmoo plot that illustrates a limiting path at frequencies. Debug exposed large number of cases of long RC paths as the primary root-cause. This observation led us to modify derates and heuristics to expose these problems and fixed the design.

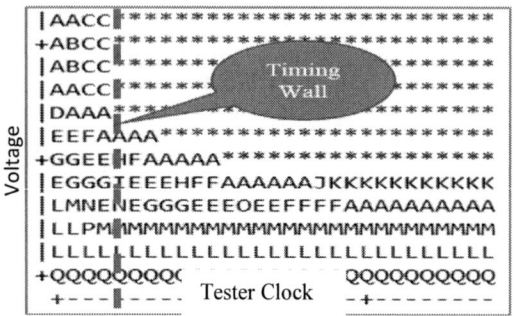

Figure 9 Shmoo Plot showing timing wall

Example2: Xeon® employs a latch based high frequency, multi-cycle transparency Ring bus that's challenging to converge. Despite robust design methodology, Silicon frequency didn't meet product targets in the largest die configuration. Design analysis highlighted latch based long multi-cycle paths as the limiters and pointed to logic imbalance between high & low clock phases as the root cause. Duty cycle corrector circuit was used to increase low clock phase as a circuit knob and was eventually productized. Figure 10 shows voltage improvement across process speed with different duty cycle settings.

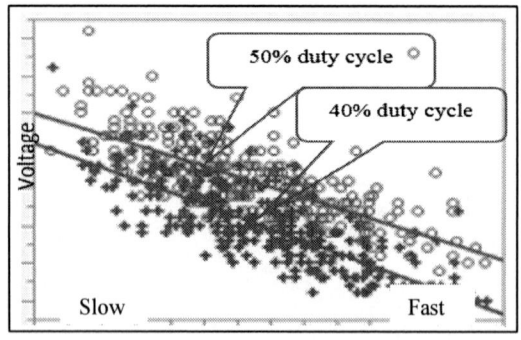

Figure 10 Voltage improvement across process range

Example3: Level shifters are critical to achieve high power efficiency by allocating block specific voltage & frequencies. Figure 11 shows the maximum voltage delta tolerable across level shifters based on early collateral. Early silicon was slower, which required higher voltage delta than allowable, thus violating functional limits. These problems were mitigated by programming on-die power/voltage management circuitry.

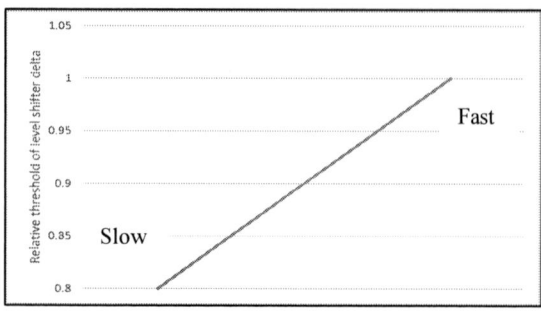

Figure 11 Relative Level shifter delta across process range

V. Conclusions

Using the methodologies and circuit knobs mentioned in the paper, we successfully productized 14nm server earlier in process maturity ramp. This paper demonstrates the need to develop circuit/process co-optimization as a means to provide quality as well as time-to-market. We expect such methods will become critical for all future nodes for early product ramps. While the exact heuristics will be process/circuit dependent, list of heuristics mentioned here are very generic and are applicable to various SoCs.

Acknowledgement

The authors thank the Intel Broadwell server worldwide teams for their creativity, passion, and dedication in bringing this product to market.

References:

[1] Nalamalpu A, et al., "Broadwell: A Family of IA 14nm Processors", VLSI Symposium 2015, pp. C314-C315.

[2] Natarajan S, A 14nm Logic Technology Featuring 2nd Generation FinFET., IEDM 2014, pp71-73

[3] E.A. Burton, et al., "FIVR–Fully .. " IEEE Applied Power Electronics Conf., pp. 432-439, 2014

A Dual-Axis MEMS Vibratory Gyroscope ASIC with 0.0061°/s/√Hz Noise Floor over 480 Hz Bandwidth

Zhichao Tan, Khiem Nguyen, Jeff Yan,

Howard Samuels, Shane Keating, Paul Crocker and Bill Clark

Analog Devices, Inc. Wilmington, MA, USA

zhichao.tan@analog.com

Abstract— This paper presents a high performance dual-axis (pitch and roll) MEMS vibratory gyroscope readout ASIC which converts angular rate information to digital output. Two signal processing chains surrounding the MEMS sensor are implemented, namely the drive channel and the sense channel. The drive channel drives the sensor to resonate at its resonant frequency, which produces a velocity of the sensor disc to generate the Coriolis force during angular rotation. The sense channel employs a low noise trans-impedance amplifier (TIA) followed by a demodulator which down-converts the angular rate input signal from the resonant frequency to baseband. Two switched-capacitor (SC) 2-1 MASH delta-sigma ADCs convert the input angular rate from the pitch and roll arises to digital output. The reference of the ADC is also demodulated from the sensor output to cancel out supply voltage dependence. The whole ASIC including the high-voltage MEMS sensor driver, digital-filter, on-chip regulator and temperature sensor is fabricated in a 0.18µm CMOS technology with an area of 7.3 mm². The design achieves a noise floor of 0.0032°/s/√Hz and 0.0061°/s/√Hz in full-scale input ranges of 500°/s and 2000°/s, respectively, over a 480Hz signal bandwidth. The bias instability is measured as 2.5°/h at input range of 500°/s. The whole ASIC consumes 7mA from a 3V supply.

Keywords— *MEMS Gyroscope; Dual Axis; Sense Channel; Drive Channel; 2-1 MASH Delta Sigma ADC*

I. INTRODUCTION

Gyroscopes which measure angular rate are widely used in motion-sensing applications such as automotive safety, gaming, navigation and so on [1]. High performance gyroscopes are particularly important for many new industry applications, such as electronic stability control (ESC) systems which improve vehicle safety [2]. Another good example is the optical image stabilization (OIS) system, in which a high-performance gyroscope detects the motion of the camera and compensates for movement while the picture is taken [3].

There are two primary signal-processing subsystems in a MEMS gyroscope readout ASIC, namely the drive and sense channels. The drive channel drives sustained oscillation of the sensor disc at resonance thereby creating velocity, which combines with rotation rate to generate Coriolis acceleration. The sense channel senses the Coriolis acceleration centered at the resonant frequency, demodulates that to baseband to extract the rotation rate information, and finally converts that to either

analog or digital output [4~6]. The drive channel is usually implemented in closed-loop to form sustain an oscillation of fixed amplitude. The sense channel can either be open or closed loop. In the case of closed-loop topology, electronic force feedback is used to null Coriolis acceleration. An open-loop topology, on the other hand, directly converts motion-induced capacitor changes to voltage or current for further processing. Open-loop sense channel topology is more popular due to its stability, robustness and straight forward implementation.

This paper describes a dual-axis (pitch and roll) MEMS vibratory gyroscope readout ASIC. The readout ASIC includes complete signal chains of the drive channel, sense channels, and other signal processing channels. An on-chip charge pump is also implemented to create a high-voltage supply for the sensor driving signal. The high voltage has two benefits: to drive the MEMS sensor disc with larger amplitudes, and to yield larger displacement sensing signals in both the drive and sense channels which lead to high signal to noise ratio (SNR). A regulator and temperature sensor are also integrated to give the analog building blocks better power supply rejection ratio (PSRR) and temperature monitoring capability.

This paper is organized as follows: Section II describes the gyroscope ASIC design and some important circuit building-block implementations. Section III presents measurement results. The paper will be concluded in Section IV.

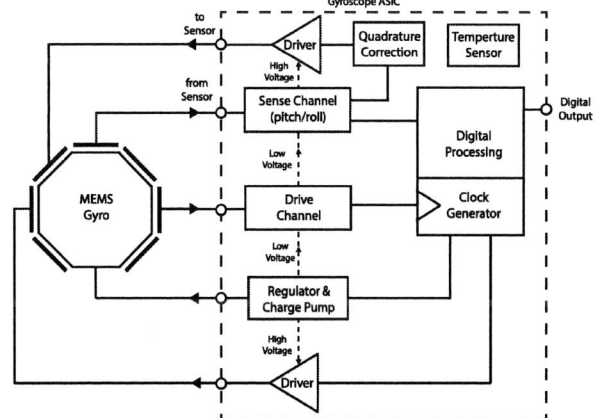

Fig. 1. Gyroscope ASIC top level diagram

II. ASIC DESIGN

Fig. 1. Shows a top-level diagram of the proposed gyroscope ASIC. The ASIC mainly consists of one drive channel, which drives the sensor at its resonant frequency, and two sense channels for pitch and roll sensing, which convert physical rotation information into digital output. Detailed system and circuit design are presented below.

A. Sense Channel Design

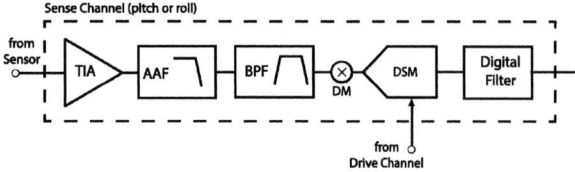

Fig. 2. Sense channel and drive channel detailed diagram

Fig. 2 shows a block diagram of the sense channel. There are two identical such sense channels in the proposed ASIC for the pitch and roll axes. Each sense loop contains a trans-impedance amplifier (TIA) to convert a displacement-induced current from the sensor to a voltage. To convert such small currents, the input impedance needs to be in the hundreds of MΩ. Such impedance is achieved by using PMOS devices in weak inversion region. This saves lots of area compared with standard solution.

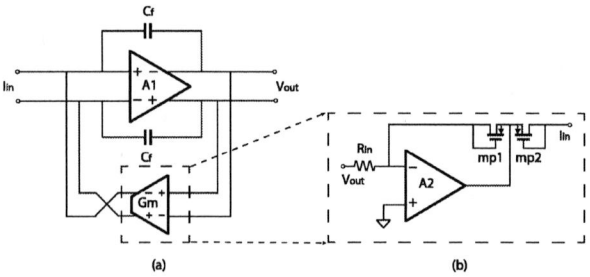

Fig. 3. Transimpedance amplifier (TIA) used in the design.

Fig. 3 shows the diagram of the trans-impedance amplifier used in the design. The AC signal gain is set to be $1/(sC_f)$, which is around 22 MΩ. The low-frequency components pass through the Gm cell, setting the DC bias for amplifier OTA_1. To balance the DC handling capability and noise performance, a MOS pseudo-resistor current divider [7] is used as depicted in Fig. 2(b). The advantage of this gm structure is high linearity and rail-to-rail swing due to negative feedback. Including the resistor R_{in}, the equivalent transconductance is:

$$g_{m,dc} = \frac{I_{out}}{V_{in}} = \frac{1}{R_{in} \dfrac{W_{mp1}}{W_{mp2}}} \tag{1}$$

If R_{in} = 8 MΩ, W_{mp1}/W_{mp2} =100, $G_{m,dc}$ = 1/800 MΩ. However, the noise from R_{in} is attenuated by $(1/100)^2$, with an equivalent R_{noise} of 80 GΩ. Thus the noise is dominated by mp2 which operates in either subthreshold or drain-well diode mode depending on the input low-frequency current direction. Choosing a long length, its noise becomes not dominant. The

capacitor C_f can be selected from either 100fF or 200fF depending on the input range setting.

An anti-aliasing filter (AAF) follows the TIA to filter out the unwanted out-of-band high-frequency signals that might alias back in to baseband by the sampling action of the subsequent band-pass filter. The anti-aliasing filter is a continuous-time, low-pass filter. Its cut-off frequency is determined by the pole formed by a resistor and a capacitor. Signal gain is also added at this stage to relax the switch-capacitor sampling noise (KT/C) requirement of the following band-pass filter.

The subsequent band-pass filter (BPF) is designed to filter out unwanted signals outside the band of interest close to the resonant frequency. Thus, the center frequency of the band-pass filter is at f0, or approximately 72 kHz. The BPF is a four-pole switched-capacitor bi-quad circuit. The sampling frequency is 6.91 MHz, which is 96 times the center frequency. The gain of the BPF can be trimmed for different input range selections.

A demodulator (DM) is placed between the band-pass filter and the ADC to demodulate the high-frequency input signal down to base band. Demodulation to baseband prior to the ADC eases the requirements for the ADC. The ADC is designed using a switched-capacitor (SC) MASH 2-1 delta-sigma modulator. To suppress the quantization noise down to the required level, 3rd-order noise shaping is required. Compared with a single loop design, the 2-1 MASH topology is chosen for its higher stability and larger input range. A folded-cascade amplifier ensures sufficient DC gain of the integrator, which minimizes leakage of quantization noise of the 1st stage. The input of the second stage is taken from the output of 2nd integrator. This arrange benefits from fully feed-forward structure of the 1st stage, in which the loop filter only processes the quantization noise [8]. The sampling clock for the ADC is 144 kHz, which is twice the resonant frequency. The oversampling ratio (OSR) is 128, which leads to an input bandwidth of 562.5 Hz. Fig. 4 shows the diagram of the 2-1 MASH delta-sigma modulator used in this design. Both the 1st stage and 2nd stage modulators have 1-bit quantizers. The digital combination logic and digital filter are also integrated to process output of the delta-sigma modulator.

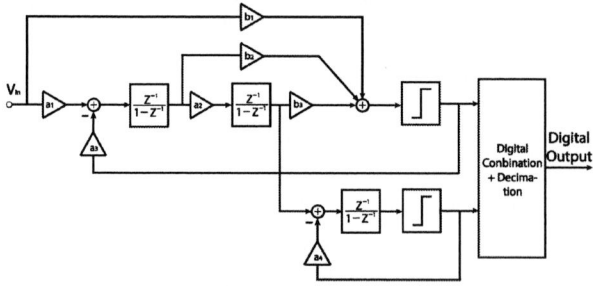

Fig. 4. 2-1 MASH Delta-Sigma Modulator.

B. Drive Channel Design

Fig.5 (a) shows the detailed diagram of the drive channel in the proposed ASIC. In the drive channel, an on-chip phase-lock loop (PLL) produces a clock signal which is level shifted to 31V. This is sufficient to electrostatically actuate the sensor mass. The 31V is generated on-chip by a high-voltage charge pump. The high-voltage clock signal then drives the MEMS sensor mass, whose motion is detected by a set of variable capacitors, which yields a displacement charge when biased at 31V. That charge creates an output current from the sensor which is sensed by the trans-impedance amplifier (TIA) in the drive channel. The TIA converts this charge into a sinusoidal voltage, and passes it to a 2nd-order band-pass filter to filter out unwanted out-of-band signal. The band-pass filtered output sinusoidal signal is then converted to a digital clock signal through a comparator which has the same frequency as the drive stimulus, but a phase which is determined largely by the relative spacing separation between the drive frequency and the resonant frequency of the sensor. Using this signal as the reference to the same PLL, the driving clock signal from the PLL and the motion of the sensor form a loop which settles at the resonant frequency of structure at approximately 72 kHz (f_0). The PLL produces clocks for entire ASIC, including both the analog and digital sections. The highest clock produced by PLL is around 27 MHz, which is used in the digital section.

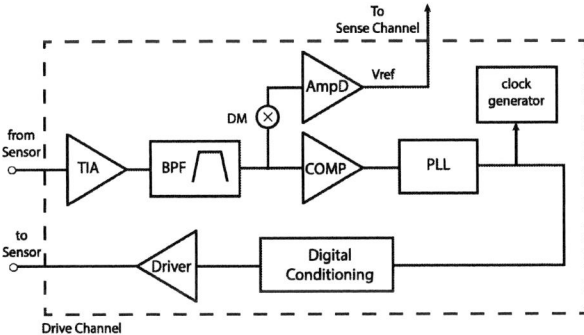

Fig. 5. Detailed drive loop diagram.

The drive loop also provides the reference voltage for the on-chip analog to digital converter (ADC) in sense channel. As shown in Fig. 5, the output of the band pass filter is also picked up by a block called amplitude detector (AmpD) which captures the amplitude of the input sinusoidal signal and converts it to a DC voltage with a preset gain. This DC voltage will be used as the reference voltage of the ADC. Compared to using a static voltage reference for the ADC, using the sensor drive amplitude can improve the sensitivity accuracy of the gyroscope during transient events. This signal can also be used to monitor whether the resonator amplitude is within prescribed limits.

C. Quadrature Correction Loop

The demodulation process discussed above yields the in-phase signal content at the resonator frequency, which contains the desired input rotation rate information. A quadrature demodulation of the same signal yields an error signal, Quadrature error, resulting from non-ideal motion of the sensor resonator with respect to the Coriolis sense detectors. This error signal can be many times larger than the desired in-phase signal. Thus, it is desirable to cancel this error at source.

Shown in Fig. 6, a quadrature correction loop is included to track and correct the quadrature error. It samples the quadrature error at the output of the bandpass filter in the sense loop, and accumulates it on an integrator. The integrator output is amplified by a high-voltage amplifier and fed back to the sensor to correct the quadrature error via quadrature correction electrodes on the MEMS sensor.

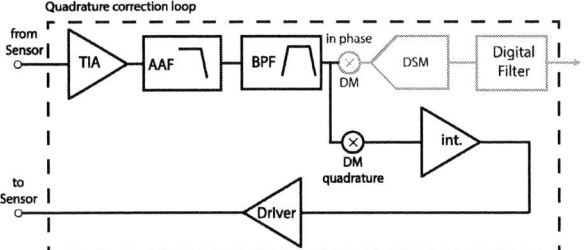

Fig. 6. Quadrature correction loop.

III. MEASUREMENT RESULTS

The proposed dual-axis gyroscope ASIC is fabricated in 0.18μm CMOS IC technology. The ASIC includes all necessary function blocks to give the digital output of the input angular rate. Fig. 7 shows a chip photo of the implemented ASIC, which occupies an area of 7.3 mm². The ASIC can work with a supply voltage from 2.7V to 5V. An integrated LDO supplies 1.8V to all the major analog and digital building blocks. The whole ASIC draws 7 mA during measurement mode. For the measurement ranges of 500°/s and 2000°/s, the ASIC achieves noise floor of 0.0032°/s/√Hz and 0.0061°/s/√Hz, respectively, in an output signal bandwidth of 480 Hz.

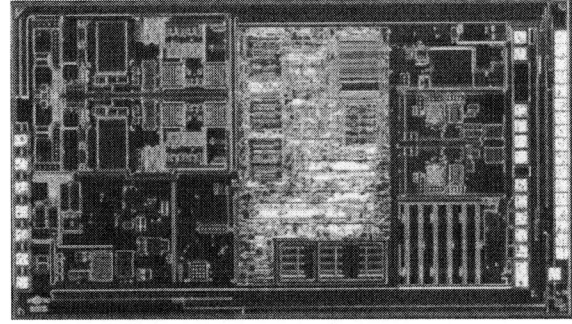

Fig. 7. Micrograph of the prototype gyroscope ASIC.

Fig. 8 shows the noise floor of the gyroscope ASIC at 2000°/s mode. Fig. 9 shows the measured Root Allan Variance (RAV) of the proposed design. It achieves 2.5°/h stability in the 500°/s mode. A performance summary and comparison with other recently published state-of-the-art works are presented at table 1.

IV. CONCLUSION

A dual-axis (pitch and roll) MEMS vibratory gyroscope readout ASIC has been presented. There are several signal channels surrounding the MEMS sensor. The sense channel converts a small output current from the sensor into voltage, and passes it to an anti-aliasing filter and a band pass filter. The output of the band pass filter is demodulated to baseband and fed to an analog-to-digital converter for digitization. The drive channel is a closed-loop signal path around sensor, which drives the sensor at its resonant frequency, which is around 72 kHz. The motion of the sensor creates a velocity of the sensor disc to generate a Coriolis force during angular rotation. The design achieves noise floor of 0.0032°/s/√Hz and 0.0061°/s/√Hz in the full-scale input range of 500°/s and 2000°/s, respectively, over the signal bandwidth of 480Hz. The bias instability is measured as 2.5°/h at input range of 500°/s. The whole ASIC consumes 7mA from a single 3V supply, and occupies an area of 7.3 mm^2.

Table 1. Performance summary and comparison with other state-of-the-arts

	This work		[3] JSSC 2016	[4] ISSCC 2015	[5] ISSCC 2011	[6] JSSC 2011
Supply (V)	3		3	3	3	3
Current (mA)	7		8.8	0.85	6.1	2.2
No. of axis	2		3	3	3	2
Signal BW (Hz)	480		80	520	50	160
Full scale (°/s)	500	2000	630	2000	2000	1000
Noise floor (°/s/√Hz)	0.0032	0.0061	0.0038	0.007	0.03	0.028/ 0.032 (X/Y)
Bias instability (°/h)	2.5	5.5	1.2	N/A	N/A	18/22 (X/Y)

ACKNOWLEDGMENT

The authors would like to thank their colleagues from the High Performance Inertial sensor group at Analog Devices Inc. (both in Wilmington and Greensboro) for their help during design, layout and chip evaluations.

REFERENCES

[1] J.Marek, "MEMS for automotive and consumer electronics," in IEEE Int.Solid-State Circuits Conf. (ISSCC) Dig. Tech. Papers, Feb. 2010, pp. 9–17.

[2] Masten, Michael K. "Inertially stabilized platforms for optical imaging systems." IEEE Control Systems 28, no. 1 (2008): 47-64.

[3] Balachandran, Ganesh K., Vladimir P. Petkov, Thomas Mayer, and Thorsten Balslink. "A 3-axis gyroscope for electronic stability control with continuous self-test." IEEE Journal of Solid-State Circuits 51, no. 1 (2016): 177-186.

[4] C. Ezekwe, W. Geiger, and T. Ohms, "27.3 A 3-axis open-loop gyroscope with demodulation phase error correction," in Proc. IEEE Int. Solid- State Circuits Conf., 2015, pp. 1–3.

[5] Prandi, Luciano, et al. "A low-power 3-axis digital-output MEMS gyroscope with single drive and multiplexed angular rate readout." Solid-State Circuits Conference Digest of Technical Papers (ISSCC), 2011 IEEE International. IEEE, 2011.

[6] L. Aaltonen et al., "A 2.2 mA 4.3 mm ASIC for a 1000 /s 2 – Axis capacitive micro-gyroscope," IEEE J. Solid-State Circuits, vol. 46, no. 7, pp. 1682–1692, 2011.

[7] F.Gozzini, G.Ferrari and M. Sampietro, "Linear transconductor with rail-to-rail input swing for very large time constant applications", Electron.Lett.,vol.42,no.19,pp 1069-1070,Sep.2006.

[8] J. Silva, U. Moon, J. Steensgaard, and G. Temes, "Wideband low distortion delta-sigma ADC topology," Electron. Lett., vol. 37, no. 12, pp. 737–738, Jun. 2001.

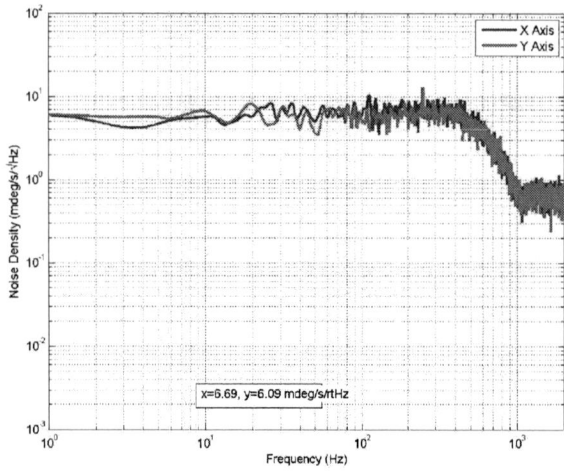

Fig. 8. Measured Noise floor of proposed gyroscope ASIC.

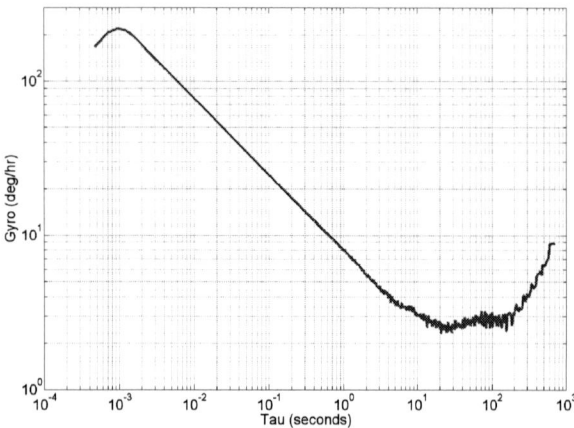

Fig. 9. Measured Bias instablity of propose gyroscope ASIC

Chaos, Deterministic Non-Periodic Flow, for Chip-Package-Board Interactive PUF

Noriyuki Miura, Masanori Takahashi, Kazuki Nagatomo, and Makoto Nagata

Kobe University, Japan
miura@cs.kobe-u.ac.jp

Abstract—This paper presents a new concept of PUF based on a chaotic behavior. Chaos is essentially not a random phenomenon but a deterministic non-periodic flow which can be utilized to extract reproducible unique ID entropy. The strong parametric sensitivity of the chaos guarantees ID variety and unclonability over unpredictable manufacturing variations. An undesirable disturbance due to dynamic parametric fluctuations can be removed by circuit-level feedback compensation together with mixed-signal post-processing for stable and long ID reproduction. The proposed chaos PUF is silicon-prototyped in 0.18µm CMOS for the proof of concept. A compact 750µm² chaos oscillator PUF successfully reproduces 512-bit unique ID with >98% stability. The proposed chaos PUF integrated with an inductive-coupling coil demonstrates chip-package-board interactive PUF for the first time with extended ID traceability including not only chip fabrication but also package and board assembly steps for advanced secure supply chain.

I. INTRODUCTION

Worldwide today, trillions of semiconductor-based computing devices are pervasively distributed everywhere, playing a crucial role to manage, process, and control critical information in essential fields. Counterfeiting of such device hardware is one of the most serious security threats against semiconductor industries, which may cause destructing the foundation of the supply chain [1]. A semiconductor-based Si Physically Unclonable Function (Si PUF) [2-5] is one best-known countermeasure against this hardware counterfeiting. Unpredictable manufacturing variations in the semiconductor process are amplified by utilizing e.g. memory meta-stability [3], matched delay line arbiter [4], or ring oscillators [5] to generate a physically-unclonable unique ID of an IC chip.

One technical challenge lies in extending ID traceability across different fabrication steps in the supply chain. All above-mentioned Si PUFs are only effective to trace just IC chip IDs. In order to enhance the security level of the supply chain, extended traceability including package and board assembly steps is needed as de-capsulated chip recycling or package/board-level manipulation could often be low-cost attack scenarios of the malicious counterfeiting [1] (Fig.1 (a)). Separate identifications of PUFs independently-integrated on the chip, package, and board are not secure enough. Such simple orthogonal combination can be bypassed as chopping and reconfiguring the essential parts of the PUF cores are

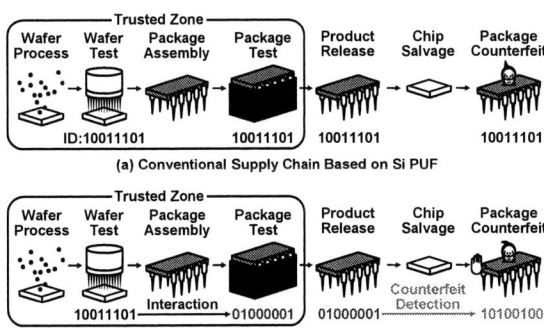

(a) Conventional Supply Chain Based on Si PUF

(b) Proposed Supply Chain Based on Chip-Package-Board Interactive PUF

Fig.1 Supply chain security based on (a) conventional Si PUF and (b) proposed chip-package-board interactive PUF.

relatively easy for the malicious attackers [1]. The key to enhance the security level is introducing mutual and thus complex chip-package-board interaction into PUF for reinforcing its physical unclonability, namely chip-package-board interactive PUF for a secure supply chain (Fig.1 (b)).

A possible approach for introducing such interaction is to implement the PUF circuit with wired connections between the chip and the package/board. However, its hardware cost penalty is non-negligible. Wire bonding (Fig.2 (a)) is common and inexpensive while the associated on-chip layout area penalty is high. A dedicated area for the bonding pad and IO/ESD circuits completely occupy ~100,00µm²/pin on-chip layout area resources. Since this area for the wire bonding hardly scales with the process scaling, the penalty becomes more and more significant in the scaled technology. The area penalty of 100,00µm²/pin is equivalent to ~1k GE (Gate Equivalent of 2NAND) in 0.18µm CMOS and is raised up to 21k GE in 28nm. Similar penalty is burdened with area-bump packaging. An advanced assembly technique such as Thru-Silicon Via (TSV) [6] may help to suppress this area penalty (Fig.2 (b)). However, an additional and expensive fabrication step is needed, finally resulting in huge cost increase.

In this paper, a complete circuit solution fully-compatible with a standard manufacturing flow is proposed for a low-cost counterfeiting countermeasure with the extended ID traceability (Fig.2 (c)). Instead of wired connections, wireless connectivity is introduced in the chip-package-board

978-1-5386-3179-9/17 $31.00 © 2017 IEEE

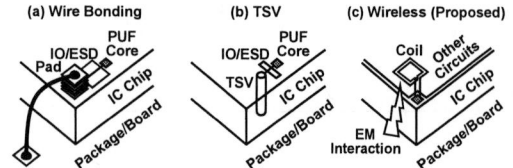

Fig.2 Chip-package-board interactive PUF by using (a) wire bonding, (b) TSV, and (c) wireless connectivity.

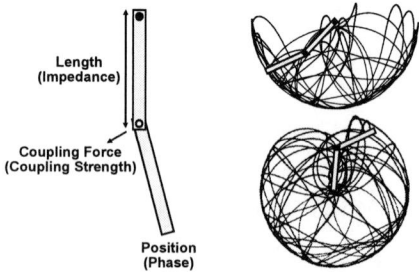

Fig.3 Conceptual illustration of chaos PUF in double pendulum analogy.

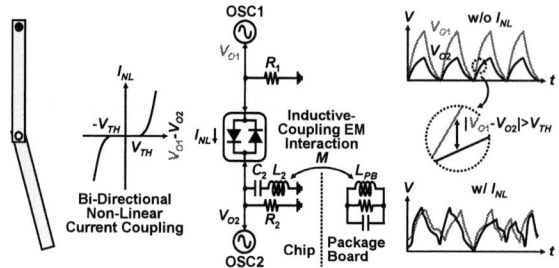

Fig.4 Simplified circuit schematic of chaos PUF.

interactive PUF. The Electro-Magnetic (EM) interaction through the coil inductively-coupled across the chip, package, and board aggressively includes the package/board assembly form with their manufacturing variations into the PUF ID generation. The on-chip area penalty is significantly suppressed because (1) the coil is drawn by using existing on-chip IC metal interconnects of only one metal layer, (2) the coil can be placed anywhere over other on-chip core circuits, (3) No ESD protection is needed because the coil is protected under an existing passivation layer of IC and thus no possible physical contact with an assembly machine or a human, and (4) the dedicated part is only the PUF core circuit but it is anyway needed in all other possible wired options.

For the PUF circuit to extract entropy from the chip-package-board assembly variations, chaotic oscillation is used as the EM interaction becomes stronger at a high-frequency signaling regime. The rest of the paper describes how to apply chaos to PUF in Section II. The detail circuit implementation is explained in Section III. Experimental results demonstrated using a 0.18μm CMOS test chip will be presented in Section IV. Finally, concluding remarks will be drawn in Section V.

II. CHAOS PUF

Chaos is an well-known physical phenomenon ever-utilized for True Random Number Generator (TRNG) in the IC field [7-9]. However, as Dr. Lorenz entitled to his legendary paper in 1963 [10], chaos is a "deterministic non-periodic flow" and is not essentially random. The randomness is observed only because the chaotic circuit in TRNG is designed to be extremely sensitive to its initial condition and run-time noise. By removing such dynamic random disturbance, a deterministic so that reproducible, and also non-periodic thus unique identity can be extracted from the chaotic behavior.

A double pendulum, the most well-known physical chaotic system, is used as an analogy to explain the concept of the chaos PUF. As illustrated in Fig.3, the trajectory of the double

pendulum strongly depends on its physical parameters such as each pendulum length, coupling force, and their initial position. Since the chaos provides non-periodic flow, the trajectory becomes clearly distinguishable with slight difference in the length and the coupling force even if the motion started from completely the same initial position (Fig.3). A dynamic variation or drift of each parameter could disturb the trajectory however without it the trajectory becomes identical repeatedly at every trials because the chaotic behavior is essentially deterministic by its governing differential equations. This double pendulum can be electrically implemented in IC by using coupled electric oscillators. The electrical load impedance, coupling strength, and phase of the oscillators can be seen like pendulum length, coupling force, and position. The electrical oscillation signal is regarded as the trajectory and can be used to generate reproducible and unique PUF ID.

III. CIRCUIT DESIGN

The proposed chaos PUF is implemented in a micro-scale electric IC. Fig.4 depicts a simplified circuit schematic of our proposed hybrid coupled chaos oscillators for the chip-package-board interactive PUF. The circuit topology is based on classic coupled chaos oscillators [11] with bi-directional non-linear current conduction (denoted as I_{NL} in Fig.4). I_{NL} can be simply implemented with two cross-coupled diodes. First, suppose two electric oscillators OSC1 and OSC2 free-running without I_{NL}, the output oscillation signals V_{O1} and V_{O2} become complete periodic waveforms (top right in Fig.4). However, by a load capacitance C_2 and resistance R_2 ($<R_1$), the slew rate and voltage swing of V_{O2} is reduced. This discrepancy compared to V_{O1} generates transient voltage difference $|V_{O1}-V_{O2}|$ to be larger than the diode threshold voltage V_{TH}. By coupling V_{O1} and V_{O2} through the cross-coupled diodes, the activated non-linear current I_{NL} introduces the chaotic behavior (bottom right in Fig.4). The additional inductance L_2 by the on-chip coil explicitly modulates the impedance of C_2 at the V_{O2} transitions. L_2 is inductively coupled through mutual inductance M with a parasitic inductance L_{PB} in any package/board assembly-related materials (e.g. lead frame, glue, or mold resin). This EM interaction between L_2 and L_{PB} finally includes the package/board assembly conditions into the chaotic behavior. The relatively large manufacturing variations in the package and board assembly produce unique chaotic oscillation waveforms for the PUF ID generation.

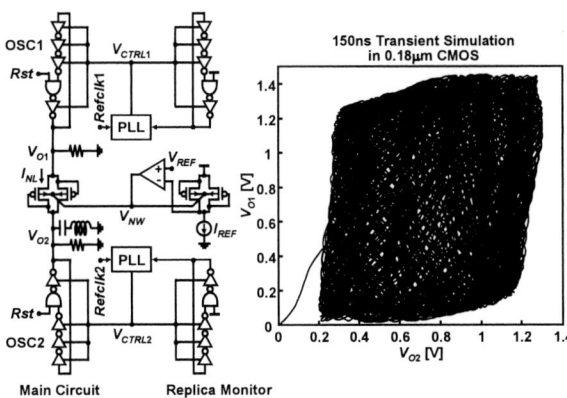

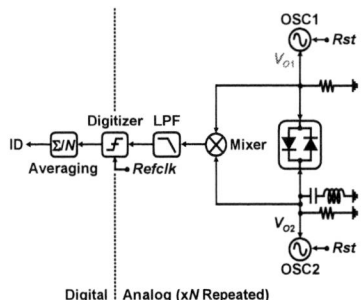

Fig.5 Detailed circuit schematic with replica feedback compensation and simulated chaos attractor plot.

Fig.5 depicts the detailed circuit schematic. The oscillator core (OSC1, OSC2) is a simple gated ring oscillator triggered by the reset pulse *Rst*. The non-linear cross-coupled diodes are simply composed of diode-connected PMOSs. All the circuits are completely standard CMOS compatible. A circuit-level technical challenge is on how to compensate for the dynamic disturbance in the chaotic behavior which deteriorates the reproducibility of the chaos PUF. A slow disturbance due to the temperature and supply voltage drift is compensated by using a replica feedback (Fig.5). A simple constant frequency feedback similar to PLL is performed on the oscillator through the control voltages $V_{CTRL1,2}$. Another critical parameter to govern the chaotic behavior is V_{TH} of the diode which is similarly compensated by a replica constant current feedback through the body biasing. The proposed circuit is designed in 0.18μm standard CMOS. The center oscillation frequency is adjusted at ~1.2GHz. Fig.5 presents the simulated V_{O1} vs. V_{O2} plot. A clear chaos attractor is successfully drawn as a graphical representation of the ID entropy source such like the trajectory in the double pendulum analogy.

A digital unique ID code is produced from the chaos attractor by a simple mixed-signal post processing (Fig.6). An analog mixer multiplies V_{O1} and V_{O2} to convert the 2D entropy of the chaos attractor information into 1D time-sequential information which is low-pass filtered to remove unnecessary noise components and digitized by a reference clock *Refclk* at ~300MHz. The digital codes are repeatedly captured by many-times negating and asserting the trigger signal *Rst*. Finally in a digital domain, a temporal majority voting is performed to decide the final output ID. This averaging process removes instantaneous run-time dynamic disturbance and enhances the reproducibility of the PUF ID generation. This feedback compensation and the averaging post-processing only remove the dynamic disturbance. Other static manufacturing variations remain for producing the unique interactive PUF ID.

IV. TEST-CHIP MEASUREMENT

A test chip was fabricated in 0.18μm CMOS (Fig.7 (a)). A 300μm-side 4-turns on-chip coil is integrated for the EM interaction. Through the near-field inductive-coupling, the coil

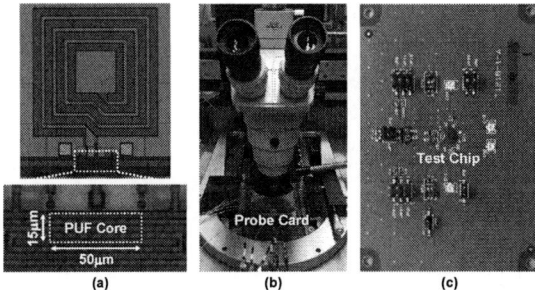

Fig.6 Mixed-signal post-processing flow for digital ID generation.

Fig.7 (a) Test-chip die photo, test setup (b) before, and (c) after assembly.

covers the interactive range of 0.5mm in vertical which is long enough to interact with the package materials and the board in a Chip-on-Board (CoB) assembly. This range can be freely adjusted by changing the coil size depending on the assembly option. The coil size is large in the footprint however again it can be placed anywhere over other circuits and does not occupy the 300μm-side complete layout area resources. The hardware cost is only metal interconnect tracks required for the coil wire. Furthermore, by employing an XY-coil layout style [12], the coil can be hidden in the sea of logic interconnects. The PUF core circuit only occupies an active layout area of 750μm² which is <0.1k GE in 0.18μm CMOS.

The test chip was evaluated in two steps before and after assembly for the proof-of-concept. First, the bare chip was tested before assembly with a probe card (Fig.7 (b)) and then tested after the CoB assembly (Fig.7 (c)). The test chip was directly die-bonded on the board ground plane with epoxy glue and was resin molded to complete packaging. The on-chip coil interacts with the ground plane, glue, and mold. Both samples were tested through the same ID generation flow. The digital averaging process was performed on an off-chip PC.

Fig.8 presents a snapshot of the chaos attractor measured before the assembly. A clear and stable chaos attractor was successfully observed with the aid of the circuit-level feedback. It was also confirmed that the averaging process in the oscilloscope further stabilized the attractor, which implies the effectiveness of the post process. After the assembly, a stable yet different attractor was observed, which partially proves the proposed concept. In order to see the feasibility of the proposed PUF, following three quantitative evaluations were conducted to the final output digital ID code.

978-1-5386-3179-9/17 $31.00 © 2017 IEEE

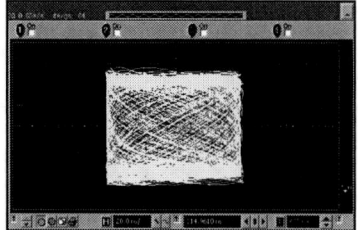

Fig.8 Snapshot of chaos attractor before assembly.

1) Uniqueness before and after assembly: First, an output ID code from the same sample before and after assembly were measured. A 1024-bit code was obtained by the averaging process over 64 ID-generation trials. The codes before and after were XORed for comparison. Fig.9 (a) presents the barcode plot of the XORed ID where the black and white stripe denotes digital bit "1" and "0". The ID difference can be clearly observed. The bottom graph presents an average Hamming Distance (HD) per bit vs. the number of bit sequence. It soon converges to 0.5, which confirms the code uniqueness before and after assembly, and guarantees unpredictability so that unclonability of ID after assembly even if the attacker knows the chip ID before assembly.

2) Uniqueness between different samples: Multiple measurement samples were fabricated and assembled. The test chip fabrication and assembly steps are completely identical. The ID codes were obtained under the same measurement condition and the ID generation flow. Fig.9 (b) presents the measured data. The uniqueness between two samples were clearly observed both graphically and quantitatively.

3) Reproducibility in the same sample: Finally, the reproducibility of the proposed PUF was evaluated. The same sample after assembly was multiple-times measured to evaluate the similarity and stability in its ID sequence. Fig.9 (c) presents the measurement results. The barcode plot clearly exhibits good reproducibility at the beginning of the generated ID sequence. The reproducibility is gradually degraded according to the duration time as this chaos PUF accumulates its disturbance effect in nature. However, as plotted also in Fig.9 (c), BER is <2% up to 512-bit ID sequence.

V. CONCLUSION

This paper presents a low-cost PUF-based counterfeiting countermeasure with an extended traceability. An inductive-coupling introduces wireless EM interaction between the chip and the package/board into PUF. The wireless scheme greatly suppresses on-chip hardware cost as otherwise required in other direct wired solutions. A chaos phenomenon can be used for PUF by exploiting its physical nature of deterministic non-periodicity. The chaos PUF has a good synergy with this wireless approach as both has strong characteristics at high frequencies. The feasibility of the chaos-based interactive PUF is demonstrated for advanced secure supply chain.

ACKNOWLEDGMENT

This work is supported by STARC Feasibility Study (FS) program in collaboration with Socionext Inc. and Sony Corp.

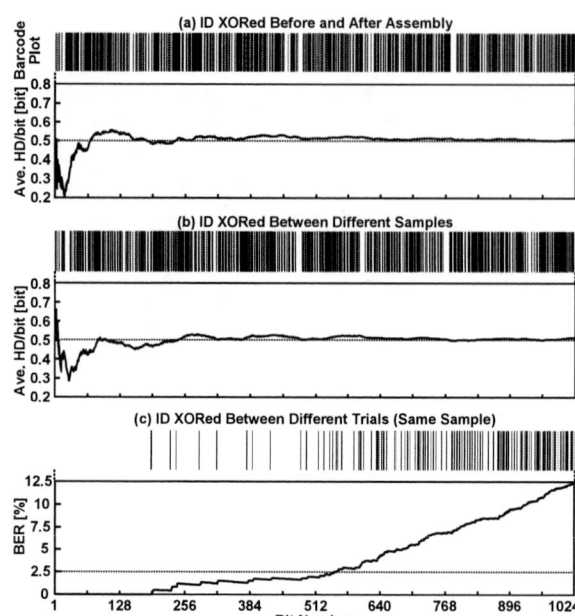

Fig.9 Barcode plot of XORed ID and average HD/bit of (a) same sample before and after assembly, (b) different samples, (c) measured BER depending on 1024bit sequence.

REFERENCES

[1] J. Villasenor, *et al.*, "The Hidden Dangers of Chop-Shop Electronics," *Spectrum*, Vol.50, No.10, pp.41-45, Oct. 2013.

[2] K. Lofstrom, *et al.*, "IC Identification Circuit Using Device Mismatch," *ISSCC Dig. Tech. Papers*, pp.372-373, Feb. 2000.

[3] Y. Su, *et al.*, "A 1.6pJ/bit 96% Stable Chip-ID Generating Circuit using Process Variations," *ISSCC Dig. Tech. Papers*, pp.406-407, Feb. 2007.

[4] J. W. Lee, *et al.*, "A Technique to Build a Secret Key in Integrated Circuits for Identification and Authentication Applications," *Symp. on VLSI Cir. Dig. Tech. Papers*, pp.176-179, June 2004.

[5] B. Gassend, *et al.*, "Silicon Physical Random Functions," *Proc. of CCS*, pp.148-160, Nov. 2002.

[6] K. Vaidyanthan, *et al.*, "Building Trusted ICs using Split Fabrication," *Proc. of HOST*, pp.1-6, May 2014.

[7] G. M. Bernstein, *et al.*, "Secure Random Number Generation Using Chaotic Circuits," *IEEE Trans. on Circuits and Systems*, Vol.37, No.9, pp.1157-1164, Sept. 1990.

[8] S. Callegari, *et al.*, "ADC Based True Random Number Generator for Cryptographic Applications Exploiting Nonlinear Signal Processing and Chaos," *IEEE Trans. on Signal Processing*, Vol.53, No.2, pp.793-805, Feb. 2005.

[9] F. Pareschi, *et al.*, "A Fast Chaos-based True Random Number Generator for Cryptographic Applications," *Proc. of ESSCIRC*, pp.130-133, Sept. 2006.

[10] E. N. Lorenz, "Deterministic Nonperiodic Flow," *Journal of the Atmospheric Sciences*, Vol.20, No.2, pp.130-141, Jan. 1963.

[11] M. Shinriki, *et al.*, "A Simultaneous Asynchronous Oscillator with Both Nonlinear Positive and Negative Conductances Connected in Series," *Proc. of IEEE*, Vol.67, No.2, pp.322-324, Feb. 1979.

[12] M. Saito, *et al.*, "An Extended XY Coil for Noise Reduction in Inductive-coupling Link," *A-SSCC Dig. Tech. Papers*, pp.305-308, Nov. 2009.

A 93μW 11Mbps Wireless Vital Signs Monitoring SoC with 3-Lead ECG, Bio-impedance, and Body Temperature

Yuxuan Luo, Kok-Hin Teng, Yongfu Li, Wei Mao, Chun-Huat Heng

Department of ECE,
National University of Singapore, Singapore
Email: yluo@u.nus.edu, elehch@nus.edu.sg

Yong Lian

Department of EECS,
York University, Canada

Abstract—This paper presents an integrated wireless multiple sensors System-on-Chip (SoC) for healthcare including 3-lead ECG, bio-impedance (Bio-Z), and body temperature. To allow continuous and real-time monitoring, the SoC has included a multi-channel reconfigurable QPSK/BFSK transmitter (TX) to accommodate different power-constraint conditions. Fabricated in 130nm CMOS technology with a total die area of 6.25mm², our SoC only consumes 93μW with 1% duty-cycling protocol in 0.9V supply.

Keywords—*healthcare-IoE multiple sensors SoC; real-time monitoring; bio-Z sensor; body temperature sensor; duty-cycling transmitter*

I. INTRODUCTION

Prevention-oriented patient specific healthcare has recently gained momentum due to the advance in technology. Under the context of the healthcare, various vital signs monitoring applications based on sensors and Bluetooth Low Energy (BLE) has emerged and aimed for commercial interest [1]. Nevertheless, the achievable data rate and power, as well as its slower turn on/off time still limits its energy efficiency.

To improve data transmission power efficiency, SoC with proprietary wireless transmission has recently been reported to demonstrate ECG, EEG, and EMG readouts at Medical Implant Communication Service (MICS) band with low power [2]. However, the single-channel transmission and data rate of 200kbps limits the achievable performance.

This paper presents a wireless vital signs monitoring SoC which includes 3-lead ECG, IQ Bio-Z for respiration, and body temperature readouts. A multi-channel reconfigurable QPSK/BFSK TX is employed for continuous real-time monitoring. With injection locking techniques, faster settling time can be achieved, which can lead to higher duty cycling ratio and better energy efficiency.

II. SYSTEM ARCHITECTURE

Fig.1 shows the system block diagram for our SoC. It includes a 2-channel ECG readout channel for 3-lead ECG monitoring, a 2-channel Bio-Z readout channel for in-phase and quadrature-phase (IQ) components monitoring and a thermistor-based temperature sensor for body temperature monitoring. The 2-channel ECG and 2-channel Bio-Z readout analog signals are sampled and quantized through two 12-bit SAR ADCs respectively in a time-interleaving manner to save the cost of silicon area. The SAR ADC is designed with a segmented split capacitor for both area and power saving.

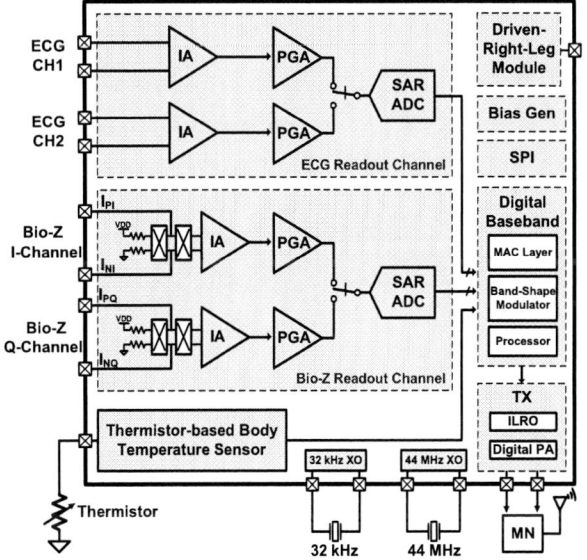

Fig. 1. Proposed architecture of the wireless vital signs monitoring SoC.

The body temperature sensor performs a direct temperature to digital conversion with noise shaping to achieve high temperature reading resolution and accuracy.

The digital baseband (BB) module will pack the collected data to implement the Media Access Control (MAC) layer before passing to the TX. For transmission, each data packet consists of 1200 bits of raw data. The raw data will contain 12 samples of 2-channel ECG, 2-channel Bio-Z, and temperature reading with each sample in 16-bit format. This will total up to 960 bits. The remaining 240 bits are used for other auxiliary data. The assembled data packet will then go through the digital BB processor to implement the proprietary MAC layer. Forward error correction (FEC) and cyclic redundancy check (CRC) are included to improve the transmission robustness. The final packet will consist of 1980 bits. Either BFSK or QPSK modulation can be chosen for the TX. This allows the trade-off of TX energy efficiency with the simplification of receiver design.

III. CIRCUITS IMPLEMENTATION

Each channel of the ECG analog front-end consists of an instrumental amplifier (IA) and programmable gain amplifier (PGA). A voltage-source-based early-demodulation technique

978-1-5386-3179-9/17 $31.00 © 2017 IEEE

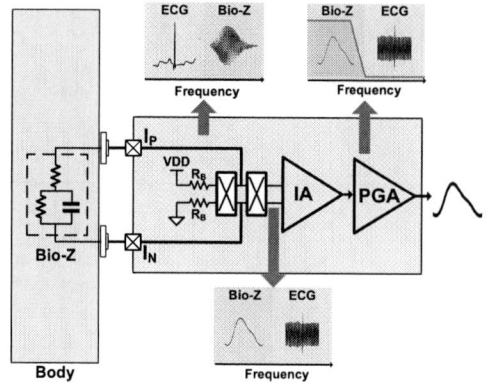

Fig. 2. Illustration of the early-demodulation Bio-Z readout channel.

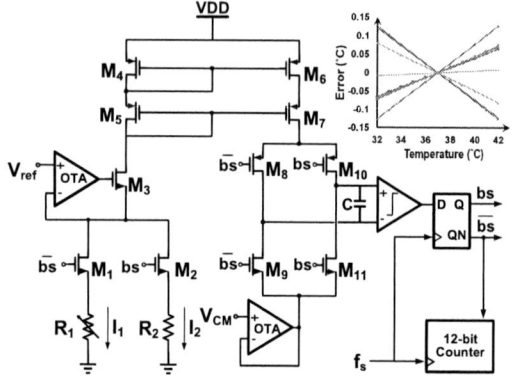

Fig. 3. Proposed H-bridge body temperature sensor schematic diagram.

is proposed for the Bio-Z readout channel as illustrated in Fig.2. It is consisted of two chopper stages followed by the gain and filtering stages. The current limit resistor R_B restricts the amount of current injected into the human body. The Bio-Z signal is first modulated to high frequency at 16 kHz through the first chopper stage. By using the second chopper at the same frequency, the modulated Bio-Z signal is then down-converted back to the baseband while the DC components such as the offset voltage and the coupled ECG signals are modulated to the high-frequency. The gain and filtering stage designs are similar to the ECG readout channels except for its lower signal bandwidth and higher gain. By exploiting its band-pass characteristic (0.5Hz~50Hz), only the low-frequency Bio-Z ($<0.25\Omega$) AC components tracking the respiration will be amplified while the ECG signal at high frequency will be suppressed. Compared to the conventional current-source-based modulation technique [3], our design does not require additional DC cancellation servo loop and wide amplifier bandwidth.

Fig.3 shows our proposed H-bridge body temperature sensing architecture by adopting duty-cycle modulation concept [4]. On-chip polysilicon resistor (R_2) with negligible temperature coefficient (TC) over 32~42°C is used to generate the constant current I_2 as a reference. Off-chip thermistor (R_1)

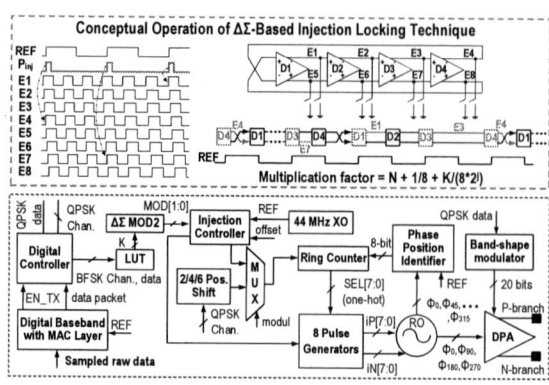

Fig. 4. $\Delta\Sigma$-based injection locking mechanism and TX architecture.

with negative TC is used to produce the temperature dependent current I_1. I_1 and I_2 can be chosen to charge or discharge the capacitor based on the comparator decision. The H-bridge structure of the switch allows the capacitor voltage to be sent directly to the differential input of a comparator without the need of any reference voltage. The comparator output will then determine the current path chosen to charge or discharge the capacitor. The resulting comparator output bit-stream (BS) will contain noise-shaped temperature information. After low-pass filtering by a 12-bit counter, high resolution temperature reading down to 0.003 °C can be extracted from the BS. From the measurement across 8 samples, the architecture can achieve error of 0.13°C with 1-point calibration at 37°C.

The proposed TX architecture is illustrated in Fig.4. It will take in the encoded data from the MAC layer and perform the desired modulation for the actual transmission. The data packet is fed into a digital control module where the modulation type and transmission channel will be selected. The multi-channel band-shaped QPSK modulation is obtained using sequential injection locking. The amount of position shift to ring counter will determine the resulting carrier frequency. For multi-channel BFSK modulation, $\Delta\Sigma$-based injection locking is proposed. The idea is best illustrated in the upper figure of Fig.4. By randomly determine the delay cell within the ring oscillator that get injected, a fractional VCO period can be obtained which allows us to synthesize any average frequency with resolution down to 1.3kHz. A digital power amplifier (DPA) is implemented in order to control both the phase and amplitude information of the modulated signal.

In order to improve the energy efficiency of the TX, duty-cycling protocol can be applied. As mentioned earlier, at the digital baseband, a buffer will accumulate 12 samples of data at a given sampling rate of 256 Hz. Hence, it will take 46.88 ms to refresh the data in the digital buffer. If 11 Mbps QPSK modulation is employed, only 180 μs will be required to fully transmit the 1980 bits of data. As the adopted injection locking has a fast turn-on time of 0.8 μs, a duty cycling ratio of 1% can be achieved. On the other hand, if simple BFSK receiver is desired, BFSK with data rate of 0.55 Mbps can be employed. This will result in duty cycling ratio of 7.7%.

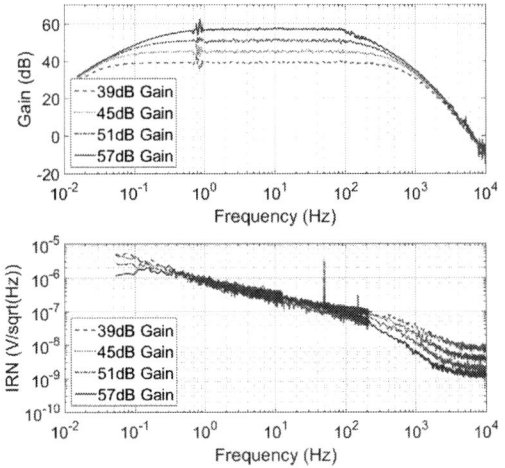

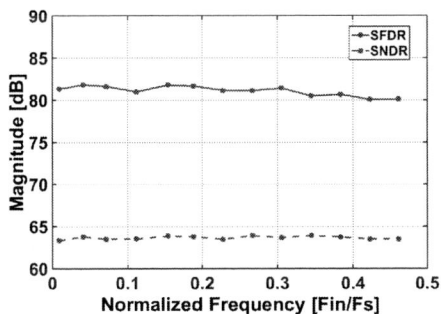

Fig. 5. Measured frequency response and input referred noise of ECG analog front-end.

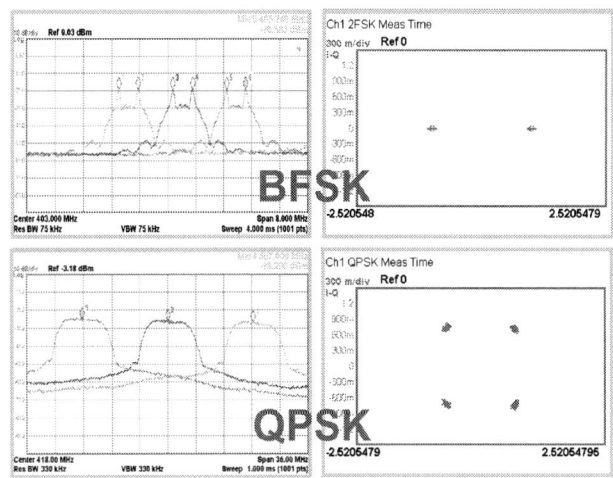

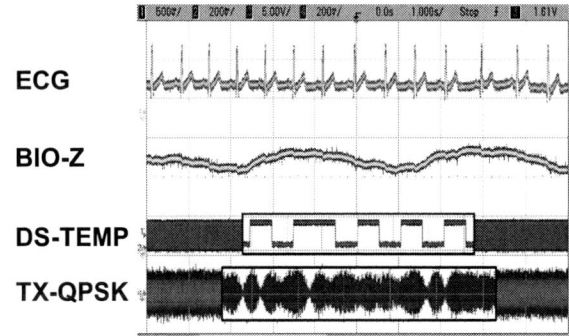

Fig. 7. The measurement results of 3-channel QPSK/BPSK with EVM of 5.38% and 3.70% at data rate of 11/0.55 Mbps respectively.

Fig. 6. Measurement results of SAR ADC (Fs = 256 Hz) SFDR and SNDR.

Fig. 8. Measured raw data of the multiple sensor and TX-QPSK outputs.

IV. EXPERIMENTAL RESULTS

As illustrated in Fig.5, the ECG AFE can cover gain from 39-57dB in 6dB step. Within 0.5∼150Hz band of interest, the measured input referred noise of 2.43μVrms and noise efficiency factor of 2.79 can be obtained. The measured SNDR and SFDR performance of the SAR ADC is shown in Fig.6. The overall signal quality exceeds an ENOB of 10 bit, which is sufficient for the targeted application.

The measurement results for the TX is shown in Fig.7. For 3-channel QPSK modulation, the data rate of 11 Mbps can be achieved with EVM of 5.38%. As for 3-channel BFSK modulation, the data rate is 0.55 Mbps and the EVM is 3.7%. The measured raw data of ECG signal, Bio-Z based respiration and body temperature on a healthy human subject are shown in Fig.8.

The left figure in Fig.9 shows the measured average power consumption of the system under duty-cycling protocol. The system power consumption linearly decreases with the decreasing duty-cycling ratio. When the TX is 100% turned ON, the system consumes an average power of 3.9 mW. As discussed earlier, with 1% duty cycling ratio, the overall power

consumption is reduced to 93 μW as illustrated. The power breakdown diagram is also shown in Fig.9.

The die photo is shown in Fig.10. The proposed SoC, fabricated in 130nm CMOS technology, occupied a total area of 6.25 mm^2. It should be pointed out that three channels of SAR ADCs are implemented in this design where one of the ADCs is designed for testing purpose.

A performance comparison is presented in Table I. Due to the low bandwidth requirement for the amplifiers in the early-demodulation Bio-Z readout channels, it demonstrates the lowest power consumption compared to the other designs. As shown, we provide the most comprehensive vital signs monitoring, covering ECG, Bio-Z and temperature while supporting multi-channel wireless transmission capability. To date, this work demonstrates the highest level of integration for wireless vital signs monitoring. The reported power including the RF is also the 2nd lowest. It should be pointed out that the reported lowest power of [2] is with the transmission of processed data at much lower data rate. For continuous ECG transmission, their power consumption of 397 μW is 4 times higher than ours. In addition, our SoC provides more vital signs sensing as well as multi-channel support.

TABLE I
PERFORMANCE COMPARISON

		This Work	M. Konijnenburg ISSCC'16	N.V. Helleputte ISSCC'14	F. Zhang ISSCC'12	W.P. Chan JSSC'Nov,14
Sensors		ECG, Bio-Z, Body Temperature	ECG, Bio-Z, GSR, PPG	ECG, Bio-Z	ECG, EMG, EEG	Pressure, Oxygen, Temeprature
Wireless Transmission		Yes	No	No	Yes	No
Supply Voltage		0.9 V	1.2/1.8 V	1.2/1.8 V	1.2/1.0/0.5 V	1.8/3.3 V
Total Chip Power		93 μW (QPSK)++ 322 μW (BFSK)++	520 μW	1845 μW	19 μW+	166 μW
Technology		130 nm	180 nm	180 nm	130nm	180 nm
Chip Area		6.25 mm²	37.7 mm²	49 mm²	8.25 mm²	5.6 mm²
AFE	Noise (0.5-150Hz)	2.43 μVrms	0.6/1 μVrms	0.6/1 μVrms	N. A.	26/36 μVrms @ 10 Hz
	Gain	39-57 dB	N. A.	N. A.	40-78 dB	N. A.
	ECG Power/Ch.	2.28 μW	<50 μW	<50 μW	4.8 μW	-
	Bio-Z Power/Ch.	9.80 μW	117 μW	58 μW	-	-
ADC		12b SAR	15b $\Delta\Sigma$	15b $\Delta\Sigma$ / 12b SAR	8b SAR	10b SAR
TS	Type	Resistor	-	-	-	BJT
	Error (32-42 range)	0.13	-	-	-	0.16 - (-0.05)°C*
	Power	2 μW	-	-	-	36 μW
Channels		>3	-	-	1	-
Data Rate		0.55 Mbps (BFSK) 11 Mbps (QPSK)	-	-	0.2 Mbps	-
Power (100% ON)		3.02 mW (BFSK) 3.53 mW (QPSK)	-	-	0.16 mW	-
Pout		-15 dBm	-	-	-18.5 dBm	-
Energy Efficiency		5.49 nJ/b (BFSK) 0.32 nJ/b (QPSK)	-	-	0.8 nJ/b	-
TX Band		401-406 MHz	-	-	402/433 MHz	-
Modulation		QPSK/BFSK	-	-	BFSK	-

+ TX duty cycled at 0.013%. Data compression is required to minimize the transmission data rate. ++ TX duty cycled at 1%/8% for QPSK/BFSK respectively. * Estimated from the measured temperature error of 9 samples, 1-point calibration at 35°C.

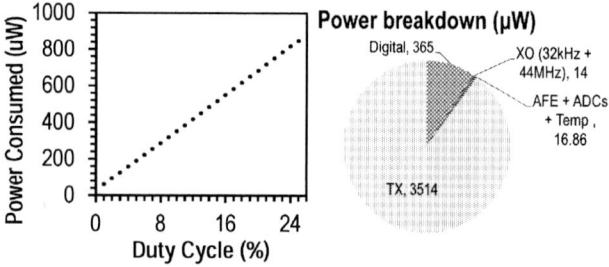

Fig. 9. Average power consumption in system-level (QPSK mode) through duty cycling the TX and power breakdown of the SoC.

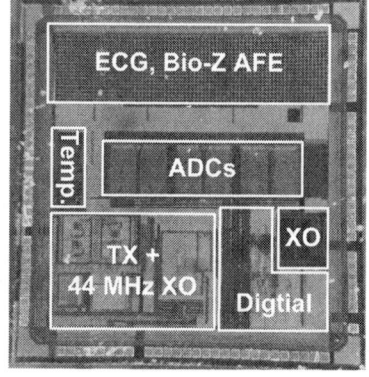

Fig. 10. Die micrograph.

V. CONCLUSIONS

This paper presents an integrated wireless multiple sensors SoC including 3-lead ECG, Bio-Z, and body temperature. The early demodulation Bio-Z readout channel reduces the bandwidth requirement and thus the amplifier power. The proposed H-bridge body temperature sensor interface achieves high resolution through inherent noise-shaping. The multi-channel reconfigurable TX can trade off energy efficiency with receiver design. Overall, low power of 93 μW has been demonstrated.

VI. ACKNOWLEDGEMENTS

This work is funded by National Research Foundation of Singapore (NRF) grant NRF-CRP8-2011-01.

REFERENCES

[1] G. Devita and et al., "A 5mW multi-standard Bluetooth LE/IEEE 802.15. 6 SoC for WBAN applications," ESSCIRC, pp. 283-286, 2014.
[2] F. Zhang and et al., "A batteryless 19μW MICS/ISM-band energy harvesting body area sensor node SoC," ISSCC, pp. 298-300, 2012.
[3] N. Van Helleputte and et al., "A multi-parameter signal-acquisition SoC for connected personal health applications," ISSCC, pp. 314-315, 2014.
[4] A. Heidary and et al., "A BJT-based CMOS temperature sensor with a 3.6 pJK 2-resolution FoM," ISSCC, pp. 224-225, 2014.
[5] M. Konijnenburg and et al., "A battery-powered efficient multi-sensor acquisition system with simultaneous ECG, BIO-Z, GSR, and PPG," ISSCC, pp. 480-481, 2016.
[6] W.P. Chan and et al., "A monolithically integrated pressure/oxygen/temperature sensing SoC for multimodality intracranial neuromonitoring," IEEE JSSC, pp. 2449-2461, Nov. 2014.

S3-4 (2090)

A 16-Channel TDM Analog Front-end with Enhanced System CMRR for Wearable Dry EEG Recording

Tao Tang[1,2], Wang Ling Goh[1], Lei Yao[2], Yuan Gao[2]

[1]School of Electrical and Electronic Engineering, Nanyang Technological University, Singapore
[2]Institute of Microelectronics, A*STAR, Singapore
ttang005@e.ntu.edu.sg

Abstract—This paper presents a 16-channel analog front-end (AFE) for wearable dry EEG recording. A novel AFE architecture that combines time division multiplexing (TDM) and chopping stabilization (CS) to improve the system common mode rejection ratio (CMRR) and the input-referred noise is proposed. With TDM, the reference electrode is connected to single channel input during its time slot to avoid the CMRR degradation caused by input impedance imbalance. Furthermore, the 2nd stage programmable gain amplifier (PGA) is shared among multiple channels, leading to significant reduction in both power consumption and chip area. In addition, input impedance boosting loop and DC servo-loop are included to boost the input impedance and cancel the input DC offset. Implemented in a 0.18-μm CMOS process, the proposed AFE shows 0.63 μV$_{rms}$ input-referred noise integrated from 0.5 Hz to 100 Hz and 560 MΩ input impedance at 50 Hz. The measured AFE intrinsic CMRR is 89 dB and the system CMRR is 82 dB, which is a 40 dB improvement compared to the conventional architecture. The AFE chip consumes merely 1.5 μW per channel under the 1 V supply voltage.

Keywords— *EEG recording; time division multiplexing; chopper stabilization; common mode rejection ratio*

I. INTRODUCTION

Electroencephalogram (EEG) is an electrophysiological method to record brain electrical activities [1]. It has been recognized as an essential clinical tool for diagnosis and monitoring of brain disorders such as epilepsy, sleep apnea and trauma. Due to the inherent small signal amplitude, EEG signal acquisition usually requires sophisticated electrode setup procedure and bulky high precision signal recording equipment [2]. Thus, the current EEG monitoring practice is mostly limited to hospital environment.

Wearable EEG recording is an emerging solution that allows continuous EEG monitoring outside hospital environment. Dry electrode is preferred in wearable EEG system for the convenient of test setup without the need for skin preparation and conductive gel. However, dry electrode does bring a few design challenges to the recording circuit development. First of all, dry electrode has much higher skin/electrode impedance as compared to the conventional wet

*This work is supported by Agency for Science, Technology and Research (A*STAR), Singapore, under BMRC Grant IAF311022.

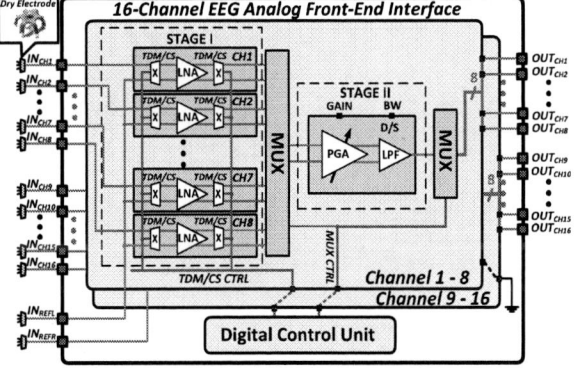

Fig. 1 Block diagram of the proposed 16-channel TDM/CS EEG AFE.

electrode. Therefore, the input impedance of readout circuit needs to be boosted to match the dry electrode [3]. Furthermore, EEG signal is susceptible to the common-mode interferences coupled from the AC power line or artifacts generated by the body movement. Hence, high common mode rejection ratio (CMRR) is required for the recording circuit. A few CMRR enhancement techniques have been introduced in the literatures, where most had focussed on the improvement of amplifier's intrinsic CMRR [4]. Conversely, the system level CMRR is more critical for overall system performance especially for the multichannel acquisition system [5].

In this paper, a 16-channel AFE for dry electrode EEG recording is presented. Time Division Multiplexing (TDM) is combined with chopping stabilization (CS) to improve the system-level CMRR and the AFE input-referred noise. Techniques such as tunable impedance boosting loop and DC-servo loop are also included to boost the AFE input impedance and to cancel the input DC offset. The overall AFE performance has been optimized for wearable multi-channel EEG recording.

II. SYSTEM ARCHITECTURE AND CIRCUIT BLOCKS

A. System Architecture

The proposed 16-channel EEG system architecture is shown in Fig 1. It comprises two identical AFE cells, each having 8 fully differential EEG recording channels. In every channel, a two-stage amplifier structure is adopted with 30-dB

978-1-5386-3179-9/17 $31.00 © 2017 IEEE

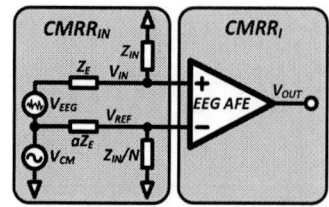

Fig. 2. Equivalent circuit model for a single channel EEG AFE in a multichannel recording system. The negative inputs of all channles are tied together and connected to the reference electrode.

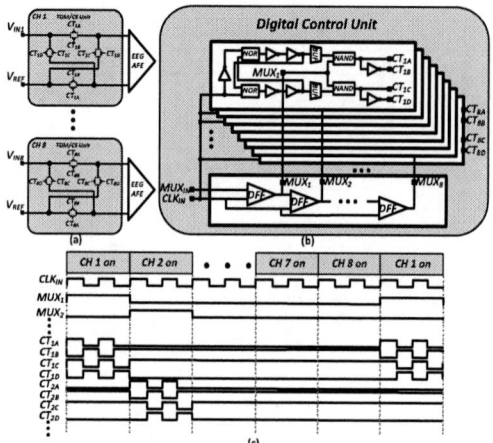

Fig. 3. (a) Proposed TDM/CS unit, (b) digital unit for control signal generation and (c) Timing scheme of TDM/CS switches .

fixed gain in the 1st stage capacitive coupled instrumentation amplifier (CCIA) and a 25/30-dB programmable gain amplifier (PGA) in the 2nd stage. One common reference electrode is shared by all working electrodes within the cell. A TDM/CS scheme is proposed to enhance the system CMRR and noise performance. Different from the scheme presented in [6], which shares the 1st stage CS-CCIA among 2 channels, this design allows the sharing of 2nd stage PGA with all the channels in the cell. The output of the 1st stages CCIA is connected to the 2nd stage PGA sequentially through a MUX synchronized with the TDM/CS control signal. In this design, by controlling the switches in the TDM/CS block, the reference electrode is only connected to one channel during the corresponding signal acquisition time slot. Thus, the proposed scheme is scalable to accommodate more channels without affecting system CMRR performance.

B. CMRR analysis

The system total CMRR ($CMRR_T$) depends on both the input interface CMRR ($CMRR_{IN}$) and the AFE intrinsic CMRR ($CMRR_I$) as illustrated in Fig. 2. $CMRR_T$ is represented as:

$$CMRR_T = \left(CMRR_{IN}^{-1} + CMRR_I^{-1}\right)^{-1} . \quad (1)$$

In a conventional multi-channel recording system, the reference input is shared by all the channels, therefore the input impedance seen by each brunch is Z_{IN}/N. This impedance

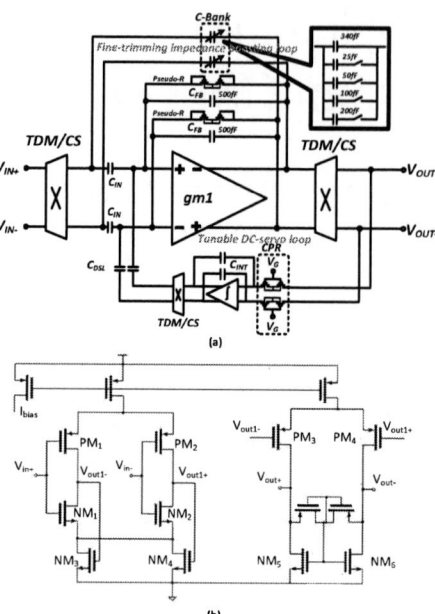

Fig. 4. (a) Schmatic of the TDM/CS-CCIA, (b) Schmatic of the OTA.

imbalance will degrade both the $CMRR_{IN}$ and $CMRR_T$, the $CMRR_{IN}$ for multi-channel system is derived as [5]:

$$CMRR_{IN} = \frac{1 + 2\left(\left|Z_{IN}/Z_E\right|\right) + N\alpha}{2\left(N\alpha - 1\right)} \quad (2)$$

where N is the number of channels that shared the same reference electrode and α is the mismatch coefficient for the impedance of reference electrode and working electrode.

Hence, although $CMRR_I$ of a well-designed fully differential amplifier can be higher than 90 dB [5], $CMRR_T$ is eventually limited by $CMRR_{IN}$ for multichannel AFE. Considering the case of $N = 2$ and $\alpha = 1$ (no impedance mismatch between reference and working electrodes), $CMRR_{IN}$ will be 60 dB under the condition of $Z_{IN}/Z_E = 1000$. The electrode/skin impedance is even higher for dry electrodes (Hundreds of $K\Omega$ to $M\Omega$), leading to even lower system $CMRR_T$. Hence, ideally, the number of channels that share the reference electrode should be reduced to one.

C. TDM/CS Technique

The TDM/CS switches schematic is shown in Fig. 3(a). The number of the switches have been minimized to avoid excessive charge injection and clock feed-through from the switches. As shown in Fig. 3(b), CLK_{IN} and MUX_{IN} can generate all the sequential MUX signals (MUX_1 to MUX_8). CLK_{IN} and each sequential MUX signal can generate the four local control signals for the TDM/CS unit. Therefore, only two control signals (CLK_{IN} and MUX_{IN}) are required in the digital control unit to generate all the control signals. The timing scheme of the control signals are illustrated in Fig. 3 (c). As can be seen, each channel MUX signal is only ON for 1/8 of the MUX_{IN} clock cycle.

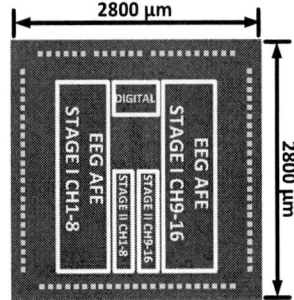

Fig. 5. Die photo of the proposed AFE.

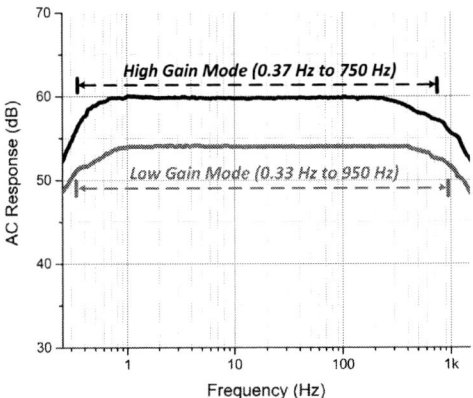

Fig. 6. Measured gains of the proposed AFE.

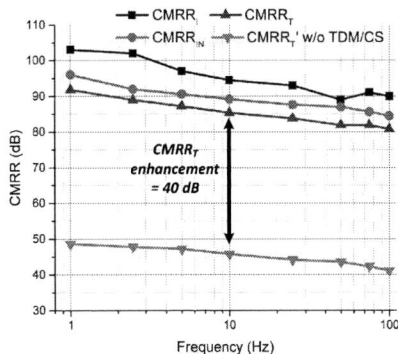

Fig. 7. Mearsured $CMRR_I$ and calculated $CMRR_{IN}$ and $CMRR_T$ based on Eq. (1) – (2) with electrode-skin impedance of 500 KΩ and α=1.05 for 5% electrode impedance mismatch.

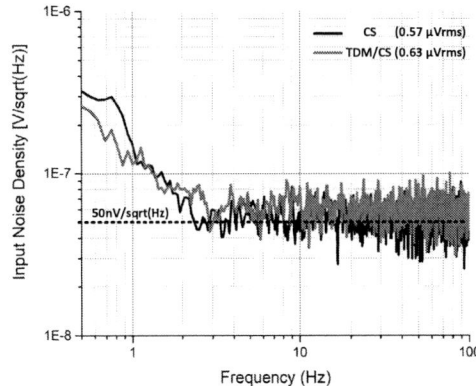

Fig. 8. Measured input-referred noise of the proposed AFE.

D. CS-CCIA Circuit

The schematic of the proposed CS-CCIA is shown in Fig. 4(a). CS technique is adopted to reduce the low frequency flicker noise. To compensate the input impedance drop due to chopping, a positive feedback path for input impedance boosting is implemented with a 4-bit fine trimming impedance boosting loop (FIBL). The FIBL provides 16 values of the feedback capacitance to avoid capacitor mismatch due to chip fabrication. The bank consists of a default capacitor of 340 fF and 4-bit trimming capacitor (25 fF, 50 fF, 100 fF and 200 fF). On the other hand, the input offset voltage can be up-modulated by chopping. Therefore, a tunable DC-servo loop (TDSL) is designed to cancel the input DC offset. The response time of the TDSL is determined by the time constant of the coupling capacitor C_{INT} and the tunable pseudo-resistor (TPR). The impedance of the TPR is controlled by the gate voltage V_G. During reset, V_G is controlled in such a way that it is ramped up from *GND* to *VDD* at a controlled speed so that the TDSL achieves a fast cancellation of input DC offset without any glitch injection.

The amplifier core is a fully differential inverter based OTA shown in Fig. 4 (b). The current-reuse technique is employed at the input stage to achieve the required noise performance with minimum current consumption. The two input transistor pairs (PM_{1-2}, NM_{1-2}) are biased in weak inversion region to maximize the current efficiency. Transistors NM_3 and NM_4 provide common-mode feedback to fix the first stage output DC level. In the second stage, PMOS transistors are chosen as the input pair to reduce the noise.

III. MEASUREMENT RESULTS

The proposed AFE chip has been designed and fabricated in a standard commercial 0.18-μm 1P6M CMOS process. The chip photo is given in Fig. 5. The chip area of the overall active circuits and each channel is 3.6 mm² and 0.23 mm² respectively. The current consumption for each channel is 1.5 μA under 1-V supply voltage. The chopping clock is selected as 4 kHz and the TDM clock frequency is 250 Hz. The measured voltage gain of two gain settings are 60/54 dB as shown in Fig. 6. The low/high cut-off frequencies are 0.37/750 Hz and 0.33/950 Hz, respectively, for the two gain modes. The measured intrinsic CMRR of the proposed design is presented in Fig. 7, the $CMRR_I$ is 89 dB at 50 Hz. Assuming 500 KΩ electrode-skin impedance and 5% impedance mismatch (α=1.05) between two electrodes, the $CMRR_{IN}$ is 87 dB. Under conventional scheme of reference electrode sharing by all the channels, the calculated $CMRR_T$ is 40 dB whereas the $CMRR_T$ with TDM/CS technique is 82 dB, which is enhanced by about 40 dB. The measured input-referred noise density is shown in

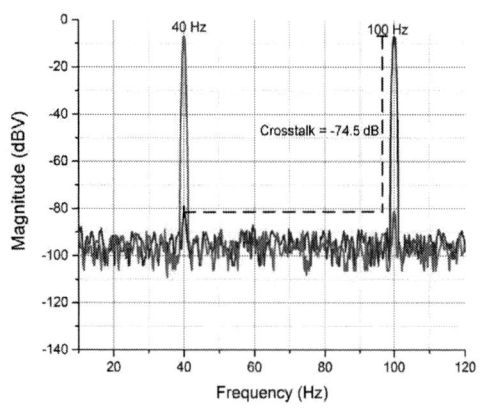

Fig. 9. Measured crosstalk for two channels of the proposed AFE.

Fig. 8. The black line indicates the noise density when the chopper is ON and TDM is OFF. This is equivalent to the setting for single channel chopped amplifier. The measured input-referred noise floor is only 50 nV/sqrt(Hz) at 100 Hz and the integrated input referred noise is 0.57 μV$_{rms}$ across 0.5 Hz to 100 Hz band. The red line represents the setting when both chopping and TDM signals are ON, the integrated noise is slightly increased to 0.63 μV$_{rms}$ in the same frequency band. This indicates negligible impact of the TDM scheme to the overall circuit noise performance. The AFE crosstalk performance is measured as shown in Fig. 9. Two sinusoid input signals (40 Hz and 100 Hz) with same magnitude are pumped into channel 1 and 3, respectively. The measured output magnitude of channel 1 and the leakage signal from channel 3 are -8.1 dBV and -82.6 dBV respectively, indicating -74.5 dBV channel crosstalk. Similar results are shown for the leakage from channel 1 to channel 3.

The AFE measured results are summarized in Table I and compared with other state-of-the-art multichannel EEG AFE designs. The proposed design achieved a low power consumption per channel of only 1.5 μW. The integrated noise is 0.63 μV$_{rms}$ from 0.5Hz to 100 Hz, which is the lowest compared to the others. Highest input impedance of 560 MΩ at 50 Hz is achieved with the FIBL. In additional to the 89 dB AFE intrinsic CMRR$_I$, the system-level CMRR$_T$ which is a parameter not reported by the other papers, is 82 dB at 50Hz and maintained above 80 dB across the frequency range up to 100 Hz.

IV. CONCLUSIONS

In this paper, a 16-channel EEG AFE chip is designed and fabricated in 0.18 μm CMOS technology. A combination of chopping stabilization and time division multiplexing is

TABLE I. COMPARISON WITH STATE-OF-THE-ARTS

Parameters	[3]	[4]	[6]	[7]	This work
Technology	0.18 μm	0.18 μm	0.18 μm	65 nm	0.18 μm
No. of Channel	15	8	16	8	16
Supply Voltage (V)	1.8	0.9	1.8	1	1
Power per channel (μW)	104.4	2.75	1.62	0.87	1.5
Gain (dB)	70	53.5 - 65	40	59.1-78	54/60
Input-referred noise (μV$_{rms}$) (0.5 to 100Hz)	0.65	1.06	0.9	0.81	0.63
Input Impedance @50Hz (MΩ)	100	NA	500	40	560
Max. EDO rejection (mV)	350	rail to rail	240	50	200
CMRR$_I$ @50Hz (dB)	78	78	97	84	89
CMRR$_T$ @50Hz (dB)*	NA	NA	NA	NA	82
Crosstalk	NA	NA	-56	NA	-74.5

*CMRR$_T$ is calculated based on Eq. (1) – (2) with electrode-skin impedance of 500 KΩ and α=1.05 for 5% mismatch for dry electrodes.

proposed to achieve high system CMRR, low input noise and low power consumption. The proposed 16-channel AFE achieves well-balanced performances and it is suitable for wearable dry electrode EEG recording applications.

REFERENCES

[1] C. Guger, *et al.* "Methods for Noninvasive Electroencephalogram Detection," *Introduction to Neural Engineering for Motor Rehabilitation*, pp. 137-154, 2013.

[2] N. Birbaumer, *et al.* "The thought-translation device (TTD): neurobehavioral mechanisms and clinical outcome," *IEEE Trans. Neural Rehabil. Syst. Eng.*, vol. 11, no. 2, pp. 120-123, 2003.

[3] J. Xu, *et al.* "A 60nV/Hz 15-channel digital active electrode system for portable biopotential signal acquisition," *IEEE Int. Solid-State Circuits Conf. Dig. Tech. Papers (ISSCC)*, pp. 424-425, 2014.

[4] D. Valle, *et al.* "Low-Power, 8-Channel EEG Recorder and Seizure Detector ASIC for a Subdermal Implantable System," *IEEE Transcation on Biomedical Circuits and Systems*, vol. 10, no. 6, pp. 1058-1067, 2016.

[5] K.A. Ng, and Y.P. Xu. "A Low-power, high CMRR neural amplifier system employing CMOS inverter-based OTAs With CMFB through supply rails," *IEEE J. Solid-State Circuits*, vol. 51, no. 3, pp. 724-737, 2016.

[6] MAB. Altaf, C. Zhang, and J. Yoo. "A 16-channel patient-specific seizure onset and termination detection soc with impedance-adaptive transcranial electrical stimulator," *IEEE J. of Solid-State Circuits*, vol. 50, no. 11, pp. 2728-2740, 2015.

[7] C.Y. Wu, and C.S. Ho, "An 8-channel chopper-stabilized analog front-end amplifier for EEG acquisition in 65-nm CMOS," *IEEE Asian Solid-State Circuits Conf. (A-SSCC)*, pp. 1-4, 2015.

An Area-Efficient Amplifier-Less Digitally-Controlled Li-Ion Battery Charger in 0.35-μm CMOS

Sheng-Ying Lin and Tsung-Hsien Lin

Graduate Institute of Electronics Engineering and Department of Electrical Engineering,
National Taiwan University, Taipei, Taiwan

Abstract - In this paper, an amplifier-less digitally-controlled lithium-ion (Li-Ion) battery charger is presented. Owing to the digitally-controlled technique, the proposed charger eliminates all analog circuits and reduces the size of the power transistor. Hence, the proposed charger features a simple circuit structure and a small chip area. Additionally, this charger provides essential operations including constant-current (CC) and constant-voltage (CV) modes. The proposed charger is implemented in the TSMC 0.35-μm CMOS process with a 5-V power supply. The output voltage is 4.2 V, and the charging current in CC mode is 600 mA with an area of only 0.768 mm^2.

Keywords – battery charger, constant-current mode, constant-voltage mode, power management.

I. INTRODUCTION

In recent years, portable devices have enabled many emerging applications. These devices usually require lightweight, small form factor, and rechargeable features for portability and continuing usage. In terms of the battery requirement, high energy density, high cycle life, high voltage, and no memory effect are characteristic of lithium-ion (Li-Ion) batteries. Hence, Li-Ion batteries are the most suitable battery for these devices. However, over discharging and over charging may damage the physical structure of Li-Ion batteries, resulting in reduced battery life or even explosion. Therefore, charging Li-Ion batteries safely and completely is an important issue.

The most popular charging method for Li-Ion batteries is constant-current constant-voltage (CC-CV) charging scheme [1], [2], as shown in Fig. 1. For the CC-CV charging scheme, there are three conventional topologies: switching-mode power supplies [4], [6], charge-pump converters [3], and low-dropout (LDO) voltage regulators [2], [5]. Because two complex interconnected loops are employed for CC mode and CV mode, all these three topologies need operational amplifiers; hence, increasing the design complexity. In addition, switching-mode power supplies and charge-pump converters usually require extra inductors and capacitors, respectively. To eliminate all external components, LDO-based chargers provide an attractive solution for Li-Ion battery chargers. However, LDO-based chargers need large size of power transistors since these analog LDOs should minimize the headroom voltage for power transistors.

Up to now, LDO-based battery chargers are analog-based design. To circumvent design issues of the analog-based approach, a digitally-controlled battery charger topology is proposed in this work. The digital control scheme allows fully

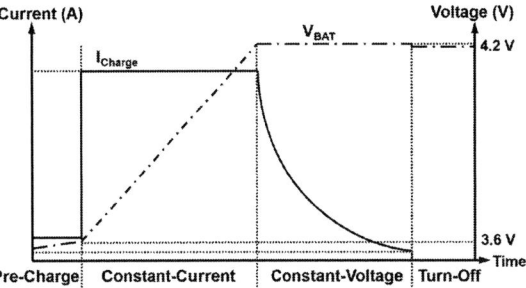

Fig. 1. Li-Ion battery CC-CV charging sequence.

turning on-and-off the power transistors with rail-to-rail gate-to-source voltage of power transistors. Therefore, smaller size of power transistors can be employed, which reduces the chip area. Furthermore, the proposed digitally-controlled charger eliminates all analog circuits by using the comparator-based control, reducing the design complexity. Besides, the proposed digitally-controlled charger has better process scalability.

This paper describes a new amplifier-less digital-controlled battery charger architecture, which achieves 600-mA constant charging current with an area of only 0.768 mm^2. This paper is organized as follows: Section II presents the proposed digitally-controlled battery charger and describes the implementation of CC and CV modes. Measurement results are provided in Section III, and finally Section IV concludes this paper.

II. PROPOSED DIGITAL-CONTROLLED BATTERY CHARGER

Fig. 2 shows the architecture of the proposed digitally-controlled Li-Ion battery charger. The proposed charger consists of a constant-voltage control logic (CVCL), a constant-current control logic (CCCL), comparators, a pre-charge detector (PCD), and array of power transistors. R_{BIR} and C_{BAT} are used to model the battery, and emulate the built-in resistance of the battery pack system. The on/off states of power transistors, $M_P[1:64]$, are determined by PCD, CCCL, and CVCL along the charging process.

The proposed charger adopts the CC-CV charging scheme. In the beginning, $V_{BAT} < V_{REF}$, CCCL dominates, and the battery is charged with constant current. When V_{BAT} reaches V_{REF}, CVCL takes over the control, and regulates V_{BAT} at the regulated voltage. The PCD, composed of comparators and digital logic, detects whether the battery should be pre-charged or not. The following describes the operation in the CC and CV modes in detail.

S3-5 (2169)

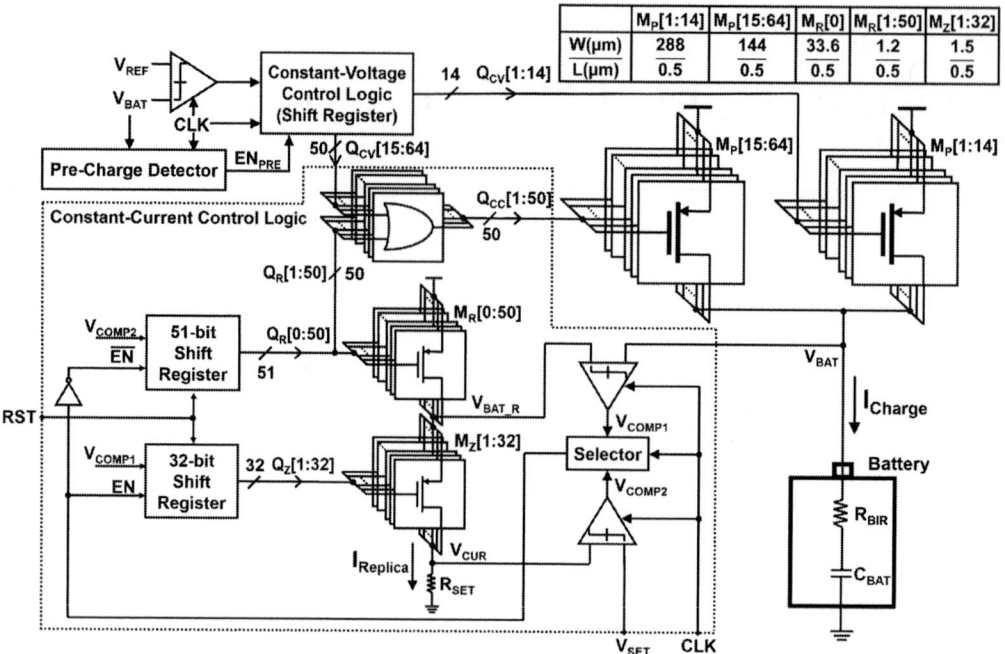

	$M_P[1:14]$	$M_P[15:64]$	$M_R[0]$	$M_R[1:50]$	$M_Z[1:32]$
W(μm)	288	144	33.6	1.2	1.5
L(μm)	0.5	0.5	0.5	0.5	0.5

Fig. 2. Proposed architecture of the digitally-controlled Li-Ion battery charger.

A. Operation in the Constant-Current Mode

As shown in Fig. 2, CVCL consists of shift registers. In the beginning of the charging process, $Q_{CV}[1:64]$ are set to 'H', so all power transistors $M_P[1:64]$ are turned off. Since V_{BAT} is lower than V_{REF}, $Q_{CV}[1:64]$ drop to 'L' one by one. When $Q_{CV}[1:64]$ are 'L,' power transistors $M_P[1:14]$ are turned on and the rest of power transistors $M_P[15:64]$ are controlled by CCCL, for the CC mode operation.

CCCL comprises a replica circuit of the main circuit. The replica current and the charging current have a ratio determined by the ratio of $M_R[1:50]$ to $M_P[15:64]$. Fig. 3(a) shows the operation block diagram of CCCL. At first, CCCL compares V_{BAT_R} with V_{BAT} and V_{CUR} with V_{SET}. If $V_{CUR} > V_{SET}$ and $V_{BAT\_R} > V_{BAT}$, the drain voltage of M_R is higher than the drain voltage of M_P, while replica current is larger than expected. This means the charging current is too high so that $Q_R[x]$ changes from 'L' to 'H,' turning off $M_R[x]$ and $M_P[14+x]$ to decrease the charging current. If $V_{CUR} < V_{SET}$ and $V_{BAT\_R} < V_{BAT}$, by contrast, this means the charging current is too low so that $Q_R[x]$ changes from 'H' to 'L,' turning on $M_R[x]$ and $M_P[14+x]$ to increase the charging current. These two situations discussed above regulate the charging current without ambiguity, so they are both viewed as current regulation mode. On the other hand, if $V_{BAT\_R} < V_{BAT}$, $V_{CUR} > V_{SET}$ or $V_{BAT\_R} > V_{BAT}$, $V_{CUR} < V_{SET}$, CCCL moves to the evaluation mode. In these cases, whether the charging current is too large or small is to be further evaluated. Hence, CCCL changes the number of turned-on $M_Z[1:32]$ to change V_{BAT_R}, V_{CUR} and regulates the charging current until moving to the current regulation mode. Fig. 3(b) shows the timing diagram of the CC mode.

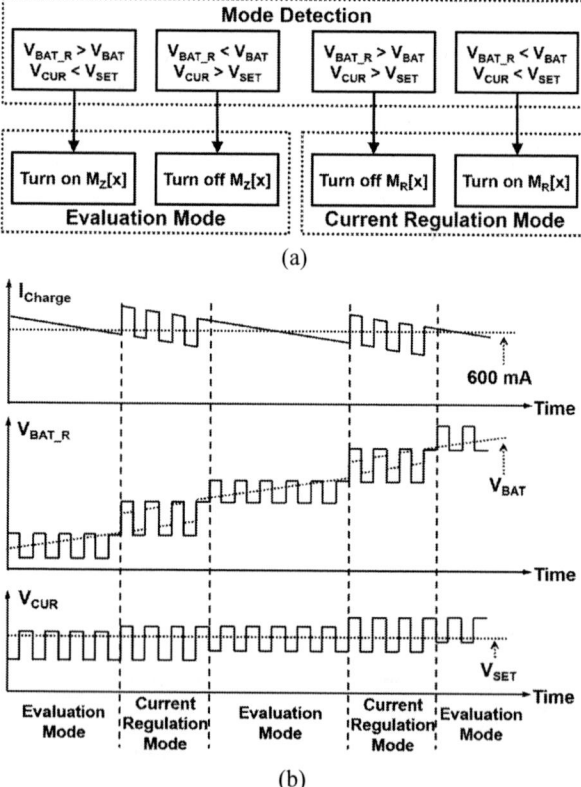

Fig. 3. (a) The operation block diagram, and (b) the timing diagram of CC mode.

978-1-5386-3179-9/17 $31.00 © 2017 IEEE

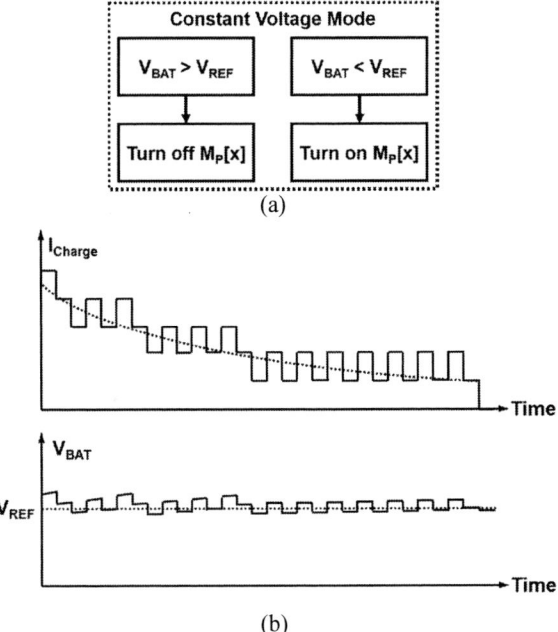

(a)

(b)

Fig. 4. (a) The operation block diagram, and (b) the timing diagram of CV mode.

B. Operation in the Constant-Voltage Mode

With the increasing of the battery voltage, if $V_{BAT} > V_{REF}$, CVCL changes $Q_{CV}[x]$ from 'L' to 'H.' $M_P[15:64]$ are no longer controlled by CCCL but by CVCL through the OR gates. Finally, the charger moves to the CV mode.

Fig. 4(a) shows the operation block diagram of the CV mode. If $V_{BAT} > V_{REF}$, CVCL turns off cells of $M_P[x]$ to gradually reduce the charging current. If $V_{BAT} < V_{REF}$, CVCL turns on cells of $M_P[x]$ to increase the charging current. The charging cycle ends when the last power transistor $M_P[1]$ is turned off. Fig. 4(b) shows the timing diagram of CV mode.

C. Dynamic Latched Comparator

Fig. 5 shows the circuit schematic of the dynamic latched comparator. While CLK is 'L,' the comparator is reset by $S_1 \sim S_4$, M_7, while V_{IN1}, V_{IN2} pass to C_1, C_2 through transmission gates. When CLK changes to 'H,' $S_1 \sim S_4$ are turned off, and C_1, C_2 store the input voltage. The comparator starts to compare the stored voltage with M_1, M_2, and then the latch brings one output to 'H' while the other to 'L.'

D. Shift Register

Fig. 6 shows the circuit schematic of a K-bit shift register. At first, all K bits are set to 'H' to turn off pMOS. If COMP_OUT is 'H,' all K bits are shifted toward right to increase the number of turned-on transistors. If comparator output is 'L,' on the other hand, all K bits are shifted toward left to decrease the number of turned-on transistors.

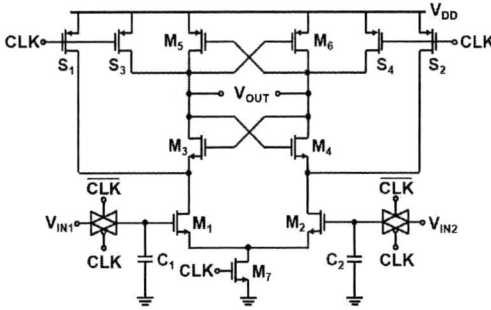

Fig. 5. Schematic of the dynamic latch comparator.

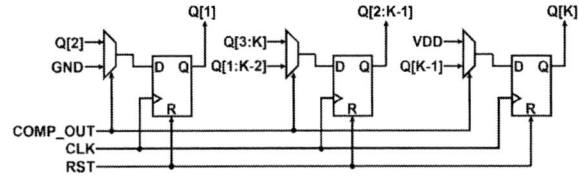

Fig. 6. Schematic of the K-bit shift register.

III. MEASUREMENT RESULTS

The proposed Li-Ion battery charger has been fabricated in a TSMC 0.35-μm CMOS process. A micrograph of the digitally-controlled charger is shown in Fig. 7. The total power transistors are sized to provide a 600-mA charging current in the CC mode and occupy only about 0.252 mm². The core area of the overall circuit is 0.768 mm².

To observe the charging process in a short time, C_{BAT} of 1 mF and R_{BIR} of 100 mΩ are used to model a battery. Fig. 8 shows the measurement results with the emulated battery. The operation frequency is 2.5 MHz. In pre-charge mode, the proposed charger turns on one unit of power transistors to charge the emulated battery with about 30 mA. After the charging mode is switched to the CC mode, the charging current is set to 600 mA. When the emulated battery voltage achieves 4.2 V, the charging mode is switched to the CV mode. The number of turned-on power transistors decreases gradually, so does the charging current. Finally, when the last turned-on power transistor is turned off, the charger stops charging.

Next, the proposed charger is applied to charge an actual 3400 mAh Li-ion battery, as shown in Fig. 7. Fig. 9 shows the measured charging profile of the CC and CV modes. Fig. 10 plots the measured charging current ripple in the CC mode. As shown in Fig. 10, the operation frequency is 1 Hz, and the charging current is 600 mA with a small ripple of 7.4 mA in the CC mode.

The proposed battery charger is compared with the previous works in Table I. This comparison shows that the proposed Li-Ion battery charger architecture is the most area-efficient without any external components. In addition, because of the digitally-controlled technique, the proposed battery charger has better process scalability.

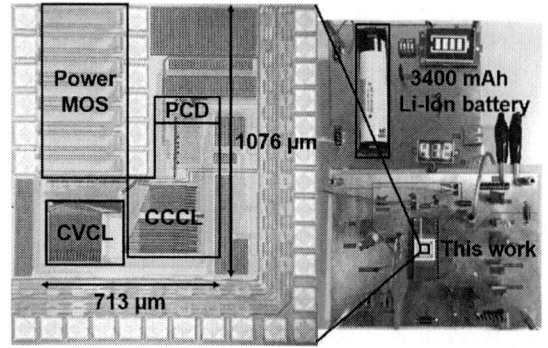

Fig. 7. Chip micrograph of the proposed digitally-controlled charger and the testing board photo.

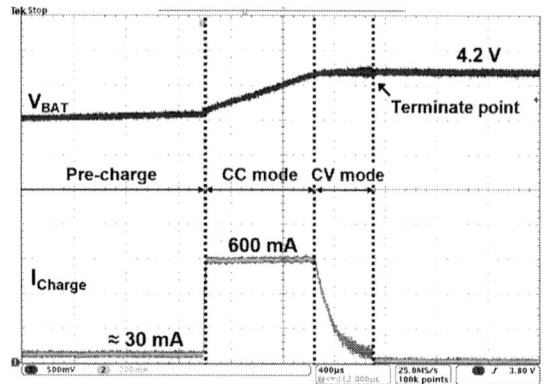

Fig. 8. Charging profile with the emulated battery of C_{BAT} of 1 mF and R_{BIR} of 100 mΩ.

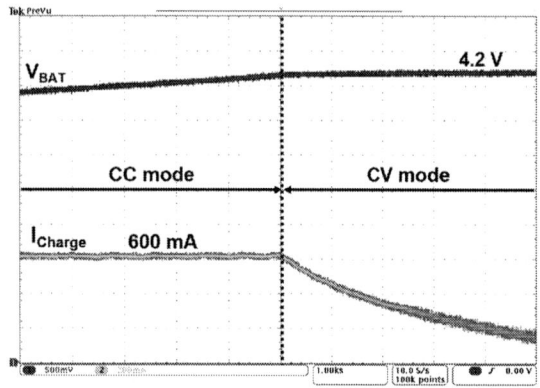

Fig. 9. Charging profile with an actual 3400mAh Li-ion battery.

IV. CONCLUSIONS

This paper presents an area-efficient, amplifier-less Li-Ion battery charger with a novel digitally-controlled architecture. The proposed Li-Ion battery charger eliminates all analog circuits and reduces the size of power transistors. Additionally, the nature of digital control suggests that the proposed charger has better process scalability. The output voltage is 4.2 V in the CV mode, and the charging current in the CC mode is 600 mA.

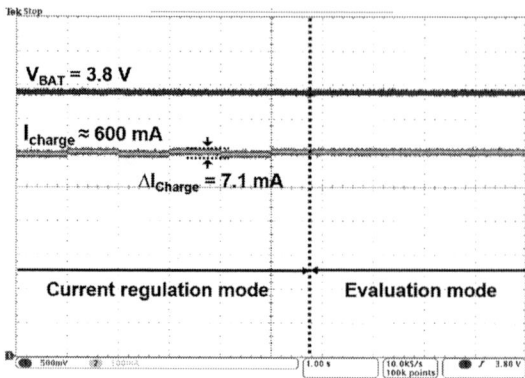

Fig. 10. Charging profile in the CC mode with an actual Li-Ion battery.

TABLE I. LI-ION BATTERY CHARGERS COMPARISON TABLE

	[2]	[3]	[4]	[5]	This work
Technology	*	TSMC 0.35-µm	TSMC 0.18-µm	TSMC 0.35-µm	TSMC 0.35-µm
Topology	Switching-based + LDO	Charge pump	Switching-based	LDO	Digitally-controlled
External component	Inductor	Capacitor	Inductor	None	None
Input voltage	5 V	5 V	5~10 V	4.5~6.5 V	5 V
Charging current	800 mA	700 mA	900 mA	500 mA	600 mA
Operation frequency	N/A	10 MHz	2.2 MHz	None	1 Hz
Efficiency	83 %	67.89 %	86 %	N/A	79 %
Chip area	*	1.96 mm²	1.6 mm²	1.1 mm²	0.77 mm²

*Discrete components

ACKNOWLEDGEMENT

The authors thank the National Chip Implementation Center for fabricating this chip. This work is supported in part by the MOST, Taiwan.

REFERENCES

[1] D. Linden and T. B. Reddy, *Handbook of Batteries*. New York: Mc-Graw-Hill, 2002, ch. 35.

[2] M. Chen and G. A. Rincón-Mora, "Accurate, compact, and power-efficient Li-Ion battery charger circuit," *IEEE Trans. Circuits Syst. II, Exp. Briefs*, vol. 53, no. 11, pp. 1180–1184, Nov. 2006.

[3] Y.-S. Hwang, S.-C. Wang, F.-C. Yang, and J.-J. Chen, "New compact CMOS Li-Ion battery charger using charge-pump technique for portable applications," *IEEE Trans. Circuits Syst. I, Reg. Papers*, vol. 54, no. 4, pp. 705–712, Apr. 2007.

[4] R. Pagano, M. Baker, and R. E. Radke, "A 0.18-µm monolithic Li-Ion battery charger for wireless devices based on partial current sensing and adaptive reference voltage," *IEEE J. Solid-State Circuit*, vol. 47, no. 6, pp. 1355-1368, Jun. 2012.

[5] C.-H. Lin, C.-Y. Hsieh, and K.-H. Chen, "A Li-Ion battery charger with smooth control circuit and built-in resistance compensator for achieving stable and fast charging," *IEEE Trans. Circuits Syst. I Reg. Papers*, vol. 57, no. 2, pp. 506-517, Feb. 2010.

[6] T.-C. Huang, R.-H. Peng, T.-W. Tsai, K.-H. Chen, C.-L. Wey, "Fast charging and high efficiency switching-based charger with continuous built-in resistance detection and automatic energy deliver control for portable electronics," *IEEE J. Solid-State Circuit*, vol. 49, no. 7, pp. 1580-1594, Jul. 2014.

A 0.5V BJT-Based CMOS Thermal Sensor in 10-nm FinFET Technology

Da Shin Lin[1] and Hao Ping Hong[2]

1 MediaTek, HsinChu, Taiwan

2 MediaTek USA, Woburn, MA USA

ds.lin@mediatek.com

Abstract

A 0.5-V BJT-based thermal sensor design is first demonstrated in a 10-nm CMOS. A charge-pump technique is proposed for operating with a digital core supply voltage as low as 0.5V without the restriction on forward junction bias (~0.7V). A switched-capacitor integrator loop is presented for process-insensitive voltage-to-frequency conversion. This thermal sensor achieves an RMS resolution of ±0.173°C, resolution FoM of $4\,nJ\,°C^2$, and 3σ-spread accuracy of ±2.8°C at 85°C. The area is $0.01\,mm^2$ and consumes 125uW from a 0.5-V supply. To the best of author's knowledge, this is the first BJT-based thermal sensor able to operate in sub-0.7V core supply voltage. The proposed low-voltage and small-area thermal sensor is easily to be integrated inside the CPUs, GPUs, and application process (AP) chips for real-time monitoring of hot spots.

Keywords—Low-voltage thermal sensor; Low-voltage bandgap; thermal sensor; AP hot spot; cellular phone hot spot

Introduction

Modern mobile devices require a fast application processor (AP). Its highest operating speed is generally limited by the thermal issues. Hence an accurate temperature sensing is essential to maximize the speed of the AP. In the 10-nm FinFET process, the self-heating effect deteriorates the temperature increasing in a small area. It is necessary to place the thermal sensors operating with a sub-1V digital core supply voltage near these local hot spots for chip reliability.

Horng [2] demonstrated a sub-1V resistor-based sensor. However, the resistor-type sensor is seldom used for mass production because of the incomplete model of the temperature coefficient of the resistor in the design phase. Sönmez [4] reports a small-sized sensor at 40nm, but his implementation requires a specific component, ETF (electrothermal filter), which is not a standard product from a foundry that is not flexible for mass production. Additionally, regarding its current consumption of 2.3mA, the number of sensors in an AP with multi-cores CPUs and GPUs is usually ten or more; therefore, the total current consumption is not appropriate for the demand of low power for a mobile device application.

Instead of using a resistor or an ETF, the BJT is more robust for mass production, and its temperature coefficient model is more accurate in the design phase. However, it is challenging to design a BJT-based thermal sensor with a digital core supply voltage as low as 0.5V.

This work demonstrates a sub-0.7V BJT-based thermal sensor low-voltage bandgap design with a ping-pong charge-pump circuit in 0.5-V core supply voltage. And, the proposed tracking loop that is insensitive to process variations for the voltage-to-frequency (V2F) conversion. The DC PSRR is calibrated by a DC supply calibration table, and the AC PSRR is solved by an integrator and long-time averaging frequency meter.

Thermal Sensor Architecture

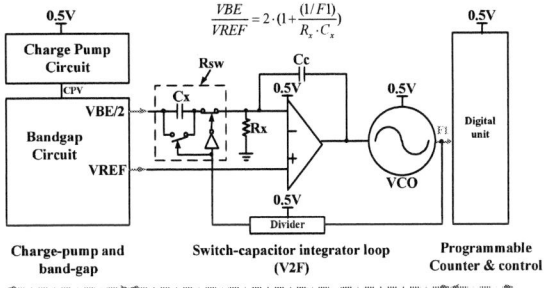

Fig. 1 Block diagram of proposed thermal sensor

Fig. 1 shows a block diagram of the proposed thermal sensor. It is composed of a low-voltage bandgap with a charge-pump circuit, a voltage-to-frequency converter, and a digital logic unit. The bandgap circuit builds a temperature-independent reference voltage, VREF, and a temperature-dependent voltage, VBE. The ratio of VBE/2 to VREF is converted into a frequency by the switched-capacitor integrator loop. In the steady state, the integrator loop forces the ratio of (Rsw+Rx)/Rx equal to the ratio of VBE/2/VREF through the frequency F1, and it is proportional to the temperature. Rsw is a switched capacitor resistor that is equal to 1/(F1·Cx). As a result, the ratio of VBE/2/VREF is equal to 1+1/(F1·Rx·Cx), which is proportional to the temperature. As in

$$\text{VBE} \propto Temperature, \frac{\frac{VBE}{2}}{VREF} = \frac{R_X + R_{SW}}{R_X}$$
$$= \left[1 + \left(\frac{1}{F1}\right)\left(\frac{1}{R_X C_X}\right)\right] \propto Temperature. \quad (1)$$

The ratio of VBE/2/VREF is proportional to temperature, so the 1/F1 is proportional to temperature. F1 is calculated by a frequency meter inside the digital unit. The digital unit includes a programmable counter with an average range of 600ns to 2.5ms and a 3-wire control interface.

The proposed charge-pump bandgap and ac-couple VCO are for dealing with such low core supply voltage. The switched-capacitor integrator loop is to convert the sensing temperature information to an oscillation frequency, and it also helps in dealing with noisy core supply power. The embedded frequency meter is to reduce the high-frequency

978-1-5386-3179-9/17 $31.00 © 2017 IEEE

signal routing spread in the chip top for power saving. A 3-wire control unit is integrated into the thermal sensor macro for interface reduction, with the more efficient thermal calibration.

Low-Voltage Thermal Sensor Techniques

A. Charge-Pump Bandgap

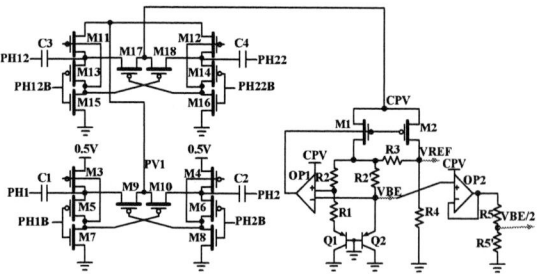

Fig. 2 Charge-pump circuit and bandgap.

Fig. 2 shows the proposed low-voltage bandgap with a cascade charge-pump circuit. The low-voltage bandgap builds a tunable bandgap reference voltage of ~0.15V, which is not restricted to the 1.2V of the conventional bandgap. The proportional to absolute temperature (PTAT) current of Q1, Q2 is summed with the negative temperature coefficient current of R3, which produces the temperature-independent current of M1. M2 current is mirrored from M1, and flows through a temperature independent resistor, R4 for the generation of the bandgap voltage of VREF.

In order to enhance the operation margin, the cascade charge pump is proposed, which pumps a core supply voltage of 0.5V to $\approx$ 1.2V (CPV) to obtain more headroom for the bandgap circuit, which is composed of IO devices Mx (x=3–18) and capacitors Cx (x=1–4). PH1, PH2, PH12, and PH22 are non-overlapping clocks. PHBx (x=1, 2, 12, 22) are out-of-phase clocks of PHx (x=1, 2, 12, 22). The drain node of M3 is preset to the supply voltage. Once PH1 goes from low to high, the voltage of the drain of M3 will ideally be twice the supply voltage. C1, M3, and M9 and C2, M4, and M10 are for ping-pong charging on node PV1. To pump the extremely low supply voltage of 0.5V, the cascade stage of Mn (n=11~18), C3, and C4 is used.

The bandgap circuit is a composite of PNP Q1 and Q2, IO devices M1 and M2, resistors R1–R4, and OP1 and OP2. Due to the finite gain of OP1 and OP2, which use IO devices. The chopping circuit is used to deal with the offset error. The mismatch between M1 and M2 must also be considered.

B. Switched-Capacitor Integrator Loop

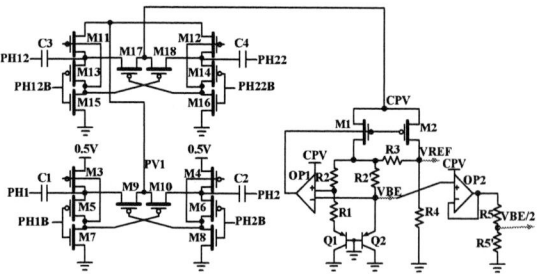

Fig. 3 Switched-capacitor integrator loop

Fig. 3 shows the proposed switched-capacitor integrator loop, which converts voltage, Vc, to an oscillation frequency, F1. The negative feedback loop ensures that the equality of (1) is established and independent of device process corners.

This loop consisted of a switch-capacitor resistor (Rsw), an integrator, an ac-coupled VCO, and a feedback divider. This loop tracking behavior is described as follows. When the temperature increases, VBE voltage will decrease, and the current will flow from Cc to Rsw which results in increasing the control voltage of VCO, Vc, as well as VCO output frequency. Then, the higher oscillation frequency leads to a lower value of switched-capacitor resistor (Rsw). In steady state, the closed loop will force the resistor ratio of Rs and Rsw to be equal to the voltage ratio of VBE/2 and VREF. In this design, F1 $\approx$ 210MHz, and the loop bandwidth (f_{3dB}) $\approx$ 250kHz at 30°C.

In this loop, several design considerations are needed for accurate temperature sensing. First, the OP3/OP4 offset error needs to be considered. Chopping technique is used to deal with this offset voltage. Second, in extremely low supply voltage (~0.5V) the VCO voltage swing would be very low. Hence, the ac-coupled connection is employed for a reliable voltage level shifting. Third, for the variation of Rx and Cx, based on (1), the VCO clock period is derived as,

$$T_x = \frac{1}{F1} = R_x \cdot C_x \left(\frac{VBE}{2 \cdot VREF} - 1 \right). \qquad (2)$$

(2), shows that the inverse of the VCO frequency is proportional to Rx·Cx times the ratio of VBE/2 and VREF. The variation of RxCx will impact the results of the VCO frequency. Therefore, the VCO output frequency is not only affected by temperature but also by the process variation of Rx·Cx . In order to eliminate the process variation of Rx·Cx, the VCO frequency or period (T_{30C}) is recorded at 30°C and the other temperature can be accurately obtained by the normalization to T_{30C} as in (3).

$$\frac{T_x}{T_{30°C}} = \frac{R_x C_x \left(\frac{VBE_{x°C}}{VREF} - 1 \right)}{R_x C_x \left(\frac{VBE_{30°C}}{VREF} - 1 \right)}. \qquad (3)$$

The first order variation of the resistor (Rx) and the capacitor (Cx) can cancel each other out. Thus, the proposed switched-capacitor integrator loop is able to sense an accurate temperature which is independent of the variation of MOS devices, resistors, and capacitors.

Calibration

For a thermal sensor, an accurate calibration is always important. In addition to a stable temperature environment, the extra heat inside the chip needed to minimize. This is a challenge for the thermal sensor, if the digital core power as our supply voltage, which suffers a lot of leakage power from other digital macros in the core power domain. Thus, it is almost impossible to do the thermal calibration precisely in the digital core power domain with package-level testing.

Instead of using package-level testing, a wafer-level testing is adopted in this design because the whole wafer is a good thermal conductor which is better for heat dissipation. Usually the wafer would have over a thousand dies on it; when the thermal calibration is done in one die, the substrate of the other thousand dies will help with the heat dissipation. The thermal distribution at the thermal calibration in the wafer-level testing calibration is confirmed by using a computational fluid dynamics analysis. Fig. 4 shows that the temperature spread error is less than ±0.5°C when there is 800mW leakage from digital core power.

In order to minimize the testing cost, the one-point calibration at 30°C is applied. The calibration not only records the accurate temperature, but also eliminates the process variation of resistor and capacitor.

due to by leveraging the advantage of using an advanced fine-line 10nm process technology.

The low-voltage thermal sensor performance summary and comparison with the state-of-the-arts are shown in Table 1. Compared to prior works [1], [5], [6], [7], all of which are BJT-based sensors, this work is the first sub-0.7V BJT-based sensor. Compared to [1] and [6], this work shows a better FoM and with a lower supply voltage. Moreover, this work shows a smaller area and lower supply voltage compared to [5] and [7].

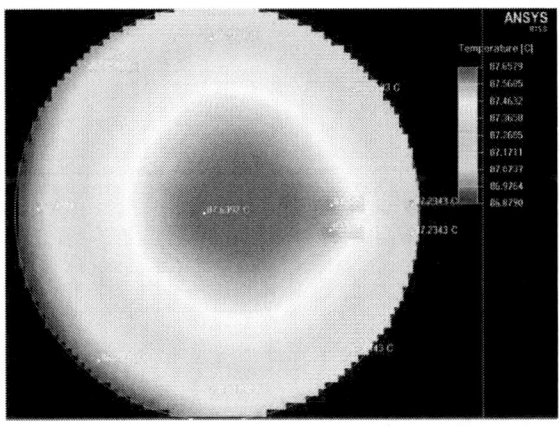

Fig. 4 Wafer-level testing. Assuming 8-site testing, each site has 100mW power, and the total is 800mW. The error from core power leakage < ±0.5°C.

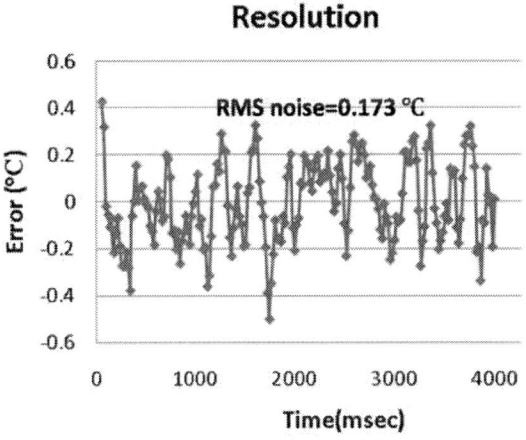

Fig. 5 Measured resolution at 30°C during 4000ms

Measurement Results

In cellular phone applications, when the IC temperature exceeds the upper limit, the chip throttles and reduces the operating speed for cooling down. Usually, the experience number of temperature of throttles is 85°C. Lower temperatures of 0°C or less are not a concern in our application.

Fig. 5 shows that the resolution of the proposed sensor is 0.179°C (RMS) over 4000ms at 30°C. The spread of the sensor accuracy is ±2.8°C (±3σ) at 85°C with 1-point calibration at 30°C for 20 samples, as shown in Fig. 6. The measured temperature range from 30°C to 85°C is based on the system requirement. This work adopts the wafer-level testing. This sensor consumes 250uA from the 0.5-V supply.

Fig. 7 shows a die photo, and the area of this sensor is 0.01 mm². Most of the area is from the charge-pump and the bandgap circuit, due to its large number of capacitors for voltage pumping and storage. The voltage to frequency converter (V2F) area is dominated by the integrator capacitor ~10pF. For the digital logic unit, the area is quite small (one sixteenth of total area),

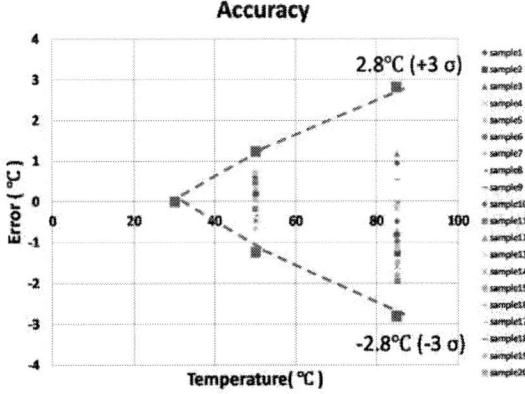

Fig. 6 Measured temperature error ±2.8°C (3 sigma) at 85°C with 1-point calibration at 30°C for 20 samples

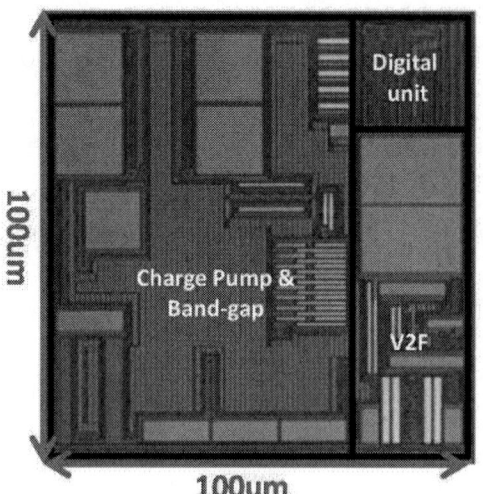

Fig. 7 Die microphotograph

Conclusion

This work demonstrates the first sub-0.7V BJT-based thermal sensor. This sensor combines the robustness of a BJT device with ultra-low-voltage operation, meaning the sensor can be fully integrated in digital core power domain to allow ball count saving and more accurate measurement of the hot spot. How to calibrate the thermal sensor in the digital core power domain with extra heat from core power leakage of 100mW is also demonstrated.

Acknowledgment

The authors would like to thank CM Chen, CS Chao, KD Chen, Alfred Tsai, Kidd Chen, MediaTek for technique discussion.

Table.1 Comparison table

	ISSCC12 [1]	ISSCC14 [5]	JSSCC15 [6]	VLSI16 [7]	This work
Technology	32nm	700nm	14nm	160nm	10nm
Type	BJT	BJT	BJT	BJT	BJT
Chip area	$0.02mm^2$	$0.8mm^2$	$0.0087mm^2$	$0.16mm^2$	$0.01mm^2$
Supply voltage	1.4V	2.9V	1.35V	1.5V	0.5V
Current	2700uA	55uA	-	4.6uA	250uA
Conversion time (tm)	50u~ 500usec	2.2msec	0.02msec	5msec	5u~ 2.5msec
Spread (3σ)	±4.5°C	±0.15°C	±1.65°C	±0.05°C	±2.8°C
Calibration method	1-point	1-point	2-point	1-point	1-point
Resolution (conversion time, tm)	0.19°C (500uS)	0.003°C (2.2mS)	0.5°C (0.02mS)	0.015°C (5mS)	0.173°C (1mS)
FOM ($nJ°C^2$) +	68	0.0036	5.7	0.0078	4

+ FOM = Energy/Conversion * Resolution^2

References

[1] J. Shor, et al. "Ratiometric BJT-based thermal sensor in 32nm and 22nm technologies," in IEEE ISSCC Dig. Tech. Papers, pp. 210–212, Feb 2012.

[2] Jaw-Juinn Horng, et al. "A 0.7V resistive sensor with temperature/voltage detection function in 16nm FinFET technologies," in IEEE VLSI Circuit Symp. Papers, pp. 54–55, June 2014.

[3] Xiao Pu, et al. "A ±0.4°C accurate high-speed remote junction temperature sensor with digital beta correction and series-resistance cancellation in 65nm CMOS," in IEEE VLSI Circuit Symp. Papers, pp. 214–215, June 2013.

[4] Uğur Sönmez, Fabio Sebastiano, and Kofi A.A. Makinwa "1650um² thermal-diffusivity sensors with inaccuracies down to ±0.75°C in 40nm CMOS," in IEEE ISSCC Dig. Tech. Papers, pp. 206 –207, Feb 2016.

[5] Ali Heidary et al. "A BJT-based CMOS temperature sensor with a 3.6pJoC2 resolution FoM," in IEEE ISSCC Dig, pp. 224–225. 2014.

[6] Takao Oshita et al. "Compact BJT-based thermal sensor for processor applications in a 14 nm tri-gate CMOS process," IEEE J. Solid-State Circuits, l. vol. 50, iss. 3, pp. 799–807, 2015.

[7] Bahman Yousefzadeh et al. "A BJT-based temperature-to-digital converter with ±60mK (3σ) inaccuracy from -70°C to 125°C in 160nm CMOS," in IEEE VLSI Circuit Symp. Papers, pp. 1–2, June 2016.

[8] K. Makinwa, "Online temperature sensor survey," Tech. Univ. Delft, The Netherlands. [Online]. Available: http://ei.ewi.tudelft.nl/docs/TSensor_survey.xls

[9] J. Shor, K. Luria, and D. Zilberman, "Ratiometric BJT-based thermal sensor in 32 nm and 22 nm technologies," in IEEE ISSCC Dig. Tech. Papers, 2010, p. 210.

[10] J. Shor and K. Luria, "Evolution of thermal sensors in Intel processors from90 nm to 22 nm," in Proc. IEEE 27th Convention Electr. Electron. Eng. Israel, Nov. 2012.

[11] H. Banba, et al. "A CMOS bandgap reference circuit with sub-1-V operation," IEEE J. Solid-State Circuits, vol. 34, no. 5, pp. 670–674, May 1999.

[12] S. Paek, W. Shin, J. Lee, H. Kim, J. S. Park, and L. S. Kim "All-digital hybrid temperature sensor network for dense thermal monitoring," in IEEE ISSCC Dig. Tech. Papers, 2013, p. 260.

An Ultra-low Power 169-nA 32.768-kHz Fractional-N PLL

Chun-Yu Lin, Tun-Ju Wang, Tzu-Hsuan Liu, and Tsung-Hsien Lin

Graduate Institute of Electronics Engineering and Department of Electrical Engineering,
National Taiwan University, Taipei, Taiwan

Abstract - This paper presents an ultra-low power 32.768-kHz fractional-N phase-locked loop (PLL). Several circuit techniques are adopted to facilitate low-power operation. A duty-cycled control scheme is proposed to turn off the charge pump intermittently for energy saving. In the VCO, a near-off switch is applied to implement a large resistor, leading to a lower power consumption and smaller chip area. Furthermore, the digital power consumption is reduced by operating the digital circuits at a lower supply voltage generated from an on-chip voltage regulator. This PLL is fabricated in the TSMC 180-nm CMOS process with a core area of 0.116 mm². Its current consumption is 169 nA from a 1-V supply. The measured peak-to-peak cycle-to-cycle jitter is 529.4 ns.

I. Introduction

Recently, 32.768-kHz clock generator has found more and more applications for timekeeping functions such as maintaining calendar time, scheduling system events, and time-stamping measured data. Furthermore, it enables longer battery life as it can facilitate a system to enter the deep sleep mode most of the time and wake up periodically to perform certain actions. Such operations are widely applied in mobile phones, wearable devices, and IoT systems.

Although crystal oscillators play a dominant role in clock generation ([1], [2]), due to the increasing demand for small footprint in portable devices, other alternatives e.g. MEMS-based oscillator, have emerged to take the place of conventional quartz crystal oscillators. MEMS-based oscillators have the advantage of enabling smaller, lower power IoT and wearable applications [3]. Typically, crystal oscillators have the frequency stability in the range of ±30 ppm, while the frequency stability of MEMS-based oscillators is in the range of ±100 ppm. Consequently, it requires a fractional-N PLL with a temperature sensor to compensate the effect of temperature variation, as shown in Fig. 1 [4].

Other than the form factor, power consumption is another critical concern for the oscillator design [5]. It goes without saying that low-power circuit architecture is the key to enable long battery lifetime for portable devices. Although [4] demonstrates the first 32-kHz low-power MEMS-based oscillator, the PLL designed in this work dissipates considerable amount of power in the whole system. It accounts 290-nA current consumption, which is larger than 30% of the current of the whole chip. Further reduction of the PLL power consumption is important.

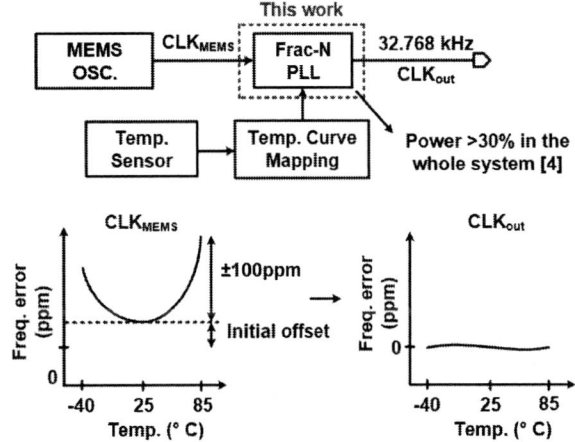

Fig. 1. Application scenario of the proposed ultra-low-power PLL.

In this paper, an ultra-low power 169-nA 32.768-kHz fractional-N PLL is presented. This paper is organized as follows. The proposed PLL architecture is given in Section II, and Section III presents the circuit designs. The measurement results and conclusion are given in Section IV and V, respectively.

II. System Architecture

Fig. 2 shows the architecture of the proposed fractional-N PLL. The reference clock (F_{REF}) is fed to a dual-modulus (N/(N+1); N = 16 in this design) pre-divider through which its frequency is divided down to 32.768 kHz. The dual-modulus pre-divider is controlled by a first-order 18-bit digital delta-sigma modulator (DSM) which determines the required division ratio to generate the target output frequency. Digital code can be adjusted to compensate for temperature variation.

Although the target frequency is already available at the output of the pre-divider, the quantization noise from the DSM will contribute large deterministic jitter. Therefore, an integer-N PLL with a narrow loop bandwidth of 1 kHz is applied after the output of the pre-divider to suppress this noise. The final output (F_{OUT}) is derived from the integer divider in the feedback path. The nominal frequency of the proposed VCO is designed to be 262.144 kHz, 8 times the target 32.768 kHz, and the division ratio of divider is 8. Several techniques are

978-1-5386-3179-9/17 $31.00 © 2017 IEEE

employed in this integer-N PLL design to lower the power consumption. These techniques are described in the next section.

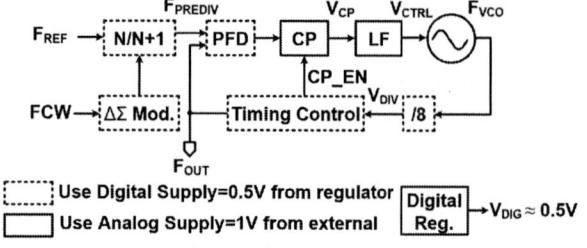

Fig. 2. Architecture of the proposed ultra-low power fractional-N PLL.

III. CIRCUIT IMPLEMENTATION

The main circuit techniques adopted in this work to achieve low-power operation are: (1) a duty-cycled charge pump (CP) to reduce the energy waste when the CP is inactive; (2) a near-off switch to implement a large resistor in the VCO, leading to a lower power consumption and smaller chip area; (3) an on-chip voltage regulator to generate lower supply voltage (~0.5 V) for digital circuits, reducing the digital power consumption. These design techniques as well as the key building blocks of the proposed PLL are discussed as follows.

A. Duty-cycled Charge Pump

Fig. 3 shows the schematic of the proposed CP. MOS switches, M_1 and M_2, allow pumping charges into or out of the loop filter according to the phase error signal UPB and DN generated from the PFD. The CP current is designed to be 16 nA. When M_1 and M_2 are both off, the currents are steered away to M_3 and M_4 instead of being switched off. This CP topology eliminates the current glitches. Furthermore, since the CP is always on, the circuit switching speed is not compromised as there is not start-up issue for the current sources. Although, keeping the CP continuously on eliminates the start-up time issue, this is at the cost of continuous power consumption.

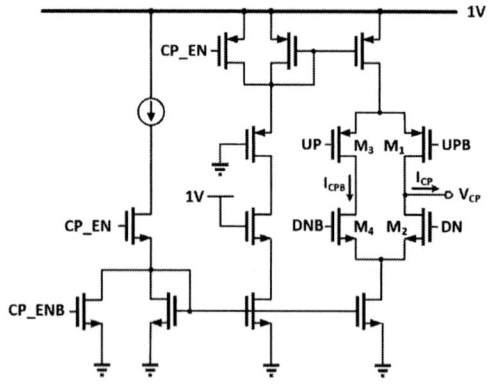

Fig. 3. Schematic of the proposed charge pump.

In order to further reduce the CP power consumption, a duty-cycled control scheme, as shown in Fig. 4, is proposed in this work. Inserted in the feedback path, a timing control circuit is applied to switch the CP on only when necessary.

To ensure a proper operation, the CP should be completely turned on and settled before the feedback signal arrives at the PFD input, but not necessarily the whole time. The control circuit implementation is described below. Three D-flip-flops which are clocked by the VCO output are used to delay the output signal from the divider by 1, 2, and 3 VCO clock cycles, generating S_1, S_2, and S_3, respectively. Signal S_2 is sent to the PFD for detection; while S_1 and S_3 are taken to produce the control signal CP_EN of the CP. When S_1 is high and S_3 is low, CP_EN is set to high and enables the CP. In other conditions, CP_EN is set to low and disables the CP. In this manner, the CP is turned on just one VCO clock cycle before the feedback signal arrives at the PFD input, and is turned off after another VCO clock cycle ends. That is, the CP is only operational for two VCO periods during 1 PFD comparison cycle.

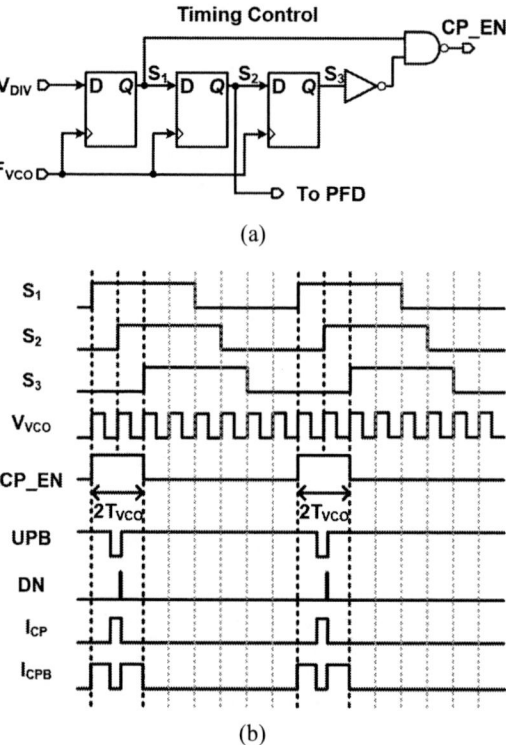

Fig. 4. (a) CP control circuit; (b) Timing diagram of the proposed CP control.

Since the division ratio is 8, and thereby $T_{PREDIV} = 8 \times T_{VCO}$, the CP is off for 75% of the time in one PFD comparison cycle. Thus, the average CP power consumption is reduced by roughly the same amount. Here, the nominal frequency of the VCO is designed to be 262.144 kHz. In the locked state, the phase error between the two inputs of the PFD is very small and falls into this range of two VCO cycles. In this design, the

CP start-up time is much less than 1 VCO period, and the scheme works well without any issue. In short, with the proposed duty-cycled control technique, the energy wasted while the CP is idling is considerably reduced.

B. VCO

The schematic of VCO is depicted in Fig. 5(a). It consists of a Schmitt trigger, inverters and a RC network. The Schmitt trigger provides the upper threshold (V_{T+}) and lower threshold (V_{T-}) for the RC network to switch between charging and discharging states. When V_{out} is pulled up, V_{inv} is pulled down to discharge V_x until V_x is lower than V_{T-}, and vice versa. The operation timing waveforms of proposed VCO are illustrated in Fig. 5(b), where $t_1 = RC \times \ln[(V_{DD,VCO}-V_{T-})/(V_{DD,VCO}-V_{T+})]$, $t_2 = RC \times \ln(V_{T+}/V_{T-})$, and the oscillation frequency can be estimated from (1).

$$f = \left\{ RC\ ln \left[\left(\frac{V_{T+}}{V_{T-}}\right) \left(\frac{V_{DD,VCO}-V_{T-}}{V_{DD,VCO}-V_{T+}}\right) \right] \right\}^{-1} \quad (1)$$

In (1), $V_{DD,VCO}$ is the supply voltage of the Schmitt trigger and inverters. The nominal frequency of the proposed VCO is designed to be 262.144 kHz.

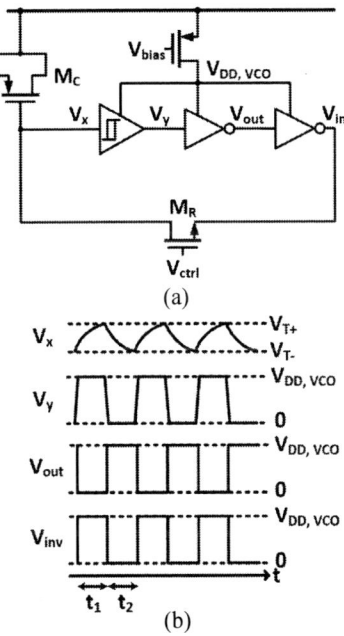

(a)

(b)

Fig. 5. (a) VCO schematic. (b) Operation timing waveforms of proposed VCO.

As shown in Fig. 5(a), the load capacitor is implemented with a MOS capacitor M_C. A large resistor is required in the VCO. However, it would exhibit large parasitic capacitance, causing extra current consumption. In order to achieve a low-power VCO, a near-off switch M_R is used to implement the large resistor (pseudo-resistor). The drain current of the MOS operating in the subthreshold region is given by (2).

$$I_D = I_{D0}\frac{W}{L}e^{V_{gs}/(nV_t)}\left(1 - e^{-V_{ds}/V_t}\right) \quad (2)$$

In (2), I_{D0} is a characteristic current, n is a slope factor, and $V_t = kT/q$. This pseudo resistor implementation also leads to a smaller chip area. Although a near-off switch by itself suffers from PVT variation, the loop will lock it to a proper state. Finally, V_{CTRL} controls the equivalent resistance of M_R and also the rate of charging or discharging the MOS capacitor, such that the F_{VCO} in the locked PLL is 8 times the F_{PREDIV}.

C. Voltage Regulator

The dynamic power consumption of a CMOS logic circuit is estimated by CV^2f, where C is the effective load capacitance, V is the supply voltage, and f is the frequency of operation. Hence, an effective power reduction strategy for the digital blocks is to reduce the supply voltage while balanced against the impact on the logic speed. Since the circuit speed is low (less than a few hundred kHz) in this work, reducing supply voltage of digital circuits for power saving is utilized for low power design.

A voltage regulator is applied here to generate a lower supply voltage for digital blocks. The schematic of the adopted voltage regulator is shown in Fig. 6 [4]. A pair of cascaded PMOS and NMOS (M_P and M_N) plays the role of a simplified replica of basic digital cells. The reference voltage of new supply level for the digital blocks, V_{REF}, is then generated with a constant bias current flowing into M_P and M_N; and $V_{DIG} \approx V_{REF}$ with the unity gain buffer. In the proposed PLL, V_{DIG} is roughly 0.5 V. The digital power consumption is then reduced by this lower supply voltage generated from the proposed on-chip voltage regulator.

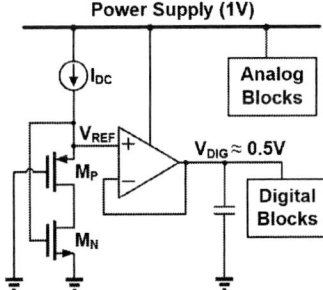

Fig. 6. Schematic of the proposed voltage regulator.

IV. MEASUREMENT RESULTS

The proposed PLL is fabricated in the TSMC 180-nm CMOS process. Fig. 7 shows the chip micrograph. The core area is 0.398×0.292 mm². The reference clock for this work is chosen to be 525 kHz, and the division ratio of the dual-modulus pre-divider is 16/17, providing an output frequency of 32.768 kHz. As shown in Fig. 8, the current consumption of the analogs blocks, including CP, VCO, voltage regulator and level shifter, is 91 nA. Without the duty-cycled control technique for the CP, it becomes 128 nA. For digital blocks, including pre-divider, DSM, PFD, divider, and timing control circuit, the current consumption is 78 nA. The total current consumption of this proposed PLL is 169 nA from a 1-V supply voltage.

Fig. 9 shows the measured output spectrum of this PLL at 32.768 kHz output. The eye diagram is shown in Fig. 10 and the peak-to-peak period jitter is 482.2 ns. Fig. 11 shows the histogram of this proposed PLL. The peak-to-peak cycle-to-cycle jitter is measured as 529.4 ns. Table 1 summarizes the performance of this work and compares this design to other similar works.

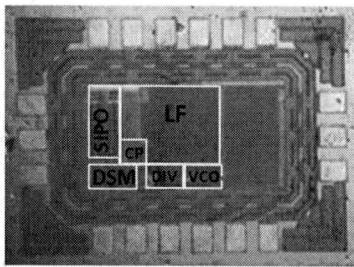

Fig. 7. Chip micrograph.

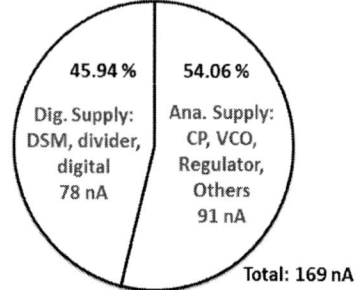

Fig. 8. Current consumption pie chart of the proposed PLL.

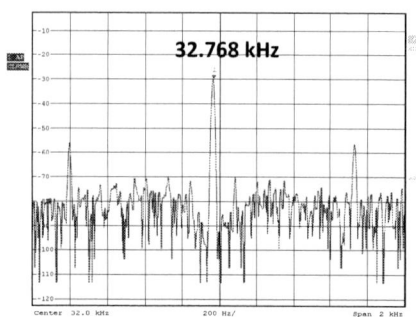

Fig. 9. Output spectrum of the proposed PLL.

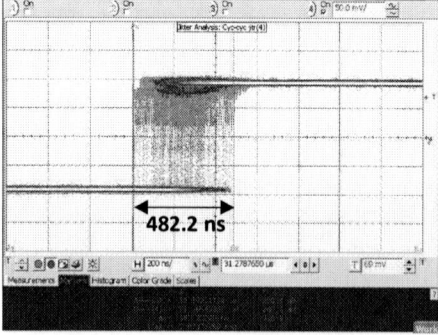

Fig. 10. Eye diagram of the proposed PLL.

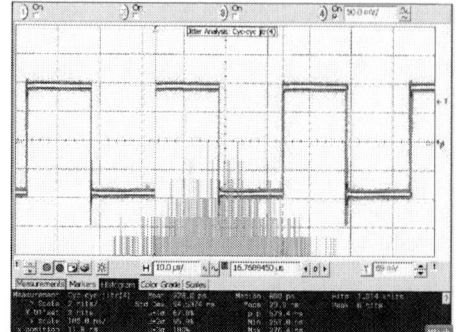

Fig. 11. Histogram of the proposed PLL.

Table I. Performance Comparison.

	[1]	[2]	[4]	[6]	**This work**
Tech. (nm)	N/A	N/A	180	60	180
Type	Crystal oscillator	Crystal oscillator	PLL	Relaxation oscillator	PLL
Supply (V)	1.2~5.5	1.2~4.5	1.2~4.5	1.6~3.2	1
Output freq. (kHz)	32.768	32.768	32.768	32.768	32.768
Current consumption (nA)	2000 max	300 typ 500 max	290	2800	169
Area (mm²)	N/A	N/A	N/A	0.048	0.116

V. CONCLUSION

This paper presents an ultra-low power PLL. A duty-cycled control technique is proposed, significantly reducing the power consumption of the CP. Also, a near-off switch is applied to implement a large resistor in the VCO, leading to a lower power consumption and smaller chip area. Furthermore, the digital power consumption is reduced by operating the digital circuits at a lower supply voltage generated from an on-chip voltage regulator. This 32.768-kHz PLL is fabricated in a TSMC 180-nm CMOS technology with core area of 0.116 mm². It consumes 169 nW from a 1-V supply. This is considerable lower than prior works.

ACKNOWLEDGMENT

The authors thank the CIC, Taiwan, for chip fabrication. This work is support in part by MOST, Taiwan.

REFERENCES

[1] Datasheet, Epson SG-3050BC [Online]. Available: https://support.eps-on.biz/td/api/doc_check.php?dl=brief_SG-3050BC&lang=ja
[2] Micro Crystal OV-7604-C7, Datasheet Farnell [Online]. Available: http://www.farnell.com/datasheets/14223.pdf
[3] M A. Partridge and H. C. Lee, "We know that MEMS is replacing quartz.But why? And why now?," in *Proc. 2013 Eur. Frequency and TimeForum & Int. Frequency Control Symp.*, Jul. 2013, pp. 411–416.
[4] S. Zaliasl et al., "A 3 ppm 1.5 x 0.8 mm² 1.0 µA 32.768 kHz MEMS-based oscillator," *IEEE J. Solid-State Circuits*, vol. 50, no. 1, pp. 1–12, Jan. 2015.
[5] A. Burdett, "Ultra-Low-Power Wireless Systems: Energy-Efficient Radios for the Internet of Things," in *IEEE Solid-State Circuits Mag.*, June 2015, pp. 18–28.
[6] K.-J. Hsiao, "A 32.4ppm/°C 3.2-1.6V self-chopped relaxation oscillator with adaptive supply generation," *Dig. Symp. VLSI Circuits*, pp. 14-15, June. 2012.

A 10kHz-BW 93.7dB-SNR Chopped $\Delta\Sigma$ ADC with 30V Input CM Range and 115dB CMRR at 10kHz

Long Xu, Johan H. Huijsing, Kofi A.A. Makinwa

Electronic Instrumentation Laboratory
Delft University of Technology, Delft, The Netherlands
Email: L.Xu-1@tudelft.nl

Abstract— **This paper presents a 15-bit $\Delta\Sigma$ ADC with 10kHz-BW which can handle 30V CM voltages with high AC CMRR (in excess of 115dB at 10kHz) while operating from a 1.8V supply. An HV capacitively-coupled chopper at its input enables the accurate sampling of input signals beyond the supply rails. Chopping is used to mitigate the ADC's offset and to enhance its CMRR, especially at high frequencies.**

Keywords—beyond-the-rails; chopping; CMRR; $\Delta\Sigma$ ADC

I. INTRODUCTION

For power management in many electronics systems, the supply current is often measured by inserting a shunt resistor between the power supply and the load as shown in Fig. 1, a technique called high-side current sensing. Compared to low-side current sensing, high-side current sensing has the following advantages: 1. no extra resistor is added to the ground path; 2. high currents caused by accidental shorts to ground can be detected. However, the small differential voltage drop across the shunt resistor must then be read out in the presence of large common-mode (CM) voltages. As a result, such systems typically consist of high voltage (HV) instrumentation amplifiers (IAs) [1-2] that translate differential signals down to a low-voltage domain, where they can be digitized by a conventional ADC. In applications where large CM voltage transients may occur, e.g. in motor current monitoring, the IA must have a high AC CMRR, while in battery monitoring applications, a bandwidth of several kHz may be required to track current peaks.

In previous work [3], an HV ADC was presented with a beyond-the-rails capability enabled by a capacitively-coupled HV chopper. Although a wide CMVR was demonstrated, it suffered from limited bandwidth and AC CMRR. This paper presents an improved design, which achieves 15-bit resolution and 10 kHz BW over a -0.7V to 29V input CM voltage range (CMVR) while operating from a single 1.8V supply. A new chopping scheme results in a minimum CMRR of 115dB over the full 10 kHz BW. Since the proposed ADC itself has wide input CMVR, it can directly digitize the voltage drop across a high-side shunt, eliminating the need for HV IAs and significantly reducing system complexity and power consumption (Fig. 1).

This paper is organized as follows. Section II introduces the overall architecture of the ADC. Section III describes the operation of the HV chopper. Experimental results are presented in Section IV and Section V concludes the paper.

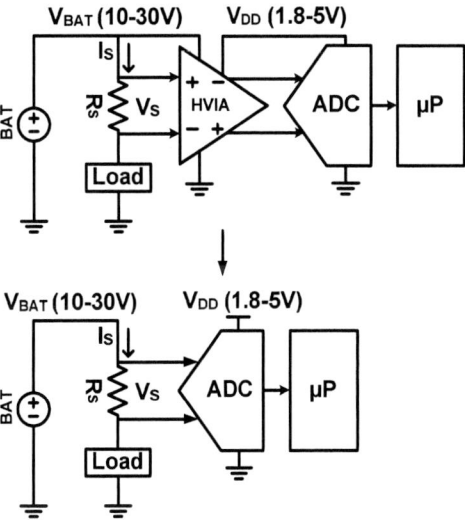

Fig. 1. High-side current sensing systems

II. ADC ARCHITECTURE AND IMPLEMENTATION

Fig. 2(a) shows the block diagram of the ADC. It is based on a single-bit 3rd-order $\Delta\Sigma$ Modulator, which employs an energy-efficient feedforward topology. Two pairs of capacitors $C_{S1,2}$ and $C_{DAC1,2}$ are used for input sampling and DAC feedback respectively. This arrangement allows the ADC's input CM voltage to be different from that of the reference, as required for high-side current sensing.

The differential input signal is sampled by the HV chopper CH_{HVIN} on two sampling capacitors $C_{S1,2}$, which simultaneously block the input CM voltage. $C_{S1,2}$ are HV fringe capacitors that can withstand CM voltages of up to 70V. To ensure good matching and gain accuracy, HV fringe capacitors are also used in the DAC feedback branch. Two diodes $D_{A,B}$ are placed after $C_{S1,2}$ to protect the rest of the ADC from large input CM transients.

All integrators are built around folded-cascode OTAs. The 1st integrator achieves 82dB DC gain and draws 230µA. Setting $C_{S1,2}$ = 2.5pF ensures that the ADC is thermal noise limited. The 2nd and 3rd integrators are scaled down by 8x to improve power efficiency. The comparator consists of a pre-amplifier and a dynamic latch. The outputs of three integrators are summed by a passive SC adder.

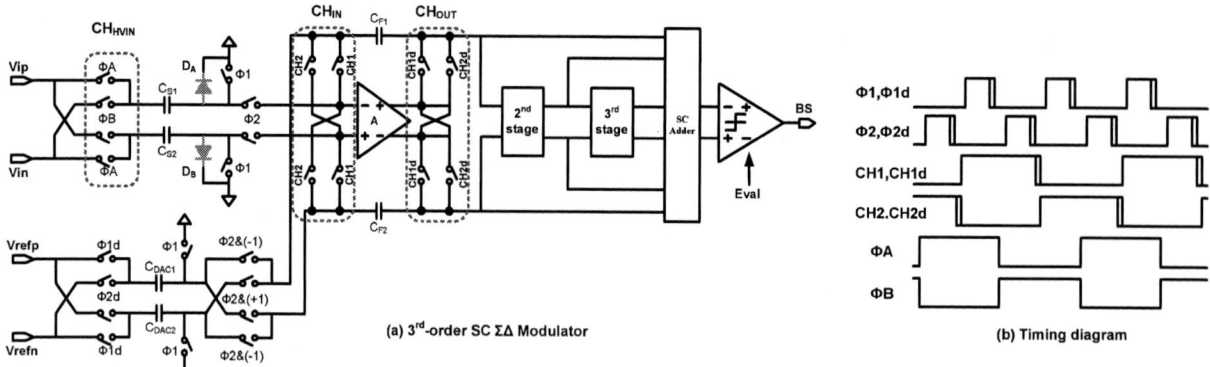

Fig. 2. Simplified ΔΣ modulator schematic and its timing diagram

The 1st integrator employs a chopping scheme to mitigate its $1/f$ noise and offset [4]. A pair of choppers (CH$_{IN}$, CH$_{OUT}$), controlled by chopping clocks CH1,2 and CH1d,2d, periodically swaps the position of the integration capacitors $C_{F1,2}$. As shown in Fig. 2, the correct polarity of the input sampling branch is maintained by the clocks ΦA and ΦB that control the HV input chopper. The chopping frequency is thus at half the modulator's sampling frequency. To minimize intermodulation between the chopping clock and the modulator's quantization noise, the chopping transitions occur in the non-overlapping phase between Φ1 and Φ2. The delayed clock CH1d,2d ensures that CH$_{IN}$ turns off slightly before CH$_{OUT}$. This ensures that signal-dependent charge from CH$_{OUT}$ is not injected into the virtual ground. The clock network is extensively shielded and balanced to avoid noise coupling.

The advantage of this chopping scheme is that it swaps the position of the sampling capacitors, thus mitigating their mismatch and improving the ADC's high-frequency CMRR. The detailed operation of the 1st integrator is shown in Fig. 3. In one integration cycle (CH1 is high), the input signal V_{ip} is sampled on C_{S1} (V_{in} is sampled on C_{S2}) during Φ1. During Φ2, the HV chopper CH$_{HVIN}$ reverses the input and thus a positive charge packet $C_{S1} \cdot (V_{ip}-V_{in})$ is transferred from C_{S1} to the C_{F1} (a charge packet $C_{S2} \cdot (V_{in}-V_{ip})$ is transferred from C_{S2} to the C_{F2}). After 1st integration is done, the position of C_{F1} and C_{F2} is interchanged with the help of CH$_{IN}$ and CH$_{OUT}$. In the next integration cycle (CH1 is low), the sampling network keeps cross-connected and V_{ip} is sampled on the C_{S2} (V_{in} is sampled on the C_{S1}). During Φ2, a positive charge packet $C_{S2} \cdot (V_{ip}-V_{in})$ is transferred to C_{F1} (a charge packet $C_{S1} \cdot (V_{in}-V_{ip})$ is transferred from C_{S2} to the C_{F2}). Assuming $C_{S1}= C_S+\Delta/2$ and $C_{S1}= C_S-\Delta/2$, the total positive charge transferred to C_{F1} after two integration cycles will be $2 \cdot C_S \cdot (V_{ip}-V_{in})$. Consequently, the sampling capacitor mismatch is averaged out by chopping.

III. HV INPUT CHOPPER

Fig. 4 shows the schematic of the capacitively-coupled HV chopper, which is similar to that in [3]. It consists of four sampling switches MN$_{1-4}$, one dynamic latch MN$_{5,6}$, three coupling capacitors C_{1-3} and a minimum selector MN$_{S1,2}$.

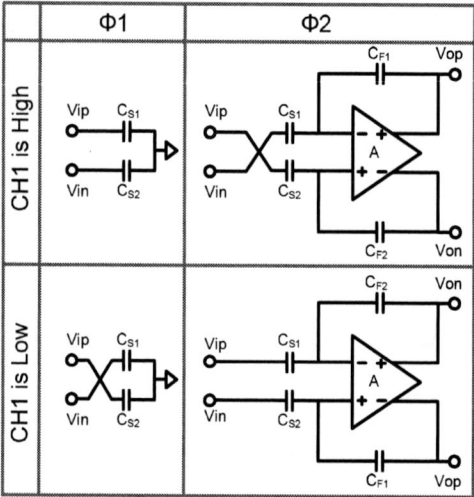

Fig. 3. The operation of the 1st integrator

All the transistors in the HV chopper are 1.8V NMOS devices. They are isolated by a semi-floating HV N-well (HVNW), which forms two back-to-back parasitic diodes D$_{par1}$, and D$_{par2}$ with the local P-well (LPW) and the P-substrate (PSUB), respectively. The LPW is connected to the input CM voltage via the minimum selector MN$_{S1,2}$. When the input CM voltage rises, the potential of the HVNW will follow the input CM voltage through D$_{par1}$. Meanwhile, D$_{par2}$ is reverse-biased. Its breakdown voltage determines the maximum input CM voltage that can be applied to the ADC. When the input CM voltage drops below ground, the HV PNP device connected to the HVNW will turn on and the potential of HVNW will be clamped at 1.8V. It ensures that D$_{par2}$ is always reversed-biased to prevent potential latch-up. Meanwhile, D$_{par1}$ is also reverse-biased and so its breakdown voltage also limits the minimum CM voltage. In this technology, the breakdown voltages of D$_{par1}$ and D$_{par2}$ are 45V and 30V respectively. As such, the HV chopper switches can withstand -30V to 45V input CMVR. In this design, the ADC's input CMVR is limited to -0.7V to 29V by the ESD diodes of the HV input pads.

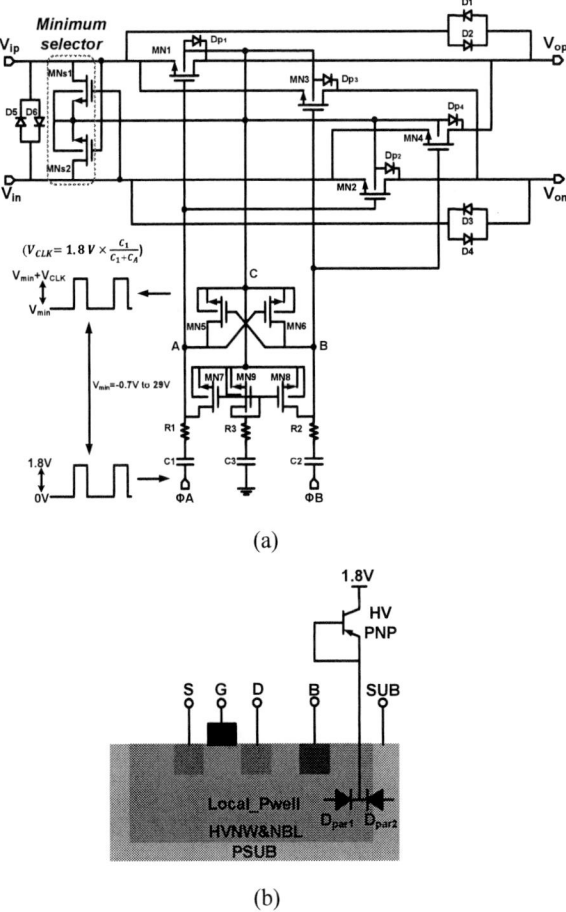

(a)

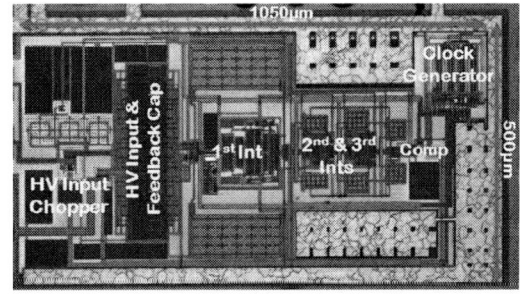

(b)

Fig. 4. Schematic of the HV input chopper (a) and a cross-section of the NMOS devices (b)

The clock signals ΦA, ΦB are capacitively-coupled to the gates of four sampling switches MN_{1-4} via two HV capacitors $C_{1,2}$. A minimum selector $MNS_{1,2}$ connected between two input terminals V_{ip} and V_{in} selects the lowest input voltage. Its output (node C) is tied to the reference of the clock level shifter comprising coupling capacitors $C_{1,2}$ and a latch $MN_{5,6}$. As a result, the coupled clocks are always superimposed on V_{min} (the lower of V_{ip} and V_{in}), which minimizes the leakage current of MN_{1-4} in the presence of bidirectional input voltages. It should be noted that the coupled clock signals at the switches' gates will be slightly attenuated by the parasitic capacitors at nodes A, B. Compared to [3], the minimum selector employs native NMOS transistors, rather than standard NMOS, thus exploiting their lower threshold voltages to extend its operation range.

IV. EXPERIMENTAL RESULTS

The $\Delta\Sigma$ modulator is fabricated in a 0.18µm HV BCD CMOS process and occupies 0.53mm² (Fig. 5). The chip draws 372µA from a 1.8V supply: 290µA by the three integrators & comparator, and 82µA by the clock generator.

Fig. 6 shows the output spectrum with a 500mV$_{p-p}$ differential input signal (the ADC's reference voltage is 600mV) that is superimposed on a 10V CM voltage. When chopping is enabled, the $1/f$ noise and 2nd-order distortion are efficiently suppressed. Fig. 7 (top) shows the measured SNR/SNDR versus input amplitude with chopping on. The ADC achieves 93.7dB peak SNR and 89dB SNDR over 10kHz bandwidth at a sampling frequency Fs = 5MHz. For small DC inputs, and chopping on, small idle tones were observed due to residual cross-coupling between high-frequency quantization noise and chopping clocks at Fs/2. As shown in Fig. 7 (bottom), the SNDR varies by only 0.3dB over the entire input CMVR, demonstrating the excellent linearity of the HV chopper.

Measurements on 12 samples (Fig. 8) show that chopping reduces the ADC's maximum offset from 2mV to 110µV. This relatively large residual offset is due to the relatively high chopping frequency (2.5MHz). The minimum CMRR at DC is 131dB with or without chopping. As shown in Fig. 9, however, chopping improves the CMRR at high frequencies. At 10 kHz, the CMRR improves from 97dB to 115dB. The ADC operates from a 1.5V to 2V supply, and its PSRR at DC and 50Hz are 95dB and 90dB respectively. In Table I, the ADC's performance is summarized and compared to similar prior art. It achieves the highest AC CMRR, 40x more BW than [3, 5], as well as a beyond-the-rails capability.

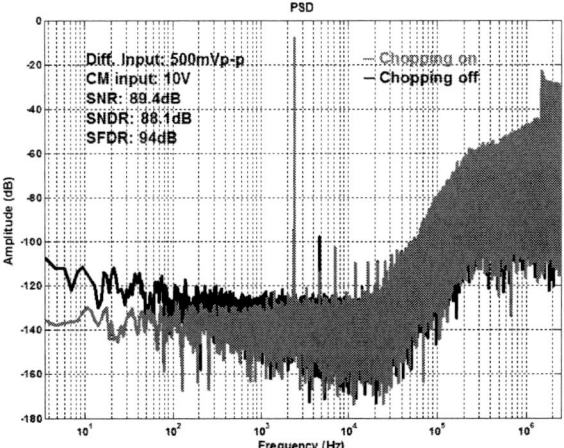

Fig. 5. Chip micrograph

Fig. 6. Measured output spectrum of the $\Delta\Sigma$ modulator under 10V input CM voltage

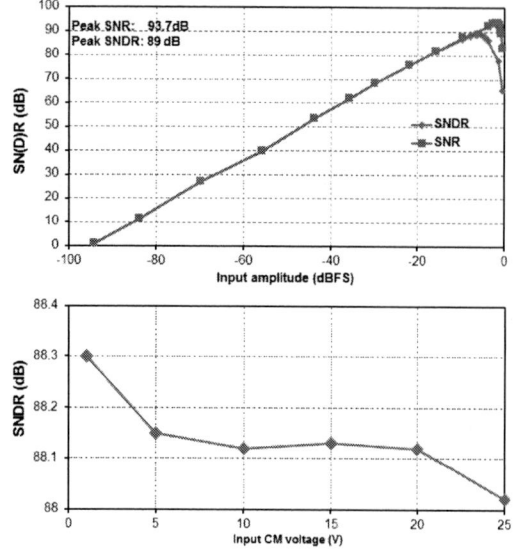

Fig. 7. SNR & SNDR versus input amplitude (top) and SNDR over input CM range (bottom)

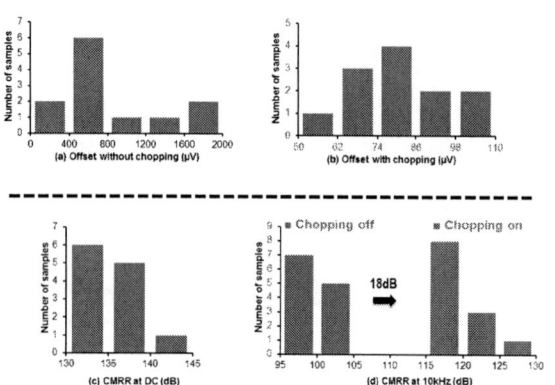

Fig. 8. Histograms (12 samples) of the measured offset and the CMRR at DC & 10kHz

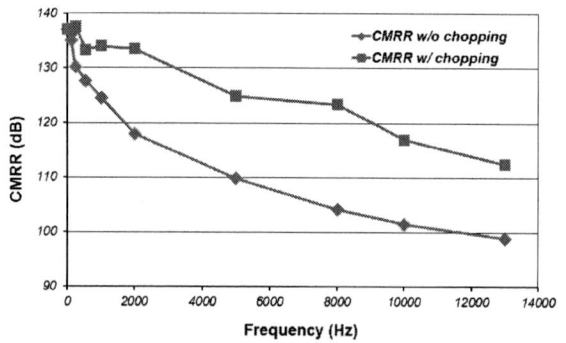

Fig. 9. Measured CMRR versus input frequency (20V_{p-p} sinewave is applied for CMRR measurement)

Table. 1. Performance summary and comparison

	This work	[3]	[5]	[6]
Supply voltage	1.5V-2V	5V	2.7V-5.5V	2.7V-60V
Architecture	ΔΣADC	ΔΣADC	IA+ADC	IA
Output signal	Digital	Digital	Digital	Voltage
Input CM range	-0.7V~29V	±30V	0V~60V	2.7V-60V
Input DM range	1V$_{P-P}$	4.4V$_{p-p}$	440mV$_{p-p}$	500mV$_{p-p}$
SNR	93.7dB	110.1dB	74dB	--
SNDR	89dB	100.6dB	--	--
SFDR	94dB	100.8dB	--	--
BW	10kHz	100Hz	250Hz*	140kHz
Offset	110µV	8µV	500µV	200µV
CMRR (@DC)	131dB	140dB	120dB	125dB
CMRR (@10kHz)	115dB	<65dB	98dB	68dB
Chip area	0.53mm²	0.8mm²	--	--
Power	670µW	505µW	4.32mW**	1.485mW
FOM***	165.5dB	163dB	124.6dB	--

* Bandwidth of the entire signal chain including current-sense amplifier and ADC
** Include the power of current-sense amplifier, ADC, voltage reference and I²C bus
*** FOM=SNR+10log(BW/Power)

V. CONCLUSION

A 15-bit ΔΣADC with a wide input CM range has been presented. Its CMVR is 16x larger than its supply voltage and it achieves a high AC CMRR (115dB at 10kHz). Also, the ADC obtains 93.7dB SNR and 89dB SNDR over 10kHz BW. Due to its beyond-the-rails capability and high accuracy, the ADC avoids the use of HV IAs for level shifting and pre-amplification, which greatly reduces the power consumption and silicon area of the resulting current sensing system.

ACKNOWLEDGMENT

The authors would like to thank Michiel, Zuyao, Lukasz, Sining, Yixuan, Burak for their support during the process of tape-out and measurement, and Sofics for providing HV ESD solution.

REFERENCES

[1] Q. Fan, J. H. Huijsing, and K. A. A. Makinwa, "A Capacitively Coupled Chopper Instrumentation Amplifier with a ±30V Common-Mode Range, 160dB CMRR and 5µV Offset," *ISSCC Dig. Tech. Papers*, pp. 374-375, Feb. 2012.

[2] J.F. Witte, J. H. Huijsing, and K. A. A. Makinwa, "A Current-Feedback Instrumentation Amplifier with 5µV Offset for Bidirectional High-Side Current-Sensing," *ISSCC Dig. Tech. Papers*, pp. 74-75, Feb. 2008.

[3] Long Xu, B. Gonen, Q. Fan, J. H. Huijsing, and K. A. A. Makinwa, "A 110dB SNR ADC with ±30V Input Common-Mode Range and 8µV offset for Current Sensing Applications," *ISSCC Dig. Tech. Papers*, pp. 90-91, Feb. 2015.

[4] Chris Binan Wang, "A 20-bit 25-kHz Delta–Sigma A/D Converter Utilizing a Frequency-Shaped Chopper Stabilization Scheme," *IEEE J. Solid-State Circuits*, vol. 36, no. 3, pp. 566–569, Mar. 2001.

[5] Maxim, MAX9612 Data Sheet. Accessed on Aug. 21, 2016.

[6] Linear, LT6119 Data Sheet. Accessed on Sep 21, 2016

S4-4 (2143)

An Energy-Efficient Self-Charged Crystal Oscillator with a Quadrature-Phase Shifter Technique

Wei-Sung Chang, Dai-En Jhou, Yu-Hong Yang, and Tai-Cheng Lee

Graduate Institute of Electronics Engineering and Department of Electrical Engineering
National Taiwan University, Taipei, Taiwan, 10617
Email: tlee@ntu.edu.tw

Abstract—An energy-efficient self-charged crystal oscillator (SCXO) employing a quadrature-phase shifter is proposed to provide wide-range pulse injection timing for power consumption reduction. The passive resistors of quadrature-phase shifter can be shared by the startup circuit to save area consumption. The double-edge extractor and the low-power comparator are added to reduce the power consumption. The proposed 32.768-kHz energy-efficient SCXO, fabricated in a 40-nm CMOS technology, consumes 34.3 nW under a 1 V supply.

Keywords—*Self-charged crystal oscillator (SCXO), Quadrature-phase shifter, Double-edge extractor, Low-power comparator, 32.768-kHz*

I. INTRODUCTION

A 32.768-kHz crystal (XTAL) is widely used to generate the real-time clock for the wearable device. Because of the limited power battery, there is a growing interest in ultra-low power crystal oscillator (XO). To achieve sub-μW power consumption, the conventional Pierce oscillator is a common architecture for XO [1]. In recent years, the DLL-assisted XO [2] and the self-charged crystal oscillator (SCXO) in [3] proved that power consumption can be greatly reduced to the nW scale by charging crystal with narrow voltage pulses. However, the higher supply voltage and the larger chip area are inevitable to realize the former technique because of the separation of multiple power domains.

Theoretically, the power consumption of dynamic circuits can be derived as

$$\frac{1}{2} V_{DD}^2 \cdot C_L \cdot f \ , \qquad (1)$$

where V_{DD} is the supply voltage, C_L means the switched capacitance and f means the switched frequency. Typically, stepping down the supply voltage is the most efficient way to reduce power consumption [3]-[4]. For example, if the V_{DD} is 4 times lower than the regular voltage, the dynamic energy can be reduced 16 times. However, additional power domain increases complexity and cost of power management IC (PMIC).

In this work, we improve the conventional SCXO under a regular supply (1V). The SCXO with quadrature-phase shifter is proposed to control the charging timing without any additional resistor. Because the passive feedback resistor of the start-up circuit is free while the start-up circuit is turned off, we can separate it into two resistors for the proposed phase shifter.

In addition, the edge extractor can be improved to reduce half power by the double-edge operation. Finally, the proposed low-power comparator is presented to extract clock edges instead of the clock slicer in [3]. This paper is organized as follows. In Section II, the conventional and the proposed SCXO architectures are introduced. The detailed scheme of each circuit block is shown in Section III with the design consideration. The measurement results of the prototype fabricated in a 40 nm CMOS technology is demonstrated in Section IV with the performance summary and comparison with previous works. Finally, Section V shows the conclusion of this work.

II. CRYSTAL OSCILLATOR ARCHITECTURE

A. Conventional Self-Charged Crystal Oscillator

Fig. 1 shows the architecture of the conventional SCXO [2]. The SCXO directly charges C_{par} and XTAL periodically to replenish the energy loss in XO to sustain oscillation. The charging switches, whose source node are connected to supply voltage, are driven by the pulses which are boosted up to four times of the supply voltage because of extremely low power supply voltage. The pulse booster enables the charging transistor to be fully turned on for energy transfer from V_{DD} to crystal. The AC-coupled slicers extract positive clock edges and the proceeding pulse boosters up-convert control pulses to charge the XTAL to in the 0° to 90° region. In this design, the injection timing is determined by the threshold of device and has large variation from chips and chips.

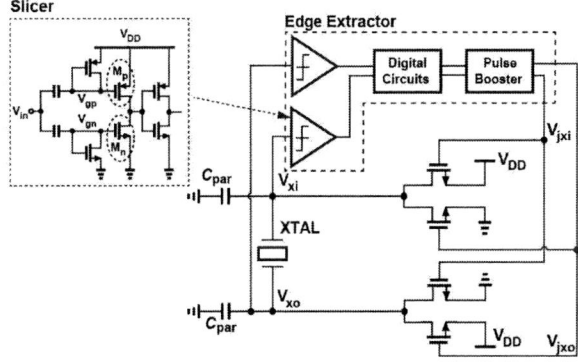

Fig. 1. The architecture of the conventional SCXO [2].

978-1-5386-3179-9/17 $31.00 © 2017 IEEE   53

B. Proposed Self-Charged Crystal Oscillator

Fig. 2 shows the proposed SCXO architecture. The proposed SCXO is composed of a quadrature-phase shifter, start-up, comparator, double edge extractor and self-charge circuit to generate a 32.768-kHz clock output. The large R for start-up circuit can be reconstructed to a quadrature-phase shifter in the self-charged mode. The charging switches for energy transfer from V_{DD} and GND to crystal are driven by V_{jxi} and V_{jxo} pulses. The double edge extractor can produce V_{jxi} and V_{jxo} pulses.

Fig. 3 shows the timing diagram of the proposed SCXO. In start-up mode, the ST signal is "low" to form a positive feedback amplifier with a large bias resistor and meet the Barkhausen criteria of oscillation. Once the XTAL oscillates stably, the ST signal is switched to "high" to form a self-charged oscillator. Furthermore, the power can be minimized when the charging injection timing is at 90 degree of 32.768-kHz signal, which is the peak voltage in crystal output node. Quadrature-phase pulse injection provides wide-range injection timing while keeping the minimum hardware cost and power consumption. Compared to the device threshold variation in [3], the variation of passive components in quadrature-phase shifter is easier to measured and calibrated.

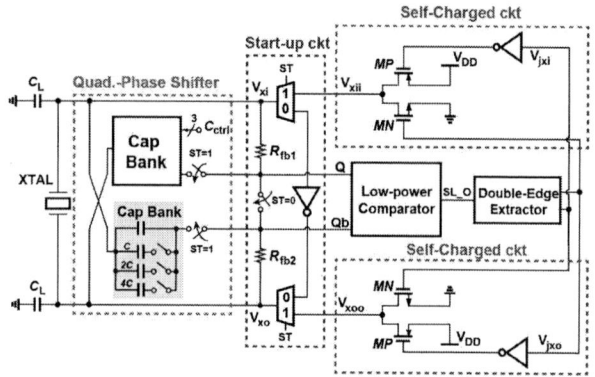

Fig. 2. The proposed self-charged crystal oscillator architecture.

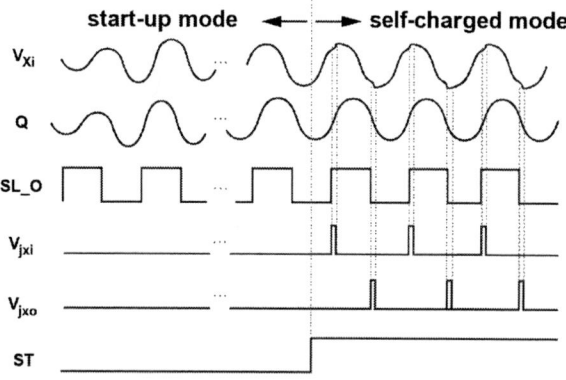

Fig. 3. The timing diagram of the proposed SCXO.

III. CIRCUIT IMPLEMENTATION

A. Quadrature-Phase Shifter

In self-charged mode, the bias resistor is separated to two and connected to cap banks respectively. The phase of both nodes in XTAL are assumed as 0 and 180 degree. Shown in Fig. 4, V_{xi} and V_{xo} correspond to the output nodes of the XTAL. Therefore, the phases between R and C bank will be 90 and 270 degree when the pole frequency of RC is designed to meet the oscillation frequency of XTAL. In this work, R is chosen as 33 MΩ and C is chosen as 147fF. The quadrature-phase shifter only increases area slightly because of reusing the large bias resistor in the start-up circuit. A 3-b capacitor bank circuit is designed for experiment and calibration.

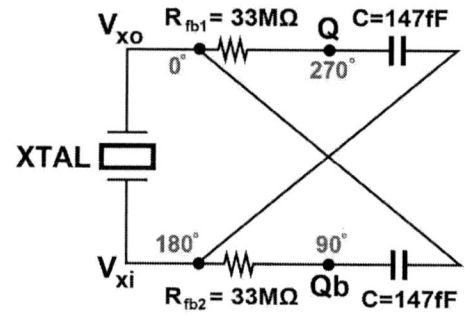

Fig. 4. The block diagram of the quadrature phase shifter.

B. Low-Power Comparator

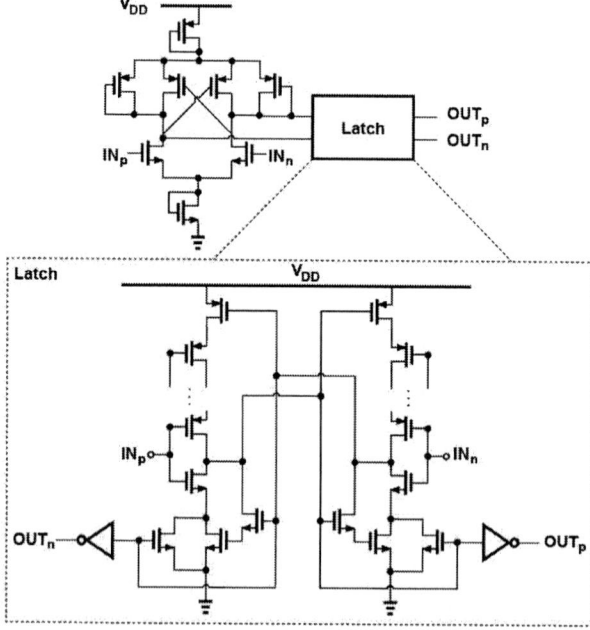

Fig. 5. The block diagram of the low-power comparator.

Because the extremely slow slope of the sinusoidal output of XTAL typically yields a steady current of the following stage circuit, low power sinusoidal-to-square comparator design becomes very challenging. In order to achieve the nanowatts power consumption, the current and voltage swing of pre-amplifier is limited by PMOS and NMOS diodes as shown in Fig. 5. The output of pre-amplifier is sharper and square-like waveform but not rail-to-rail due to limited by diodes. The following latch is shown in Fig. 5. Multiple cross-coupled pairs are inserted in each node to reduce the short current from V_{DD} to GND during the transition time. The output of the latch is designed to be rail-to-rail square wave.

C. Double-Edge Extractor

For power saving, only one delay chain with the AND and NOR gate extract both positive and negative edges from the comparator output to produce a narrow width pulse. In the double-edge extractor, the pulse width is generated on the delay of rising and falling edge. It is very important to have the same pulse width for both sides to balance charge and discharge time for XTAL. Therefore, both edges of the delay unit are mainly delayed by an inverter with PMOS diode stage as shown in Fig. 6. Since both delays are determined by the same type device, the difference caused by PVT influence can be minimized.

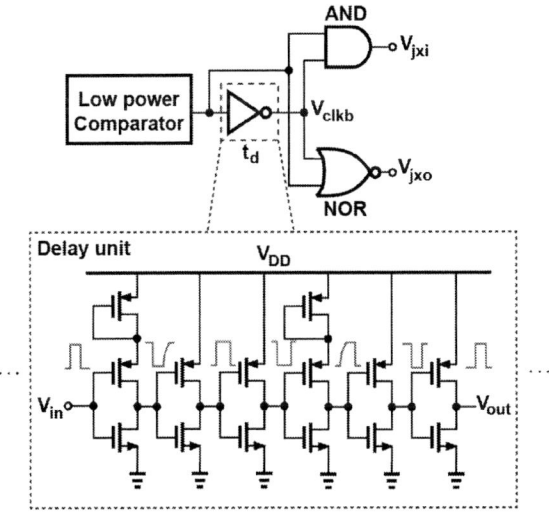

Fig. 6. The block diagram of the double edge extractor.

IV. MEASUREMENT RESULTS

The proposed SCXO has been fabricated in a 40nm CMOS technology which occupies 0.045 mm². The chip micrograph is shown in Fig. 7. It can be noted that the passive resistor is separated into R_{fb1} and R_{fb2}, which can be shared with startup circuit and RC-filter circuit. An off-chip model (Fork-2060) is used as the XTAL with the loading capacitor C_L (25 pF).

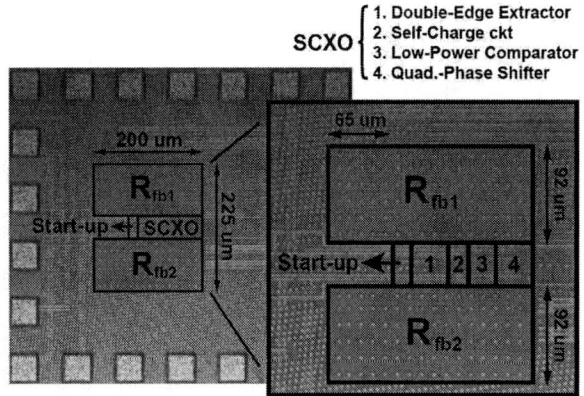

Fig. 7. Die photo.

Fig. 8 shows the measured output waveform and the two triggering pulses of the proposed SCXO under 1-V supply voltage. This is the most energy-efficient case when the outputs of SCXO are triggered at the peak of the swing.

It can be also seen that the mechanism of power charging is triggered by the short pulses which are V_{jxi} and V_{jxo}. By using the proposed delay circuit, the triggering pulses can be extracted by both the positive and negative clock edges through the low-power comparator. Furthermore, the difference of the pulse widths between differential triggering pulses is less than 10% from the measurement result. In this design, it is enough to tolerate for the successful charging of the XTAL because of the smaller charging transistors which are insensitive to the mismatch of the power-charging timing.

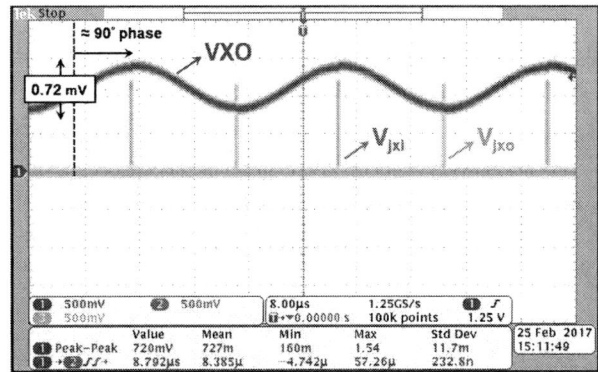

Fig. 8. The measured XO output and triggering pulses waveform.

The power distribution of each block is shown in Fig. 9. It can be observed that the energy of the self-charged circuit is the dominant term compared with other blocks.

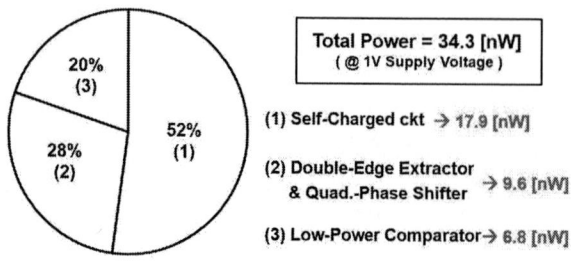

Fig. 9. The measured power distribution of each block
with the most energy-efficient case.

Thanks to the proposed 3-bit RC-filter circuit, the direct-charging timing can be controlled without any additional resistor. The dynamic range of this RC-filter circuit is roughly equal to 7 us with a step size of about 1 us in this work. The measurement results versus different cases of charging timing are shown in Fig. 10 (a) and (b).

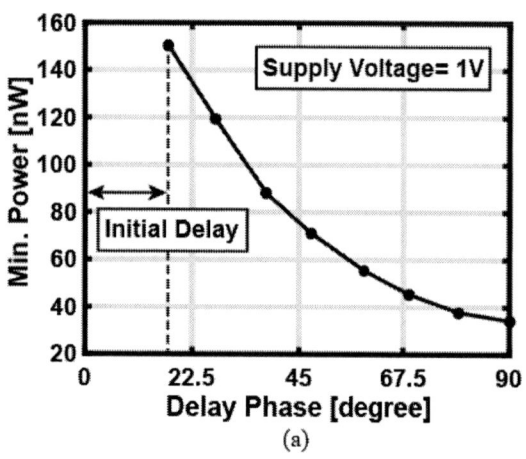

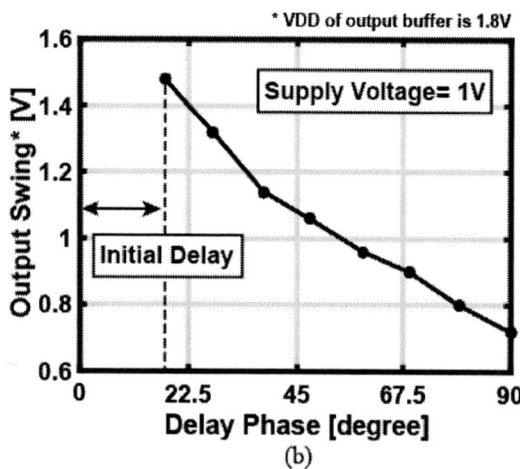

Fig. 10. The measurement results with different charging delay phase.
(a) Minimum Power (b) Output Swing

Table I. Performance summary and comparison table.

	This work	ISSCC '12 [2]	ISSCC '14 [3]	Electronics Letters '16 [4]
Technology	40 nm	0.18 µm	28 nm	0.18 µm
Frequency [Hz]	32.768 k	32.768k	32.768 k	32.768 k
Supply Voltage [V]	1~1.32	0.92~1.8	0.15~0.5	0.25~0.5
Power [nW] (*:Min.)	34.3* (@90° delay) 71.2 (@47.5°delay) 150.4 (@17.5°delay)	5.58*	1.89*	2.89*
Area [mm²]	0.045	0.3	0.03	0.027
Freq. Var with V$_{DD}$	16 ppm@ 1~1.32V	N/A	30 ppm@ 0.15~0.5V	N/A

Table I summarizes the performance of the proposed SCXO and shows the comparison with the other type XOs. The total power consumption with the different phase delay of 90°, 47.5°, and 17.5° are 34.3, 71.2, and 150.4 nW, respectively. If the phase delay is 90°, which is at the peak of the output swing, the most energy-efficient power consumption is achieved.

V. CONCLUSION

The proposed SCXO, which consuming about nano-watts under 1V supply voltage is presented. The proposed double edge extractor can save about half power of digital logic block. It also reduces the power consumption by the low-power comparator instead of the conventional clock slicer to achieve the stable oscillation under the low swing conditions. While the start-up circuit turns off, the passive feedback resistors can be reused to control the charging timing to increase the power efficiency. The total power of the proposed SCXO is smaller than 35 nW with the most energy-efficient case.

ACKNOWLEDGMENT

The authors would like to thank Nuvoton Technology and Ministry of Science and Technology (MOST) under Contract 103-2221-E-002-266-MY3 for supporting this research.

REFERENCES

[1] W. Thommen, "An Improved Low Power Crystal Oscillator," *in Proc. IEEE ESSCIRC*, Sept. 1999, pp. 146-149.

[2] Dongmin Yoon, Dennis Sylvester, David Blaauw, "A 5.58nW 32.768kHz DLL-Assisted XO for Real-Time Clocks in Wireless Sensing Applications," *in ISSCC. Dig. Tech. Papers*, Feb 2012, pp. 366-367.

[3] Keng-Jan Hsiao, "A 1.89nW/0.15V Self-Charged XO for Real-Time Clock Generation," in *ISSCC. Dig. Tech. Papers*, Feb. 2014, pp. 298-299.

[4] Shang-Chi Wu, Tai-Cheng Lee, "Ultra-low-power one-pin crystal oscillator with self-charged technique, " *in IEEE Electronics Letters*, Feb. 2016, pp. 325-327.

An Area-Efficient Capacitively-Coupled Sensor Readout Circuit with Current-Splitting OTA and FIR-DAC

Chih-Chan Tu[1], Feng-Wen Lee[1,2], Han-Chun Chen[1], Yu-Kai Wang[1,2] and Tsung-Hsien Lin[1]

[1]National Taiwan University, Taipei, Taiwan
[2]Mediatek, Hsinchu, Taiwan

Abstract - An area-efficient CTDSM for sensor applications is presented. The proposed capacitively-coupled CTDSM combines the functions of precision amplifier and DSM to achieve hardware efficiency. The 1-bit quantizer with FIR-DAC improves the linearity without using DEM in multi-bit architectures or large-size analog filters. The proposed current-splitting OTA saves the capacitor area in the first integrator stage. Fabricated in a 0.18-μm CMOS, this chip draws 70 μA (?) from a 1.8-V supply, and achieves 12-bit resolution under 2-kHz bandwidth at ±40-mV input range. The circuit area is only 0.2 mm².

I. INTRODUCTION

Precision sensors such as e-compasses or altimeters requires medium-to-high resolution (>12 bits) and accuracy, while maintaining good energy efficiency for portability in the future IoT era. Such sensor elements are usually in the form of a bridge topology, and need to interface with readout circuits with high input impedance.

Conventionally, the sensing element is followed by a precision instrumentation amplifier (IA). A high-resolution ADC, usually a DSM, then digitizes the amplified signal (e.g. Fig. 1). A current-feedback IA (CFIA) with a discrete-time delta-sigma modulator (DTDSM) is proposed in [1], and a capacitively-coupled IA (CCIA) with a continuous-time delta-sigma modulator (CTDSM) is realized in [2]. These approaches implement two different circuits separately; both architectures are quite complex and occupy considerable area.

Alternatively, interfacing with the sensor directly with an ADC but without IA offers a viable solution to reduce the hardware complexity. Without the IA, the design challenge is to maintain sufficient SNDR and energy efficiency for practical applications. The sampling operation of the discrete-time approach [3, 4] results in noise aliasing issues, and leads to larger input-referred noise. The continuous-time approach with Gm-C CTDSM [5] achieves similar noise performance compared with that of a CFIA+DTDSM [1], while provides high input impedance. However, this design imposes two new design issues. First, an area-consuming, a passive LPF is required after the DAC to improve the linearity. Second, the large time constant requirement in the CTDSM loop makes the capacitor of the integrator very large. In [6], the multi-bit DAC alleviate the need of a large LPF, with the price of requiring complex digital circuits with DEM technique. The VCO-based loop filter mitigates the time-constant issue in a CTDSM, with the trade-off of needing to address the PVT issue.

In this work, a capacitively-coupled sensor readout circuit is proposed. It is essentially a second-order CTDSM with an FIR-DAC, which embed the CCIA operation. This paper is organized as follow. In Section II, the proposed architecture is introduced. Circuit implementation is described in Section III. The experimental data is presented in Section IV.

II. CIRCUIT ARCHITECTURE

Fig.1 shows an example of a conventional sensor readout system, which consists a precision IA (can be a CFIA or a CCIA), and a high-resolution DSM. As mentioned previously, such topology incurs complex design and large chip area.

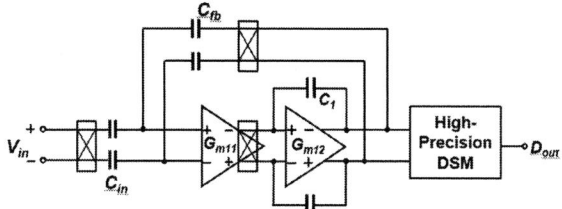

Fig. 1 Conventional sensor readout system, composing an IA and a high-resolution ADC.

The proposed sensor readout system is shown in Fig. 2. It is in effect a second-order CTDSM. Comparing with Fig. 1, it can be viewed as merging the quantizer into the CCIA loop [6]. The capacitively-coupled topology is more energy-efficient than that of the CFIA or Gm-C CTDSM counterparts. The 1-bit quantizer with the FIR DAC makes the DSM operation more linear without using large-area LPF [5]. The original two-stage opamp, G_{m11} and G_{m12}, and C_1 serve as the first integrator of the CTDSM. G_{m21} and G_{m22} are added before the quantizer to form a feedback loop for one more order of noise shaping; this is termed as the "inner loop" in this work. An attenuation block with a gain α (α < 1) is inserted before G_{m21}, to limit the signal swing of G_{m21} to suppress the nonlinearity. The reference voltage of DAC2 is also scaled with α.

Choppers are used to modulate the input signal V_{in} and feedback signal V_{fb} to mitigate the flicker noise; the demodulator is implemented at G_{m11} output. The input full swing range is designed to be ±40 mV. The amplification ratio C_{in}/C_{fb} is set to 10 to suppress the noise contribution from the reference voltage and FIR-DAC.

In the DSM, a 1-bit quantizer is chosen for its inherent linearity. However, the 1-bit feedback DAC causes large signal jumps at its output (V_{fb}), which in turn causes severe non-linearity in the G_{m1} stage. To mitigate this problem, prior designs use large passive analog filters (200-pF on-chip in [3] and 10-nF off-chip in [4]) in the feedback path at the cost of large area. In this work, a semi-digital design strategy employing a 21-tap FIR filter is adopted, as shown in Fig. 3(a). The 1-bit output signal (D_{OUT}) is first fed into a delay chain composed of D-FFs, and then controls the DAC. The DAC is realized with logic gates, reference voltages (V_{ref} and 0,5 V_{ref}), and resistors. The impulse response h[n] of the FIR filter is a bell-shape behavior, as illustrated in Fig. 3(b). It smooths the signal at V_{fb} to improve overall linearity. Here, using two reference voltages, V_{ref} and $V_{ref}/2$, makes the realization of the coefficients easier. The total R used in this design is 1.3 MΩ.

978-1-5386-3179-9/17 $31.00 © 2017 IEEE

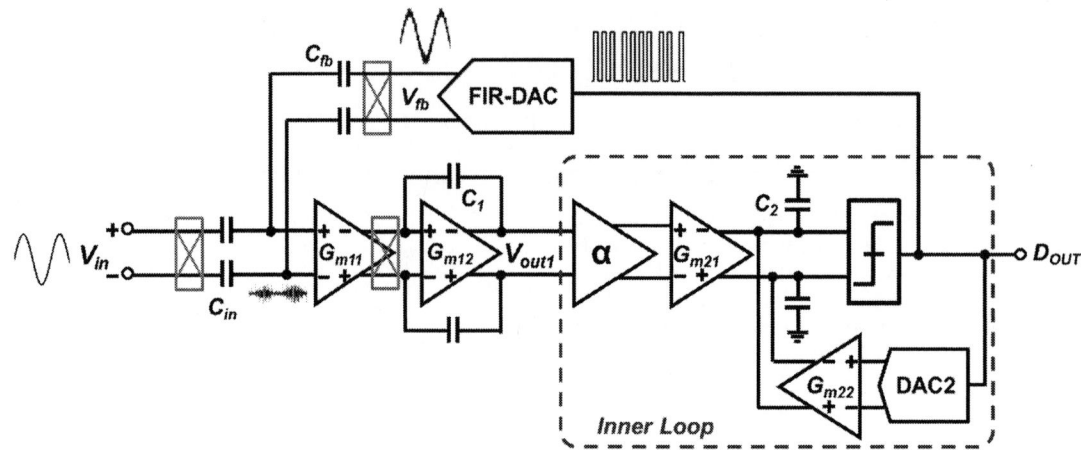

Fig. 2 Architecture of the proposed sensor readout system.

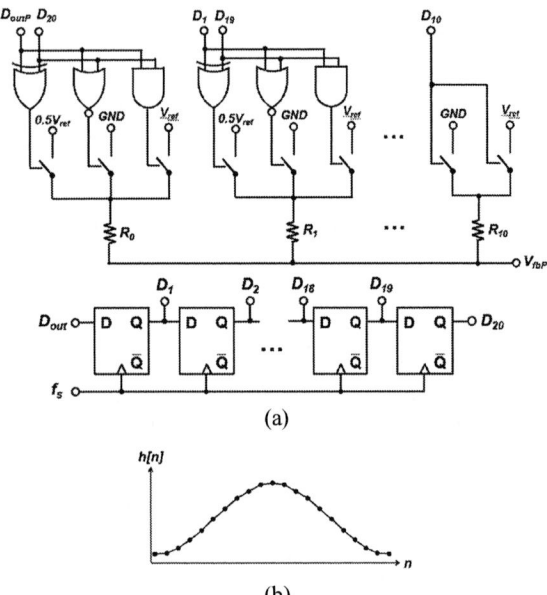

(a)

(b)

Fig. 3 (a) Schematic of the FIR-DAC (only half side is shown). (b) Impulse response of the FIR-DAC.

The matching and absolute accuracy of the resistors are not critical, since the output digital code must go through all delays to pass the DAC. The overall output noise of the FIR-DAC is contributed by the reference voltages and the resistors. Assuming noise from the references is not dominating, the worst case noise with all resistors connecting in series (1.3 MΩ) is approximately 144 nV/√Hz.

The loop gain of the proposed CTDSM can be written as

$$L(s) = \frac{G_{m1}}{sC_1} \times \frac{1}{1 + s\frac{C_2}{G_{m2}}} \times H_{FIR}(s) \times \frac{C_{fb}}{C_{in}} \quad (1)$$

In (1), $H_{FIR}(s)$ has a DC gain of 1. The 3-dB low pass corner is designed at 200 kHz to provide enough attenuation of quantization error at V_{fb} under a 5-MHz sampling frequency. To maintain loop stability, the unity-gain frequency of $L(s)$, which is determined by $(G_{m1}/sC_1)*(C_{fb}/C_{in})$, is designed to be

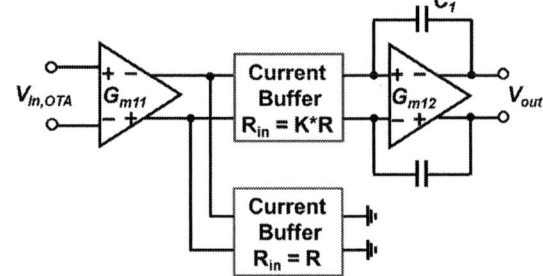

Fig. 4 Proposed current-splitting integrator stage.

small at 80 kHz, to avoid other non-dominant poles near 200 kHz due to the FIR filter. The second term represents the frequency response of the inner loop ($\alpha = 0.1$ and $V_{REF,DAC2} = 0.1 \times V_{REF} = 0.1$ V). With $G_{m2} = 3$ µS and $C_2 = 1$ pF, the resulted 3-dB corner of 500 kHz is sufficiently high and is beyond the stability concern.

Designing a small G_{m1}/sC_1 is a challenging task considering area efficiency. To keep low input-referred noise of the first integrator at approximately 25 nV/√Hz, it requires 100 µS of G_{m1}, which in term mandates a 20-pF capacitance for C_1. More stringent input-referred noise requirement makes this problem even worse. Adding another voltage divider after the first integrator (before the inner loop) helps to suppress the loop gain. However, the signal swing at the first integrator output must be carefully optimized.

Fig. 4 shows the proposed method to suppress the bandwidth of the first-stage integrator. Two current buffers are employed at the output of G_{m11}, with different input impedances K×R and R. As a result, only 1/(1+K) portion of the output current from G_{m11} is integrated in C_1; the other K/(1+K) portion is bypassed to ground. By setting N = 9, the effective bandwidth is suppressed by 10 times, and the capacitor size requirement is relaxed by the same order. The design of G_{m11} and the buffers is called current splitting OTA in this paper. This OTA with the current diversion technique is termed as current-splitting OTA in this paper. The main drawback of the proposed technique is additional noise contribution from the buffer stage and G_{m12}. Since the signal is attenuated by 1/(1+K) times, the noise of G_{m12} needs to be optimized. The design details are described in the next section.

The overall input-referred noise PSD of the DSM is

$$S_{VIN}(f) = S_{Gm11}(f)\left(1 + \frac{C_{fb} + C_p}{C_{in}}\right) + S_{FIR}(f)\left(\frac{C_{fb}}{C_{in}}\right)^2 \quad (2)$$

In this design, the main noise comes from the G_{m11} of the first-stage integrator. Furthermore, due to the parasitic capacitance C_p at the virtual ground node, the input-referred noise of G_{m11} is slightly elevated. The input-referred noise contributed from the FIR-DAC is about 14 nV/√Hz, which is not the major noise source in this design.

III. CIRCUIT IMPLEMENTATION

Fig. 5 shows the design of the proposed current-splitting OTA G_{m11}, together with G_{m12} and C_1. It consists of the main gm stage (M_{0P} with g_{m0}), and two current buffer (folded-cascode) stages. The output buffer stage with M_1 and M_L conducts current I, while the other bypassing buffer stage conducts K×I current. The main G_{m11} stage conducts M×K×I current. At a signal frequency close to the unity-gain frequency of the integrator, the input impedance of the current buffers are $1/g_m$, with a ratio of K between these two buffers. Therefore, only 1/(1+K) of the signal current from g_{m0} will flow into C_1; the effective bandwidth is reduced by 1/(1+K) times.

There are two additional main noise contributors: M_1 and M_L in the buffer stage. The noise power from M_L is $(1+K)^2$ times larger when referred to g_{m0} input, since the signal current is attenuated by (1+K) times at OTA output. The noise from M_1 is no longer suppressed by large output impedance from M_{0N} and M_{0P}, and is K^2 times larger when referred to g_{m0} input. Assume that $g_{m1} = g_{mL} = g_{m0}/MK$, the overall input-referred noise of the G_{m11} stage is thus

$$S_{Gm11}(f) \approx \frac{16kT\gamma}{g_{m0}}\left(1 + \frac{K}{M}\right) \quad (3)$$

Choosing a large M mitigates the additional noise contributions from the buffer stage. However, this also increases the voltage swing before the current buffer stage, and gives rise to non-linearity or intermodulation components.

In this design, M = 5 and K = 9 are chosen. The overall input-referred noise of G_{m11} is approximately 40 nV/√Hz under M*K*I = 5 μA and g_{m0} = 100 μS. C_1 is chosen to be 2 pF to obtain the desired loop BW of 80 kHz.

Fig. 6 shows the design of the inner loop. G_{m21} and G_{m22} are degenerated with resistors R_2 = 300 kΩ; the input stage of G_{m21} is attenuated with capacitive dividers. The arrangements are to suppress the non-linearity from the inner loop. The capacitive attenuator relaxes the driving capability required for G_{m12}. However, since the capacitors are inherently high-pass units, another set of choppers are added before C_{att} and the G_{m12} output to make sure the low-frequency signal can pass the system. The DAC2 is implemented with a set of switches controlled by output digital code. V_{REF2P} and V_{REF2N} are set to be 900 mV ± 50 mV, which corresponds to the 10× attenuation of V_{REF} (= 1 V) in the FIR-DAC. Pseudo-resistors are used to bias the input node of G_{m21}.

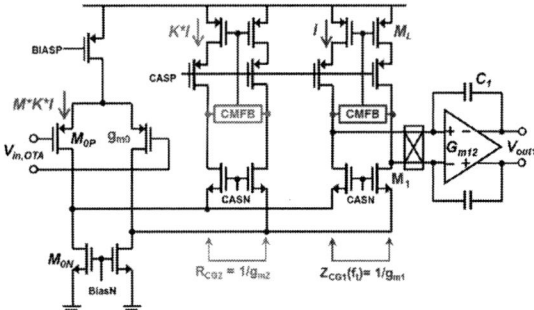

Fig. 5 The proposed current-splitting OTA

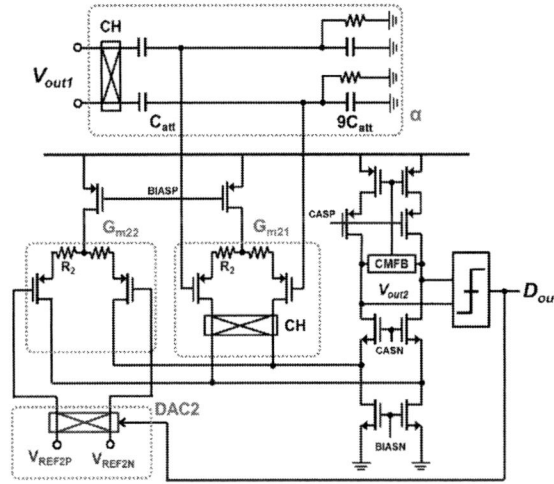

Fig. 6 Schematic of the inner loop.

Fig. 7 Chip Photo.

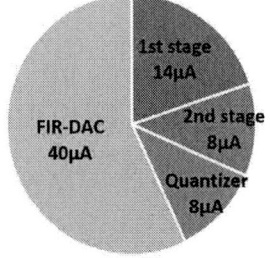

Fig. 8 Measured current distribution.

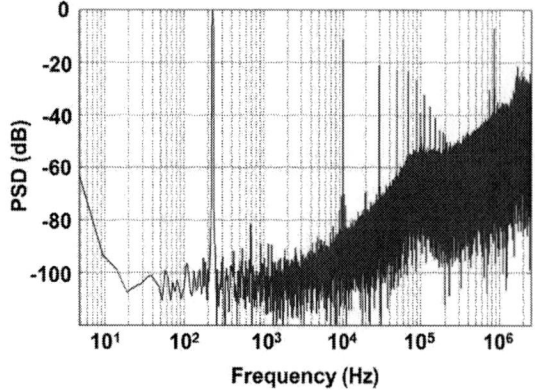

Fig.9 Measured 2^{20}-point FFT of this chip.

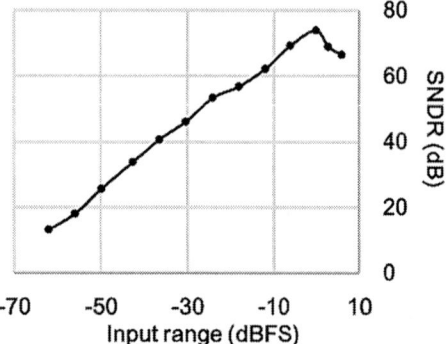

Fig. 10 Dynamic range measurement.

IV. MEASUREMENT RESULTS

The prototype chip is implemented in the TSMC 0.18-μm process. The chip photo is shown in Fig. 7; the core circuit area is 0.2 mm^2. The total current consumption is 70 μA under a 1.8-V supply. Among all currents, the loop filter draws 22 μA; the quantizer and FIR-DAC consumes 8 μA and 40 μA, respectively. Most power is consumed in the D-flip-flops for the delay-chain, which operate at 5-MHz sampling frequency.

Fig. 9 shows the measure output 2^{20}-point FFT result, with a 229-Hz, 80-mV$_{pp}$ input signal. The chopping frequency is set to 10 kHz and the sampling frequency is 5 MHz. The measured SNR/SNDR under 2-kHz bandwidth is 76.14 dB and 75.1 dB, respectively; the SFDR is 81.5 dB. The input-referred offset is 50 μV, which can be further improved if slow choppers are used [1, 5].

Table I shows the comparison with other high-resolution sensor readout circuits. This prototype achieves better area efficiency compared to conventional IA + DSM approaches [1] [2]. Also, the proposed approaches address the issues from the prior arts by trying to integrate them together. Compared to discrete-time approaches [4], this architecture shows improved noise performance. The measured results of 12-bit resolution under 2-kHz bandwidth and < -80-dB THD suggests that this work can be potentially applied to many sensor readout applications, such like electronic compasses.

V. ACKNOWLEDGEMENT

The authors wish to thank CIC, Taiwan for chip fabrication. This work is supported by MOST, Taiwan.

VI. REFERENCES

[1] R. Wu et al.,"A 20-b ±40-mV range read-out IC with 50-nV offset and 0.04% gain error for bridge transducers," *IEEE J. Solid-State Circuits*, vol. 47, no. 9, pp. 2152–2163, Sep. 2012.

[2] H. Jiang et al., "An Energy-Efficient 3.7nV/√Hz Bridge-Readout IC with a Stable Bridge Offset Compensation Scheme" in *Proc. IEEE Symp. on VLSI Circuits*, Feb. 2017, pp. 171–173.

[3] B. Yousefzadeh et al., "A compact sensor readout circuit with combined temperature, capacitance and voltage sensing functionality," in *Proc. IEEE Symp. on VLSI Circuits*, Jun. 2017, pp. 78–79.

[4] E. Bonizzoni et al., "An incremental ADC sensor interface with input switch-less integrator featuring 220-nVRMS resolution with ±30-mV input range," in *Proc. IEEE ESSCIRC*, Sep. 2012, pp. 389–392.

[5] G. Singh et al., "A 20bit continuous-Time ΣΔ modulator with a Gm-C integrator, 120dB CMRR and 15 ppm INL," in *Proc. IEEE ESSCIRC*, Sep. 2012, pp. 385–388.

[6] C.-C. Tu et al., "A 0.06mm^2 ± 50mV range -82dB THD chopper VCO-based sensor readout circuit in 40nm CMOS," in *Proc. IEEE Symp. on VLSI Circuits*, Jun. 2017, pp. 84–85.

Table I. Performance Summary and Comparison

	Process	Architecture	Supply	f_s	Input	BW	Noise	SNR	Linearity	Power	FoMs	Area
[1] JSSC '12	0.7μm	CFIA+ DTDSM	5V	10 kHz	80 mV$_{pp}$	6 Hz	16 nV /√Hz	117.2 dB*	5 ppm INL	1.35mW	153.7 dB	6mm^2
[5] ESSCIRC '12	0.7μm	Gm-C CTDSM	5V	5 MHz	100 mV$_{pp}$	2 kHz	20 nV /√Hz	91.9 dB	-102dB THD	1.2mW	151.2 dB	6mm^2
[2] ISSCC '17	0.18μm	CCIA+ CTDSM	1.8 V	2 MHz	20 mV$_{pp}$	2 kHz	3.7 nV /√Hz	92.6 dB*	28 ppm INL	2.16 mW	152.3 dB	0.73 mm^2
[3] VLSI '17	0.18μm	SAR+ DTDSM	1.8 V	35 kHz	1.8 V$_{pp}$	10 Hz	3800 nV /√Hz	94.5 dB	0.5% Error	8.3 μW	155.3 dB	0.33 mm^2
[6] VLSI '17	0.04μm	VCO-CTDSM	1.2 V	1 MHz	100 mV$_{pp}$	2 kHz	142 nV /√Hz	74.6 dB	-82 dB THD	21 μW	154.7 dB	0.06 mm^2
This Work	0.18μm	CC-CTDSM	1.8 V	5 MHz	80 mV$_{pp}$	2 kHz	98 nV /√Hz	76.14dB	-81.5dB THD	126 μW	148.1 dB	0.2 mm^2

* DC SNR is related to AC SNR by a ratio of 2√2. FoMs = SNR + 10×log(BW/Power)

S5-1 (2049)

25 fJ/bit, 5Mb/s, 0.3V True Random Number Generator With Capacitively-Coupled Chaos System and Dual-Edge Sampling Scheme

Anh Tuan Do and Xin Liu

Institute of Microelectronics, Agency for Science, Technology and Research (A*STAR), Singapore,
Email: doat@ime.a-star.edu.sg; liux@ime.a-star.edu.sg

Abstract—This paper proposes a low-power, small form-factor all-digital RNG utilizing a concept of capacitive coupling between two ROs to amplify jitter and a dual-edge sampling scheme to increase the data rate. It is the most useful in battery-powered IoT applications where both energy and silicon area are critical constraints. The capacitive coupling effect with all digital circuit allows our design to operate reliably over a wide range of supply voltages. It consumes only 1.85 pJ/bit at 1V supply and 67 Mb/s, which scales down to 25 fJ/bit at 0.3 V, 5 Mb/s. Its overall FOM is 1.7×10^4, which is more than $3\times$ better than existing designs.

I. INTRODUCTION

Low-power, robust True Random Number Generators are critical to ensuring the security of wireless communication devices, ranging from IoT devices to satellite modem applications [1]. Although quantum processes are preferable sources of entropy, circuits to harvest them are usually bulky and power hungry [2]. As a result, several CMOS-based TRNGs have been proposed based on device noise [3], meta-stability of the cross-coupled pair [4] and oscillator jitter [1, 5-6]. Among them, analog-based designs are not readily suitable for integrating with nano-scale digital circuits [5] while the meta-stability-based design requires calibration to eliminate circuit biased due to inherent process variations [4]. [1, 5-6] utilize jitter noise in all-digital ring oscillator (RO) structures to provide more efficient and low-cost solutions for IoT applications. The conventional RO-based RNG (Fig. 1a) utilizes a DFF to sample random data from a fast RO (f_1), using a slow RO (f_2) as a clock. To ensure good randomness of the output stream, the jitter of the slow RO (i.e. clock) must be comparable to the oscillation period of the fast RO (i.e. data) [6]. Thus f_1 is much higher than f_2. Consequently, the total power consumption of the system is not optimized, mainly due to the high-speed f_1. Another drawback of these designs is their ability to demonstrate reliable operation at low

voltages. For example, in [5], the data rate may degrade quickly as the counter needs to count the number of cycles-to-collapse of the 3-edge RO. [1] reduces power consumption by proposing a chaos system with transmission gate (Tx-gate) coupling [Fig. 1b]. As shown in Fig. 1b, the flow RO f_2 is replaced by two ROs (f_{21} and f_{22}). These two ROs coupled with each other by a transmission-gate (TX-gate) to amplify the intrinsic jitter of f_{21} so that the high-speed RO f_1 can be slowed down, thus power is saved, as illustrated in Fig. 1b. Nevertheless, the coupling effect of the Tx-gate in [1] rapidly diminishes when the supply voltage falls below the threshold voltage of the transistors.

This paper improves the energy efficiency of the existing design using a capacitive coupling concept and a dual-edge sampling scheme. The all-digital design offers high bit rates with low-power consumption, even in subthreshold operating condition. It is most suitable for IoT applications where both power supply and silicon real estate are limited resources.

II. PROPOSED DESIGN

We propose a TRNG consists of one fast RO (f_1) followed by a divide-by-two frequency divider, two slow ROs (f_{21} and f_{22}) and two DFFs (Fig. 2). Unlike [1], f_{21} and f_{22} are coupled by a capacitor to form a chaos system so that its jitter is amplified. We call it a jittered clock. In CMOS implementation, this coupling capacitor can be MIM cap or MOS-cap with a capacitance of a few femtoFarad so its area overhead is quite small. In general, $f_1 \gg f_{21} > f_{22}$. f_{21} and f_{22} are capacitively-coupled to create a chaos system to amplify jitter noise of f_{21}. The capacitive coupling is more efficient than existing diode or Tx-gate coupling because the coupling effect between its two terminals is always present, even at low supply voltage. Thus it performs better when V_{DD} scales down. For example, the measurement result in Fig. 3

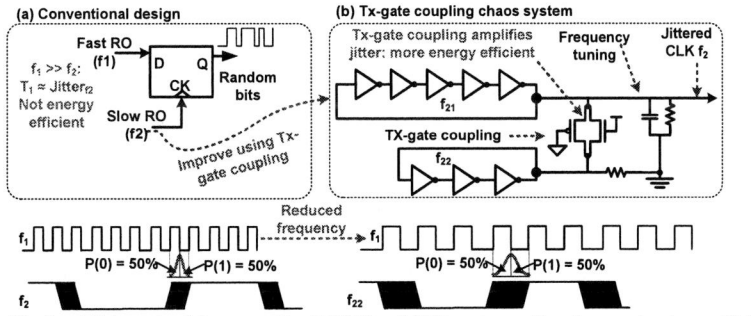

Fig. 1. a) Schematic of the conventional TRNG and (b) Tx-gate coupling chaos system to amplify jitter.

This work was supported by A*STAR, Singapore EDB and Singapore National Research Foundation (NRF) under the grant No. S15-1321-IAF OSTIn-SIAG, NRF2014SAS-SRP001-057.

978-1-5386-3179-9/17 $31.00 © 2017 IEEE

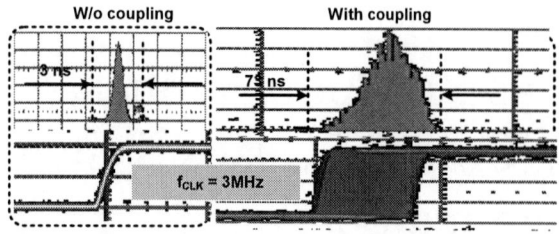

Fig. 3. The measured jitter of the ROs with and without the coupling effect at 0.4 V.

confirms that with the coupling effect, the jitter of the ROs is amplified by more than 24×, even at 0.4 V supply voltage.

With a jittered clock, the whole design's operation is as follow: the output of RO f_1 is fed to a divide-by-two frequency divider to obtain a 50% duty cycle (T_D). Output of the frequency divider (i.e. F) has a frequency of $f_1/2$ and is connected to the D input of the first DFF, which is sampled by a jittered clock (i.e. CLK at the output of f_{21}) to produce XQ. With the assumption that jitter of f_{21} is amplified enough to be comparable to the period of f_1, a random sequence is sampled by the DFF. In addition, the inverted version of F and CLK (i.e. FB and CLKB) are fed to the D and CK inputs of the second DFF, respectively, as shown in Fig. 2 to produce XP. Subsequently, XQ and XP are XOR-ed to obtain XQC. Finally, XQC is passed through a 4b-linear-feedback-shift-register to enhance its uniformity (XQCE).

Since CLK is a jittered signal, statistically CLKB contains an equal amount of jitter. Therefore CLKB can also be used as a sampler with random arrival time. Note that both DFF are positive-edge triggered and they are latched at the alternative edges of the CLK, as illustrated in Fig. 4. Thus, if XQ is random due to jitter of CLK, XP is also random due to jitter of CLKB. In other words, XQ and XP should have similar randomness. Furthermore, we can assume that individual XQ and XP outputs are independent of each other since the exact rising edges of the clocks are determined by the amplified jitter noise of the chaotic system. Since XQ and XP are sampled at alternative clock edges, XOR-ing them create an output stream with a double data rate as follow:

$$XQC_{2n} = XQ_n \oplus XP_n \qquad (1a)$$
$$XQC_{2n+1} = XQ_{n+1} \oplus XP_n \qquad (1b)$$
$$XQC_{2n+2} = XQ_{n+1} \oplus XP_{n+1} \qquad (1c)$$

as shown in the timing diagram in Fig. 4. By sharing the ROs to create F, FB, CLK, CLKB our design doubles the output data rate with minimal area and energy overhead. The purpose of the pairing of F-CLK and FB-CLKB are two folds: (1) XQ may be biased (i.e. statistically more "0" than 1" in the long run) due to the non-50% duty cycle of F (Fig. 4). Note that this may still happen even though we already use a divide-by-2 to obtain 50% T_D at the output of the frequency divider. This is because of the non-ideality at the rising and falling edge of the signal F. However, the duty cycles of F and FB compensate and thus the bias in the output streams of XQ and XP also compensate each other but not correlated (Fig. 4). Note that XQ ≠ ~XP due to the randomness at each sampling point. (2) Using CLK and CLKB results in XQ and XP to be latched at the alternative timing (Fig. 4). By XOR-

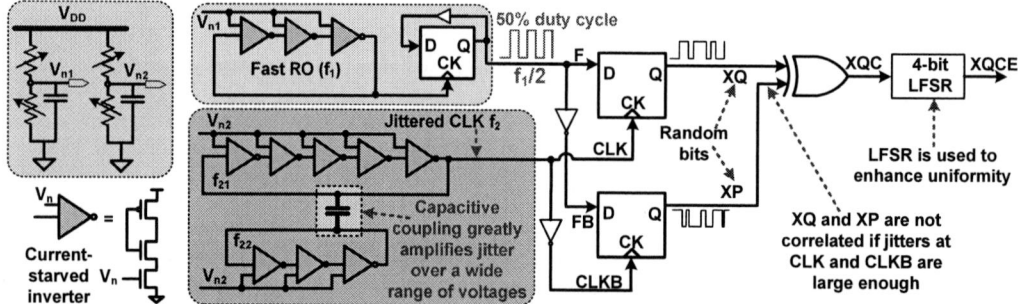

Fig. 2 Proposed design using capacitive coupling and dual-edge sampling to double the data rate. LFSR is used to enhance the uniformity.

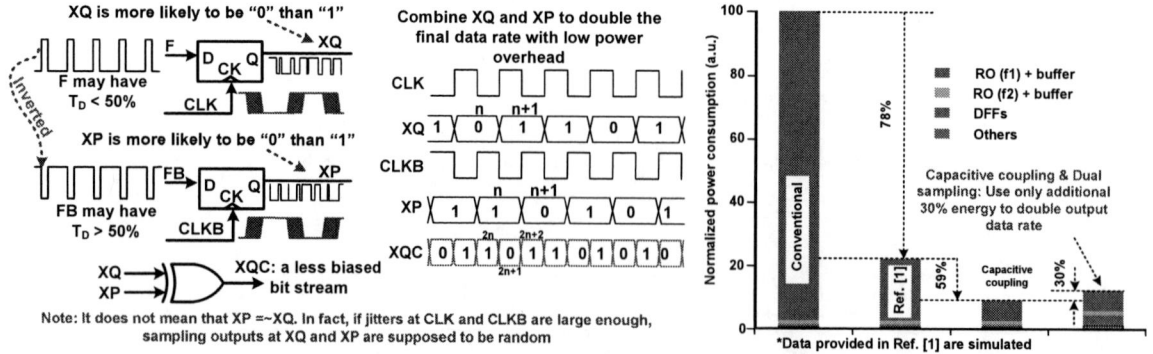

Fig. 4 The dual-edge sampling scheme mitigates the bias in the output stream (b) Example timing diagram.

Fig. 5 Simulated power breakdown of the proposed design in comparison with the conventional design.

ing them the design automatically generates a third bit-stream (XQC) at both clock edges which doubles the data rate. Finally, XQC is passed through a 4-b linear feedback shift register (LFSR) to enhance its uniformity (XQCE). The sharing scheme doubles the data rate with only 30% power overhead (compared to itself, with single-edge sampling), as shown in Fig.5. Equivalently, it reduces 91% power when compared to the conventional design. The main concern when sharing ROs is the correlation between XQ and XQB, which will be evaluated later (Fig. 9).

In this implementation, a current-starved inverter is used to realize the ROs, as shown in Fig. 2. A biasing voltage V_{ni} is used to tune the delay of the inverter and thus the frequency of the ROs. This implementation is more area and energy efficient than using the number of inverter stages. The relative speed between f_1 vs f_{21}/f_{22} is controlled by biasing voltage V_{n1} and V_{n2}. V_{n1} controls the fast RO f_1 while V_{n2} controls f_{21} and f_{22}. f_{22} is made faster than f_{21} by using less number of delay stages, also shown in Fig. 2. In this work, V_{n1} and V_{n2} are generated off-chip to provide testing and characterization flexibility across a wide operating voltage range. However, it can be implemented on-chip with similar structure.

III. TEST CHIP MEASUREMENT RESULTS

The proposed TRNG has been implemented in 65 nm CMOS, occupying 560 um^2. Its chip microphotograph is shown in Fig. 6. Fig. 7 illustrates our testing strategy and the measured minimum V_{n1}/V_{n2} ratio required to obtain good output stream. Note that given a fixed V_{n2}, the higher V_{n1}/V_{n2} the faster f_1 compared to f_{21}/f_{22} and thus, in theory, the quality

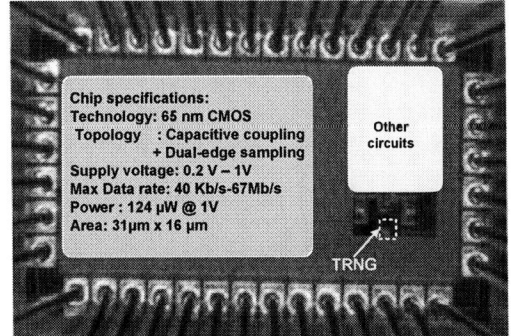

Fig. 6 Microphotograph of the test chip implemented in 65nm CMOS and chip specification.

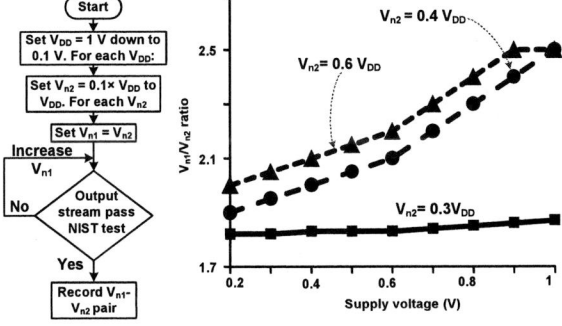

Fig. 7 Testing strategy and the measured minimum Vn1/Vn2 ratio plot vs V_{DD}.

TABLE I: NIST TEST RESULTS

Test name	All 18Mb, V_{DD}=0.2→1 V		First 2Mb, V_{DD}=0.2 V		Last 2Mb, V_{DD}= 1V		PASS?
	P-value	Pass Rate	P-value	Pass Rate	P-value	Pass rate	
Frequency	0.9	99%	0.83	95%	0.71	100%	passed
BlockFreq.	0.81	99%	0.21	100%	0.32	100%	passed
AccumulativeSum	0.19	98%	0.53	95%	0.013	100%	passed
Runs	0.73	100%	0.96	100%	0.36	95%	passed
LongestRun	0.43	98%	0.025	95%	0.53	100%	passed
Rank	0.02	99%	0.83	100%	0.89	95%	passed
FFT	0.65	100%	0.21	100%	0.56	100%	passed
NonOver.Temp.	0.27	98%	0.36	100%	0.57	95%	passed
Over.Temp.	0.40	100%	0.52	100%	0.72	100%	passed
Approximate.Entr.	0.85	99%	0.35	100%	0.15	95%	passed
Serial	0.35	98%	0.36	100%	0.37	100%	passed
LinearComlexity	0.25	100%	0.12	100%	0.21	95%	passed
Universal	0.46	97%	0.16	100%	0.18	100%	passed
Rand.Excursion	0.21	100%	N.A*		N.A*		passed
Rand.Exc.Variant	0.34	100%	N.A*		N.A*		passed

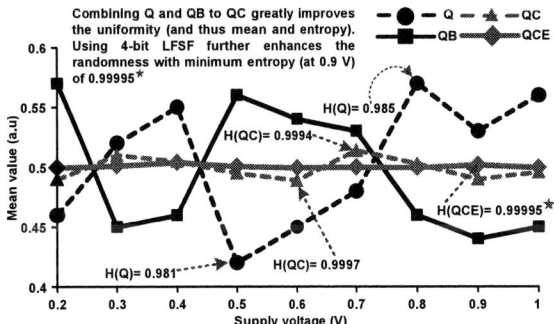

Fig. 8 Means and the corresponding entropies (H) of the output streams at different supply voltages. Mean = P(X=1) = PX. H(X) =-[PXlog2(Px) + (1-PXlog2)(1-Px)].

of the output stream is better, at the cost of more power consumption. As can be seen from Fig. 7, minimum V_{n1}/V_{n2} ratio depends on both V_{DD} and V_{n2}. In general, the higher V_{DD} or V_{n2} (i.e. faster f_{21}/f_{22}), a higher V_{n1}/V_{n2} ratio is required. Since it is not possible to characterize all V_{n1}/V_{n2} pair, for the sake of data clarity, we set $V_{n1} = 2\times V_{n2}$ and $V_{n2} = 0.3 V_{DD}$ before extensively analyzing the randomness of the output stream at different voltages. This choice allows us to scale V_{DD} without adjusting the V_{n1}/V_{n2} ratio (Fig. 7). At each supply voltage condition (i.e. 0.2 V, 0.3 V to 1V), 2Mb of data are collected at node XQCE and padded into an ASCII file, resulting in a total of 18Mb for NIST test [7]. As shown in Table I, the whole 18Mb bit stream passed all 15 tests, proving that the collected data has a high probability of being random across the tested voltage range. Data collected from each voltage (i.e. 2Mb) are also evaluated individually. They all 13 passed NIST tests, except for RandomExcursion and RandomExcursionVariant tests due to the shortage of data. Results from 0.2 V and 1 V are also shown in Table I as examples.

Due to non-ideality of the circuits, there are biases in XQ and XP, thus their expectation (i.e. mean) is not perfectly 50%. Measured results confirm that E(XQ) and E(XP) tends to deviate from 0.5 in opposite directions (Solid and dashed black lines in Fig. 8) which agrees with our hypothesis that if the duty cycle of F is statistically less than 50%, that of FB is more than 50%. These deviations also change with V_{DD}, as illustrated in Fig. 8. XOR-ing them makes XQC much less biased. For example, the worst case entropy of XQ and XP is 0.981/bit at 0.5V but that for XQC is 0.9994/bit at 0.7 V. After passing through the LFSR, XQCE has the worst-case entropy of 0.99995 across the voltage range (Fig. 8). Thus the

TABLE II: PERFORMANCE COMPARISON

	HOST '14 [1]	ISSCC '16 [2]	ISSCC '14 [5]	VLSI Symp.'10 [4]	ISSCC '08 [3]	Trans. Comp.' 03 [6]	This work
Technology	45 nm	0.15 µm	28 nm	45 nm	45 nm	0.18 µm	65 nm
Architecture	Diode coupling	Light detector	Jitter In 3-edge RO	Meta-stability of Cross-coupled pair	Electron-trap noise	Oscillator jitter	Capacitive coupling; Dual Sampling
V_{DD} (V)	1	> 2 V	1	0.28-1.35	1	1.8	0.2 - 1
Bit rate (Mb/s)	127	128	23.16	15-2400*	2	10	0.05 (0.2V)-67 (1V)***
Area (F^2)	46x10^3	49×10^6	478×10^3	1.9×10^6	19×10^3	493×10^3	132×10^3
NIST pass	11/15	N.A	**15/15**	**15/15**	N.A	N.A	**15/15**
Calibration needed	No	N.A	No	Yes	No	No	**No******
Energy efficiency (pJ/bit)	8.6	N.A	23	0.3@0.28V** 3@1V**	950	230	**0.025@0.3V/ 1.85@1V**
FOM#	3.6×10^3 (simulated)	N.A	2.43	4.8×10^3	1.26	1	**17.2×10^3 @0.3V, 5MHz**

* Extracted from Fig. 5 of ref [4]   *** The maximum bit rate of our design at 1 V is limited by the testing board

Estimated from Fig. 5 of ref [4]   ** V_{n1} & V_{n2} are generated off-chip for characterization. It can be build on-chip with similar structure

\# $FOM = \frac{Data\ Rate\ (Mb/s)}{Energy.per.bit\ (pJ)*Area(F^2)}$ (Normalized to [6])

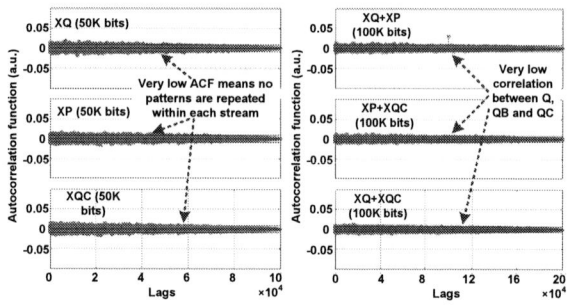

Fig. 9 Auto Correlation Function of the output bitstreams collected at 0.3 V.

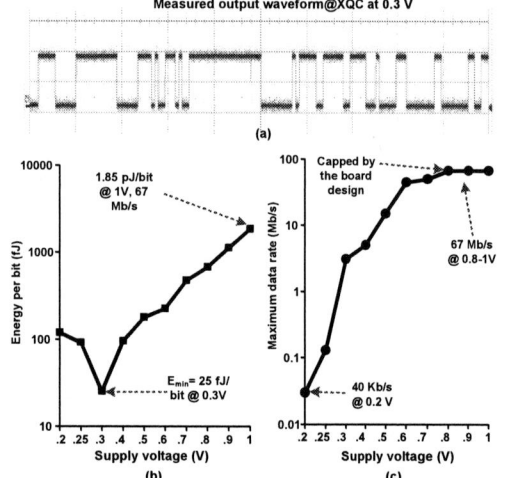

Fig. 10   (a) Measured output waveform. (b) Energy consumption per bit and (c) max operating frequency vs V_{DD}

proposed scheme has successfully mitigated the non-ideality of the circuit. In fact, as a random source, it should not output a perfect entropy of 1 (i.e. P(x=1) = P(x=0) = 0.5).

Next, we investigate if there are any obvious patterns that are repeated within each bit stream using the Auto Correlated Function (ACF) in Fig. 9. The very low ACFs on the left of Fig. 9 indicate that no obvious pattern is found within each of XQ, XP, and XQC. We also need to access the cross-correlations between XQ, XP and XQC. By padding these stream together (e.g. 50Kb from Q followed by 50Kb from XQB), low ACFs on the right of Fig. 9 confirms that there is an extremely low correlation between them and thus sharing the ROs minimally affects the total entropy of the system.

Fig. 10 (a) shows the measured output waveforms at 0.3 V while the energy per bit and maximum bit rate across different voltages are shown in Fig. 10(b) and Fig. 10(c), respectively. The test chip operates properly from 0.2 V to 1 V and achieves the optimum energy efficient point at 0.3 V where it consumes only 25 fJ/bit while having a data rate of 5Mb/s. From 0.8 V to 1 V, its maximum data rate is 67 Mb/s, which is capped by the test board capability. Table II compares the measurement results with existing state of the arts. It is the most energy-efficient design which has 3× better energy/bit than the lowest reported prior art to-date [4].

IV.   CONCLUSION

A RO-based RNG utilizing capacitive coupling technique and dual-edge sampling has been implemented in 65nm CMOS process. Our measurement results confirm that the design offers good randomness, high data rate output steam with ultra low power consumption over a wide voltage range. The collected data passed the widely used NIST random test with all sub-tests. The proposed design offers an all-digital solution for IoT applications and consumes only 0.025pJ/bit at 0.3V, 5Mb/s. Its overall is FOM of 1.72×10^4, more than 3× better than existing designs.

REFERENCES

[1] S. N. Dhanuskodi, et. al., "A Chaotic Ring oscillator based Random Number Generator," IEEE International Symposium on Hardware-Oriented Security and Trust (HOST), 2014, pp. 160-165.

[2] N. Massari et al., "16.3 A 16x16 pixels SPAD-based 128-Mb/s quantum random number generator with -74dB light rejection ratio and -6.7ppm/C bias sensitivity on temperature," Digest of Technical Papers, IEEE International Solid-State Circuits Conference, 2016, pp. 292-293.

[3] M. Matsumoto, et. al., "1200 µm2 Physical Random-Number Generators Based on SiN MOSFET for Secure Smart-Card Application," Digest of Technical Papers, IEEE International Solid-State Circuits Conference, 2008, pp. 414-624

[4] S. Srinivasan et. al., "2.4GHz 7mW all-digital PVT-variation tolerant True Random Number Generator in 45nm CMOS," VLSI Circuits Symp., Honolulu, 2010, pp. 203-204.

[5] K. Yang, et. al., "16.3 A 23Mb/s 23pJ/b fully synthesized true-random-number generator in 28nm and 65nm CMOS," Digest of Technical Papers, IEEE International Solid-State Circuits Conference, 2014, pp. 280-281.

[6] M. Bucci, et al., "A High Speed Oscillator Based True Random Number Generator for Cryptographic Applications on a Smart Card IC," IEEE Trans. Computers, vol. 52, no. 4, pp. 403-409, 2003.

[7] http://csrc.nist.gov/groups/ST/toolkit/rng/documentation_software.html

978-1-5386-3179-9/17 $31.00 © 2017 IEEE

A 1.25pJ/bit 0.048mm^2 AES Core with DPA Resistance for IoT Devices

Shengshuo Lu[1], Zhengya Zhang[1], Marios Papaefthymiou[1,2]

[1] University of Michigan, Ann Arbor, MI, USA. [2] University of California, Irvine, CA, USA.

Abstract—An AES core designed for low-cost and energy-efficient IoT security applications is fabricated in a 65nm CMOS technology. A novel Dual-Rail Flush Logic (DRFL) with switching-independent power profile is used to yield intrinsic resistance against Differential Power Analysis (DPA) attacks with minimum area and energy consumption. Measurement results show that this 0.048mm^2 core achieves energy consumption as low as 1.25pJ/bit while providing at least 2604x higher DPA resistance over its conventional CMOS counterpart, marking the smallest, most energy-efficient and most secure full-datapath AES core.

Keywords—Advanced Encryption Standard, Differential Power Analysis, Dual-Rail Flush Logic, Instrinsic DPA Resistance.

I. INTRODUCTION

The Advanced Encryption Standard (AES) is a widely-used algorithm for symmetric cryptography. Although AES is extremely difficult to break in theory, AES chips can be subject to side-channel attacks that use information such as the chip's power profile to reveal the secret key stored in the chip. Differential Power Analysis (DPA) is one of the most effective side-channel attacks [1]. In a DPA attack, a switching-dependent power profile of the chip is generated through monitoring its power supply current. This profile is then correlated with a switching behavior model to reveal the secret key [2-5].

Previous AES chips report 66x to 2500x DPA resistance over DPA-unprotected cores [2-5]. One defense mechanism is to augment an unprotected core with countermeasure circuits that scramble its supply voltage and current [2, 3]. This approach is not amenable to voltage scaling, however, because the scrambled supply voltage is limited to a certain minimum level, and no work reports on its performance under voltage scaling. Another defense is to use intrinsically DPA-resistant logic gates that exhibit constant energy consumption during operation and hide the impact of switching activity from the power trace [4, 5]. This approach typically suffers from high area overheads.

This paper presents a voltage-scalable full-datapath 128-bit AES chip with intrinsic DPA resistance that is suitable for Internet-of-Things (IoT) applications thanks to its energy-efficient operation and small die area. Compared to previous DPA-protected cores [2-5], this chip is the smallest, most energy-efficient, and most DPA-resistant.

This work was supported in part by NSF under grants No. CCF 1320027 and No. CCF 1161505.

II. DUAL-RAIL FLUSH LOGIC AND ARCHITECTURE

Logic gates designed to be intrinsically DPA-resistant have the potential to provide superior resilience to DPA attacks. However, published designs fall short of fulfilling this potential while at the same time suffering from 2x to 4x high area overheads [4, 5]. This paper proposes Dual-Rail Flush Logic (DRFL), a logic family whose gate structure and architecture are designed and optimized to achieve a switching-independent energy profile. With balanced gate topology and pipeline flushing mechanism, DRFL provides robust and intrinsic resistance to DPA attacks with minimal area overhead.

DRFL is a derivative of static dual-rail CMOS logic [6]. Fig. 1 shows a DRFL XOR gate. When inputs A (A_b) and B (B_b) present valid complementary logic values, the gate is in evaluation mode, and output Y (Y_b) presents valid complementary values. When all inputs are set to the same value, the gate is in precharge mode, and both outputs present the opposite values. During consecutive cycles in its operation, the gate alternates between evaluation mode and precharge mode. For example, when the XOR gate in Fig. 1 changes from precharge mode to evaluation mode, one of the outputs goes high, with the other output remaining low, depending on the input values. When the gate changes from evaluation mode back to precharge mode, the high output node is discharged low, and the gate consumes the same amount of energy as during evaluation. Therefore, irrespective of input values, the energy consumption of each individual DRFL gate remains the same throughout its operation.

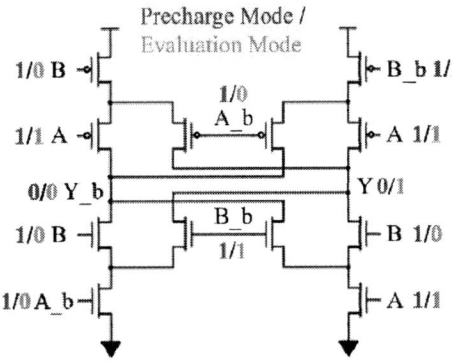

Fig. 1. DRFL XOR gate. Input and output values are shown for precharge and evaluation mode.

978-1-5386-3179-9/17 $31.00 © 2017 IEEE

One of the advantages of static dual-rail gate is that it fully utilizes the benefit of complementary inputs, yielding intrinsic DPA resistance with only modest area overheads over static CMOS. For example, a relatively complex function like XOR, which is the function used most often in AES, requires 10 transistors when implemented as a single-rail static CMOS gate. A dual-rail implementation of XOR uses 12 transistors by sharing transistors in the pull-up and pull-down branches as shown in Fig. 1, only incurring a 20% overhead. In general, as functions become more complex, more transistor sharing is possible, and dual-rail implementations use much less than twice as many devices as their single-rail counterparts. For this reason, DRFL designs eventually achieve much better area efficiency than other intrinsic solutions [4, 5]. The area of the DRFL AES core in this paper is only 50% higher than its CMOS counterpart. By comparison, the WDDL-based design in [5] has 4x the area of its CMOS counterpart, because it uses CMOS gates and inverters to mimic the functionality of dual-rail logic gate. The BBL-based design in [4] has twice the area of its CMOS counterpart, because each BBL gate has a boost structure for charge recovery.

In addition to balanced gate design, pipeline flushing is essential for DRFL to achieve superior DPA resistance. To ensure correct operation and pipeline flushing, in cascades of DRFL gates, adjacent gates precharge to opposite values, as shown in Fig. 2, with the gates precharging to 1/0 denoted as P/N gates. P gates must connect to N gates, and N gates must connect to P gates. To that end, buffers must be inserted as necessary to balance paths and ensure the proper polarity of connected gates. Due to the regular and balanced topology of the AES datapath, the number of balancing buffers is relatively small. Moreover, unlike WDDL [5], which introduces two inverters at the output of each gate, DRFL avoids inverters thanks to the alternating P/N architecture, eliminating the area and energy consumption overhead associated with inverters and offsetting in part the buffer overheads.

After each pipeline stage, a pair of CMOS flip-flops is used to store logic values. Each flip-flop pair alternates between precharge mode (both flip-flops store the same value) and evaluation mode (flip-flops store complementary values). Hence, the flip-flops consume the same amount of energy when alternating between precharge and evaluation, ensuring switching-independent energy consumption.

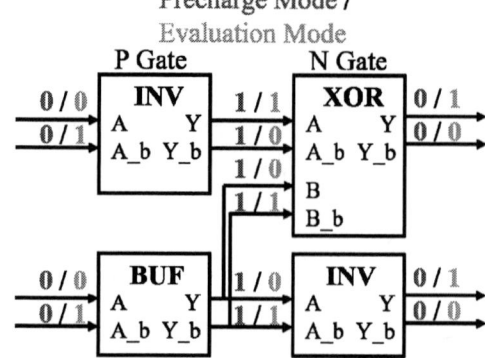

Fig. 2. Cascade of DRFL gates.

In DRFL pipelines, evaluation data and precharge data propagate in an interleaved manner, as shown in Fig. 3. During consecutive cycles, each pipeline stage alternates between evaluation mode and precharge mode.

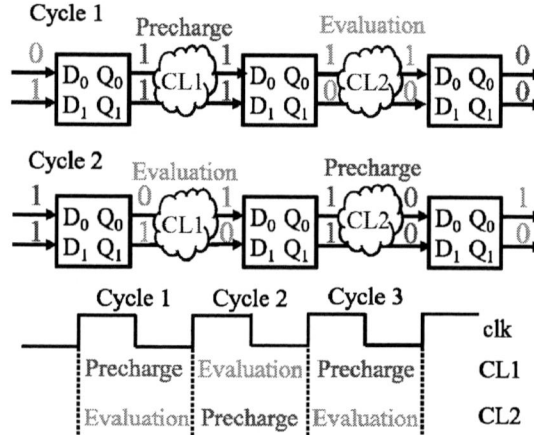

Fig. 3. Pipeline of DRFL gates, and interleaving of precharge and evaluation mode.

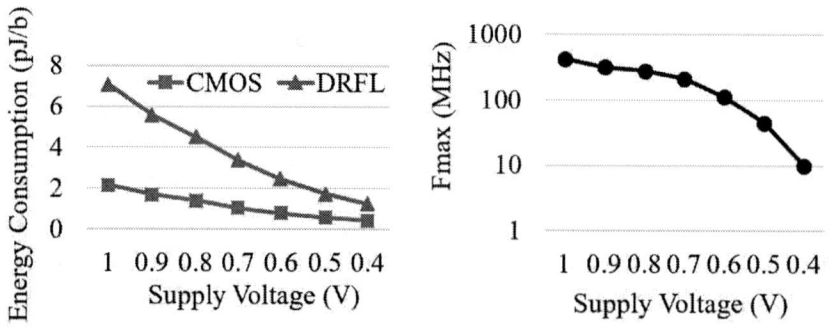

Fig. 4. Measured frequency vs. supply voltage.

III. EXPERIMENTAL EVALUATION

The DPA-resistant AES core has been fabricated in a 65nm CMOS process. Its standard CMOS counterpart has been included on the same die. The two cores have the same RTL specification from [7], architecture, and target frequency.

Leveraging its CMOS underpinnings, the DRFL core functions correctly across a wide voltage range, as shown in Fig. 4. At the nominal supply of 1V, the core achieves a clock frequency of 430MHz and consumes 7.09pJ/bit. With a near-threshold supply of 0.4V, the core operates at 10MHz and consumes 1.25pJ/bit.

DPA attacks are performed on both cores at the nominal voltage, the weakest operating point for DPA attacks. (As energy consumption decreases with voltage scaling, the chip's power profile is masked by noise, and DPA attacks require more traces [8].) Fig. 5 shows Measurements to Disclosure (MTD) for the standard CMOS core. MTD of a byte in the key is the number of measurements needed for the correlation of the correct key value to surpass the correlation of all other 255 values [2]. In the CMOS core, the key byte that requires the least number of power traces to be disclosed (a.k.a. the 1st key byte) is revealed relatively soon, as its correlation crosses the maximum correlation among all other 255 values after 768 measurements and continues to increase with the number of measurements. Fig. 6 illustrates the hardness of disclosing the key in the DRFL core. Even after 2 million measurements, the normalized correlation value of the correct key remains relatively stable and is indistinguishable from the other key hypotheses.

Measurement results from the two cores are illustrated in Table I. Both cores attain a maximum frequency of 430MHz at the nominal 1V supply level. The DRFL core has half the throughput of its CMOS counterpart, since its pipelines are in precharge mode every other cycle. The DRFL core is 50% larger than the CMOS core, due to the overheads of dual-rail logic and balancing buffers. Measured at 1V and 0.4V, the DRFL core consumes 2.3x and 1.8x more energy, respectively, than the CMOS core. However, the DRFL core exhibits at least 2604x higher DPA resistance than the CMOS core. Die photo of the two cores is shown in Fig. 7.

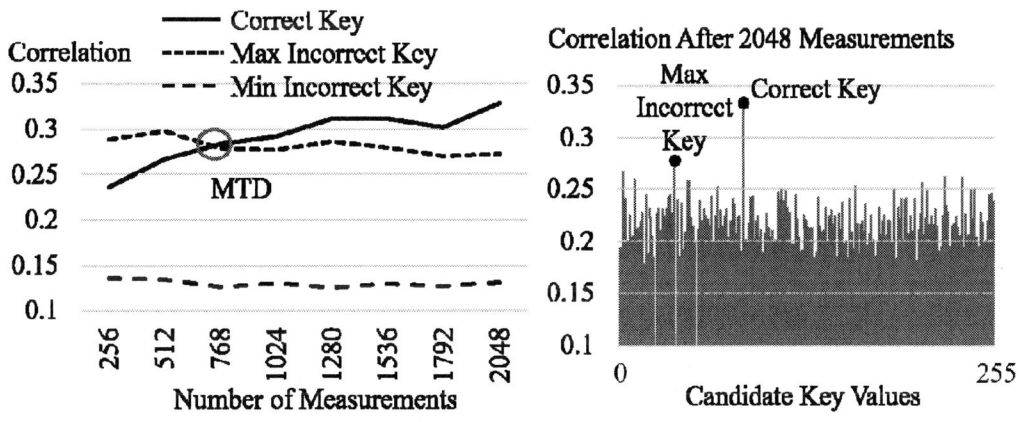

Fig. 5. DPA attack on standard CMOS AES.

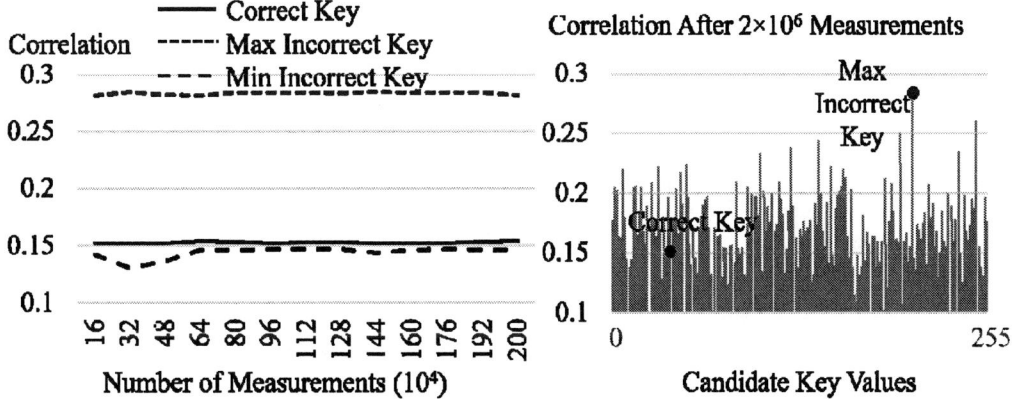

Fig. 6. DPA attack on DRFL AES core.

978-1-5386-3179-9/17 $31.00 © 2017 IEEE

Fig. 7. Die photo.

Table II compares the area, performance, energy efficiency, and DPA resistance of the DRFL core and other published AES cores [2-5, 9]. The designs in [2-5] are DPA resistant with MTD of 1st key byte ranging from 66x to 2500x compared to

an unprotected AES core. The design in [9] achieves superior throughput and energy efficiency, but is not DPA resistant. Consuming 1.25pJ/bit, the 0.048mm² DRFL core achieves superior balance between area, energy efficiency and DPA resistance.

REFERENCES

[1] Kocher, Paul, Joshua Jaffe, and Benjamin Jun. "Differential power analysis." In Annual International Cryptology Conference, pp. 388-397. Springer Berlin Heidelberg, 1999.

[2] Tokunaga, Carlos, and David Blaauw. "Secure AES engine with a local switched-capacitor current equalizer." In International Solid-State Circuits Conference-Digest of Technical Papers, pp. 64-65. IEEE, 2009.

[3] Liu, Po-Chun, Ju-Hung Hsiao, Hsie-Chia Chang, and Chen-Yi Lee. "A 2.97 Gb/s DPA-resistant AES engine with self-generated random sequence." In ESSCIRC, 2011 Proceedings of the, pp. 71-74. IEEE, 2011.

[4] Lu, Shengshuo, Zhengya Zhang, and Marios Papaefthymiou. "1.32 GHz high-throughput charge-recovery AES core with resistance to DPA attacks." In VLSI Circuits, 2015 Symposium on, pp. C246-C247. IEEE, 2015.

[5] Hwang, David D., Kris Tiri, Alireza Hodjat, B-C. Lai, Shenglin Yang, Patrick Schaumont, and Ingrid Verbauwhede. "AES-Based Security Coprocessor IC in 0.18um CMOS With Resistance to Differential Power Analysis Side-Channel Attacks." IEEE Journal of Solid-State Circuits 41, no. 4 (2006): 781-792.

[6] Weste, N. H. E., and D. Harris. "Array subsystems." CMOS VLSI design: a circuits and systems perspective (4th edition)(M. Goldstein and M. Suarez-Rivas, eds.) (2011).

[7] Canright, David. "A very compact S-box for AES." In International Workshop on Cryptographic Hardware and Embedded Systems, pp. 441-455. Springer Berlin Heidelberg, 2005.

[8] Haider, Syed Imtiaz, and Leyla Nazhandali. "Utilizing sub-threshold technology for the creation of secure circuits." In International Symposium on Circuits and Systems, pp. 3182-3185. IEEE, 2008.

[9] Mathew, Sanu K., Farhana Sheikh, Michael Kounavis, Shay Gueron, Amit Agarwal, Steven K. Hsu, Himanshu Kaul, Mark A. Anders, and Ram K. Krishnamurthy. "53 Gbps Native GF(2^4)² Composite-Field AES-Encrypt/Decrypt Accelerator for Content-Protection in 45 nm High-Performance Microprocessors." IEEE Journal of Solid-State Circuits 46, no. 4 (2011): 767-776.

TABLE I. DRFL AND CMOS DESIGNS CHARACTERISTICS

	DRFL		CMOS	
Technology	65nm			
Area (mm²)	0.048		0.032	
Supply Voltage (V)	1.0	0.4	1.0	0.4
Frequency (MHz)	430	10	430	10
Throughput (Gb/s)	2.752	0.064	5.504	0.128
Power (mW)	19.5	0.080	11.8	0.056
Energy Efficiency (pJ/b)	7.09	1.25	2.14	0.44
DPA resistance	2×10⁶ measurements without cracking any key		768 measurements to disclosure of first key byte	
Key bytes disclosed (out of 16 keys bytes)	0		16	
DPA resistance comparison	At least 2604x			

TABLE II. COMPARISON WITH PREVIOUSLY PUBLISHED AES CHIPS.

	DRFL		[2]	[3]	[4]	[5]	[9]	
Technology	65nm		130nm	90nm	65nm	180nm	45nm	
Area(mm²)	0.048		0.364	0.104	0.291	2.45	0.026	
Supply Voltage (V)	1	0.4	1.2	1	0.41	1.8	1.1	0.36
Maximum Frequency (MHz)	430	10	110	255	1320	85.5	2100	Not reported
Maximum Throughput (Gb/s)	2.752	0.064	1.28	2.97	16.9	0.99	26.5	Not reported
Power (mW)	19.5	0.08	44.34 (100MHz)	7.10 (200MHz)	98	200 (50MHz)	62.5	Not reported
Energy Efficiency (pJ/b)	7.09	1.25	38.10	3.04	5.79	345	2.358	0.455
DPA Resistance	2604x		2500x	1086x	720x	66x	DPA unprotected	

978-1-5386-3179-9/17 $31.00 © 2017 IEEE

S5-3 (2218)

A $0.40\,\mathrm{pJ/cycle}$ $981\,\mu\mathrm{m}^2$ Voltage Scalable Digital Frequency Generator for SoC Clocking

Martin Cochet[*†], Sylvain Clerc[†], Guénolé Lallement[†*], Fady Abouzeid[†], Philippe Roche[†], Jean-Luc Autran[*]
* Aix-Marseille University & CNRS, IM2NP (UMR 7334), Marseille, France
† STMicroelectronics, Crolles, France.

Abstract—This work presents an area-efficient voltage and frequency scalable clock generator for low-power digital SoC clocking. Named Direct Digital Sampling and Synthesis (DDSS), the open-loop generator implemented in $28\,\mathrm{nm}$ FD-SOI operates from $0.45\,\mathrm{V}$ to $1.1\,\mathrm{V}$ with measured jitter from 1.7% to 5.1% UI. Its low power consumption of $0.40\,\mathrm{pJ/cycle}$ at $57\,\mathrm{MHz}$ $0.5\,\mathrm{V}$ combined with the ability to perform fast frequency changes makes this circuit an alternative to PLLs for fast Dynamic Voltage and Frequency Scaling (DVFS) strategies in low power SoCs.

I. INTRODUCTION

The last decade has seen a trend in taking advantage of digital logic downscaling to port analog building blocks into digital designs. This has enabled area savings and voltage downscaling for elements such as power monitors, temperature sensors as well as Phase Locked Loops (PLLs) [1]–[5]. The All Digital PLLs now replace the traditional LC oscillators and charge pumps with ring oscillators and digital loop filters.

These circuits offer better area and power performance, at the expense of slightly higher jitter values, which is less critical for clocking than for RF applications. However, due to their closed loop nature, the digital PLLs face the same lock time restrictions as their analog counterparts. Some instant switching strategies added to the PLLs come at the price of added area and power consumption [5].

Moreover, power management strategies rely heavily on fine grain frequency scaling [6]. This fine granularity applies both in space and time, requiring a clock generator with low area and instant switching capability respectively. To limit its power overhead, the clock generator also needs a low leakage current, in clock gating mode, and a wide voltage scalability with output frequency matching that of the digital logic it clocks.

The open-loop principle of Direct Digital Sampling and Synthesis (DDSS) [7] offers an alternative to PLLs, trading off the phase locking for instant switching. The previously published DDSS, however, suffered from limited voltage and frequency scalability ($0.6\,\mathrm{V}$ minimum voltage, $574\,\mathrm{MHz}$ maximum frequency at $0.9\,\mathrm{V}$), as well as a complicated calibration mechanism, limiting its practical use.

The proposed design, implemented in 28nm FD-SOI, improves on the DDSS principle by using a phase selection approach to the fractional division unit, rather than delay lines. Compared to [7], this method allows for a simpler calibration-free design, offering a 14x reduction in area down to $981\,\mu\mathrm{m}^2$, as well as extended voltage operating range (down to $0.45\,\mathrm{V}$), 6.5x reduction in power consumption at Vmin and a maximum frequency on par with digital clocking requirements ($879\,\mathrm{MHz}$ vs $574\,\mathrm{MHz}$ at $0.9\,\mathrm{V}$). This makes the phase selection based DDSS a good candidate for low power voltage scalable Glob-

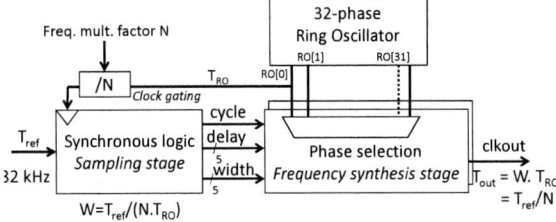

Fig. 1. Principle of operation of the DDSS clock generator.

ally Asynchronous, Locally Synchronous (GALS) SoCs.

II. CIRCUIT ARCHITECTURE

A. Direct Digital Sampling and Synthesis

Fig. 1 illustrates the phase selection principle of operation of the phase-selection based DDSS. A low frequency clock reference of period T_{ref} is provided (for example from an off-chip quartz). First, in the *sampling stage*, the reference frequency is compared with the period T_{RO} of an internal free-running ring oscillator via a simple counter, which produces a digital output W proportional to T_{ref}/T_{RO}. Thanks to a configurable clock gating, the first stage counter is incremented by one every N cycles of T_{RO}, resulting in $W = T_{ref}/(N.T_{RO})$.

Then, in the *synthesis stage*, a phase selector operates the fractional frequency division of the RO by a programmable factor proportional to the first stage output W, ie $T_{out} = T_{RO}.W = T_{ref}/N$. By using the same RO reference for the sampling and synthesis stages, this feed-forward design guarantees that the output frequency is N times that of the reference, independently of the exact RO frequency.

The feed-forward topology also provides a change in W and T_{out} when N or T_{ref} are changed, after only one reference cycle, compared to several cycles re-locking for PLLs [1]–[4].

Contrary to delay line types of fractional division [5], [7], the phase selection method does not require any specific calibration, as the sum of the 32 phases delay is by construction equal to one period T_{RO} of the ring oscillator. The only timing constraint is that T_{RO} must be larger than the setup time of the synchronous logic stage. But as this logic is very simple this setup constraint is low ($0.61\,\mathrm{ns}$ at $0.9\,\mathrm{V}$) and can be safely margined. The calibration-free operation offers a drastic reduction of area compared to [7] which required 2720 latches and logic for PVT dependent configuration.

Last, thanks to a selection of both the rising and falling edge of the generated clock, the width of the output pulse and hence its duty cycle can be controlled.

B. Oscillator design

Fig. 2 presents the schematics and layout of the 32-phase ring oscillator. It is based on a conventional cross coupled

978-1-5386-3179-9/17 $31.00 © 2017 IEEE   69

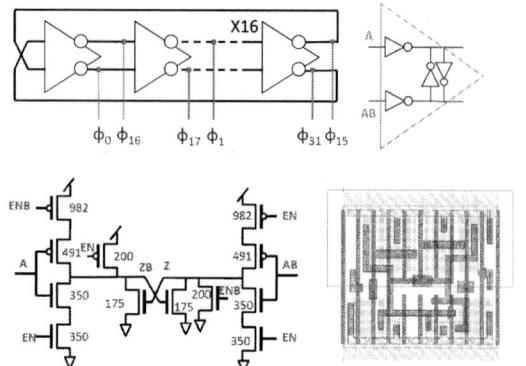

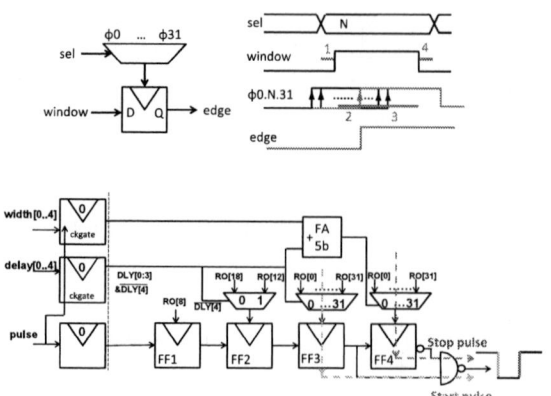

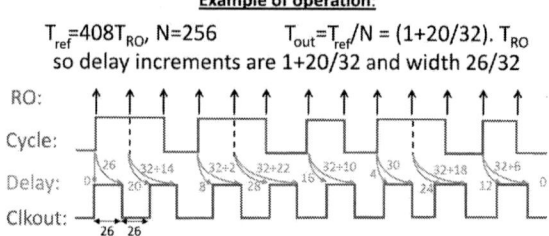

Fig. 2. Top: principle of the 32-phase RO. Bottom: schematics and layout of a differential inverter stage

Example of operation:

$T_{ref}=408T_{RO}$, N=256 $T_{out}=T_{ref}/N = (1+20/32)$. T_{RO}

so delay increments are 1+20/32 and width 26/32

Fig. 3. Example of the DDSS operation.

Fig. 4. a) Top left, principle of flop based phase selection, b) Top-right resulting timing constraints c) bottom, full schematic of a Phase Selection Unit (PSU).

inverter pair topology. The transistor-level design is optimized as the inverter pair is reduced to minimum sized NMOS only to reduce area and power. An enable command is also added at each stage to enable ring gating for power savings in idle mode. This command is added on each of the 16 stages to avoid phase imbalance. Last, the design is laid out in order to allow abutting between stages and with standard cell logic without area overhead.

C. Phase selection principle

Fig. 3 illustrates the general principle, with the commands cycle and delay sent to the phase selection block. On each cycle the rising and falling edge delay values are incremented by one step (20/32 in the example). When the increment overflows, the cycle command is set to 0 for one cycle and no pulse is processed.

Fig. 4 presents the details of the phase selection operation. The general principle, illustrated in sub-figure 4.a) consists in using a flip-flop to propagate a selected phase at its rising edge. The conceptual timing of the selection, presented in sub-figure 4.b), is first to set the phase multiplexer selection, then to enable the edge capture by setting the "window" D input of the flip-flop to 1. However, as illustrated on the timing diagram, some margin must be set between the selection: (1) for multiplexer setting before the window is enabled, (2) and (3) are the timings the window needs to be enabled before and after the desired edge is selected and (4) for disabling the window before the multiplexer command is changed.

Because of these constraints, the phase selection cannot be performed in a single cycle: the full window has to cover the phases Φ_0 to Φ_{31} plus the margins (1-4), as illustrated in Fig.4b). For this reason, two Phase Selection Units (PSUs) run in parallel, each operating over 2 cycles. The first half of

the first cycle is used to guarantee constraints (1) and (2), one cycle for the actual phase selection and another half cycle for constraints (3) and (4). The sub-figure 4.c) presents the details at gate level of the implementation. The first two flip-flops FF1 and FF2 guarantee the margin (1) and (2), while the FF3 selects the rising edge and FF4 the falling edge. Moreover, the only cells affecting the output jitter are the 32:1 MUX and the FF3, which limits mismatch impact of the two parallel PSUs. This design is very compact and easy to implement, requiring only 44 standard cells per PSU, compared to over 200 per delay element in [7].

D. Digital flow and simulations

The highly digital and very compact nature of the DDSS take benefit of the digital flow for quick design iterations, to explore different design strategies and cells sizings as well as timing verification and simulations. The trade-off being that the automated P&R does not ideally match timing between paths, causing added deterministic jitter.

The size (13k transistors total) and digital behavior of the circuit makes full SPICE simulations on the extracted netlist possible. This allows for simulation of the estimated output deterministic jitter due to delay mismatch between the gates.

Fig. 5 shows the results of the simulation. In green cross the nominal run across three corners shows the effect of P&R mismatch only, while the box plot shows the spread of 25 Monte Carlo runs. This simulation predicts the level of jitter expected and demonstrates that at low voltage its dominant contributor is the random variation rather than P&R mismatch, which validates the digital flow approach.

III. MEASURED PERFORMANCES

A. Testchip implementation

Fig. 6 presents the full test vehicle view, details of the DDSS layout and test harness. 16 chips have been fabricated, packaged and measured. The testchip integrates frequency dividers to allow validation of functionality even at GHz frequency range where standard digital IOs cannot transmit the generated clock off chip directly. The circuit is designed in a 28nm FD-SOI Regular Voltage Threshold (RVT) process to minimize leakage in idle mode for low power applications. The total DDSS area is $981\,\mu m^2$ and can be placed inside of digital logic with no guard area overhead.

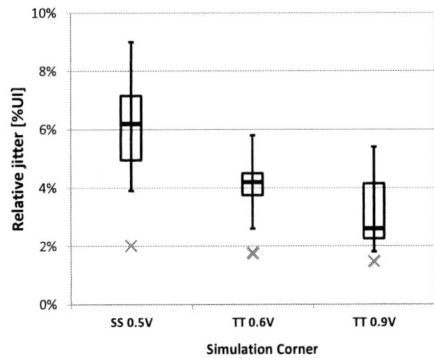

Fig. 5. Extracted Spice simulations of the full design across corners. Green cross: nominal run, box plot: result of 25 Monte Carlo runs.

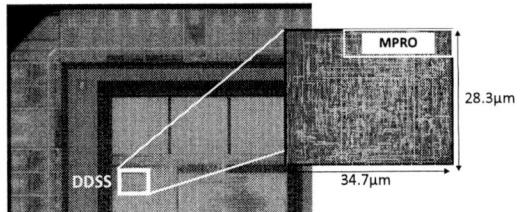

Fig. 6. View of the test chip and DDSS layout.

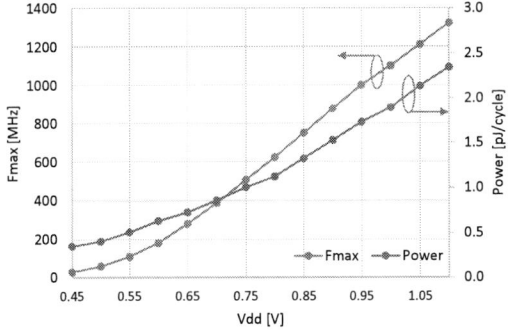

Fig. 7. Maximum output frequency and energy consumption of the clock across 0.45-1.1V range. Median values at room temperature across 16 dice.

B. Maximum frequency and power

Fig. 7 shows the measured maximum generated frequency and power consumption at Fmax of the 16 chips across 0.45 V to 1.1 V supply. The circuit achieves an energy efficiency of 0.45 pJ/cycle at 57 MHz 0.5 V and 1.53 pJ/cycle at 879 MHz 0.9 V.

Table I (median value measured across 16 dice) shows the DDSS can be Reverse Body Biased (RBB) at the same time as the core when it is power gated for 6x to 11x leakage reduction, down to 10 nW at 0.5 V 1.5 V RBB, enabling Internet of Things type duty-cycled operations.

C. Duty cycle control

As described previously, the phase selection approach makes it possible to control the pulse width. This is implemented in practice by 7 settings, the first 4 control the width from 1/8th to 4/8th, while the last 3 invert the output of the first

TABLE I. LEAKAGE POWER IN POWER OFF MODE AT 0.5V AND 1V SUPPLY AND DIFFERENT BIAS LEVELS.

Vdd \ Vbb	no BB	0.5 V RBB	1 V RBB	1.5 V RBB
0.5 V	0.11 µW	0.04 µW	0.02 µW	0.01 µW
1 V	0.60 µW	0.25 µW	0.16 µW	0.10 µW

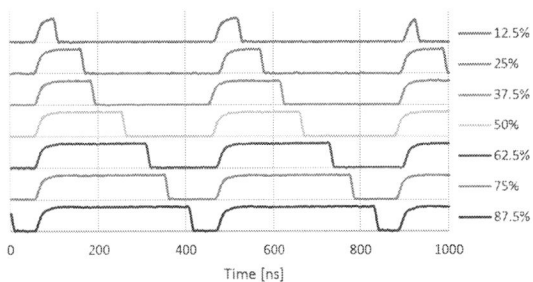

Fig. 8. Measured clock output with the seven available duty cycle settings.

3. This sets 7 settings between 1/8th to 7/8th. This feature is illustrated in the measured data of fig. 8. This feature is useful to offset clock tree unbalance at lower voltages and can be used in low power pulse based latch logic [8].

D. Jitter measurement

First, it is important to note that the jitter constraints are different for SoC clocking than for RF or data recovery applications. The relative peak to peak period jitter is the main metric and directly translates to a frequency penalty, or extra margining in the digital logic the DDSS clocks. For example a 6% UI jitter corresponds to only a 3% Fmax degradation in the clocked logic.

Fig. 9 presents the rms and peak-to-peak jitter measurements at 0.5 V across the 16 dice, as well as a capture of a jitter histogram, illustrating the superposition of the deterministic component, from phase selection paths mismatch, with the random jitter from supply and components noise. As the deterministic jitter is dependent on the phase increment values, the values are measured for the 16 dice and two different output frequencies (20 MHz and 35 MHz) to demonstrate measurements are not made on a best case. The median measured peak-to-peak jitter value is 1.47 ns and 2.01 ns at 35 MHz and 20 MHz, ie 5.1% and 4.0% UI respectively.

Due to IO bandwidth, the jitter at 0.9 V is measured at 100 MHz. Median value of pk-pk and RMS jitter is 167 ps and 20.7 ps, ie 1.7% UI and 0.21% UI respectively.

E. SoC level performance

This jitter performance has to be put in perspective with the full SoC power budget. As an illustration, the test-chip also includes a low power ARM M0+ core [9] operating at the same voltage and clocked by the DDSS. On a Dhrystone testbench the M0+ consumes 0.94 pJ/cycle at 0.5 V. Hence, when compared to an ideal clock, the DDSS with a 5.1% UI jitter requires an increase in frequency margin of 2.6%. From measured data, this corresponds to a 1.6 mV increase in core voltage for margining, which in turn increases the core energy by only +1.1% ie. by 0.01 pJ/cycle. So, for low power cores, the benefit of the energy efficiency improvement in the clock generator far outweighs the penalty in jitter performance compared to some conventional PLLs [4], [5].

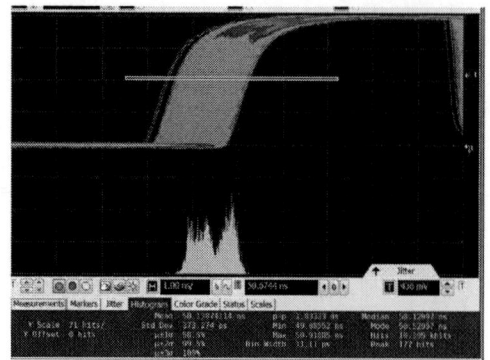

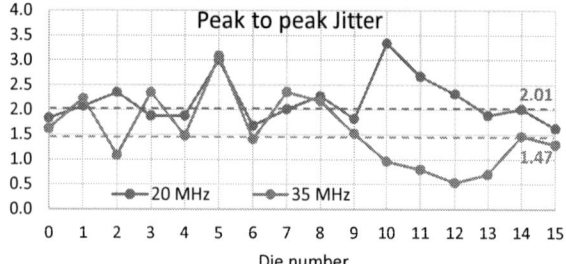

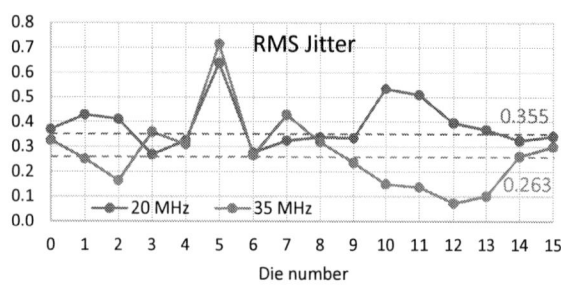

Fig. 9. Top: details of the measurements on die 0 at 20MHz. Bottom: Jitter measurements on 16 dice at 0.5V with 20MHz and 35MHz output frequency.

F. Comparison with the state of the art

Table II summarizes the performance of the proposed DDSS clock generators compared with previous DDSS and state of the art PLLs. This work combines a wide voltage range with frequencies compatible with digital logic clocking (unlike [2]), the best reported area and an excellent maximum energy efficiency of 0.40 pJ/cycle.

IV. CONCLUSION

This paper proposes a novel implementation of the Direct Digital Sampling and Synthesis (DDSS) frequency generation circuit. This all-digital approach, compared to conventional PLLs, offers instant frequency scaling. By using a phase selection approach, this circuit offers an extremely compact implementation ($981 \, \mu m^2$) and low power across a full 0.45 V–1.1 V operating range, with 0.40 pJ/cycle at 0.5 V. This combined with its satisfying jitter performance (5.1% UI at 35 MHz 0.5 V) demonstrates the possibility for important system-level energy savings from single clock domain ultra low-power systems to large GALS SoCs.

ACKNOWLEDGMENT

The authors would like to thank their colleagues from STMicroelectronics Janit Kumar, Amit Patel, Manohara Mr and David Bonciani for tests setup and Dominique Zamora from Maya Technology for chip top level design.

REFERENCES

[1] C. C. Chung et al., "A 0.52/1 V fast lock-in ADPLL for supporting dynamic voltage and frequency scaling," IEEE Transactions on Very Large Scale Integration (VLSI) Systems, vol. 24, no. 1, pp. 408–412, Jan 2016.

[2] Y. Ho et al., "A near-threshold 480 MHz 78uW all-digital PLL with a bootstrapped DCO," Solid-State Circuits, IEEE Journal of, vol. 48, no. 11, pp. 2805–2814, 2013.

[3] Y. W. Chen and H. C. Hong, "A fast-locking all-digital phase locked loop in 90nm CMOS for gigascale systems," in 2014 IEEE International Symposium on Circuits and Systems (ISCAS), June 2014, pp. 1134–1137.

[4] W. Deng et al., "A fully synthesizable all-digital pll with interpolative phase coupled oscillator, current-output dac, and fine-resolution digital varactor using gated edge injection technique," IEEE Journal of Solid-State Circuits, vol. 50, no. 1, pp. 68–80, Jan 2015.

[5] A. Elkholy et al., "A 20-to-1000MHz +/-14ps peak-to-peak jitter re-configurable multi-output all-digital clock generator using open-loop fractional dividers in 65nm CMOS," in 2014 IEEE International Solid-State Circuits Conference Digest of Technical Papers (ISSCC), Feb 2014, pp. 272–273.

[6] I. Miro-Panades et al., "A fine-grain variation-aware dynamic vdd - hopping avfs architecture on a 32 nm gals mpsoc," IEEE Journal of Solid-State Circuits, vol. 49, no. 7, pp. 1475–1486, July 2014.

[7] M. Cochet et al., "A 28nm fd-soi standard cell 0.6-1.2v open-loop frequency multiplier for low power soc clocking," in 2016 IEEE International Symposium on Circuits and Systems (ISCAS), May 2016, pp. 1206–1209.

[8] W. Jin et al., "A 0.35v 1.3pj/cycle 20mhz 8-bit 8-tap fir core based on wide-pulsed-latch pipelines," in 2016 IEEE Asian Solid-State Circuits Conference (A-SSCC), Nov 2016, pp. 129–132.

[9] G. Lallement et al., "A 2.7pJ/cycle 16MHz SoC with 4.3nW Power-Off ARM Cortex-M0+ Core in 28nm FD-SOI," IEEE European Solid-State Circuits Conference (ESSCIRC), 2017.

TABLE II.   COMPARISON OF THE PROPOSED ARCHITECTURE WITH STATE OF THE ART ALL-DIGITAL CLOCK MULTIPLIERS

	This work		ISCAS 2016 [7]	TCAS 2016 [1]		JSSC 2013 [2]	ISSCC 2014 [5]	ISCAS 2014 [3]		JSSC 2015 [4]
Technology	28nm FD-SOI RVT		28nm FD-SOI LVT	90nm CMOS		90nm CMOS	65nm CMOS	90nm CMOS		65nm CMOS
Design	Open-loop multi-phase		Open-loop DL	All-digital PLL						
Instant switching	1 cycle		1 cycle	2 cycles		No	Yes	7 cycles		No
Automated P&R	Yes (except RO)		Yes	No		No	No	No		Yes
Duty-cycle control	Yes		No	No		No	No	No		No
Supply [V]	0.45-1.1		0.6-1.2	0.52-1		0.25-0.5	0.9	1.2		0.8
Area [μm^2]	981		14,000	65,000		57,000	120,000	64,600		6,600
Normalized area	1		14.3	6.4		5.6	22.7	6.3		1.2
	0.5V	0.9V	0.6V	0.9V	0.52V	1V	0.25V	0.5V		
Fmax [MHz]	57	879	93	574	120	600	48	480	1000	3000	6000	1410
E/cycle [pJ]	0.40	1.53	1.53	4.75	0.31	8.4	0.062	0.163	3.2	NA	9.24	0.87
Jitter pk-pk [% UI]	5.1	1.7[b]	2.7	5.6[a]	1.9[c]	4.9[c]	3.0	2.9	2.7	12.0	5.1	1.51[c]

[a] Simulated value.   [b] At 100 MHz.   [c] Only rms value reported, peak to peak jitter estimated as 6 times the rms.

A 10-GHz Multi-purpose Reconfigurable Built-in Self-Test Circuit for High-Speed Links

Myungguk Lee, Seungho Han, Jae-Yoon Sim, Hong-June Park, and Byungsub Kim

Dept. of Electrical Engineering, Pohang University of Science and Technology (POSTECH)

Pohang, Korea

byungsub@postech.ac.kr

Abstract—**This paper presents a multi-purpose built-in self-test circuit to reduce large design overhead in preparing various tests of high-speed links. The proposed circuit can be configured as a pattern generator, a pseudo-random bit sequence generator, a scrambler, a descrambler, or a snapshot, all of which are frequently used in various link tests but also require significant design effort. To reduce the large design overhead, we efficiently implemented the five aforementioned functions in a single design by sharing their common structure. A test chip was fabricated in 28-nm CMOS technology, and the five target functions were successfully verified at 10Gb/s consuming 9.27 mW. To demonstrate the usefulness of the proposed circuit, a horizontal eye and a pulse-response were also measured *in-situ* by utilizing the proposed circuits which were configured to conduct various different functions.**

Keywords—**built-in self-test, testing of a high-speed link, a linear-feedback shift register, productivity of design**

I. INTRODUCTION

Testing of a high-speed link is increasingly becoming problematic. High-speed instruments like a bit-error-rate tester (BERT) are often not affordable. Off-chip measurement may greatly differ from the actual link behavior due to the parasitic effects. Because a built-in self-test (BIST) circuit can tackle both the cost and the accuracy problems of tests of a high-speed link [1]-[3], it is becoming an irreplaceable element in development of a high-speed link.

However, a BIST for a high-speed link significantly increases the design overhead. Fig. 1 illustrates a typical link with a BIST utilizing five frequently-used blocks: a pseudo-random bit sequence (PRBS) generator [4], a pattern generator, a scrambler, a descrambler, and a snapshot [1]. Because these blocks must reliably operate at high speed, they must be carefully designed. Especially, to be functional at ultra-high speed of tens of Gb/s [4], layouts of such blocks need extreme cares. Since a BIST usually takes much larger area than the link core (Fig. 2) [1], designing the target five different blocks often requires much more workload than the link core.

To reduce the design overhead related with testing, we propose a 10-GHz multi-purpose BIST circuit which can be configured to conduct five functions of a pattern generator, a PRBS generator, a scrambler, a descrambler, and a snapshot. By sharing their common structure, the five target functions are efficiently combined into the proposed circuit. A test chip was fabricated in 28-nm CMOS technology. The target five functions of the proposed circuit were verified with 10 GHz clock consuming 9.27mW. To demonstrate usefulness of the

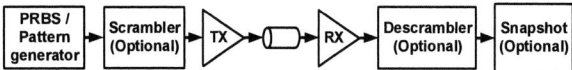

Fig. 1. An example block diagram of a link exploiting a BIST.

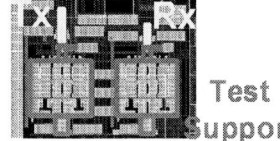

Fig. 2. An example layout of a BIST circuit and a serial link core.

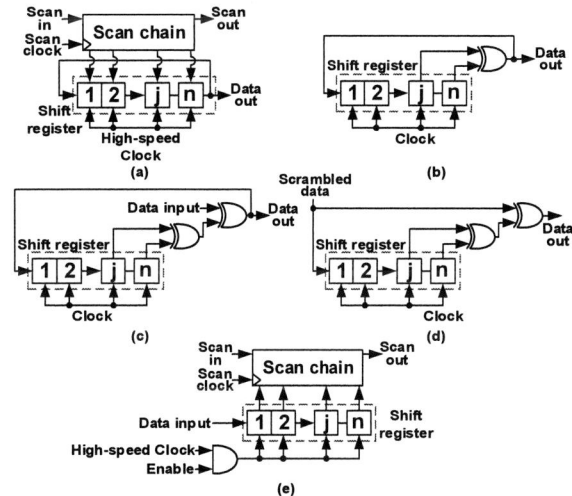

Fig. 3. Block diagrams of (a) a pattern generator, (b) a PRBS generator, (c) a scrambler, (d) a descrambler, and (e) a snapshot.

proposed circuits, we also measured a horizontal eye size and a pulse *in-situ* using the proposed circuits.

II. BACKGROUND

In this section, the five target functions are explained as background.

A. A Pattern generator

A pattern generator is a high-speed rotational shift register (Fig. 3 (a)) which produces a repetitive bit sequence by rotating a seed pattern. The seed can be loaded by a scan chain when the high-frequency clock is disabled. This circuit is often used to test the response of a link to a user-defined pattern (Fig. 1).

B. A Pseudo-Random Bit Sequence (PRBS) Generator

A PRBS-N generator is a high-speed linear-feedback shift register (LFSR) which generates a PRBS with a length of 2^N-1

(Fig. 3 (b)). A PRBS generator is widely used as a data source in bit-error-rate (BER) test of a link.

C. Scrambler

A scrambler is a high-speed block which randomizes the input data sequence by adding (or multiplying) a PRBS [5]. A typical scrambler is implemented using a LFSR (Fig. 3. (c)), and is widely used at the transmitter (Fig. 1) for many purposes including for better clock-recovery or better security.

D. Descrambler

At the receiver, a descrambler recovers the original data from the scrambled data by subtracting (or dividing) the same PRBS used for scrambling (Fig. 1) [5]. A descrambler is typically implemented using a high-speed LFSR as in Fig. 3 (d).

E. Snapshot

A snapshot is a circuit which allows users to monitor a part of a high-speed bit stream though a slow scan I/O [1]. A snapshot typically consists of a scan chain and a clock-gated high-speed shift register (Fig. 3. (e)). A snapshot can pause the high-speed bit stream by disabling the high-speed clock so that the paused part of the bit stream can be reliably accessed by the slow scan chain. A snapshot is frequently used in the receiver (Fig. 1) to perform various *in-situ* measurements [1]-[3].

III. THE PROPOSED CIRCUIT

We propose a P²SDS circuit which can be configured to conduct five target functions of a pattern generator, a PRBS generator, a scrambler, a descrambler, and a snapshot. P²SDS is designed based on the observation that the five target blocks in Fig. 3 share the structural similarity: all of them have high-speed shift registers and some of them have scan chains. Because the five target blocks differ only by slight difference in the connectivity among elements, they can be seamlessly integrated into one P²SDS design by making the connectivity logically re-configurable. Therefore, P²SDS can greatly reduce the design overhead to implement the five target functions.

The schematic diagram of the P²SDS is depicted in Fig. 4 (a). P²SDS consists of a clock-gated high-speed data path (H-path) and a slow scan path. The H-path is a reconfigurable shift register consisting of high-speed master-slave D flip-flops (H-flops) whose values can be written and read in parallel by the scan path when the high-speed clock HCLK is disabled. The H-flop has the additional data input S and control input LOAD for the parallel write operation (Fig. 4 (c)). With proper setting of the multiplexer, two XOR gates, and two switches, the logical connectivity of the H-path can be configured the same as any of the target five blocks in Fig. 3. The high-frequency clock buffers and clock-gates of the H-path are carefully designed to minimize the clock skew. The clock can be disabled at any time for reliable data interchange between the scan path and the H-path. The scan path exploits a trivial design in which two multiplexers of the scan flops (Fig. 4(b)) select whether the H-path will be read or written in parallel by the scan path. Two non-overlapping scan clocks are used to prevent race conditions systematically.

In this work, we designed a 31-bit P²SDS, but the length of P²SDS is not necessarily restricted to 31 bit.

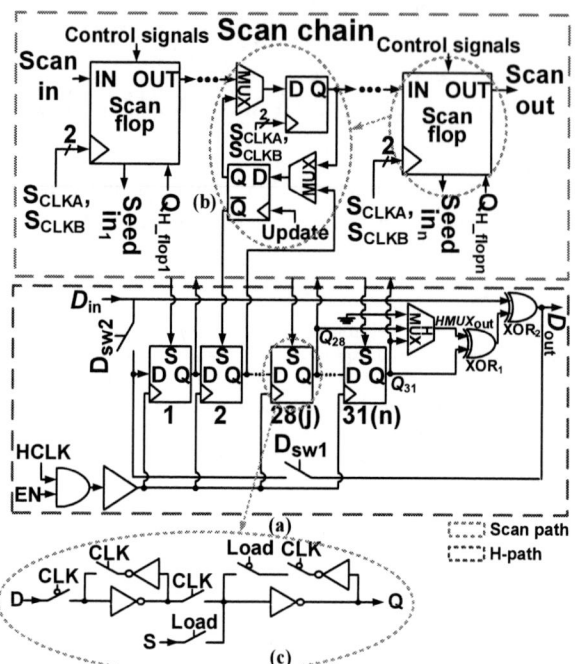

Fig. 4. Schematic diagrams of (a) a P²SDS, (b) a scan-flop, and (c) a H-flop

A. Pattern generator

A P²SDS works as a pattern generator if we set $D_{in}=HMUX_{out}=D_{SW2}=$'0', and $D_{SW1}=$'1'. In this mode, a user-defined 31-bit seed pattern can be loaded by the scan path when HCLK is disabled, and the P²SDS rotates the seed pattern producing the repetitive sequence when HCLK is enabled.

B. Pseudo-random bit seuqence generator

If we set $D_{in}=$'0', $HMUX_{out}=Q_{28}$, $D_{SW1}=$'1', and $D_{SW2}=$'0', then the P²SDS work as a PRBS-31 because $Q_{28} \oplus Q_{31}$ is fed back to the first data input of the shift register. When HCLK is disabled, the seed of the PRBS can be loaded by the scan path.

C. Scrambler

When $HMUX_{out}$ is Q_{28}, D_{SW1} is '1', and D_{SW2} is '0', the H-path of the P²SDS is configured the same as the scrambler in Fig. 3 (c). Therefore, the P²SDS works as scrambler.

D. Descrambler

When $HMUX_{out}$ is Q_{28}, D_{SW1} is '0', D_{SW2} is '1' and D_{in} are the scrambled data, the connectivity of the P²SDS is identical to the one of the descrambler in Fig. 3 (d), and therefore, it operates as a descrambler.

E. Snapshot

A P²SDS operates as a snapshot if we set $D_{SW1}=$'0' and $D_{SW2}=$'1'. In this mode, the H-path is configured as a simple shift register. When HCLK is enabled, the bit stream of D_{in} passes through the shift register at very high speed. When HCLK is disabled, 31 consecutive bits of D_{in} are immediately held in the shift register of the H-path. Users can read these 31 bits via the scan chain through slow I/Os. Since this clock disabling and scan reading can be also conducted when the P²SDS plays one of other four target functions (a pattern generator, a PRBS generator, a scrambler, and a descrambler), P^2SDS can conduct one of four other operations and snapshot operation at the same time.

IV. EXPERIMENTS AND APPLICATION

To prove the concept of P^2SDS and to demonstrate its usefulness, a test chip was fabricated in 28-nm CMOS technology and tested at 10 Gb/s with 1 V supply voltage.

A. Verification of the target five functions of P^2SDS

The test circuit in Fig. 5 is implemented and tested to verify the five target functions of P^2SDS.

1) Pattern generator / Snapshot

The operations of P^2SDS as a pattern generator and a snapshot were verified by comparing the data sequence transmitted by the pattern generator and the captured snapshot sequence at the receiver in Fig. 5. P^2SDS_2 and P^2SDS_5 were configured as a pattern generator and a snapshot, respectively, while P^2SDS_1, P^2SDS_3, and P^2SDS_4 were configured as a simple buffer passing the input to the output intact by setting the multiplexers of their H-paths to select Q_{31}s so that their XOR_1 outputs were always '0'. D_{out} values of P^2SDS_1 and P^2SDS_3 were '0' since their D_{in} = '0.' Therefore, the data sequence generated by P^2SDS_2 must be captured intact by P^2SDS_5. Fig. 6 shows the sequence transmitted by P^2SDS_2 and the sequence received and captured by P^2SDS_5 in experiment at 10Gb/s. The identical sequences prove that the proposed circuit successfully worked as a pattern generator and a snapshot.

2) Scrambler / Descrambler

The scrambling and descrambling functions of P^2SDS were verified by comparing the bit sequences of the original data, the scrambled data, and the descrambled data. P^2SDS_1, P^2SDS_2, P^2SDS_4, and P^2SDS_5 of Fig. 5 were configured as a pattern generator, a scrambler, a descrambler and a snapshot, respectively. The P^2SDS_3 output is set to '0' in the same way as before. In the experiment, a user-defined bit sequence was generated by P^2SDS_1, and then scrambled and transmitted by P^2SDS_2. The scrambled data were descrambled by P^2SDS_4, and then captured by P^2SDS_5. Fig. 7 show the experimental results measured at 10Gb/s. The scrambled data in Fig. 7 (b) were captured by P^2SDS_2 like the snapshot operation. In Fig. 7, the original data completely differ from the scrambled data but are identical to the descrambled data, verifying the scrambling and descrambling operations of P^2SDS.

3) PRBS generator

The PRBS operation of P^2SDS was verified by using a descrambler as a PRBS-checker. If a PRBS is descrambled, then the descrambled sequence must be a consecutive '0s' [6], [7]. To use this property in verification of PRBS, P^2SDS_2, P^2SDS_4, and P^2SDS_5 of Fig. 5 were configured as a PRBS-31 generator, a descrambler, and a snapshot, respectively. The outputs of P^2SDS_1 and P^2SDS_3 were set to '0' as before. Fig. 8 shows the measured results at 10 Gb/s. The sequence captured by P^2SDS_5 contains only '0's (Fig. 8 (b)) while the pattern of P^2SDS_2 looks random (Fig. 8 (a)), verifying the PRBS operation. To prove that the Fig. 8 (b) result were not caused by some unexpected disconnection of logical paths, we also performed the same test with error-injection by configuring P^2SDS_3 as a pattern

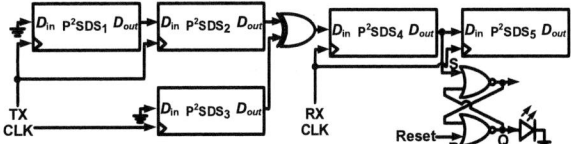

Fig. 5. A test circuit to prove the target five functions of the proposed circuit.

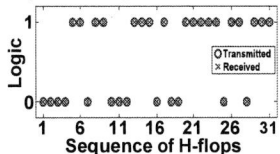

Fig. 6. A transmitted user-defined sequence from a pattern generator and the received sequence at a snapshot.

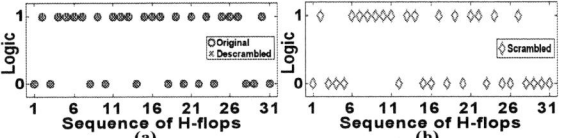

Fig. 7. (a) A transmitted original sequence from a pattern generator and the descrambled sequence captured by snapshot and (b) the scrambled sequence.

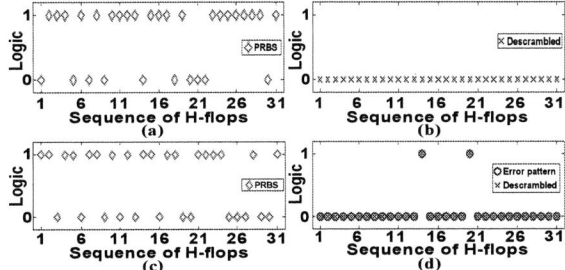

Fig. 8. (a) A PRBS (b) the descrambled sequence captured by a snapshot, (c) a PRBS, and (d) the injected error pattern and the descrambled sequence with errors.

generator containing two bit errors. When two bit errors (Fig. 8 (d)) were added in the PRBS (Fig. 8 (c)) generated by P^2SDS_2, the exact two bit errors appeared at the PRBS-checker's outputs (Fig. 8 (d)) which were originally all '0's.

B. Applications of P^2SDS

The proposed circuits can be utilized in various tests of high-speed links. As examples of the interesting and useful applications of P^2SDS, we demonstrate that a horizontal eye size and a pulse response can be measured in-situ utilizing P^2SDS circuits configured in various ways. Applications of P^2SDS are not limited to these examples.

1) In-situ horizontal eye measurement

The horizontal eye size can be conveniently measured in-situ by utilizing P^2SDS circuits. As demonstrated in Fig. 8, bit errors in an incoming PRBS sequence can be detected by a descrambler which is working as a PRBS-checker [6], [7]. Therefore, the horizontal eye size can be measured in-situ by examining bit errors for various clock phases using two P^2SDSs configured as a PRBS generator and a PRBS-checker.

In demonstration, the horizontal eye size observed by P^2SDS_4 in Fig. 5 was measured in-situ at 10Gb/s. For measurement, P^2SDS_2 and P^2SDS_4 were configured as a PRBS generator and a descrambler (a PRBS-checker), respectively, while P^2SDS_1 and P^2SDS_3 output were set to '0'. As demonstrated in Fig. 8 (d), if the PRBS received by P^2SDS_4 has errors, then the output bit stream of P^2SDS_4 must have '1's at the position of the bit errors. Easily to monitor the error detection by P^2SDS_4 through a slow I/O, we used an S-R latch which is connected to the output of P^2SDS_4 and a light emitting diode (LED) as in Fig. 5. If any bit error is detected by P^2SDS_4, then the latch output Q will become '1' which will be held until the latch is reset. Otherwise, Q will remain '0'. By using this scheme, the bit error can be easily monitored by the LED light (Fig. 9(b)) at DC without any problematic high-speed data connection with an expensive BERT. By checking errors in this way for more than 10^{10} bits at 10Gb/s while sweeping RX CLK phases with a 10-ps interval, the horizontal eye size were measured *in-situ*. The measured eye width is 0.7 unit-interval (UI) (Fig 9 (a)).

2) In-situ pulse response measurement

The proposed circuits can be used to measure a pulse response *in-situ* like a time-domain reflectometer (TDR) scope [1]. Fig. 10 (a) shows a test circuit to demonstrate the pulse response measurement. The test circuit consists of two P^2SDS circuits, a differential buffer, a current mode logic (CML) driver, a RC-dominant metal wire with length of 130 μm and width of 0.05 μm, and a slicer with the reference voltage which is set by an external digital-to-analog converter. P^2SDS_1 and P^2SDS_2 were configured as a pattern generator and a snapshot, respectively. Repetitive 1-UI-wide pulses with a 31-UI cycle were transmitted by P^2SDS_1 to the slicer through the metal wire. The received bits were sliced and captured by the snapshot while sweeping the phase of the receiver clock RX CLK and the reference voltage V_{ref}. The voltage of the pulse response at a specific RX CLK phase corresponds to the V_{ref} value which makes almost the same numbers of '1' and '0' appeared in the sliced bit stream at the RX CLK phase [1]. The pulse response measured *in-situ* at 10Gb/s is plotted in Fig. 10 (b).

V.   CONCLUSION

We propose a multi-purpose reconfigurable BIST circuit named P^2SDS to reduce the design overhead to prepare tests of a high-speed link. The proposed P^2SDS circuit can be configured as a pattern generator, a PRBS generator, a scrambler, a descrambler, and a snapshot, which are widely used in various tests of high-speed links and require large design effort. P^2SDS can reduce such design overhead in preparing tests of a serial link since P^2SDS can provide the five aforementioned functions just by changing configuration. By using the proposed circuits, various interesting tests of high-speed links can be done very easily and conveniently. A test chip to prove the concept of P^2SDS and to demonstrate its interesting applications was fabricated in in 28-nm CMOS technology. In the experiment, the five target functions of P^2SDS were successfully verified at 10GHz clock. As demonstrations of useful applications of P^2SDS, a horizontal eye size and a pulse response were measured *in-situ* at 10Gb/s without expensive high-speed test equipment. The proposed circuit worked at 10 GHz and consumed 9.27 mW. The area occupied by a P^2SDS circuit is 141 μm × 13.1 μm (Fig. 11).

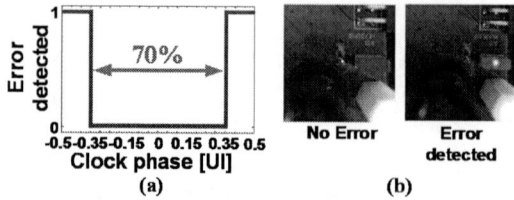

Fig. 9. (a) A measured horizontal eye size and (b) a photograph of the LED which indicates bit error detection during the experiment.

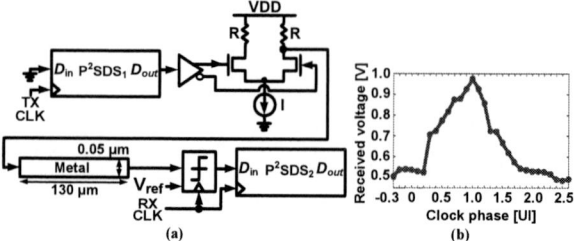

Fig. 10 (a) A test circuit to measure a pulse response *in-situ* and (b) the measured pulse response at 10Gb/s.

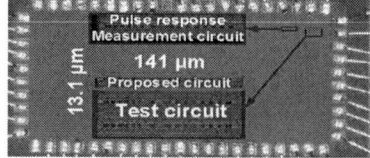

Fig. 11. The micrograph of the fabricated chip.

ACKNOWLEDGMENT

This work was supported by the National Research Foundation (NRF) of the MSIP, Korea, under Grant 2015R1A2A2A09001553, and by Samsung POSTECH Research Center (SPRC) funded by Samsung Electronics. Authors appreciate IDEC for tool support.

REFERENCES

[1] B. Kim, and V. Stojanovic, "An Energy-Efficient Equalized Transceiver for RC-Dominant Channels," IEEE Journal of Solid-State Circuits, vol. 45, no. 6, pp. 1186-1197, June 2010.

[2] B. Kim, and V. Stojanovic, "A 4Gb/s/ch 356fJ/b 10mm equalized on-chip interconnect with nonlinear charge-injecting transmit filter and transimpedance receiver in 90nm CMOS," in IEEE International Solid-State Circuits Conference Digest of Technical Papers, Feb. 2009, vol. 978, pp. 66-67.

[3] M. Choi et al., "An FFE TX with 3.8x Eye Improvement by Automatic Impedance Adaptation for Universal Compatibility with Arbitrary Channel and RX Impedances," IEEE Symposium on VLSI Circuits, Jun. 2017.

[4] T. O. Dickson et al., "An 80-Gb/s 2^{31}-1 pseudorandom binary sequence generator in SiGe BiCMOS technology," IEEE Journal of Solid-State Circuits, vol. 40, no. 12, pp. 2735-2745, Dec. 2005.

[5] J. E. Savage, "Some Simple Self-Synchronizing Digital Data Scramblers," The Bell System Technical Journal, vol. 46, no. 2, pp. 449-487, Feb. 1967.

[6] R. Westcott, "Testing digital data transmission systems," U.K. Patent no. 1 281 390, 1972.

[7] O. Kromat, U. Langmann, G. Hanke, and W. J. Hillery, "A 10-Gb/s Silicon Bipolar IC for PRBS Testing," IEEE Journal of Solid-State Circuits, vol. 33, no. 1, pp. 76-85, Jan. 1998.

A 56Gbps PAM-4 Optical Receiver Front-end

Kuan-Lin Fu, and Shen-Iuan Liu

Graduate Institute of Electronics Engineering & Department of Electrical Engineering
National Taiwan University, Taipei, Taiwan 10617, R.O.C.
E-mail: lsi@ntu.edu.tw

Abstract—A 56Gbps PAM-4 optical receiver front-end is presented. In order to reduce the input-referred current noise of the receiver front-end, the shunt feedback resistor R_F of the TIA is enlarged. And, the equalizer is inserted to boost the high-frequency gain and extend the bandwidth. The AGC amplifier using the proposed dB-linear VGAs is further to lower the noise. This PAM-4 optical receiver front-end is fabricated in a 40 nm CMOS process. Its power is 120mW from a supply of 1.2V excluding the output buffers. Its core area is 0.21mm².

I. INTRODUCTION

In the ultra-high-speed wireline communications, four-level pulse amplitude modulation (PAM-4) signaling becomes popular. This is because that the PAM-4 signaling needs a lower bandwidth than the non-return-to-zero (NRZ) one. For an optical receiver front-end, a low-noise transimpedance amplifier (TIA) [1-5] is required. In general, a low-noise TIA has a tradeoff among its gain, bandwidth, and input-referred current noise. When the shunt feedback resistor of a TIA is increased, its gain and input-referred current noise are improved, but its bandwidth is decreased. For a NRZ signaling, a limiting amplifier is widely used. However, the limited linear range of the limiting amplifier is not suitable for a PAM-4 signaling. It is because the limiting amplifier will distort the four-level linearity of a PAM-4 signaling. Thus, an automatic gain control (AGC) amplifier [6-8] is used to maintain the signal swing of a PAM-4 signaling.

To maintain a uniform loop transient response and settling time, a dB-linear gain characteristic of a variable gain amplifier (VGA) is required in an AGC amplifier. A Gilbert cell [6-8] uses an exponential generator to realize a VGA with the dB-linear gain. So, the AGC amplifier [6-8] is composed of a Gilbert cell as a VGA, several constant gain amplifiers (CGAs) and an exponential generator. Assume the gain range of an AGC amplifier is constant. It leads to the VGA realized by a Gilbert cell to have a low gain. Thus, the input-referred noise of this kind of AGC amplifiers becomes poor.

In this paper, two TIAs with an equalizer is adopted. In addition, the AGC amplifier adopts five proposed dB-linear VGAs without any CGA. Since the proposed VGA has a finite gain, thus the input-referred noise is reduced. A 56Gbps PAM-4 optical receiver front-end is presented by using the TIAs with an equalizer and the AGC using the proposed VGAs.

II. CIRCUIT DESCRIPTION

A. PAM-4 Receiver Front-end

Fig. 1 shows a PAM-4 optical receiver front-end, which is composed of two low-noise shunt-feedback TIAs with an equalizer, an AGC amplifier, an offset canceling network, and

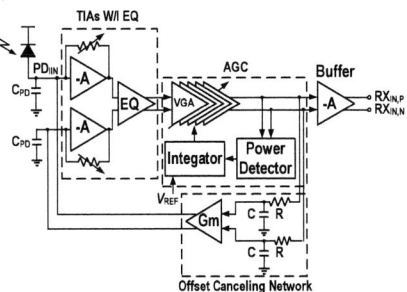

Fig. 1. A 56Gbps PAM-4 optical receiver front-end

a buffer. The AGC amplifier is composed of five dB-linear VGAs, a power detector, and an integrator.

B. Low-Noise Shunt-Feedback Transimpedance Amplifier and Equalizer

Fig. 2 shows two TIAs with an equalizer. The upper TIA is major and the lower one is dummy, which reduces the common-mode noises. Two TIAs are realized by two inverter-based amplifiers, M_1-M_2 and M_{1D}-M_{2D}, two shunt feedback resistors R_F, two triode NMOS transistors, M_3 and M_4. The control voltage V_{con} adjusts M_3 and M_4 parallel with R_F to alter the gain of the TIAs. Two NMOS transistors, M_0 and M_{0D}, serves the transconductance stage of the offset canceling network, which cancels the dc offset. Because M_0 and M_{0D} contribute to the noises less than other devices, they are ignored in the following noise analysis. The equalizer is composed of M_5-M_7, two inductors L_{L1} and two resistors R_{L1}. The input-referred current noise of the TIAs with the equalizer is derived as

$$\overline{I_{n,in,TIA}^2} = [\frac{g_m r_o /(g_m r_o +1)}{s^2(\frac{R_F C_{in,t}}{(1+g_m r_o)\omega_o})+s(\frac{\omega_o R_F C_{in,t}+1}{(1+g_m r_o)\omega_o})+1}]^2(\frac{4kT}{R_F})$$

$$+[\frac{(1+sR_F C_{in,t})(g_m r_o \omega_o)/R_F C_{in,t}}{s^2+s(\omega_o + \frac{1}{R_F C_{in,t}})+(g_m r_o +1)\frac{\omega_o}{R_F C_{in,t}}}]^2(\frac{4kT\Upsilon}{g_m R_F^2}) \quad (1)$$

$$+(\frac{Z_{L1}^2}{g_{m5}^2\{[(1+g_{m7}r_{o7})r_{o5}+r_{o7}]\parallel Z_{L1}\}^2 R_F^2})(4kT\Upsilon g_{m5}+\frac{4kT}{R_{L1}})$$

where $g_m = g_{m,M1} + g_{m,M2}$, $g_{m,M1}$ and $g_{m,M2}$ are the transconductances of M_1-M_{1D} and M_2-M_{2D}, respectively. $r_o = r_{o1} \parallel r_{o2}$, r_{o1} and r_{o2} are the output resistances of M_1-M_{1D} and M_2-M_{2D}, respectively. $\omega_o = 1/(r_o C_{LA})$ is the dominant pole frequency of the inverter-based amplifier and C_{LA} is the output capacitance of the inverter-based amplifier; $C_{in,t}$ is the total input capacitance of the TIA including the photodiode capacitance C_{PD}, the TIA input capacitance $C_{in,tia}$ and other parasitic capacitances. The coefficient γ is 2/3 for a long channel length device. $Z_{L1} = (R_1 + sL_1) \parallel (1/sC_{L1})$ represents the output impedance of the equalizer and C_{L1} is the output capacitance of the equalizer. The simulated and calculated input-referred current noises of two TIAs with the equalizer

978-1-5386-3179-9/17 $31.00 © 2017 IEEE

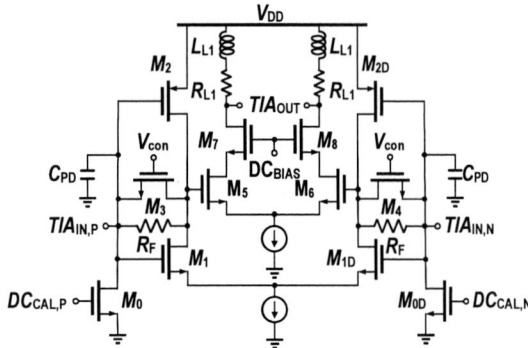

Fig. 2 Two TIAs with an equalizer.

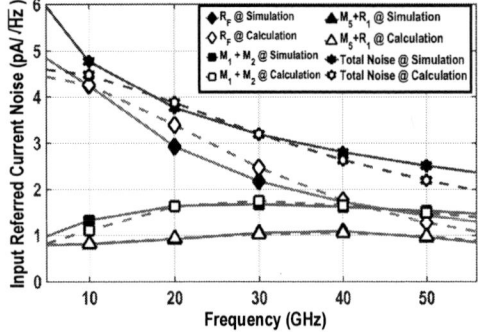

Fig. 3 Simulated input-referred current noise of the TIAs and the equalizer.

are shown in Fig. 3. The total input-referred current noise is 3.2pA/Hz$^{1/2}$ at 30GHz. R_F contributes by 2.46pA/Hz$^{1/2}$ at 30 GHz. M_1 and M_2 contribute by 1.78pA/Hz$^{1/2}$ at 30GHz. Thus, R_F dominates the input-referred current noise while the noise frequency is less than 30 GHz. When R_F is increased, both the input-referred current noise and the bandwidth of the TIA are reduced. To extend the bandwidth, an equalizer is cascaded with TIAs to boost the high-frequency gain.

C. dB-Linear VGA

Fig. 4 shows the proposal dB-linear VGA. The currents of the current sources M_{13} and M_{14} are 7mA and 1mA, respectively. Assume the differential pair, M_9 and M_{10}, operates in the saturation region. Another differential pair, M_{11} and M_{12}, operates in the subthreshold region. One may wonder that the speed of a subthreshold device is very limited. For a 40nm CMOS process, Fig. 5 shows the simulated unity current gain frequency of M_{11} or M_{12}. When $I_{D,11,12}$ changes from 0.32mA to 1.4mA, the simulated unity current gain frequency ranges from 25GHz to 75GHz. It can offer the bandwidth requirement of 14GHz for the 56Gbps data. The gain of this VGA is defined as

$$A_{VGA} \triangleq \frac{VGA_{OUT,P} - VGA_{OUT,N}}{VGA_{IN,P} - VGA_{IN,N}} = (g_{m,M9,10} - g_{m,M11,12}) \cdot Z_{L2} \quad (2)$$

where $g_{m,M9,10}$ and $g_{m,M11,12}$ are the transconductances of M_9-M_{10} and M_{11}-M_{12}, respectively. $Z_{L2}=(R_2+sL_2)||r_{ds9,10}||r_{ds11,12}$. $r_{ds9,10}$ and $r_{ds11,12}$ are the channel resistances of M_9-M_{10} and M_{11}-M_{12}, respectively. Assume M_{15} operates in the triode region, its channel resistance is given as

$$R_{ds,15} \cong 1/[K_n \frac{W}{L}(V_{CTRL} - V_{S,15} - V_{TH,15})] \quad (3)$$

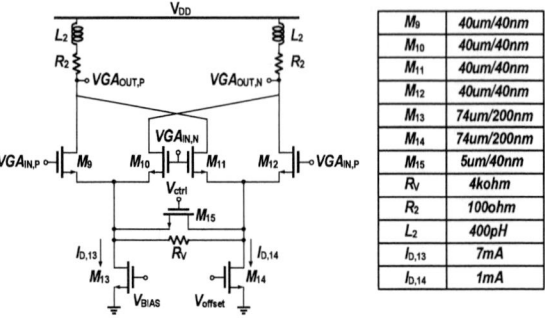

Fig. 4 Proposed dB-linear VGA.

M_9	40um/40nm
M_{10}	40um/40nm
M_{11}	40um/40nm
M_{12}	40um/40nm
M_{13}	74um/200nm
M_{14}	74um/200nm
M_{15}	5um/40nm
R_V	4kohm
R_2	100ohm
L_2	400pH
$I_{D,13}$	7mA
$I_{D,14}$	1mA

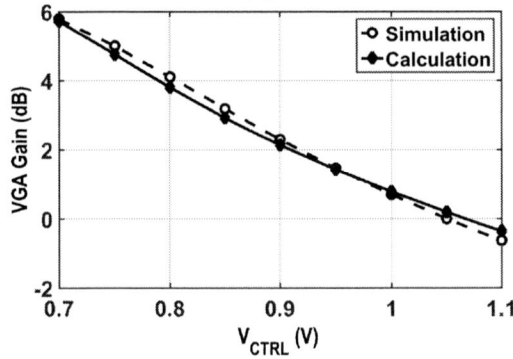

Fig. 5 Simulated unity current gain frequency of $M_{11,12}$ versus $I_{D,11,12}$.

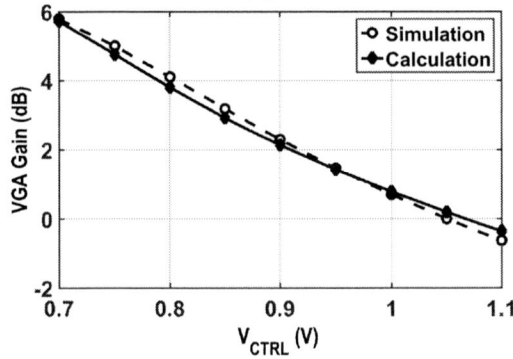

Fig. 6 Simulated VGA gain versus V_{CTRL}.

Since M_{11} and M_{12} operates in the subthreshold region, eq. (2) can be rewritten as

$$A_{VGA} = [g_{m,M9,10} - (\frac{I_{D0,11,12}}{nV_T})(\frac{W}{L})_{11,12} e^{\frac{V_{GS9,10} - 2I_{D11,12}(R_{ds,15}||R_V)}{nV_T}}] \cdot Z_{L2} \quad (4)$$

where n>1 is the subthreshold slope factor, and $V_T=kT/q$. $I_{D0,11,12}$ is the drain current while $V_{GS,11,12} = V_{TH,11,12}$. $V_{TH,11,12}$ is threshold voltage of M_{11}-M_{12}. K =1.38 × 10^{-23} V^2/Hz. T is Kelvin temperature. q=1.6× 10^{-19} C. According to eq. (4), the gain of the proposed VGA is exponentially controlled by the control voltage V_{CTRL}. So, it is a dB-linear VGA. Fig. 6 shows the simulated and calculated VGA gains versus V_{CTRL}. The simulated VGA gain ranges from -0.6dB to 5.78dB when V_{CTRL} changes from 0.7 V to 1.1 V.

D. Automatic Gain Control (AGC) Amplifier

The AGC amplifier is composed of five dB-linear VGAs, a power detector, and an integrator. Fig. 7 shows the simulated input-referred voltage noises of the proposed AGC and the conventional one. The conventional AGC amplifier is composed of a Gilbert cell VGA and four CGAs. Both AGC amplifiers have the same gain range of 30dB. The simulated input-referred voltage noise of the proposed AGC amplifier

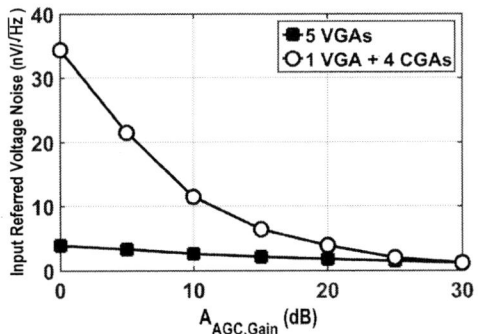

Fig. 7 Simulated input-referred voltage noises of the conventional AGC amplifier and the proposed one

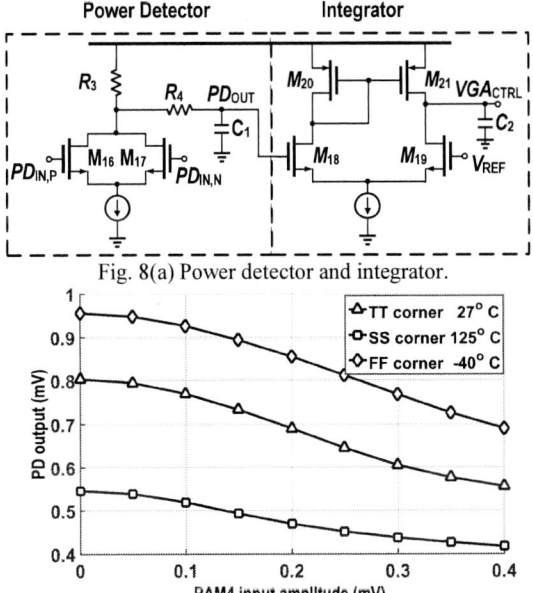

Fig. 8(a) Power detector and integrator.

Fig. 8(b) Simulated output voltage of power detector versus PAM-4 input amplitude.

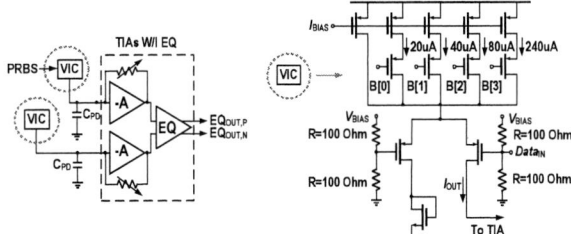

Fig. 9 Voltage-to-current convertor

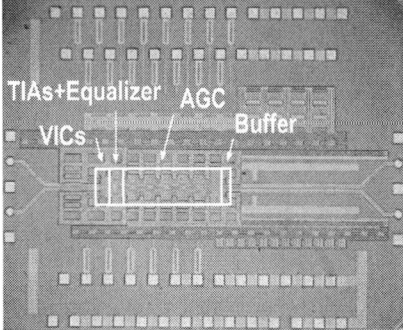

Fig. 10 Die photo.

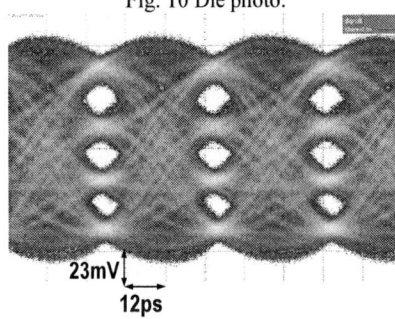

Fig. 11 Measured eye diagram.

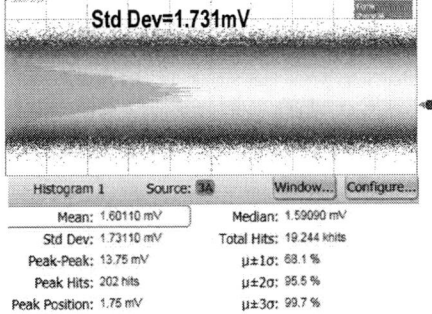

Fig. 12 Measured output noise.

ranges from $1.11\text{nV/Hz}^{1/2}$ to $3.57\text{nV/Hz}^{1/2}$ when its gain changes from 0dB to 30dB, respectively. The simulated input-referred voltage noise of the conventional AGC amplifier ranges from $1.11\text{nV/Hz}^{1/2}$ and $34.38\text{nV/Hz}^{1/2}$ when its gain changes from 0dB to 30dB, respectively. The input-referred voltage noise of the proposed AGC amplifier is also low even the gain of the AGC amplifier is low. Fig. 8(a) shows the power detector and the integrator for the AGC amplifier. The power detector is used to convert the output power of the AGC amplifier into an output voltage PD_{out}. The difference between PD_{out} and an external reference voltage V_{REF} is converted and integrated by the integrator and the capacitor C_2. The simulated transfer curve of the power detector is shown in Fig. 8(b). The output voltage varies from 550mV to 800mV at the typical corner while the PAM-4 input amplitude changes from 0.4V to 0.1V. A first-order RC low-pass filter is realized by R_4 and C_1. This filter has the -3dB bandwidth of 159 MHz, which is used to reduce the ripple of PD_{out}.

E. Voltage-to-Current Convertor

Fig. 9 shows a voltage-to-current convertor (VIC) which mimics a photodiode. The output current of the VIC is digitally

controlled by a 4-bit code with a current step I_{LSB} of 20uA. The upper VIC is connected to a PRBS and the lower one is dummy.

III. EXPERIMENTAL RESULTS

A 56Gbps PAM-4 optical receiver front-end is fabricated in a 40nm CMOS process. Its die photo is shown in Fig. 10. The core area of this PAM-4 optical receiver front-end is 0.21mm^2. The total power is 120mW when the supply voltage is 1.2V. Fig. 11 shows the measured eye diagrams with an input 56Gbps PRBS of 2^7-1. Fig. 12 shows the integrated output noise by the oscilloscope. Since the oscilloscope contributes to a noise of 0.634mV_{rms}, it should be de-embedded. Then the integrated input-referred noise of the PAM-4 optical receiver

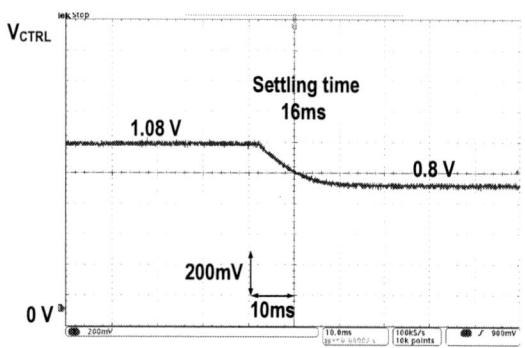

Fig. 13. Measured V_{CTRL} for the AGC.

front-end is calculated as

$$I_{n,in,tot} = \frac{2\sqrt{(1.731mV)^2 - (0.634mV)^2}}{1.257k\Omega} = 2.56uA_{rms} \quad (5)$$

The calculated input-referred rms noise current is 2.56uA. The average input-referred noise density per bandwidth is calculated as

$$I_{n,in,avg} = \frac{I_{n,in,tot}}{\sqrt{BW}} = 20.23\,pA/\sqrt{Hz} \quad (6)$$

where the -3dB bandwidth BW is 16GHz. The average input-referred noise density per bandwidth is 20.23pA/ Hz$^{1/2}$. Fig. 13 shows measured transient response of the optical receiver front-end. When the amplitude of a PAM-4 PRBS is 150mV, the control voltage V$_{CTRL}$ starts from 1.08 V and settles to 0.8V. The measured settling time is 16ms. The gain range of this PAM-4 optical receiver front-end is 38~68dBΩ. Its energy efficiency is 2.14pJ/bit.

IV. CONCLUSIONS

A 56Gbps PAM-4 optical receiver front-end is fabricated in a 40nm CMOS process. In order to reduce the input-referred current noise of the receiver front-end, the shunt feedback resistors R_F of the TIA is enlarged. The equalizer is inserted to boost the high-frequency gain and extend the bandwidth. The integrated input-referred rms noise of this PAM-4 optical

receiver front-end is 2.56uA. The AGC amplifier using the proposed dB-linear VGAs is further to lower the noise. Table I shows the performance summary of this optical receiver front-end and other works. By using the TIAs with the equalizer and the AGC amplifier, this PAM-4 optical receiver front-end achieves a high data rate of 56Gbps and a low input-referred rms noise current.

ACKNOWLEDGMENT

This chip is fabricated through the TSMC University Shuttle Program. The authors would like to thank Ministry of Science and Technology, Taiwan, and Donation Grant FD105012.

REFERENCES

[1] T. Takemoto, H. Yamashita, T. Yazaki, N. Chujo, Y. Lee, and Y. Matsuoka, "A 4 X 25-to-28 Gb/s 4.9 mW/Gb/s -9.7 dBm High-Sensitivity Optical Receiver Based on 65 nm CMOS for Board-to-Board interconnects", *IEEE ISSCC Dig.* Tech. Papers, pp. 118-119, Feb. 2013.

[2] J. Y. Jiang, P. C. Chiang, H. W. Hung, C.L. Lin, T. Yoon, and J. Lee, "100 Gb/s Ethernet Chipsets in 65 nm CMOS Technology", *IEEE ISSCC Dig.* Tech. Papers, pp. 120-121, Feb. 2013.

[3] A. Cevrero, et. al., "A 64 Gb/s 1.4 pJ/b NRZ Optical-Receiver Data-Path in 14 m CMOS FinFET", *IEEE ISSCC Dig.* Tech. Papers, pp. 482-483, Feb. 2017.

[4] S. H. Huang and W. Z. Chen, "A 25 Gb/s 1.13 pJ/b -10.8 dBm Input Sensitivity Optical Receiver in 40 nm CMOS", *IEEE J. Solid-State Circuits*, vol. 52, pp 747-756, March 2017.

[5] D. Li, G. Minoia, M. Repossi, D. Baldi, E. Temporiti, A. Mazzanti, and F. Svelto, "A Low-Noise Design Technique for High-Speed CMOS Optical Receivers," *IEEE J. Solid-State Circuits*, vol. 49, no. 6, pp 1437-1447, June 2014.

[6] C. Liu, Y. P. Yan, W. L. Goh, Y. Z. Xiong, L. J. Zhang, and M. Madihian, "A 5-Gb/s Automatic Gain Control Amplifier with Temperature Compensation", *IEEE J. Solid-State Circuits*, vol. 47, no. 6, pp 1323-1333, June 2012.

[7] S. Ray and M. M. Hella, "A 10Gb/s Inductorless AGC Amplifier with 40dB Linear Variable Gain Control in 0.13 um CMOS " *IEEE J. Solid-State Circuits*, vol. 51, no. 6, pp 440-456, Feb. 2016.

[8] C. F. Liao and S. I. Liu, "A 10 Gb/s CMOS AGC amplifier with 35 dB Dynamic Range for 10 Gb Ethernet", *IEEE ISSCC Dig.* Tech. Papers, pp. 2092-2101, Feb. 2006.

TABLE I PERFORMANCE SUMMARY

	[1]	[2]	[3]	[4]	[5]	[6]	[7]	[8]	Our
Technology	65 nm CMOS	65 nm CMOS	14-nm FinFET	40nm CMOS	65 nm CMOS	130 nm SiGe	130 nm CMOS	180 nm CMOS	40 nm CMOS
Data Type	NRZ	NRZ	NRZ	NRZ	NRZ	NRZ	NRZ	NRZ	PAM-4
Data rate (Gbps)	25-28	25	64	25	25	5	10	10	56
TIA	Yes	Yes	Yes	Yes	Yes	No	No	No	Yes
AGC	No	No	No	No	No	Yes	Yes	Yes	Yes
Gain control	N/A	N/A	N/A	N/A	N/A	dB-linear	dB-linear	dB-linear	dB-linear
PRBS	2^9-1	2^7-1	2^7-1	2^{15}-1	2^{31}-1	2^{15}-1	2^{31}-1	2^7-1	2^7-1
Input-referred rms noise current (uArms)	2.6	4.2	N/A	N/A	1.8	N/A	N/A	N/A	2.56
Supply (V)	1.1/3.3	1.2	0.9/1.15	1.2	1.1/1.8	1.2	1.2	1.2	1.2
Photodiode capacitance (fF)	N/A	150	69	100	160	N/A	N/A	N/A	100
Energy Eff. (pJ/bit)	4.91	2.76	1.4	1.13	3.72	14.4	5.0	5.4	2.14
Power (mW)	137.5	69	89.6	28.25	93	72	50	54	120
Core Area (mm²)	0.32	1.04	0.028	0.007	0.42	1	1.07	1.32	0.21

A Low-Power PAM4 Receiver Using 1/4-Rate Sampling Decoder with Adaptive Variable-Gain Rectification

Guang Zhu, Quan Pan[#], John Zhuang[*], Charlie Zhi[*], and C. Patrick Yue

Department of Electronic and Computer Engineering, the Hong Kong University of Science and Technology, Hong Kong
[#]Etopus, Sunnyvale, CA, USA   [*]Brite Semiconductor, Shanghai, China

Abstract—This paper presents a 1/4-rate PAM4 receiver employing a sampling decoder with an adaptive variable-gain rectifier (AVGR) to achieve a bit efficiency of 1.38 pJ/bit. By concurrently performing gain adaptation and amplitude rectification for decoding the least significant bit (LSB), the proposed decoder greatly reduces power consumption compared with the conventional full-rate topology using three comparators. The sense amplifier (SA) in the AVGR is designed to have large gain and wide programmability in order to accommodate wide PAM4 input dynamic range (DR) while using a single constant reference voltage for decoding. Experimental results verify that the receiver IC, implemented in SMIC 28-nm high-k CMOS process, can receive and decoder a 24-Gb/s 190-mV$_{pp}$ PAM4 signal at a BER of 10^{-11}.

Keywords—PAM4 receiver decoder; LSB decoding; adaptive variable-gain rectifier.

I. INTRODUCTION

The rapid growth of mobile internet data traffic has driven the speed of serial links to increase significantly with improved power efficiency during the past decades. However, the trend is hard to maintain for NRZ signaling due to the bandwidth (BW) limitation from both the channel and IC process. PAM4 signaling is attracting more and more attention for its doubled BW efficiency [1–5]. However, PAM4 receivers face new design challenges in terms of the decoding scheme, input sensitivity, and linearity requirement [6]. ADC-based PAM4 receiver architecture is widely adopted because of its design flexibility and portability for process migration. The reported bit efficiency is around 10 pJ/bit [5]. In contrast, analog PAM4 receiver design can achieve a better efficiency at 4 pJ/bit as reported in [3]. The decoder in an analog PAM4 receiver is one of the most critical blocks in determining the overall receiver power consumption and input sensitivity since it needs to handle different input amplitudes and can be very power hungry if the decoder worked at full rate. In [1-2], the decoder employs three comparators with manually tuned reference levels control to accommodate different input amplitudes. The receiver in [3] uses a variable gain amplifier (VGA) with adaptive gain control loop so that the reference level of the decoder can be fixed. However, the high power consumption of the decoder still need to be addressed.

To alleviate the above design issues, a novel decoder architecture utilizing an adaptive variable-gain rectifier (AVGR) in a 1/4-rate receiver is proposed. In this design, only two voltage-mode comparators are needed for decoding to further save power. For different input amplitudes, the AVGR adaptively tunes the gain of its sense amplifier (SA), and correctly performs the decoding function.

II. RECEIVER DECODER ARCHITECTURE

Figure 1 presents the comparison between three different PAM4 receiver architectures. A continuous time linear equalizer (CTLE) is used to pre-condition the input PAM4 signal. As shown in Fig. 1(a), the decoder based on three comparators is the most common topology. Each comparator has different reference levels for decoding. To deal with different input amplitudes, a VGA is introduced before the decoder. Considering the design robustness and speed, power-hungry CML-type comparators are usually employed in full-rate decoders [1]. On the other hand, the comparators in a 1/4-rate decoder, as shown in Fig. 1(b), are clocked at much lower speed and hence are based on voltage-mode topology. It should be mentioned that the chip area also benefits from adopting 1/4-rate architecture since inductors required in full-rate VGA for shunt peaking can be avoided. Although the number of comparators is fourfold, the total power is actually halved because a full-rate CML-type comparator consumes up to 8 times the power of a 1/4-rate voltage-mode comparator [2]. The subsequent thermometer-to-binary (T2B) logic can also operate at lower speed and hence consume less power.

The VGA function is not necessary for decoding the most significant bit (MSB) of the PAM4 signal due to the high gain of the comparator. If the four logic levels of PAM4 signal are gray coded, the least significant bit (LSB) can be decoded by distinguishing two amplitudes of the PAM4 signal. Fig. 1(c) shows the proposed decoder architecture where the MSB decoding path includes a comparator CMP1 while the LSB decoding path consists of an AVGR and a comparator CMP2. The VGA is replaced by a variable-gain SA and the gain adaptation block adaptively tunes its gain for different amplitudes. The SA output is rectified before the CMP2 where a reference voltage V$_{ref}$ is used to compare with two rectified voltage levels corresponding to two amplitudes of the PAM4 signal for LSB decoding. By doing so, the proposed decoder saves one comparator and the T2B logic is also eliminated. For binary coding, an XNOR gate is required.

III. BUILDING BLOCKS

A. CTLE

Figure 2(a) shows the circuit diagram of the CTLE. Shunt peaking is employed to enhance the BW and the area of the customized inductor is 40×40 μm^2. The source degeneration resistor R$_S$ is controlled to tune the equalization and the simulated equalization ability is 5.2 dB at 10 GHz.

Due to the multi-level signaling and the associated transitions, PAM4 signals are more sensitive than NRZ signals

S6-2 (2129)

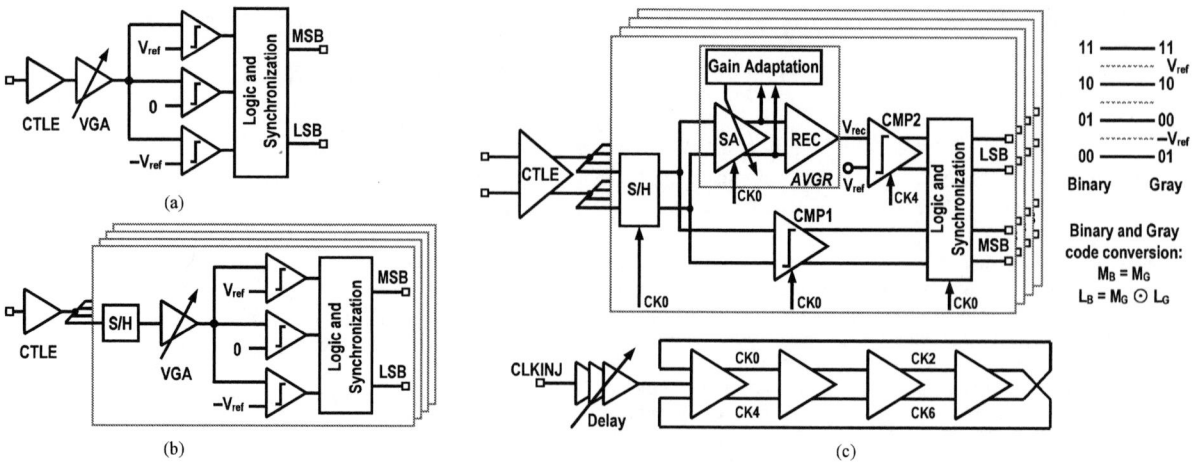

Fig. 1. Comparison of PAM4 receiver (a) a full-rate three-comparator-based topology with full-rate VGA, (b) a 1/4-rate three-comparator-based topology with 1/4-rate VGA, and (c) the proposed 1/4-rate topology with AVGR utilizing two comparators.

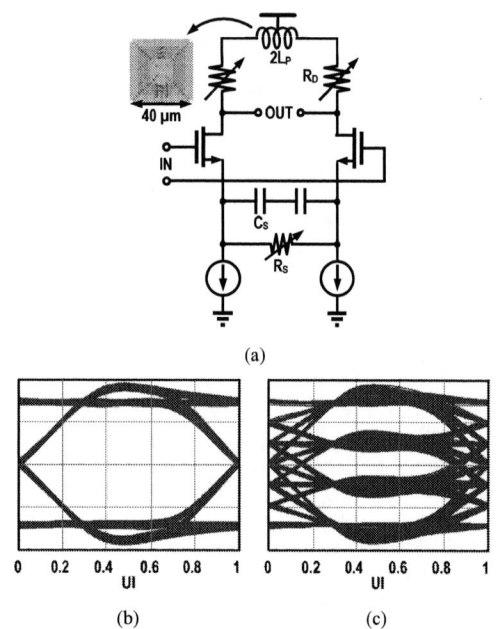

Fig. 2. (a) CTLE circuit diagram, (b) and (c) are the simulated NRZ and PAM4 outputs of CTLE when it is with 2-dB over-peaking.

to over-peaking in the equalization. As illustrated in Fig. 2(b) and 2(c), the PAM4 and NRZ signals at the CTLE outputs are compared with 2-dB over-peaking. The eye opening of the NRZ eye diagram is barely degraded by the over-peaking; as a result, the recovered clock can sample the data at the 0.5 UI instance to achieve optimal data slicing. However, for the PAM4 signals, the optimal sampling point is obviously shifted away from the middle point of the data eye due to the level squeezing caused by the over-peaking. Therefore, the CTLE or any equalization for PAM4 signal requires additional care to ensure the data slicing occurs at the optimal instance.

B. AVGR

As shown in Fig. 1(c), the AVGR includes three sub-blocks: gain adaptation block, variable-gain SA, and rectifier. The sampled differential signal is amplified by the SA and then rectified to a single-ended signal V_{rec} which is proportional to the input amplitude. For a PAM4 signal, there are two amplitudes and one is three times of the other one. Therefore, V_{rec} also has two voltage levels (V_{max0}, and V_{max1}) and a reference voltage V_{ref} between them helps recover the LSB in the CMP2. When the receiver starts working, the gain of the VGR is adaptively adjusted according to the input amplitude by the gain adaptation block resulting in a fixed V_{max1}, so that V_{ref} is easy to be decided if the DC component (rectifier output when input is zero) of V_{rec} is as small as possible.

Figure 3 shows the circuit diagrams of the two main blocks of the AVGR: the variable-gain SA and the rectifier. The SA is modified from a strong arm latch (SAL) which is widely used in a clocked comparator. The gain of the SA is programmable by adopting both switchable capacitance loading and switchable input transistor size. Using a linear SAL instead of a differential pair as the SA has the following advantages [7]: 1) The SA has a high gain and large output swing due to the inherent latch; 2) There is no static current in the SA and the rectifier, so the power is small; 3) The DC component of the rectifier's output is also small since the conduction time of the rectifier is very short when the differential input is zero. The transfer characteristic of the VGR is obtained by maintaining the gain of the SA and sweeping the input amplitude of a periodic square waveform. Fig. 4 shows the simulation result with the input amplitude sweeping from 20 to 90 mV$_{pp}$. As shown, the transfer curve is quite linear and the gain is up to 17 dB. For the PAM4 input with an amplitude of 90 mV$_{pp}$, the two levels of V_{rec} are 300 mV and 730 mV, so the margin of V_{ref} is large. The DC component of the rectifier's output is around 60 mV (the intersection point of the extrapolated curve and vertical axis in Fig. 4). When the input amplitude of the VGR is 300 mV$_{pp}$, the gain of the SA is set to 7 dB so that V_{max1} of

978-1-5386-3179-9/17 $31.00 © 2017 IEEE   82

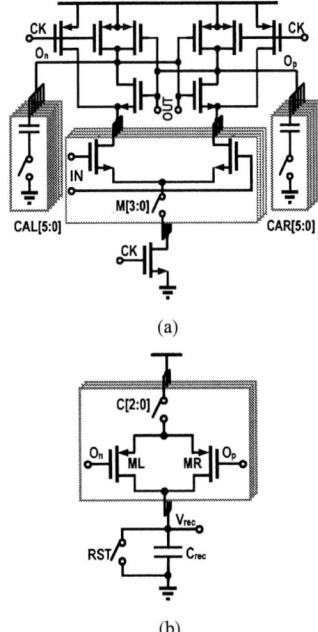

(a)

(b)

Fig. 3. (a) Variable-gain sense amplifier and (b) rectifier.

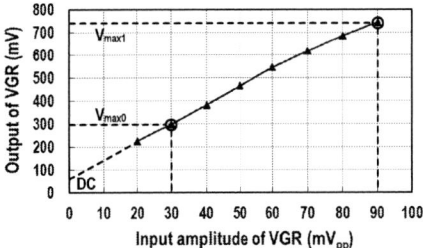

Fig. 4. Simulated transfer characteristic of variable-gain rectifier.

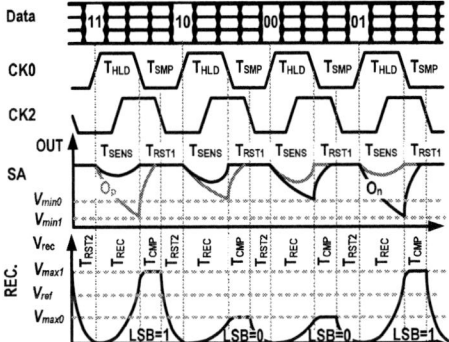

Fig. 5. Timing diagram of variable-gain rectifier for four PAM4 levels.

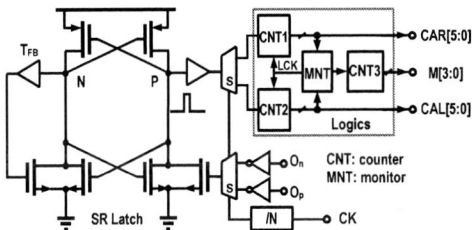

Fig. 6. Diagram of gain adaptation.

the rectifier's output is the approximately same to that in Fig. 4. Therefore, the programmable gain range of the VGR is 10 dB.

Figure 5 is the timing diagram of the VGR for the four gray-coded PAM4 levels in Fig. 1(c). After sampling, the SA output makes transistors in the rectifier conduct for part of the holding time T_{HLD} and the current is integrated on the capacitor C_{rec}. Once T_{HLD} is over, the SA outputs reset to high and V_{rec} holds for the comparison with V_{ref} in the CMP2. Before T_{HLD} starts again, V_{rec} discharges to ground. As Fig. 5 shows, V_{rec} is a series of pulses with two voltage levels V_{max0} and V_{max1} which are corresponding to LSB = 0 and LSB = 1, respectively.

C. Gain Adaptation

For the PAM4 signal, the gain adaptation is highly demanded due to the linearity issue and decoder design. The rectifier with adjustable gain is designed to overcome the effect of process variation and the setting is fixed after an initial calibration. Fig. 5 shows that V_{rec} is determined by the SA output, so the gain adaptation is used to regulate the SA output so that the minimum of the SA output (V_{min1}) is the same for different input amplitudes. The SA output is monitored by the gain adaptation block whose outputs digitally control the gain of the SA. The adaptation diagram is shown in Fig. 6 and a self-resetting SR latch works as the interface between the SA and digital logics. The SR latch is triggered if V_{min1} is smaller than its trigger voltage V_{tg}. After it is triggered, the SR latch produces a pulse to drive the counters (CNT1,2) in digital logics to increase, and then resets. The CNT1,2 outputs which control the capacitor arrays of the SA are monitored by the block MNT. Once CNT1/CNT2 reaches its maximum, both of them reset and the transistor size decreases. The increase of the CNT1,2 and the decrease of the CNT3 leads to a smaller gain

of the SA and a higher V_{min1}. As V_{min1} reaches V_{tg}, CNT1,2 will keep for a certain time and the MNT produces a signal to lock the CNT1,2 then the adaptation process ends. The SA output is determined by V_{tg} of the SR latch. According to simulations, V_{tg} has a design margin of 200 mV which is enough to cover the process, voltage, and temperature (PVT) variation. The SR latch is driven alternately by the SA output O_p/O_n; therefore, the mismatch effect of the SA is eliminated.

IV. MEASUREMENT RESULTS

The proposed receiver is fabricated in SMIC 28-nm CMOS process and Fig. 7(a) shows the chip photograph. The core part occupies an area of 0.15×0.16 mm^2. The chip consumes 33 mW from a 1 V supply. As shown in Fig. 1(c), the 1/4-rate clock signals are generated by a four-stage injection locked ring oscillator (ILRO) for de-skewing the forwarded clock under source-synchronous operation. Fig. 7(b) shows the power consumption breakdown. The 1/4-rate architecture requires the clock to drive a larger loading. Hence, more than half of the total power is consumed by clock distribution with multi-stage buffers. The 1/4-rate decoder inherently performs

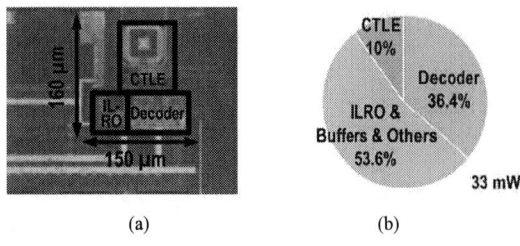

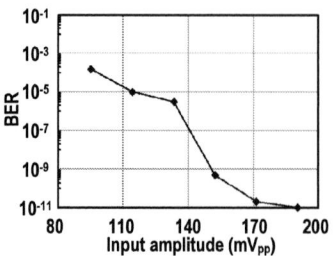

Fig. 9. Measured BER under different input amplitudes.

(a)   (b)

Fig. 7. (a) Chip photograph and (b) power consumption breakdown.

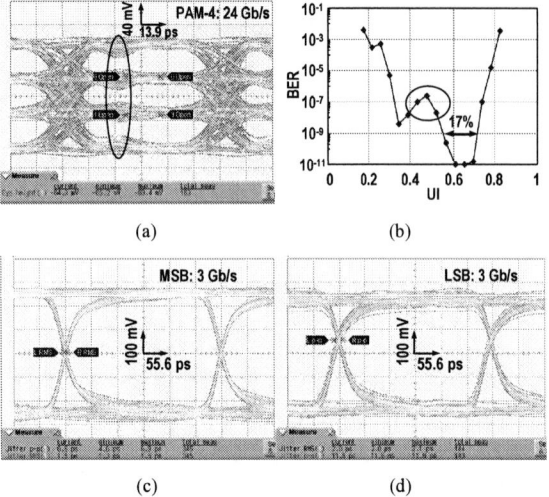

(a)   (b)

(c)   (d)

Fig. 8. (a) 24-Gb/s 190-mV$_{pp}$ PAM4 input and (b) measured BER bathtub curve when with over-peaking, (c) recoverd 3-Gb/s MSB and (d) LSB.

1:8 de-multiplexing which eliminates the need for a dedicated de-mux in full-rate topology.

The PAM4 signal is generated by a two-channel pattern generator and a power combiner. Before going to the receiver, the signal is attenuated by 10 dB. The loss of the attenuators and cables is ~3dB at 6 GHz. Fig. 8(a) shows the 24-Gb/s 190-mV$_{pp}$ PAM4 input and the ripple in the eye diagram mainly results from the imperfect frequency response of the transmitter in the pattern generator. Fig. 8(b) shows the measured BER bathtub curve. As discussed in III.A, the spike in the curve is induced by the over-peaking in the CTLE which in turn leads to BER degradation when sampling at the 0.5 UI instance. The BER performance between 0 to 0.5 UI is worse than that between 0.5 to 1.0 UI because of the residue ripple in the input eye diagram. Fig. 8(c) and 8(d) show the eye diagrams of the recovered MSB and LSB with measured RMS jitters of 1.5 ps and 2.0 ps, respectively. The BER performance of the receiver under different input amplitudes is also measured and Fig. 9 shows a BER of 10^{-11} is achieved when the input amplitude increases to 190 mV$_{pp}$. The reference voltage does not change when the amplitude varies because of the adaptation. The performance of this work is summarized and compared with similar works in Table I. This receiver can perform decoding adaptively under a good power efficiency.

TABLE I. Performance Summary of PAM4 Receiver Works

	[1]	[3]	[4]	**This work**
Function	CTLE+ Decoder +CDR	CTLE+ DFE+CDR +Decoder	DFE+ Decoder+ Clock buffer	CTLE+ Decoder +ILRO
Clocking	Full rate	1/2-rate	1/4-rate	1/4 rate
Gain Adaptation	No	Yes	No	Yes
Data Rate (Gb/s)	56	40-56	32	24
Bit Efficiency (pJ/bit)	7.5	4.11	0.55	1.38
Eye Width of BER 10^{-9} @ Input Amp. (mV$_{pp}$)	NA	13% @300	17% @300	17% @190
CMOS Process Node	40nm	16nm	65nm	28nm

V. CONCLUSION

A 24-Gb/s PAM4 receiver in 28-nm technology has been reported. With the proposed AVGR, the receiver achieves a bit power efficiency of 1.38-pJ/bit while adaptively decoding PAM4 signals with different amplitudes.

ACKNOWLEDGMENT

This work was sponsored in part by the Research Grant Council of Hong Kong SAR Government, China, under project No. 16201815. The authors also would like to thank SMIC for chip fabrication.

REFERENCES

[1] J. Lee *et al.*, "56Gb/s PAM4 and NRZ SerDes transceiver in 40nm CMOS," *IEEE Symp. VLSI Circuits Dig. Tech. Papers*, Jun. 2015, pp. 118−119.

[2] T. Toifl *et al.*, "A 22-Gb/s PAM4 receiver in 90-nm CMOS SOI technology," *IEEE J. Solid-State Circuits*, vol. 41, no. 4, pp. 954−965, Apr. 2006.

[3] J. Im *et al.*, "A 40-to-56Gb/s PAM-4 receiver with 10-tap direct decision-feedback equalization in 16nm FinFET," *IEEE Int. Solid-State Circuit Conf. Dig. Tech. Papers*, Feb. 2017, pp. 114−115.

[4] O. Elhadidy *et al.*, "A 32 Gb/s 0.55 mW/Gbps PAM4 1-FIR 2-IIR tap DFE receiver in 65-nm CMOS," *IEEE Symp. VLSI Circuits Dig. Tech. Papers*, Jun. 2015, pp. 224−225.

[5] D. Cui *et al.*, "A 320mW 32Gb/s 8b ADC-based PAM4 analog front-end with programmable gain control and analog peaking in 28nm CMOS," *IEEE Int. Solid-State Circuit Conf. Dig. Tech. Papers*, Feb. 2016, pp. 58−59.

[6] H. Zhang *et al.*, "PAM4 signaling for 56G serial link applications: A tutorial," *DeisgnCon*, Jan. 2016.

[7] B. Razavi, "The StrongARM latch [a circuit for all sessions]," *IEEE Solid-State Circuits Mag.*, vol.7, no. 2, pp. 12−17, Jun. 2015.

S6-3 (2050)

A 82 mW 28 Gb/s PAM-4 Digital Sequence Decoder with built-in Error correction in 28nm FDSOI

Masum Hossain[1], Aurangozeb[1], AKM Delwar Hossain[1], Maruf Mohammad[2]

[1]University of Alberta, Edmonton, Canada, [2]Qualcomm Atheros

masum@ualberta.ca

Abstract— This paper describes an energy-efficient PAM-4 digital receiver based on sequence detection. This scheme takes advantage of the ISI in the channel to reconstruct the time domain 5-bit sequence including MSB and LSB. The architecture also enables built-in error correction with very low latency. This concept is demonstrated with prototype implemented in a 28nm FDSOI CMOS using only 18-data comparators. Consuming only 82mW at 28 Gb/s, the receiver is capable of compensating 27 dB channel loss with BER lower than 10^{-8}. With error correction, the receiver can achieve BER lower than 10^{-10}.

Keywords— *MLSD; Sequence DFE; PAM-4; Error correction and SNR Optimized Receiver.*

I. INTRODUCTION (*Heading 1*)

As the bandwidth in data centers continues to grow, we are moving to higher modulation (i.e. PAM-4) schemes to improve spectrum efficiency. Since the channel characteristics are not improving to keep up with the data rate, this seems to be a viable way of increasing data rate. However, this spectrum efficiency comes at the cost of signal-to-noise ratio (SNR).

In PAM-4, two bits of information are encoded into four levels. Therefore, compared to NRZ signaling for the same transmit swing MSB's amplitude is reduced to 2/3 and LSB's amplitude is reduced by 1/3. As a result, symbol eye height is reduced by a factor of 3 and that translates to 9 dB SNR penalty. This 9 dB SNR penalty manifests itself in many ways – first, the degradation of SNR makes it significantly more challenging to achieve low BER, therefore most standards adopt forward error correction (FEC) encoding and decoding that costs additional area, power and significant latency. Second, residue ISI of the MSB and crosstalk has much bigger impact on the LSB that further degrades voltage and timing margin.

Conventional approach to receiver equalization is CTLE and DFE. CTLE response is difficult to control over PVT variation – therefore, ISI cancellation is often insufficient. DFE is more effective in post-ISI cancellation. However, canceling the 1st post-cursor ISI requires feedback loop latency to be less than 1 UI – this has proven to be challenging especially at high data rates. Therefore, 1st post-cursor tap is often implemented in loop unrolled manner. Although this technique alleviates the latency issue, a number of comparators significantly increases. In the case of NRZ signaling, for N tap loop unroll DFE we need 2^N number of comparators. Unfortunately, in the case of PAM-4 this penalty is much higher – for N tap PAM-4 loop unroll DFE we require 3×4^N. Therefore, 1 tap loop unroll DFE

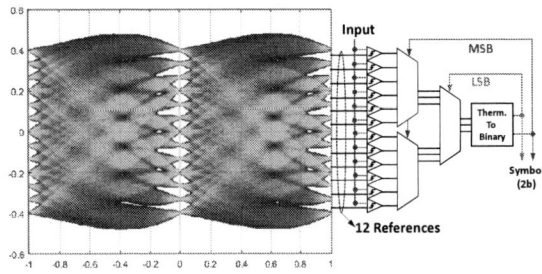

Fig. 1. PAM-4 1-tap loop unrolled DFE structure in conventional implementation using 12 comparators

requires 12 comparators as shown in Fig. 1. As a result, use of DFE in PAM-4 receiver is very limited.

To address these challenges, most of the PAM-4 receivers are adopting ADC based front-end with advanced digital equalization [1,2]. Since an ADC alone with reasonable resolution (6 bits) consumes around 10 pJ/bit, when DSP [1-6] is included power consumption easily exceeds 15 pJ/bit. Therefore, an SNR-optimized approach is attractive especially if the receiver energy efficiency can be improved to less than 5 pJ/bit.

II. SNR OPTIMIZED PAM-4 DECODING

Conventional PAM-4 DFE is challenging –both previous MSBs and LSBs can influence current symbol's decision process. Therefore, any error in LSB can propagate to MSB selection comparator causing error. As a result, DFE in PAM-4 receiver is less effective.

Energy distribution in PAM-4 signaling is better understood from the decomposed MSB and LSB single bit responses shown in Fig. 2. Note that most of the signal energy is concentrated within three taps – 1st pre-, main and 1st post-cursor taps. After sorting them according to the energy in each tap we find the following sequence - h^{MSB}_0 h^{MSB}_{+1} h^{LSB}_0 h^{MSB}_{-1} h^{LSB}_{+1} h^{LSB}_{-1}. Interestingly, energy in pre-cursor ISI of MSB is comparable to the main cursor of the LSB, h^{LSB}_0. In conventional approach, this pre-cursor is equalized from transmitter side but due to maximum swing limitation equalization is mainly attenuates low-frequency component and as a result the SNR degrades (Fig. 2(c)). In this example with pre-cursor equalization, LSB SNR penalty can be as high as15 dB (9 dB for PAM-4 signaling and 6 dB to correct pre-cursor ISI). Although Tx and Rx equalization can reduce the pre- and post-cursor ISI introduced by the channel, signal amplitude reduces significantly.

978-1-5386-3179-9/17 $31.00 © 2017 IEEE

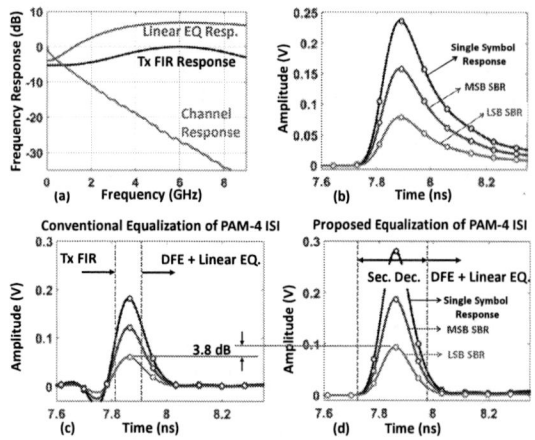

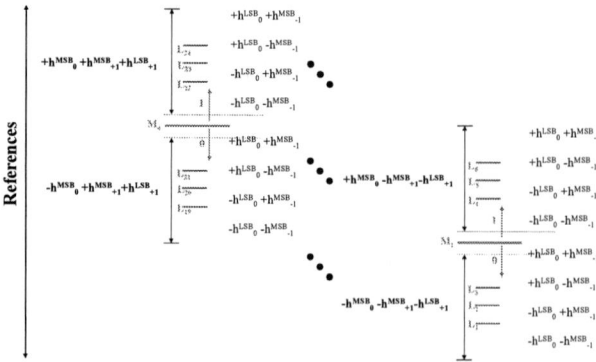

Fig. 4. Reference generation for PAM-4 sequence detection based on channel impulse response

Fig. 2. (a) Frequency response of the channel, Tx FIR and Rx LEQ (b) Channel single symbolresponse and decomposed SBR (c) Equalized SBR using Tx FIR based pre-cusrosr cancellation, and (d) Using sequence decoding.

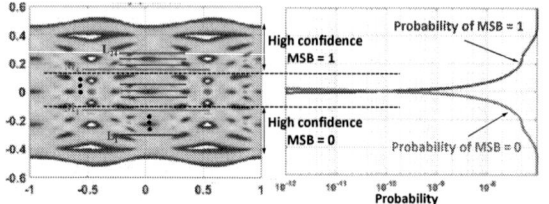

Fig. 3. An example of input eye with high confidence MSB detection and corresponding MSB PDF

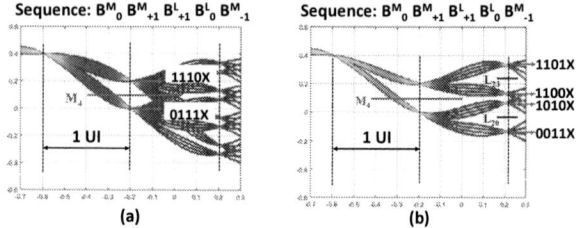

Fig. 5. Partial response eye with reference level for sequence detection (a) MSB detection and (b) LSB detection with 1 post-cursor ISI and no pre-cursor

A. High confidence MSB detection

Compared to traditional NRZ signaling, MSB and LSB suffers 3 and 9 dB penalty respectively. Although MSB may have better SNR, symbol SNR in PAM-4 is limited by LSB SNR. In other words, it is still possible to have symbol error even if the MSB is detected correctly. Therefore, it is beneficial to decode MSB and LSB separately rather than decoding the symbol directly. For better understanding we plot the probability of MSB=1 and MSB=0 in the signal space with the eye diagram in Fig. 3 that shows that it is possible to define regions where we can confidently (lower than BER = 10^{-12}) decode MSBs even when ISI is not corrected.

To detect 'high confidence' MSB =1, a threshold is placed where the probability of MSB = 0 is less than 10^{-12} along with additional noise margin. Similarly, a threshold is set on the lower signal space where if the signal is below that level we can confidently decide that the MSB = 0. This high confidence MSB detection provides several advantages: first, these decisions in high confidence region are not affected by DFE. Therefore, these decisions are not affected by LSB error and error propagation is truncated. Second, these MSBs are error free, they can be used to detect and correct previously decoded MSBs and LSBs. Note that outside these two 'high confidence' regions MSB can be either 0 or 1 and requires DFE to decide between them. Never the less, these high confidence regions cover 66% of the signal space providing sufficient improvement in mitigation of error propagation as well as error correction.

In a SNR-optimized approach, rather than reducing these ISI components, we prefer to decode them as a sequence of bits. To enable that, this work introduces several techniques: first, instead of reducing pre-cursor ISI, we use the ISI in a constructive way to decode sequence. Specifically, the MSB pre-cursor energy that is comparable to the main of the LSB should be used for detection. Second, to mitigate DFE error propagation, we use high confidence MSB detection and sequence DFE with redundancy. Finally, pre-cursor based error correction is used to improve raw BER of the system. The potential SNR benefits of the proposed approach visible in the LSB SBR that shows 3.8 dB larger amplitude (Fig. 2(d)).

Considering both MSB and LSB, three taps translate to 6-bit sequence and its decoding requires significant hardware complexity. Since pre-cursor LSB component has the smallest energy, 5-bit sequence is a good compromise between hardware complexity and SNR. Therefore, each received sample is decoded to a 5-bit sequence: $B^M_0 B^M_{+1} B^L_{+1} B^L_0 B^M_{-1}$, where the subscript indicates bit location and superscript indicates MSB or LSB. Here B^M_0 and B^L_0 are current MSB and LSB respectively. Similarly, $B^M_{+1} B^L_{+1}$ are previous MSB and LSB. More importantly, B^M_{-1} is the next MSB detected based on the precursor ISI of MSB. Therefore, the proposed approach helps in two ways: first, proposed approached have a higher SNR as shown in the SBR and second, the decoded future MSB is used for error correction.

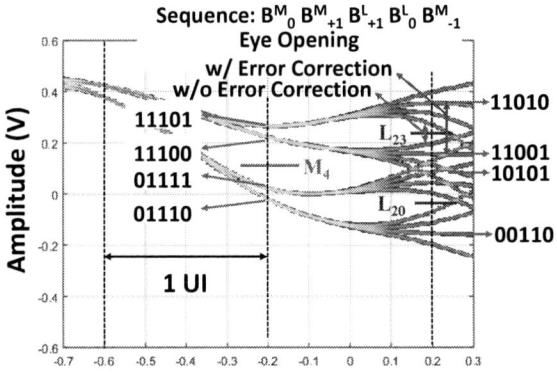

Fig. 6. PAM-4 partial response eye with precursor ISI based sequence detection example

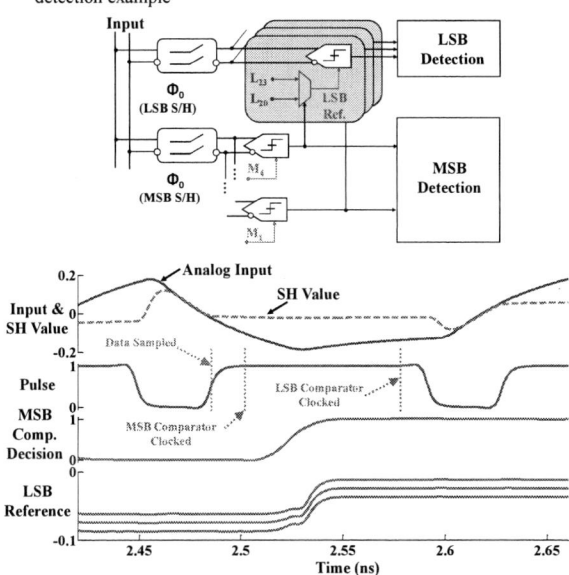

Fig. 7. Two step MSB- LSB detection scheme with timing diagram. Simulation results to show reference settling for LSB detection

B. Separate decoding for MSB and LSB

MSB and LSB are decoded as part of 5-bit sequence. A straightforward implementation would require 32 comparators. Inspired by the two-step ADC, sequential detection of MSB and LSB can reduce required number of comparators.

MSB is decoded as part of partial sequence $B^M_0 B^M_{+1} B^L_{+1}$ from the 5bit sequence. First, we generate four likely set of sequences and then the most likely sequence is selected based on previously decoded MSB and LSB ($B^M_{+1} B^L_{+1}$) similar to DFE. To generate the likely set of sequences, 4 MSB comparator references are placed between the $h^{MSB}_0 h^{MSB}_{+1}$ h^{LSB}_{+1} tap combinations as shown in the Fig. 4. For example, M_4 comparator is placed between bottom of 111XX and top of 011XX references to maximize the noise margin. Here the last two don't care bits (X) represent current LSB (B^L_0) and next MSB (B^M_{-1}). Starting with a simplified case of no pre-cursor ISI ($h_{-1}=0$), Fig. 5a shows the detection of MSB. Once the MSB is known, we require only one additional comparator to

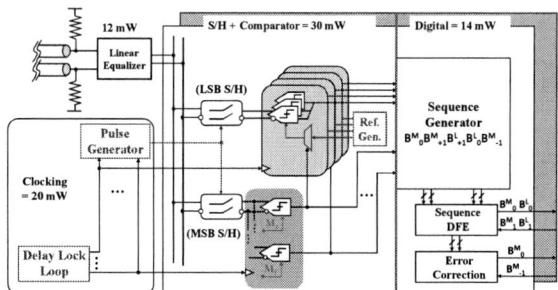

Fig. 8. Complete two way time interleaved 28Gb/s PAM-4 receiver with builtin error correcction and power consumption berakdown

decode LSB. However, relying on MSB detection creates timing constraint. To avoid that timing constraint, LSB's are also decoded speculatively by placing a single comparator each selected sequence set. One example is shown in Fig. 5(b) one comparator is placed between 1101 and 1100 to detect LSB speculatively assuming previous MSB and LSBs are also '1' and '0' respectively. Compared to conventional loop unroll DFE, here number of comparators reduces from 12 to 8 saving 33% hardware.

C. Pre-Cursor based Error Correction

The hardware efficient sequence DFE described in the previous section can compensate low to moderate loss channels with excellent power efficiency. However, as channel loss increases, precursor ISI becomes a major concern. In this work proposes a more SNR-optimized approach by using the precursor ISI to detect the next symbol. To do that we place two additional comparators above and below the LSB detection comparators at:

$$\pm h^{MSB}_0 \pm h^{MSB}_{+1} \pm h^{LSB}_{+1} + h^{LSB}_0 - h^{MSB}_{-1}$$
$$\pm h^{MSB}_0 \pm h^{MSB}_{+1} \pm h^{LSB}_{+1} - h^{LSB}_0 + h^{MSB}_{-1} \qquad (1)$$

Here ± indicate these two references will appear in 8 possible set of sequences. One example is shown in Fig. 6. Note that the region between top and bottom comparator is where current LSB and next MSB are negatively correlated – meaning current LSB is simply an inversion of next MSB. Therefore, in the next sample time when the MSB will be decoded, it can be used to detect the error by comparing predicted MSB and actual decoded MSB. Although the error is detected, it remains unclear in which symbol the error has happened. This is resolved using high confidence MSBs – since the probability of their error is lower than 10^{-12}, we can consider those MSB decisions to be correct. After the sequence DFE, two sequences are generated – one of them is DFE selected sequence and the other one is 'most likely' alternative. These two sequences are generated with opposite next MSB (B^M_{-1}), therefore can be used to select the correct sequence in the next bit period when MSB is decoded.

The effectiveness of this error correction depends on two factors: first, we need high confidence MSBs in the sequence and this can be done from the transmitter side through encoding. Second, although it would be preferred to correct the errors before feedback, the hardware complexity significantly increases. Instead, error correction is placed in the forward path after DFE that results in significantly simplified hardware with

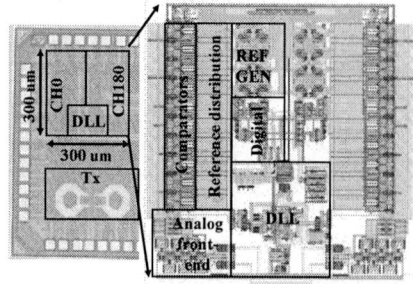

Fig. 9. Implemented prototype in ST 28nm FDSOI

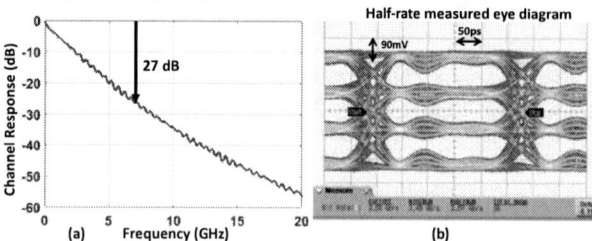

Fig. 10. (a) Channel response (b) measured half-rate eye diagram of recovered data

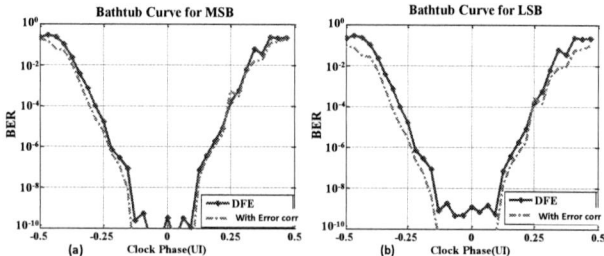

Fig. 11. Measured BER bathtub of (a) MSB & (b) LSB bit for 2^7-1 pattern

manageable latency. Note that in sequence DFE, sequences are generated such that in most cases LSB error propagation can be avoided through redundancy. On the other hand, probability of MSB error propagation is low due to higher SNR.

III. IMPLEMENTATION & MEASUREMENT

The sequential MSB and LSB detection provides power and area benefits, but the timing constrain requires careful consideration. For two-step conversion, separate sample and hold are used for MSB and LSB comparators. First, sampled value is compared with MSB comparators (M_1 to M_4). The output of these comparators are used to generate 4 possible MSB sequences. These outputs are also used to sub-range the possible LSB sequences. From 8 possible LSB references we need to consider only four. By careful mapping, it is possible to use the raw comparator output to drive the reference selection mux. The LSB reference must be settled before the comparator can be triggered. Therefore, to meet the timing, MSB comparator decision time and Mux selection time together should be less than 1.5 UI. In 28 nm FDSOI comparator takes about 35 ps to resolve and reference selection mux takes another 35 ps – total 70 ps can comfortably support 28Gb/s PAM-4. Note that sample and

TABLE I. Comparison with state-of-art

	Shafik ISSCC '15 [2]	Frans VLSI '16 [3]	Cui ISSCC '16 [1]	Rylov ISSCC '16 [4]	Im ISSCC '17 [5]	This Work
Technology	65 nm CMOS	16 nm FinFET	28 nm CMOS	32 nm CMOS	16 nm FinFET	28 nm FDSOI
Data Rate (Gb/s)	10 NRZ	56 PAM-4	32 PAM-4	25 NRZ	40-to-56 PAM-4	28 PAM-4
Equalization	Analog FFE ADC Based DFE & FFE	CTLE ADC Based DFE & FFE	CTLE ADC Based DSP	CTLE ADC Based DFE & FFE	CTLE 10-TAP FFE ($h_1,...,h_{10}$)	CTLE ADC Based Sequence DFE
ADC Architecture	32x TI SAR ADC	32x TI SAR ADC	32x TI SAR ADC	4x Flash ADC	N/A	2x Flash ADC
Resolution	6-bit	8-bit	8-bit	5-bit	N/A	5-bit sequence
Channel Loss Equalization	36.4 dB @ 5 GHz	25 dB @ 14 GHz	32 dB @ 8 GHz	40 dB @ 12 GHz	10 dB @ 14 GHz	27 dB @ 7 GHz
Power (mW)	79 (w/o DSP) 87 (w/ DSP)	410 (w/o DSP) —	320	453	230	82 @ 27 dB
FOM (pJ/bit)	8.7	7.32	10	18.12	4.11	2.93
BER	<10⁻¹⁰	<10⁻⁸	--	--	<10⁻¹²	<10⁻¹⁰

hold value is held for 1.5 UI after 0.5 UI tracking time as shown in Fig. 7. Therefore, half-rate architecture is chosen to meet the timing constraint and simplify clocking. The highly digital receiver architecture shown in Fig. 8 includes a front-end with both passive and active equalization. Passive equalization can accommodate large swing with excellent linearity whereas active equalization provides high-frequency gain. Keeping the power consumption low (12 mW), the analog front-end needs to provide only 6 to 10 dB programmable boost. This is sufficient to keep the ISI within a single pre and post-tap that can be corrected using PAM-4 sequence DFE. MSB comparator references are placed such that their outputs can be used to generate four possible 3bit sequences, $B^M_0 B^M_{+1} B^L_{+1}$, where B^M_0 is the current MSB bit and B^M_{+1}, B^L_{+1} are previous symbol's MSB and LSB respectively.

The implemented prototype shown in Fig. 9 occupies only 300 um x 300 um in 28nm FDSOI. It also includes transmitter and DLL for clock generation. The detail layout of the receiver shows two-way time-interleaved architecture with 18 comparators. Received signal after 5-bit sequence DFE goes through pre-cursor based error correction. The implemented prototype allows visibility to both with and without error correction (Fig. 10). Without error correction, symbol error rate is limited to 10⁻⁸ only as set by the LSB BER. With error correction LSB, error improves to better than 10⁻¹⁰ as shown in the Fig. 11. When compared to other existing PAM-4 receiver, the proposed work achieves highest energy efficiency by eliminating the ADC and power and area hungry DSP. This highly digital architecture is portable and can be scalable to decode longer sequence at higher channel loss.

REFERENCES

[1] D. Cui et al., "3.2 A 320mW 32Gb/s 8b ADC-based PAM-4 analog front-end with programmable gain control and analog peaking in 28nm CMOS," ISSCC 2016.

[2] A. Shafik et al., "3.6 A 10Gb/s hybrid ADC-based receiver with embedded 3-tap analog FFE and dynamically-enabled digital equalization in 65nm CMOS," ISSCC, 2015.

[3] Y. Frans et al., "A 56Gb/s PAM4 wireline transceiver using a 32-way time-interleaved SAR ADC in 16nm FinFET," VLSI 2016.

[4] S. Rylov et al., "3.1 A 25Gb/s ADC-based serial line receiver in 32nm CMOS SOI," ISSCC 2016.

[5] J. Im et al., "6.3 A 40-to-56Gb/s PAM-4 receiver with 10-tap direct decision-feedback equalization in 16nm FinFET," ISSCC, 2017.

[6] Aurangozeb et. al., "Channel Adaptive ADC and TDC for 28 Gb/s PAM-4 Digital Receiver ," CICC, 2017.

A 51Gb/s, 320mW, PAM4 CDR with Baud-Rate Sampling for High-Speed Optical Interconnects

Nan Qi[1,2], Yuhang Kang[3], Qipeng Lin[3], Jianxu Ma[4], Jingbo Shi[2], Bozhi Yin[2], Chang Liu[2],
Rui Bai[4], Shang Hu[2], Juncheng Wang[2], Jiangbing Du[5], Lin Ma[5], Zuyuan He[5], Ming Liu[3],
Feng Zhang[3,*], Patrick Yin Chiang[2,6]

[1]Institute of Semiconductors, Chinese Academy of Sciences, Beijing, China
[2]State Key Laboratory of ASIC and Systems, Fudan University, Shanghai, China
[3]State Key Laboratory of Microelectronics Devices and Integrated Technology,
Institute of Microelectronics, Chinese Academy of Sciences, Beijing, China.
[4]PhotonIC Technologies, Shanghai, China. [5]Shanghai Jiao Tong University, Shanghai, China.
[6]Oregon State University, Corvallis, OR, USA. *E-mail: zhangfeng_ime@ime.ac.cn

Abstract—A quarter-rate 51Gb/s PAM4 CDR with decoded dual-NRZ outputs is presented for 400GbE optical transceivers. A baud-rate data-only sampling PD with zero-crossing integrating front-end is proposed to minimize the clock generation and distribution overhead, as well as improving the noise resilience under PAM4 low-SNR inputs. Measurement results show the CDR features 1.08ps RMS clock jitter, 3.4×10⁻⁹ PAM4 data recovery BER, and 6.27pJ/bit power efficiency.

I. INTRODUCTION

With the explosive growth of the mobile computing and cloud services, optical interconnects with higher throughput is demanded. When NRZ data-rate goes beyond 40Gb/s, it is the electrical-optical (E/O) conversion bandwidth (BW) that limits the transceiver's speed. Instead of further increasing the data-rate, the 4-level Pulse-Amplitude Modulation (PAM4) can double the throughput without additional wavelengths or higher BW. Meanwhile, for single-channel data-rate above 25Gb/s, a CDR is preferred in both the transmitter and receiver, extending their electrical reach and suppressing jitter. Compared to cooper wirelines, the fiber channel loss is much smaller. Therefore, a TIA with CTLE, or 1-2 taps DFE are typically integrated to provide equalization prior to the CDR. Illustrated in Fig. 1, this work focuses on the PAM4 CDR with the full-rate outputting capability, which can be used as a repeater for both the E/O and O/E conversion. Input signal conditioning is not concerned in this work.

Design challenges of PAM4 CDR include (1) quantization of the high-speed multi-level signal, (2) clock/data alignment based on the PAM4 signal with multiple transition crossings, (3) low recovery error, low timing jitter with low power consumption. In this work, we demonstrate a PAM4 CDR prototype that attempts to address the clocking complexity and related power consumption by utilizing baud-rate sampling with an integrating front-end, thereby minimizing the clock generation and distribution overhead. Measured bit-error rate (BER) across a realistic ~10dB loss channel at Nyquist verifies the feasibility of this baud-rate CDR concept for 51Gb/s (or 25.5GBaud/s) PAM-4 operation.

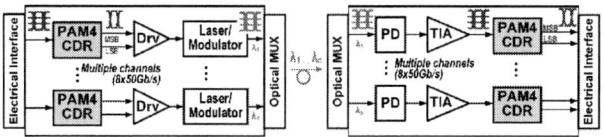

Fig. 1. PAM4 CDR in a 400GbE optical transceiver

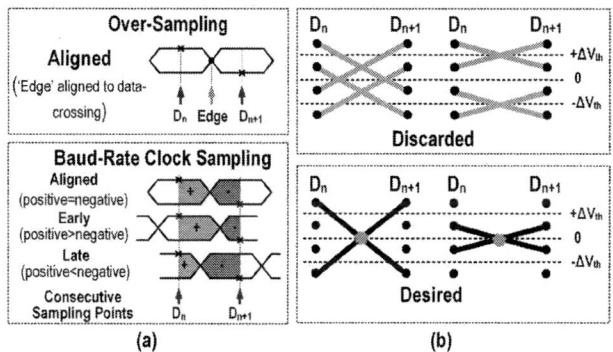

Fig. 2. Principle of the Baud-rate Sampling, (a) over-sampling vs. Baud-rate sampling and (b) transition-selection.

II. SYSTEM-LEVEL DESIGN

A. PAM4 Baud-Rate Sampling and Transition Selection

Traditional over-sampling CDR employs two consecutive (90° apart) clock phases for each data-bit. When phase-0 is aligned to the zero-crossing, phase-90 samples at the UI center with the highest SNR (Fig. 2a). Since this alignment tries to converge to a single crossing point, the recovered clock jitter and BER rely largely on the input signal quality, especially for PAM4 signals with reduced SNR. Moreover, to distribute additional clock phases draws large power and chip area. The proposed Baud-rate sampling utilizes the same clock phase to sample and align the clock, eliminating the phase at zero-crossings. The 'Edge' path accumulates charge within a predefined time-window, and evaluates the differential voltage polarity. When data-bit toggles, the integrated differential voltage (shadow-area of the eye) should be zero if the sampling happens at the mid-point in each UI. A positive non-zero voltage implies an earlier clock arrival than data, while vice versa. The PLL tunes the clock phase, trying to eliminate the non-zero voltage value. Another benefit is that, when the input PAM4 signal features slow rising/falling, the sampling point would be slightly shifted (due to the non-uniform shaped shadow area), to a point that the data is better settled, thereby optimizing the SNR (Fig 2a).

Fig 2b depicts the transition selection for PAM4 signals. There are multiple crossing regions (on the time-axis) in PAM4 transitions, which are data-pattern dependent and non-uniformly distributed [1]. This work selects the symmetric-crossing region with the highest transition-occurrence probability, preventing the recovered clock from wandering across all crossing regions, which otherwise would introduce large timing jitter.

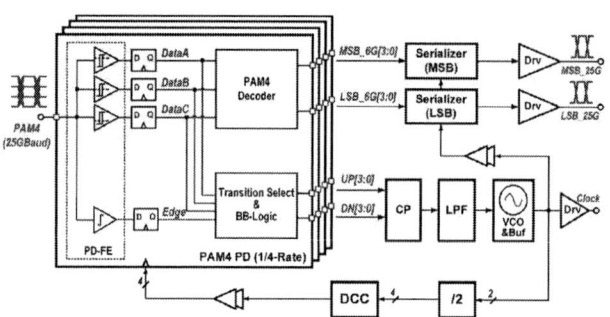

Fig. 3. Block diagram of the proposed PAM4 CDR.

B. Architecture

Fig. 3 shows the block diagram of the proposed PAM4 CDR. A quarter-rate phase detector (PD) down samples the 25GBaud/s PAM4 input into four 6.25G lanes. Each PD lane consists of a front-end (PD-FE), DFF re-timing stages, a PAM4 decoder and a digital logic module. The PD-FE employs 3 tunable-threshold slicers to quantize the 4-level amplitudes into a 3-bit thermal-code to the decoder, which then translates it into binary 2-bit outputs, namely the MSB (2x weight) and the LSB (1x weight). These quarter-rate data streams are serialized into 25Gb/s and sent out by voltage-mode drivers with 1-tap pre-emphasis. During the multiplexing, instead of retiming the data in each 2:1 MUX, a quadrature phase scheme is utilized to keep the phase relationship between two input data streams [3]. Working in parallel with the slicers, an integrator is employed in the PD-FE, which extracts the early/late information of the sampling, and sends to the digital logic module, where specific PAM4 data transitions are picked-up for clock-data alignment, with UP/DN generated to shift through the PLL. An on-chip VCO generates differential 12.5GHz clocks, which gets divided into 4-phase 6.25GHz for PD sampling. The clock phase is dynamically tuned by the PLL, comprising of a PAM4 PD, a charge pump (CP), a low-pass filter (LPF), a /2 divider and a Duty-Cycle Corrector (DCC). The loop bandwidth is designed tunable from 5MHz to 15MHz, depending on the CP current and the RC corner of the LPF.

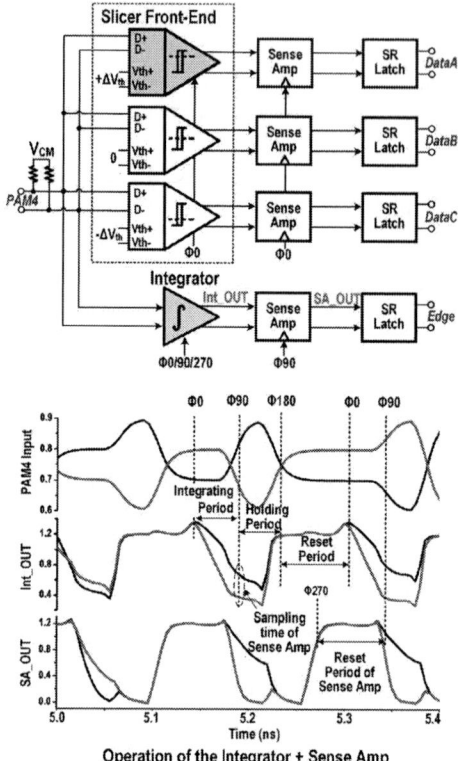

Fig. 4. Schematic of the PD front-end and its operation principle

III. Circuits Implementations

A. Phase Detector (PD)

A PAM4 PD plays the most important role in the CDR, which detects the phase difference between the PAM4 data and the recovered clock, generating UP/DN signals for phase locking (Fig. 3). The PD's core building block is its front-end, which composes of three data-paths and an edge-path in parallel (Fig. 4). Each slice in the PD-FE has a dedicated threshold, which is realized by applying a pre-determined DC-bias at the output node. The slice outputs are firstly amplified in a sense amplifier (SA), and then receive the regeneration in a SR-latch. The recovered data is derived in the Thermometer-code format as DataA, DataB and DataC outputs. Different to the data-path, the edge-path employs an integrator in the first stage, integrating the differential voltage value within a 1/4 clock period window. As explained in Fig. 2, a positive non-zero voltage implies earlier clock arrival than the data, while a negative integrated voltage indicating 'late'.

Fig. 4 describes the operation principle in detail. An input PAM4 signal is transformed into differential currents and then gets integrated within a 1/4 clock period window. Then the integrating result is held for another 1/4 clock period, during which the SA senses the voltage difference for the next stage, and amplifies it to a sufficient high swing for a fast regeneration in the SR latch. Fig. 4b depicts the detailed operation in transient voltages, in which the complete clock period is split into 3 parts, namely the integrating, holding and reset periods. The edge path evaluates early or late by adopting the Baud-rate sampling technique described in section II. All the slicer front-ends and integrators employ current DACs for input offset calibration.

B. Transition Selection

As described in Fig. 2, only the symmetric transitions are utilized for the clock/data alignment in the proposed PAM4 CDR. Transitions select is performed based on the initial thermometer code outputs from the PD slicers. Depicted in Fig. 5, A<2:0> and B<2:0> represent for the PD outputs of the past and current Baud respectively. The transition-select circuit recognizes the transitions between pattern '111' and '000', and enables the UP and DN control signals only when valid transitions are detected. Retiming stages are inserted to prevent from any decision error introduced by the mis-alignment between the select signal and the initial UP/DN outputs from the Bang-bang (BB) logic.

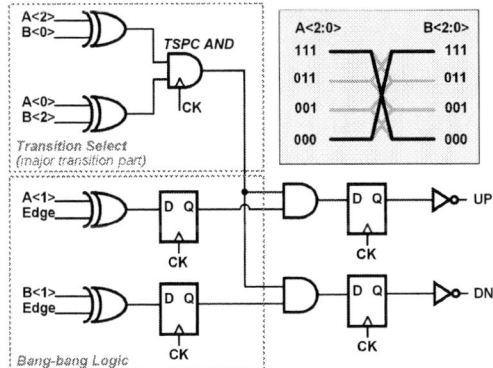

Fig. 5. Transitions-select for PAM4 operation

C. Serializer and Output Driver

Figure 6 shows the block diagram of the CDR's dual data output paths, which are used to interface with laser/modulator drivers. The recovered quarter-rate 6.25Gb/s data firstly receives a 4:1 multiplexing, and then get transmitted through voltage-mode SST drivers. Considering to save power in the final 2:1 MUX, the original 5-latches topology is replaced with a direct feed-through one, in which quadrature phase 6.25GHz clocks are utilized to generate 180° phase-shift at 12.5Gb/s, thereby eliminating the additional half-UI delay latches.

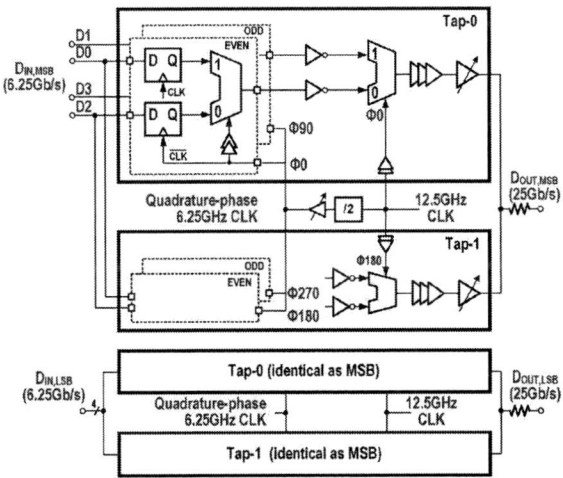

Fig. 6. Schematic of the serializer for interfacing between the CDR to laser/modulator divers.

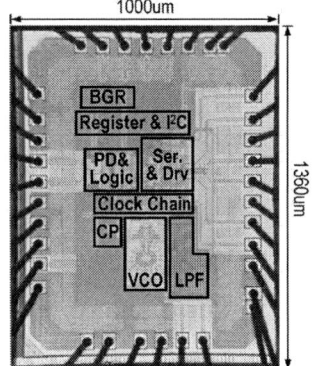

Fig. 7. Micro-photo of the proposed PAM4 CDR

IV. MEASUREMENT RESULTS

Implemented in 65nm CMOS, the proposed CDR occupies 0.45mm^2 core area in a 1.36 mm^2 test-chip (fig. 7). The CDR is tested by wire bonding to a lossy PCB, with the PAM4 signal applied at the input. To measure the BER, a 25.5G Baud/s PAM4 PRBS-9 signal is fed into the CDR, experiencing cable and PCB trace loss. The CDR then recovers the data/clock and generates two 25Gb/s NRZ signals, respectively MSB and LSB, that are both independently assessed using a 30G BERT.

A. Recovered Clock

Regarding to the clock recovery, a 2-channel 64GSs AWG is used to generate a 51Gb/s PAM4 data input, and an 80GSs real-time oscilloscope is used to analyze the recovered 12.55GHz clock, of which the jitter is measured as 1.08ps$_{RMS}$ and 8.4ps$_{PP}$ respectively. The closed-loop phase-noise is measured as -93dbc/Hz, -112dBc/Hz and -122dBc/Hz at 1kHz, 8.1MHz and 120MHz offset frequency respectively (fig. 8).

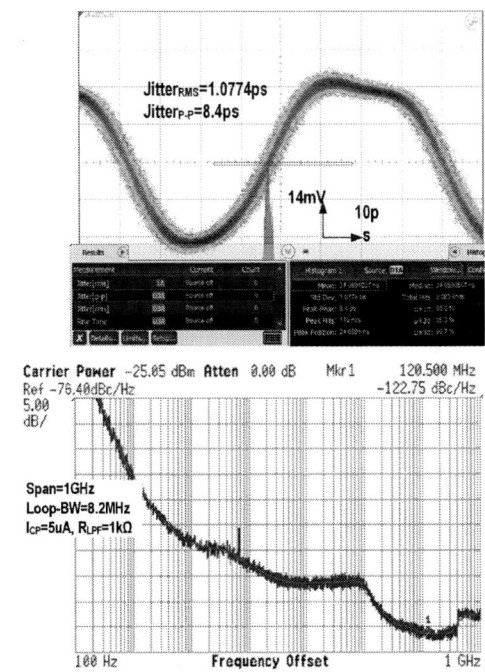

Fig. 8. Measured recovered clock: clock jitter and phase noise

B. Serializer Outputs

Shown in Fig. 9 are the measured 25G-NRZ eye diagrams of the individual MSB and LSB channels. With an 8.0dB pre-emphasis boost enabled to overcome the PCB-channel loss, the measured jitter of the MSB/LSB 25G channels is 1.7ps/11.5ps (rms, pk-pk). Besides, a '0-1-2-3' fixed pattern PAM4 signal is fed into the CDR, and the MSB/LSB NRZ outputs are checked against the input, which verifies the function at 25GBaud/s.

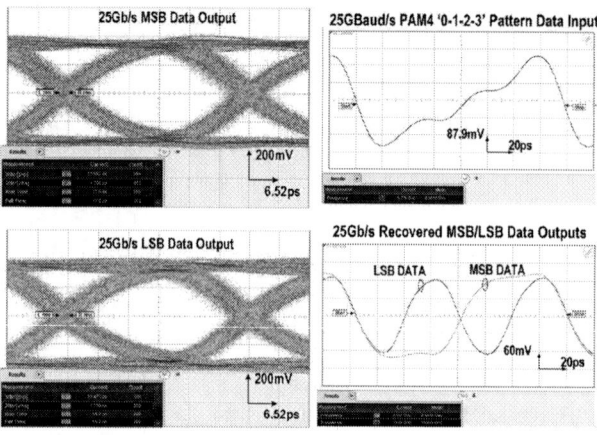

Fig. 9. Measured MSB and LSB output eye-diagrams, as well as the recovered data pattern against the corresponding PAM4 input

C. BER

The BER of the CDR is measured by inputting a 25GBaud/s PAM4 data, while observing at the driver outputs as LSB and MSB data streams. The BER is measured through the above NRZ streams respectively. The recovered BER is error-free for the MSB and 3.4e⁻⁹ for the LSB, respectively (fig. 10). It is believed that the input channel-added high-frequency loss and jitter prevent the LSB data recovery from complete error-free in this design, since no input CTLE is integrated in this work.

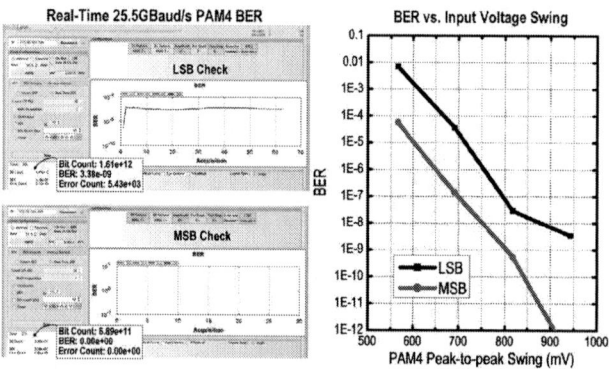

Fig. 10. Measured IRR and statistic auto-calibration results

The performance of this PAM4-CDR is summarized in Table-I. The measured power consumption is 321mW in a standard digital 65nm-CMOS technology. This work is compared with previously published PAM4-CDRs from major conference/journal papers, of which only one prior work [2] has been previously published above 50Gbps-PAM4.

TABLE I. PERFORMANCE SUMMARY AND COMPARISON

	This Work	[2] JSSC2015	[3] JSSC2006	[4] JSSC2015
Data Rate	51Gb/s	56 Gb/s	22 Gb/s	40Gb/s
Modulation	PAM4	PAM4	PAM4	NRZ
DEMUX Ratio	1:4	1:1	1:8	1:4
Jitter of Rec. Clock	1.07ps-rms 8.4ps (pk-pk) at 12.55GHz	532fs,rms; 2.0ps, pk-pk at 28GHz	1.64ps,rms; 13.3ps,pk-pk; at 1.38GHz	N/A
Input Sensitivity	940mV$_{diff}$ (PAM4 output from AWG)	23 mVpp, at BER=10⁻¹²	N/A	N/A
Power Diss. (mW)	Driver: 136mW Phase-Detector: 56.4mW Reference Gen.: 39.6mW VCO+Clk-Buf: 85.2mW I2C: 3.3mW 321mW (total)	420 mW	228 mW	630 mW (RX)
Chip Area	1.36X1.0 mm²	1.6 x 1.0mm²	1.0mm²	2.7 x1.2 mm² (4 lanes)
Technology	65nm Digital CMOS	40nm Digital CMOS	90 nm SOI CMOS	28 nm CMOS
Supply Voltage	1.2 V	1.2 V	1.1 V	N/A

V. CONCLUSION

A quarter-rate 51Gb/s PAM4 CDR with decoded dual-NRZ outputs is presented for 400GbE optical transceivers. A baud-rate data-only sampling PD with zero-crossing integrating front-end is proposed to minimize the clock generation and distribution overhead, as well as improving the noise resilience under PAM4 low-SNR inputs. Measurement results show the CDR features 1.08ps RMS clock jitter, 3.4×10⁻⁹ PAM4 data recovery BER, and 6.27pJ/bit power efficiency.

VI. ACKNOWLEDGEMENTS

This work is supported in part by China's CAS Pioneer Hundred Talents Program, 1000-Talents Foreign Experts program, Shanghai STCSM Science Foundation (15511103103); the National Natural Science Foundation of China under Grant 61534002, 61474025, 61474134, an industry grant from PhotonIC Technologies, Shanghai, and software donations from Integrand Software and Scientific Analog corporations.

REFERENCES

1. J.Zerbe et al., "Equalization and Clock Recovery for a 2.5–10-Gb/s 2-PAM/4-PAM Backplane Transceiver Cell," in IEEE Journal of Solid-State Circuits, vol. 38, no. 12, pp. 2121-2130, Dec 2003.

2. J. Lee et al., "Design of 56 Gb/s NRZ and PAM4 SerDes Transceivers in CMOS Technologies," IEEE J. Solid-State Circuits, vol. 50, no.9, pp. 2061–2073, Sep. 2015.

3. T. Toifl et al., "A 22 Gb/s PAM-4 receiver in 90 nm CMOS SOI technology," IEEE J. Solid-State Circuits, vol. 41, no. 4, pp. 954–965, Apr. 2006.

4. R. Navid et al., "A 40 Gb/s Serial Link Transceiver in 28 nm CMOS Technology," in IEEE J. Solid-State Circuits, vol. 50, no. 4, pp. 814-827, April 2015.

A 15-μW, 103-fs step, 5-bit Capacitor-DAC-based Constant-Slope Digital-to-Time Converter in 28nm CMOS

Peng Chen, Feifei Zhang, Zhirui Zong*, Hao Zheng, Teerachot Siriburanon, Robert Bogdan Staszewski

University College Dublin, Ireland   *Delft University of Technology, the Netherlands

Abstract—This paper proposes a power-efficient capacitor-array-based digital-to-time converter (DTC) using a constant-slope approach. Fringe-capacitor-based digital-to-analog converter (C-DAC) array is used to regulate starting supply voltage of the constant slope fed to a fixed threshold comparator. The proposed DTC consumes only 15 μW from a 1V supply, while achieving fine resolution of 103 fs when running at 40 MHz. The measured INL and DNL are 0.73/0.35 LSB within a 5-bit range. The DTC achieves the best figure-of-merit of 8.5 fJ among state-of-the-art when normalizing the product of power and INL to the product of input frequency and range.

Keywords—digital-to-time converter (DTC); low-power; low voltage; power-efficient; capacitor-based DAC (C-DAC); constant slope; high resolution; INL; PLL.

I. INTRODUCTION

CMOS scaling and configurability of all-digital phase-locked loops (ADPLL) have spurred significant development in its architectures and building blocks. Traditionally, to achieve low in-band phase noise, a fine-resolution time-to-digital converter (TDC) with at least one oscillator period of dynamic range is required [1]. With such a wide range, it is usually one of the most power-hungry building blocks in the ADPLL. Moreover, high linearity is essential in avoiding significant in-band spurious tones. Recent publications show significant interests in the use of a *digital-to-time converter* (DTC) in ADPLLs, e.g., a reference delay technique to reduce the TDC detection range [2, 14], dithering reference phases to suppress spurs in near integer-*N* channels [3], *etc.* Recent publications also adopt DTC for the use in subsampling PLLs or ADPLLs to achieve fractional-*N* operation, which require high resolution with wide dynamic range [4, 5]. The dynamic range of DTC can be relaxed using multi-phase outputs from DCO divider [5]. Furthermore, it can be significantly reduced with an assistance of phase interpolator (PI) in the feedback path (e.g., quadrature divider and 4-bit PI can reduce required dynamic range of a 5 GHz oscillator from 200 ps to 3 ps) [5]. Consequently, the design of low-power DTC with fine resolution and good linearity becomes increasingly important for modern frequency synthesizers.

A conventional DTC is based on a delay of inverter/buffer [2]. However, this approach suffers from limited resolution and high-power consumption due to a large number of delay cells. Recent developments of high-resolution DTCs usually address two main components, *i.e.*, input/output buffers and delay generation part. The input buffer is used to drive the delay generation circuit and the output buffer is used to drive the output load. The delay generation part is digitally controlled to regulate the desired time delay. When comparing different DTC

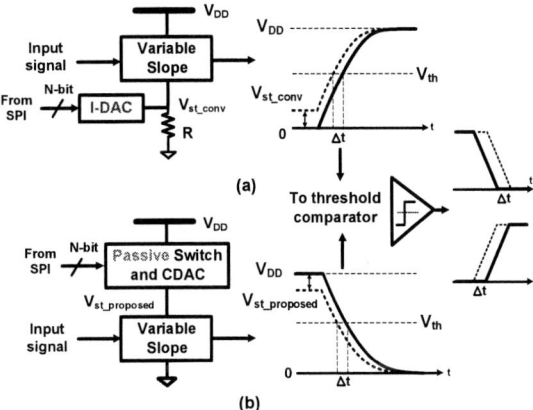

Fig.1. Conceptual diagram of the delay generation circuit in (a) conventional constant-slope DTC using I-DAC [8], (b) proposed passive C-DAC-based constant-slope DTC.

architectures, the main differences are in the delay generation circuits. One popular method to use is a *variable-slope* method, in which the delay is regulated by a variable voltage ramp which drives a threshold comparator. The slope of the voltage ramp is tuned through tunable capacitances and resistances. Even though fine resolution <500 fs can be achieved [4, 7], conventional DTCs based on the *variable-slope* method suffer from poor linearity due to nonlinear relationship between propagation delay and input ramp time [8]. This problem can be suppressed by exploiting a *constant-slope* method, which generates voltage ramps with different starting voltages. As shown in Fig. 1(a), the slope is maintained and it ideally produces a linear relationship for a constant delay after a threshold comparator. Despite the significant improvements in resolution and linearity, the main power contributor in [8] is the current DAC which consumes over 1 mW.

We propose a DTC featuring high resolution and high linearity using a passive DAC in the constant-slope architecture. By exploiting a capacitor-based DAC (C-DAC) as part of the ramp generator, the power consumption is significantly reduced while maintaining comparable performance in terms of resolution and linearity. In this work, a fringe capacitive array is used as the DAC for regulating the supply of ramp generator with high resolution and good matching. The DAC output further drives the threshold comparator to convert the starting voltage experiencing the constant slope drop into a variable delay with extremely fine resolution, as shown in Fig. 1(b). Similar to the case when C-DAC is applied in Successive

978-1-5386-3179-9/17 $31.00 © 2017 IEEE

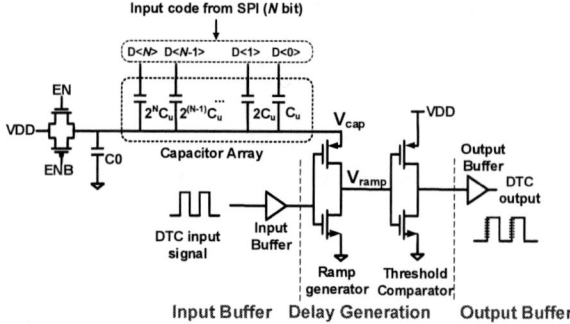

Fig. 2. Proposed power-efficient C-DAC-based digital-to-time converter (DTC)

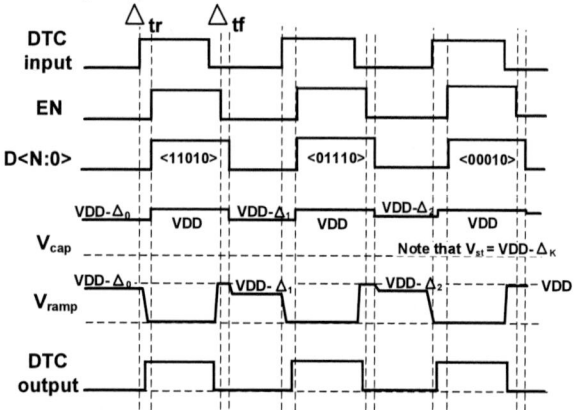

Fig.3. Timing Diagram of the proposed passive C-DAC-based constant-slope DTC at all the corresponding nodes

Approximation Register Analog-to-Digital Converter (SAR-ADC) [9,10], due to its mostly passive structure, it consumes significantly lower power when compared to any of previous DTC architectures while achieving high resolution with good linearity due to good matching of the layout geometry [10].

II. DESIGN OF PROPOSED DIGITAL-TO-TIME CONVERTER (DTC)

Thanks to the constant-slope and charge sharing techniques, the proposed DTC architecture is best suited for ultra-low power (ULP) and low-voltage (LV) environments. In a constant-slope DTC, the delay generation part is implemented using a ramp generator and a threshold comparator, in which the latter is realized based on a simple inverter. In contrast, in [8] the ramp generator is composed of a Current-DAC (I-DAC), a resistor, a pulse generator and a current mirror, as shown in Fig. 1(a), ultimately consuming considerable of power.

In this work, a C-DAC-based constant-slope ramp generator is proposed to modulate the power supply directly, as shown in Fig. 1(b). The output of ramp generator drives the threshold comparator to generate the time delay corresponding to how much the initial voltage deviates from the ideal supply (V_{DD}). Here we use a straightforward CMOS inverter due to its compact size and low power consumption. The detailed schematic of the proposed DTC is shown in Fig. 2. Since the capacitor array and

switch are passive, the inverter cell is the only active power-consuming block in the ramp generator. Instead of using the I-DAC and the resistor to generate the start voltage of ramp generator, as in Fig. 1(a), the proposed design lets the ramp start around the supply voltage (V_{DD}) and then regulates the initial voltage (V_{cap}) through the charge re-distribution technique from the fringe-capacitor array. Energy consumed by switching the input digital code and the enable/disable signal for transmission gate are negligible and the main power consumption only flows from the input of the transmission gate to the ramp inverter, which is also known as the charge-sharing technique. The timing diagram of the proposed DTC is shown in Fig. 2 and it operates as follows:

Specific Timing Considerations: Compared to the conventional DTCs, there is one extra control signal whose timing should be taken care of here, i.e., the control signal EN for the transmission gate. Since it affects the next cycle operation, together with the DTC delay control word, D<N:0>, the rising edge EN should come *after* the DTC input's rising edge. Δt_r is the time to allow the ramp inverter output, V_{ramp}, to drop to ground, thus settling before EN rises. After the transmission gate is enabled, V_{cap} is preset to V_{DD}. After the transmission gate is turned off, the charge stored on the C-DAC capacitor array is $(C_0 + K*C_u)*V_{DD}$ where, C_0 is a fixed capacitance, C_u is the unit capacitance of C-DAC and K represents the number of such chosen units. Δt_f is reserved for the transmission gate to be disabled completely. After D<N:0> is reset to zero, the stored charge is shared over $C_0 + C_a$ (C_a is the *total* capacitance of the capacitor array, i.e., K=max) and V_{cap} drops to $V_{DD} - \Delta_K$, which is the initial voltage for the ramp. When the next DTC input arrives, the falling edge of V_{ramp} drops from $V_{DD} - \Delta_K$ to ground.

III. CIRCUIT IMPLEMENTATION

The proposed DTC contains two main blocks: the input/output buffers and the delay generation block, which consists of the ramp generator and the threshold comparator. In this design, the input buffer is sized relatively large since it is critical for phase noise contribution. On the other hand, the proposed ramp generator is made up of the transmission gate, the capacitor array and the ramp inverter. Their circuit designs are discussed below.

1) C-DAC-based Ramp Generator

The size of transmission gate is carefully chosen to minimize its on-resistance and to reduce settling time for V_{cap} so that it does not limit the operating speed of the DTC.

The implementation of capacitor array of small unit values is important to achieve high power efficiency and high accuracy. In this work, similar to the use in [9], a custom-designed metal-oxide-metal (MOM) capacitor of a small value is used, as shown in Fig. 4. In this design, the unit capacitance is laid out using Metal-2 to Metal-7 with 0.05 μm width and 7.2 μm length. The layout extraction reveals a unit capacitance of 6.3fF. The fixed capacitor (C_0) is built as a MOM capacitor from the PDK with a value of 9.43 pF. The capacitors occupy majority of the core area. The power consumption and area could potentially be reduced with smaller capacitors since it is the ratio between the total capacitance of the capacitor array C_a over the constant capacitance of C_0 that determines the V_{cap} variation range, rather

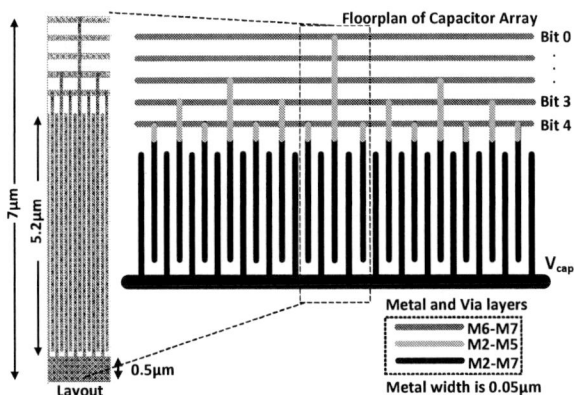

Fig.4. Floorplan and layout of the capacitor array as part of the ramp generator in the proposed DTC

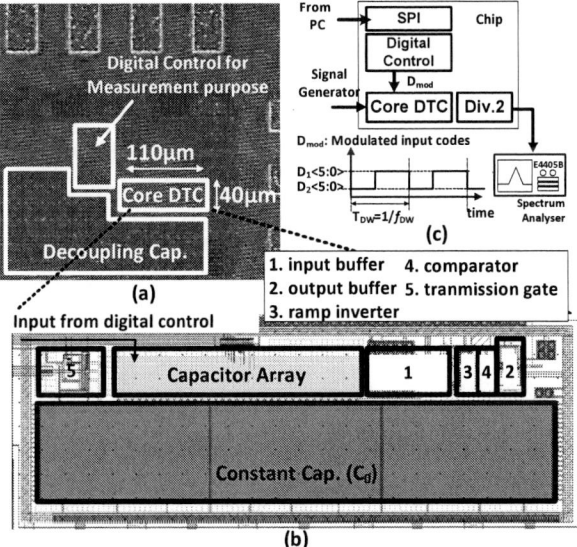

Fig.5. (a) Chip micrograph of the proposed DTC and power consumption of each block (b) layout of core area of the proposed DTC (c) measurement setup

than the absolute capacitor value. The voltage variation range for the V_{cap} is from the minimum initial voltage ($V_{st,min}$) to V_{DD}, whereas $V_{st,min}$ can be determined as:

$$V_{st,min} = \frac{C_O}{C_a + C_O} V_{DD} \qquad (1)$$

In the layout, Metal-1 and Metal-2 are not used in C_0 in order to reduce the substrate interference. The ramp inverter size should not be overly large as to avoid large self-parasitic capacitance. Without a cascade structure in the ramp inverter, the proposed DTC is more suitable for low supply voltage. In the layout, guard-ring is used for noise isolation.

2) Threshold Comparator

The size of the inverter is optimized to slow down the input ramp and to cover the desired dynamic range using a relatively large size. The threshold voltage (V_{th}) of this comparator can be adjusted through sizing of PMOS and NMOS in order to ensure

enough margin from $V_{st,min}$. Moreover, the chosen comparator size can provide enough driving capability to its following output buffer.

IV. EXPERIMENTAL RESULTS

To demonstrate the effectiveness of the proposed DTC idea, it has been implemented in TSMC 28-nm LP CMOS. Fig. 5 shows the die microphotograph and chip core layout. The effective DTC area is 0.004mm². When operating at 1.0V supply, it consumes only 15μW while running at 40MHz. Since the supply is common, power breakdown of each building block is derived from simulations. The input buffer dominates more than half of the total power. The transmission gate consumes 7% of power due to one inverter generating the complementary control signal ENB. Ramp inverter and comparator both consume 11% of power.

To measure the DTC nonlinearity, a sensitive frequency domain method was adopted from [12], and is shown in Fig. 5(c). A digital control block is provided to generate periodic digital control codes modulating the DTC delay. The spurious tone level then corresponds to the resolution. Similar to the method in [12], the DTC output is divided by 2, yielding a 20MHz square wave as the output. The spectrum is measured using Agilent E4405B ESA-E. The relation between the spur level and the relative time can be expressed as:

$$spur_h(f_{Div} \pm f_{DW}) = 20 \log_{10}\left(\frac{\tau_h}{T_{CK}}\right) \text{ [dBc]} \qquad (2)$$

In which, τ_h is the delay difference between two different control codes $D_1<4:0>$ and $D_2<4:0>$; T_{CK} is the period of DTC input clock; f_{Div} is the fundamental frequency of the divided

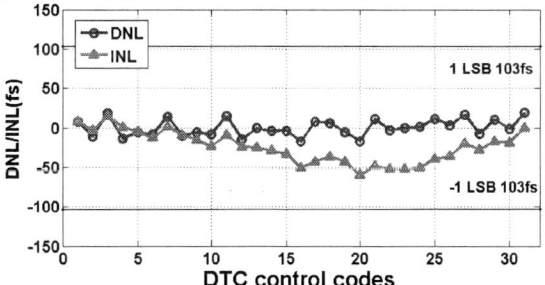

Fig. 6. Measured DNL/INL

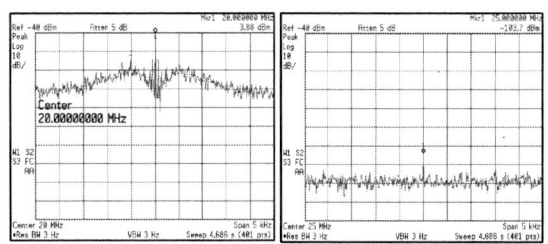

Fig. 7. Measured output spectrum when input frequency is 40MHz and the DTC is being modulated by 1LSB code difference

DTC output; f_{DW} is the frequency of the code waveform. As shown in Fig. 7, referring to the fundamental frequency at 20 MHz, the relative spur level at 5 MHz offset frequency is -107.6 dBc when the DTC control codes are modulated with one

TABLE I: Performance Comparison with The-State-of-the-Art Digital-To-Time Converter (DTC)

	[6] ISSCC11	[7] ISSCC11	[4] ESSCIRC14	[8] JSSC15	[13] JSSC16	[11] VLSI06	[2] ISSCC14	This work
Method	Variable Slope	Variable Slope	Variable Slope	Constant Slope (I-DAC)	Interpolation	Variable Threshold	Buffer-based	Constant Slope (C-DAC)
Technology (nm)	65	65	28	65	28	90	40	28
Supply Voltage (V)	1.1	1.2	0.9	1.2	1.1	1.0	1	1.0
Resolution (fs)	4700	241-330	550	19	244	1000	21500	103
Number of bits	5.3	10	10	10	11	6	6	5
INL (fs)	1900	3000	990	64	1200	3200	67600	75
Power (mW)	>0.22@ 48MHz	2.2@ 40MHz	0.5@ 40MHz	0.8+1.0@ 55MHz	19.8@2GHz	N/A	0.0137@ 32MHz	**0.015@ 40MHz**
FoM[1](fJ)	46.8	541.0	22.0	107.7	23.8	NA	21.4	**8.5**

[1] FoM=Power*INL/(Freq*Range), only core DTC is considered

LSB step difference. This corresponds to a 104.2 fs delay step. The measured DNL and INL are 0.35 and 0.73 LSB with 103 fs resolution, respectively, as shown in Fig. 6.

The measured phase noise of the divided DTC output shows an integrated jitter of 1209 fs, which is limited by the input buffer, as per post-layout simulations. Avoiding the use of cascode MOSFETs in the proposed DTC makes it more suitable to a low supply voltage. It is verified through measurements that when operating at 0.76 V, the DTC consumes only 8.6 μA (i.e., 6.5 μW).

The key performance metrics of the proposed DTC are compared in Table I with state-of-the-art DTCs. It can be observed that the constant-slope DTCs can achieve fine resolution and excellent linearity. In this work, the proposed DTC maintains such performance while consuming significantly less power. To fairly compare the proposed work with other publications, the INL of the DTC is normalized to the product of input frequency and range as a figure-of-merit (FoM). Note that in [8], INL is normalized to the range but power consumption and input frequency have not been included. As a result, the proposed DTC achieves an FoM of 8.5fJ, which appears to be the best to our knowledge. As the very first demonstration of the proposed idea, the range only covers 5 bits, thus limiting its applications to a few cases, such as dithering and PI assisted synthesizer [5]. However, the C-DAC range can be easily extended to more bits, e.g. 8 bits, thus showing good potential for widened use in modern low-power frequency generation.

V. Conclusions

In this paper, we have proposed a high-resolution and ultra-low-power DTC. When operating at 1V, it achieves 103fs resolution at 15μW drained power. Benefiting from the constant slope operation, the DTC achieves 75fs INL. The proposed DTC makes use of a simple inverter structure with scalable passive capacitor array as a DAC. It can offer high resolution with good linearity while consuming extremely low power which is suitable for the Internet-of-Things (IoT) applications.

Acknowledgment

The authors acknowledge Science Foundation Ireland (SFI) and MCCI for their support, and TSMC University Shuttle for chip fabrication.

References

[1] R. B. Staszewski et. al., "All-Digital TX Frequency Synthesizer and Discrete-Time Receiver for Bluetooth Radio in 130-nm CMOS," IEEE JSSC, pp 2278-2291, Dec. 2004.

[2] V. Chillara et. al., "An 860 μW 2.1-to-2.7 GHz All-Digital PLL-Based Frequency Modulator with a DTC-Assisted Snapshot TDC for WPAN (Bluetooth Smart and ZigBee) Applications", IEEE ISSCC, 2014, pp. 172-173.

[3] G. Marzin, et al., "A 20 Mb/s Phase Modulator Based on a 3.6 GHz Digital PLL With -36 dB EVM at 5 mW Power", IEEE JSSC, pp. 2974-2988, Dec. 2012.

[4] N. Markulic, et al., "A 10-bit, 550-fs step digital-to-time converter in 28 nm CMOS," IEEE ESSCIRC, 2014, pp. 79-82.

[5] A. Narayanan, et al., "A Fractional-N Sub-Sampling PLL using a Pipelined Phase-Interpolator with an FoM of -250dB," IEEE JSSC, pp. 1630-1640, Jul. 2016.

[6] N. Pavlovic and J. Bergervoet, "A 5.3 GHz digital-to-time-converter-based fractional-N all-digital PLL," IEEE ISSCC, 2011, pp. 54-56.

[7] D. Tasca, et al., "A 2.9-to-4.0 GHz fractional-N digital PLL with bang-bang phase detector and 560 fsrms integrated jitter at 4.5 mW power," IEEE ISSCC, 2011, pp. 2745-2758.

[8] Z. Ru, et al., "A High-Linearity Digital-to-Time Converter Technique: Constant-Slope Charging," IEEE JSSC, pp. 1412-1423, Jun. 2015.

[9] P. Harpe, et al., "A 26 μW 8 bit 10 MS/s Asynchronous SAR ADC for Low Energy Radios," IEEE JSSC, pp. 1585-1595, Jul. 2011.

[10] M. Saberi, et al., "Analysis of Power Consumption and Linearity in Capacitive Digital-to-Analog Converters Used in Succesive Approximation ADCs," IEEE TCAS-I, pp. 1736-1748, Aug. 2011.

[11] K. Inagaki, et al., "A 1-ps resolution on-chip sampling oscilloscope with 64:1 tunable sampling range based on ramp waveform division scheme," IEEE VLSI-C, 2006, pp. 61-62.

[12] C. Palattella, et al., "A sensitive method to measure the integral nonlinearity of a digital-to-time converter, based on phase modulation," IEEE TCAS-II, pp. 741-745, Aug. 2015.

[13] S. Sievert, et al. "A 2GHz 244 fs-Resolution 1.2 ps-Peak-INL Edge Interpolator-Based Digital-to-Time Converter in 28nm CMOS," IEEE JSSC, pp. 2992-3004, Dec. 2016.

[14] Y. Wu, et al., "A 3.5–6.8GHz wide-bandwidth DTC-assisted fractional-N all-digital PLL with a MASH ΔΣ TDC for low in-band phase noise," IEEE ESSCIRC, 2016, pp. 209-212.

A 173–200 GHz Quadrature Voltage-Controlled Oscillator in 130 nm SiGe BiCMOS

Paul Stärke, Vincent Rieß, Corrado Carta and Frank Ellinger
Chair for Circuit Design and Network Theory
Technische Universität Dresden
01069 Dresden, Germany

Abstract—**This work presents a fundamental-mode voltage-controlled oscillator (VCO) realized in a 130 nm SiGe BiCMOS process for operation up to 200 GHz. The implemented topology is derived from a feedback phase shifter design, which includes a wideband passive polyphase filter based on 90° and 180° transmission line couplers and a variable-gain active combiner. The fabricated circuit achieves a very large frequency tuning range of 173–200 GHz with a phase noise down to −85 dBc/Hz at 10 MHz offset around 188 GHz. As an unique feature this topology allows the simultaneous generation of 90° quadrature signals, which are amplified by an additional single-stage output buffer. The power level of both output signals over a wide bandwidth is −3 dBm each, with a total dc power consumption of 27.7 mW for the oscillator core and 10.3 mW per buffer.**

Index Terms—**voltage-controlled oscillator (VCO), BiCMOS, SiGe, phase shifter, hybrid, balun, coupler, splitter, quadrature**

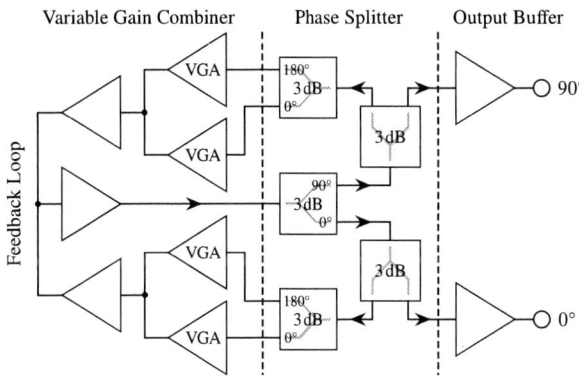

Fig. 1. Block diagram of the proposed VCO with signal flow direction

I. INTRODUCTION

For many radar, communication and measurement systems the frequency synthesis with low noise and spurious content is important for an optimal performance. For this purpose voltage-controlled oscillators (VCO) in combination with a phase-locked loop and an fixed high-quality reference are used. In the mm-wave region this approach becomes more and more difficult as the available gain per transistor decreases, while the overall losses increase.

Common VCO designs, like the Colpitts or cross-coupled topology, rely on some form of tunable reactance, e.g. an LC resonator with varactor [1] or an actively-tuned inductor [2]. While the achievable phase noise performance is usually good, the frequency range is limited by the tuning range of the involved components. As a possible solution to this problem, this work presents an approach for realizing a fundamental oscillator with a very large frequency tuning range. The design is based on positive feedback over a wideband active phase shifter, realized with passive coupled line structures.

The circuit is implemented in the most advanced SiGe BiCMOS process of IHP, the SG13G2, which offers HBTs with a maximum frequency of oscillation of up to 500 GHz. The RF back-end of the process enables the realization of low-loss passive structures, including MIM capacitors and wide transmission lines.

II. OSCILLATOR DESIGN

A. Principle of Operation

The basic principle of operation of the presented circuit relies on connecting the output of a phase shifter to its input

and tune the loop phase to achieve positive feedback at the desired oscillation frequency. The corresponding block diagram is shown in Fig. 1. For this particular design it is necessary to generate four signals with a 90° shift between each other to allow a full phase shift of 360°. A continuous tuning range can be achieved then by weighting and combining the paths accordingly. However, dividing the source signal by four causes a loss of at least 6 dB, which must be compensated. An additional loss of 3 dB is necessary to decouple the oscillator core from the output buffer. To allow oscillation the loop gain must exceed unity, therefore additional amplification of at least 9 dB is needed.

At first the feedback signal is divided by a 90° hybrid, where two power splitters then forward half of the energy of each path to an output buffer, while the other half is divided again by a 180° hybrid. The following active variable-gain combiner (VGC), consisting of three stages with four variable-gain amplifiers (VGA) and three buffer amplifiers, superposes the weighted phase components and forwards it to the input again to close the loop. As an additional feature this oscillator is able to generate a pair of signals in quadrature with the power split after the 90° hybrid.

B. Circuit Implementation

1) Phase Generation: The most important components for this VCO topology are the wideband hybrids. For the 90° coupler the commonly used design of Lange is implemented [3]. The advantage is a straightforward solution with very good phase accuracy and amplitude balance, as well as a low area

978-1-5386-3179-9/17 $31.00 © 2017 IEEE

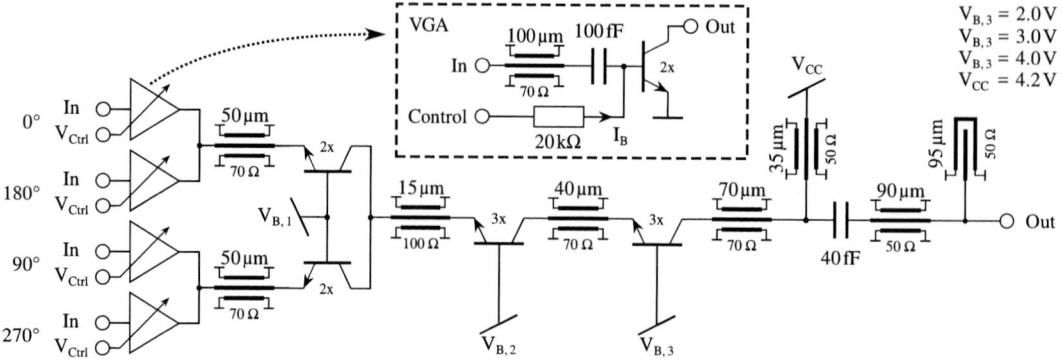

Fig. 2. Schematic of the variable-gain combiner with variable-gain amplifiers, common base buffers and matching networks – the bias voltages $V_{B,1...3}$ are generated by an additional dc network and decoupled with large on-chip capacitors

requirement. The final design is numerically optimized by EM simulations and shows an insertion loss of around −4 dB per path, with an imbalance in amplitude below 0.1 dB and in phase below 2° between 140 GHz and 220 GHz.

More challenging is the realization of a 180° coupler. Common approaches include Marchand baluns and transformers. However, both show disadvantages in regard of balance, area requirement and losses. A very interesting alternative is presented by Cho using three coupled lines [4]. This approach is adapted to the used process and optimized numerically by EM simulations as well. The final design shows an insertion loss of around −3.7 dB per path, while the bandwidth and imbalances are comparable with the Lange coupler.

For the power splitter a simple Wilkinson divider is used with an insertion loss of −3.6 dB per path. The total loss of the phase generation network is therefore 11.3 dB, which is only 2.3 dB more than for the ideal case.

2) Variable-Gain Combiner: For the active VGC a dc coupled transistor stack is implemented as shown in Fig. 2. The variable-gain capability is realized by tuning the base current of the common-emitter (CE) VGA stages with the control voltages $V_{\text{Ctrl},0...270}$. The inputs of the CE stages and the subsequent common-base (CB) stages are matched with series transmission line elements to increase the total available gain, while the output is connected with a wideband network to the Lange–coupler.

For the performance of the VGC it is beneficial if the total bias current only varies slightly over the target phase shift φ. Under the assumption of a linear dc current gain, this can be achieved if the base currents $I_{\text{B},0...270}$, which are linearly dependent on the control voltages $V_{\text{Ctrl},0...270}$ due to the large series resistance, follow a squared sinusoidal dependency:

$$I_{\text{B},0} = \frac{I_{\text{B,max}}}{4}(1+\cos\varphi)^2 \qquad I_{\text{B},90} = \frac{I_{\text{B,max}}}{4}(1+\sin\varphi)^2$$
$$\text{(1a)} \hspace{4cm} \text{(1c)}$$

$$I_{\text{B},180} = \frac{I_{\text{B,max}}}{4}(1-\cos\varphi)^2 \qquad I_{\text{B},270} = \frac{I_{\text{B,max}}}{4}(1-\sin\varphi)^2$$
$$\text{(1b)} \hspace{4cm} \text{(1d)}$$

$I_{\text{B,max}}$ is the maximum base current for a single CE stage and is set to result in a total supply current of $I_{\text{CC}} = 6.6$ mA. Fig. 3

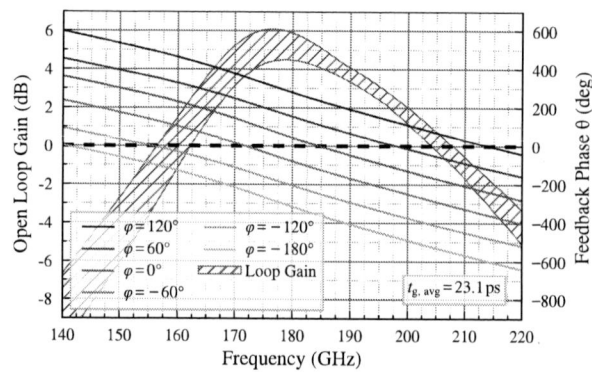

Fig. 3. Simulated envelope of the open loop gain and the corresponding feedback phase θ for an phase shift φ between −180° and 120°

shows the simulation results of the full open loop, including EM models for all passive structures, at different values of φ. The loop gain varies slightly by 1.6 dB with a peak between 4.4 dB and 6.0 dB and an unity bandwidth B_{unity} between 160 GHz and 205 GHz. The group delay around the center is almost independent of φ, with a value of

$$t_{\text{g}} = -\frac{d\theta_{\text{rad}}}{d\omega} = -\frac{1}{360°}\frac{d\theta_{\text{deg}}}{df} \approx 23.1\,\text{ps}. \qquad (2)$$

To ensure that the oscillation condition is always satisfied for exactly one frequency, B_{unity} must lie within a full 360° change of the feedback phase θ. This correlation can be approximated, under the assumption of a constant group delay, as

$$B_{\text{unity}} \leq B_{360} \approx t_{\text{g}}^{-1} = 43.3\,\text{GHz}. \qquad (3)$$

3) Output Buffer: For the output buffers a simple cascode stage with approximately 9 dB gain and up to −3 dBm output power is implemented. Due to the Wilkinson divider the feedback to the oscillator core is very small. It is therefore possible to independently operate the two buffers, including turning one off completely.

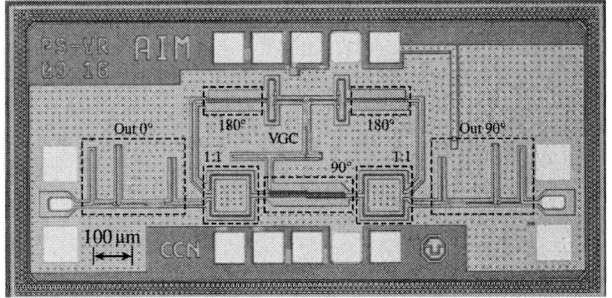

Fig. 4. Photograph of the fabricated chip with a size of 0.65 mm × 1.44 mm – the passive and active functional blocks are highlighted

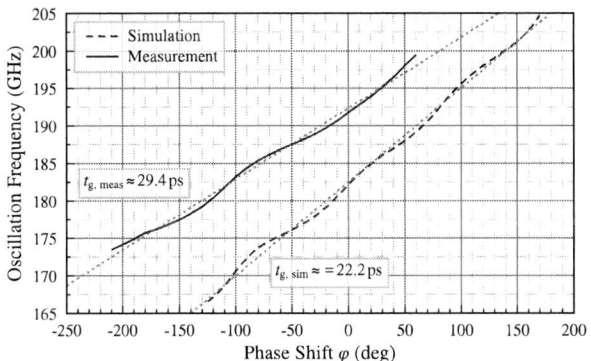

Fig. 5. Simulated and measured oscillation frequency over the phase shift φ – the group delay of the oscillator core is derived from a linear fit

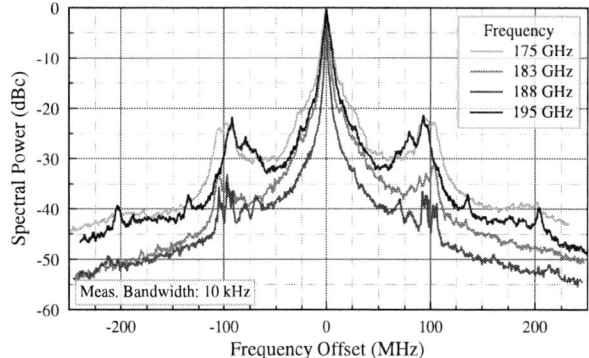

Fig. 6. Measured spectrum, characterized with 10 kHz resolution bandwidth over 500 MHz, for 175 GHz, 183 GHz, 188 GHz and 195 GHz

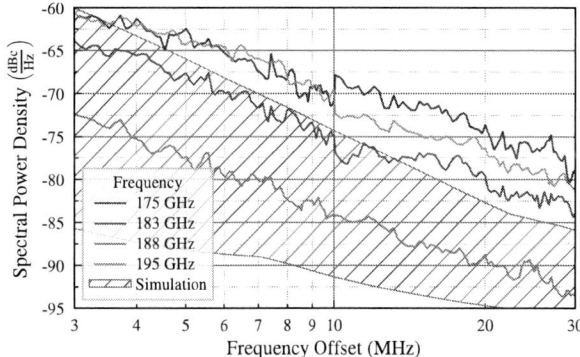

Fig. 7. Simulated envelope of the phase noise and measurement results for 175 GHz, 183 GHz, 188 GHz and 195 GHz within 3 MHz to 30 MHz

III. EXPERIMENTAL RESULTS

The fabricated chip has a size of 0.65 mm × 1.44 mm = 0.94 mm², including the pads with a pitch of 100 µm, while the oscillator core itself occupies only 0.24 mm². A chip photograph is given in Fig. 4.

The measured operating point is 6.6 mA at 4.2 V supply for the oscillator core and 4.3 mA at 2.4 V supply per output buffer (second buffer turned off), resulting in a total power consumption of 27.7 mW + 10.3 mW = 38 mW.

The following frequency measurements are performed with a spectrum analyzer and a G-band subharmonic mixer, using the 14th harmonic for the down-conversion. For the power measurement a wideband waveguide calorimeter is used. The insertion loss of the contact probe and the waveguide feed are deembedded with a corresponding set of s-parameters.

The simulated and measured oscillation frequency versus the phase shift is shown in Fig. 5. The simulated frequency range of 40 GHz is close to the unity bandwidth of the open loop. In the measurement the achieved bandwidth is reduced to 27 GHz, which is likely caused by a slightly lower loop gain.

As the group delay and the frequency tuning range are correlated according to (3), the group delay can also be approximated from the performance of the closed loop as well. For the simulation a value of 22.2 ps is calculated, which agrees with the previous result of 23.1 ps from (2). However, in the measurement a significant larger value of 29.4 ps is observed.

The difference of 6.8 ps corresponds to an additional electrical length of over 1 mm (dielectric SiO_2 with $\varepsilon_r - 3.9$). An inaccuracy in the passive component model is therefore unlikely. This would also imply a significant impact on the phase generation network, which is not observed. The actual cause is not completely understood, but one possibility, evaluated by simulations, is a slower performance of the transistors, which would increase the capacitive delay for each stage.

The measured spectrum for four frequency settings is shown in Fig. 6. The spurious peaks around 100 MHz offset are caused by the down-conversion process. Due to the comparably high amount of phase noise the spectrum analyzer is not able to eliminate all higher-order mixing products completely.

The measured phase noise at the same four frequency points is shown in Fig. 7. For comparison the simulated envelop over the complete tuning range is included as well. At 10 MHz offset the phase noise reaches a value down to −85 dBc/Hz. However, close to the frequency limits, the noise characteristic worsens by up to 15 dB, because the loop gain reaches unity and the losses become dominant. In general the measured values are about 5 dB above the simulations.

The measured and simulated power levels at the 0° output are shown in Fig. 8. Up to 195 GHz the values are nearly constant around −3 dBm, indicating that the buffer is always in saturation and independent of the oscillator state. Near the

TABLE I
COMPARISON OF RECENT OSCILLATOR DESIGNS FOR MM-WAVE APPLICATIONS

Source	[This]	[1]	[2]	[5]	[6]	[7]	[8]
Technology	130 nm SiGe	130 nm SiGe	130 nm SiGe	120 nm SiGe	65 nm CMOS	32 nm CMOS	250 nm InP
Topology	Phase Shifter	Cross-Coupled	Push-Push	Push-Push	Push-Push	Cross-Coupled	Cross-Coupled
Frequency Range	173 to 200 GHz	184 to 197 GHz	198 to 205 GHz	309 to 330 GHz	206 to 220 GHz	205 to 213 GHz	248 to 262 GHz
Relative Bandwidth[a]	14.5 %	6.4 %	3.5 %	6.6 %	6.6 %	3.8 %	2.3 %
Phase Noise[b]	−85 dBc/Hz	−91 dBc/Hz	−103 dBc/Hz	−78 dBc/Hz	−81 dBc/Hz	−98 dBc/Hz	−88 dBc/Hz
Power Consumption	38 mW	16 mW	30 mW	63 mW	43 mW	42 mW	85 mW
Output Power	−3.0 dBm	−0.8 dBm	−7.2 dBm	−13.3 dBm	−1.6 dBm	−13.5 dBm	2.9 dBm
Power Efficiency[c]	1.32 %	5.13 %	0.64 %	0.07 %	1.60 %	0.11 %	2.29 %
Figure of Merit[d]	−157 dB	−160 dB	−158 dB	−133 dB	−146 dB	−146 dB	−155 dB

[a] $B_{\text{rel}} = \frac{\Delta f_{\text{osc}}}{f_{\text{center}}}$ [b] $\mathscr{L}(f_{\text{Noise}})$ at 10 MHz [c] $\eta_{\text{dc}} = \frac{P_{\text{out}}}{P_{\text{dc}}}$ [d] $FoM_{\text{T}} = \mathscr{L}(f_{\text{Noise}}) \left(\frac{f_{\text{Noise}}}{10 \Delta f_{\text{osc}}}\right)^2 \eta_{\text{dc}}$

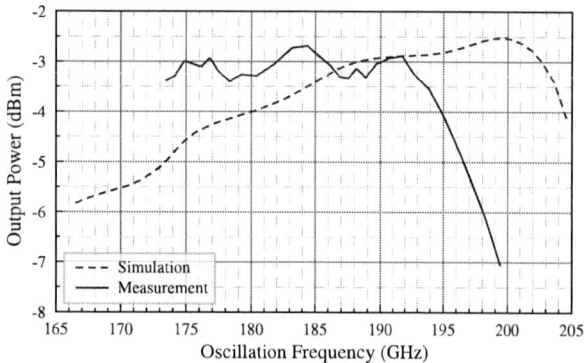

Fig. 8. Simulation and measurement results of the power level at the 0° output over the oscillation frequency – interconnect losses are deembedded

upper frequency limit at 200 GHz they decrease rapidly. This is likely caused by a diminishing loop gain, as well as some mismatch at the output stage. To allow a comparison with different designs, the oscillator was operated single-ended by turning off the 90° buffer to reduce the power consumption.

To compare different oscillator performances it is beneficial to use a figure of merit (FoM). A common definition is FoM_{T} [2], which weights the phase noise against the frequency tuning range and dc efficiency and yields a value of −157 dB (lower values are better). Table I gives an overview of recent VCO designs and their performance. Compared to the state-of-the-art the proposed VCO design performs very well with high output power, a good efficiency and an excellent tuning range.

IV. CONCLUSION

A voltage-controlled oscillator was presented using a topology based on a feedback phase shifter design. To the authors best knowledge, this is first example of a successfully realization operating beyond 100 GHz. While the phase noise characteristic, with −85 dBc/Hz at 10 MHz offset, is comparable to other designs, the main feature of this approach is its very large frequency tuning range between 173 GHz and 200 GHz

corresponding to a relative bandwidth of 14.5 %. By increasing the loop gain and reducing the group delay even higher values seem achievable.

As additional feature this oscillator allows the simultaneous generation of quadrature signals with identical amplitude and symmetric output characteristic. The single-ended output power is −3 dBm with a dc power consumption of 38 mW, resulting in a power efficiency of 1.32 %.

ACKNOWLEDGEMENT

This work has been supported by the German Research Foundation (DFG) within the frame of the projects Automatic Impedance Matching (AIM) and 3D-LommID, as well as by the German Federal Ministry of Education and Research (BMBF) within the frame of the project FAST-SPOT of the Zwanzig20 program FAST.

The authors also thank the Center for Information Services and High Performance Computing (ZIH) at Technische Universität Dresden for generous allocations of compute resources.

REFERENCES

[1] D. Fritsche, S. Li, N. Joram, C. Carta, and F. Ellinger, "Design and Characterization of a 190-GHz Voltage-Controlled Oscillator," in *2016 46th European Microwave Conference (EuMC)*, Oct. 2016, pp. 493–496.

[2] P. Y. Chiang, O. Momeni, and P. Heydari, "A 200-GHz Inductively Tuned VCO With -7-dBm Output Power in 130-nm SiGe BiCMOS," *IEEE Trans. Microw. Theory Tech.*, vol. 61, no. 10, pp. 3666–3673, Oct. 2013.

[3] J. Lange, "Interdigitated Strip-Line Quadrature Hybrid," in *1969 G-MTT International Microwave Symposium*, May 1969, pp. 10–13.

[4] C. Cho and K. C. Gupta, "A new design procedure for single-layer and two-layer three-line baluns," *IEEE Trans. Microw. Theory Tech.*, vol. 46, no. 12, pp. 2514–2519, Dec. 1998.

[5] S. P. Voinigescu *et al.*, "A Study of SiGe HBT Signal Sources in the 220-330-GHz Range," *IEEE J. Solid-State Circuits*, vol. 48, no. 9, pp. 2011–2021, Sep. 2013.

[6] P. H. Chiang *et al.*, "A 206-220 GHz CMOS VCO using body-bias technique for frequency tuning," in *2015 IEEE MTT-S International Microwave Symposium*, May 2015, pp. 1–3.

[7] Z. Wang, P. Y. Chiang, P. Nazari, C. C. Wang, Z. Chen, and P. Heydari, "A CMOS 210-GHz Fundamental Transceiver With OOK Modulation," *IEEE J. Solid-State Circuits*, vol. 49, no. 3, pp. 564–580, Mar. 2014.

[8] J. Yun, N. Kim, D. Yoon, H. Kim, S. Jeon, and J. S. Rieh, "A 248-262 GHz InP HBT VCO with Interesting Tuning Behavior," *IEEE Microw. Wirel. Compon. Lett.*, vol. 24, no. 8, pp. 560–562, Aug. 2014.

A 67 GHz Dual Injection Quadrature VCO with -182.9 dBc/Hz FOM in 90-nm CMOS

Cuei-Ling Hsieh, Hong-Shen Chen, Hou-Ru Pan, and Jenny Yi-Chun Liu

Department of Electrical Engineering
National Tsing Hua University, Taiwan
Email: concertohsieh@gmail.com, s102061809@m102.nthu.edu.tw, s101323013@mail1.ncnu.edu.tw, and jennyliu@gapp.nthu.edu.tw

Abstract—An ultra low phase noise millimeter-wave (mm-wave) quadrature voltage-controlled oscillator (QVCO) is proposed. By introducing an additional coupling path and using uneven sized cross-coupled oscillators with a fully symmetrical layout, the phase noise and the amplitude/phase mismatches are minimized. Implemented in 90-nm CMOS process, the 67-GHz compact QVCO achieves 0.46 degree and 0.47 dB phase and amplitude mismatches, respectively, and a phase noise of -105.8 dBc/Hz at 1-MHz offset, resulting in a figure-of-merit of -183 dBc/Hz.

Keywords—CMOS, millimeter-wave, QVCO, phase noise, V-band.

I. INTRODUCTION

The demands for multi-Gb/s wireless communication have increased rapidly in recent years. The unlicensed frequency spectrum around 60 GHz provides opportunities to satisfy such demands due to its wide bandwidth. To fully utilize the bandwidth, complex modulation schemes are adopted that requires the generation of low phase noise quadrature LOs. Millimeter-wave phased array transceivers also benefit from quadrature LO generation. However, the generation of low phase noise quadrature signals is challenging. Several solutions are proposed. A fundamental oscillator with hybrids or polyphase filters can be used at the cost of large power consumption from the buffers required to compensate for the passive loss especially at mm-wave frequencies [8]-[9]. An oscillator at twice the target frequency ($2f_{osc}$) followed by a divider produces quadrature outputs; however, designing oscillator and divider at $2f_{osc}$ is challenging especially with f_{osc} close to f_{max} ($f_{max} \sim 135$ GHz in this work), and it is power hungry and occupies a large area due to the increased number of inductors [10]. Similarly, an oscillator at a fraction of the target frequency (f_{osc}/N) with an injection locked multiplier is another possibility with a wide tuning range, but it consumes a large area and has poor phase error due to the imbalanced structure [1].

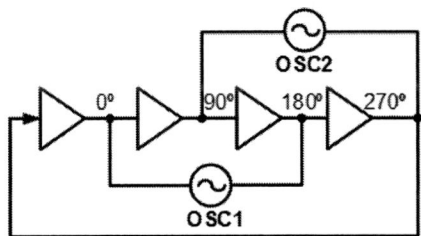

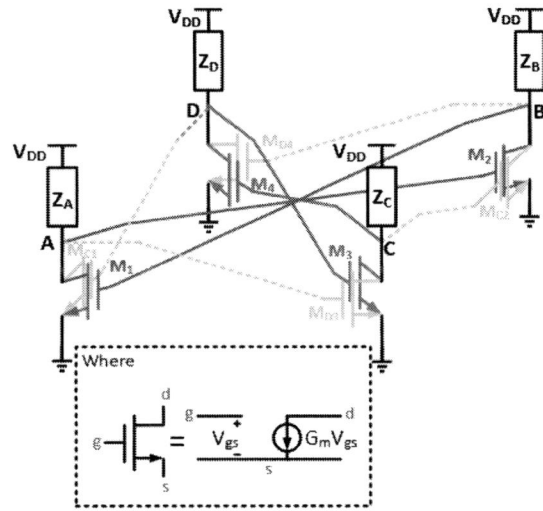

Fig. 1. Block diagram and circuit schematic of conventional P-QVCO.

Coupling two identical VCOs through a coupling network is straightforward. Various coupling means have been proposed including parallel coupling (P-QVCOs), serial coupling (S-QVCOs), in-phase injection coupling [2], and transformers-based coupling [3]-[5]. Considering the limited headroom, the loss and design complexity of transformers, we present a low cost QVCO based on conventional P-QVCO with a new dual injection technique and uneven cross-coupled pair selection. With the above mentioned techniques and a symmetrical ring layout, the proposed QVCO achieves low phase noise and quadrature imbalance. Section II discusses the P-QVCOs and the details of the proposed QVCO. Measurement results are presented in Section III. Conclusions are drawn in Section IV.

II. PROPOSED QVCO DESIGN AND ANALYSIS

A. Proposed Oscillator Architecture

Parallel-QVCO is one of the most common quadrature oscillator architectures. Its block diagram and corresponding circuit schematic are shown in Fig. 1. Two cross-coupled VCOs (M_1/M_2 and M_3/M_4) are coupled through a coupling network consisting of four transistors (M_{C1}, M_{C2}, M_{D3}, and M_{D4}) to form quadrature signals. Without the loss of generality, assume the transconductances of M_1-M_4 are equal to G_{mo} and the

978-1-5386-3179-9/17 $31.00 © 2017 IEEE

transconductances of M_{D1}, M_{D2}, M_{C3}, and M_{C4} are equal to G_{mi}. The load impedances Z_A-Z_D have an impedance of Z_L, which is the equivalent impendence of the loaded RLC tank. The oscillation frequency ω_0 of this P-QVCO is

$$\omega_{osc1} = -\left(\frac{1}{2RC} \times \frac{G_{mi}}{G_{mo}}\right) \pm \sqrt{\left(\frac{1}{4C^2R^2}\left(\frac{G_{mi}}{G_{mo}}\right)^2 + \frac{1}{LC}\right)} \quad (1)$$

,where $\omega_0 = \sqrt{\frac{1}{LC}}$ is the self-resonant frequency of the tank. This equation shows that the oscillation frequency deviates from the self-resonance of the tank.

In comparison to P-QVCO, the block diagram of the proposed QVCO is shown in Fig. 2. From Fig. 2, compared to P-QVCO, the proposed architecture introduces one extra coupling path in the reverse direction to the original coupling path that strengthens the coupling between two VCOs, which provides a reverse anti-phase injection, and facilitates a symmetrical implementation. The circuit schematic and the layout placement of the MOS transistors are shown in Fig. 3 and Fig. 5. With the assumption that the transconductances of M_{C1}, M_{C2}, M_{D3}, and M_{D4} are equal to G_{mr}, the oscillation frequency of proposed QVCO is:

$$\omega_{osc2} = \left(\frac{1}{2RC} \times \frac{G_{mi}-G_{mr}}{G_{mo}}\right) \pm \sqrt{\left(\frac{1}{4C^2R^2}\left(\frac{G_{mi}-G_{mr}}{G_{mo}}\right)^2 + \frac{1}{LC}\right)} \quad (2)$$

Compared to (1), by careful design, $G_{mi} - G_{mr}$ can be minimized to reduce the deviation of the oscillation frequency of the QVCO from the self-resonant frequency of the tank.

Being fully symmetrical, two switching pairs (M_1/M_2 and M_3/M_4) and their load inductors (L_D, 37 pH) form the switching quadruplet. The width of the switching transistors is 5 μm with the minimum length to guarantee oscillation startup. Two VCOs are coupled by two differential paths, comprised of two parallel coupling transistor sets, $M_{C1}/M_{C2}/M_{D3}/M_{D4}$ and $M_{C3}/M_{C4}/M_{D1}/M_{D2}$. The widths of these two sets of coupling transistors are 5 μm and 35 μm, respectively. There exists an optimum size ratio for oscillation frequency, phase noise, power dissipation, output power, and quadrature accuracy. As denoted in the schematic, M_{Ci} and M_{Di} inject differential signals to their quadrature counterpart, M_i, (i = 1, 2, 3, 4), e.g. Q_+ and Q_- inject I_+. To simplify signal routing, M_{Ci}, M_{Di}, and M_i share one current source controlled by V_{ctrl}. More importantly, this arrangement shows better immunity to component mismatches and further relaxes the tradeoff between phase noise and phase error [6].

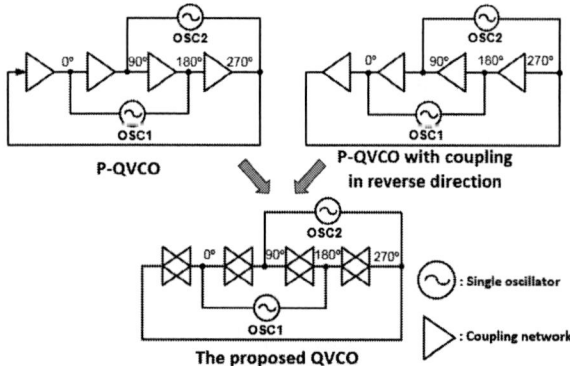

Fig. 2. Block diagrams of P-QVCO and the proposed QVCO.

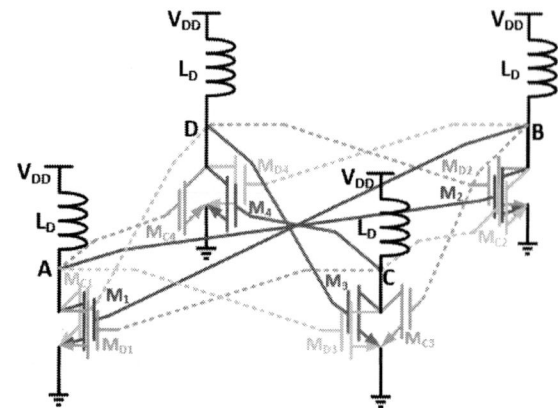

Fig. 3. Schematic of the proposed QVCO.

B. Phasor Analysis

Proposed in this work, an extra coupling network is designed to strengthen the coupling of two VCOs and introduce an opposite phase shift with anti-phase injection. The schematic and phasor of the proposed QVCO are shown in Fig. 4. In Fig. 4, I_i (i= 1, 2, 3, 4) represents the drain current of each MOS transistor in two cross coupled oscillators, and I_{Ci} and I_{Di}, (i= 1, 2, 3, 4) are the drain current of coupling MOS transistors. I_{ZA} is the total current flowing through the impendence Z_A. The Phase noise is minimized when the oscillation frequency is the same as the self-resonant frequency, ω_0, of the tank. However, in P-QVCO, the oscillation frequency deviates from the resonant frequency of the tank that inevitably degrades the phase noise. From (1), the frequency deviation is $\Delta\omega = \frac{\omega_0}{2Q_{tank}}\tan^{-1}\frac{I_{CP}}{I_{SW}}$, where Q_{tank} is the quality factor of the tank, and I_{CP} and I_{SW} are the drain currents of the coupling and switching transistors, respectively. This induces the oscillation frequency of P-QVCO varied from ω_0 to ω_{osc1}. With the proposed dual coupling, the total coupling current I_{CP} is formed by two differential currents, $I_{CP} = I_C + I_D$, where I_C and I_D are the drain currents of the differential coupling transistors, respectively. The angle between the phasors of I_{CP} and I_{SW} can be minimized such that the oscillation frequency is moved closer to the resonance of the tank. As a result, the oscillation frequency of proposed circuit will now be ω_{osc2}, as shown in Fig. 4. Compared with ω_{osc1}, ω_{osc2} has higher tank quality factor, which implied that the phase noise will be better in the proposed QVCO.

C. Implementation

As the quadrature errors are mainly due to the mismatches of the tanks, the interconnects, and the coupling networks in practical implementation, a symmetrical ring layout topology is proposed as shown in Fig. 5. Four identical sections are placed at the four sides of the ring. Each section consists of one switching transistor (M_i), and two coupling transistors (M_{Ci} and

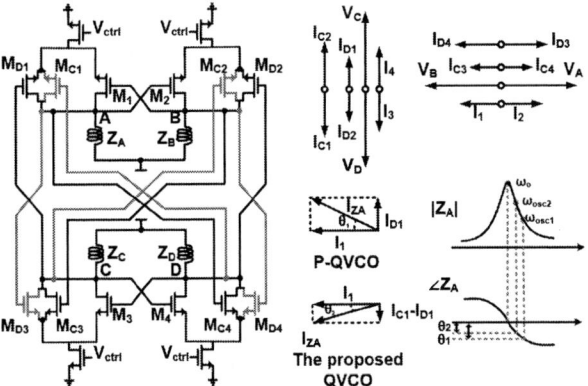

Fig. 4. Phasor diagram of the proposed QVCO.

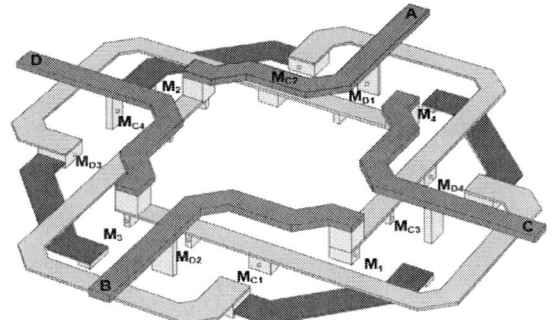

Fig. 5. Layout planning of the proposed QVCO.

M_{Di}). Their gates are shorted on top of the transistors while the drains are routed independently to the other three sides. Inductors are placed at four opposite sides to minimize the unwanted coupling. This arrangement avoids the cross connections commonly seen in P-QVCOs that lower the symmetry of the circuit and increase the parasitic capacitances.

This work is varactorless as varactor usually deteriorates the overall quality factor of the tank. Rather, the bias of the tail current source, V_{ctrl}, is adjusted to vary the junction capacitance of the cross-coupled oscillators, hence the oscillation frequency. However, the achievable tuning range is limited by the tail current since the junction capacitance will not vary too much when the V_{GS} of MOS transistor is larger than V_{th}, the threshold voltage of MOS transistor.

III. MEASUREMENT RESULTS

Two chips are designed for measurement purpose. Both of them are implemented in a 90-nm CMOS technology. As shown in Fig. 7, one chip contains only the proposed QVCO while the other chip integrates the proposed QVCO with a downconversion mixer, which is mainly for time domain measurement due to the limited speed of the oscilloscope. The frequency domain and time domain testing setups are shown in Fig. 6. The QVCO draws 76 mA from a 1.2-V supply. Fig. 8 shows the spectrum and phase noise measurement results. The QVCO achieves a phase noise of -105.8 dBc/Hz at 1-MHz offset and -118.6 dBc/Hz at 10-MHz offset. The $1/f^3$ corner is around

700 kHz. By introducing an extra anti-phase coupling, the tank current is in phase with the transconductor current. This effectively reduces the phase noise to a level comparable or better than single LC VCOs. The output power is -6.3 dBm. By varying the control voltage (V_{ctrl}) from 0.4 V to 1.2 V, the oscillation frequency changes from 67 GHz to 71.5 GHz. The minimum power to start oscillation is 24 mW. For time domain measurement, the quadrature outputs are downconverted to 8 GHz rather than the range of MHz due to the area constraint while smaller LC serving as filter can be easily implemented at high frequency. The quadrature outputs are not perfect sinusoidal waveform because of the downconverted mixer induces many non-linear terms in the downconverted outputs. Quadrature outputs are shown in Fig. 9(a). Measurements from ten chips show the quadrature outputs have phase and amplitude mismatches below 0.46° and 0.47 dB, respectively, including the mismatches from the mixer and testing environment. The low phase noise and quadrature accuracy result from the dual coupling architecture that can be implemented in a fully symmetrical fashion. The phase noise and FOM under different V_{ctrl} are shown in Fig. 9(b).

Table I compares this work with the state-of-the-art millimeter-wave CMOS QVCOs. This work shows more than 8-dB improvement in phase noise compared to the prior art, which results in the lowest reported FOM of -182.9 dBc/Hz at 1-MHz offset.

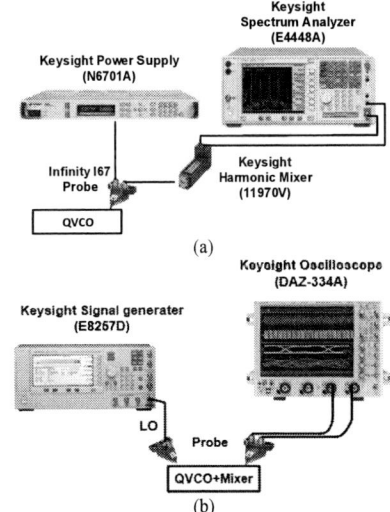

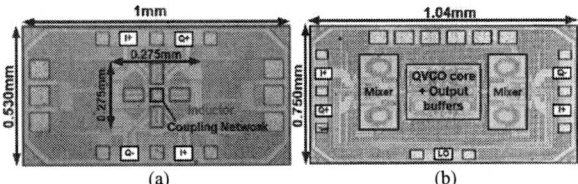

Fig. 6. Test setups for (a) spectrum and (b) time domain measurement.

Fig. 7. Chip micrograph of (a) QVCO only for frequency domain measurement, (b) QVCO and downconversion mixer for time domain measurement.

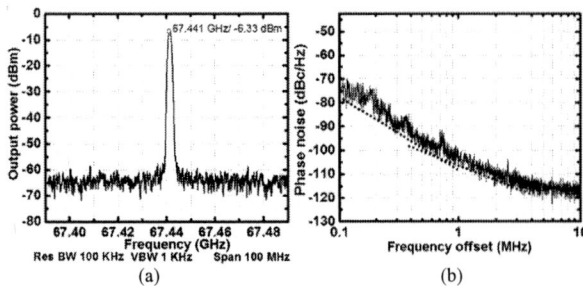

Fig. 8. Measured (a) spectrum and (b) phase noise performance when $V_{ctrl} = 1.2$ V.

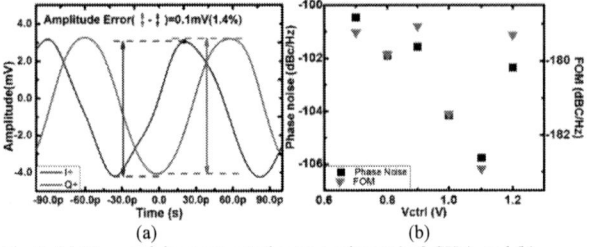

Fig. 9. (a) Measured downconverted output voltages (at 8 GHz), and (b) measured performance vs. V_{ctrl}.

IV. CONCLUSION

A novel millimeter-wave QVCO with bi-directional coupling structure is proposed to improve the phase noise and minimize the phase and amplitude mismatches of the quadrature outputs. This structure facilitates a fully symmetrical layout. Fabricated in a 90-nm CMOS process, the compact QVCO achieves 0.46° and 0.47 dB minimum phase and amplitude mismatches, respectively, and a superior phase noise of -105.8 dBc/Hz at 1-MHz offset, resulting in a figure-of-merit of -183 dBc/Hz at 1.2-V supply.

ACKNOWLEDGMENT

The authors thank the Ministry of Science and Technology, Taiwan (NSC 104-2221-E-007-105) for financial support, National Chip Implementation Center and National Nano Device Laboratories for chip fabrication and measurement.

REFERENCES

[1] Z. Huang *et al.*, "A 70.5-to-85.5GHz 65nm phase-locked loop with passive scaling of loop filter," in *ISSCC Dig. Tech. Papers*, Feb. 2015, pp. 448-449.

[2] X. Yi *et al.*, "A 57.9-to-68.3GHz 24.6mW frequency synthesizer with in-phase injection-coupled QVCO in 65nm CMOS", *IEEE J. Solid-State Circuits*, vol. 49, no. 2, pp. 347-359, Feb. 2014.

[3] M. Vigilante *et al.*, "Analysis and design of an E-Band transformer-coupled low-noise quadrature VCO in 28-nm CMOS", *IEEE Trans. Microwave Theory Techniques*, vol. 64, no. 4, pp. 1122-1132, Apr. 2016.

[4] U. Decanis *et al.*, "A mm-wave quadrature VCO based on magnetically coupled resonators," in *ISSCC Dig. Tech. Papers*, Feb. 2011, pp. 280-281.

[5] V. Szortyka *et al.*, "A 42 mW 200 fs-Jitter 60 GHz sub-sampling PLL in 40 nm CMOS," *IEEE J. Solid-State Circuits*, vol. 50, no. 9, pp. 2025-2036, Sep. 2015.

[6] A. Mazzanti *et al.*, "On the amplitude and phase errors of quadrature LC-tank CMOS oscillators," *IEEE J. Solid-State Circuits*, vol. 41, no. 6, pp. 1305-1313, Jun. 2006.

[7] A. Moroni *et al.*, "Analysis and design of a 54 GHz distributed "hybrid" wave oscillator array with quadrature outputs," *IEEE J. Solid-State Circuits*, vol. 49, no. 5, pp. 1158-1172, May 2014.

[8] D. Zhao *et al.*, "A 60-GHz outphasing transmitter in 40-nm CMOS," *IEEE J. Solid-State Circuits*, vol. 47, no. 12, pp. 3172-3183, Dec. 2012.

[9] T. Zhang *et al.*, "A 55~70GHz two-stage tunable polyphase filter with feedback control for quadrature generation with <2° and <0.32dB phase/amplitude imbalance in 28nm CMOS process," in *European Solid-State Circuits Conference (ESSCIRC)*, Sep. 2015, pp. 60-63.

[10] W. Volkaerts *et al.*, "118 GHz fundamental VCO with 7.8% tuning range in 65 nm CMOS," in *IEEE Radio Frequency Integrated Circuits Symposium (RFIC)*, Jun. 2011, pp. 1-4.

TABLE I
PERFORMANCE COMPARISON WITH THE STATE-OF-THE-ART CMOS QUADRATURE MM-WAVE OSCILLATORS.

Reference	This Work		[1]	[2]	[3]	[4]	[5]	[7]
Frequency (GHz)	67.4		73.5	62.7	72.7	58.2	60	54.2
Technology (nm)	90		65	65	28	65	40	65
Supply voltage (V)	1.2		1.2/1	1.2	0.7	1	0.9	1.2
DC power (mW)	91@ V_{ctrl}=1.2V	59@ V_{ctrl}=0.7V	47.3	11.4	35.6	22	14	36
PN @ 1 MHz (dBc/Hz)	-105.8	-101	-94.6[a]	-95	-97.7[b]	-97	-94.5	-78[c]
FOM @ 1 MHz (dBc/Hz)	-182.9	-179.8	-175.2[a]	-179.7	-179.4	-179	-177	-157[c]
Amp. mismatch	3.8%	2.4%	n/a	0.9dB	1dB	1dB	n/a	0.1dB
Phase mismatch (deg)	0.46	1.03	n/a	0.7	1.5	1.5	5	2
Tuning range (GHz)	4.5		14	10.4	4.7	4.4	9.5	4.6
Core area (mm²)	0.076		0.16[c]	0.039	0.031	0.075	0.053[c]	0.055[c]

[a] Phase noise of PLL
[b] Estimated from 10-MHz result
[c] Estimated from figure

$$\text{FOM} = PN - 20\log_{10}\left(\frac{f_0}{\Delta f}\right) + 10\log_{10}\left(\frac{P_{DC}}{1\text{ mW}}\right)$$

A 350-mV 2.4-GHz Quadrature Oscillator with Nearly Instantaneous Start-Up Using Series LC Tanks

Yue Chen[*], Masoud Babaie[*], and Robert Bogdan Staszewski[*†]

[*]Department of Microelectronics, Delft University of Technology, 2628 CD, Delft, The Netherlands
[†]School of Electrical and Electronic Engineering, University College Dublin, Dublin 4, Ireland

Abstract—We propose a ring-based quadrature LC-tank oscillator for Internet-of-Things (IoT) that can operate under a 350-mV supply voltage of energy harvesters. The oscillator is based on a series LC tank, with additional control circuitry to realize a nearly instantaneous start-up of one/two RF cycles to facilitate a deeply duty-cycled burst-mode operation of IoT. Fabricated in TSMC 28nm CMOS, the prototype consumes less than 1.3mW from 350mV supply and 0.07mm² in area, while achieving phase noise of better than -118dBc/Hz @3MHz. Frequency tuning range is 2.24~2.61GHz.

Keywords—*LC oscillator; quadrature oscillator; ring oscillator (RO); series LC tank; low supply; fast start-up.*

I. INTRODUCTION

Emerging applications, such as Internet-of-Things (IoT) and wireless sensor networks (WSN), are attracting an enormous research interest in ultra-low-power (ULP) and ultra-low-voltage (ULV) transceivers. For these applications, the system's lifetime tends to be severely limited by the power consumption of RF transceivers and battery capacity. To extend the lifetime without bulky batteries, the power consumption of the transceivers should be drastically minimized. At the same time, energy harvesting techniques are maturing to dramatically increase the system lifetime, and even realize batteryless operation. However, the output voltage level of energy harvesters is typically low [1,2], while power efficiency of state-of-the-art boost converters at such low input voltages is limited. Consequently, it becomes highly desirable for IoT systems to be powered directly by the low harvested voltage, which can be as low as 0.4V [1,2]. On the other hand, IoT/WSN systems widely use heavily duty cycled operation to improve their energy (i.e., averaged power) efficiency, e.g. in burst mode operation [3]. Hence, it appears highly beneficial to achieve a fast start-up capability to minimize the duty-cycle ratio and to reduce the overall energy consumption. Moreover, the silicon area should be kept as small as possible to reduce the cost.

To address the aforementioned issues, we propose an RF oscillator comprising a ring of four tiny series LC tanks with a nearly instantaneous startup capability. This structure combines advantages of conventional inverter-based ring oscillators (RO) with those of traditional LC oscillators. The supply voltage can go to as low as 350mV, which is one of the lowest ever reported at RF frequencies. The nearly instantaneous start-up ability competes with that in the conventional ring oscillators,

but here the phase noise performance is superior due to the use of resonating LC tanks. The occupied area is tiny as compared to the traditional LC oscillators, while providing quadrature outputs intrinsically.

II. LOW-SUPPLY FAST START-UP QUADRATURE OSCILLATOR

The ring-based oscillator structure is chosen here to generate quadrature outputs directly at the fundamental frequency. This avoids the need to double the oscillation frequency followed by a ÷2 frequency divider, both of which consume additional power. Conventional inverter-based ROs, even those specifically designed for ultra-low supply voltage through internally boosting the supply voltage [4], could not operate at such high frequencies (i.e. ~2.4GHz) under the targeted supply voltage (i.e. 350mV), which is already lower than the threshold voltage (V_{th}) of transistors in this 28-nm technology. Hence, to guarantee oscillation, some kind of passive voltage gain is needed to boost up the voltage swing at the input of inverter stages directly.

A. Oscillator Core

The passive voltage gain can be implemented by employing a series LC tank, as shown in Fig. 1. Both tank options boost the input voltage swing by about Q_{res} times at the resonant frequency, where Q_{res} is the quality factor of the tank. However, for the Fig. 1(a) configuration, the inductor will be in parallel with an input parasitic capacitance of the next-stage inverter. This will not only short-circuit to ground the next inverter's dc input but it would also degrade the voltage boosting effect. Simulations show that this degradation can be as large as 40%. In contrast, the input parasitic capacitance of the next stage in the Fig. 1(b) configuration is absorbed into the tank capacitance. Hence, this type of configuration is adopted in this design. As shown in Fig. 2, the tank output voltage is amplified by Q_{res} times and shifted -90° at the resonant frequency f_{res}. Together with the 180° phase shift coming from the inverter, the total phase shift of one basic unit at the resonant frequency is 90°. To meet the Barkhausen criterion, four such units are arranged in a ring to form the oscillator core, as shown in Fig. 3 [5]. Thus, the oscillator would intrinsically produce the quadrature outputs.

In contrast to the parallel LC tank, a voltage source (i.e., of low impedance) is here required to drive the tank. A simple inverter driven by a sufficiently large input signal, as shown in

The project is supported in part by the China Scholarship Council under Grant 201406280031; in part by the European Research Council under Grant 307624. TSMC university shuttle program is acknowledged for chip fabrication.

978-1-5386-3179-9/17 $31.00 © 2017 IEEE

Fig. 1, provides the required low driving impedance while simultaneously acting as a high impedance isolator to the preceding stage [5]. The voltage waveform generated by the inverter at the tank input is a near square wave, jumping rapidly between the supply rails. The tank acts as a low-pass LC filter and suppresses higher harmonic components of the pseudo-square wave, consequently generating a sinusoidal waveform with a boosted magnitude of $\alpha_V \times V_{dd} \times Q_{res}$ at the input of the subsequent inverter stage. Note that α_V is the oscillator voltage efficiency factor, and defined as a ratio of the fundamental component of the tank input voltage to dc supply V_{dd} ($2/\pi$ for ideal square wave). Even though the inverter has a supply voltage lower than V_{th}, due to tank's voltage gain, its gate terminal is driven by the large-enough voltage swing to flip the inverter.

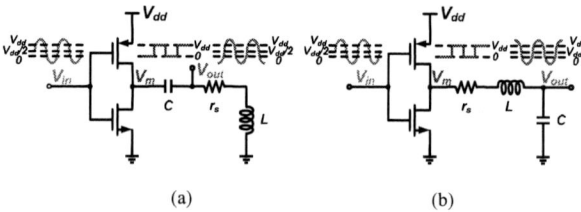

Fig. 1. Basic unit of the proposed oscillator with: (a) inductor connected to ground, (b) capacitor connected to ground.

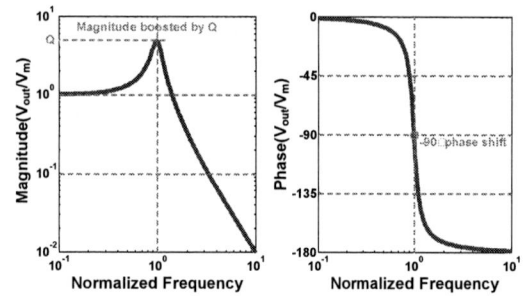

Fig. 2. Transfer function of series LC tank in Fig. 1(b).

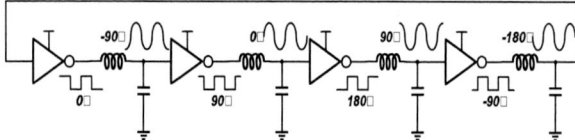

Fig. 3. Structure of oscillator core.

When the inverter delay is negligible compared to the phase shift of the series LC tank, the oscillation frequency f_{osc} of the proposed oscillator is identical to the tank's resonant frequency f_{res} [5].

$$f_{osc} = f_{res} = \frac{1}{2\pi\sqrt{LC}} \qquad (1)$$

However, since the supply voltage is very low, the delay of the driving inverter could not be neglected here. When the inverter delay is taken into consideration, the required phase shift from the tank would move downwards, lowering the oscillation

frequency. The voltage gain of the tank would also degrade when f_{osc} deviates from f_{res}. A smaller voltage swing at the inverter's input terminal increases the inverter delay, thus lowering f_{osc} even further. Assuming the phase shift corresponding to the inverter delay is ϕ, the oscillation frequency could be calculated by

$$f_{osc} = f_{res} \cdot \left(\sqrt{1 + \left(\frac{\tan\phi}{2Q_{res}}\right)^2} - \frac{\tan\phi}{2Q_{res}} \right) \qquad (2)$$

Equation (2) clearly shows that f_{osc} will be lower than f_{res}, since the value of the outer bracket term is smaller than 1, and the difference between these two frequencies grows with ϕ. Q_{res} also has an influence on the oscillating frequency. A higher Q_{res} increases the $f_{res} - f_{osc}$ difference. This is in accordance with the fact that the rate of phase variation of a series LC tank is proportional to Q_{res} near the resonant frequency. By using larger inverters, the above problem could be alleviated because the inverters run faster. Furthermore, up-sizing the inverters also results in the waveforms becoming more 'squarish'. Consequently, the time when neither NMOS nor PMOS transistors operate in the deep triode region is very small compared to the oscillation period. This way, the inverters are working as highly efficient class-D amplifiers and their contribution to the total oscillator's phase noise could be negligible. This is in contrast to traditional LC oscillators, where phase noise contribution from active devices is γ (the channel noise factor of MOS transistors) times that from the tank losses, and does not improve with technology scaling. Here, the phase noise from the active devices can be made negligible by employing 'better switches' (lower turn-on resistance and higher current capability) in the advanced technologies [6].

The power consumption of the series LC oscillator could be calculated as in (3)

$$P = 4 \cdot V_{dd} \cdot I_{dc} = 4 \cdot V_{dd} \cdot \frac{V_{dd} \cdot \alpha_V}{r_s \cdot \alpha_I} = \frac{4 \cdot V_{dd}^2}{r_s} \cdot \frac{\alpha_V}{\alpha_I} \qquad (3)$$

where the current efficiency factor α_I is defined as the ratio of the fundamental component of tank current to dc current I_{dc} of each stage. From analysis, the ideal value of α_I is π. As can be gathered from (3), the power consumption is inversely proportional to the series resistance r_s of inductor L. To reduce power, a higher r_s is needed, leading to larger inductance under certain Q. However, the large inductance, combined with the large input parasitic capacitance of the driving inverter, may limit the highest oscillation frequency and the tuning range. Hence, there is a trade-off between the contribution of inverter to the phase noise and the oscillator's power consumption. By considering the aforementioned remarks, the inductor implemented in this oscillator is about 9.88nH with a Q around 12 at 2.5GHz. A coarse tuning of the oscillation frequency is done through a 5-bit switched-capacitor bank, while the fine tuning is achieved with MOS varactors.

B. Instantaneous Start-Up Control

A significant drawback of the Fig. 3 oscillator topology is its inability to start-up automatically. The inductor in each stage becomes electrical short at dc. Thus, the circuit has a solid dc feedback, preventing it from self-oscillation [5]. Moreover, as mentioned in Section I, a short start-up time is crucial in extremely low duty-cycle burst-mode systems. For inverter-based ROs, the start-up time could be reduced to a few oscillation periods through presetting voltage of internal nodes [3]. However, little attention has been paid in literature to LC oscillators, which generally take up many RF cycles to settle.

Since the start-up solution in [5] would anyway not work at ULV supplies, here switches within the oscillator are added to enable fast start-up. As shown in Fig. 4(b), switches S_{p1}~S_{p8}, implemented with NMOS transistors, are used to preset the corresponding oscillator nodes to their initial sleep-mode values. Unlike in conventional ROs, here the phase relationship between each stage in the sleep mode is very different from that in the oscillation mode. As depicted in Fig. 4(a), the four stages of the oscillator could be divided into 2 groups: stages 1&3, and stages 2&4. In the oscillation mode, the two stages of each group are in anti-phase, leading to a pseudo-differential operation, while the 90° phase shift exists between the two groups. However, in the sleep mode, the two stages of each group are forced to be in-phase due to the electrical short of the inductor. This difference makes it difficult to directly preset the node voltages while ensuring both fast oscillation start-up and low idle power in the sleep mode. For example, assume the first (vi1&vr1) and third (vi3&vr3) stages are preset to *low* and *high* voltage, respectively, according to the expected quadrature phase relationship in the oscillation mode. Stage4 (vi4&vr4) should then be preset to a low voltage. Otherwise, the first falling edge from stage3 at start-up would cause no action to stage4, preventing the propagation of the oscillation signal. However, both the input and output of the driving inverter of the first stage are now biased at a low voltage in the idle state, causing large sleep-mode supply current flowing through the PMOS transistor of the stage1 inverter to ground. To overcome this problem, a set of NMOS switches S_{i1}~S_{i4} are inserted between the output of the driving inverters and the input of their tanks. These transistors are off in the sleep mode, thus isolating the two originally connected nodes. Hence, they could be preset to different voltages to simultaneously satisfy the requirements for fast start-up and low idle power. In the oscillation mode, these transistors are on, and could be simply lumped into the series resistance of the inductor. When the oscillator gets triggered, switches S_{i1} and S_{i3} generate voltage jumps at opposite directions at the tank input of the first and third stages respectively, in order to achieve the fast oscillation start-up. This would also result in a fixed phase relation between the oscillator's ports after start-up, which is beneficial for reducing the PLL's locking time. Dummy switches S_{i2} and S_{i4} are added for symmetry. An enable transistor is also added to each driving inverter to further reduce the leakage current in the sleep mode.

The digital logic realizing the aforementioned fast start-up and its corresponding timing sequence are also shown in Fig. 4(b). When the rising edge of 'trigger' signal arrives, the enable signal ('en') and the preset control signal ('p_ctrl') become high and low respectively, making the oscillator ready for start-up. Then a pulse is generated on the trigger control signal ('tri_ctrl') to close the switches S_{i1} and S_{i3} for a short period. During this period, the corresponding tank inputs are

pulled up and down from their presetted levels in the sleep mode, respectively, generating the voltage jumps needed for the start-up. At the same time, the NMOS switches S_{i1}~S_{i4} are also turned on, enabling the normal propagation of the oscillation signal in the ring.

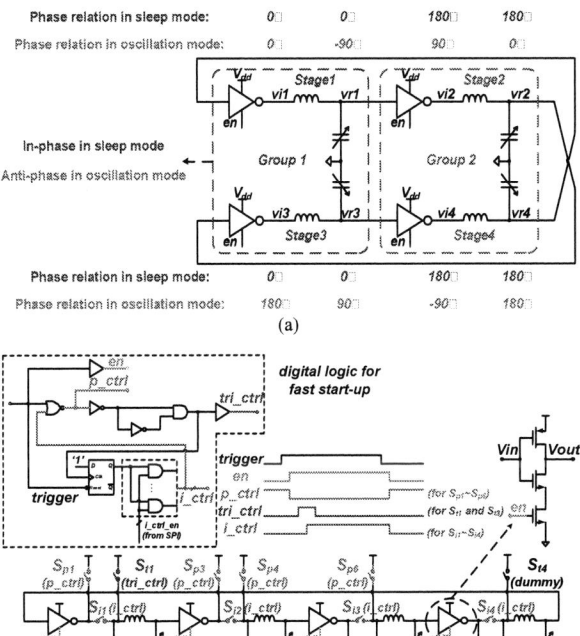

(a)

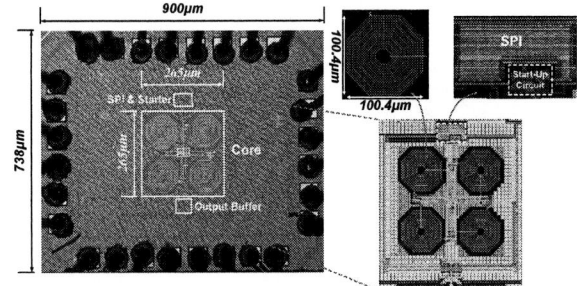

(b)

Fig. 4. (a) Phase relationship of the oscillator ports in the two modes, (b) oscillator structure with additional circuits for fast start-up.

III. IMPLEMENTATION AND MEASUREMENT RESULT

The proposed oscillator is implemented in TSMC 28nm CMOS. The chip micrograph is shown in Fig. 5 together with layout details. The inductor is implemented in the ultra-thick top metal. The core area is 0.07mm², while the start-up circuit occupies a negligible area. Under a 350mV supply voltage, the power consumption of the oscillator is ≤1.3mW.

Fig. 5. Chip micrograph and layout details of proposed oscillator.

The oscillation frequency could be tuned from 2.24GHz to 2.61GHz. Fig. 6(a) shows the measured phase noise at the maximum and minimum oscillation frequencies. At 3MHz offset from the carrier, the measured values are -118.2dBc/Hz and -120.7dBc/Hz, respectively, while $1/f^3$ corner frequency varies from 700kHz at 2.24GHz to 1.5MHz at 2.61GHz. The relatively high corner frequency is a result of the class-D operation of the driving inverter [6]. Fig. 6(b) illustrates the measured oscillator phase noise and its corresponding Figure-of-Merit (FoM) at 3MHz offset from the carrier frequency across the tuning range (TR).

Fig. 7 shows the settling behavior of the oscillator at start-up. The oscillation frequency settles in about 1~2 RF cycles after the start-up trigger, which is comparable to that of fast start-up inverter based ROs. The oscillator also shows fixed start-up waveform during repeated measurements.

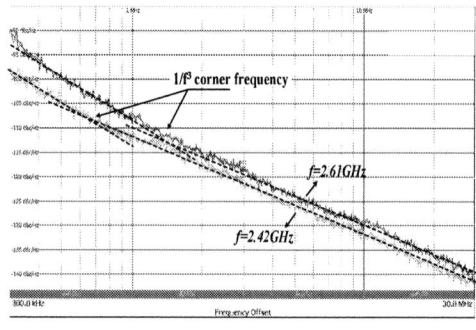

(a)

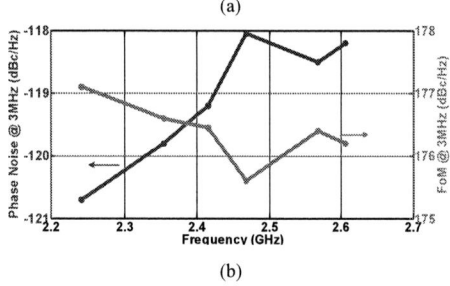

(b)

Fig. 6. (a) Measured phase noise at 2.24GHz and 2.61GHz, (b) measured phase noise and its corresponding FoM at 3MHz offset across TR.

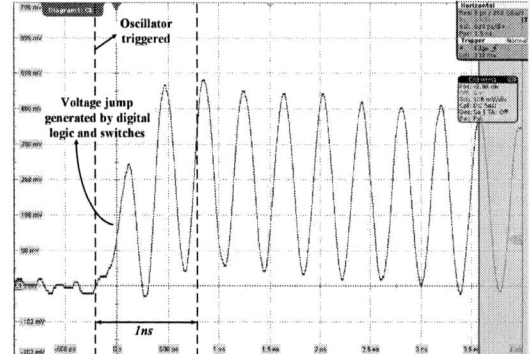

Fig. 7. Measured frequency settling behavior of the oscillator at start-up.

The performance of the proposed 350mV supply LC ring oscillator is summarized and compared with state-of-the-art in Table I. When contrasted with the previously designed series

LC oscillator [5], the proposed oscillator features much lower supply voltage and power consumption. When compared with other low-voltage supply LC oscillators [8]-[9], the supply voltage and power consumption of the proposed oscillator are among the lowest, while providing quadrature outputs with an area that is at least 5x smaller. The proposed oscillator also has fast start-up ability, which is not addressed in other works.

TABLE I. PERFORMANCE SUMMARY AND STATE-OF-ART COMPARISON

	This Work	[5]	[7]	[8]	[9]
Technique(nm)	28	40	65	65	65
Architecture	Series LC	Series LC	Ring	XFMR Feedback	Class-D
Area(mm²)	0.07	0.0063	0.003	0.5†	0.4
Supply(V)	0.35	1.1-0.9	0.7-1.0	0.3	0.35‡
f_{osc}(GHz)	2.24-2.61	2.66-4.97	1.7-3.47	2.32-2.6	4.4-5.4
TR(%)	15.3	60.5	68.5	11.4	20.4
PN@3MHz* (dBc/Hz)	-120.1- -118.7	-120.5	-109.9- -108.2	-122.4	-133.3- -135.2
Power(mW)	1.3	16-8.5	0.65-2.5	0.6	2.6-2.3
FoM(dB)	177.1- 175.6	171.3- 176.9	166.9- 165.6	182.9	194.4- 194.8
Quadrature	Yes	Yes	No**	No	Yes
Start-Up time(ns)	0.5~1	–	–	–	–

* Normalized to 3MH offset assuming 20dB/decade slope. ** 35 output phases

† Estimated from layout. ‡ Gate terminal of cross-coupled pair biased at 0.4V.

IV. SUMMARY

To benefit IoT nodes powered by energy harvesters and operating at very low duty cycles, an ultra-low supply voltage quadrature oscillator based on a series LC tank is proposed. Additional switches and digital logic are introduced for fast start-up. Compared with prior art, the proposed oscillator features a small area, low supply voltage and power consumption, and less than 1ns start-up time.

REFERENCE

[1] S. Bandyopadhyay and A. P. Chandrakasan, "Platform architecture for solar, thermal, and vibration energy combining with MPPT and single inductor," *IEEE JSSC*, vol. 47, no. 9, pp. 2199–2215, Sep. 2012.

[2] M. Babaie et al., "A fully integrated Bluetooth low-energy transmitter in 28 nm CMOS with 36% system efficiency at 3 dBm," *IEEE JSSC*, vol. 51, no. 7, pp. 1547–1565, Jul. 2016.

[3] S. Drago, D. M. W. Leenaerts, B. Nauta, F. Sebastiano, K. A. A. Makinwa, and L. J. Breems, "A 200µA duty-cycled PLL for Wireless Sensor Nodes in 65nm CMOS," *IEEE JSSC*, vol. 45, no. 7, pp. 1305–1315, Jul. 2010.

[4] Y. Ho, Y.-S. Yang, C. Chang, and C. Su, "A near-threshold 480 MHz 78 uW all-digital PLL with a bootstrapped DCO," *IEEE JSSC*, vol. 48, no. 11, pp. 2805–2814, Nov. 2013.

[5] M. Tohidian, S. A.-R. Ahmadi-Mehr, and R. B. Staszewski, "A tiny quadrature oscillator using low-Q series LC tanks," *IEEE Microw. Wireless Compon. Lett.*, vol. 25, no. 8, pp. 520–522, Aug. 2015.

[6] L. Fanori and P. Andreani, "A 2.5-to-3.3 GHz CMOS Class-D VCO," in *IEEE ISSCC*, pp. 346–347, Feb. 2013.

[7] J. Yin, P.-I. Mak, F. Maloberti, and R. P. Martins, "A 0.003mm² 1.7-to-3.5GHz dual mode time-interleaved ring-VCO achiving 90-to-150kHz $1/f^3$ phase-noise corner," in *IEEE ISSCC*, pp. 48-49, Feb. 2016.

[8] F. Zhang, K. Wang, J. Koo, Y. Miyahara, and B. Otis, "A 1.6mW 300mV-Supply 2.4GHz Receiver with -94dBm Sensitivity for Energy-Harvesting Applications," in *IEEE ISSCC*, pp. 456-457, Feb. 2013.

[9] A. G. Roy, S. Dey, J. B. Goins, T. S. Fiez, and K. Mayaram, "350 mV, 5 GHz Class-D enhanced swing differential and quadrature VCOs in 65 nm CMOS," *IEEE JSSC*, vol. 50, no. 8, pp. 1833–1847, Aug. 2015.

On-Chip Spur and Phase Noise Cancellation Techniques

Yi-An Li[1], Monte Mar[2], Borivoje Nikolić[1], Ali M. Niknejad[1]

[1] Berkeley Wireless Research Center (BWRC), University of California, Berkeley, CA

[2] Solid State Electronics Development, The Boeing Company

Abstract—Two techniques for spur and phase noise cancellation have been proposed. A fully integrated design achieves a measured spur cancellation of 15dB at 250MHz and 750MHz offset as well as phase noise cancellation from 4MHz to 200MHz offset with maximum 25dB cancellation depth for a 1-GHz clock. The proposed ideas have been verified through a fabricated 65nm CMOS prototype with power consumption of 11mW from a supply voltage of 1.2V.

Index Terms—Spur, phase noise, cancellation, notches, and delay-line discriminator.

I. INTRODUCTION

Spur and phase noise are two of the most critical specifications that can ultimately limit the performance of communication system. For example, spurs resulting from device mismatches in the LO generation circuit could cause a transmitter to fail its spectral mask requirement. On the receiver side, blocker induced reciprocal mixing of phase noise is the ultimate sensitivity limit, even for a perfectly linear receive chain. As a result, we tend to spare more power budget on clock sources to meet the worse case corners and scenarios, which might, in fact, rarely happen. Therefore, it would be beneficial to have a post-process module cascaded after the clock source and turned active only when needed. Our goals are to generate notches against far-out spurs, and to produce high-pass filtering on the phase noise of the clock to suppress close-in phase noise, as conceptually shown in Fig. 1. In this way, we can relax the specifications of the clock source and achieve lower power design to potentially extend battery life considerably.

II. PROPOSED IDEA

A. Spur Cancellation

To simplify illustrations, we can first consider a special case that a clock has spurs with offset frequency of half carrier frequency, as shown in Fig. 2. The spurs affect the clock by fluctuating clock edges periodically. Thus, if we can delay this clock by half of the jitter period (i.e. one clock period in this case), and interpolate it with the original clock, then the periodic jitter can be cancelled out. Such an operation in the time domain is an average on the past and present phase, which can be described in frequency domain as

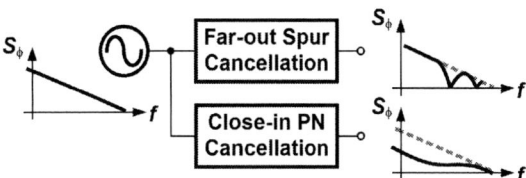

Fig. 1. Spur and phase noise cancellation goals.

$$|\Phi_{\text{out}}(s)| = \frac{|\Phi_{\text{in}}(s)\cdot(1 + e^{-sT})|}{2}$$
$$= |\Phi_{\text{in}}(s)|\cdot|\cos \pi fT| \qquad (1)$$

, where $\Phi_{\text{in}}(s)$ is input phase noise and T is the delay time. We notice that this transfer function create notches at the offset frequencies of $1/2T$, $3/2T...$, etc. By programming the time delay, we can line up the notch frequencies on top of the spurs to reject them. Note that such rejection not only applies to spurs, but also applies to phase noise.

Although it is possible to push the notches into low offset frequency with large delay, for close-in phase noise cancellation, the notch bandwidth (which is proportional to $1/T$) shrinks and therefore has little impact on the integrated phase noise. In addition, the noise accumulated by the long delay line also limits the notch depth. Nevertheless, it is still effective for far-out spurs that only require short delay. The issues above can be circumvented if we can change $1+e^{-sT}$ into $1-e^{-sT}$. In this way, the transfer function becomes $|\sin \pi fT|$, which has its first notch locating at dc and is more suitable for close-in phase noise cancellation. To realize such a transfer function, we will introduce a delay line discriminator method in next section.

B. Phase Noise Cancellation

In order to perform phase noise cancellation, the first step is to extract the phase noise information from the clock source, and then apply it back to the original clock with opposite polarity [1]. The delay line discriminator is a good candidate to serve this function and is used widely in spectrum analyzers with high sensitivity for phase noise measurement [2]. When a clock, $\cos(\omega t+\phi_{\text{in}}(t))$, passes through a delay line of delay T_1, a frequency dependent phase shift is then applied on its spectrum. By comparing the phase difference between the two ends of delay line,

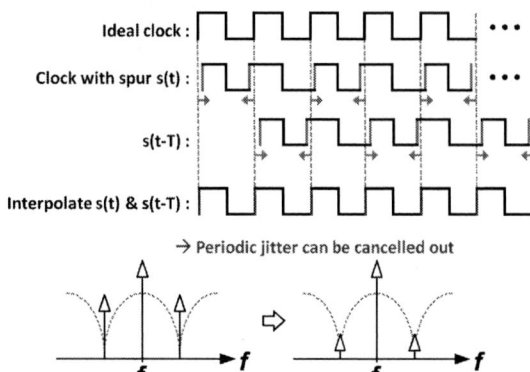

Fig. 2. Delay-and-interpolate spur cancellation technique.

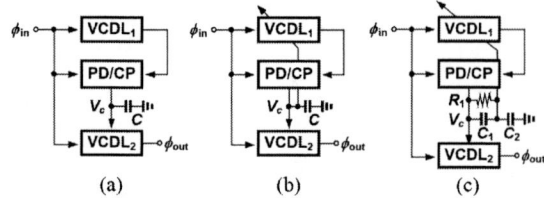

Fig. 3. Evolutions of phase noise cancellation architecture: (a) basic design (b) with DLL to avoid V_C from saturating, and (c) using series C_1 and C_2 to break the trade-off on loop capacitance value selection.

the resulting signal at the output of the phase detector (PD) is given by

$$\Delta\Phi_{PD}(s) = \Phi_{in}(s)\cdot(1\text{-}e^{-sT_1})$$
$$\cong \Phi_{in}(s)\cdot sT_1. \quad (2)$$

This signal is a differentiated version of input phase information and is down-converted into baseband. We can then recover such a signal by an integrator and then feed-forward it to modulate a voltage control delay line (VCDL) where the same clock is passing through. In this way, the baseband phase noise information can be up-converted back to the clock frequency to cancel out the original phase noise. The operations above can be realized in a simplified circuit shown in Fig. 3(a), which incorporates a VCDL and a PD for phase comparison, a charge pump (CP), a capacitor as an integrator, and another VCDL at output for phase noise cancellation. In this circuit, the output phase noise can be expressed as

$$|\Phi_{out}(s)| \cong \left| -\Phi_{in}(s)\cdot sT_1 \cdot \frac{K_P K_D}{sC} + \Phi_{in}(s)\cdot e^{-sT_2} \right| \quad (3)$$

, where K_P, T_2, and K_D are the gain of PD/CP, the delay of VCDL$_2$, and the gain of VCDL$_2$, respectively. When the cancellation condition is met (i.e. $T_1 K_P K_D/C = 1$), it can be further reduced to

$$|\Phi_{out}(s)| \cong |\Phi_{in}(s)\cdot(1\text{-}e^{-sT_2})|$$
$$= |\Phi_{in}(s)| \cdot |2\sin \pi f T_2|. \quad (4)$$

In this way, the close-in phase noise can be cancelled out largely by the high-pass filtering on original phase noise. Note that, although (2) and (4) have the same form, they are totally different, since the signal in (2) is at the baseband, whereas the signal in (4) is at clock frequency.

III. NON-IDEALITY CONSIDERATIONS

Although the result in (4) looks promising theoretically, there are some possible issues in circuit realization. Therefore, it is worthwhile to analyse how they impact the performance and how to work around them.

A. Mismatch

First, since we perform the cancellation in the analog domain, the cancellation depends on the device matching and would be limited by mismatches and variations over PVT. Fortunately, for >20dB cancellation depth (Fig. 4), the mismatch only needs to be controlled within 10%, which is not difficult to achieve with careful design and layout. Furthermore, calibration knobs can be added, if needed, to track over PVT.

B. DC Balancing Loop

Second, due to the nature of integrator, any small constant phase offset will finally pump the control voltage (V_C) into saturation. Since we already have PD, CP, and VCDL, we can mitigate this issue by looping them into a delay-locked loop (DLL), as shown in Fig. 3(b), such that V_C would settle to a proper voltage level by the feedback. Rather than introduce another CP and variable-gain amplifier (VGA) [5] for the dc balancing loop, this sharing method shows zero power and minimum area penalty. It is worth noting that the DLL locks the phases at two ends of the delay line within its bandwidth, which in turn wipes out low frequency phase information and makes the phase noise cancellation at low offset frequency malfunction, as

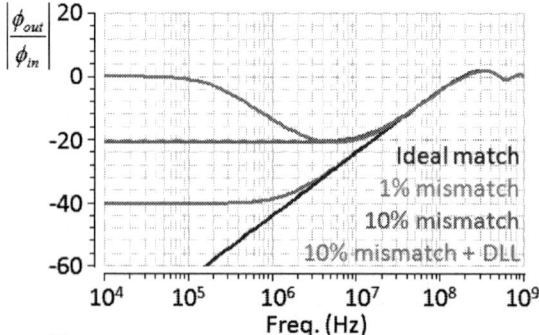

Fig.4. Non-ideal effects on phase noise cancellation.

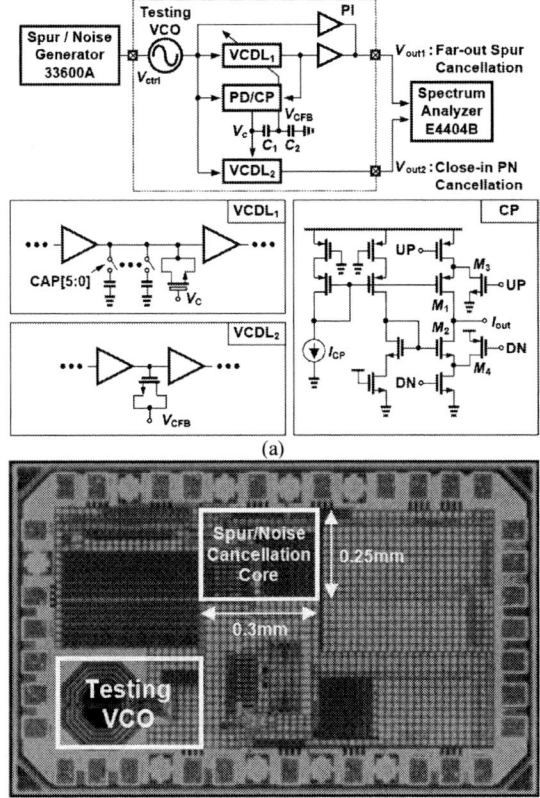

(a)

(b)

Fig.5. (a) Overall architecture and testing setup. (b) Test chip die photo.

can be shown by the green curve in Fig. 4. Such an impact might be less critical within a PLL, where the close-in phase noise will be cleaned by the phase noise of external reference clock. Nevertheless, we still prefer to keep the DLL bandwidth as small as possible in this design.

IV. CIRCUIT IMPLEMENTATION

According to the analysis above, we have to meet the cancellation condition and to keep low DLL bandwidth simultaneously. However, it would lead to a trade-off on capacitance value. To break this trade-off, we proposed a new loop filter composed of series capacitors C_1 and C_2, shown in Fig 3(c). In this way, the DLL loop sees a large C_2 to achieve small bandwidth, and the feed forward path sees a capacitance of $C_1C_2/(C_1+C_2)$ that can still meet the cancellation conditions by this extra freedom. Also, in order to define the dc voltage and compensate any possible leakage with minimum effect on the transfer

function, a tuneable large resistor R_1 (from 100k to 1MΩ) is connected in shunt with C_1.

The final architecture is shown in Fig. 5(a), which combines both spur and phase noise cancellation techniques with a shared delay line. The phase interpolator (PI) for spur cancellation can be easily implemented by directly shorting the two inverters' outputs that are driven by the clocks to be interpolated, since the DLL has aligned them nominally. Note that in this test chip, the output of the two techniques are separated for testing purpose, and can be implemented in cascade by feeding V_{out1} into the input of VCDL$_2$. The delay line of VCDL$_1$ can cover the delay range from 1.25nsec to 2.75nsec by 6-bit capacitor banks switching and +/−70psec by varactor tuning. To minimize possible leakage on the internal node between C_1 and C_2, we use thick oxide varactor in VCDL$_1$. The resulting low tuning sensitivity is not an issue, since small varactor gain is desired to minimize DLL bandwidth. The output VCDL$_2$ is also made of the inverter chain loaded by normal varactors, but with opposite varactor polarity and much larger tuning gain to achieve cancellation. In addition to DLL bandwidth, the output resistance of CP also limit the cancellation at low offset frequency. Therefore, long channels are selected for the M_1/M_2 in CP to keep high output resistance, and M_3/M_4 are added to help driving the extra capacitance loading (Fig. 5(a)).

V. EXPERIMENTAL RESULT

This chip has been fabricated in 65nm CMOS technology, which occupies $0.3 \times 0.25 \text{mm}^2$ core area (Fig. 5(b)). The circuit consumes a total power of 11mW, of which 5mW dissipates in the delay line, 2mW in the PD/CP, and 4mW in the output VCDL. Fig. 5(a) shows the measurement setup. The spur and white noise waveforms are generated externally by 33600A waveform generator and then modulate the control line of a 1-GHz testing voltage-controlled oscillator (VCO) on the chip to mimic a low power VCO with relaxed performance. The

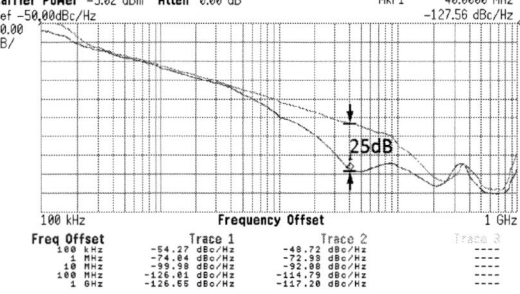

Fig.6. Phase noise is cancelled from 4MHz to 200MHz offset with maximum cancellation of 25dB on a 1-GHz clock.

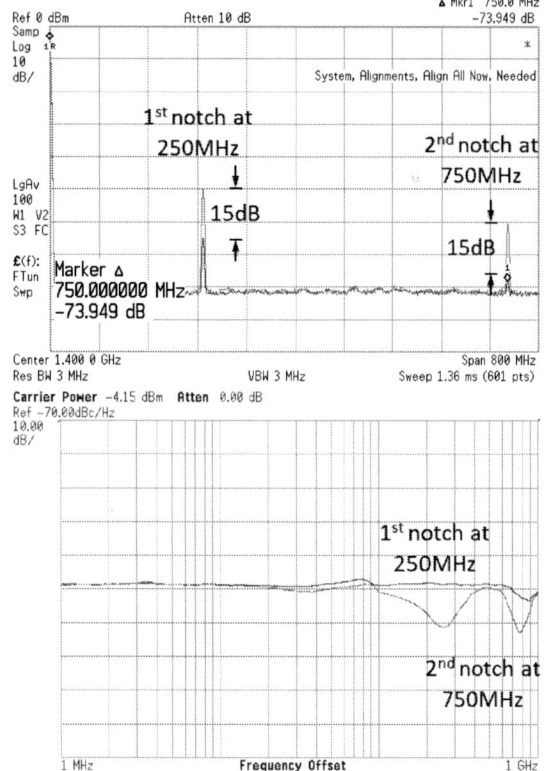

Fig.7. (a) Far-out spurs at 250MHz and 750MHz offset frequencies can be cancelled up to 15dB by the notches. (Blue curve). (b) Far-out noise can also be attenuated by notches. (Red curve).

TABLE I
PERFORMANCE SUMMARY

	TMTT 2015[3]	RFIC 2012[4]	IMS 2016[6]	This work
Frequency	1.5GHz	5GHz	10GHz	1GHz
Delay Line Type	Off-chip FBAR	On-chip inverter	On-chip LC + Off-chip SAW	On-chip inverter
Phase Noise Cancellation BW	1k ~ 2MHz	100k ~ 20MHz	100k ~ 10MHz	4M ~ 200MHz
Max Phase Noise Cancellation Depth	40dB	12.5dB	15.5dB	25dB
Far-out Spur Cancellation	N/A	N/A	N/A	15dB
Power Consumption (excluding VCO)	340mW	20.9mW	102mW	11mW
Core Area	1.8×1.2 mm²	0.38×0.32 mm²	1.68×1.5 mm²	0.3×0.25 mm²
Technology	130nm CMOS	90nm CMOS	65nm CMOS	65nm CMOS

phase noise cancellation with delay-line-discriminator method is demonstrated in Fig. 6. The cancellation applies from 4MHz to 200MHz offset frequency and achieves a maximum of 25dB cancellation at 40MHz offset frequency. On the other hand, the delay-and-interpolate method for spur cancellation is also verified. Fig. 7(a)

shows the results of spur cancellation. In this design, the delay line is set to be 2nsec, which generates notches at 250MHz ($1/2T$) and 750MHz ($3/2T$) offset frequency. The spurs at those two offset frequencies can be rejected by 15dB (from -56 to -71dBc). Such rejections also apply to phase noise. To felicitate the observation at high offset frequencies, we use external clock with broadband noise as testing clock instead of on-chip VCO. As shown in Fig. 7(b), two notches at 250MHz and 750MHz are clearly shown on phase noise plot after applying the technique. Table I summarizes and compares our work with recent publications.

VI. CONCLUSION

Two novel techniques to generate transfer functions of $|\cos \pi fT|$ and $|\sin \pi fT|$ for spur and phase noise cancellation has been proposed and verified in the test chip. For spur cancellation, the delay-and-interpolate technique can generate notches that are able to reduce the spur level by 15dB. For phase noise cancellation, a delay-line-discriminator and feed-forward technique can cancel the phase noise on the offset frequency from 4MHz to 200MHz with maximum cancellation depth of 25dB. This cancellation module can be cascaded after the clock source targeted for low power with relax the specifications and turned active only when high performance scenarios occur, allowing extended battery life by optimum power and performance trade-off management .

ACKNOWLEDGMENT

This work is funded in part by DARPA RFFPGA. The authors wish to acknowledge the TSMC University Shuttle Program for chip fabrication.

REFERENCES

[1] F. Aflatouni, et al."Design methodology and architectures to reduce the semiconductor laser phase noise using electrical feed-forward schemes", *IEEE TMTT*, vol. 58, no. 11, 2010.

[2] D. Scherer "The art of phase noise measurement" [Online] Available:http://www.hparchive.com/seminar_notes/Scher er_Art_of_PN_measurement.pdf

[3] A. Imani and H. Hashemi, "An FBAR/CMOS Frequency / Phase Discriminator and Phase Noise Reduction System," *IEEE Trans. Microwave Theory & Tech.*, vol. 63, no. 5, pp. 1658-1665, May 2015.

[4] S. Min, et al., "A 90nm CMOS 5GHz Ring Oscillator PLL with Delay-Discriminator Based Active Phase Noise Cancellation," *IEEE RFIC*, pp. 173-176, June 2012.

[5] S. Hao, et al., "A 10 GHz delay line frequency discriminator and PD/CP based CMOS phase noise measurement circuit with -138.6 dBc/Hz sensitivity at 1 MHz offset," *IEEE RFIC*, pp. 63–66, 2015.

[6] S. Hao, et al., "A 10 GHz Phase Noise Filter with 10.6 dB Phase Noise Suppression and -116 dBc/Hz Sensitivity at 1 MHz Offset," *IEEE IMS*, May 2016.

S10-1 (2057)

A Single-Inductor Triple-Input-Triple-Output (SITITO) Energy Harvesting Interface with Cycle-by-Cycle Source Tracking and Adaptive Peak-Inductor-Current Control

Chi-Wei Liu, Ming-Jie Chung, Hui-Hsuan Lee, Pei-Chun Liao, and Po-Hung Chen
Institute of Electronics
National Chiao Tung University
Hsinchu, Taiwan

Abstract—In this paper, a single-inductor triple-input-triple-output (SITITO) buck-boost converter with cycle-by-cycle source tracking (CCST) is developed for multi-source energy harvesting. The proposed CCST is capable of harvesting power from the PV and TEG simultaneously, and automatically selects the appropriate source according to the maximum power point (MPP) of the transducers in each switching cycle. The converter provides two regulated outputs for the analog ($V_{OUT1} = 1.2$ V) and the digital ($V_{OUT2} = 0.4$-0.6 V) circuits, while storing the excess energy in a storage capacitor (C_{OUT3}). An adaptive on-time (AOT) circuit with an adaptive peak-inductor-current (APIC) control is developed for dynamically adjusting the on-time based on the input voltage (V_{IN}) to obtain high conversion efficiency under different V_{IN}. The proposed converter achieves a peak efficiency of 84.4% with an available input voltage of 0.05 V to 0.8 V. Compared to the conventional constant peak-inductor-current (CPIC) control, the proposed APIC improves 11 % efficiency at low V_{IN}.

Keywords—dc-dc converter, multi-source energy harvesting, source tracking, adaptive peak-inductor-current control

I. INTRODUCTION

Ambient energy harvesting has been adopted in various promising applications such as wireless sensors, wearable devices, and the internet of things (IoT). Conventional single-source energy harvesting systems fail to operate adequately, when the ambient energy is insufficient. To develop a self-sustaining system, a multi-source energy harvesting interface that includes different energy transducers such as a photovoltaic cell (PV) and a thermoelectric generator (TEG) is critical.

Recent work in [1] presents a single-inductor multiple-input-dual-output power converter harvesting energy from different sources; however, the sources are selected manually. The converter needs to cover a wide input voltage range with high conversion efficiency to handle different types of energy transducers. The conventional constant-on-time (COT) limits the conversion efficiency, when the input voltage varies. An adaptive-on-time (AOT) with a constant peak-inductor-current (CPIC) control demonstrated in [2] obtains a high conversion efficiency for different input voltages. However, the efficiency is still low at low V_{IN} and it has been used for a single-source energy harvesting system.

This paper presents a single-inductor triple-input-triple-output (SITITO) buck-boost converter with cycle-by-cycle source tracking (CCST) for PV and TEG hybrid energy harvesting system. The target input voltage ranges from 0.05 V to 0.8 V for accommodating the output voltages of the PV and TEG. The converter generates two regulated outputs and store the excessive energy to the storage capacitor (C_{OUT3}). The first output voltage (V_{OUT1}) is regulated at 1.2 V for analog and RF circuit, while the second output (V_{OUT2}) is regulated between 0.4 V to 0.6 V for energy-efficient digital circuits. The proposed cycle-by-cycle source tracking (CCST) detects input and output conditions in each cycle and automatically select an appropriate source according to maximum power point (MPP) of energy transducers. Thus, the converter is capable of continuously extracting the maximum power from both energy transducers. To obtain a high conversion efficiency under wide input voltage range (0.05-0.8 V), an AOT with adaptive peak-inductor-current (APIC) control is developed for dynamically adjusting the on-time based on the input voltages.

The paper is organized as follows. The system architecture and the operational sequences of the converter are presented in Section II. Section III describes the circuit implementation of sub-blocks. The measurement results and a comparison of the state-of-the-art power converters are discussed in Section IV. Finally, conclusions are drawn in Section V.

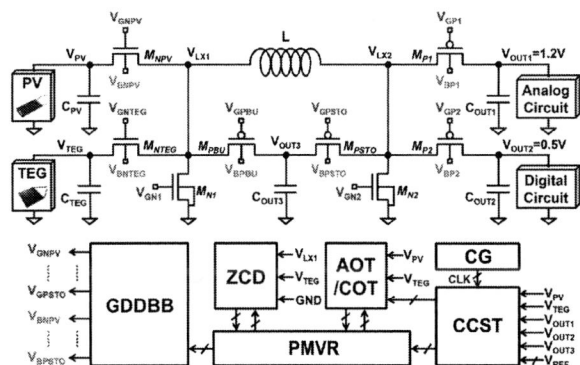

Fig. 1. Block diagram of proposed SITITO buck-boost converter.

978-1-5386-3179-9/17 $31.00 © 2017 IEEE

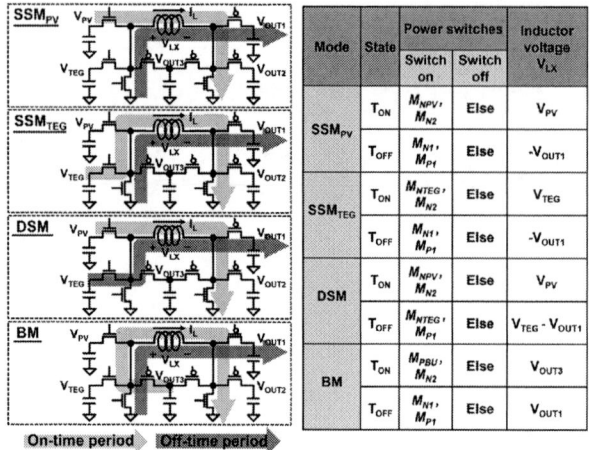

Fig. 2. Power flow in each operation mode.

II. SYSTEM ARCHITECTURE

Fig. 1 shows the block diagram of the proposed SITITO buck-boost converter. It consists of a dual-input triple-output converter core, the clock generator (CG), the cycle-by-cycle source tracking (CCST), the power management and voltage regulation (PMVR), the gate driver and dynamic body bias (GDDBB) and the adaptive on-time (AOT) /constant on-time (COT) circuit. The clock generator (CG) generates the clock signals for the system. The proposed CCST detects the conditions of energy transducers and outputs. The PMVR determines the power flow of the converter core to regulate two of the outputs (V_{OUT1}, V_{OUT2}) and maintain two energy transducers operating at MPP (V_{PV}, V_{TEG}). The AOT/COT circuits generate appropriate on-time signals for the power transistors to obtain high conversion efficiency under wide input voltage range.

To extract the maximum power from transducers while regulating output voltages, the converter is operating at single-source mode (SSM$_{PV}$, SSM$_{TEG}$), dual-source mode (DSM), and a backup mode (BM) according to input and output conditions. Fig. 2 illustrates the power flow in each mode when charging V_{OUT1}. The SSM extracts power from the PV (SSM$_{PV}$) or TEG (SSM$_{TEG}$), whereas the DSM extracts power from both the PV and TEG in a switching period. The BM transfers energy from the storage capacitor to the loads, when the harvested energy is lower than the load requirement. On the other hand, when the energy is higher than the load requirement, the excessive energy is stored in C_{OUT3}.

Fig. 3 illustrates the timing diagram of CCST in the proposed converter. The inductor stores current and delivers it to one of the outputs; the priority sequence is V_{OUT1}, V_{OUT2}, and V_{OUT3}. The converter is operating under discontinuous conduction mode (DCM) and regulate the output voltage using pulse-skip modulation (PSM). The on time in each operating mode is determined by AOT/COT circuit. The ZCD with an offset cancellation controls the off-time of the power transistors to avoid a reverse inductor current. The relationship between the

output (V_{PV}, V_{TEG}) and MPP (V_{PVMPP}, V_{TEGMPP}) of energy transducer is also illustrated in the table in Fig. 3. The power is extracted from energy transducer once the output voltage is higher the corresponding MPP.

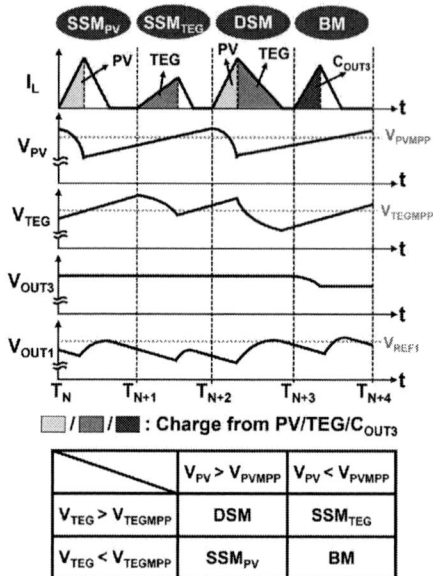

	$V_{PV} > V_{PVMPP}$	$V_{PV} < V_{PVMPP}$
$V_{TEG} > V_{TEGMPP}$	DSM	SSM$_{TEG}$
$V_{TEG} < V_{TEGMPP}$	SSM$_{PV}$	BM

Fig. 3. Timing diagram of CCST illustrating each operating mode of proposed converter.

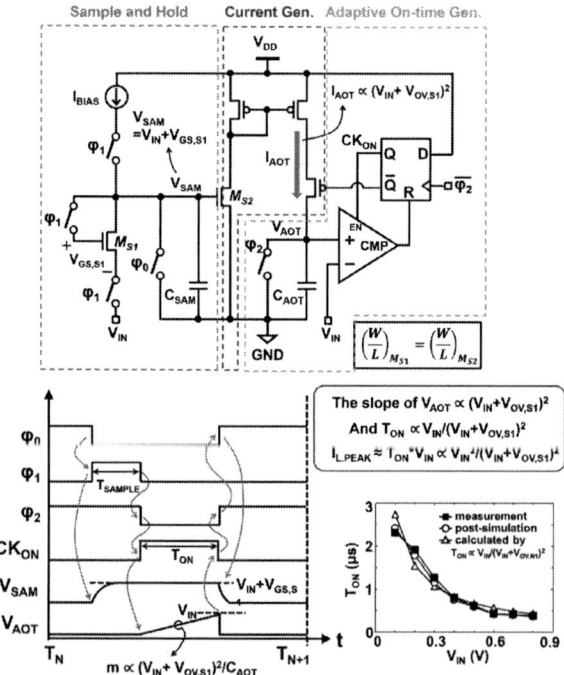

Fig. 4. Circuit schematic and timing diagram of proposed APIC AOT circuit.

978-1-5386-3179-9/17 $31.00 © 2017 IEEE

III. ADAPTIVE PEAK INDUCTOR CURRENT CONTROL

The converter employs APIC technique when the energy is harvested from transducers (SSM_{PV}, SSM_{TEG}, and DSM) to increase the conversion efficiency under different V_{IN}. The APIC covers the input voltage ranging from 0.05-0.8 V. To obtain high conversion efficiency in backup mode, COT technique is employed when the converter extracts energy from C_{OUT3}.

Fig. 4 shows the circuit schematic and the timing diagram of the proposed APIC AOT circuit. The sample and hold generates a voltage (V_{SAM}) related to V_{IN} while current generator produces a modulated current I_{AOT}, that is proportional to $(V_{IN}+V_{OV,S1})^2$. The capacitor C_{AOT} is charged by I_{AOT} and the comparator compares V_{AOT} and V_{IN} to generate the on time signal. As a result, the on time is proportional to $V_{IN}/(V_{IN}+V_{OV,S1})^2$, which reduces as V_{IN} increases.

The proposed APIC AOT circuit can adaptively adjust the peak inductor current according to input voltages (V_{PV} or V_{TEG}) to improve the conversion efficiency under different input voltages. The measurement results demonstrate the generated on time is close to our design.

IV. MEASUREMENT RESULT

The proposed SITITO converter with CCST and APIC has been fabricated in 180 nm CMOS process. The system is demonstrated using a 40x40x4 mm³ TEG and a 40x20x2 mm³ PV module. Fig. 5 shows the chip micrograph with total chip area of 2.9 mm² and active area of 1.8 mm². Fig. 6 shows the measured dependences of conversion efficiency on V_{IN} by the proposed APIC and the conventional CPIC. Compared to the conventional CPIC, the proposed APIC achieves a higher conversion efficiency over a wide input voltage range and an efficiency improvement of 11% is obtained at low V_{IN}.

Fig. 7 shows the measured waveforms under different light illuminations and temperature differences. The converter automatically determines operation modes in each cycle and operates at SSM_{PV}, SSM_{TEG}, and DSM according to input and output conditions. The PV and TEG transfer power more frequently with the increases in the light and temperature difference, respectively. The measured waveform also demonstrates that V_{PV} and V_{TEG} always remain around MPP and the converter extracts the maximum power from both the energy transducers. Furthermore, the converter is capable of harvesting both PV and TEG in one clock period without any external control signals, which is suitable for multi-source energy harvesting system. Due to limited test channels, the measured waveforms at BM is not included in Fig. 7.

Fig. 8 shows the measured dependences of the power conversion efficiency on the output power (P_{OUT}) at SSM, DSM, and BM. SSM at 0.6V is for PV operation while SSM at 0.3 V is for TEG operation. The converter achieves a peak efficiency of 84.4 % at 2mW while covering the output power ranging from 20 μW to 4 mW.

Table I draws a comparison with recently published energy harvesting interface ICs [1-5]. The proposed converter manages two different energy sources at MPP automatically, while simultaneously harvesting both energies during one switching period. The proposed APIC technique enables a wide input voltage range with a high efficiency. The proposed SITITO converter successfully produces three output voltages with a small inductor of 4.7 μH.

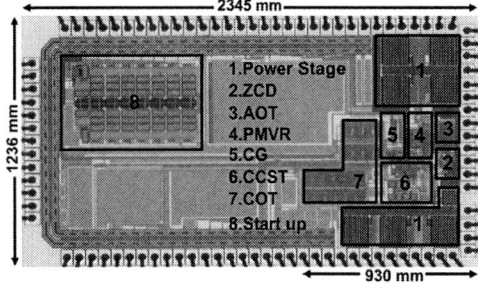

Fig. 5. Chip micrograph.

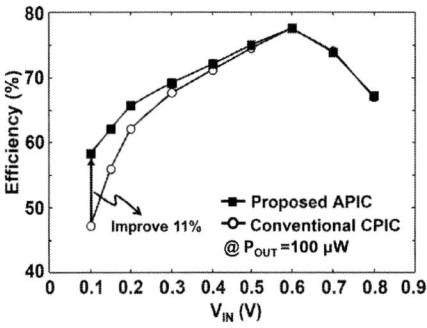

Fig. 6. Measured dependences of efficiency on V_{IN}.

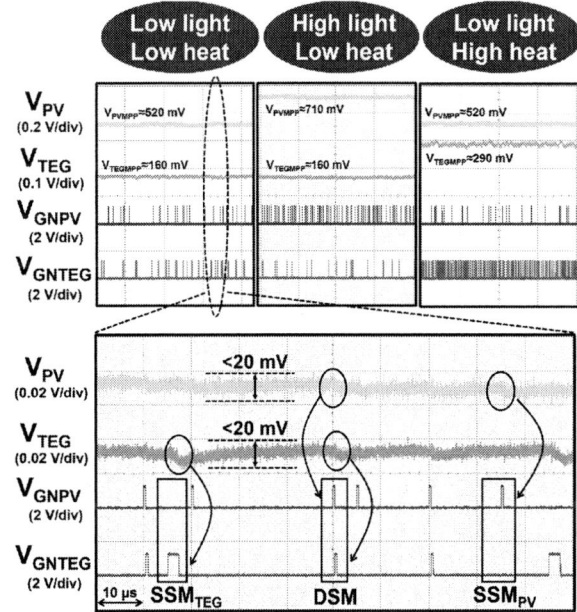

Fig. 7. Measured waveform under different light intensity and temperature differences

TABLE I. PERFORMANCE COMPARISON

	[1] JSSC 12	[2] VLSI 14	[3] JSSC 15	[4] JSSC 15	[5] ISSCC 16	This work
Technology	350 nm CMOS	130 nm CMOS	130 nm CMOS	180 nm CMOS	350 nm CMOS	180 nm CMOS
Input voltage	0.02 V-5 V	0.38 V-3.3 V	0.01 V-0.3 V	-	0.03 V-3.6 V	0.05 V-0.8 V
Regulate mechanism	PWM	PFM	-	PFM&PSM	Hysteretic	PSM
I_L control due to V_{IN}	-	AOT(CPIC)	AOT(APIC)*	-	AOT(CPIC)	AOT(APIC)
Output power	9μW-10mW	1.2μW-100mW	<22mW	1μW-10mW	-	20μW-4mW
Peak efficiency	83%	85%**	83%	83%	85%	84%
Inductor	22 μH	-	10 μH	10 μH	22 μH	4.7 μH
Renewable source	PV&TEG&PZT	PV	PV/TEG	PV	PV	PV&TEG
Source selection	Manual	-	-	-	-	Automatic

* APIC due to the LS resistance from V_{IN}-to-V_{SS}
** 2-stage power conversion

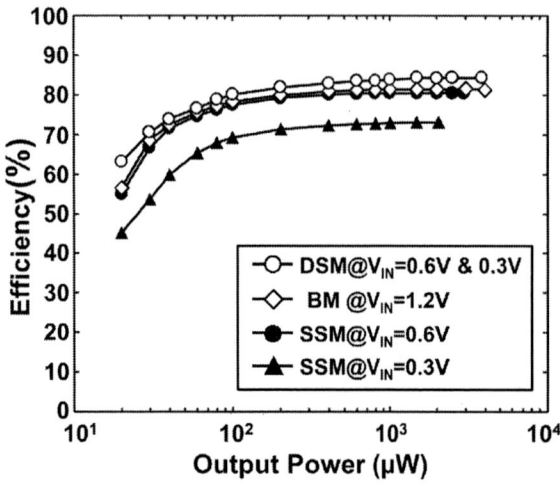

Fig. 8. Measured dependences of power conversion efficiency on output power under different operation modes.

V. CONCLUSION

This paper presents a single-inductor triple-input triple-output (SITITO) energy harvesting interface, capable of simultaneously converting the energy from two energy transducers. The proposed CCST senses the input and output conditions and selects the appropriate source according to MPP of each energy transducer. The operating mode is determined cycle by cycle to make both transducers operating at MPP. The proposed converter achieves a maximum conversion efficiency of 84.4 % with output power ranging from 20 μW to 4 mW. Compared to the conventional CPIC, the proposed APIC improves 11 % conversion efficiency at low V_{IN}. To the best of our knowledge, this is the first energy harvesting interface that is capable of harvesting energy from multiple energy transducers simultaneously.

ACKNOWLEDGMENT

This work was partially supported by MOST, R.O.C. (103-2221-E-009-199-MY3, 105-2221-E-009-180), and MoE ATU Program under NCTU BETRC. We would like to thank the National Chip Implementation Center (CIC) for chip fabrication.

REFERENCES

[1] S. Bandyopadhyay and A. P. Chandrakasan, "Platform Architecture for Solar, Thermal, and Vibration Energy Combining with MPPT and Single Inductor," *IEEE J. Solid-State Circuits (JSSC)*, vol. 47, no. 9, pp. 2199-2215, Sept. 2012.

[2] A. Shrivastava, Y. K. Ramadass, S. Khanna, S. Bartling and B. H. Calhoun, "A 1.2μW SIMO Energy Harvesting and Power Management Unit with Constant Peak Inductor Current Control Achieving 83–92% Efficiency Across Wide Input and Output Voltages," *IEEE Symp. on VLSI Circuits*, pp. 1-2, Jun. 2014.

[3] A. Shrivastava, N. E. Roberts, O. U. Khan, D. D. Wentzloff and B. H. Calhoun, "A 10 mV-Input Boost Converter With Inductor Peak Current Control and Zero Detection for Thermoelectric and Solar Energy Harvesting With 220 mV Cold-Start and −14.5 dBm, 915 MHz RF Kick-Start," *IEEE J. Solid-State Circuits*, vol. 50, no. 8, pp. 1820-1832, Aug. 2015.

[4] G. Yu, K. W. R. Chew, Z. C. Sun, H. Tang and L. Siek, "A 400 nW Single-Inductor Dual-Input–Tri-Output DC–DC Buck–Boost Converter With Maximum Power Point Tracking for Indoor Photovoltaic Energy Harvesting," *IEEE J. Solid-State Circuits*, vol. 50, no. 11, pp. 2758-2772, Nov. 2015.

[5] Y. Lu, S. Yao, B. Shao and P. Brokaw, "A 200nA single-inductor dual-input-triple-output (DITO) converter with two-stage charging and process-limit cold-start voltage for photovoltaic and thermoelectric energy harvesting," *IEEE International Solid-State Circuits Conference (ISSCC) Dig. Tech. Papers*, pp. 368-369, Feb. 2016.

[6] X. Zhang, P. Chen, Y. Okuma, K. Ishida, Y. Ryu, K. Watanabe, T. Sakurai, and M. Takamiya, "A 0.6 V Input CCM/DCM Operating Digital Buck Converter in 40 nm CMOS," *IEEE J. Solid-State Circuits (JSSC)*, vol. 49, no. 11, pp. 2377-2386, Nov. 2014.

[7] R. Damodaran Prabha and G. A. Rincón-Mora, "0.18-μm Light-Harvesting Battery-Assisted Charger–Supply CMOS System," *IEEE Transactions on Power Electronics (TPE)*, vol. 31, no. 4, pp. 2950-2958, Apr. 2016.

978-1-5386-3179-9/17 $31.00 © 2017 IEEE

An 88% Efficiency MPPT for PV Energy Harvesting System with Novel Switch Width Modulation for Output Power 100nW to 0.3mW

Karim Rawy, Taegeun Yoo and Tony T. Kim

School of EEE, Nanyang Technological University, Singapore

Karim002@e.ntu.edu.sg

Abstract— This paper presents a novel three-dimensional maximum power point tracking (3D-MPPT) system for ultra-low power (ULP) solar energy harvesting systems (EHS) for internet of things (IoT) smart nodes. The proposed 3D-MPPT utilizes a gate-source voltage (V_{gs}) dependent switch width modulation (SWM) technique for improving power efficiency (PE) at standby (<1 μA) and heavy (>300μA) load scenarios, and eliminating the gate driver and conduction loss trade-off for a reconfigurable switched capacitor charge pump (SCCP). The proposed SWM technique modulates the SC transistors size in proportion to the load condition, input voltage and V_{gs} applied. The tested chip, fabricated in 65-nm CMOS technology, can harvest from 0.35 V and provides a regulated output voltage at 1 V with peak efficiency of 88% at 200 μW and PE > 60% at 100 nW.

I. INTRODUCTION

Solar energy harvesting (SEH) using photovoltaic cells (PVC) become an attractive solution to power the IoT smart nodes including outdoor and indoor wearable electronics and wireless sensor networks for self-sustaining and prolonged lifetime systems [1-4]. The need of an efficient DC-DC up-conversion along with an MPPT circuit rises due to the limited harvested power level and its environmental condition dependency [4-7]. Many SEH solutions based on hill-climbing MPPT methods deliver hundreds of μW with >70% efficiency. However, delivering hundreds of nW during light load scenarios for low duty functions within IoT sensor nodes with high efficiency is a non-trivial task [1]. Fig. 1 shows an example of a wireless sensor node operation scenario for IoT applications, presented in [8]. Conventional MPPT systems achieve a considerable PE during normal and heavy load conditions (> 50μW). However, their PE is deteriorated to 30-40% at ultra-low load conditions (<1μW) such as standby and light duty functions which occupy more than 50% of the overall operation time [1, 8]. Thus, it degrades the overall operation efficiency, especially at low duty functions within IoT smart nodes and implantable sensors (Fig. 1).

Conversion ratio modulation (CRM) or/and switching frequency (f_{sw}) modulation (SFM) techniques have been reported in literature to track the maximum power point (MPP) [6]. However, using fixed transistor size for the DC-DC converter limits the PE in a wide power level range [1]. It stems from mainly two facts: (i) heavy load requires large transistors to avoid conduction loss while (ii) standby scenarios (<1μA) need small transistors to avoid gate driver loss.

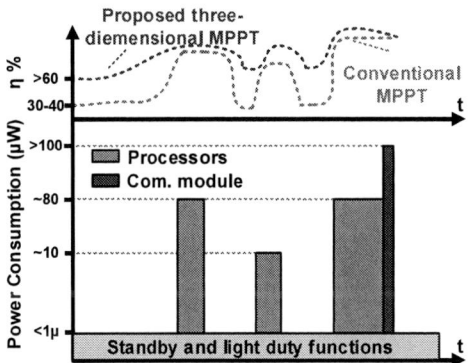

Fig. 1. Wireless sensor operation scenario and advantage of the proposed 3D-MPPT.

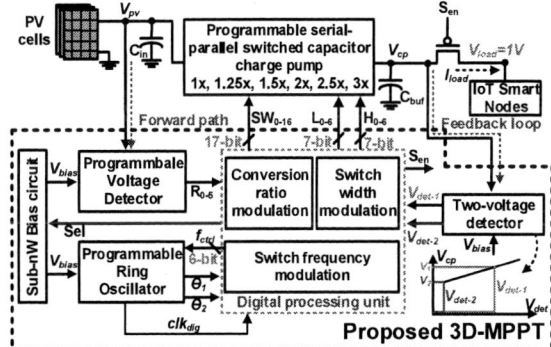

Fig. 2. System diagram of the proposed 3D-MPPT.

In this work, we propose a fully-integrated 3D-MPPT (Fig. 2) with a novel V_{gs} dependent SWM technique that proportionally optimize the transistors size of a reconfigurable SCCP with the load condition and the input PVC voltage along with the V_{gs} applied per switch. Since the proposed technique eliminates the aforementioned trade-off, the MPP is attained by a three-dimensional tracking process, namely SWM, CRM and SFM. It also efficiently cover a wide input range at various load scenarios using a programmable voltage detectors and an ultra-low power digital processing unit (DPU), and avoiding power hungry voltage or current sensors. The proposed 3D-MPPT enhances the PE by at least 20% in the sub-μW load power compared with the conventional MPPT designs.

978-1-5386-3179-9/17 $31.00 © 2017 IEEE

II. PROPOSED 3D-MPPT METHOD

The proposed 3D-MPPT introduces a hill-climbing technique including SWM, CRM and SFM to enhance the PE across a wide load range as shown in Fig.1. The basic idea is to proportionally modulate the SCCP transistors size with I_{load} and V_{pv}. More precisely, the gate driver and conduction power losses depend on the SCCP transistors size at a given V_{gs} and f_{sw} due to the SC converter operation characteristics as follows:

$$P_g \propto C_{sw} V_{gs}^2 f_{sw} \qquad (1)$$

where P_g is the gate driver loss and C_{sw} is the transistor gate capacitance. Hence, at ultra-light load where f_{sw} is reduced by the SFM, small transistors will reduce C_{sw}, minimizes P_g and thus enhances the PE. Utilizing this information for MPPT, the proposed SWM technique optimizes the transistors sizes (i.e. C_{sw}) in proportion to f_{sw}, which is an indicator for the load state (Fig. 3). Due to the quadratic relation between P_g and V_{gs} in (1), PE can be further improved by additional proportional optimization using V_{gs} applied. The SC transistors are divided into two types: (i) low side transistors (P_{LS}) with $V_{gs} \approx V_{pv}$ and (ii) high side transistors (P_{HS}) with $V_{gs} \approx V_{cp}$, as shown in Fig.4. In principle, the higher is the available V_{pv}, the lower is the conversion ratio (C_r). Therefore, C_r can be an indicator to the V_{pv} (i.e. V_{gs}) applied on P_{LS} transistors, and modulates their sizes accordingly besides f_{sw}. However, the P_{HS} size is proportionally modulated only with f_{sw}, since their V_{gs} is constant and identical to V_{cp}. Thus, SWM is triggered simultaneously with the variations in V_{pv} (case-3) or I_{load} (case-2), or both (case-1), as shown in Fig. 3. The SWM technique along with CRM and SFM will be discussed in the next section.

III. PROPOSED 3D-MPPT ARCHITECTURE AND CIRCUIT IMPLEMENTATION

The proposed 3D-MPPT system contains five main blocks (Fig.2), namely, the DPU where the proposed 3D-MPPT algorithm is executed, a programmable ring oscillator (PROSC), programmable input voltage detectors (PIVD) that detect V_{pv} value, two output voltage detectors to regulate V_{cp},

and finally an ULP bias circuit for generating bias voltage (V_{bias}) for the voltage detectors and the PROSC.

To accommodate a wide input range, a reconfigurable serial-parallel (SP) SCCP is implemented (Fig. 4). CRM sets the C_r by a forward path (FWP) realized by the PIVD (Fig.2). By manipulating the SCCP configuration using SW_{0-16}, the SP SCCP can boost V_{pv} with six different C_r values: 1X, 1.25X, 1.5X, 2X, 2.5X and 3X for a regulated 1 V output voltage (Fig. 2).

The DPU depicted in Fig.5 shows the CRM, SFM and SWM control blocks to realize the proposed 3D-MPPT algorithm. The SFM adopts an event-time driven technique realized by a closed-feedback loop (Fig.2). Following the timing diagram in Fig. 6 (a), at a heavy load scenario, f_{sw} increases by one step with each V_{det-2} negative edge (event driven) indicating a V_{cp} droop (i.e. $V_{cp} < V_2$). However, f_{sw} can also increase '$\Delta t_L/t_d$' steps if the V_{det-2} low time (Δt_L) exceeds a predefined time (t_d) (time driven). Likewise, at a light load condition, f_{sw} decreases by one step with each V_{det-1} positive pulse (i.e. $V_{cp} > V_1$). However, if V_{det-1} high time (Δt_H) exceeds t_d, f_{sw} decreases by '$\Delta t_H/t_d$' steps. Δt_L and Δt_H are calculated by the Δt_L and Δt_H digital counters (Fig.5) while the edge detector

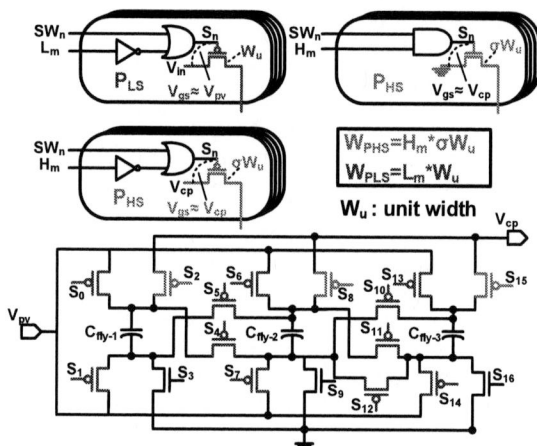

Fig. 4. Circuit schematic of the switched capacitor charge pump and its transistors different types and width.

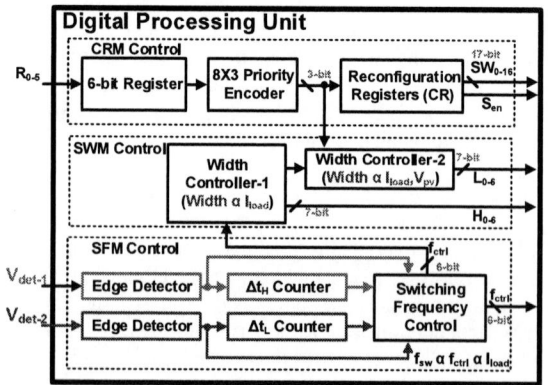

Fig. 5. Digital processing unit (DPU) with SFM, CRM and SWM control block diagram

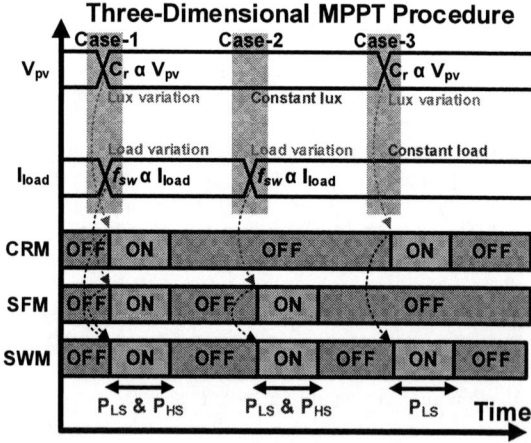

Fig. 3. The proposed three-dimensional MPPT operation

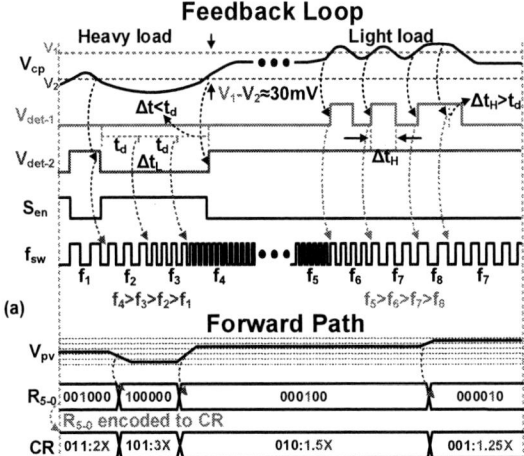

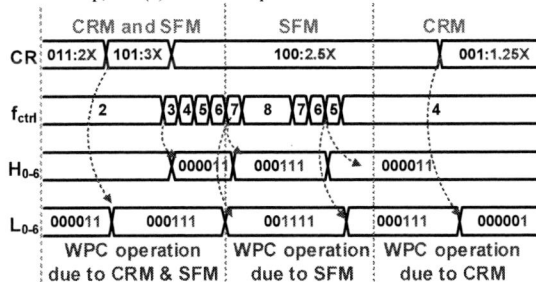

Fig.6. Timing diagram for the digital processing unit (a) with the feedback loop, and (b) the forward path.

Fig. 7. Timing diagram of the proposed switch width modulation technique.

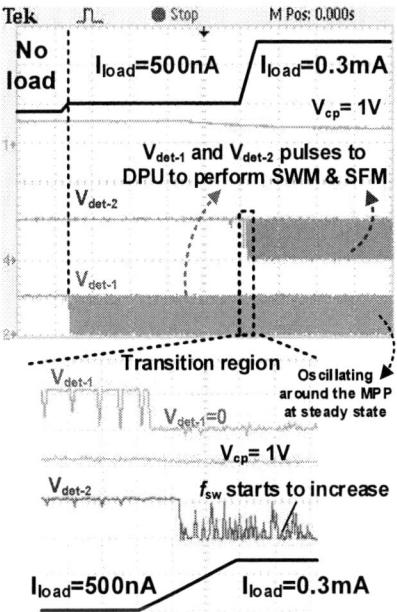

Fig. 8. Transient response of the proposed 3D-MPPT at a 0.3mA load step.

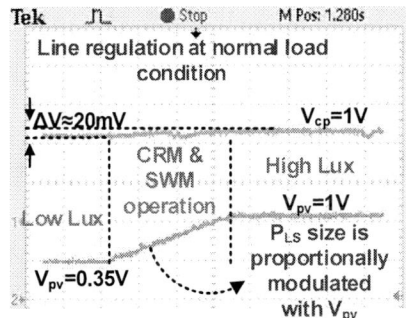

Fig. 9. Regulation performance of the proposed 3D-MPPT.

blocks sense the V_{det-1} and V_{det-2} pulses to increase or decrease f_{sw} accordingly. The f_{sw} control block sets the PROSC output frequency using f_{ctrl} to maintain the regulation and achieve the MPP across the load range. While the CRM control block sets the SCCP C_r by a 6-bit digital word (R_{0-5}) received from the PIVD, which is encoded and stored in a 3-bit reconfiguration register (CR), as explained by the timing diagram in Fig. 6 (b).

The Proposed SWM control (Fig. 5) optimizes P_{HS} and P_{LS} sizes using H_{0-6} and L_{0-6} control signals respectively. L_{0-6} value is set using width controller-2 which uses CR and f_{ctrl} values to proportionally modulate P_{LS} sizes with I_{load} and V_{pv}. Hence, minimizing the P_g losses resulting from unnecessarily large transistor size (i.e. C_{sw}) and the applied variable voltage swing (i.e. V_{pv}). On the other hand, H_{0-6} value, defined by width controller-1, modulates P_{HS} sizes in proportion to I_{load} only, since they have a constant voltage swing (i.e. V_{cp}). The timing diagram shown in Fig. 7 explains the proposed SWM technique along with the SFM and CRM. L_{0-6} is updated with the CR and f_{ctrl} variations, while H_{0-6} is updated only with f_{ctrl}. Fig. 4 illustrates the SP SCCP schematic along with P_{HS} and P_{LS} configuration. The generic size for P_{LS} and P_{HS} (i.e. W_{PHS} and W_{PLS}) is defined by H_{0-6} and L_{0-6} in Fig. 4.

The IC encloses all the tracking algorithms together with auxiliary circuits including PROSC and the sub-nW bias circuit (Fig. 2). The PROSC is designed to generate f_{sw} from 19 KHz

to 16 MHz by altering f_{ctrl} digital value. The sub-nW bias circuit provides multiple V_{bias} which are sequentially connected to the PIVD by the 'Sel' digital signal from the DPU.

IV. MEASUREMENT RESULTS

The proposed 3D-MPPT IC was fabricated in 65-nm CMOS process. The implemented system harvests from 0.35 V to 1 V, and provides a regulated output voltage at 1 V with a <2 nW control circuit and an ultra-low power DPU.

The transient performance of the proposed 3D-MPPT is presented in Fig. 8. When the load varies abruptly from 500 nA (light condition) to 0.3 mA (heavy condition), the DPU receives $V_{det-1/2}$ pulses so to modulate the f_{sw} (SWM) and proportionally modulate the SCCP transistors size (SWM) to track the MPP. Likewise, as V_{pv} varies from 0.35 V (low light irradiance) to 1 V (high light irradiance), as shown in Fig. 9, the DPU modulates the SCCP conversion ratio (CRM) and its transistor sizes in proportion to the applied V_{pv} (i.e. V_{gs}) to achieve the MPP. Initially, V_{det-1} tends to decrease f_{sw}, then during the transition region, V_{det-2} starts its pulses to increase f_{sw}, while

978-1-5386-3179-9/17 $31.00 © 2017 IEEE

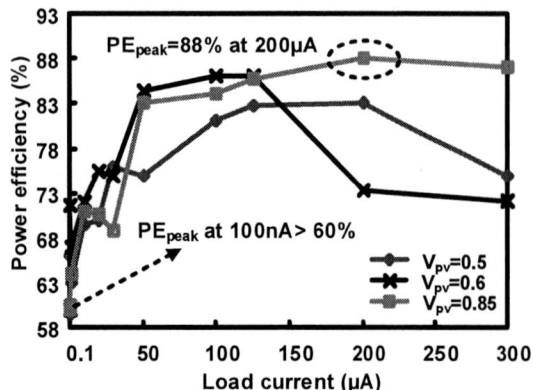

Fig. 10. Power efficiency at different V_{pv} across the load range.

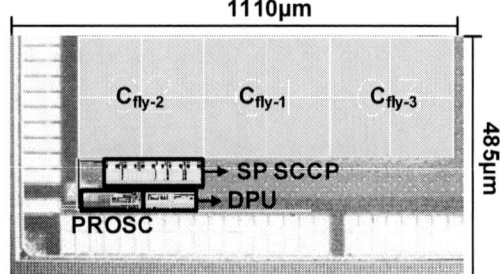

Fig. 11. Die photo of the proposed EHS system.

Table I: Comparison with prior art

Parameters	[6] ISSCC'16	[4] ISSCC'14	[7] ISSCC'14	This work
Technology	0.18μm	0.18μm	0.13μm	**65nm**
Fully integrated	Yes	Yes	No	**Yes**
Conversion ratios	Reconf.	Reconf.	Single	**Reconf.**
MPPT technique	SFM and CRM	SFM	No MPPT	**SFM, CRM and SWM**
Regulation	Yes	Yes	No	**Yes**
Output voltage	1.8V	2.2V-5.2V	0.619V	**1V**
Input range	0.5V-1.8V	0.14V-0.5V	0.15V-0.45V	**0.35V-1V**
Frequency range	NA	70Hz-19MHz	250KHz	**19KHz-16MHz**
Output power range	<34.8μW	5nW-5μW	<60μW**	**100nW-300μW**
Peak Efficiency	72%	50% @ 0.45V	35% @ 0.18V 72.5% @ 0.45V	**83% @ 0.5V 88% @ 0.85V**
Efficiency @ <1 μA	NA	40% @ 100nA**	<10% @ 1μA**	**60.5% @ 100nA**
Cold start-up	Yes	Yes	Yes	**Yes**

** extracted from the measurement results

V_{det-1} turns '0', indicating an abrupt voltage droop due to a heavy load condition (Fig. 8). Finally, at steady state, $V_{det-1/2}$ oscillates around the f_{sw} corresponds to the MPP. The PE of the proposed 3D-MPPT is recorded in Fig. 10 at different V_{pv} across the load range. The tested chip achieved a peak efficiency of 88% at 200 μA (heavy load), and a PE > 60% at 100nA (standby and light load condition). The results show that the proposed SWM technique improves the PE by at least 20% at light load conditions (<1 μA), and by 10% at heavy load conditions (> 100 μA), compared with the prior state-of-the-art

works. This improvement is due to the proportional optimization of the SCCP transistor size with I_{load}, along with the SFM and CRM, eliminating the trade-off between the conduction and gate driver power losses.

The chip has an area of 0.538mm² with three MIM on-chip flying capacitors as shown in Fig. 11. Table I shows the comparison with prior art EHS. The suggested 3D-MPPT improves the PE by at least 20% as compared with the MPPT design in [4]. Hence, the proposed 3D-MPPT with the novel SWM technique improves the PE at various load scenarios for EHS within IoT smart nodes and implantable sensors.

V. CONCLUSION

The proposed 3D-MPPT was implemented with ultra-low power digital and analog circuits. The fabricated IC shows expected results regarding the regulation and transient performance. The novel tracking algorithm eliminates the trade-off between the conduction power loss at heavy load and gate driver power loss at standby and light load conditions, showing a high PE across a wide load range compared with previous works. It successfully achieves a peak PE of 88% at 200 μA, and a PE >60% at ultra-light load. The input harvesting range is extending from 0.35 V to 1 V with a regulated output voltage at 1 V. The implemented EHS is fully integrated without the need of external reference voltage or passive off-chip components, making it suitable for ultra-low power applications within IoT, implantable and wearable sensor networks.

REFERENCES

[1] I. Lee, W. Lim, A. Teran, J. Phillips, D. Sylvester, and D. Blaauw, "21.4 A >78%-efficient light harvester over 100-to-100klux with reconfigurable PV-cell network and MPPT circuit," in *2016 IEEE International Solid-State Circuits Conference (ISSCC)*, 2016, pp. 370-371.

[2] J. Leicht, M. Amayreh, C. Moranz, D. Maurath, T. Hehn, and Y. Manoli, "20.6 Electromagnetic vibration energy harvester interface IC with conduction-angle-controlled maximum-power-point tracking and harvesting efficiencies of up to 90%," in *Solid- State Circuits Conference - (ISSCC), 2015 IEEE International*, 2015, pp. 1-3.

[3] K. Rawy, F. K. George, D. Maurath, and T. T. Kim, "A Time-based Self-Adaptive Energy-Harvesting MPPT with 5.1-μW Power Consumption and a Wide Tracking Range of 10-μA to 1-mA " in *ESSCIRC Conference 2016: 42nd European Solid-State Circuits Conference*, 2016, pp. 503-506.

[4] J. Wanyeong, O. Sechang, B. Suyoung, L. Yoonmyung, D. Sylvester, and D. Blaauw, "23.3 A 3nW fully integrated energy harvester based on self-oscillating switched-capacitor DC-DC converter," in *Solid-State Circuits Conference Digest of Technical Papers (ISSCC), 2014 IEEE International*, 2014, pp. 398-399.

[5] J. Goeppert and Y. Manoli, "Fully Integrated Startup at 70 mV of Boost Converters for Thermoelectric Energy Harvesting," *IEEE Journal of Solid-State Circuits*, vol. 51, pp. 1716-1726, 2016.

[6] X. Liu and E. Sanchez-Sinencio, "21.1 A single-cycle MPPT charge-pump energy harvester using a thyristor-based VCO without storage capacitor," in *2016 IEEE International Solid-State Circuits Conference (ISSCC)*, 2016, pp. 364-365.

[7] K. Jungmoon, P. K. T. Mok, and K. Chulwoo, "23.1 A 0.15V-input energy-harvesting charge pump with switching body biasing and adaptive dead-time for efficiency improvement," in *Solid-State Circuits Conference Digest of Technical Papers (ISSCC), 2014 IEEE International*, 2014, pp. 394-395.

[8] Y. Lee, Y. Kim, D. Yoon, D. Blaauw, and D. Sylvester, "Circuit and system design guidelines for ultra-low power sensor nodes," in *DAC Design Automation Conference 2012*, 2012, pp. 1037-1042.

A DVS-Based Burst Mode with Automatic Entrance Point Control Technique in DC-DC Boost Converter for Wearable Devices and IoT Applications

Chiao-Hung Cheng[1], Li-Chi Lin[1], Jian-He Lin[1], Ke-Horng Chen[1], Ying-Hsi Lin[2], Jian-Ru Lin[2], and Tsung-Yen Tsai[2]

[1]Institute of Electrical Engineering, National Chiao Tung University, Hsinchu Taiwan, [2]Realtek Semiconductor Corp., Hsinchu, Taiwan

Abstract—the proposed dynamic voltage scaling (DVS) based burst mode links the DVS technique in a system on a chip (SoC) and the burst mode operation of the DC-DC converter for further efficiency improving. Conventional burst mode has only one entrance transition point (ETP) between the pulse width modulation (PWM) and the burst mode, so the voltage ripple is high at low DVS voltage while the efficiency is low at high DVS voltage. The proposed DVS-based burst mode uses the automatic entrance point control (AEPC) technique to decide multiple ETP values corresponding to the voltage identification (VID) code from the SoC. The quality enhancement technique deliberately adjusts the burst reference voltage to further reduce the output ripple with acceptable loss of efficiency. The tested DC-DC boost converter with the DVS-based burst mode technique is fabricated in 0.18μm CMOS process. Measurement results show that the efficiency is higher than 85% when the output voltage varies from 1.8V to 3.2V (controlled by the DVS) and load current ranges from 0.1mA to 140mA, with peak efficiency 94%.

I. INTRODUCTION

The development of electronics in the Internet of Things (IoT) and wearable electronics has become an important trend in consumer electronics. Due to the limited battery capacity, it is necessary to extend the usage time of these electronics. Thus, the well-known dynamic voltage scaling (DVS) technique adjusts the supplying voltage V_{OUT} to V_{OUT1}-V_{OUTn} corresponding to each operation performance point (OPP) as shown in Fig. 1(a) where the microprocessor schedule decides suitable OPPs corresponding to different tasks [1]. In general, task-based DVS technique can reduce power loss by dynamically adjusting the supplying voltage V_{OUT} for different tasks. However, the power converter does not have changed entrance transition point (ETP) corresponding to different V_{OUT} when it transits between the pulse width modulation (PWM) and the burst mode [2]. Although the load-dependent switching frequency in the burst mode reduces switching power loss, the power converter cannot find its optimum operation among the output voltage ripple percentage ($\Delta V_{OUT}/V_{OUT}$) and the power saving. In Fig. 1(b), the ETP of conventional burst mode is set at a fixed loading current point for decreasing design complexity. If the V_{OUT}=1.8V, the ETP is too early so the $\Delta V_{OUT}/V_{OUT}$ is too large. Contrarily, if the V_{OUT}=3.2V, the ETP is too late to cause the deterioration of power conversion efficiency.

In conventional designs, there is no relationship between the ETP and the DVS technique in the system-on-a-chip (SoC) applications. This loose connection between ETP and DVS cannot find an optimum operation between efficiency and performance of IoT devices. Therefore, the automatic entrance point control (AEPC) technique in this paper decides the ETP according to the output voltage ripple percentage ($\Delta V_{OUT}/V_{OUT}$) as depicted in Fig.

2(a). The DVS-based burst mode uses multiple ETP values set by 8-bit voltage identification (VID) code (VID[0:7]) to find the trade-off between efficiency and ΔV_{OUT}. That is, smaller V_{OUT} can tolerate smaller ΔV_{OUT}. The ETP should be late. On contrarily, larger V_{OUT} can tolerate larger ΔV_{OUT}. The ETP should be early for power saving. Consequently, the power saving performance becomes more apparent at higher V_{OUT} as illustrated in Fig. 2(b) where the efficiency is raised when the V_{OUT} is 3.2V with an early ETP.

In general, the task-based DVS technique of the microprocessor decides the V_{OUT} while the power converter roughly adjusts the ETP corresponding to each task. However, different task is constituted by versatile instructions including low and high performances. In other words, the power converter needs to fine-tune the ETP value for different instructions. In this paper, the proposed DVS-based burst mode technique includes more instruction information from the microprocessor to improve the power saving ability. A 3-bit $V_{quality}[0:2]$ is used to fine-tune the performance between the efficiency and the ΔV_{OUT}.

The paper is organized as follows. Section II discusses the design concept of the proposed DVS-based burst mode with the AEPC technique and illustrates the system operation. It also presents the reduction of ΔV_{OUT}, efficiency, and transient response. Section III shows the circuit implementations, and experimental results are shown in Section IV. Finally, conclusions are made in Section V.

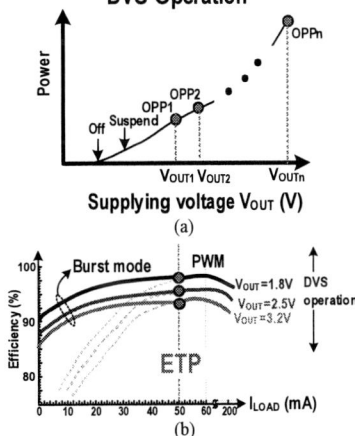

Fig. 1. (a) Different OPPs are set by the DVS technique in the microprocessor. (b) The efficiency degradation at different V_{OUT} due to DVS technique because one ETP is set in conventional burst mode.

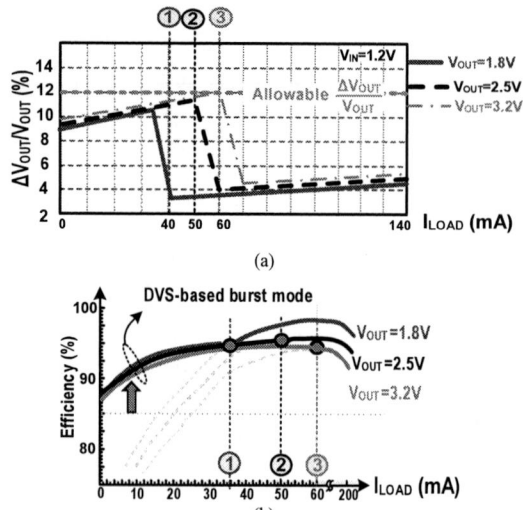

(a)

(b)

Fig. 2. (a) The decision of the ETP is based on the output voltage ripple percentage ($\Delta V_{OUT}/V_{OUT}$). (b) Different ETP results in different ΔV_{OUT}.

II. PROPOSED DVS-BASED BURST MODE CONTROL DC-DC BOOST CONVERTER

For wearable and IoT electronics, the boost converter in Fig. 3 is used to get a higher V_{OUT} since the capacity and volume of input voltage V_{BAT} are usually limited. The power management (PM) master controlled by the microprocessor in the SoC implements the DVS technique by two signals VID[0:7] and $V_{quality}[0:2]$. VID[0:7] code sets dynamic scaling output voltage V_{OUT} and multiple ETP values while $V_{quality}[0:2]$ fine-tunes the ΔV_{OUT} for different instructions.

In order to fit the DVS-based burst mode to the peak current mode control, the current sensor detects the inductor current signal V_S and generates a bias to control the weighted current sources in the dynamic current scaling (DCS) circuit for tuning the desired ΔV_{OUT}. The output control signal V_{DCS} of the DCS circuit changes the original reference voltage V_{REF} to the reference voltage $V_{REF(Burst)}$ in the DVS-based burst controller corresponding to the versatile instructions of each task. The VID[0:7] of the DVS technique in the SoC can vary its supplying voltage of operation while it also sets the ETP of the DVS-based burst mode of the boost converter. Besides, $V_{quality}[0:2]$ controlled by the versatile instructions of each task has trade-off between efficiency and supplying performance.

Fig. 4 shows the PWM and burst mode controller. PWM mode operation ensures reduced ΔV_{OUT} and higher driving capability while the burst mode operation reduces much power loss according to DVS technique and the versatile instructions of each task. Owing to the peak current mode control, the output of the error amplifier V_{EA} controls the peak inductor current level of the current-mode boost converter. Thus, different VID code (VID[0:7]), which is sent from the SoC, can set its own ETP by the V_{REF_VID} to decide when the controller transits from the PWM to the burst mode.

Once the V_{EA} is higher than V_{REF_VID}, the reference voltage V_{REF} used by the PWM is changed to the adaptive reference voltage $V_{REF(Burst)}$ by the V_{DCS}. A 3-bit $V_{quality}[0:2]$ further adaptively tunes the value of V_{DCS} in the DCS circuit for controlling the ΔV_{OUT}. Owing to the change of $V_{REF(Burst)}$ according to the loading current and VID code, the ETP entrance can be adapted to the different IoT applications in Fig. 1 by the AEPC technique for finding trade-off between efficiency and performance.

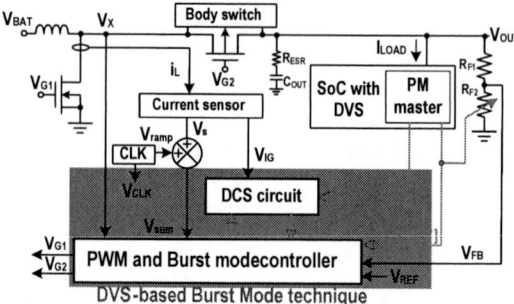

Fig. 3. Boost converter with the proposed DVS-based burst mode technique.

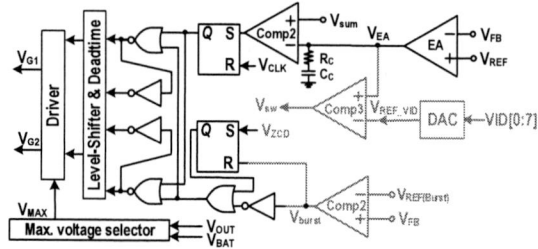

Fig. 4. PWM and burst mode controller.

III. CIRCUIT IMPLEMENTATION

Fig. 5 shows the circuit implementation of current sensor and DCS circuit. The low-side current sensor uses the transistor M_7 to scale down the inductor current. The derived inductor current signal is V_S. Besides, it converts the current information to a bias to the weighted current source in the DCS circuit to change the value of I_{DCS}. In the DCS circuit, the $V_{quality}[0:2]$ controls the weighted current sources implemented by transistors $M_{W0} - M_{W7}$. The ΔV_{OUT} can be fine-tuned by the 3-bit $V_{quality}[0:2]$ according to the loading control and DVS-based reference voltage $V_{REF(Burst)}$.

Once the ETP value is detected by the comparator 'Comp3' in Fig. 4, the V_{SW} is set high to trigger the DCS circuit. If the I_{LOAD} is smaller than the setting value V_{REF_VID} which is controlled by the VID[0:7], the DVS-based burst mode starts to reduce power loss. Meanwhile, the ΔV_{OUT} is also controlled by the 3-bit $V_{quality}[0:2]$, so the ΔV_{OUT} is reduced by increasing the value of $V_{REF(Burst)}$ as shown in Fig. 6. Owing to increasing $V_{REF(Burst)}$, more burst pulses are generated to reduce the ΔV_{OUT}. In contrast, ΔV_{OUT} can be slightly increased by decreasing the value of $V_{REF(Burst)}$ to reduce the number burst pulses.

Consequently, the switching frequency can be slightly reduced for reducing switching power loss and increasing efficiency.

IV. EXPERIMENTAL RESULTS

The boost converter with the DVS-based burst mode and the AEPC technique is verified by converting input voltage of 1.2V to 1.8V to dynamic scaling output voltage from 1.8V to 3.2V. The switching frequency of the converter is 1.5MHz, the output capacitor is 4.7μF, and the inductor is 2.2μH. An 8-bit VID[0:7] code is used to have 5mV for each voltage step. Besides, the SoC system requests loading current ranging from 0.1mA to 100mA.

At low V_{OUT}=1.8V, it is suitable to set a lower ETP value for meeting the requirement of allowable $\Delta V_{OUT}/V_{OUT}$. At I_{LOAD}=30mA, the boost converter in Fig. 7(a) operates at the PWM mode without the DVS-based burst mode. Thus, the efficiency is sacrificed although its ΔV_{OUT} is 6mV. High switching power loss is not acceptable. In contrast, the ETP in Fig. 7(b) is dynamically set at 40mA by the proposed AEPC technique. As a result, the reduced switching frequency increases efficiency about 5% at a little cost of increasing ΔV_{OUT} (around 22mV). $\Delta V_{OUT}/V_{OUT}$ is 12% and acceptable by the IoT electronics at light loads.

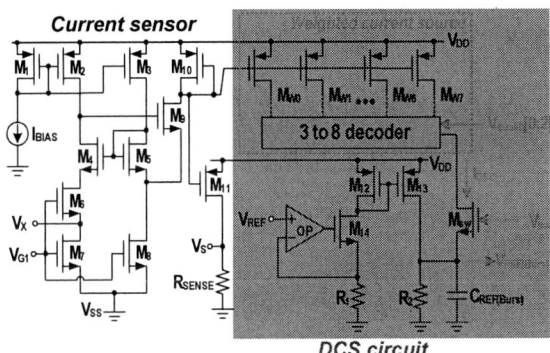

Current sensor

DCS circuit

Fig. 5. Current sensor and the DCS circuit.

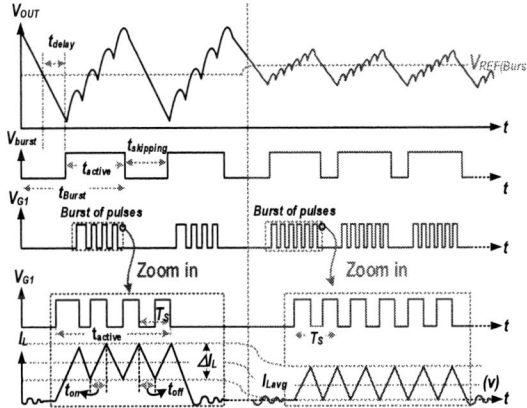

Fig. 6. ΔV_{OUT} is also reduced by increasing the value of $V_{REF(Burst)}$.

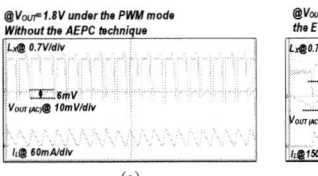 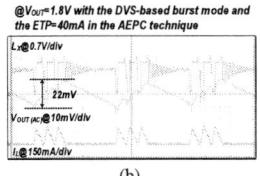

Fig. 7. (a) The ΔV_{OUT} is small but the efficiency is sacrificed due to the PWM operation. (b) The AEPC technique can set the ETP to have the burst operation according to the VID code where ETP is low for meeting allowable $\Delta V_{OUT}/V_{OUT}$.

To avoid data loss at transmission or storage on IoT, small supplying voltage ripple is demanded even in the burst mode operation. The quality-enhanced control signal $V_{quality}[0:2]$ is set by different instructions in the task to get low voltage ripple for ensuring high performance operation of the SoC. Fig. 8 shows the measured waveforms at the I_{LOAD}=45mA where the DVS-based burst mode works since the V_{OUT}=3.2V has a higher ETP=60mA. The quality-enhanced function for supplying voltage ripple reduction by setting $V_{quality}[0:2]$ also works. Fig. 8(a) shows a large ΔV_{OUT} of 58mV ($\Delta V_{OUT}/V_{OUT}$=18%) since $V_{quality}[0:2]$ is set at a low value. It is larger than the allowable 12%. As a result, it easily degrades the SoC performance or causes data error in access, communication or cloud storage for IoT applications.

Fortunately, the output voltage quality can be enhanced by increasing the value of $V_{quality}[0:2]$. Due to the setting of Fig. 8 (b) and (c), the ΔV_{OUT} are reduced gradually to about 25mV and 13mV, respectively, corresponding to $\Delta V_{OUT}/V_{OUT}$ of 8% and 4% separately due to a large value of $V_{quality}[0:2]$. Finally, Fig. 8(d) shows that the output voltage ripple can be reduced to less than 10mV while $V_{quality}[0:2]$ is at its highest value. However, the power converter seems to work at the PWM mode, so it is not suitable to have a highest $V_{quality}$ value because the efficiency will be greatly decreased. The quality-enhancement control has the trade-off between efficiency and output voltage ripple in the proposed DVS-based burst mode. In other words, the task-based DVS technique can fit to different instructions in the SoC. It does not increase the cost of the system and achieves smart power management of wearable devices.

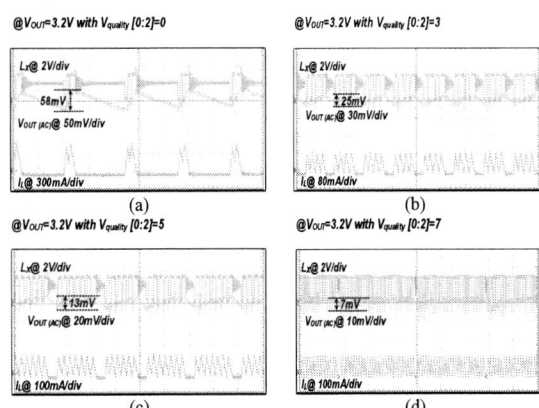

Fig. 8. The quality enhanced output voltage by setting $V_{quality}[0:2]$ when V_{OUT}=3.2V, ETP=60mA, and I_{LOAD}=45mA. (a)$V_{quality}[0:2]$=0. (b) $V_{quality}[0:2]$=3. (c) $V_{quality}[0:2]$=5. (d) $V_{quality}[0:2]$=7.

Fig. 9 demonstrates the transient behavior between the PWM and the DVS-based burst mode. The output voltage ripple is increased from 8mV to 28mV when load current steps down from 75mA to 60mA where ETP=60mA, $V_{quality}[0:2]=3$, and $V_{OUT}=3.2V$. The $\Delta V_{OUT}/V_{OUT}$ is kept smaller than 12%, which is acceptable for the IoT devices.

Experimental results of efficiency and voltage ripple versus load current from 0.1mA to 140mA are illustrated in Fig. 10 and 11, respectively. Conventional burst mode in Fig. 10 shows the three different DVS activities corresponding to three output voltages 1.8V, 2.5V and 3.2V where the ETP has a fixed value of 50mA. It is obvious to see the efficiency degradation, which is below 80% at light loads, when the V_{OUT} is 3.2V in Fig. 10(a). Moreover, the $\Delta V_{OUT}/V_{OUT}$ is larger than 12% when the V_{OUT} is 1.8V.

On the other hand, Fig. 11 shows the measured statistic results when the DVS-based burst mode is implemented. The measured efficiency of these conditions is higher than 85% at light loads and the peak efficiency is 94%. The maximum $\Delta V_{OUT}/V_{OUT}$ is smaller than 12%. In conclusion, the V_{OUT} is changed by the DCS technique while the efficiency and the $\Delta V_{OUT}/V_{OUT}$ can be considered at the same time by the proposed DVS-based burst mode. Fig. 12 shows the measured $\Delta V_{OUT}/V_{OUT}$ with the setting $V_{quality}[0:2]=5$ in the DVS-based burst mode. Obviously, the efficiency is only sacrificed a little bit (about 1%) while the $\Delta V_{OUT}/V_{OUT}$ is further reduced to smaller than 10% by the quality enhancement technique. High quality supplying voltage is still provided in the burst mode if the instruction sets need it.

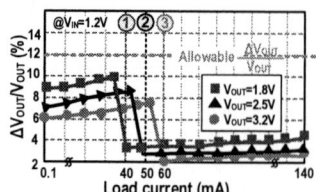

Fig. 12. $\Delta V_{OUT}/V_{OUT}$ versus load current w/i $V_{quality}[0:2]=5$. High quality supplying voltage is still provided in the burst mode.

Fig. 13 shows the chip micrograph and the prototype of the proposed boost converter. Table. I summarizes the performance of the developed DVS-based burst mode boost converter and shows the comparison with state-of-the-arts. The proposed AEPC technique achieves high efficiency over a wide load range corresponding to different dynamic voltage scaling values.

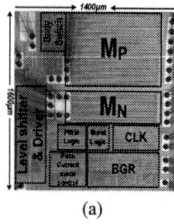 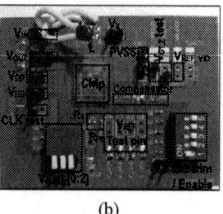

(a)   (b)

Fig. 13. (a) Chip micrograph. (b) Prototype of the proposed DVS-based boost DC-DC converter.

Table. I: Specifications and comparison with prior arts.

	This work	[3]	[4]
Topology	Boost	Boost	Boost
Control scheme	DVFS-based burst	Burst	Burst
L, C_OUT	2.2μF/4.7μF	2.2μF/10μF	6.8μF/4.7μF
F_SW	1.5MHz	1.25MHz	2MHz
Input voltage	1.2V~2.8V	0.7V~5.5V	3V~40V
Output voltage	1.8V~3.2V	1.8V~5.5V	48V
Output load	0.1m~140mA	10m~100mA	5m~135mA
Output voltage ripple	<10mV	<100mV	<15mV
Peak efficiency	94%@0.09A	93%@0.09A	90%@0.135A

V. CONCLUSIONS

The proposed DVS-based burst mode boost DC-DC converter with the AEPC technique is fabricated in 0.18μm CMOS process. In this paper, the link of the DVS technique of the SoC and the burst mode of the DC-DC converter further enhances the conversion efficiency. Due to different instructions in the task of the DVS technique, the proposed quality-enhanced control can get the trade-off between the efficiency and the voltage ripple in the DVS-burst mode control. Measurement results show that overall efficiency is higher than 85% from 0.1mA to 140mA and the peak efficiency is 94% at the load current of 90mA.

REFERENCES

[1] Brian Zimmer, et al., "A RISC-V Vector Processor With Simultaneous-Switching Switched-Capacitor DC–DC Converters in 28 nm FDSOI ", *IEEE Journal of Solid-State Circuits*, pp. 930-942, Apr. 2016.

[2] Ferran Reverter, et al., "Optimal Inductor Current in Boost DC/DC Converters Operating in Burst Mode Under Light-Load Conditions", *IEEE Transactions on Power Electronics*, pp. 15-20, Jan. 2016.

[3] *TPS61099 Datasheet*, " TPS61099x Synchronous Boost Converter with Ultra-Low Quiescent Current", Texas Instruments, Dallas, TX, USA, 2016.

[4] *LT8330 Datasheet*, " Low IQ Boost/SEPIC/Inverting Converter with 1A, 60V Switch", Linear Technology, Milpitas, CA, USA, 2015.

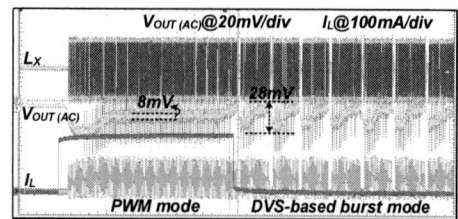

Fig. 9. During the boundary of the PWM and the DVS-based burst mode, the power saving is achieved by the APEC technique again.

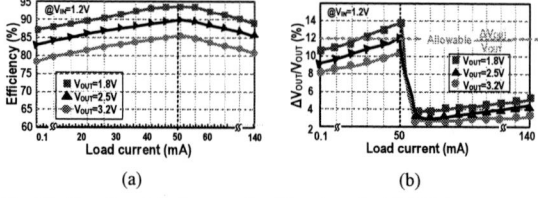

(a)   (b)

Fig. 10. Experimental of conventional burst mode. (a) Efficiency versus load current. (b) $\Delta V_{OUT}/V_{OUT}$ versus load current.

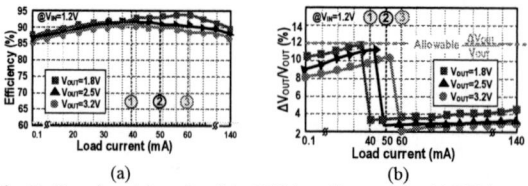

(a)   (b)

Fig. 11. Experimental results of the DVS-based burst mode. (a) Efficiency versus load current. (b) $\Delta V_{OUT}/V_{OUT}$ versus load current.

A Wide Load and Voltage Range Switched-Capacitor DC-DC Converter with Load-Dependent Configurability for DVS Implementation in Miniature Sensors

Hassan Saif, Yongmin Lee, Minsun Kim, Hyeonji Lee, Muhammad Bilawal Khan and Yoonmyung Lee

Department of Electrical and Computer Engineering, Sungkyunkwan University, Suwon, Korea

Email: hsaif194@skku.edu, yoonmyung@skku.edu

Abstract—A new switched capacitor (SC) converter for powering miniature sensor systems with wide load current and output voltage ranges is proposed in this paper. By adopting a multiple-ratio SC stage and reconfigurable stage interconnect scheme, the proposed converter offers fine granularity of conversion ratios, which improves efficiency for light load operation. The multiple-leaf structure for switch size control further improves light load efficiency, while parallel stage configuration improves heavy load efficiency. Under light load (500 nA), the converter maintains an efficiency of greater than 62% with a peak efficiency of 78% over a wide voltage range (400 mV to 1.6 V). Furthermore, with an active state load of 200 µA, it maintains an efficiency of greater than 64% with a peak efficency of 72% over an output voltage range of 700 mV to 1.6 V. The effective step resolution of the proposed converter is 16 mV from a 4 V battery.

Keywords— switched-capacitor converter; DC-DC converter; wide load and voltage range; miniature sensors; DVS

I. INTRODUCTION

Achieving energy-efficient operation in battery-powered miniature sensor systems [1] is a key design goal for maximizing lifetime. In such systems, battery voltage, typically 3.4-4.2 V for high energy density Li batteries, should be down-converted to circuit operation voltage, which can range from 1.6 V down to 0.4 V, for energy-efficient near-threshold operation of digital circuits. At this voltage range, efficient conversion is required not only for 100 s of µA load current in active mode but also for sub-µA load current for standby mode since such systems can spend most of the time and consume a significant portion of the total energy in standby mode.

A switched capacitor (SC) converter is an attractive solution for such miniature systems since it can be fully integrated and operate with relatively high efficiency for µA-order load current compared with inductor-based schemes. However, due to its topology-dependent nature, it can only provide a limited number of voltage conversion ratios, which limits implementation of dynamic voltage scaling (DVS) for efficient operation and adaptive voltage regulation with battery voltage degradation. Moreover, maintaining high efficiency for both active and standby mode operation is a nontrivial challenge, especially when the output voltage is low, i.e., when the conversion ratio is high.

A number of SC topologies for use in such systems have been recently proposed [2-7]. A few multi-stage SC converters [2-4] provide good efficiency over a fairly wide output voltage range, but the efficiency at near threshold voltages (0.4-0.7 V)

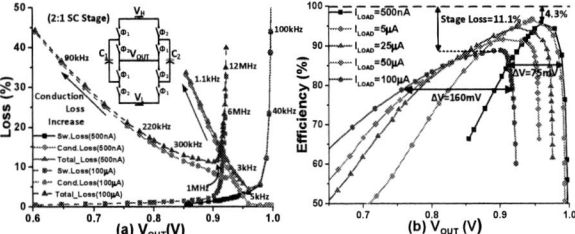

Fig. 1 (a) Loss contribution in a single stage 2:1 SC converter in light (500nA) and heavy (100µA) load conditions (b) conversion efficiency trend with different I_{LOAD}

drastically degrades to less than 40%, especially with loads less than 1 µA. Single-stage SC converters [5-7] can provide high efficiencies but only at a few fixed conversion ratios.

In this paper, a new SC converter with a load-dependent topology configurability is proposed to achieve high efficiency over wide load current and output voltage ranges. By adopting a multi-ratio, multi-leaf stage structure, fine conversion ratio resolution is achieved with fewer cascaded stages, minimizing conduction and switching loss. Reconfigurable stage interconnect (RSI) allows the selection of the optimal stage connection depending on the load current level for a target output voltage.

II. LOSS ANALYSIS IN SC CONVERTERS

In an SC converter, the contribution of conduction and switching losses can drastically vary with load conditions. Fig. 1 shows such variation with simulation results for a single-stage 2:1 SC converter when the no-load output voltage (V_{NL}) is 1.0 V.

In the light load condition (I_{LOAD}=500 nA) in Fig. 1(a), the total loss is minimal when the output voltage (V_O) is ~0.96 V, which is the optimal operation voltage (V_{opt}). To increase V_O above V_{opt}, the operating frequency must be increased exponentially, which also increases the switching loss exponentially. On other hand, to decrease V_O below V_{opt}, the frequency must be reduced. Since the converter is operating at a low frequency (less than 5 kHz), the voltage ripple for the flying capacitors rapidly increases as the frequency is lowered. Such an increase in ripple causes a sharp increase in conduction loss as V_O is lowered below V_{opt}. This results in a narrow convex-shaped total loss curve for light-load SC converter operation.

Under a heavy-load condition (I_{LOAD}=100 µA), in the same manner, switching loss increases exponentially as V_O is

978-1-5386-3179-9/17 $31.00 © 2017 IEEE

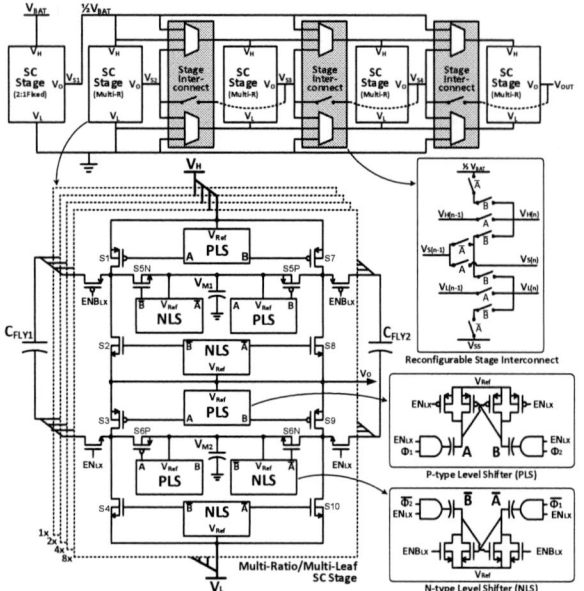

Fig. 2. Structure of the proposed SC converter

increased above V_{opt}. However, when V_O is decreased below V_{opt}, the voltage ripple increases at a slower rate than in the light-load scenario since it is operating at a higher frequency (1 MHz). This results in a slower increase in conduction loss and a wider convex-shaped total loss curve.

The trends discussed above result in the conversion efficiency curves shown in Fig. 1(b). Two observations can be made from this plot:

1) The efficiency curve around V_{opt} is narrower with a light load compared with a heavy load. For example, the voltage range (ΔV) that exhibits 10% lower efficiency than peak efficiency is 160 mV for 100 μA load and 75 mV for 500 nA load. This implies that many of these efficiency curves with different V_{NL} are required to maintain high efficiency over a range of output voltages with light-load current. Therefore, achieving fine-resolution by cascading many SC stages is required for efficient voltage conversion under light load.

2) The loss per stage around V_{opt} is significantly larger with a heavy load compared with a light load (11.1% vs 4.3%). This implies that if more SC stages are cascaded to generate many conversion ratios to achieve a wide output voltage range, the overall efficiency will drop significantly under heavy-load conditions. Therefore, the number of cascaded stages should be minimized.

Considering these observations, the following strategies can be applied to multi-stage SC converters. For efficient light-load operation, fine resolution is required with cascading stages. By employing a multi-ratio stage structure that can provide multiple conversion ratios, fine voltage resolution can be achieved with fewer stages, resulting in efficiency improvement by limiting conduction loss. However, in the light-load loss curve, switching loss is a significant factor around V_{opt} due to the large switches that are sized for minimizing conduction loss in heavy-load conditions. To minimize the switching loss, binary-sized switches can be selectively activated depending on the load condition.

TABLE I. STAGE INTERCONNECTION SCHEME

Control Bits		$V_{H(n)}$	$V_{L(n)}$	$V_{S(n)}$
A_n	B_n			
0	0	$V_{S(n-1)}$	V_{SS}	$\neq V_{S(n-1)}$
0	1	$\frac{1}{2}V_{BAT}$	$V_{S(n-1)}$	$\neq V_{S(n-1)}$
1	X	$V_{H(n-1)}$	$V_{L(n-1)}$	$= V_{S(n-1)}$ (parallel)

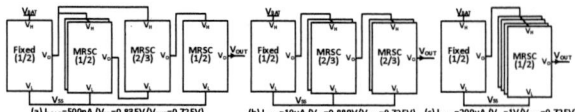

(a) I_{LOAD}=500nA (V_{NL}=0.835V/V_{OUT}=0.725V) (b) I_{LOAD}=10μA (V_{NL}=0.889V/V_{OUT}=0.725V) (c) I_{LOAD}=200μA (V_{NL}=1V/V_{OUT}=0.725V)

Fig. 3. Example optimal configurations for various load conditions with target V_{OUT} of 0.725V

For heavy-load operations, fewer stages are required. Therefore, parallel connection of unused stages can improve efficiency by increasing the flying capacitance per stage and decreasing conduction loss by effectively increasing switch sizes. Moreover, cascading loss can be reduced by providing connection to power or ground at each stage by adopting a recursive connection structure [4].

III. SYSTEM ARCHITECTURE

Fig. 2 shows the top level architecture of the proposed SC converter. Five multi-ratio SC stages are cascaded with RSIs, allowing generation of the desired output voltage (V_{OUT}) by adjusting the conversion ratio in the SC stages and the connection configuration in the RSIs.

A. V_{OUT} Generation with Load-dependent Configurability

Each SC stage can generate an output voltage equal to 1/2, 1/3 or 2/3 of the applied differential input voltage (V_H-V_L). A differential input pair is provided through RSIs as shown in Fig. 2. By setting control bits A_n and B_n, one pair among [$V_{S(n-1)}$, V_{SS}], [$\frac{1}{2}V_{BAT}$, $V_{S(n-1)}$] or [$V_{H(n-1)}$, $V_{L(n-1)}$] can be connected to [$V_{H(n)}$, $V_{L(n)}$] as summarized in Table I. When [$V_{H(n-1)}$, $V_{L(n-1)}$] is connected to [$V_{H(n)}$, $V_{L(n)}$], stage n-1 and stage n are parallel-connected.

Since the proposed converter is designed for V_{OUT} less than 2 V, the conversion ratio of the first SC stage is fixed to $\frac{1}{2}$, which generates 2 V from V_{BAT} of 4 V. This $\frac{1}{2}V_{BAT}$ is supplied as power for the rest of the stages. To minimize switching loss, clock is generated from the output of the 2nd stage, which provides a low clock swing of 1 V or 1.33 V, depending on the conversion ratio of the 2nd stage. This voltage is sufficiently higher than the threshold voltages of the switch transistors in later stages.

The multiple conversion ratios of each SC stage together with a flexible interconnection scheme yield (3×2=6) configurations per stage, excluding the parallel connection case. This results in 1×3×6³=648 configurations with five stages. However, since there are many duplicated conversion ratios, the total number of unique conversion ratios is 250, which results in an effective resolution of 16 mV for output voltage lower than $\frac{1}{2}V_{BAT}$ when V_{BAT} is 4 V. Such fine resolution is beneficial for efficient operation with light-load current.

A multi-ratio SC stage and reconfigurable stage interconnect scheme provide load-dependent configurability for optimized conversion operation. Depending on the load level, the number of cascaded stages and parallel connected stages can be adjusted to achieve optimal efficiency. Fig. 3 shows an example where the desired output voltage is 0.725 V.

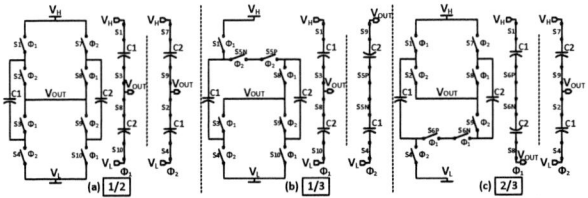

Fig. 4. Configurations of SC stage for (a) 1/2 (b) 1/3 (c) 2/3 ratio

TABLE II. SWITCH SEQUENCE

Ratio	Enable	Disable	Φ_1	Φ_2
1/2	1,2,3,4,7,8,9,10	5P, 5N, 6P, 6N	1,3,8,10	2,4,7,9
1/3	1,3,4,5P,5N,8,9,10	2, 6P, 6N, 7	1,3,8,10	4,5P,5N,9
2/3	1,2,4,6P,6N,7,8,9	3, 5P, 5N, 10	1,6P,6N,8	2,4,7,9

When the load is 500 nA, a configuration with 4 cascaded stages and a no-load voltage (V_{NL}) of 0.835 V can be used to reach 68.1% efficiency. However, when the load is 200 μA, maximizing parallel connections is desired, and a configuration with V_{NL}=1.0 V can achieve 64% efficiency. If the configuration optimized for 500 nA is used for a 200 μA load, the efficiency would drop to 27%.

B. Multi-ratio Multi-leaf SC Stage

The details of a reconfigurable SC stage are shown on the bottom-left of Fig. 2. The SC stage consists of 4 leaves where switches are binary-sized. Only the minimum-sized leaf is activated when the load is sub-μA, and larger leaves are added in parallel as the load increases. Flying capacitors are connected directly to the smallest leaf, which is always enabled, whereas flying capacitors are connected to larger leaves only when they are enabled.

Each leaf consists of 12 switch transistors. Three conversion ratios–1/2, 1/3, 2/3–can be realized, as shown in Fig. 4, by driving eight transistors with two-phase interleaved signals as summarized in Table II. The switches are driven by a capacitive level-shifter scheme, as shown in the bottom-right of Fig. 2, to minimize driving loss by reducing clock swing.

The boundary switches (S1-S4, S7-S10) have their source at stable supply (V_H, V_{OUT} or V_L), which can be used as reference for level-shifter switching. However, the middle switches (S5-S6), which are required to generate 1/3 or 2/3 ratios, lack such stable references for corresponding level shifters. To address this issue, local references (V_{M1}, V_{M2}) are implemented for level shifter reference. However, when the switches are disabled, the boundary switches can be disabled by shorting their gates to sources, whereas the middle switches' local reference voltages cannot be maintained if the switches are disabled. Therefore, instead of shorting to source, the gates of the middle switches are tied to power/ground rails, eliminating the need for local reference for disabling.

Flying capacitors are distributed with binary weight amongst consecutive stages, with the highest capacitance at the last stage, where the current is the largest.

C. Leakage-based Symmetric Clock Generator

The proposed design utilizes a leakage-based clock generator to handle the wide frequency range from kHz to MHz as shown in Fig. 5. The leakage-based delay cell originally introduced in [8] has only one voltage-controlled

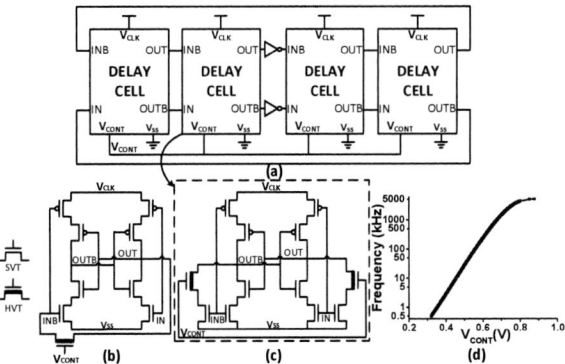

Fig.5. (a) Block diagram of leakage-based clock generator (b) Delay Cell structure in [8] (c) Modified delay cell with symmetric leakage branch (d) Frequency vs. control voltage (V_{CONT})

leakage branch (Fig. 5(b)). This causes an asymmetric falling/rising transition and uneven duty cycle, which affects the stability and limits the frequency range. To address this issue, a symmetric leakage structure is adopted as shown in Fig. 5(c). Thanks to the symmetrical structure, the duty cycle at each stage output is 50%, which helps to reduce ripple and conduction losses over the entire frequency range. This symmetry also helps increase the frequency range of the oscillator. High threshold transistors are used for the leakage control branch, which provides fine control of the leakage current, resulting in fine resolution and a wide frequency range.

The leakage control voltage (V_{CONT}) in the clock generator is also used for controlling the number of active leaves in the SC stages. As the load current increases, V_{CONT} and the corresponding frequency must be increased to transfer more charges. Therefore, by comparing V_{CONT} with pre-determined reference voltages, more leaves can be activated as the load increases.

IV. EXPERIMENTAL RESULTS

The proposed SC converter was fabricated in a 180-nm process with an area of 1.525 mm² and total on-chip MIM capacitance of 2.09 nF. A chip micrograph is shown in Fig. 6.

The measured efficiency over the target output voltage range (0.4-1.6 V) with various load conditions is presented in Fig. 7. For each loading condition, only those efficiency curves that are required to obtain the highest possible efficiency for any given output voltage are shown. With a light-load current of 500 nA, efficiency curves are narrow which validates loss analysis simulations, the fine granularity of the conversion ratios helps maintain efficiency above 60% for the entire

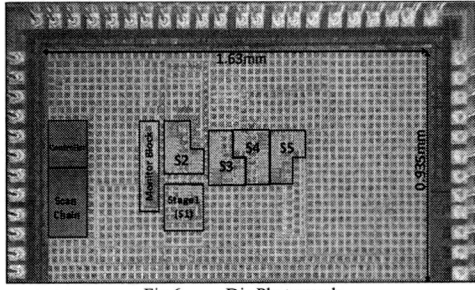

Fig.6. Die Photograph

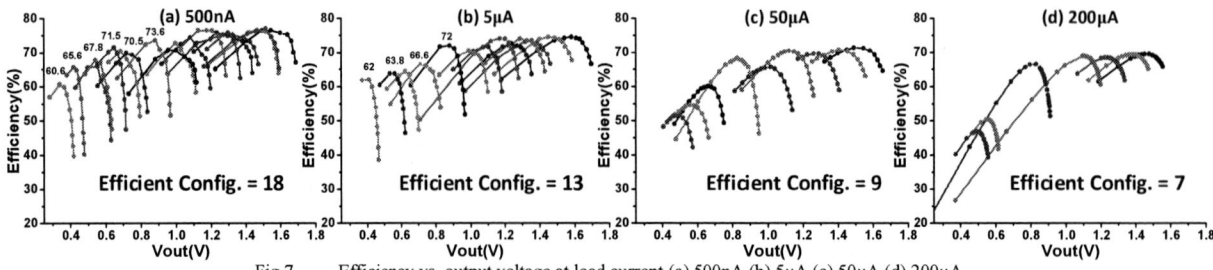

Fig.7.   Efficiency vs. output voltage at load current (a) 500nA (b) 5μA (c) 50μA (d) 200μA

voltage range and reach peak efficiency of 78% by utilizing 18 different configurations. This is a significant improvement from an earlier binary SAR-approach [2] where the light load efficiency was 45% at 2 μA. As the load is increased, the efficiency curves become wider, and fewer configurations with parallel stage connections are sufficient to cover the voltage range as predicted by loss analysis simulations (e.g. 7 configurations for 200 μA load).

The regulated output voltage and the corresponding efficiency for load currents ranging from 300 nA to 200 μA when V_{BAT} = 4 V are shown in Fig. 8(a). The proposed converter can maintain regulated voltage at V_{OUT} = 1.45 V, 1.2 V, and 0.78 V within ±16 mV with peak efficiency of 77.5%, 77.9% and 73.9%, respectively. Fig. 8(b) illustrates the significance of a multiple-leaf structure by plotting frequency, number of active leaves and efficiency for 1.2 V regulated output. The optimal frequency increases with increasing load, and as frequency increases, leaves are added in parallel gradually, which helps to maintain efficient operation over a wide load range. Table III compares the system's performance with that of the state-of-the-art SC converters.

V.   CONCLUSION

A wide voltage and current range SC converter for miniature sensors is presented in this paper. A multi-ratio SC stage and a reconfigurable interconnect scheme allow fine granularity for conversion ratio control, which improves the efficiency with light load. It also allows parallel stage configuration, improving efficiency with heavy load. The multiple-leaf approach optimizes loss over a broad load range by controlling switch sizes. The experimental results validate that the proposed SC converter can maintain greater than 62% efficiency over a target load current range of 300 nA–200 μA

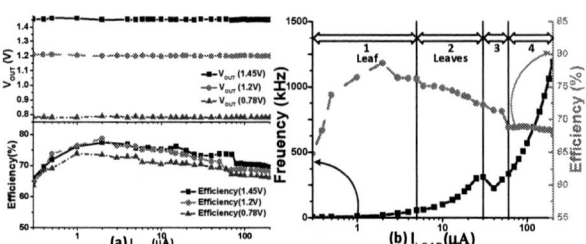

Fig.8. (a) Regulated V_{OUT} (top), Corresponding efficiency (bottom), (b) Frequency and Efficiency vs. I_{LOAD}

over wide output voltage range. Therefore, the proposed architecture provides well-regulated voltage efficiently during both the active and idle states. It allows efficient implementation of DVS for the active state and standby power reduction for the idle state, which can improve battery lifetime for miniature sensors.

ACKNOWLEDGMENT

This work was supported by the Center for Integrated Smart Sensors funded by the Ministry of Science, ICT & Future Planning as Global Frontier Project (CISS-2011-0031860)

REFERENCES

[1] H. Kim, et al., Symp-VLSI, pp. 202-203, June 2015.

[2] S. Bang, et al., ISSCC, pp. 370- 371, Feb. 2013.

[3] L. G. Salem, et al., CICC, Sep. 2014.

[4] L. G. Salem, et al., ISSCC, pp. 88–89, Feb. 2014.

[5] D. El-Damak, et al., ISSCC, pp. 374–375, Feb. 2013.

[6] H. P Le, et al., ISSCC, pp. 372–373, Feb. 2013.

[7] T. Ozaki, et al., A-SSCC, pp. 225-228, Nov. 2016.

[8] W. Jung, et al., ISSCC, pp. 398–399, Feb. 2014.

TABLE III. PERFORMANCE COMPARISON SUMMARY

	This Work	[2]	[3]	[4]	[5]	[6]	[7]
Technology	0.18μm	0.18μm	0.25μm	0.25μm	0.13μm	65nm	0.18μm
V_{IN} (V)	4	3.4 – 4.3	2.5	2.5	1.5	3 – 4	3.6 – 4.2
V_{OUT} (V)	0.4 – 1.6	0.9 – 1.5	0.1 – 2.24	0.1 – 2.18	0.4 – 1.1	1	0.98
Number of Conversion Ratios	75 for 0.4V - 1.6V 250 for 0V - 4.0V (V_{BAT} = 4.0V)	20 for 0.9V - 1.5V 128 for 0V - 4.0V (V_{BAT} = 4.0V)	45	15	4	2	1
Step Resolution	16mV @ 4.0V_{BAT}	31mV@ 4.0V_{BAT}	55mV @ 2.5V_{BAT}	N/R	N/R	N/R	N/R
Number of Stages	5	1+5	3	4	1	1	1
Capacitor Type	MIM Cap	On chip	MIM Cap	MIM cap	Ferroelectric	Bulk PMOS	On chip
Output Current Range	300nA – 200μA (667x)	3μA – 300μA (100x)	< 1.86mA	<2mA	20μA – 1mA (50x)	N/R	50nA – 1mA (20000x)
Total Capacitance	2.09nF	2.24nF	2.8nF	3nF	8nF	3.88nF	360pF
Clock frequency(Hz)	2k – 2.3M	80k – 1.7M	N/R	N/R	N/R	1M – 300 M	0.38k – 18M
Area (mm²)	1.525	1.69	4.3	4.645	0.366	0.64	0.79
η_{MAX} (%)	78	73	86	85	93	74.3	54
η@ Heavy Load	68.2% @200 μA (1.2V)	56% @200 μA (1.2V)	74%@1.86mA (1.2V)	73% @2mA (1.2V)	93% @200 μA (0.96V)	68.6% @ 0.12A (1V)	54% @ 1mA @0(.98V)
η@ Light Load	76% @500nA (1.2V) 64% @300nA (1.2V)	58% @3μA (1.2V)	N/R	N/R	73% @20 μA (0.96V)	N/R	54% @4μA (0.98V)
η@ Light Load & Low Voltage	64% @500nA (0.4V) 48% @300nA (0.4V)	45% @ 2μA (0.9V)	N/R	N/R	N/R	N/R	N/R

S10-5 (2180)

A High Efficiency and Fast Transient Digital Low-Dropout Assisted Switched-Capacitor Converter for EMI-Free Internet of Everything (IoE) Systems

Shao-Qi Chen[1], Yen-Ting Lin[1], Yu-Sheng Ma[1], Wen-Hau Yang[1], Ke-Horng Chen[1], Ying-Hsi Lin[2], Jian-Ru Lin[2], and Tsung-Yen Tsai[2]

[1]Department of Electrical and Computer Engineering, National Chiao Tung University, Hsinchu, Taiwan,
[2]Realtek Semiconductor Corporation, Hsinchu, Taiwan

Abstract—State-of-the-art closed-loop switched-capacitor (SC) designs usually regulate the output voltage by pulse frequency modulation (PFM) with the disadvantage of large electromagnetic interference (EMI). It may cause the failure of the Internet of Everything (IoE) devices. This paper proposes the hybrid digital low drop-out (DLDO) SC converter with a constant switching pulse width modulation (PWM) control for low EMI. Besides, the analog-to-digital current conversion provided by the DLDO due to its fast transient response can predict the desired ratio for the SC and reduce the dip voltage at the same time. The handshaking technique achieves smooth transition in the hybrid architecture. Experimental results show peak efficiency of 85% (closed-loop) and 95% (open loop) since the power converter has fast transient response by the DLDO prediction and reduction of switching power loss. Accurate fine tune operation by the DLDO and smooth handshaking improve load regulation from 49mV to 3mV and undershoot from 263mV to 93mV with a reduced recovery time of 200ns.

I. INTRODUCTION

Fig. 1 shows the advance of the System-in-Package (SiP) which has multiple chips in one package. The actual size can be smaller than 50mm². The Internet of Everything (IoE) devices can be miniature by the SiP technique. However, if the power management IC is designed by the inductor-based topology, the shielding technique becomes necessary for reducing electromagnetic interference (EMI). Nowadays compartment shielding (CPS) technique is used to solve this problem, but high cost and reduced yield rate cause the products less competitive. Thus, the closed-loop switched-capacitor (SC) DC-DC converters [1],[2],[4]-[6] become very popular to supply a regulated output voltage to the IoE systems compared to inductor-based converters, with the advantages of less number of external passive components and compact size in standard CMOS technology.

Conventional closed-loop SC designs use the pulse frequency modulation (PFM) technique. The EMI influence is large for IoEs since the voltage regulation is achieved by varying switching frequency. Although variable switching frequency achieves low load regulation and fast transient response, the undesired EMI may cause the failure of the IoE system since designing a filter to discard random switching noise is difficult. For low EMI effect, constant frequency pulse width modulation (PWM) is demanded while the number of ratios is also necessary to be increased for good load regulation.

There is a trilemma of design among load regulation (LR), transient response (TR), and numbers of conversion ratio (No. CR) as illustrated in Fig. 2. Recursive SC (Re-SC) converter in [2] has fast TR but poor LR and few ratio numbers. On the other hand, the Rational SC (Ra-SC) converter in [6] has

more ratio number and good LR but slow TR. The rule of thumb is larger ratio number results in better LR at the cost of longer TR time. Contrarily, less ratio numbers improve the TR but deteriorate the LR. Moreover, in case of light-to-heavy load step, the disadvantage of large dip voltage for both SC converters will cause shut down or idle time in the SoC to seriously influence the overall performance.

Therefore, this paper presents a hybrid digital low drop-out (DLDO) SC converter to achieve high performance in the trilemma, low power consumption, high conversion efficiency, and less EMI over wide load ranges for the IoE systems. The paper is organized as follows. Section II discusses the comparison among state-of-the-arts. The proposed hybrid converter is also shown to present the improvement in LR, efficiency and TR. Section III shows the circuit implementation and experimental results are shown in Section IV. Finally, conclusions are made in Section V.

II. PROPOSED DLDO SC CONVERTER

Overall consideration of LR, TR and efficiency of SC converters during transient period results in the similar disadvantages of state-of-the-arts. For example, Fig. 3 shows the conversion efficiency ratios including 1/3, 5/14 and 4/11 versus the efficiency.

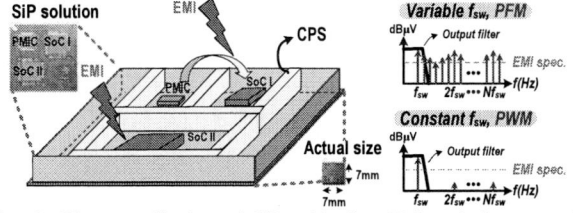

Fig. 1. The overall size of SiP with the CPS technique can be miniature but the cost is increased due to the CPS.

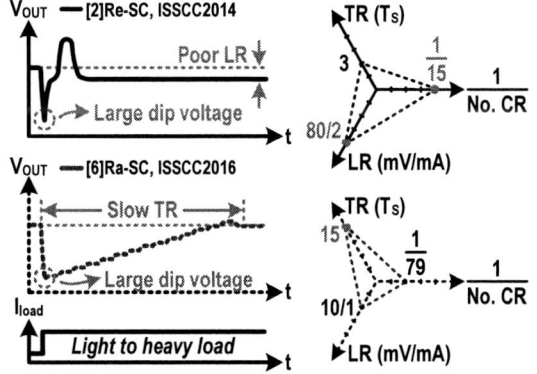

Fig. 2. Trade-off between LR, TR, and the No. of CR among state-of-the-arts.

978-1-5386-3179-9/17 $31.00 © 2017 IEEE

The peak of each curve indicates the converter at light loads. If the SC converter works at the ratio of 1/3 in the beginning, which is marked as label of '(1)' in Fig. 3, the efficiency is high. High efficiency and low LR is derived at light loads only. In case of light-to-heavy load step, the efficiency decreases and the operation point falls to the label of '(2a)' if the SC converter cannot immediately change the ratio to react to the load change. After the closed-loop operation, the ratio changes from 1/3 to either 5/14 or 4/11 which are labeled as '(3a) and '(4a)', respectively. Label of '(3a)' suffers from low efficiency while label of '(4a)' has an overvoltage problem.

Thus, the proposed hybrid DLDO SC converter in Fig. 4(a) utilizes the advantage of fast TR of the DLDO to provide sufficient energy to the output by turning on sufficient number of power switches. The number of power switches is a load-dependent coding used to rapidly change the ratio of the SC converter to a suitable ratio. Thus, the operation point moves from '(2a)' to '(3b)' in Fig. 4(b). The LR becomes low at '(3b)' compared to '(3a)' in Fig. 3(a) due to the help of DLDO. Thus, high efficiency, low LR, fast TR, and reduced dip voltage can be achieved. Compared to [3], the proposed technique is a real fully-integrated solution.

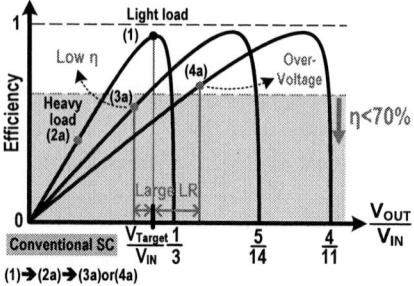

Fig. 3. Poor efficiency and transient response during the light-to-heavy load change from (1) to (2a) to (3a) or (4a).

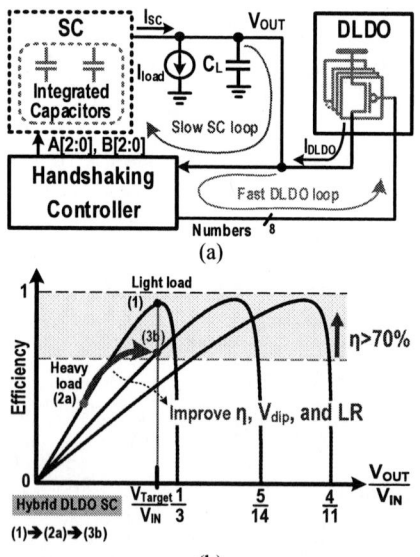

Fig. 4. (a) Structure of hybrid DLDO SC converter. (b) Improvement of efficiency and transient response in case of light-to-heavy load

change from (1) to (2a), and then (3b).

The proposed hybrid DLDO SC converter has dual loops, which are controlled by the handshaking controller, to regulate the V_{OUT} as shown in Fig. 5. The DLDO provides an additional current path once load transient is detected, where $I_{load} - I_{DLDO}$ represents the equivalent driving current from the SC converter in transient response. The SC ratio ascends at the next cycle to provide larger current to the V_{OUT}. Meanwhile, the current from the DLDO retreats gradually, which is controlled by the finite state machine (FSM) for high linearity in the conduction region. Finally, high efficiency and small offset voltage are obtained simultaneously.

For the SC converter, 3-bit controllable input variables $A_{(10)} = a_2a_1a_{0(2)}$ and $B_{(10)} = b_2b_1b_{0(2)}$ derive an output voltage V_{OUT} in (1).

$$V_{OUT} = \frac{A \cdot V_{IN}}{8 + B} \qquad (1)$$

Moreover, the coarse SC controlled by codes $A_{c(10)}(=a_2a_10)$ and $B_{c(10)}(=b_2b_10)$ provides a suitable and fast voltage level to the DLDO regulator for fast voltage regulation and high efficiency. Owing to the fast response of the DLDO regulator, the C[31:0] can be determined rapidly. Thus, $A_{c(10)}$ and $B_{c(10)}$ can be derived to meet the requirement in (2) where R_{on} is the on resistance of power MOSFET. The function of the DLDO regulator works as an analog-to-digital (ADC) current conversion to accurately predict the desired $A_{c(10)}$ and $B_{c(10)}$. A and B may be derived too early before the time at t_3. In this work, it is necessary to ensure smooth transition from the DLDO to the SC. Although the target A and B is derived at the time t_0, the transition is still done at the time t_3.

$$V_{OUT}(A(t_0), B(t_0)) < V_{Target} - I_{DLDO}(C[31:0])*R_{on} < V_{OUT}(A_c, B_c) \qquad (2)$$

Fig. 6 shows the timing diagram in case of abrupt load change. The DLDO meets the energy request of output immediately to reduce the dip voltage. Furthermore, with the handshaking between DLDO and SC, $I_{OUT} = I_{SC} + I_{DLDO}$ are well distributed and equal to I_{load} in steady state to optimize efficiency and minimize load regulation. Flow chart in Fig. 7 illustrates both transient and steady state operations.

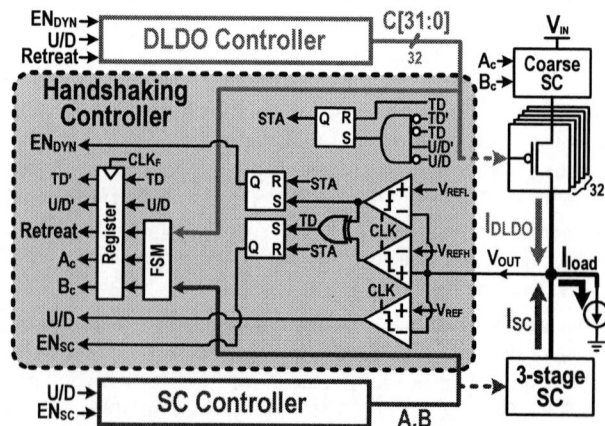

Fig. 5. Handshaking and dual-loop control of the proposed hybrid

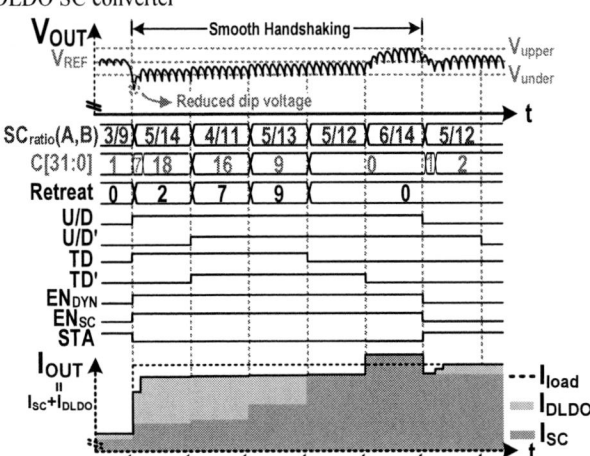

Fig. 6. Timing diagram of DLDO and SC in case of abrupt load change.

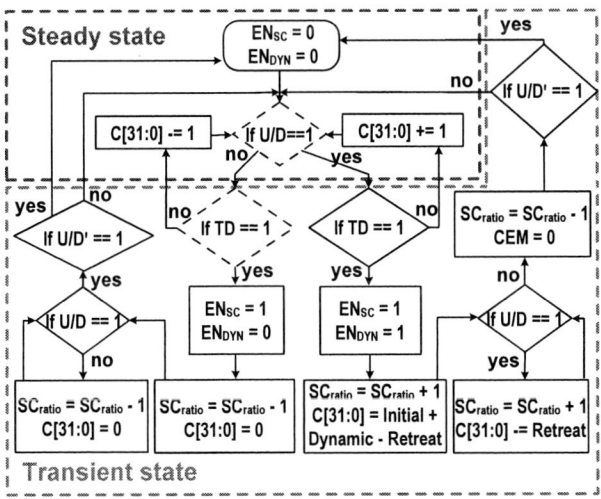

Fig. 7. Operation flow chart of overall controller.

III. CIRCUIT IMPLEMENTATION

A. SC design

Fig. 8(a) describes the structure of 3-stage SC and 2:1 SC unit. In steady state, the 3-stage SC regulates at the target voltage by fixed coarse-tuning codes A[2:0] and B[2:0] generated from the handshaking controller in Fig. 4(a). During load transient, the transient detect signal, TD, triggers the signal EN_{SC} to enable the operation of SC, which dynamically adjusts A[2:0] and B[2:0].

3-bit controllable input variables $A_{(10)}=a_2a_1a_{0(2)}$ and $B_{(10)}=b_2b_1b_{0(2)}$ select different power rails to the top plates of each 2:1 SC unit. The corresponding output voltage for different a_i and b_i, where i=0, 1, and 2, is shown in Table I.

Table I: The corresponding value for each a_i and b_i.

(a_i,b_i)	(0,1)	(1,1)	(1,0)	(0,0)
Mux,out	$-V_{OUT}$ or $2V_{DD}-V_{OUT}$	$V_{DD}-V_{OUT}$	V_{DD}	V_{SS}

The 2:1 SC unit in Fig. 8(b) is a four-phase SC converter with two input V_H and V_L, and one output $V_M=(V_H+V_L)/2$.

With four-phase interleaving operation, the output ripple of V_M is greatly reduced as depicted in Fig. 8(c).

B. Dual loop controller design

Fig. 9 shows the implementation of dual loops controller. The DLDO enhances the step resolution with the additional fine-tuning control bit array C[31:0] which is regulated on the basis of up/down (U/D) control. U/D control is the comparing result of V_{OUT} and V_{REF}. In steady state, the control signal stays at the least significant bit (LSB) C[0]. During load transient, the triggered signal 'EN_{DYN}' leads DLDO to choose the decoded C[31:0], which is decoded from the summation of signals 'Initial', 'Dynamic' and 'Retreat'. The signal 'Initial' represents an initial code. The signal 'Dynamic' is calculated from the difference between V_{OUT} and V_{REFL}. The signal 'Retreat' is derived according to the flow chart deigned in Fig. 7 to eliminate the voltage difference between two adjacent ratios of 3-stage SC.

IV. EXPERIMENTAL RESULTS

Fig. 10 shows the 8mA load step transient response. Recovery time is reduced from 1.08μs to 200ns owing to the help from the DLDO regulator. The proposed handshaking control suppresses the undershoot voltage from 263mV to 93mV and improves load regulation from 49mV to 3mV.

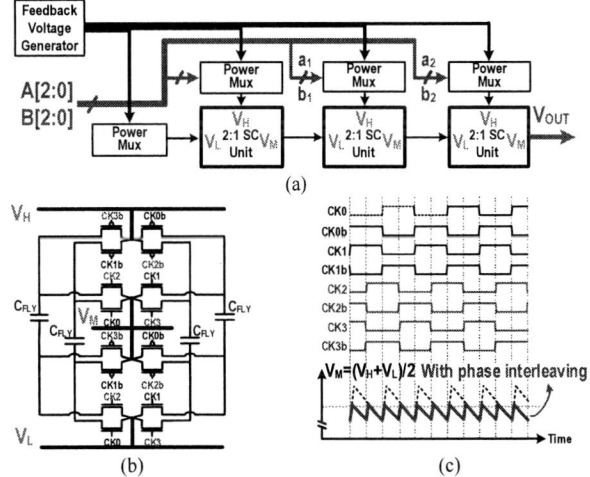

Fig. 8. Circuit Implementation. (a) 3-stage SC converter. (b) 2:1 SC unit. (c) Timing diagram of the output voltage ripple reduction.

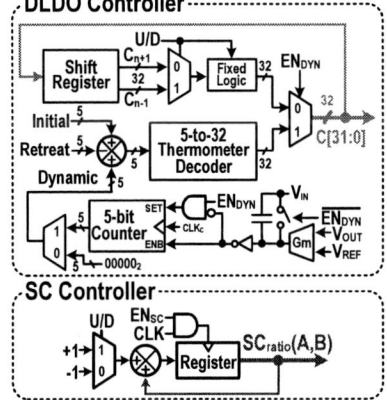

Fig. 9. Implementation of dual loop controller.

Under different loads at V_{IN}=3.3V, Fig. 11 compares the offset voltages among the measured results of the proposed hybrid DLDO SC converter and [6]. In [6], large offset voltages occur as the load current increases. The plots show that the DLDO has the advantage of smaller offset voltage.

Fig. 12 compares the conversion efficiency with state-of-the-arts when the V_{OUT} varies from 0.2V to 2.85V with a maximum load current of 20mA. The efficiency drops between the selected ratios are smoothed by the DLDO. Fig. 13 summarizes the performance of state-of-the-arts and proposed converter. Table II compares when the efficient 43 conversion ratios are selected from 79 ratios in [6]. 301 equivalent ratios enhanced by DLDO are employed with output voltage offset below a predefined resolution 11mV. Chip micrograph is shown in Fig. 14 with an active area of 8.84mm² fully integrated in 0.25μm CMOS process.

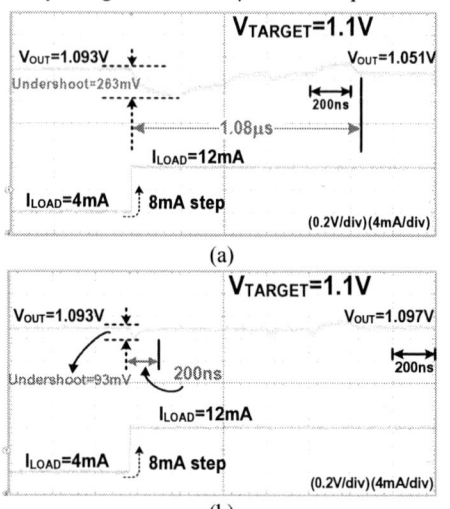

(a)

(b)

Fig. 10. Measured transient waveforms. (a) Without the help of the DLDO regulator. (b) With the help of the DLDO regulator.

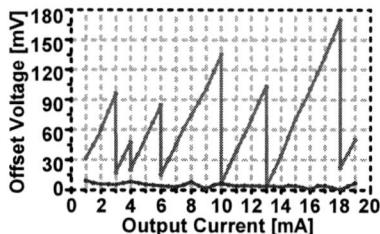

Fig. 11. Offset voltage versus output current.

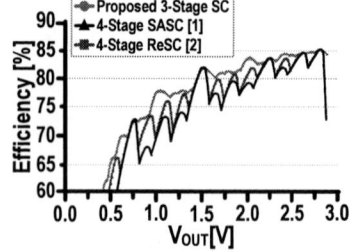

Fig. 12. Conversion efficiency versus output voltage.

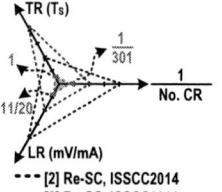

--- [2] Re-SC, ISSCC2014
--- [6] Ra-SC, ISSCC2016
--- Proposed hybrid DLDO SC

Fig. 13. Trade-off between LR, TR, and the No. of CR.

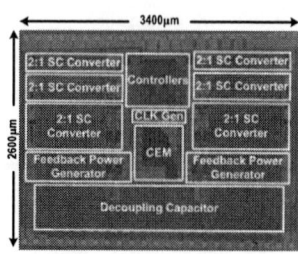

Fig. 14. Chip micrograph.

Table II: Comparison table.

	This Work	[1]	[2]	[6]
Technology	0.25μm CMOS	0.18μm CMOS	0.25μm CMOS	0.18μm CMOS
Reconfigurability Type	Rational	Binary	Binary	Rational
Number of Stages	3+2	7	4	3+2
Ratios Count	43 without CEM *301 with CEM	117	15	79
Voltage Resolution	≤11mV @V_{in}=3.3V	31.25mV @V_{in}=4V	156.25mV @V_{in}=2.5V	10~70mV @V_{in}=2V
Transient Recovery Time	200ns @ΔI_{load}=10mA	10ms @ΔI_{load}=0.05mA	**8μs @ΔI_{load}=2mA	N/A
Transient Voltage Drop	93mV @ΔI_{load}=10mA	250 mV @ΔI_{load}=0.05mA	**950mV @ΔI_{load}=2mA	N/A
Maximum Output Current	20mA	0.3mA	2mA	N/A
Input Voltage	3.3V	3.4V-4.3V	2.5V	2V
Output Voltage	0.2V-2.85V	0.9V-1.5V	0.1V-2.18V	0.47V-1.87V
Peak Efficiency	85% @ Close loop 95% @ Open loop	72% @ Close loop	85% @ Close loop	95% @ Open loop
Fully Integrated	YES	YES	YES	YES

*In this design, CEM is selected for 11mV resolution
** Transient recovery time and transient voltage drop derived from [2] measured result

V. CONCLUSIONS

The proposed hybrid DLDO SC converter promises the output voltage offset below 11mV. Smooth handshaking in the hybrid architecture improves load regulation from 49mV to 3mV and undershoot from 263mV to 93mV with a reduced recovery time of 200ns. Experimental results show peak efficiency of 85% (closed-loop) and 95% (open loop), higher than those of prior arts due to the DLDO prediction and reduction of switching power loss.

REFERENCES

[1] S. Bang et al., "A Fully Integrated Successive-Approximation Switched-Capacitor DC-DC Converter with 31mV Output Voltage Resolution," *ISSCC Dig. Tech Papers*, pp. 370-371, Feb. 2013.

[2] L. G. Salem et al., "An 85%-Efficiency Fully Integrated 15-Ratio Recursive Switched-Capacitor DC-DC Converter with 0.1-to-2.2V Output Voltage Range," *ISSCC Dig. Tech Papers*, pp. 88-89, Feb. 2014.

[3] L. Cheng et al., "A 30MHz Hybrid Buck Converter with 36mV Droop and 125ns 1% Settling Time for a 1.25A/2ns Load Transient," *ISSCC Dig. Tech Papers*, pp. 188-189, Feb. 2017.

[4] H.–P. Le et al., "A Sub-ns Response Fully-Integrated Battery-Connected Switched-Capacitor Voltage Regulator Delivering 0.19W/mm2 at 73% Efficiency," *ISSCC Dig. Tech Papers*, pp. 372-373, Feb. 2013.

[5] J. Jiang et al., "A 2-/3-Phase Fully Integrated Switched-Capacitor DC-DC Converter in Bulk CMOS for Energy-Efficient Digital Circuits with 14% Efficiency Improvement," *ISSCC Dig. Tech Papers*, pp. 366-367, Feb. 2015.

[6] W. Jung et al., "A Rational-Conversion-Ratio Switched-Capacitor DC-DC Converter Using Negative-Output Feedback," *ISSCC Dig. Tech Papers*, pp. 218-219, Feb. 2015.

A CMOS Time of Flight (TOF) Depth Image Sensor with In-Pixel Background Cancellation and Sensitivity Improvement Using Phase Shifting Readout Technique

Ting Liao, Nien-An Lee, and Chih-Cheng Hsieh*

Department of Electrical Engineering,
National Tsing Hua University, Hsinchu, Taiwan
Email: *cchsieh@ee.nthu.edu.tw

Abstract—This paper presents a CMOS time-of-flight (TOF) image sensor with in-pixel background light cancellation for outdoor depth imaging application. The using of P+/N_well diode with proposed polarity switching integration and phase-shift readout (PSR) technique achieves the in-pixel background cancellation capability and sensitivity improvement. Moreover, the PSR also suppresses the column fixed-pattern-noise (FPN) without need of extra frame capturing. A prototype TOF sensing chip with a 64x64 array has been fabricated in TSMC standard 0.18μm CMOS process and verified. The pixel pitch is 20μmx20μm with a fill-factor of 33%. The achieved depth measurement range is 0.75 to 7.5 meters with a linearity error below 1.1%. The measured relative precision is 4.2% at a 7.5-meter target distance; and the background light suppression capability is up to 180k lux without saturation.

Keywords — Depth imaging, CMOS image sensor, Time-of-flight imaging, Background cancellation, Fixed-pattern noise suppression, 3-D camera.

I. INTRODUCTION

Range imaging is one of the most popular application of smart image sensing system. The increasing need of three-dimensional (3-D) range imaging has been created by consumer electronics applications such as gaming control, robotic vison, and gesture recognition. For above applications, the development of depth camera capable of capturing 3-D range information in real time with high accuracy under background illumination becomes very critical.

Continuous-wave-modulation time-of-flight (TOF) is a preferred method for 3-D range imaging due to its less complex data processing requirement and no geometric constraints. Fig. 1 shows the operation principle of TOF sensing system. The 3-D range imaging is obtained by detection of the phase difference of light emitter and reflection. The depth information L can be determined by the four-phase signal integration and modulation as shown in the following equations [1]:

$$L = \frac{c\theta}{2\pi f_{mod}}$$

$$\theta = \frac{\pi}{2} \times \left(1 - \frac{V(0) - V(\pi)}{|V(0) - V(\pi)| + |V(\pi/2) - V(3\pi/2)|}\right) \quad (1)$$

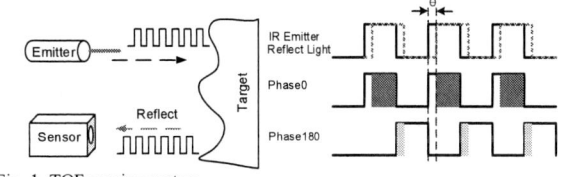

Fig. 1. TOF sensing system

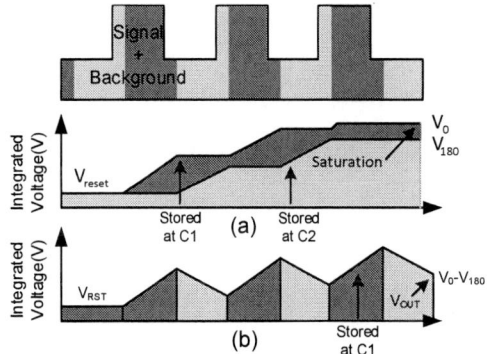

Fig. 2. Operation timing and integrated voltage waveforms of (a) 2-storage-nodes operation (b) 1-storage-node with in-pixel subtraction operation

where c is the speed of light; f_{mod} is the modulation frequency; and $V(0) \sim V(3\pi/2)$ are the integrated signals of different phases.

The demodulation operation of multiple-phase TOF sensor requires the corresponding storage nodes for signal accumulations at the specific time windows of different phases. Therefore, the limited well capacity of pixel is usually saturated by background light (BGL) during accumulation, which degrades the achievable performance. Fig. 2 shows the integrated voltage waveforms of two-phase accumulation operation as an example. Fig. 2(a) shows the storage nodes (C1 and C2) for both phases (phase 0 and 180 degrees) are saturated by BGL after certain cycles of accumulation. On the contrast, Fig. 2(b) shows only the difference of in-phase (0 degree) and out-phase (180 degrees) signals are accumulated (on C1) with in-pixel background cancellation, since the shot noise and readout noise are inversely proportional to the square of accumulation cycles, we can achieve a better SNR performance

978-1-5386-3179-9/17 $31.00 © 2017 IEEE

by signal integration of more accumulation cycles without saturation.

Several sensing devices and techniques have been developed for TOF sensing and BGL suppression. In SPAD-based (SPAD: single-photon-avalanche-diode) TOF sensing system [2], the depth information is calculated directly by sensing the light traveling time between source and object; and the BGL suppression is implemented with temporal-spatial correlation. However, SPAD-based system suffers from the high-voltage operation and a quite-large pixel size. The use of pinned photo diode with multiple transfer gates [3-4] has demonstrated a high performance in a small pixel size. However, the in-pixel BGL suppression cannot be implemented in these structures due to the one-way charge transfer characteristic of transfer gate.

In this work, we propose a new TOF sensing structure with in-pixel BGL suppression by using polarity switching integration and phase-shift-readout (PSR) techniques to improve the SNR and column-FPN performance. The rest of this paper is organized as follows. Section II describes the architecture and circuit implementation. Section III presents the measurement results. Finally, the conclusion is drawn in Section IV.

II. CIRCUIT DESCRIPTION

A. Proposed TOF imager arichitecture

Fig. 3 shows the chip architecture of the implemented TOF depth image sensor. The prototype consists of a 64x64 TOF pixel array, column-wise 10-bit single slope analog to digital converters (SS-ADC), row and column control circuits, TX delay control and driver circuits, and the peripheral bias and array-shared ramp generators. The SS-ADC is implemented with a comparator, a 10-bit counter, output latches, and a ramping reference from the global on-chip generator. The conversion speed of SS-ADC is 7kS/s for video-rate depth imaging application. The TX delay control can adjust the phase difference between the modulation window and light emitter triggering edge through a tunable voltage bias. The analog supply voltage is 3.3V for signal dynamic range consideration, while the digital supply is 1.8V for power reduction.

B. Circuit structure and operation

Fig. 4 shows the schematic of pixel circuit. The sensing diode is implemented with P+ on N-well structure for anode (P+) and cathode (N-well) switching capability. The demodulation circuit consists of an in-pixel OPAMP with a negative feedback integration-cap C_{INT} as a capacitive trans-impedance amplifier (CTIA) for I-to-V conversion and constant photodiode bias providing. Four switches controlled by Φ_{TX0} and Φ_{TX180} are used to change the photo-generated signal polarity corresponding to the light emission timing ($\Phi_{emitter}$ in Fig. 5) for polarity switching integration and in-pixel BGL suppression operation. The switch coupling effects are minimized by matching sizing and careful layout.

Fig. 5 illustrates the timing diagram of TOF pixel operation. First, Φ_{PIX_RST} is used to reset pixel at the beginning of every

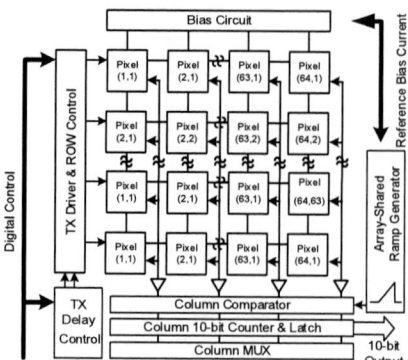

Fig. 3. Architecture of the implemented TOF imaging sensor

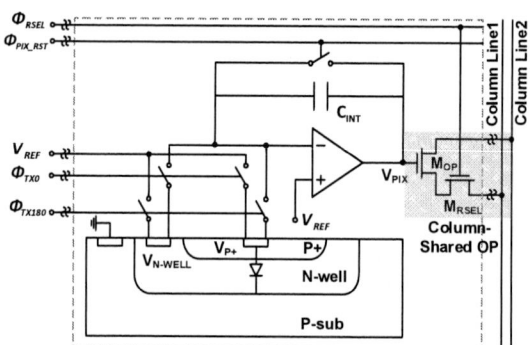

Fig. 4. The proposed TOF sensing pixel

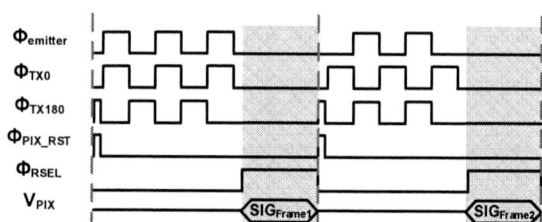

Fig. 5. Operation timing of TOF pixel

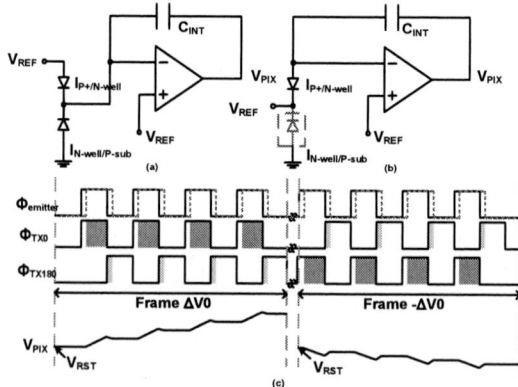

Fig. 6. Pixel operation when (a) Φ_{TX0} = 1 (b) Φ_{TX180} = 1 (c) The timing diagram of the proposed PSR technique.

frame. The following signal demodulation and readout operations will be described in detail in the next section. After certain demodulation cycles, the demodulation switches (controlled by Φ_{TX0} and Φ_{TX180}) are both off to stop the signal integration. The integrated voltage on C_{INT} is then read out through a unity-gain column-shared OPAMP [5] (instead of the conventional source follower) with better linearity and less gain loss. Finally, the readout signals are converted to digital data by the column-wise SS-ADC and MUX-out in serial.

C. In-pixel BGL suppresion with polarity switching integration

When $\Phi_{TX0} = 1$, the anode (P+) is connected to the input of CTIA and the cathode (N-well) is connected to the reference voltage V_{ref}. The photo-generated current will flow into CTIA to charge C_{INT} and convert to voltage (V_{PIX}). On the contrary, when $\Phi_{TX180} = 1$, the diode connection is reversed (P+ to V_{ref} and N_well to CTIA input) to discharge C_{INT}. By steering the diode (i.e. photocurrent) polarity, only the signal difference of Φ_{TX0} and Φ_{TX180} are integrated on C_{INT} to perform the in-pixel signal subtraction, that is, the BGL suppression. The integrated voltage V_{PIX} can be expressed as following equation:

$$V_{PIX} = V_{RST} + \frac{I_{PD0}}{C_{INT}} \cdot t_0 - \frac{I_{PD180}}{C_{INT}} \cdot t_{180} \quad (2)$$

D. Phase-shift readout (PSR) technique

The using of P+/N_well diode ($PD_{P+/N\text{-}well}$) comes with an inherent parasitic diode N_well/Psub ($PD_{N\text{-}well/Psub}$), which may cause an unbalance response issue between Φ_{TX0} and Φ_{TX180} as depicted in Fig. 6. When $\Phi_{TX0} = 1$ as shown in Fig. 6(a), both photo currents from $I_{P+/N\text{-}well}$ and $I_{N\text{-}well/Psub}$ are summed up and integrated. While when $\Phi_{TX180} = 1$ as shown in Fig. 6(b), only the photo current $I_{P+/N\text{-}well}$ is integrated through CTIA, and induce an unbalance response. To solve the mentioned issue, two continuous frames ($\Delta V0$ and $-\Delta V0$) readout with phase-shift readout (PSR) is proposed as shown in Fig. 6(c). In the frame $\Delta V0$, Φ_{TX0} is in phase with $\Phi_{emitter}$ and the output is expressed as $V_{FFRAME,\Delta V0}$ in (3). In the frame $-\Delta V0$, the Φ_{TX180} is changed to be in phase with $\Phi_{emitter}$ and the output is expressed as $V_{FFRAME,-\Delta V0}$ in (4).

$$
\begin{aligned}
&V_{FRAME,\Delta V0} \\
&= V_{RST} + \frac{I_{p+\_Nwell} + I_{Nwell\_Psub}}{C_{INT}} \cdot t_0 - \frac{I_{p+Nwell}}{C_{INT}} \cdot t_{180} \quad (3)
\end{aligned}
$$

$$
\begin{aligned}
&V_{FRAME,-\Delta V0} \\
&= V_{RST} - \frac{I_{p+Nwell}}{C_{INT}} \cdot t_0 + \frac{I_{p+Nwell} + I_{Nwell Psub}}{C_{INT}} \cdot t_{180} \quad (4)
\end{aligned}
$$

The effective demodulation signal $\Delta V0$ for depth calculation, defined as the voltage difference of $V_{FFRAME,\Delta V0}$ and $V_{FFRAME,-\Delta V0}$, is shown in (5).

$$
\begin{aligned}
\Delta V(0) &= V_{FRAME,\Delta V0} - V_{FRAME,-\Delta V0} \\
&= \frac{2 \cdot I_{p+Nwell} + I_{Nwell Psub}}{C_{INT}} \cdot (t_0 - t_{180}) \quad (5)
\end{aligned}
$$

Fig7. Micrograph of the TOF imaging chip

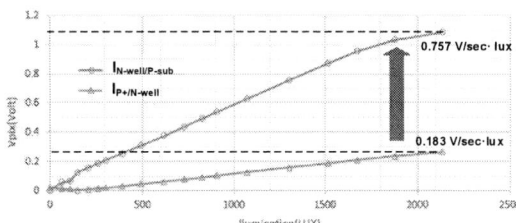

Fig 8. Lux transfer curve of $PD_{P+/N\text{-}well}$ and $PD_{N\text{-}well/Psub}$

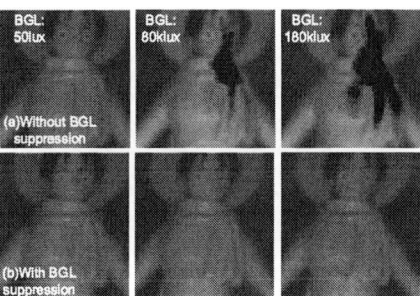

Fig 9. Captured images wo/wi BGL suppression

It shows $I_{P+/N\text{-}well} \cdot t_{180}$ in (3) and $I_{P+/N\text{-}well} \cdot t_0$ in (4) are cancelled out in (5) by applying the PSR operation. Moreover, the photo-response of t_0 and t_{180} are balanced with a bonus photocurrent contribution ($I_{N\text{-}well/Psub}$) from the parasitic diode. Since the demodulation signal is a differential output with a correlated subtraction, the column-FPN induced from the non-ideal effects of readout path including the CTIA reset level variation, the column-shared OPAMP offset, and the comparator offset in SS-ADC can be eliminated.

III. Measurement Result

A prototype chip with 64×64 pixel array and 3.3V operation has been fabricated in TSMC standard 0.18μm CMOS process with MIM capacitors. The pixel pitch is 20μm with a fill-factor of 33%.

A. Two-dimensional (2-D) Imaging

Fig 8 shows the measured lux transfer curve of the implemented pixel. The sensitivity of $PD_{N\text{-}well/Psub}$ and $PD_{P+/N\text{-}}$

well is 0.757 V/lux·sec and 0.183 V/lux·sec, respectively. With the proposed pixel structure and polarity switching integration technique, the photo sensitivity ($PD_{P+/N-well} + PD_{N-well/Psub}$) is improved by 5 times compared to that with $PD_{P+/N-well}$ only. Fig. 9 shows the captured 2-D images with a background spotlight illumination range of 50-180k lux. Fig 9(a) shows the image is saturated after 80k lux without BGL suppression. On the contrary, the captured images in Fig. 9(b) are normal up to 180k lux by applying the proposed in-pixel BGL suppression.

B. Depth Imaging

The depth (3-D) imaging operation has been tested using a light source of 850nm IR-LED and a modulation frequency of 2.5 MHz to achieve an equivalent frame rate of 26 frame/sec. Fig. 10(a) shows the measured distances versus the actual depth ranging from 0.75m to 7.5m. The maximum non-linearity, defined as the maximum error of each data point normalized to its actual distance, is 1.1% measured from 0.75m to 7.5m. Fig. 10(b) shows the relative precision, defined as the RMS error normalized to the maximum distance, is 4.2% at 7.5m with a measured readout noise of 0.86 mV. Fig. 11 shows the captured 2-D and corresponding depth images of a doll body placed at a distance of 0.75m.

Table I summarizes the performance comparison of the state-of-the-art TOF sensors. In non-pinned-photodiode based TOF sensor, this work shows the smallest pixel size and the best BGL suppression with a competitive precision performance.

IV. CONCLUSION

A 64x64 pixel CMOS TOF depth imaging sensor has been carefully designed and tested. The proposed TOF pixel with PSR technique has demonstrated a good performance with in-pixel BGL suppression up to 180k lux, a 5-times sensitivity improvement of 0.94 V/lux.sec, and column-FPN elimination. The detectable distance range is 0.75~7.5m with a non-linearity of 1.1% and the achieved relative precision is 4.2% at 7.5m.

ACKNOWLEDGMENT

The authors thank National Chip Implementation Center (CIC) Taiwan for fabrication of the test chip. This research is particularly supported by National Science Council, Taiwan under contract number 104-2221-E-007-103-MY3, and MOST 105-2218-E-007-006

REFERENCES

[1] J. Cho, J. Choi, S.-J. Kim, J. Shin, S. Park, J. D. K. Kim, and E. Yoon, "A 5.9μm-pixel 2D/3D image sensor with background supression over 100klx," IEEE Symp. VLSI Circuits Dig., pp. 6–7, Jun. 2013.

[2] C. Niclass, M. Soga, H. Matsubara, S. Kato, and M. Kagami, "A 100-m range 10-frame/s 340×96-pixel time-of-flight depth sensor in 0.18-μm CMOS," IEEE J. Solid-State Circuits, vol. 48, no. 2, pp. 559–572, Feb. 2013.

[3] Y. M. Wang, I. Ovsiannikov, S. Byun, T. Lee, Y. Lee, G. Waligorski, H. Wang, S. Lee, D. Min, Y. Park, T. Kim, C. Choi, G. Han, and E. R. Fossum, "Compact ambient light cancellation design and optimization for 3D time-of-flight image sensors," Intl. Image Sensor Workshop, vol. 2, Jun. 2013.

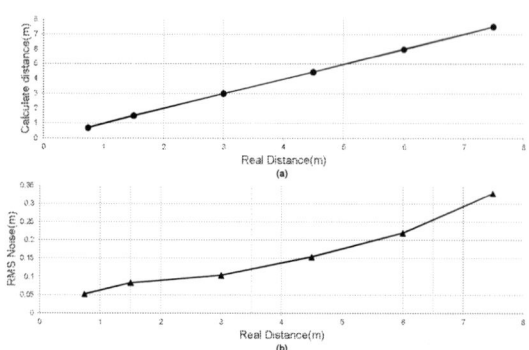

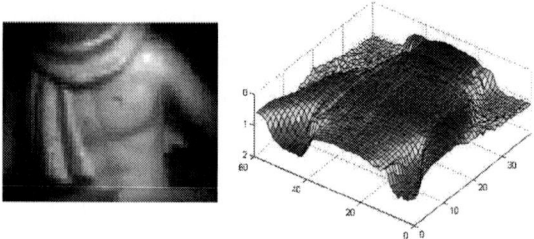

Fig. 10. (a) Measured distance (b) RMS noise versus actual depth

Fig. 11. The captured 2-D and 3-D images

TABLE I.   COMPARISON TABLE

	Zach JSSC 2010[7]	Perenzoni JSSC 2011[6]	Niclass JSSC 2013[2]	This work
Technology	0.6um CMOS	0.18um CMOS	0.18um CMOS	0.18um CMOS
BGL scheme	Pixel-level charge subtraction	Pixel-level CDS	Temporal-spatial correlation	Pixel-level voltage subtraction
PIX pitch	125um	29.1um	25um	20um
Fill factor	66%	34%	70%	33%
Resolution	16x16	160x120	340x96	64x64
Frame rate	16	27.8~80	10	26
BGL range	150k	n.a.	80k	180k
	0.1~3.2m	1~4.5m	10~100m	0.75~7.5m
Relative precision	1.3%	11%	0.1%	4.2%
Non-linearity	3.3%	unknown	0.37%	1.1%

[4] T. Oggier, R. Kaufmann, M. Lehmann, B. Buttgen, S. Neukom, M. Richter, M. Schweizer, P. Metzler, F. Lustenberger, and N. Blanc, "Novel pixel architecture with inherent background suppression for 3D time-of-flight imaging," Proc. SPIE, vol. 5665, no. 1, pp. 1–8, Jan. 2005.

[5] Wei-Fan Chou, Shang-Fu Yeh, Chih-Cheng Hsieh," A 143dB 1.96% FPN linear-logarithmic CMOS image sensor with threshold-voltage cancellation and tunable linear range", IEEE Sensors 2012.

[6] M. Perenzoni, N. Massari, D. Stoppa, L. Pancheri, M. Malfatti, and L. Gonzo," A 160×120-pixels range camera with in-pixel correlated double sampling and fixed-pattern noise correction," IEEE J. Solid-State Circuits, vol. 46, no. 7, pp. 1672–1681, Jul. 2011.

[7] G. Zach, M. Davidovic, and H. Zimmermann, "A 16×16 pixel distance sensor with in-pixel circuitry that tolerates 150 klx of ambient light," IEEE J. Solid-State Circuits, vol. 45, no. 7, pp. 1345–1353, Jul. 2010.

An Element-Matched Band-Pass Delta-Sigma ADC for Ultrasound Imaging

Michele D'Urbino[*‡], Chao Chen[*], Zhao Chen[*], Zu-Yao Chang[*], Jacco Ponte[†], Boris Lippe[†] and Michiel Pertijs[*]

[*]Electronic Instrumentation Laboratory, Delft University of Technology, Delft, The Netherlands
[†]Oldelft, Delft, The Netherlands   [‡]now with Caeleste CVBA, Mechelen, Belgium
michele.durbino@gmail.com

Abstract—**This work presents a compact ADC architecture capable of digitizing the signals received by every individual element of a 2D ultrasound transducer array. An element-matched layout of 150 μm × 150 μm is realized by exploiting each piezo-electric transducer element not only as the signal source, but also as the electro-mechanical loop-filter of a continuous-time band-pass ΔΣ ADC, thus minimizing the required circuit blocks. The transducer's frequency response, which is inherently matched with the signal bandwidth of interest, provides noise shaping to the ADC. A prototype chip has been fabricated in a 0.18 μm CMOS technology, featuring 20 ADCs located directly underneath a 150 μm-pitch piezo-electric transducer array fabricated on top of the chip. Each ADC, clocked at 200 MHz, consumes 800 μW from a 1.8 V supply, and achieves an SNR of 47 dB in a 75% bandwidth around a center frequency of 5 MHz. Acoustic measurements show that the ADC successfully digitizes incoming echo signals.**

Keywords: Band-pass ADC, Electro-mechanical Resonator, Tracking Quantizer, Inverter-based OTA.

I. INTRODUCTION

Heart-related diseases are the most common cause of death. Echocardiography, i.e. the use of ultrasound to image the heart, is a safe and affordable imaging technique that is crucial for the effective diagnosis and treatment of heart-related diseases. This work focuses on the realization of miniature ultrasound probes that are capable of making real-time three-dimensional (3D) images of the heart. In particular, we focus on endoscope-based probes for trans-esophageal echocardiography (TEE), which are used to make high-quality cardiac images from the patient's esophagus.

In order to obtain real-time 3D images, a 2D array of transducer elements is needed, leading to a total of 1000+ elements, of which the echo signals have to be processed. In conventional (2D) ultrasound probes, the elements are connected individually to an imaging system. The number of cables that would be needed in a 3D probe, however, exceeds by far what can be accommodated by an endoscope, calling for the use of in-probe integrated circuits. Prior work focuses on analog approaches to reduce the number of cables, such as analog sub-array beamforming [1] [2]. In this work, we report an element-level ADC architecture that enables the digitization of all received signals. Thus, digital beamforming, compression and multiplexing techniques can employed to reduce the number of cables, paving the way towards miniature ultrasound probes with a fully-digital interface.

The proposed design exploits the filtering properties of the ultrasound transducer to reduce the hardware needed for the digitization of the acoustic signal. In particular, the band-pass characteristic of the transducer is used as the noise shaping element of a Continuous-Time Band-Pass ΔΣ Modulator. This paper demonstrates this concept for the first time for ultrasound applications, and shows that it enables the integration of an entire oversampled ADC underneath a 150 μm × 150 μm transducer element even in a standard 0.18 μm CMOS technology, allowing massively-parallel digitization of all received signals in a 2D transducer array. This size is substantially smaller than a recently-reported element-matched ΔΣ ADC, which employs a conventional electrical loop-filter and was implemented in 28 nm CMOS [3].

II. TRANSDUCER IMPEDANCE CHARACTERISTICS

In this work, bulk PZT piezo-electric transducer elements are employed, with a pitch of 150 μm. These transducers are integrated directly on top of the proposed application-specific integrated circuit (ASIC) using the procedure described in [1], in which one electrode of each element connects to a bond-pad of the ASIC, while the other electrode is grounded using a thin conductive layer that covers the whole array.

The electrical impedance of a transducer element, measured with an impedance analyzer, is shown in Fig. 1. The plot shows a largely capacitive behavior, with a resonance around 5 MHz due to the mechanical thickness-mode resonance of the transducer. This behavior can be captured well by an equivalent Butterworth-Van Dyke circuit model, shown in the inset of Fig. 1, in which a motional branch, consisting of R, L and C, accounts for the mechanical resonance, and a parallel capacitor C_S models the capacitance between the transducer's electrodes. The transducer impedance can be expressed as $Z(s) = \frac{LCs^2+RCs+1}{sC_S(LCs^2+RCs+1+\frac{C}{C_S})}$, which shows the capacitive baseline, the resonance in the numerator and the anti-resonance in the denominator. The target of this work is to exploit the resonance in the loop filter of a ΔΣ ADC.

The resistance R directly impacts two extremely important aspects: the intrinsic thermal noise added by the transducer and the quality factor of the resonance. The first yields the required thermal noise performance of the ADC's input stage, and the second, defined as $Q = \frac{\omega_0 L}{R} = \frac{\omega_0}{Bandwidth}$, determines the intrinsic bandwidth of the system.

III. SYSTEM-LEVEL DESIGN

A. Working Principle

The quantization noise should be suppressed by the modulator around the center frequency. Since the resonance modeled

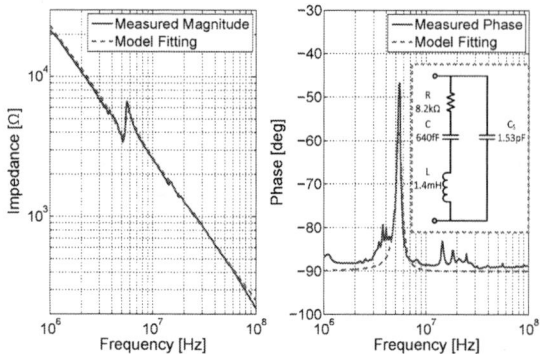

Fig. 1. Impedance of the employed PZT element (magnitude and phase).

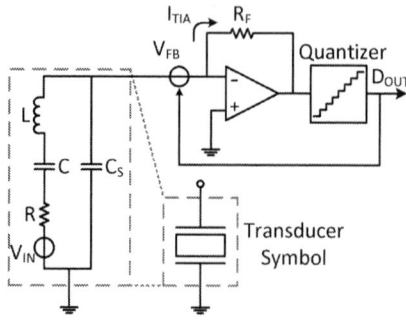

Fig. 2. Explanation of the working principle behind the element-level $\Delta\Sigma$ ADC.

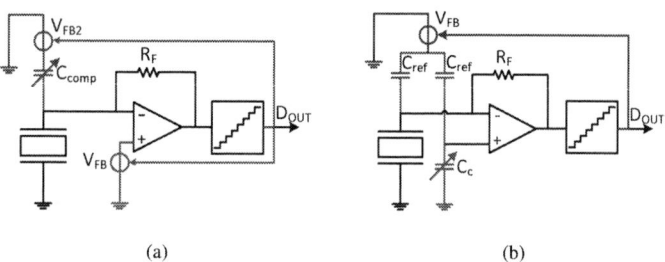

Fig. 3. ADC block diagram with (a) a compensation capacitor and (b) a capacitive bridge configuration to compensate for the effect of C_S

by the RLC branch provides a corresponding band-pass characteristic, this is chosen as the noise-shaping element of the modulator. The signal source within the transducer model is represented by an ideal voltage source V_{IN} in series with R, approximately ranging between 50 µV and 250 mV. Ideally, the summing node of the modulator should be realized at that same location, but that node in the equivalent circuit does not have a physical counterpart. Therefore, voltage feedback (V_{FB}) is applied to the available electrode of the transducer, while the resulting current is read out using a transimpedance amplifier (TIA), as shown in Fig. 2. The TIA's output is then digitized by a quantizer that controls the feedback voltage and thus closes the loop of the modulator. The current sensed by the TIA can be expressed as:

$$I_{TIA} = (V_{IN} - V_{FB})\frac{sC}{LCs^2 + RCs + 1} - V_{FB}sC_S \quad (1)$$

The first part of this equation shows the desired noise shaping behavior, but the second part, due to C_S, adds to the output a differentiated version of V_{FB}, which is undesirable. Moreover, the required floating voltage-feedback DAC is difficult to implement. Fig. 3a shows how these problems can be mitigated. First, the floating DAC is replaced by feedback to the non-inverting input of the TIA, causing the TIA's virtual ground to follow this voltage, thus effectively applying voltage feedback to the transducer. Second, an additional voltage DAC V_{FB2} and a capacitor C_{comp} compensate for the last term in (1), by providing a current $V_{FB2}sC_{comp}$ that cancels out the undesired current $V_{FB}sC_S$ coming from the transducer. A further refinement is shown in Fig. 3b, where the two feedback voltages are merged into one, and the feedback to the non-inverting input is replaced by a capacitive divider, leading to a bridge-type configuration in which the bottom capacitor C_C is made tunable and should be as close to C_S as possible for a perfect compensation.

B. Complete Element-Level Architecture

The complete element-level ADC architecture is shown in Fig. 4. Here, V_{FB} is replaced by a bank of inverters, supplied by two reference voltages V_{REF+} and V_{REF-}, each driving a pair of feedback capacitors. A programmable resistor R_C (Full scale = 1.5 kΩ, LSB = 100 Ω) was added in series with C_C (Full scale = 3.2 pF, LSB = 50 fF) in order to compensate for the transducer's high-frequency behavior [4]. C_F, added in

parallel with R_F, provides the loop filter with a dominant pole and helps to attenuate the effect of an imperfect compensation of C_S. Finally, a second stage is introduced to amplify the output of the TIA and thus relax the noise requirements of the quantizer.

Due to the propagation attenuation of the acoustic echo signals, the ADC has to be able to handle a relatively large input dynamic range: the nearest tissue produces high-amplitude echoes that reach the transducer elements sooner, while farther tissue generates weak echoes that will take a longer time to be caught by the transducers. To compensate for this, so-called time-gain compensation (TGC) is implemented by varying the reference voltages as a function of time, implemented using off-chip DACs, which could be integrated on-chip in a re-design. This ensures that the feedback voltage is always comparable with the input signal. In addition, the second-stage gain is programmable by two bits (decoded into four states) so as to maintain loop stability and to keep the voltage swing at the input of the quantizer approximately constant. The second stage gains are 9 dB, 16 dB, 24 dB and 33 dB. The second stage also introduces an additional pole, which, together with that of the first stage, helps to attenuate the first-stage noise. Both the g_m stages are implemented using current-reuse differential-to-single-ended operational transconductance amplifiers (OTA) [5], to improve the ADC's current efficiency. An element-level constant-g_m biasing block provides bias currents to the g_m stages.

A 3-bit quantizer is used in order to reach the desired SQNR requirements without having to increase the over-

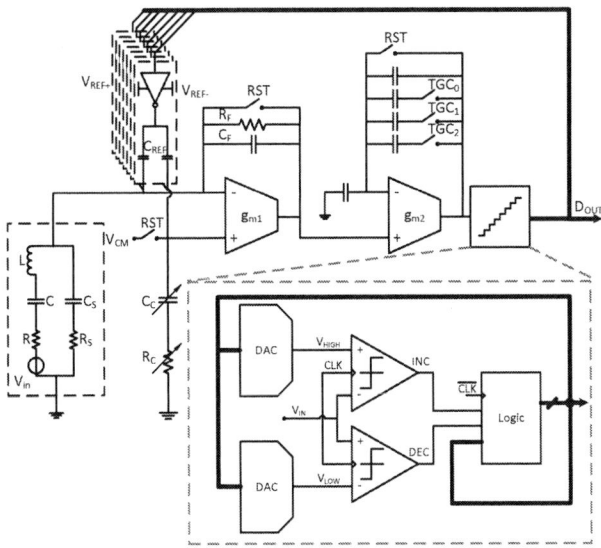

Fig. 4. Implemented ADC architecture.

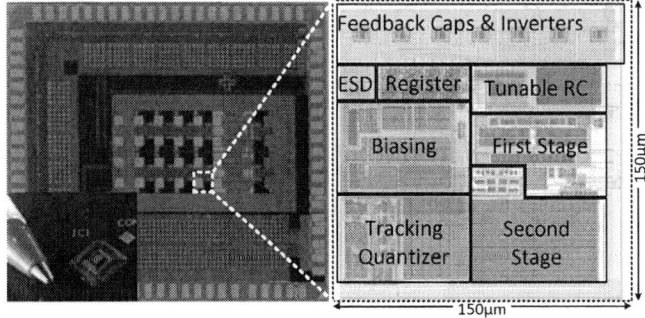

Fig. 5. Chip micrograph, element layout and photo of the prototype array built on top.

sampling ratio (OSR) and therefore the sampling frequency excessively. This quantizer, clocked at 200 MHz, employs a tracking architecture [6], depicted in the inset of Fig. 4. Two comparators compare the quantizer's input voltage with reference levels V_{HIGH} and V_{LOW}, generated by two 3-bit capacitive DACs based on the quantizer's previous output. The outputs of these comparators then determine whether the quantizer's output should be incremented, decremented, or kept as-is. While this type of quantizer is unable to track fast changing inputs, system simulations show that this is not an issue in this design, because the signal that reaches the quantizer has already been limited in bandwidth by the poles added by the two stages. This type of quantizer provides a good compromise among speed, power and die area. It also allows the ADC's output to be encoded in a 2-bit format, thus reducing the data rate compared to a full 3-bit output.

IV. CHIP-LEVEL DESIGN

A prototype with a transducer array of 5×4 elements has been realized, each of which is interfaced with a copy of the element-matched $\Delta\Sigma$ ADC described above. Every ADC produces two 200 MHz output bitstreams, coming from the tracking quantizer comparators, which are exported to an off-chip FPGA via on-chip LVDS transmitters. To reduce the chip's I/O count to a manageable number, a total of 20 LVDS transmitters and a 2:1 channel multiplexer are implemented, allowing 10 out of the 20 ADCs to be read out simultaneously. For flexibility, software-based decimation filters are employed; in a re-design, these filters could be implemented in each element or in the chip periphery. A micrograph of the chip, fabricated using TSMC 0.18 μm technology, is shown in Fig. 5, together with the layout of the 150 μm × 150 μm element-matched ADC. The inset in Fig. 5 shows the prototype chip mounted on a PCB substrate with a piezo-electric transducer array built on top using the process described in [1], before applying the top ground layer, so that the individual elements can be seen.

V. MEASUREMENT RESULTS

Fig. 6 shows the measured ADC output spectrum, with a full-scale acoustic continuous-wave input generated by an unfocused 5 MHz transmitter positioned above the chip (see Fig. 6). The desired noise shaping behavior can be clearly observed: the quantization noise is lowest in the bandwidth of interest, highlighted in green. The measured SNR is 45 dB, 47 dB, 45 dB and 41 dB for the four TGC steps (i.e. gain settings of the second stage). The lower SNR in the last TGC step is likely due to noise coming from the off-chip reference voltages V_{REF+} and V_{REF-}. Digital beamforming (on- or off-chip) could be employed in a re-design to improve the effective SNR of the image.

In order to verify that the ADC output is a good approximation of the actual signal received by the transducer, the response of the ADC to an acoustic pulse has been compared in Fig. 7 to the same element's TIA output, with the ADC feedback disabled. In the latter case, the TIA provides an amplified version of the transducer's output current. The two responses are in good agreement.

Tab. I compares this work with recently published ultrasound ADCs. As this design uses the ultrasound transducer as the loop filter for a $BP\Delta\Sigma$ converter, the area typically reserved for the reactive components in standard loop filters is saved. This allows the designed converter to be very competitive in terms of area. Furthermore, the power consumption is an order of magnitude lower than the state of the art, even if it achieves a worse SNR.

Fig. 8a compares this work with $\Delta\Sigma$ ADCs in general, as reported in [7]. The x-axis represents the area, while on the y-axis, the Walden figure of merit (FOM) is shown, defined as: $FOM_W = \frac{Power}{\frac{f_s}{OSR} 2^{\frac{SNR-1.76}{6.02}}}$ This work achieves by far the smallest area compared to designs using a similar technology. Furthermore, only two of the converters employing a smaller feature size obtain a smaller area. The Walden FOM, while being more than one decade above the state of the art, can still be considered as a competitive result, since the proposed design is not just an ADC, but also includes a TIA and a programmable-gain amplifier (PGA). Finally, Fig. 8b shows a direct comparison between area and power consumption, for ADCs with similar SNR and bandwidth, using any technology node. This work features the smallest area, as well as the lowest power consumption among these designs.

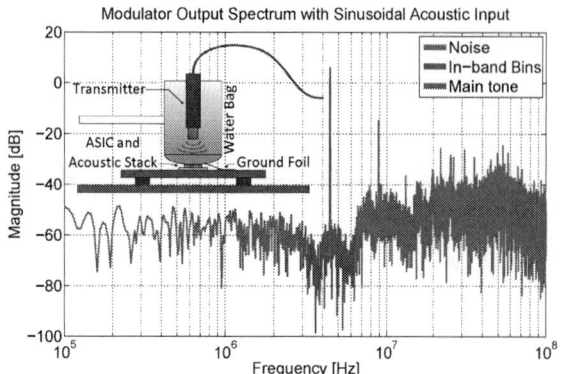

Fig. 6. FFT of the ADC's measured bitstream. The center frequency is slightly shifted because of the uncertainties linked to transducer fabrication

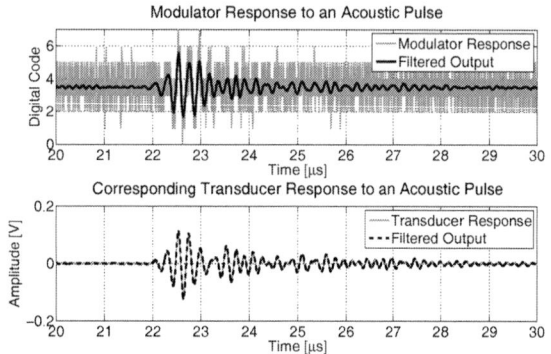

Fig. 7. Time-domain response of the ADC to an acoustic pulse compared with that of the transducer. Measured in water at a distance of 2 cm

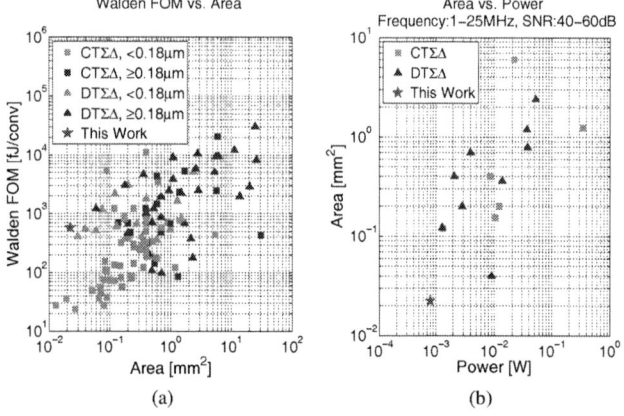

Fig. 8. Comparison of this work's FOM and Power vs. active area with designs published at ISSCC and VLSI [7]

VI. CONCLUSION

This work demonstrates that a piezo-electric ultrasound transducer can serve as the electro-mechanical loop filter of a $\Delta\Sigma$ ADC, thus inherently reducing complexity, area and power consumption compared to an implementation using a conventional electrical loop filter. We expect that this concept is not limited to the bulk PZT piezo-electric transducers

TABLE I. COMPARISON OF THIS WORK WITH SIMILAR PUBLICATIONS

	This Work	[3]	[8]	[9]	[10]
Transducer Type	PZT	CMUT	N/A	N/A	N/A
Architecture	CTBPDS	DTLPDS	SAR	CTBPDS	CTLPDS
Technology [nm]	180	28	130	65	65
No. of Channels	20	16	64	8	1
Elem. Matched [μm]	Yes (150)	Yes (250)	No	No	No
Center Freq. [MHz]	5	5	5	260	4
Bandwidth [MHz]	3.125-6.875	10	8	20	15
Area/Channel [mm^2]	0.025	0.0625	0.1	0.03	0.4
Power/Channel [mW]	0.8	17.5	6.32	13.1	6.96
SNR/Channel [dB]	47	60	48.5	54	74.6

employed in our work, but can be extended to micro-machined transducers, such as capacitive micro-machined ultrasound transducers (CMUTs). An attractive feature of the presented ADC architecture is that it does not introduce much additional circuitry compared to a traditional analog front-end (AFE), which also tends to employ a TIA and a programmable-gain stage. The only extra circuits needed for element-level digitization are the tunable RC branch, the feedback DACs and the tracking quantizer. These circuits occupy only 37% of the area and consume roughly 50% of the power. The acoustic results obtained with our prototype show that the ADC successfully digitizes acoustic echo signals, demonstrating the effectiveness of the presented architecture and making it a promising approach for in-probe digitization in future high-element-count 3D ultrasound probes.

REFERENCES

[1] C. Chen et al., "A Prototype PZT Matrix Transducer With Low-Power Integrated Receive ASIC for 3-D Transesophageal Echocardiography," *IEEE Transactions on Ultrasonics, Ferroelectrics, and Frequency Control*, vol. 63, no. 1, pp. 47–59, 2016.

[2] Y. Katsube et al., "Single-chip 3072ch 2D array IC with RX analog and all-digital TX beamformer for 3D ultrasound imaging," in *IEEE International Solid-State Circuits Conference*, 2017, pp. 458–459.

[3] M.-C. Chen et al., "A Pixel-Pitch-Matched Ultrasound Receiver for 3D Photoacoustic Imaging with Integrated Delta-Sigma Beamformer in 28nm UTBB FDSOI," in *Int. Solid-State Circuits Conference*, San Franciso, USA, 2017, pp. 456–458.

[4] M. Guan et al., "Studies on the circuit models of piezoelectric ceramics," in *Proc. International Conference on Information Acquisition*, 2004, pp. 26–31.

[5] M. Bazes, "Two novel fully complementary self-biased CMOS differential amplifiers," *IEEE Journal of Solid-State Circuits*, vol. 26, no. 2, pp. 165–168, 1991.

[6] L. Dörrer et al., "A 3-mW 74-dB SNR 2-MHz continuous-time delta-sigma ADC with a tracking ADC quantizer in 0.13-μm CMOS," *IEEE Journal of Solid-State Circuits*, vol. 40, no. 12, pp. 2416–2426, 2005.

[7] B. Murmann, "ADC Performance Survey 1997-2016," 2016. [Online]. Available: http://web.stanford.edu/ murmann/adcsurvey.html

[8] Y.-J. Kim et al., "A Single-Chip 64-Channel Ultrasound RX-Beamformer Including Analog Front-End and an LUT for Non-Uniform ADC-sample-clock Generation," *IEEE Transactions on Biomedical Circuits and Systems*, vol. 11, no. 1, pp. 87–97, 2017.

[9] J. Jeong et al., "A 260 MHz IF Sampling Bit-Stream Processing Digital Beamformer With an Integrated Array of Continuous-Time Band-Pass $\Delta\Sigma$ Modulators," *IEEE Journal of Solid-State Circuits*, pp. 1–9, 2016.

[10] Y. Zhang et al., "A Continuous-Time Delta-Sigma Modulator for Biomedical Ultrasound Beamformer Using Digital ELD Compensation and FIR Feedback," *IEEE Transactions on Circuits and Systems I: Regular Papers*, vol. 62, no. 7, pp. 1689–1698, 2015.

S11-3 (2190)

A 12.1mW, 60dB SNR, 8-Channel Beamforming Embedded SAR ADC for Ultrasound Imaging Systems

Taehoon Kim and Suhwan Kim

Department of Electrical and Computer Engineering, Seoul National University, Seoul, Korea

E-mail: taehoon.kim@analog.snu.ac.kr, suhwan@snu.ac.kr

Abstract— **A power-efficient analog beamforming embedded SAR ADC for ultrasound imaging systems is presented. It is constructed from multiple sub-beamforming SAR ADCs, which sequentially perform analog beamforming and analog-to-digital conversion for an assigned focal point on a scan-line. Power is saved because these operations are carried out in the charge domain without a summing op-amp. This is realized by a capacitor beamforming digital-to-analog converter, which acts as both an analog memory cell and a digital-to-analog converter. The proposed analog beamformer was implemented in a 0.18μm CMOS process, and its power consumption is 12.1mW. The delay resolution is 4.17ns and the maximum achievable delay is 292ns. The sampling frequency is 20MS/s and the SNR of the prototype is 60dB. The circuit occupies an active area of 2.2mm².**

Keywords— *Analog beamformer; Hybrid beamformer; Beamforming SAR ADC; ultrasound imaging systems*

I. INTRODUCTION

Ultrasound imaging systems with multiple transducer elements are widely used to observe cross-sections of the human body. Beamforming techniques are typically employed during both transmission and receiving modes to improve the image quality by increasing the SNR of the received signals. In transmission mode, a high-voltage pulse is applied to each transducer at staggered times, so that the acoustic waves from all the transducers arrive at a focal point simultaneously. In receiving mode, the echo signals received by the transducers, are sent to analog front-end circuits. These amplify and filter the signals, and then following beamformer performs a time-alignment operation, and finally sums the signals over all the channels. Since the echo signals at the input of the beamformer are correlated but the noise signals are not, the SNR increases in proportion to the square root of the number of channels.

Recently, two-dimensional transducer arrays have been introduced in advanced devices to obtain three-dimensional images [1]. These arrays contain of as many as a thousand transducer elements. This high channel count causes wiring problems, because connecting so many channels to an ultrasound machine through coaxial cables is not feasible. A common way of reducing the channel count is sub-array beamforming, shown in Fig. 1, in which an M-channel beamformer realigns and sums a subset of M channels after signal amplification and filtering. An M-channel beamformer

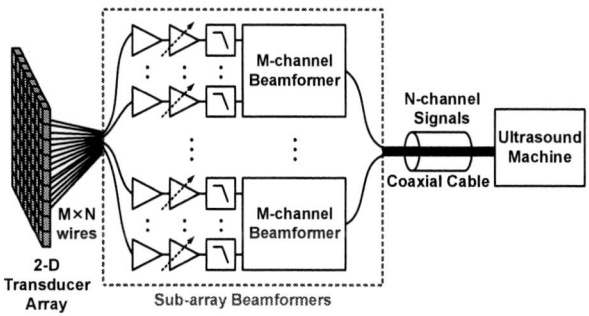

Fig. 1. Sub-array beamforming architecture for a 2D transducer array [2].

can be as implemented either a digital [3]-[5] or an analog circuit [6]-[8]. Digital beamformers (DBFs) have excellent performance, but they require large numbers of analog-to-digital converters (ADCs) which are essential for each channel. Analog beamformers (ABFs) have been suggested as a more suitable choice because of small area and low power requirements. ABFs can be implemented using analog filters or analog memory cells. However, they typically suffer from limited delay, coarse delay resolution and low SNR.

Hybrid beamformers (HBFs) offer a compromise between ABFs and DBFs [9]-[11]. HBFs are used to reduce the number of ADCs while utilizing the digital beamforming operation. The two-step architecture of an HBF also reduces total memory required in the sub-array beamformer and minimizes power dissipation. However, the ABF and the ADC are still independently located to perform each function in the HBF. Since the output signal must be voltage type in order for the ABF output to be delivered to the ADC, a power-hungry op-amp that adds the signal in the voltage domain is necessary. This can be a factor that prevents further increase in the number of channels because the power dissipation should be minimized in order not to violate the safety regulation.

To overcome the above issue, this paper proposed a new architecture. First of all, the summing op-amp is eliminated, and the capacitor digital-to-analog converter (CDAC) of a successive-approximation register (SAR) ADC is reconfigured to perform beamforming. Instead of summing the signals in the voltage domain, this is now done in the charge domain using a charge-averaging method. Following beamforming, A/D conversion is performed in the same block using a

978-1-5386-3179-9/17 $31.00 © 2017 IEEE

S11-3 (2190)

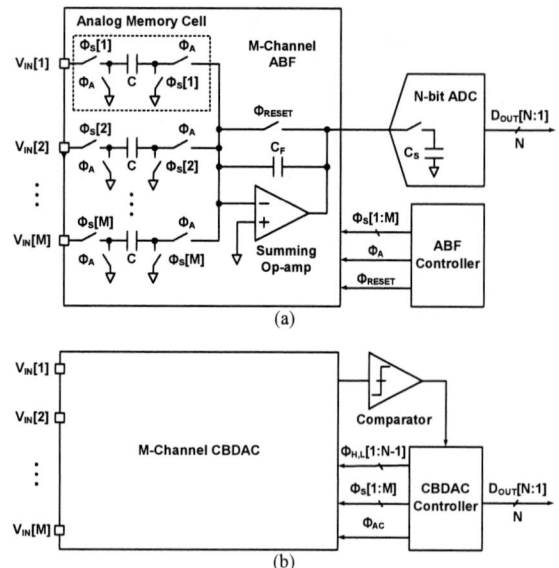

(a)

(b)

Fig. 2. (a) Established architecture of an *M*-channel ABF and an *N*-bit ADC, and (b) the proposed *M*-channel, *N*-bit BF-SAR ADC architecture.

charge-redistribution method. Performing these operations in the charge domain allows the summing op-amp to be eliminated, reducing power consumption.

II. ARCHITECTURE

The established architecture of a two-stage HBF consisting of an *M*-channel ABF and an *N*-bit ADC for processing a single focal point is shown in Fig. 2(a). The *M*-channel ABF consists of analog memory cells, a summing op-amp and a timing controller. It accepts M channels of amplified echo signals and performs beamforming in the voltage domain. In the sampling phase, M echo signals $V_{IN}[1:M]$, returning from the focal point, are sampled sequentially into the analog memory cells, controlled by the sampling signals $\Phi_S[1:M]$. In the subsequent addition phase, the ABF controller activates the addition signal Φ_A, which causes all the charges sampled by the analog memory cells to be added by the summing op-amp. The output of the ABF is sampled by the capacitor C_S and the ADC converts it to a digital code.

Fig. 2(b) shows our alternative *M*-channel beamforming embedded SAR (BF-SAR) ADC. The capacitor arrays as analog memory cells for the analog delay line is merged with the capacitor arrays in the CDAC of the typical SAR ADC. We called this modified CDAC a capacitor beamforming digital-to-analog converter (CBDAC). The proposed BF-SAR ADC consists of a CBDAC, a comparator and a CBDAC controller. During beamforming, the CBDAC acts as ABF with sample-and-hold delay lines and performs charge-mode summation. These operations controlled by the sampling clock $\Phi_S[1:M]$ and the addition and conversion signal Φ_{AC}. After beamforming, the CBDAC performs same function as the charge redistribution based CDAC of a SAR ADC and is controlled by the DAC control signals $\Phi_{H,L}[1:N-1]$. Unlike the established architecture summing the signals in the voltage

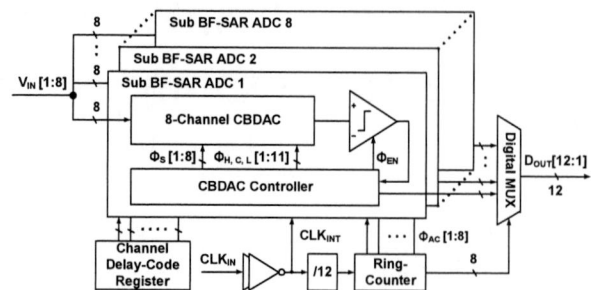

Fig. 3. Block diagram of the proposed 8-channel BF-SAR ADC.

domain, there is no need for the summing op-amp in proposed architecture.

III. CIRCUIT IMPLEMENTATION

Fig. 3(a) is a block diagram of our ADC, which consists of eight sub BF-SAR ADCs, a digital multiplexer, a clock driver, a clock divider, and an 8-bit ring-counter. The eight sub BF-SAR ADCs are time-interleaved, so that beamforming and A/D conversion are performed sequentially for multiple focal points on a scan-line. The digital multiplexer selects one of the digital outputs of the eight sub BF-SAR ADCs. The clock divider divides the operating clock by 12, and the 8-bit ring-counter takes this divided signal and generates the time-interleaved control signals $\Phi_{AC}[1:8]$.

Fig. 4 is a schematic diagram of the CBDAC, which is based on a split-capacitor array designed with a 9-bit upper array and a 3-bit lower array, with V_{CM}-based capacitor switching scheme. The first MSB capacitance of 128C is divided by four capacitances of 32C, and the second MSB capacitance of 64C is divided by two 32C capacitors to perform beamforming, where C is 26fF. The sampling phase is the same as it is in the established architecture, except that the input of the 8th channel is sampled into a binary-weighted capacitor array. In the addition and A/D phase, the CBDAC controller activates the signal Φ_{AC} which causes addition and A/D conversion to take place. The input sampling switches are off, and all the capacitors are connected to the common voltage V_{COM}, which achieves charge-mode summation of all eight channels. Then the comparator compares the voltage V_X with V_{COM} and determines the MSB of the digital output. In the second conversion cycle, the first four capacitors are connected to the top-level reference voltage V_{REFT} or the bottom-level reference voltage V_{REFB}, depending on the output of the comparator, which also decides the second MSB of the digital output. In the third conversion cycle, the next two capacitors are used to determine the third MSB of the digital output. From the next conversion cycle, the same operations with conventional SAR ADC are performed. A compensation capacitor C_C is added to the lower side of CBDAC to compensate for errors in linearity caused by parasitics and the integer value of the bridge capacitor C_B.

Fig. 5(a) is a block diagram of the part that generates the sampling clock $\Phi_S[1:8]$ in the CBDAC controller. It consists of a 7-bit up-counter and eight sampling-signal generators,

978-1-5386-3179-9/17 $31.00 © 2017 IEEE

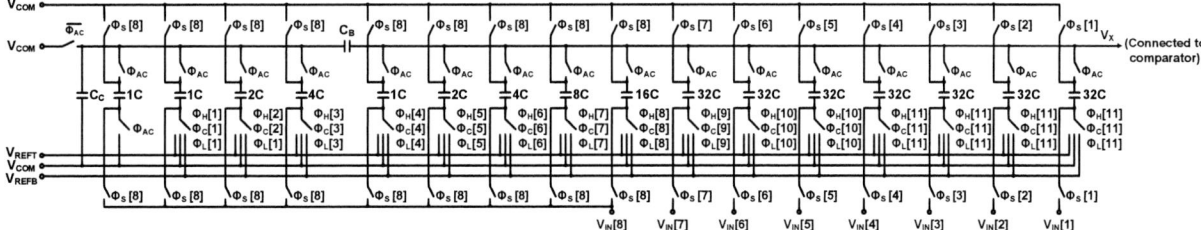

Fig. 4. Schematic diagram of the 8-channel CBDAC.

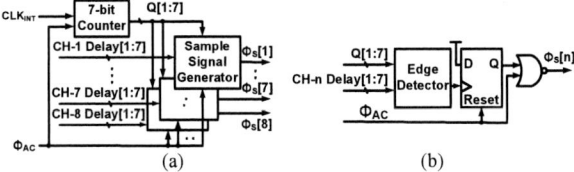

Fig. 5. Implementation of (a) the part that generates the sampling clock $\Phi_S[n]$ ($n=1,...,8$) in the CBDAC controller and (b) n^{th} sample-signal generator in Fig. 5(a).

each of which compares the output of the up-counter with the channel delay-code CH-n Delay[1:7] ($n=1,...,8$). A falling edge of the sampling clock $\Phi_S[1:8]$ is detected if the output of the up-counter is the same as the channel delay code. Fig. 5(b) is a block diagram of nth sample-signal generator in Fig. 5(a).

IV. EXPERIMENTAL RESULTS

The prototype BF-SAR ADC was implemented in a 0.18μm CMOS process. Fig. 6 shows the measurement setup and a die micrograph, in which the BF-SAR ADC occupies an active area of 2.2mm². For testability, eight analog front-ends (AFEs) [12] were included in front of the BF-SAR ADC. Our chip was implemented so that the analog inputs can be applied directly to the BF-SAR ADC or via the AFE. When validating the delay compensating function and measuring the dynamic performance of the BF-SAR ADC itself, the input signals were connected directly to its inputs. But when verifying the effect of beamforming on SNR enhancement, the inputs were applied through AFEs for adding uncorrelated input noise. During all of measurements reported in this paper, the offset mismatch from the time-interleaved architecture was cancelled by equalizing the averages of each sub BF-SAR ADC output to the average of the output of the first sub BF-SAR ADC [13].

The measured differential nonlinearity (DNL) and integral nonlinearity (INL) of the proposed BF-SAR ADC are shown in Fig. 7. The peak DNL and INL are 0.97/-0.67 LSB and 1.24/-1.37 LSB, respectively. Fig. 8 shows the output spectrum of the summation of all 8 channels when sinusoidal inputs of 1MHz were applied to the BF-SAR ADC. The SNR and SNDR were 60.8dB and 58.3dB respectively. The SNR of a total of 30 sample chips was measured to check the robustness of the design against process variations. The histogram in Fig. 9 shows a total spread of 60 to 61dB. Fig. 10 shows the change of SNR gain with the number of channels: the linear relationship verifies the beamforming function and agrees with calculated values. Table I compares this work with other beamformers, and shows that the BF-SAR ADC

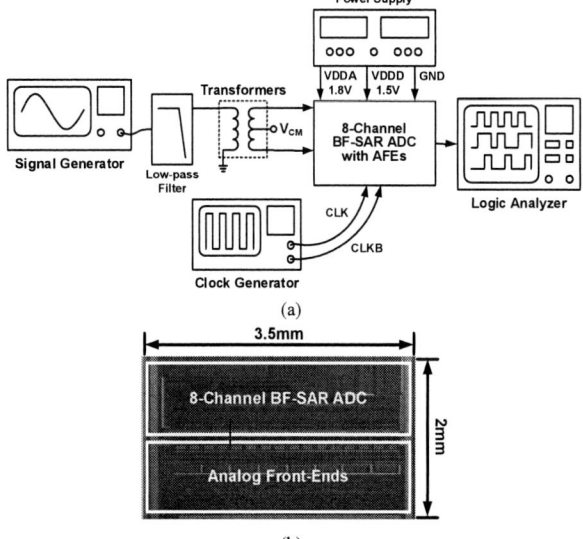

(a)

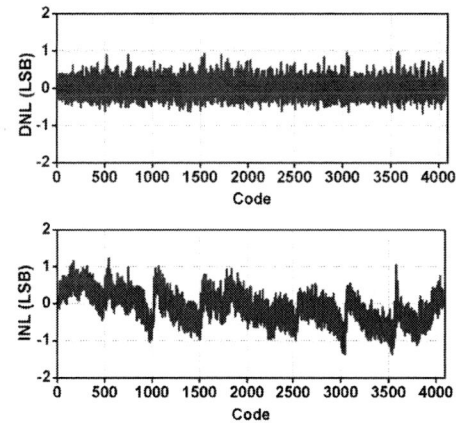

(b)

Fig. 6. (a) Measurement setup and (b) die micrograph.

Fig. 7. Measured DNL and INL.

consumes less power than other beamformers which require an additional power hungry ADC.

V. CONCLUSION

We have proposed a power-efficient beamforming embedded SAR ADC for ultrasound imaging systems. Our BF-SAR ADC performs charge-mode summation and charge-redistribution based A/D conversion by reconfiguring the

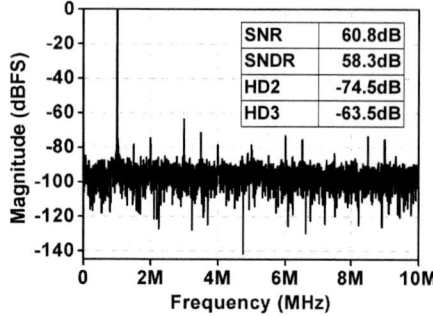

Fig. 8. Measured output spectrum of the BF-SAR ADC with a 1MHz input.

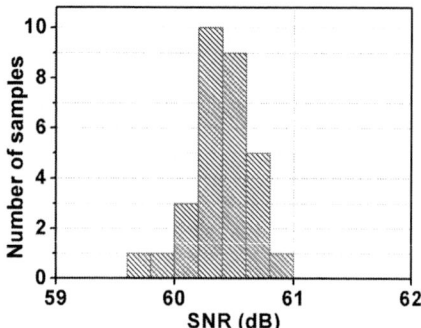

Fig. 9. Histogram showing the SNR distribution over 30 samples.

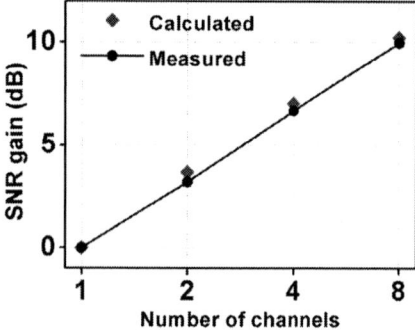

Fig. 10. Measured SNR gain with different numbers of summed channels.

CDAC of a SAR ADC. Utilizing this BF-SAR ADC, the need for a summing op-amp is eliminated, resulting in lower power consumption. The prototype was implemented in a 0.18μm CMOS process with an active area of 2.2mm² and its measured power consumption is 12.1mW. The SNR increases linearly with each doubling of the number of channels, and the proposed BF-SAR ADC achieves 60dB SNR when all 8-channels are summed. The experimental results suggest that this prototype is suitable for ultrasound imaging systems with 2D transducer arrays.

ACKNOWLEDGMENT

The authors would like to thanks to the Alpinion Medical Systems for technical discussions. This work was financially supported in part by Samsung Research Funding Center of Samsung Electronics under Project Number SRFC-IT1701-04.

TABLE I. PERFORMANCE COMPARISON WITH OTHER BEAMFORMERS

Parameter	This work	[5]	[6]	[7]	[8]	[10]
Functionality	BF-SAR ADC	DBF	ABF	ABF	ABF	HBF (ABF+DBF)
Technology (μm)	0.18	0.028	0.35	0.13	0.18	0.13
Supply (V)	1.8/1.5	1.5/1	3.3	1.2	1.8	3.3/1.2
Number of channels	8	16	8	32	16	64
Delay dynamic range	70	128	20	192	15	1280
Maximum delay (ns)	292	1067	35	1920	600	8000
Delay resolution (ns)	4.17	8.33	1.75	10	40	6.25
Power consumption (mW)	12.1	280	67	150	12.6	1140
Sampling frequency (MHz)	20	N/A	N/A	N/A	25	N/A
Active area (mm²)	2.2	0.656	0.36	16.4	1.3	19.4
SNR (dB)	60	60	N/A	N/A	60	<53

REFERENCES

[1] J.-M. Bureau et al., "A two-dimensional transducer array for real-time 3D medical ultrasound imaging," in *IEEE Ultrason. Symp.*, 1998, pp. 1065–1068.

[2] B. Savord et al., "Fully sampled matrix transducer for real time 3D ultrasonic imaging," in *IEEE Ultrason. Symp.*, Oct. 2003, pp. 945–953.

[3] B. D. Steinberg, "Digital beamforming in ultrasound," *IEEE Trans. Ultrason., Ferroelect., Freq. Contr.*, vol. 39, no. 6, pp. 716–721, Nov. 1992.

[4] J. A. Brown et al., "A digital beamformer for high-frequency annular arrays," *IEEE Trans. Ultrason., Ferroelect., Freq. Contr.*, vol. 52, no. 8, pp. 1262–1269, 2005.

[5] M-C. Chen et al., "A Pixel-Pitch-Matched Ultrasound Receiver for 3D Photoacoustic Imaging with Integrated Delta-Sigma Beamformer in 28nm UTBB FDSOI," in *Proc. IEEE Int. Solid-State Circuits Conf., Dig. Tech. Papers*, Feb. 2017, pp. 456–457.

[6] G. Gurun et al., "An analog integrated circuit beamformer for high-frequency medical ultrasound imaging," *IEEE Trans. Biomed. Circuits Syst.*, vol. 6, pp. 454–467, Oct. 2012.

[7] J. Y. Um et al., "A single-chip time-interleaved 32-channel analog beamformer for ultrasound medical imaging," in *IEEE Asian Solid-State Circuits Conf.*, Nov. 2012, pp. 193–196.

[8] S. Sharma et al., "In-probe ultrasound beamformer utilizing switched-current analog RAM," *IEEE Trans. Circuits Syst. II, Exp. Briefs*, vol. 62, no. 6, pp. 517–521, Jun. 2015.

[9] B-H. Kim et al., "Hybrid volume beamforming for 3-D ultrasound imaging using 2-D CMUT arrays," in *IEEE Ultrason. Symp.*, 2012, pp. 2246–2249.

[10] J. Y. Um et al., "An analog-digital-hybrid single-chip RX beamformer with non-uniform sampling for 2-D CMUT ultrasound imaging to achieve wide dynamic range of delay and small chip-area," in *Proc. IEEE Int. Solid-State Circuits Conf., Dig. Tech. Papers*, Feb. 2014, pp. 426–427.

[11] M. Bae et al., "A Novel Beamforming Method for Wireless Ultrasound Smart Probe," in *IEEE Ultrason. Symp.*, 2014, pp. 2185–2188.

[12] T. Kim et al., "A CMOS analog front-end for driving a high-speed SAR ADC in Low-Power Ultrasound Imaging Systems," in *IEEE Int. Conf. on System-on-Chip*, Sept. 2016, pp. 160-163.

[13] N. Le Dortz et al., "A 1.62 GS/s time-interleaved SAR ADC with digital background mismatch calibration achieving interleaving spurs below 70 dBFS," in *IEEE Int. Solid-State Circuits Conf., Dig. Tech. Papers*, Feb. 2014, pp. 386–388.

A 2.79-mW 0.5%-THD CMOS Current Driver IC for Portable Electrical Impedance Tomography System

Jaeeun Jang[*], Minseo Kim[*], Joonsung Bae[†] and Hoi-Jun Yoo[*]

E-mail: jaeeun.jang@kaist.ac.kr

[*]Department of EE, Korea Advanced Institute of Science and Technology (KAIST), Daejeon, Republic of Korea

[†]Department of EE, Kangwon National University, Chuncheon, Kangwon, Republic of Korea

Abstract— A low-power and high-accuracy current driver IC is proposed for portable electrical impedance tomography (EIT) systems. The proposed IC supports active electrode configuration and has three key features. First, high output impedance current driver is implemented with phase compensation scheme through a delay-locked loop (DLL) to significantly alleviate a phase shift and a required power consumption in the driver. Second, low-power pseudo-sine wave generator is used not only for a low-harmonic distortion but also for a flexible injection frequency utilization. Through the proposed quantization scheme, total harmonic distortion (THD) is reduced by 35%, achieving THD less than 0.5%. Third, the calibration unit is proposed to compensate injection-current amplitude variations among active electrodes. As a result, the proposed current driver IC guarantees more than 1MΩ output impedance over a wide range of injection current up to 400μA and operating frequency from 1kHz to 5MHz, only consuming 2.79mW.

Keywords—Bioimpedance circuit, current driver, EIT, low-power, phase error, wideband operation

I. INTRODUCTION

Electrical Impedance Tomography (EIT) has been actively studied for many years as an alternative medical imaging solution. The most attractive point of EIT is that it can provide long-term and continuous patient monitoring based on electrical tissue properties, resulting in harmless measurement process without any ionizing radiation and high intensity electromagnetic field [1]. Moreover, it can be implemented with small form-factor devices, such as smart belt, so that spatial constraints are dramatically reduced. The conventional EIT [2] system was limited in the bedside monitoring application, but nowadays, portable EIT application is getting more attention for real-time imaging system via its compactness [3].

In the portable EIT system, both low-power consumption and accurate measurement are key requirements since portable EIT system is powered by battery sources, low-power impedance measurement is of significant importance for a longer battery lifetime [4]. Moreover, accurate electrical impedance measurements and well-matched current injections are essential because EIT imaging is based on an inverse problem-solving algorithm which is vulnerable to electrode to electrode variations. Especially, the accurate measurements are hampered by a large number of electrodes unavoidably accompanying long cable length and stray capacitance issue [5]. Therefore, the active electrode configurations directly

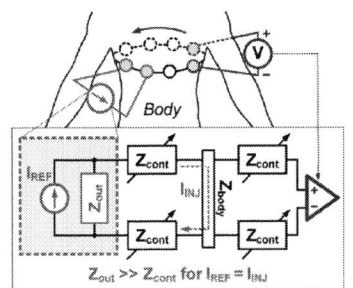

Figure. 1. Requirement of accurate current injection in EIT

integrating current driver IC on the electrode are widely applied to mitigate the accuracy issues [6]. In such a configuration, each electrode injects electrical current through its own current driver. Since the accuracy of EIT imaging is directly affected by quality of injection current, high-accuracy and well-matched current driver becomes key-building block.

Consequently, the current driver in the active electrode configuration should satisfy following requirements. First, it has to provide accurate current amplitude under the variation of load impedance. The load impedance varies from hundreds Ω to tens of kΩ with respect to the injection frequency [7]. As shown in Fig. 1, the output impedance of current driver should be high enough for accurate current injection. Second, low-harmonic and reconfigurable sinusoidal signal generator that supports wide range of injection frequency (a few kHz to few MHz) needs to be integrated for the simultaneous current injection among various electrodes [8]. Last, the current driver-to-driver mismatch should be minimized for the high image quality.

To tackle the high output impedance issue, the previous work [9] proposed output impedance boosting technique. However, it consumes large power in order to achieve high gain-bandwidth performance under large parasitic capacitances, reducing output phase shift problem [10] at the high frequency signal. In addition, the sinusoidal generator is not integrated with current driver, which leads to its limited usage for EIT application. For a sinusoidal current driver, the previous work in [3] adopts differential sinusoidal current stimulator based on wien-bridge oscillator, but it cannot cover wide range of frequencies because of its limited tuning scheme. Above all, these previous works have not considered current driver-to-driver matching issues in the active electrode configuration, so that they are not able to correspond to unpredictable current mismatch.

978-1-5386-3179-9/17 $31.00 © 2017 IEEE

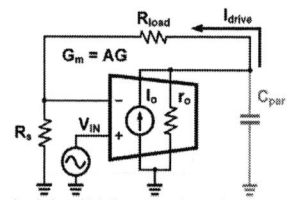

Figure. 2. Basic principle of high output impedance current driver

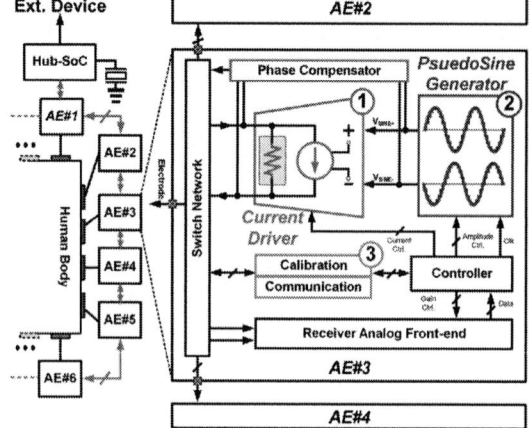

Figure. 3. Overall Architecture of IC

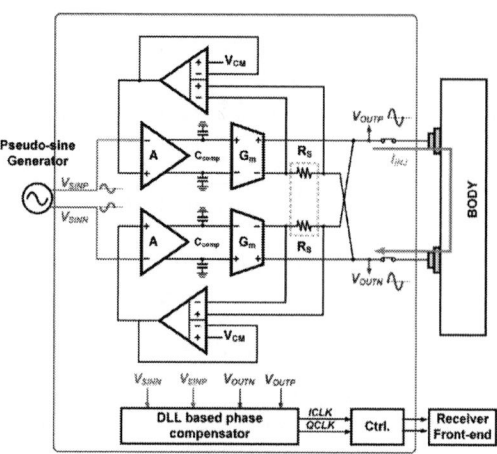

Figure. 4. Proposed current driver architecture

are connected to switch network. The signal paths are managed by the configuration of switching network. Each IC shares the same global clock received from Hub-SoC. From the global clock, pseudo-sine generator makes differential sinusoidal voltage used as input signals of current driver. From the input sinusoidal wave, the current driver generates sinusoidal injection current. The DLL based phase compensator senses input and output signals of current driver and generates the phase-synchronized I/Q clocks with respect to injection current. These clocks are forwarded to the switch network, and used in the demodulator of receiver analog front-end. Before impedance measurement, injection current from each electrode is equalized through the calibration unit. During the calibration, current paths to electrodes are turned off, and each node sequentially drive current to the reference resistor in calibration unit. The received peak amplitudes of currents are processed through the Hub-SoC and Hub-SoC delivers control code to each pseudo-sine wave generator.

III. CIRCUIT IMPLEMENTATION

A. Phase compensated current driver

Fig. 4 shows proposed current driver architecture [11]. It consists of gain stage with gain A, transconductance stage G_m, differential difference amplifier (DDA), on-chip current sensing resistor R_s with fully differential configuration and DLL based phase compensator. The gain stage provides enhanced transconductance and loop gain to increase the total output impedance of the current driver. The output current is sensed by R_s and fed back to the gain stage through the DDA. Since common mode voltages in both sides of resistor are not equal, the voltage buffer using DDA maintains output voltage level to V_{CM}. The DDA circuit is shown in Fig. 5, and V_{CM} is set to common mode voltage in the outputs of pseudo-sine wave generators.

With proposed design, a gain-bandwidth requirement remains as a bottleneck for the low-power design. Due to its feedback topology, large bandwidth and phase margin consideration are required for high frequency injection. Moreover, the parasitic capacitance of output nodes contributes

In this paper, we propose a low-power and accurate current driver IC for portable EIT system to fulfill three requirements with following key features. First, phase compensated high output impedance current driver is proposed for both low-power and accurate measurement. Second, low-harmonic pseudo-sine wave generator is integrated for low-power and flexible frequency generation. Third, the calibration unit is proposed for well-matched injection current between active electrodes.

The rest of this paper is organized as follows: In Section II, the basic principle and architecture of propose IC will be introduced. Section III shows detail circuit implementation and Section IV shows the measurement results. Finally, the conclusion will be made in Section V.

II. CURRENT DRIVER ARCHITECTURE

Fig. 2 shows basic principle of current driver. It consists of current sensing voltage feedback topology. Due to its feedback configuration, output impedance R_{out} is given by

$$R_{out} = r_o + (1 + G_m r_o) R_s$$

where G_m is transconductance, r_o is small signal output impedance, R_s is feedback current sensing resistor, and R_{load} is load resistance. According to the results, large G_m and r_o increase output impedance of current driver.

Fig. 3. shows overall architecture of current driver IC. It consists of current driver, pseudo-sine wave generator, calibration and communication unit, controller, and receiver analog front-end block. Since each electrode should support both current injection and voltage sensing, the building blocks

978-1-5386-3179-9/17 $31.00 © 2017 IEEE

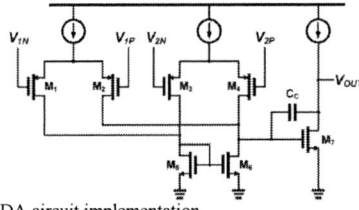

Figure. 5. DDA circuit implementation

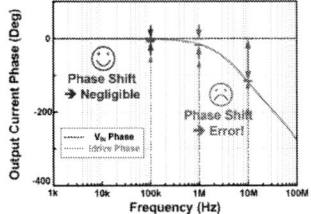

Figure. 6. Phase shift in current driver

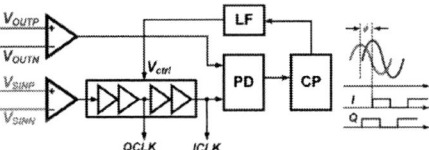

Figure. 7. Proposed DLL based phase compensator circuit.

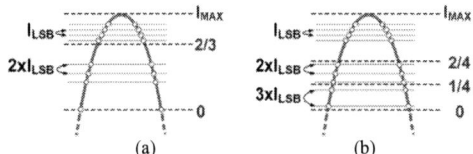

Figure. 8. Comparison between previous (a) and proposed (b) low-harmonic quantization scheme

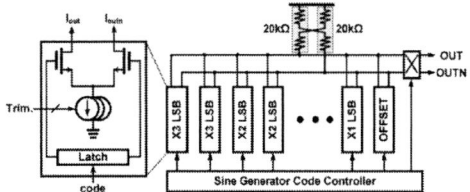

Figure. 9. Pseudo-sine generator circuit implementation

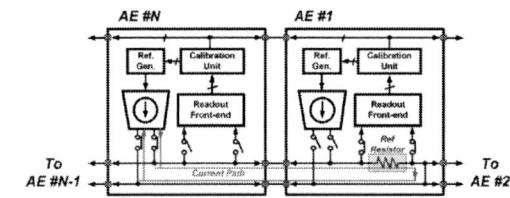

Figure. 10. Mismatch compensation scheme

to increase input-to-output phase shift in current driver. Since it varies according to real measurement environment, it is hard to estimate parasitic capacitance accurately. As shown in the Fig. 6, to avoid phase shift in higher frequency, overall power consumption of building blocks should be increased.

Fig. 7 shows proposed phase compensator circuit. It has two input signal path. V_{OUTP}, V_{OUTN} are output voltages of current driver and V_{SINP}, V_{SINN} are output voltage of pseudo-sine wave generator. Each input signal is amplified through the identical limiter block. With delay stage, delayed $V_{SINP,N}$ phase information is compared to $V_{OUTP,N}$ phase information in phase detector (PD), and its difference converted to v_{ctrl} through the charge pump (CP) and loop filter (LF). After phase locking, accurate in-phase (ICLK) and quadrature-phase (QCLK) are obtained. These clock signals are forwarded to the receiver demodulator. As a result, phase delay of injection current and demodulator is reduced to less than 1 degree in operating frequency range. The additional power consumption of phase compensator is only 120µW and phase delay from two limiter circuits is negligible.

B. Low-harmonic pseudo-sinewave generator

Fig. 8-(a) shows pseudo-sine generator with adaptive quantization scheme [12]. By utilizing adaptive quantization scheme, dominant harmonic components are surpressed at a high frequency, improving power-linearity efficiency. However, because the previous implementation quantizes signal level using only two way, its THD performance is limited. Furthermore, its output impedance is not enough for EIT measurement as sinusoidal current is directly injected through cascoded current driver.

As shown in Fig. 8-(b), the improved low-harmonic quantization scheme is proposed. The basic idea is that more LSBs (Least significant bits) are assigned in the peak region of sinewave. For the current amplitude higher than half of the peak current, quantization is fine such that its level is rounded off to the nearest unit current level. For the current amplitude less than half, each of quantization levels are rounded off to the nearest multiple of two or three of the unit current cell. With this configuration, more unit LSBs can be allocated to the peak region. According to simulation results, THD is reduced by 35%, from 0.736% to 0.48%.

Fig. 9 shows circuit implementation. Each unit LSB consists of a current source and NMOS switch pair. To resolve glitches, a latch is inserted before NMOS switches. According to code controller, each current source is turned on and off sequentially. The output currents are converted into voltage through 20kΩ on-chip resistors and forwarded to mixer to generate fully differential pseudo-sine wave. The layouts of resistors and current sources are carefully placed with common centroid configuration.

C. Current mismatch calibration unit

In Fig. 10, mismatch calibration scheme is shown. Since each node consists of active electrode configuration, injection current scale should be matched identically. Before the measurement begins, calibration unit sequentially generates test current and each test current is measured through reference resistor. The measured current amplitude are processed in the Hub-SoC, and calibration codes are fed back to each active electrode IC, so each of injection current scale can be equalized.

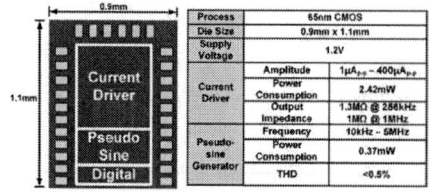

Process	65nm CMOS
Die Size	0.9mm x 1.1mm
Supply Voltage	1.2V
Current Driver	Amplitude	1µA$_{p-p}$ ~ 400µA$_{p-p}$
	Power Consumption	2.42mW
	Output Impedance	1.3MΩ @ 256kHz / 1MΩ @ 1MHz
Pseudo-sine Generator	Frequency	10kHz ~ 5MHz
	Power Consumption	0.37mW
	THD	<0.5%

Figure. 11. Chip photo and performance summary

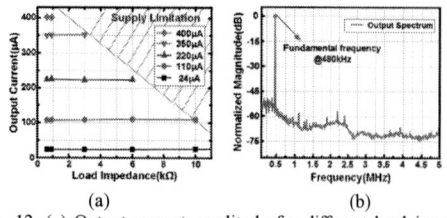

(a) (b)

Figure. 12. (a) Output current amplitude for different load impedance (b) Output spectrum of pseudo-sine wave generator

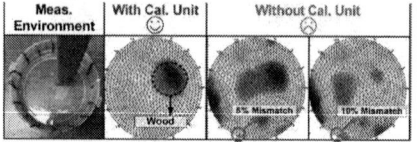

Figure. 13. EIT image acquisition results with proposed calibration unit

IV. MEASUREMENT RESULTS

Fig. 11 shows chip photograph and performance summary table. The proposed current driver IC is implemented in 65nm CMOS process and occupies 0.99mm². It uses 1.2V supply voltage. The current driver supports current up to 400µA and its power consumption is 2.42mW. The measured output impedance is beyond 1MΩ over wide frequency up to 1MHz. The current driver maintains phase error less than 1 degree in operating frequency range from 10kHz to 5MHz and pseudo-sine wave generator requires 0.37mW at maximum frequency.

Fig. 12-(a) shows measured output current peak-to-peak amplitude with different load impedance at 500kHz. The injected current amplitudes are tested from 24µA to 400µA and the measured variation of current is less than 1%. The frequency spectrum of pseudo-sine wave generator is shown in Fig. 12-(b). Thanks to adaptive quantization, low frequency harmonics are effectively reduced and measured THD distortion is less than 0.5% at maximum current injection.

Fig. 13 shows the EIT image results with calibration unit. With off-chip receiver, the proposed current driver is used for measurement. With calibration unit, clean image is obtained, but with current mismatch, distorted image is obtained.

Table. I shows the comparison table with previous current driver for EIT application. The proposed IC is the first work that fully integrates sine-wave generator in active electrode configuration.

V. CONCLUSION

In this paper, a current driver IC for portable EIT system with low-power and high accuracy is proposed. For the high output impedance current driver, DLL based phase compensator is proposed. It releases phase shift issue significantly with low-power consumption. For the low-harmonic and flexible injection frequency, pseudo-sine wave generator is adopted. Also, the calibration unit is proposed to compensate current variations among active electrodes. As a result, the proposed IC satisfying more than 1MΩ output impedance in wide frequency range of injection current while consuming 2.79mW only.

REFERENCES

[1] E. Teschner, M. Imhoff and S. leonhardt, "Electrical Impednace Tomography: The realisation of regional ventilation monitoring 2nd edition", Drager.

[2] D. Holder, Electrical Impedance Tomography: Methods, History and Applications.. Bristol, U.K.: IOP Pub., 2005.

[3] S. Hong, J. Lee, J. Bae and H. J. Yoo, "A 10.4 mW Electrical Impedance Tomography SoC for Portable Real-Time Lung Ventilation Monitoring System," in IEEE Journal of Solid-State Circuits, vol. 50, no. 11, pp. 2501-2512, Nov. 2015.

[4] H. Wi, H. Sohal, A. L. McEwan, E. J. Woo and T. I. Oh, "Multi-Frequency Electrical Impedance Tomography System With Automatic Self-Calibration for Long-Term Monitoring," in IEEE Transactions on Biomedical Circuits and Systems, vol. 8, no. 1, pp. 119-128, Feb. 2014.

[5] A. McEwan, G. Cusick and D. S. Holder, "A review of errors in multi-frequency EIT instrumentation", Physiological Measurement, vol. 28, June 2007.

[6] M. Guermandi, R. Cardu, E. Franchi Scarselli and R. Guerrieri, "Active Electrode IC for EEG and Electrical Impedance Tomography With Continuous Monitoring of Contact Impedance," in IEEE Transactions on Biomedical Circuits and Systems, vol. 9, no. 1, pp. 21-33, Feb. 2015.

[7] M. Rahal, J. M. Khor, A. Demosthenous, A. Tizzard, and R. Bayford, "A comparison study of electrodes for neonate electrical impedance tomography," Physiol. Meas., vol. 30, pp. S73–S84, 2009.

[8] R. J. Halter, A. Hartov and K. D. Paulsen, "A Broadband High-Frequency Electrical Impedance Tomography System for Breast Imaging," in IEEE Transactions on Biomedical Engineering, vol. 55, no. 2, pp. 650-659, Feb. 2008.

[9] L. Constantinou, I. F. Triantis, R. Bayford and A. Demosthenous, "High-Power CMOS Current Driver With Accurate Transconductance for Electrical Impedance Tomography," in IEEE Transactions on Biomedical Circuits and Systems, vol. 8, no. 4, pp. 575-583, Aug. 2014.

[10] A. S. Tucker, R. M. Fox, and R. J. Sadleir, "Biocompatible, high precision, wideband, improved Howland current source with lead-lag compensation," IEEE Trans. Biomed. Circuits Syst., vol. 7, no. 1, pp. 63–70, Feb. 2013.

[11] M. Kim et al., "21.2 A 1.4mΩ-sensitivity 94dB-dynamic-range electrical impedance tomography SoC and 48-channel Hub SoC for 3D lung ventilation monitoring system," 2017 IEEE International Solid-State Circuits Conference (ISSCC), San Francisco, CA, 2017, pp. 354-355.

[12] L. Yan et al., "A 13µA Analog Signal Processing IC for Accurate Recognition of Multiple Intra-Cardiac Signals," in IEEE Transactions on Biomedical Circuits and Systems, vol. 7, no. 6, pp. 785-795, Dec. 2013.

[13] L. Constantinou, R. Bayford and A. Demosthenous, "A Wideband Low-Distortion CMOS Current Driver for Tissue Impedance Analysis," in IEEE Transactions on Circuits and Systems II: Express Briefs, vol. 62, no. 2, pp. 154-158, Feb. 2015.

Table. I. Comparison table

	JSSC2015 S. Hong [3]	TBioCAS2015 M. Guermandi [6]	TBioCAS2014 L. Constantinou [9]	TCAS2015 L. Constantinou [13]	This work
Process	0.18µm CMOS	0.35µm CMOS	0.8µm CMOS	0.35µm CMOS	65nm CMOS
Architecture	Differential Current Source	Differential	Negative Feedback	Negative Feedback	Negative Feedback
Supply Voltage	1.8V	3.3V	18V	5V	1.2V
Maximum Output Current	1mA$_{p-p}$	127µA$_{p-p}$	5mA$_{p-p}$	400µA$_{p-p}$	400µA$_{p-p}$
Frequency	<200kHz	<256kHz	<1MHz	<500kHz	<5MHz
Power Consumption	-	0.844mW	-	-	2.79mW
Output Impedance	-	>665kΩ @ 100kHz	>1MΩ @ 500kHz	1MΩ @ 1MHz	1MΩ @ 1MHz
Sine-wave Generator Integration	Yes	No (Non-linear)	No	No	Yes
Configuration	-	Active Electrode	Active Electrode	Active Electrode	Active Electrode

S11-5 (2186)

0.5 and 1.5 THz Monolithic Imagers in a 65 nm CMOS Adopting a VCO-Based Signal Processing

Suna Kim, Kyoung-Yong Choi, Dae-Woong Park, Joo-Myoung Kim, Seok-Kyun Han, and Sang-Gug Lee
Department of Electrical Engineering
Korea Advanced Institute of Science and Technology (KAIST)
Daejeon, Republic of Korea
suna.kim@kaist.ac.kr

Abstract— A new architecture of monolithic THz imager adopting a VCO-based signal processing is proposed, and the enhancement of signal-to-noise ratio is analyzed. The 0.5/1.5 THz imagers, implemented in a 65 nm CMOS, show the lower noise equivalent power (NEP) than that of the detector itself, presenting the best NEP among the state-of-the-art imagers. The fully integrated THz imagers do not require external equipment, illustrating its feasibility as a portable high resolution real-time THz camera.

Keywords— *THz imager, focal-plane array, THz detector, distributed resistive self-mixing, voltage-controlled oscillator, CMOS, monolithic integration.*

I. INTRODUCTION

Interest in terahertz (THz) imagers in CMOS technologies is growing rapidly due to their advantages of room temperature operation and monolithic implementation. Many CMOS THz

imagers have been reported but most of them required lock-in amplifier (LIA) to obtain high resolution images [1]-[3], which can be an obstacle for small form-factor, a key factor in portable applications. A 1k-pixel monolithic CMOS THz camera that does not utilize the LIA showed severe performance degradation caused by the noise of readout circuits. To address these limitations, this work proposes a new architecture for a monolithic THz imager that adopts a VCO-based signal processing. While the prior readout circuits tend to increase the total system noise, the proposed THz imager achieve better noise equivalent power (NEP) than that of the detector itself. Implemented in a 65 nm CMOS, 0.5 THz 4×3 pixel and 1.5 THz 8×8 pixel imagers present 21 and 0.5 pW/√Hz NEP, respectively, which outperforms all the prior imagers.

II. THz MONOLITHIC IMAGER ADOPTING A VCO-BASED SIGNAL PROCESSING TECHNIQUE

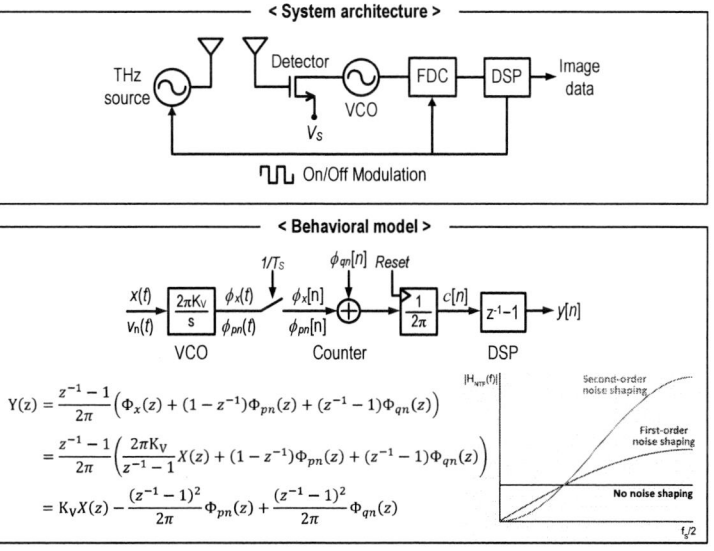

Fig. 1. System architecture of proposed single pixel THz imager and behavioral model of VCO-based signal processing.

This research was supported by the Pioneer Research Center Program through the National Research Foundation of Korea funded by the Ministry of Science, ICT & Future Planning (No. 2012-0009594).

978-1-5386-3179-9/17 $31.00 © 2017 IEEE

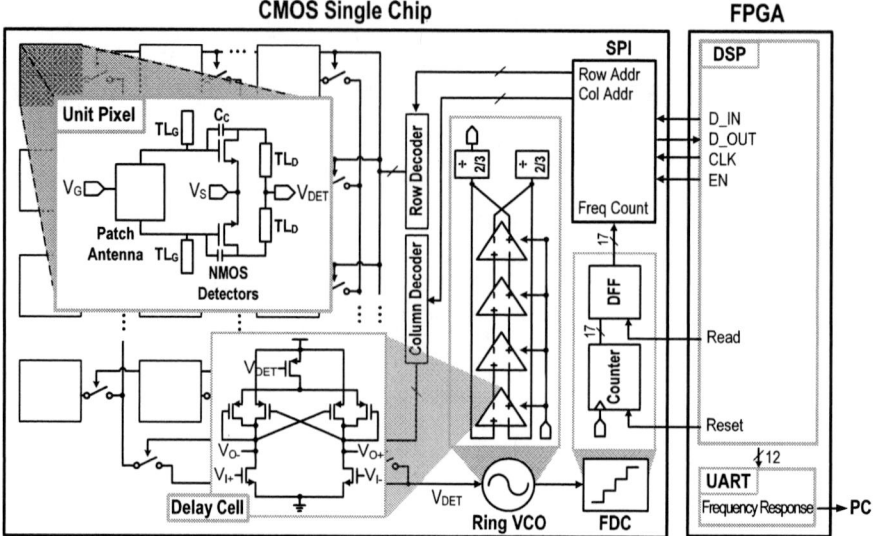

Fig. 2. Implementation details of the proposed array imagers.

This section describes the proposed imager which adopts a VCO-based signal processing technique, and analyzes its enhanced SNR characteristics. Also, the integrated structure, which reduces pixel size and enhances the radiation efficiency of the antenna, is described.

A. Proposed architecture

The architecture of the proposed single-pixel THz imager is presented in Fig. 1. The THz signal applied at the gate-source terminal generates a DC voltage drop across the drain-source terminal of the cold-FET detector. A VCO is adopted as a readout circuit, and by applying the detector output as the control voltage of the VCO, the variation in the drain-source voltage of the detector is translated into the variations in oscillation frequency. Then, the VCO output is converted into a digital number by the frequency-to-digital converter (FDC). The DSP block generates a clock to on/off modulate the THz source, and the frequency difference of the on/off states is detected as a final image data.

The behavioral model of the proposed imager is depicted in Fig. 1, where $x(t)$ represents the output voltage of detector, $v_n(t)$ the input referred noise of VCO including the output noise of detector, $\phi_x(t)/\phi_{pn}(t)$ the phase signal/noise at the output of the VCO, $\phi_{qn}[n]$ the quantization noise, $c[n]$ the output of the counter (FDC), and $y[n]$ the image data. The frequency domain model of VCO can be expressed as $2\pi K_V/s$ (K_V: KVCO of the VCO), and the counter can be modeled as a switch with the sampling period of T_S and the gain of $1/2\pi$. As the quantization noise of the previous sampling period (residual phase) inherently becomes the initial phase of the next period [5], the output of the counter can be expressed as

$$c[n]=(1/2\pi)(\phi_x[n]+\phi_{pn}(nT_S)-\phi_{pn}((n-1)T_S)+\phi_{qn}[n-1]-\phi_{qn}[n]). \quad (1)$$

Lastly, because the DSP yields $y[n]=c[n]-c[n-1]$, the frequency domain expression of image data $Y(z)=(z^{-1}-1)C(z)$ can be finally represented as shown in Fig. 1. As shown by the equation in Fig. 1, the phase and quantization noises of VCO and FDC, respectively, are second-order high-pass shaped while the signal is transferred with the gain of K_V. As the phase noise is second-order high-pass shaped, the noise of detector itself is also first-order high-pass shaped, because of the first-order low-pass shaping characteristic of VCO. This means that the NEP of the whole system can be lower than that of detector itself.

B. 0.5/1.5 THz 4×3/8×8 pixel monolithic imager designs

Implementation details of the proposed imager is presented in Fig. 2. The detector pixels, VCO, FDC, row/column decoders, and SPI are all integrated into a single chip. The cross-sectional view of the unit pixel is shown in Fig. 3, where

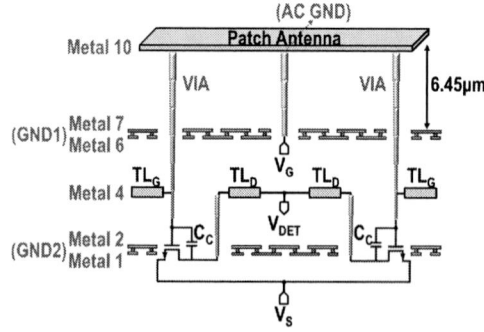

Fig. 3. Cross-sectional view of the unit pixel.

the open stubs (TL$_G$), $\lambda/4$ transmission lines (TL$_D$), and the detectors including the coupling capacitor (C$_C$) are vertically integrated under the patch antenna, minimizing the size of pixel. The patch type antenna is chosen to prevent signal loss into the Si substrate with the size of 137×184 μm^2 and 36×60 μm^2 for 0.5 and 1.5 THz designs, respectively. Metal 10 is used for the patch, and metal 6 and 7 for the ground of the patch antenna. Another ground is implemented using metal 1 and 2, and TL$_G$ and TL$_D$ are implemented as a strip line using the two ground layers and metal 4. The impedance between the antenna and detectors is matched by adjusting the feeding position of antenna and TL$_G$. TL$_D$ presents high impedance at the drain terminal of the MOSFET detector to the signal coming through C$_C$ [1]. The ground wall between the pixels is adopted only for the 1.5 THz antenna, and finally the efficiency of 0.5/1.5 THz antenna is designed to be 58%/85%. The source terminals of MOSFET detectors are biased (V_S) to set the initial control voltage of the VCO while V_G set the gate voltage to turn on the MOSFET detectors. The ring type VCO consists of four delay cells [6] and 2/3 dividers. A single VCO is used for the whole imager array, and is sequentially connected to the detector output of each pixel by the row/column decoders. The FDC counts the oscillation frequency of the VCO and transfers the digital data to SPI by the *Read* signal, and then resets the count for the next pixel. The SPI communicates with the FPGA for the frequency count and the row/column addresses. The final output data of the DSP is delivered to a PC via UART communication. Note that the VCO, FDC, row/column decoders, and SPI are all integrated below the M$_{6-7}$ ground plane. Therefore, the radiation efficiency of the antenna is

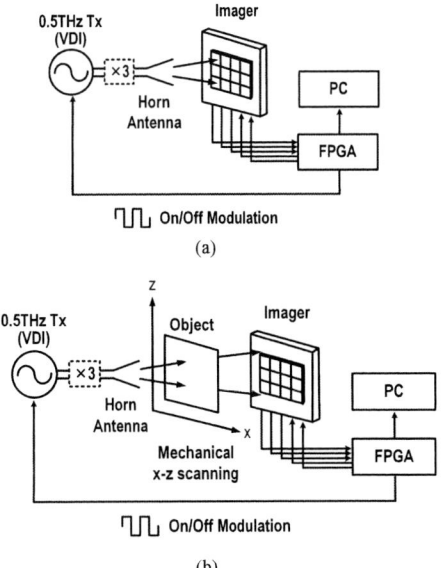

Fig. 4. Measurement setup for (a) the responsivity and (b) the imaging of the object (bottom).

enhanced as the M$_{6-7}$ presents a large ground plane.

III. MEASUREMENT RESULTS

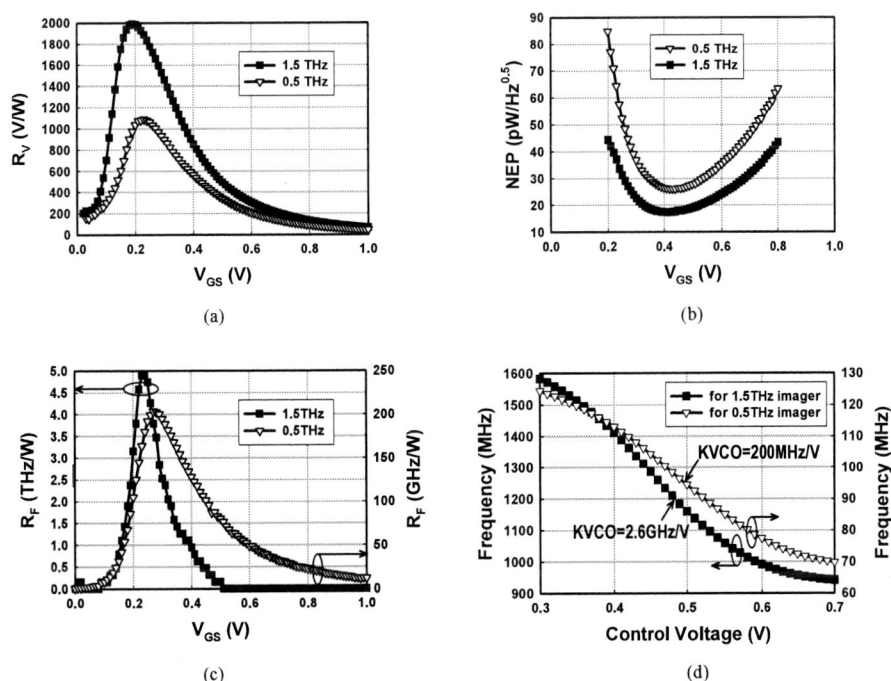

Fig. 5. Measured (a) R$_V$ and (b) NEP versus V$_{GS}$ of the FET detector, (c) RF versus V$_{GS}$ of the imager, and (d) oscillation frequency versus control voltage of the VCO.

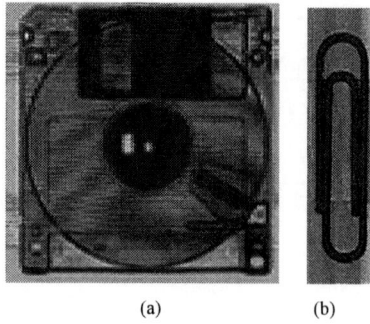

(a)   (b)

Fig. 6.   (a) 0.5 THz and (b) 1.5 THz images

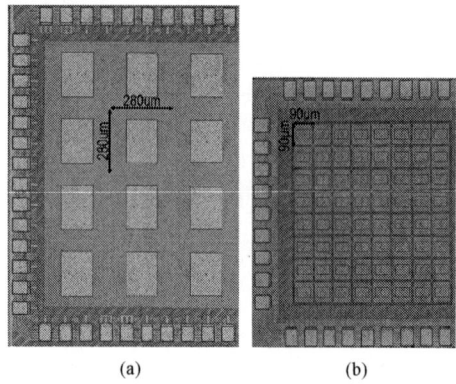

(a)   (b)

Fig. 7.   Chip micrograph of (a) 0.5 THz (980×1400 μm²) and (b) 1.5 THz (630×460 μm²) imagers.

Fig. 4 shows the measurement setup for the responsivity and the imaging. Fig. 5 shows the measured voltage responsivity (R_V) and NEP vs. V_{GS} of the unit pixel detector, and the frequency responsivity R_F(Hz/W) vs. V_{GS} of the imager and the oscillation frequency vs. control voltage of the VCO. To obtain the voltage responsivity (R_V=Output voltage/Input power), the input power into the detector is estimated using the Friis transmission equation where the transmitted power is 5.83m/40μW, transmitting antenna gain 25/25 dBi, receiving antenna gain 2.63/5.9 dBi, effective wavelength of receiving antenna 274/72 μm, and distance from transmitting to receiving antenna R 15/5 mm for 0.5/1.5 THz cases, respectively. To obtain the NEP (NEP=Noise spectral density/Responsivity), the noise spectral density from the channel thermal noise is estimated by the measured channel resistance. In Fig. 5, the maximum R_V and minimum NEP of the unit pixel detector are 1.09/2 kV/W and 26/17 pW/√Hz at 0.5/1.5 THz cases, respectively.

The transmission mode imaging is performed with mechanical x-z scanning (Fig. 4). With the 0.5 THz 4×3 pixel imager, an 8-bit resolution image of a floppy disk is obtained without the aid of an LIA (Fig. 6). With the 1.5 THz imager, single-pixel imaging is performed at the moment, because the

power of the 1.5 THz Tx is too weak to obtain an image without fixed pattern noise correction. An image of a clip is presented with a 3-bit resolution (due to the weak Tx power). A performance summary of the proposed imager in comparison with previous array imagers is presented in Table I. The system noise is measured by observing the standard deviation of output data. As expected from the analysis, in Table I, the system NEPs of the proposed imagers are 21/0.5 pW/√Hz for 0.5/1.5 THz imagers, respectively, which are lower than the NEPs of unit pixel detector alone. Fig. 7 shows the chip micrographs.

IV. CONCLUSION

A new architecture of a THz imager adopting a VCO-based signal processing technique is presented, and its enhanced SNR characteristics is analyzed. Structurally, all the circuits are located beneath the patch antenna, reducing the pixel size and enhancing the radiation efficiency of the antenna. Implemented in a 65 nm CMOS process, the NEP performance of the proposed imagers is best compare to the prior works despite the highest level of integration and the frequency of operation.

TABLE I
PERFORMANCE COMPARISON WITH PREVIOUS IMAGERS.

	This work		[3]	[4]	[7]
Technology	65 nm CMOS		130 nm CMOS	65 nm CMOS	SBD in 130 nm CMOS
$f_{operating}$ (THz)	1.5	0.5	0.82	0.86	0.28
Array size	8×8	4×3	8×8	32×32	4×4
DC Power (mW)	6.4	13	9.6	2.56	6
Responsivity	4.9 THz/W	203 GHz/W	3.4 kV/W	140 kV/W	336 V/W
System NEP (W/√Hz)	0.5 p	21 p	28 p	12 n	29 p

REFERENCES

[1] Erik Öjefors Ullrich R. Pfeiffer, Alvydas Lisauskas, and Hartmut G. Roskos, "A 0.65 THz Focal-Plane Array in a Quarter-Micron CMOS Process Technology", *IEEE J. Solid-State Circuits*, vol. 44, no. 7, July 2009

[2] Franz. Schuster et al., "A Broadband THz Imager in a Low-Cost CMOS Technology", *ISSCC Dig. Tech. Papers*, pp. 42-43, Feb. 2011.

[3] Dae Yeon Kim, Shinwoong Park, Ruonan Han, and Kenneth K. O., "820-GHz Imaging Array Using Diode-Connected NMOS Transistors in 130-nm CMOS", *Symposium on VLSI Circuits*, pp. C12-C13, June 2013.

[4] Richard Al Hadi et al., "A 1 k-Pixel Video Camera for 0.7-1.1 Terahertz Imaging Applications in 65-nm CMOS", *IEEE J. Solid-State Circuits*, vol. 47, no. 12, Dec. 2012.

[5] Jaewook Kim, Tae-Kwang Jang, Young-Gyu Yoon, and SeongHwan Cho, "Analysis and design of voltage-controlled oscillator based analog-to-digital converter," *IEEE Trans. Circuits Syst. I*, VOL. 57, NO. 1, pp. 18-30, Jan. 2010.

[6] William Shing Tak Yan and Howard Cam Luong, "A 900-MHz CMOS Low-Phase-Noise Voltage-Controlled Ring Oscillator", *IEEE Trans. Circuits Syst. II, Analog Digit. Signal Process*, vol. 48, no. 2, pp. 216–221, Feb. 2001.

[7] Ruonan Han et al., "Active Terahertz Imaging Using Shottky Diodes in CMOS: Array and 860-GHz Pixel", *IEEE J. Solid-State Circuits*, vol. 48, no. 10, Oct. 2013.

Dual-Loop 2-step ZQ Calibration for Dedicated Power Supply Voltage in LPDDR4 SDRAM

Chang-Kyo Lee, Junha Lee, Ki-Ho Kim, Jin-Seok Heo, Gil-Hoon Cha, Jin-Hyeok Baek, Dae-Sik Moon, Yoon-Joo Eom, Tae-Sung Kim, Hyunyoon Cho, Younghoon Son, Seonghwan Kim, Jong-Wook Park, Sewon Eom, Si-Hyeong Cho, Young-Ryeol Choi, Seungseob Lee, Kyoung-Soo Ha, Youngseok Kim, Bo-Tak Lim, Dae-Hee Jung, Eungsung Seo, Kyoung-Ho Kim, Yoon-Gyu Song, Youn-Sik Park, Tae-Young Oh, Seung-Jun Bae, In-Dal Song, Seok-Hun Hyun, Joon-Young Park, Hyuck-Joon Kwon, Young-Soo Sohn, Jung-Hwan Choi, Kwang-Il Park, Seong-Jin Jang

Samsung Electronics, DRAM Design Team
Hwaseong-si, Gyeonggi-do, Korea 18448
ck1980.lee@samsung.com

Abstract— **This paper presents a dual-loop 2-step ZQ calibration scheme with 20nm DRAM process to support dedicated supply voltage (VDD, VDDQ). The proposed calibration scheme maintains a target value of on-die termination (ODT) in DQ/CA regardless of the supply-voltage variations which are caused by dynamic voltage frequency switching (DVFS) and alleviates the calibration time which is increased by insertion of additional CA calibration. The offset of a comparator is averaged out by fraction-referred input switching-then-averaging scheme (FISA). And code-referred periodic ZQ update (CPZU) scheme can track the VT variation while minimizing the interference.**

Keywords—DRAM, ZQ calibration, DVFS, Offset calibration

I. INTRODUCTION

Recently the mobile era requires more high-speed operation for the data processing. Thus, many trails are continuously performed for implementation of high speed IO interfaces without the degradation of signal integrity (S/I). Generally in previous mobile applications, because of the relative short channel configurations between DRAM and SoC, the signal integrity (S/I) is not degraded severely. However, due to the increases of operating speed beyond 3.2Gbps and PKG placement depending on board level applications, the S/I degradation causes the serious limitation on system operations and give an impact on performance. Thus, a careful consideration on the impedance match between source to termination become very important factors on system optimization. In LPDDR4 SDRAM [1], the impedance of a DQ pull-down driver is calibrated to a precise external resistor while that of DQ pull-up driver is calibrated based on a replicated DQ pull-down driver through the ZQ calibration during the training sequence. In case of applying DVFS where voltage differences exist between VDD and VDDQ for low power consumption in LPDDR4 SDRAM, then the value of ODT can be shift between CA and DQ even though the ZQ calibration is executed. Because the power rail of a CA driver is VDD, while that of a DQ driver is VDDQ respectively. This ODT value shift causes the difference in ODT and driver strength.

We proposed dual-loop 2-step ZQ calibration scheme to overcome the system performance degradation by DVFS operation where a VDD-VDDQ gap exits.

 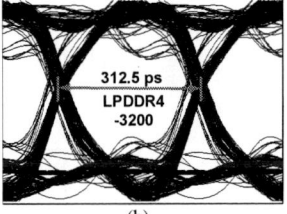

(a)   (b)

Fig. 1 Eye diagrams of (a) un-calibrated output driver and (b) impedance-calibrated output driver

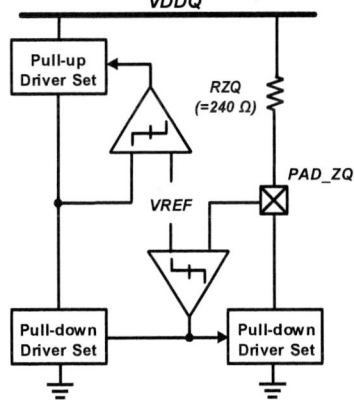

Fig. 2 Conventional LPDDR4 ZQ calibration

II. ZQ CALIBRATION

In high speed interface, output driver and on-die termination impedances should have small offset value against process, temperature, and voltage (PVT) for improving signal integrity. If there is impedance mismatch due to PVT variations, eye windows of transmitted and received data becomes smaller, and bit error rate (BER) is increased due to reflection and degradation of drivability as shown in Fig. 1. If the un-calibrated output driver is used for wire-line communication, the rising and falling slew rates are lower than the expected values and signal distortion by reflections is more serious. To avoid such S/I degradation, DRAM has a function for impedance calibration which is called as ZQ calibration. Fig. 2

shows a simplified block diagram of ZQ calibration circuit. When ZQ calibration command is issued, the pull-down driver set that is connected to an external resistor is turned on. The feedback loop controls impedance of the pull-down driver to be equal to resistance of the external resistor. Simultaneously pull-up driver calibration is also performed using a replica circuit of the pull-down driver set as a reference. Throughout the ZQ calibration, pull-down impedance is fitted to the external resistor and pull-up impedance is fitted to the calibrated pull-down impedance across process, temperature, and voltage. In case of applying dedicated supply voltage (VDD, VDDQ) to reduce power consumption, the ODT variation between DQ

and CA which driven by VDDQ, VDD respectively can be worse. Thus, the dual ZQ calibration to support the DVFS is required to enhance the energy efficiency for mobile applications.

III. PROPOSED DUAL-LOOP 2-STEP ZQ CALIBRATION

A. Dual-loop 2-step search algorithm

Fig. 3 shows block diagram of proposed dual-loop 2-step ZQ calibration architecture to support dedicated VDD, VDDQ applications. The dual-loop 2-step ZQ calibration circuit includes a pull-down driver, a replica pull-down driver, a pull-up driver for DQ and an additional pull-down driver for CA including two dynamic comparators. As the VDD-based pull-down driver is added to calibrate the ODT of CA which adopts VDD as supply voltage, the limitation of timing margin is alleviated by 2-step ZQ calibration [2], where 2-step ZQ calibration is composed of a 3-bit coarse search algorithm and a 3-bit fine search algorithm and both algorithms are based on thermometer codes to suppress the monotonicity and decision error. And the acceptable error range in this design is determined as 16 bits which is equal to 2 LSBs of a coarse step.

Fig. 4 shows the fundamental operation of the proposed dual-loop 2-step ZQ calibration. In first phase, the DQ pull-down driver is calibrated reference to a precise external resistor (RZQ) as shown in Fig. 4(a). The PAD_ZQ node is consecutively controlled by ZQ code [5:0] with discretely

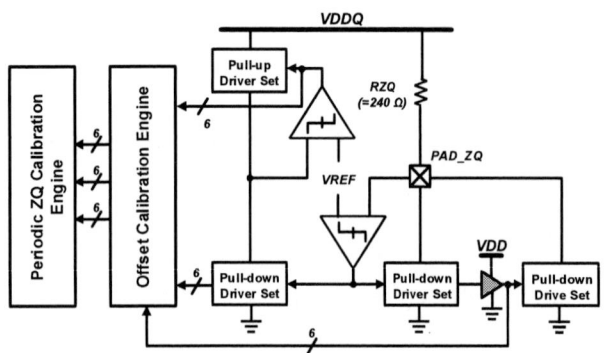

Fig. 3 Proposed dual-loop 2-step ZQ calibration

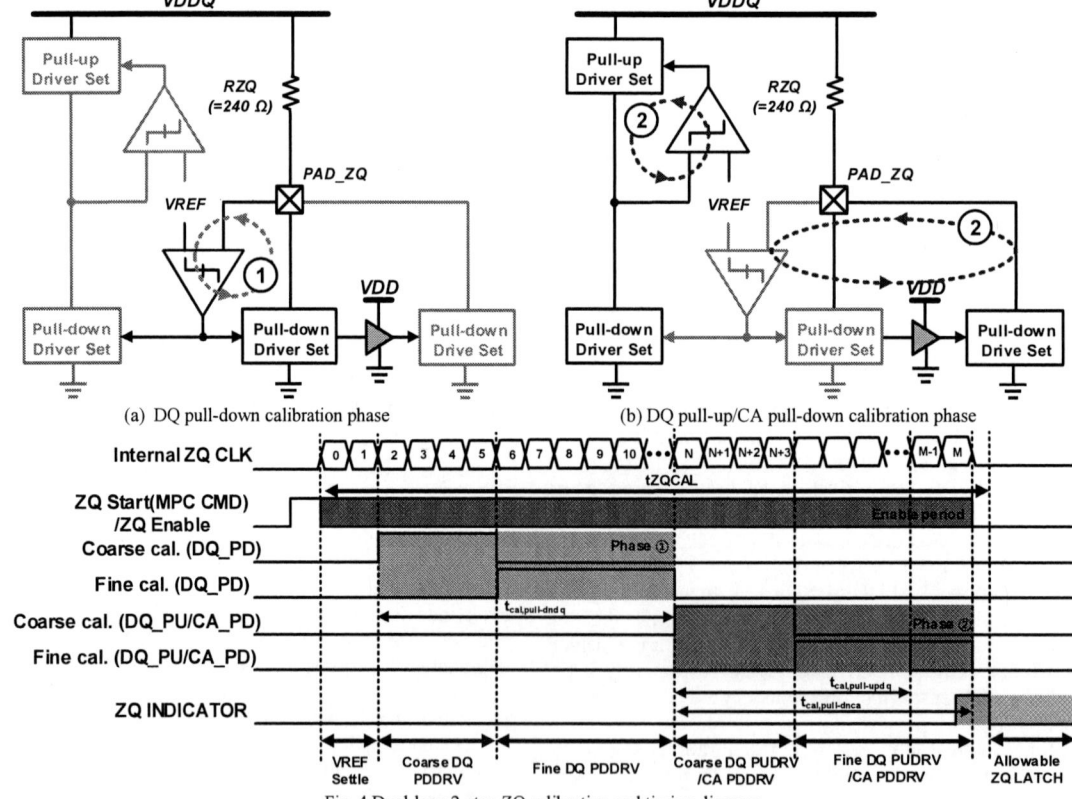

(a) DQ pull-down calibration phase   (b) DQ pull-up/CA pull-down calibration phase

Fig. 4 Dual-loop 2-step ZQ calibration and timing diagram

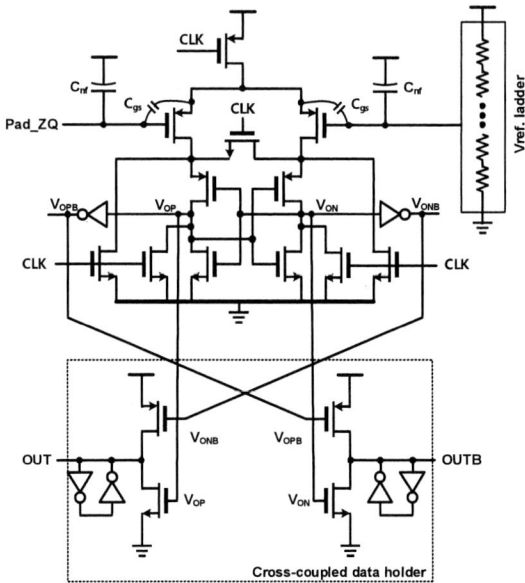

Fig. 5 Dynamic latch with cross-coupled data holder

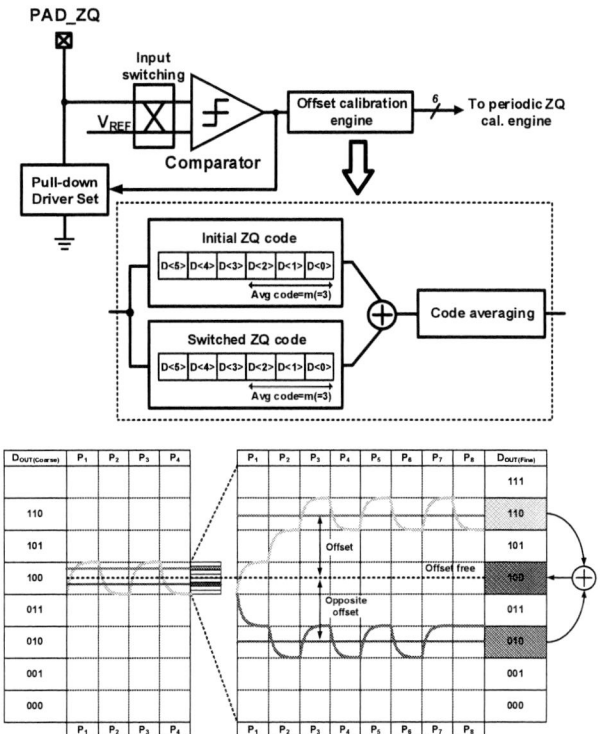

Fig. 6 Fraction-referred input switching-then-average scheme

negative feedback manner to reach the target V_{OH} (=1/3VDDQ) level and those codes are transferred to its replicated pull-down driver and ZQ calibration is done with consecutive 3-bit data toggling to minimized the error due to the noise. In second phase, both a DQ pull-up driver and a CA pull-down driver are calibrated simultaneously as shown Fig 4(b). The calibration reference of a DQ pull-up driver is a replicated DQ pull-down driver, while that of a CA pull-down driver is precise external RZQ, respectively. And the calibration sequences of both a DQ pull-up driver and a CA pull-down driver follow a DQ pull-down driver in the first phase. Using proposed dual-loop 2-step calibration, both DQ and CA pull-down drivers are commonly calibrated reference to RZQ without calibration time increase by comparator reuse and mismatch due to the calibration code transfer. The calibration is controlled by synchronous clock based on internal oscillator, while timing of calibration done can be varied depending on the driver conditions such as sensitivity to PVT and intrinsic offset. Thereby the calibration time can be written as $t_{cal,total}=t_{cal,pull-dndq}+ \max[t_{cal,pull-updq}, t_{cal,pull-dnca}]$ and that time should be within tZQCAL (<1us). Each phase also includes additional calibration period for offset cancellation. And the indicator signal which blocks the ZQ latch signal to ensure the update timing of ZQ code is generated before the all ZQ calibration is finished.

B. Dynamic latch with cross-coupled data hoder

Fig. 5 shows a dynamic latch with cross-coupled data holder (CCDH) based on a strong-arm typed dynamic latch [3]. The output data (OUT/OUTB) of CCDH isolates from the dynamic latch during reset phase. Thus, the output of dynamic latch hold its value during the reset phase and additional DFFs which store the sampled data before reset phase are not required. Thus, data processing delay to control the DAC operation can be relived. And the intrinsic kick-back effect caused by the un-balanced input network is minimized using

kick-back noise filtering capacitor (C_{nf}), where value of C_{nf} is determined considering the offset of the comparator and capacitor dividing ratio between C_{gs} and C_{nf}

C. Fraction-referred input switching-then-averaging scheme

Major error source of ZQ calibration causes from a ZQ comparator, thus the offset of a comparator should be minimized as possible. To mitigate the offset of a comparator, the proposed fraction-referred input switching-then-averaging scheme (FISA) is proposed as shown in Fig. 6. The calibration sequence of FISA is described as follow. In first phase, normal 2-step search algorithm is executed and the ZQ data codes are stored in initial ZQ code registers. In second phase, each the ZQ pad and Vref are connected with opposite direction by an input switching network and only fine code searching is executed with fixed upper 3-bits. After both 2-phase ZQ calibration, both ZQ data codes are added and then shifted by 1-bit in a code averaging circuit. By doing this, the offset of dynamic latch can be averaged out and those ZQ codes are transferred to DQ and CA, respectively. The offset of the dynamic latch is determined within the lower 3 bits range of a fine step to minimize the calibration time.

D. Code-referred periodic ZQ calibration

After initial ZQ calibration during training sequence, the ZQ calibration is executed periodically to track the VT variations which change the ODT value and driver strength depending on MPC command from SoC. However, as periodic

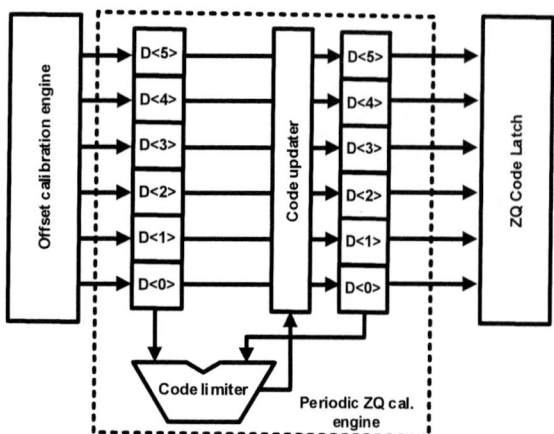

Fig. 7 Periodic ZQ calibration engine with code limiter

ZQ calibration operates in background mode irrespective of read/write operation, there can be a chance of abrupt VOH level variation in terms of ODT and driver strength by ZQ codes shift due to the internal noise and supply voltage fluctuation. Thus, in this design, the code variation is limited and a CPZU scheme based on a linear search scheme is applied as shown in Fig. 7. Comparing the code distance between the previous ZQ code and newly updated ZQ code, that code distance is beyond predefined range (3 bits in this design), the code update is not allowed to minimize the external disturbance while tracking the VT variation.

IV. MEASUREMENT RESULTS

Fig. 8 shows the chip photo of proposed dual-loop 2-step ZQ calibration circuit which is implemented with 20nm DRAM process. The ODT of DQ is calibrated in reference to VDDQ and that of CA is calibrated in reference to VDD, where the VDDQ is 0.95V and VDD is 1.1V. In case of dedicate ZQ calibration is not applied, the value of CA ODT is approximately 20% lower compare with that of DQ ODT because the ZQ code are shared by both DQ and CA. After dual ZQ calibration, the value of ODT converges on target range as shown in Fig. 9. Fig. 10 shows the measured ODT value of DQs and CA depending on temperature (-25°C~85°C). Although temperature can change the ODT value, ZQ calibration can track that variation and updates ZQ codes to recent value. According to the measured results based on RZQ/4, the min/max of DQ ODT value is 59.5 ohm and 61.6 ohm respectively and that of CA is 60.8 ohm and 62.1 ohm respectively. Those values can satisfy the LPDDR4 spec. irrespective of voltage gap between the VDDQ and VDD. Considering the feasibility of the proposed design, DVFS can be applied to enhance the power efficiency of current LPDDR4 without dynamic performance drop.

V. CONCLUSION

In this paper, the dual-loop 2-step ZQ calibration is introduced. As the additional CA pull-down calibration is

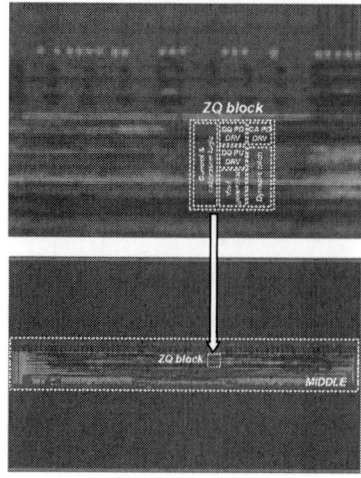

Fig. 8 Chip photo

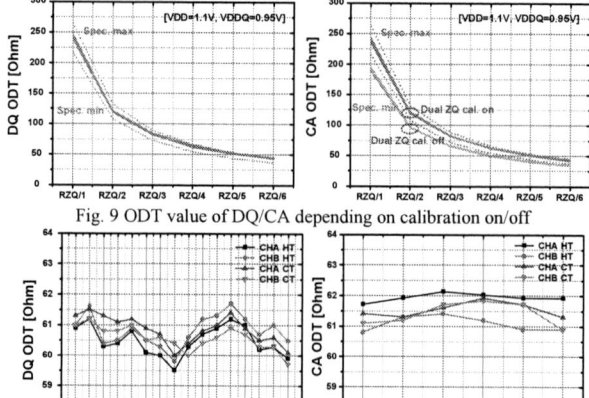

Fig. 9 ODT value of DQ/CA depending on calibration on/off

Fig. 10 ODT value depending on temp. variation

applied to normal ZQ calibration, the DVFS scheme for low power consumption can be applied without dynamic performance drop. Moreover, driver strength and ODT can be maintained without mismatch to the newest value by the FISA scheme and the CPZU scheme.

REFERENCES

[1] Tae-Young Oh, et al., "A 3.2Gb/s/pin 8Gb 1.0V LPDDR4 SDRAM with integrated ECC engine for sub-1V DRAM core operation," ISSCC. Dig. Tech. Papers, pp. 430-431, 2014

[2] Chang-Kyo Lee, et al., "A 5.0Gb/s/pin 8Gb LPDDR4X SDRAM with with Power-Isolated LVSTL and split-die architecture with 2-die ZQ calibration scheme," ISSCC. Dig. Tech. Papers, pp. 390-391, 2017

[3] J. H. Kim, et al., "Simulation and Analysis of Random Decision Errors in Clocked Comparators", IEEE TRANSACTIONS ON CIRCUITS AND SYSTEMS-I, Vol. 56, pp. 1844-1857, Aug. 2009

MLC/3LC NAND Flash SSD Cache with Asymmetric Error Reduction Huffman Coding for Tiered Hierarchical Storage

Hikaru Watanabe, Yoshiaki Deguchi and Ken Takeuchi

Department of Electrical, Electronic, and Communication Engineering, Chuo University, Tokyo, Japan

E-mail : watanabe@takeuchi-lab.org

Abstract—**Asymmetric Error Reduction Huffman Coding (AERH) is proposed for MLC/3LC NAND flash memory used as SSD cache in the tiered hierarchical storage. AERH compresses data, reduces asymmetric memory cell errors by modulating V_{TH} distribution and thus enhances reliability and endurance of memory cells. AERH decreases data-retention errors by 72.6% and 82.6% of MLC and 3LC NAND flash, respectively. In addition, this paper analyzes the compression ratio and the cost (effective bits/cell) of proposed AERH. Proposed MLC/3LC NAND flash with AERH realizes the utmost reliability of enterprise SSD cache where SLC NAND flash is commonly used.**

Keywords— NAND flash memory; compression; reliability

I. INTRODUCTION

Generally, multi-level cell (MLC: 2bits/cell) NAND flash has been used at the primary storage at data centers [1]. MLC NAND flash has high reliability and high endurance compared with triple-level cell (TLC: 3bits/cell) NAND flash, which is used in lower reliability consumer applications such as smartphones and USB memories. On the other hand, in the tiered hierarchical storage system at data centers, single-level cell (SLC: 1bit/cell) NAND flash-based SSD is used as the storage cache or Tier 0 because frequently accessed cache needs the high reliability.

This paper proposes Asymmetric Error Reduction Huffman coding (AERH), which compresses data and optimizes V_{TH} distribution of MLC/3LC NAND flash at the same time without cell area overhead. 3LC NAND flash is the three level cell that uses three states out of four states of MLC NAND flash. By applying AERH to MLC or 3LC NAND flash, the proposed SSD can be used as SSD cache or Tier 0 and thus achieves both high reliability and high capacity because of the multi-level storage, compared with conventional SLC NAND flash SSD cache. As a result, the proposed MLC/3LC NAND flash can be used as high capacity pseudo-SLC NAND flash as shown in Fig. 1.

Fig. 2 shows the measured data-retention BER of each V_{TH}-state in 1Ynm TLC NAND flash [2]. In TLC NAND flash, BER increases as the V_{TH} increases because the stronger electric field across the tunnel oxide accelerates the electron de-trap or electron leakage from the floating gate [3]. To improve the reliability of TLC NAND flash, FRBH-TLC, which monotonically reduces the population of the higher V_{TH}-state cells, has been proposed [2]. On the other hand, in MLC NAND flash, measured data-retention errors occur asymmetrically. The measured data-retention BER of "B"-state is much higher than the other V_{TH}-states as shown in Fig.3. To improve reliability of MLC NAND flash, proposed AERH-MLC reduces the population of "B"-state cells. Moreover, to further improve the reliability, AERH-3LC eliminates "B"-state cells and modulates the V_{TH} distribution. Fig. 4 shows the data-retention errors in 1Ynm MLC NAND flash. The narrow read margin between "A" and "B"-states may probably cause the high data-retention BER of "B"-state. In this paper, AERH, which compresses data and simultaneously modulates the V_{TH} distribution, is proposed to reduce this asymmetric memory cell errors of MLC NAND flash.

II. ASYMMETRIC ERROR REDUCTION HUFFMAN CODING (AERH)-MLC/3LC

AERH-MLC/3LC modifies the conventional Static Huffman coding [4]. Figs. 5-6 show the algorithm of proposed AERH-MLC/3LC. First, these compression methods analyze the information of frequencies of each data. The algorithm of counting frequencies is shown in Fig. 5. Raw data is divided into 8bit units and the frequencies of each 8bit data are counted. In Fig. 6, the frequencies of data are counted per 4bit data such as "0000" to explain the proposed algorithm clearly whereas the actual proposed SSD counts the frequencies per 8bit to simplify the computation. Second, each data is described as a leaf node and a Huffman tree is generated step-by-step from the bottom

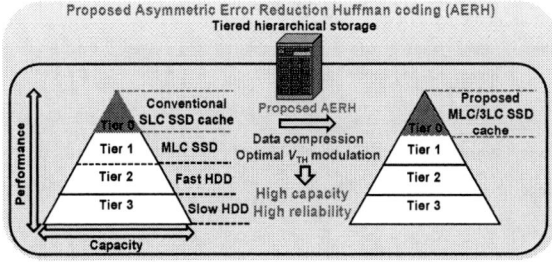

Fig. 1. Concept of this work.

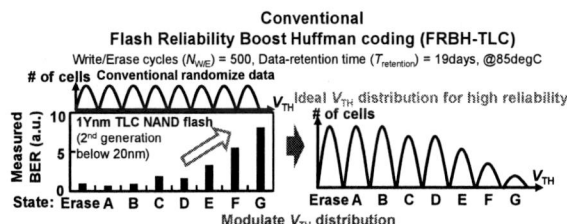

Fig. 2. Measured data-retention bit error rates (BERs) of 1Ynm TLC NAND flash [2].

most infrequent node to the top most frequent node as shown in Fig. 6. In AERH, an internal node is created with 3-4 sub-nodes, whereas in the conventional Huffman tree 2 sub-nodes are used. Increasing the sub-nodes in the proposed scheme enhances the memory cell reliability but slightly degrades the compression ratio, which will be discussed in Section IV. Then,

each V_{TH} state is assigned to each branch. To achieve both efficient data compression and high reliability, the most unreliable "B"-state is assigned to the most infrequent branches and the most reliable "Erase"-state is assigned to the most frequent branches, which is different from [2] where the more frequent branches are monotonically assigned to the lower V_{TH} state.

In case of the 3LC NAND flash operation where only 3 states are used as the memory storage to achieve the better reliability than MLC NAND flash, AERH-3LC does not assign "B"-state to any branches and thus eliminates the "B"-state. When all nodes are connected, the Huffman tree is completed. After that, AERH encodes data directly as V_{TH}-states to the assigned branches. As a result, by allocating more frequent data to more reliable V_{TH}-state, the ideal V_{TH} distribution that realizes the efficient data compression and high reliability is realized as shown in Fig. 7. Because the more frequent data are directly encoded to fewer cells, data compression is realized at the same time, which further enhances the endurance.

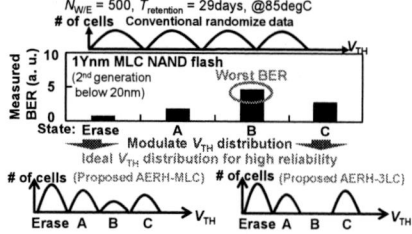

Fig. 3. Measured data-retention BER of 1Ynm MLC NAND flash. Asymmetric errors are observed, where BER of "B"-state is higher than the other V_{TH}-states. Proposed AERH modulates V_{TH} distribution to reduce the "B"-state errors.

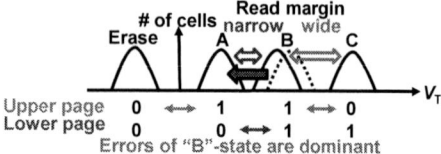

Fig. 4. V_{TH} distribution and page allocation of MLC NAND flash. Errors of "B"-state are dominant probably because read margin between "A" and "B"-state is set small to make the BER of the lower page similar to the BER of the upper page.

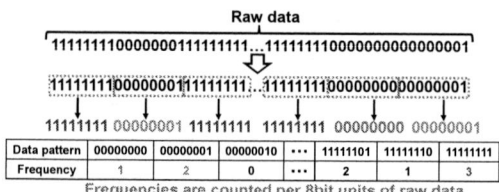

Fig. 5. Algorithm of counting frequencies. Raw data are divided into 8bit units and frequencies are counted.

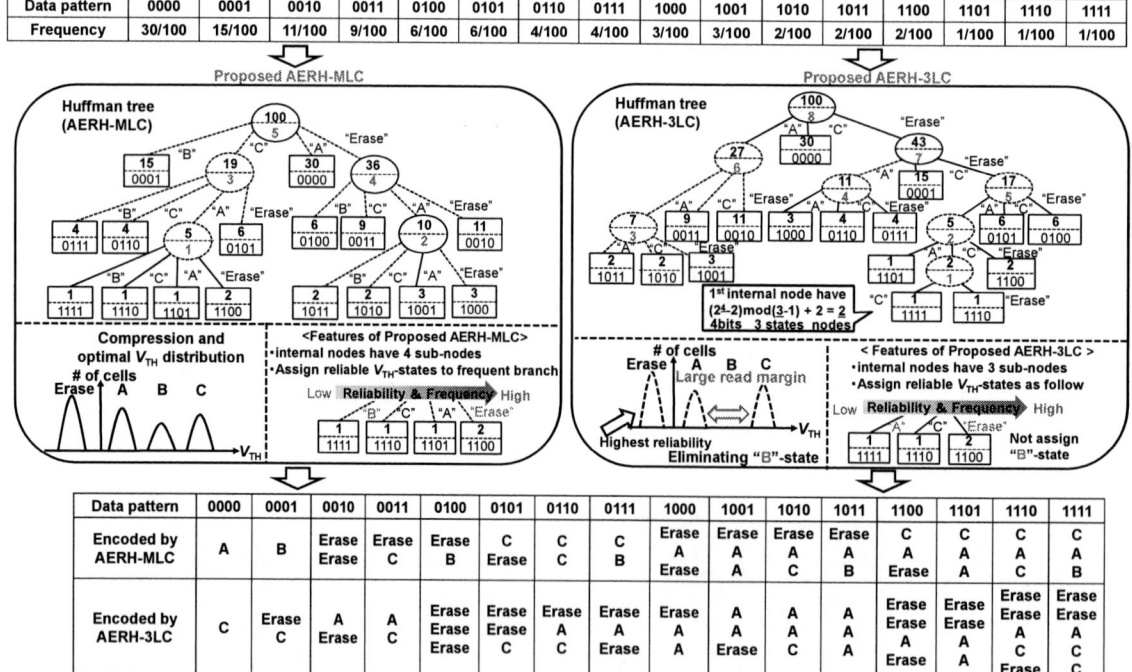

Fig. 6. Algorithm of proposed AERH for MLC (left figure) and 3LC (right figure) NAND flash. First, the frequencies of each data are counted. Second, the huffman tree is generated. Third, data are encoded direcrtly as V_{TH}-states of memory cells.

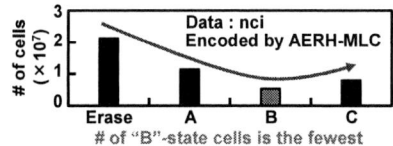

Fig. 7. V_{TH} distribution with AERH-MLC. The population of "B"-state cells is fewest, which is an ideal V_{TH} distribution for MLC NAND flash to reduce errors of "B"-state.

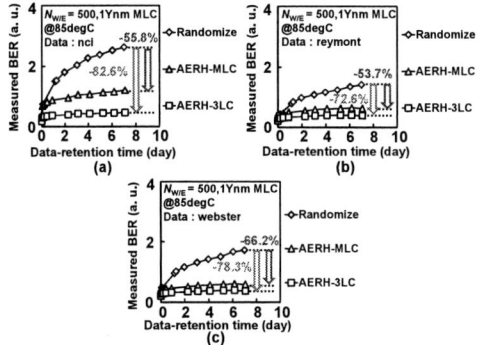

Fig. 8. Measured BER of data-retention errors for 3 pattern files [6]. AERH-MLC and 3LC decrease the BER by over 50% and 70%, respectively.

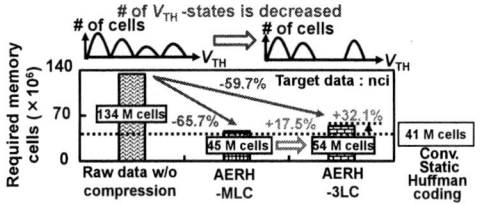

Fig. 9. Measured required memory cells to store encoded data with AERH-MLC/3LC. AERH-3LC requires 17.5% more cells than AERH-MLC because one V_{TH}-state is not used as the memory.

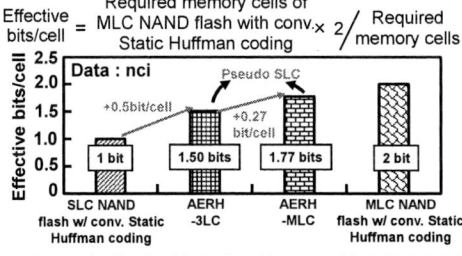

Fig. 10. Measured effective bits/cell or bit cost, with AERH-MLC/3LC. AERH-MLC/3LC achieves better bit costs than SLC NAND flash.

III. EXPERIMENTAL RESULTS OF PROPOSED AERH-MLC/3LC

Fig. 8 shows the measured BER of data-retention errors for 3 pattern files [6]. AERH-MLC/3LC decrease the BER by over 50% for all of these files. Especially, AERH-MLC decreases the data-retention errors by 66.2% for "webster" file compared with conventional randomize. Similarly, AERH-3LC decreases the data-retention errors by 82.6% for "nci" file. Moreover, AERH-MLC and 3LC decrease the memory cells required to store encoded data by 65.7% and 59.7%, respectively, as shown in Fig. 9. Meanwhile, proposed AERH-3LC increases the

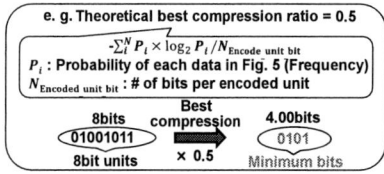

Fig. 11. Definition of theoretical best compression ratio. Theoretical best compression ratio can be calculated from frequencies of each data.

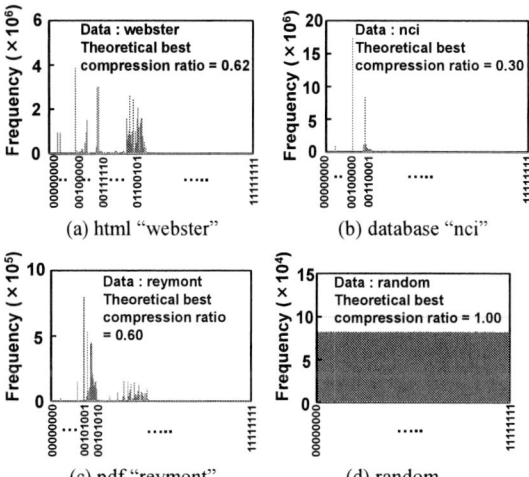

Fig. 12. Measured data frequencies of each 8bit unit of 3 Pattern files [6] and random data. Theoretical best compression ratio strongly depends on the frequency distribution.

required memory cells by 17.5% and 32.1% compared with proposed AERH-MLC and conventional Static Huffman coding, respectively, because of eliminating one V_{TH}-state. Fig. 10 shows the comparison of effective bits/cell which represents the bit cost compared with conventional Static Huffman coding that have two branches. The effective bits/cell increases by 0.5 and 0.77bit/cell compared with conventional SLC NAND flash. However, the effective bits/cell of AERH-MLC is slightly smaller than the conventional MLC NAND flash with Static Huffman coding. The reason of the slight compression ratio degradation is discussed in the next section. On the other hand, the reliability of AERH is improved significantly by over 50%.

IV. DISCUSSION OF COMPRESSION RATIO AND EFFECTIVE BITS/CELL

In this section, compression ratio is calculated as follows.

Compression ratio = Encoded data size / Raw data size (1)

Moreover, theoretical best compression ratio of Huffman coding is calculated from the frequencies of each data as shown in Fig. 11 [5]. The compression ratio strongly depends on the characteristics of data, that is, the frequencies of data. Fig. 12 shows the analyzed frequencies of each 8bit data of 3 pattern files used for the compression evaluation experiments [6] and random data file. In this analysis, "nci" file (Fig. 12 (b)) has the lowest theoretical best compression ratio because the frequencies of each 8bit data are the most unevenly distributed. Fig. 12 (d) shows the frequencies of each 8bit data of random

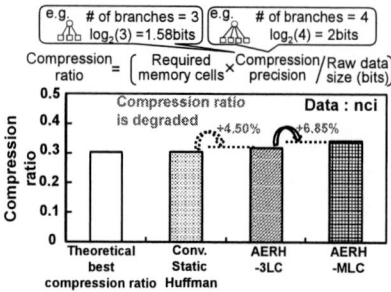

Fig. 13. Comparison of measured compression ratio for "nci" file between conventional Static Huffman coding and proposed AERH for MLC/3LC NAND flash.

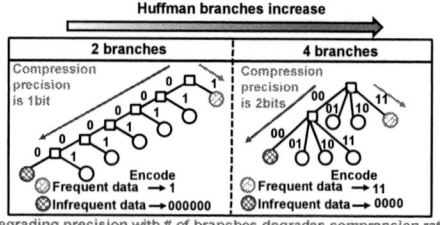

Fig. 14. Relation between Huffman branches and the compression ratio. In a larger branches, compression ratio becomes worse because the compression precision is degraded.

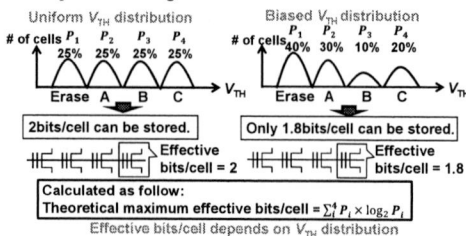

Fig. 15. Relation between effective bits/cell and V_{TH} distribution. Theoretical maximum effective bits/cell is determined by the V_{TH} distribution. Again, effective bits/cell becomes worse (smaller) when the V_{TH} distribution is unevenly distributed.

data file. Since the frequencies are evenly distributed, theoretical best compression ratio of random data file is 1.00. This means that random data file cannot be compressed by Huffman coding.

Fig. 13 shows the comparison of measured compression ratio. In this figure, the equation of compression ratio is defined so that even in AERH-MLC/3LC, where data is directly encoded as V_{TH}-states. As shown in Fig. 13, the compression ratio of the conventional Static Huffman coding with 2 branches almost achieves the theoretical best compression ratio. On the other hand, the compression ratio of AERH-3LC is 4.50% worse than conventional Static Huffman coding, and that of AERH-MLC is 6.85% worse than AERH-3LC. This result indicates that the compression ratio is degraded by increasing Huffman branches. The reason is expressed in Fig. 14. In conventional Static Huffman coding with 2 branches, the compression precision is 1bit because 1bit data, '0' or '1', is assigned to each branch. On the other hand, in the proposed 3-4 branches, the precision is degraded as the branches increase. In this example, conventional Static Huffman coding encodes the most frequent and infrequent data to "1" and "000000", respectively. On the

Table I. Summary of this work

Data		Compression ratio	Required memory cells	Effective bits/cell	Measured data-retention BER
nci	Randomize w/ conv.Static Huffman	0.304	41 M	2	Baseline
	AERH-3LC	0.319	54 M	1.50	-82.6%
	AERH-MLC	0.342	45 M	1.77	-55.8%
reymont	Randomize w/ conv.Static Huffman	0.608	16 M	2	Baseline
	AERH-3LC	0.617	21 M	1.55	-72.6%
	AERH-MLC	0.623	17 M	1.94	-53.7%
webster	Randomize w/ conv.Static Huffman	0.625	104 M	2	Baseline
	AERH-3LC	0.633	132 M	1.56	-78.3%
	AERH-MLC	0.635	105 M	1.96	-66.2%

other hand, AERH-MLC encodes the most frequent and infrequent data to "11" and "0000", respectively. Since compression ratio is improved by larger difference of encoded data length between more frequent and infrequent data, more branches degrade the compression ratio. This degradation of compression ratio slightly decreases the effective bits/cell.

In Fig. 10, effective bits/cell of AERH-MLC is 1.77bits/cell, which is lower than 2bits/cell of the conventional MLC NAND flash. Moreover, the V_{TH} distribution is modulated for high reliability, that is, reducing the population of the lower reliable "B"-state. Again, the asymmetric distribution of the V_{TH} states shown in Fig. 15 slightly degrades the compression ratio and decreases the effective bits/cell.

V. CONCLUSION

In MLC NAND flash, the data-retention errors of "B"-state is more significant than the other V_{TH}-states. Proposed AERH compresses data and modulates V_{TH} distribution simultaneously to enhance the reliability. The most unreliable "B"-state cells are reduced by AERH-MLC and are completely eliminated by AERH-3LC.

Table I shows the summary of this work. Proposed AERH-MLC and 3LC decrease the BER of data-retention errors by 55.8% and 82.6%, respectively, for "nci" file. AERH slightly decreases the effective bits/cell compared with MLC NAND flash due to the modulation of V_{TH} distribution. If compared with SLC NAND flash, AERH-MLC and 3LC have larger effective bits/cell by 0.77 and 0.5bit/cell, respectively. As a result, proposed MLC and 3LC NAND flash can be used as high capacity pseudo-SLC NAND flash of SSD cache in a tiered hierarchical storage.

VI. REFERENCE

[1] S. Tanakamaru, Y. Kitamura, S. Yamazaki, T. Tokutomi and K. Takeuchi, "Application-aware solid-state drives (SSDs) with adaptive coding," in *IEEE Symp. VLSI Circuits*, pp. 103-104, 2014.

[2] Y. Deguchi, A. Kobayashi, H. Watanabe and K. Takeuchi, "Flash Reliability Boost Huffman Coding (FRBH): Co-Optimization of Data Compression and V_{TH} Distribution Moduration to Enhance Data-Retention time by over 2900x," in *IEEE Symposium on VLSI Technology*, pp. T206-T207, 2017.

[3] J. Lee, J. Choi, D. Park and K. Kim, "Degradation of Tunnel Oxide by FN Current Stress and its Effects on Data Retention Characteristics of 90-nm NAND Flash Memory," in *IEEE International Reliability Physics Symposium*, pp. 497-501, 2003.

[4] DAVID A. HUFFMAN, "A Method for the Construction of Minimum-Redundancy Codes," *PROCEEDINGS OF THE I.R.E.*, pp. 1098-1102, 1952.

[5] C. E. SHANNON, "A Mathematical Theory of Communication," Reprinted with corrections from the Bell System Technical Journal, Vol.27, pp. 379-423, 623-656, july, October, 1948.

[6] Silesia compression corpus. Available: http://sun.aei.polsl.pl/~sdeor/index.php?page=silesia

Word-line Batch V_{TH} Modulation of TLC NAND Flash Memories for Both Write-Hot and Cold Data

Yoshiaki Deguchi and Ken Takeuchi

Department of Electrical, Electronic, and Communication Engineering, Chuo University, Tokyo, Japan
E-mail : deguchi@takeuchi-lab.org

Abstract— Word-line Batch V_{TH} Modulation (WBVM) is proposed as a comprehensive solution for both write-hot and cold data to improve the reliability of Triple Level Cell (TLC) NAND Flash memories. For write-hot data, WBVM V_{TH} score modulation decreases the program-disturb errors by 49% and enhances the endurance by 1.8-times of 2D-TLC NAND Flash. On the other hand, for write-cold data, WBVM BER score modulation decreases the data-retention errors by 36% and extends the acceptable data-retention time by 2.3-times of 2D-TLC NAND Flash. For 3D vertical TLC NAND Flash memory, the data-retention errors decrease by 13% and the acceptable data-retention time is extended by 1.4-times.

Keywords— *TLC NAND Flash memory; Reliability; Data-retention error; program-disturb error; endurance*

I. INTRODUCTION

Triple Level Cell (TLC) NAND Flash memory is used for Solid-State Drives (SSDs) due to the low bit-cost and the high capacity. However, reliability enhancement techniques are required because of the low reliability caused by the small memory cell threshold voltage (V_{TH}) margins [1]. For write-cold data where data is not programmed frequently, the higher V_{TH} state has the lower reliability due to the high electric field across the tunnel oxide [2]. To decrease the bit error rate (BER) of TLC Flash, the randomizer is implemented in the SSD controller to avoid the worst case of the data pattern as shown in Fig. 1(a) [3]. Moreover, Asymmetric Coding (AC) which modulates the V_{TH} distribution of a page has been proposed to further improve the reliability as shown in Fig. 1(b) [3]. However, these techniques do not decrease errors enough for scaled 2D or vertical 3D TLC NAND Flash memories. These techniques have been proposed based on the assumption that the write/read access unit is a page.

Recently proposed advanced high-speed programming schemes program the whole word-line, that is, three pages of the same word-line (WL) sequentially, e.g. three-step program for TLC NAND Flash [4] and High Speed Program for V-NAND [5]. For such advanced word-line unit programming, this paper proposes Word-line Batch V_{TH} Modulation (WBVM) as shown in Fig. 1(c). Proposed WBVM realizes the optimal V_{TH} distributions for both write-hot and cold data. In this paper, WBVM V_{TH} score modulation and BER score modulation are proposed for write-hot and cold data, respectively. Write-hot and cold data represent the frequently and infrequently programmed/erased data, e.g. proxy server and web server, respectively. Although the reliability enhancement techniques for read-hot and cold data have been proposed in [6], these techniques are not optimized for write-hot and cold data.

Fig. 2 compares the bit-flipping algorithm between conventional AC (Fig. 2(a)) and proposed WBVM (Fig. 2(b)). Proposed WBVM utilizes the TLC flag. Although the TLC flag has been proposed in [7], it is optimized only for the error-tolerant image identification system using the deep-learning. In this paper, WBVM makes use of the TLC flag to improve the reliability comprehensively for both write-hot and cold data. Conventional AC monotonically flips all data or not flip of a page in the code length unit. On the other hand, proposed WBVM can flexibly modulate the V_{TH} distribution in more detail because WBVM flips data in a whole word-line unit, not a page. Thus, WBVM can best optimize the reliability of both write-hot and cold by flexibly changing the algorithm of selecting the TLC flag. Fig. 3 shows the measured error patterns of the write-hot and cold data. In write-hot data, Write/Erase (W/E) stress is dominant and increases the program-disturb errors as shown in Fig. 3(a). On the other hand, in write-cold data, data-retention errors increase as shown in Fig. 3(b). Proposed WBVM realizes

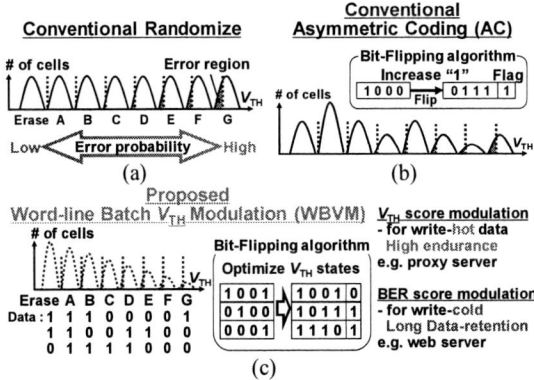

Fig. 1. Concept of this work. (a) conventional Randomize, (b) Asymmetric Coding (AC) and (c) proposed Word-line Batch V_{TH} Modulation.

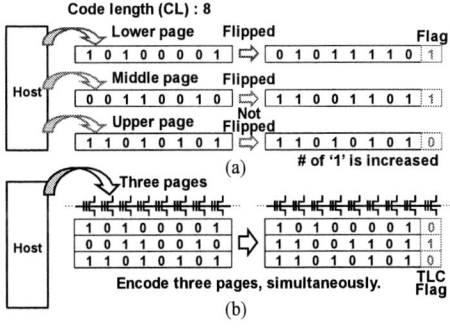

Fig. 2. Comparison of bit-flipping algorithms. (a) Conventional AC encodes data per page unit, and (b) proposed WBVM encodes data per whole word-line (WL), that is, 3 pages.

978-1-5386-3179-9/17 $31.00 © 2017 IEEE

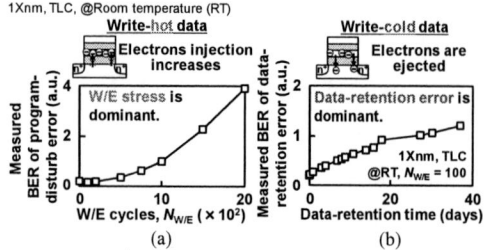

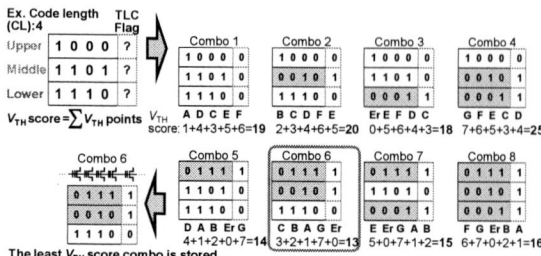

Fig. 3. Comparison of measured error patterns. (a) In write-hot data, the program-disturb error which increases V_{TH} is dominant. (b) In write-cold data, the data-retention error which decreases V_{TH} is dominant.

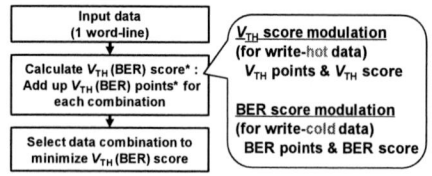

Fig. 4. Flowchart of proposed WBVM encoder. The V_{TH} or BER score is calculated to decrease the Write/Erase (W/E) stress or data-retention errors, respectively.

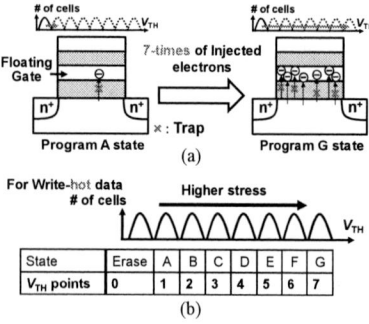

Fig. 5. Program operation in write-hot data. (a) The injected electrons in 'G' state program are about 7-times more than 'A' state program. (b) V_{TH} points are defined as the V_{TH} states.

a high endurance and a long retention for write-hot and cold data, respectively, by modulating the V_{TH} distribution differently.

II. PROPOSED WBVM V_{TH} SCORE MODULATION

Proposed algorithm of WBVM encoder is shown in Fig. 4. First, data of the whole WL is input to the proposed WBVM encoder. Second, the V_{TH} or BER score is calculated for write-hot or cold data, respectively. Finally, the optimal data combination is selected by minimizing the V_{TH} or BER score for write-hot and write-cold data, respectively. In the read of SSD, data is decoded based on the TLC flag.

In write-hot data, the gate tunnel oxide is damaged by the frequent W/E cycling. Moreover, the W/E stress is enlarged as more electrons tunnel through the floating gate as shown in Fig. 5(a) [3]. Assuming that the V_{TH} states are almost evenly distributed as V_{TH}, the V_{TH} points are linearly defined as the V_{TH} states as shown in Fig. 5(b). This linear relation may be caused by the phenomena where the damage to the tunnel oxide or the amount of traps generated in the tunnel oxide depends on the total charge tunneling through the tunnel oxide during the write/erase cycles.

Fig. 6. Algorithm of proposed WBVM V_{TH} score modulation for write-hot data. V_{TH} score is calculated based on the V_{TH} points. After that, the combination with the least V_{TH} score is selected. In this case, Combo 6 is selected and is stored to the TLC NAND Flash memory.

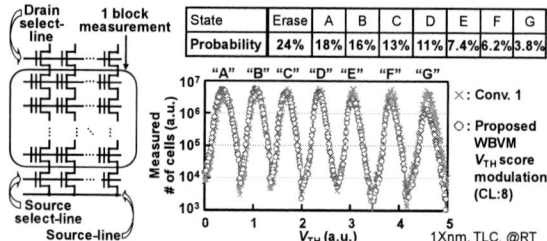

Fig. 7. Probabilities of each V_{TH} state and measured V_{TH} distribution with proposed WBVM V_{TH} score modulation.

For write-hot data, proposed WBVM V_{TH} score modulation decreases the W/E stress by minimizing the V_{TH} score which is the sum of the V_{TH} points.

Proposed bit-flipping algorithm of WBVM V_{TH} score modulation for write-hot data is shown in Fig. 6. In this figure, code length (CL) is 4 just for the plain explanation purpose. The TLC flag provides the 8 possible bit flipping combinations. When the flag bit is '1', the same page bits are flipped as shown in the gray region in Fig. 6. To select the flag, the V_{TH} scores of each possible combination are calculated based on the V_{TH} points of data. After that, the bit flipping combination with the least V_{TH} score is selected to minimize the W/E stress per code length unit. The modulated V_{TH} distribution is shown in Fig. 7. Higher V_{TH} state cells are actually decreased.

Fig. 8 indicates the evaluation method of the proposed WBVM V_{TH} score modulation for write-hot data. Program data is encoded by conventional Randomize, AC, and proposed WBVM V_{TH} score modulation. Data are sequentially programmed and erased. After that, data in memory cells are read to measure the program-disturb errors. Fig. 9 shows the measured program-disturb errors which increase with the W/E stress. The measured BERs of program-disturb error decrease by 49% and 38% compared with Conv. 1 and Conv. 2, respectively, when code length (CL) is 8. Moreover, the acceptable W/E cycle which means the endurance is extended by 1.8 and 1.7-times compared with the Conv. 1 and Conv. 2, respectively.

III. PROPOSED WBVM BER SCORE MODULTION

In write-cold data, the data-retention failure is critical because electrons (or electron-traps) are ejected from the floating gate as the data-retention time increases. Fig.10 shows the measured BER points of data-retention errors. Each V_{TH} state BER is called BER point in the proposed WBVM BER score modulation for write-cold data. The BER points abruptly

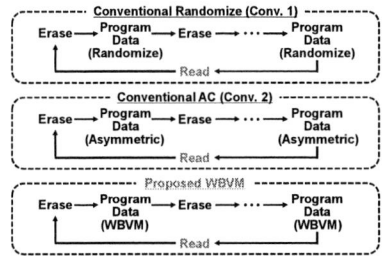

Fig. 8. Evaluation method of W/E stress for write-hot data. Encoded data are repeatedly programed/erased with conventional Randomize (conv. 1), AC (Conv. 2), and proposed WBVM V_{TH} score modulation.

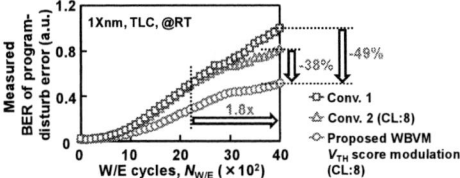

Fig. 9. Measured program-disturb errors for write-hot data. BER of program-disturb errors depends on the W/E stress and decreases by 49% and 38% compared with Conv. 1 and Conv. 2, respectively.

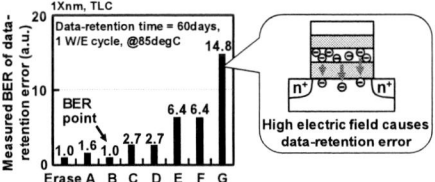

Fig. 10. Measured BER points of data-retention errors for write-cold data. The BER points represent the measured BER during the testing. The BER point of 'G' state is highest due to the highest electric field across the tunnel oxide.

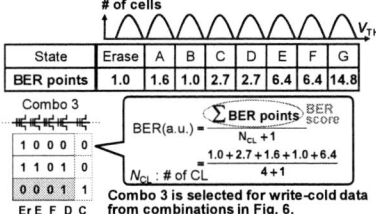

Fig. 11. Example of the BER score calculation of WBVM BER score modulation. The BER score is the sum of the BER points of data in the code length. Moreover, the BER score represents the estimated BER of the code length unit.

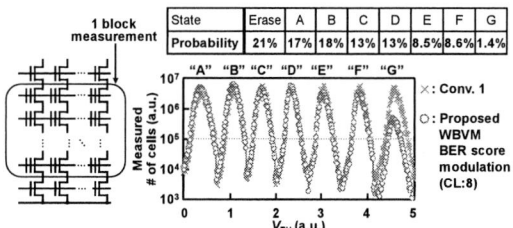

Fig. 12. Probabilities of each V_{TH} state and measured V_{TH} distribution with proposed WBVM BER score modulation.

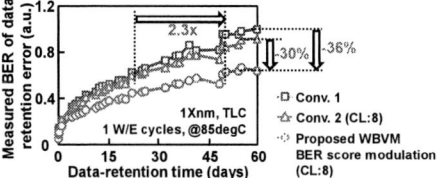

Fig. 13. Measured data-retention errors for write-cold data. Proposed WBVM BER score modulation decreases the data-retention errors by 36% and 30% compared with Conv. 1 and Conv. 2, respectively.

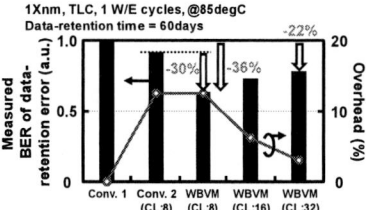

Fig. 14. Measured BER of the data-retention errors for write-cold data and the TLC Flash overhead. Proposed WBVM BER score modulation decreases the BER by over 22% even if the code length (CL) is 32 with 3.13% flag area overhead.

increase as the V_{TH} increases just like the exponential function of V_{TH} or the shape of the FN tunneling current, I_{FN}. I_{FN} is expressed as $E_{OX}^2 \exp(-B/E_{OX})$ where E_{OX} is the electric field across the tunnel oxide or is proportional to the V_{TH}, and B is the constant [8]. Hereafter, the BER score is explained as the exponential function of the V_{TH} for the simple explanation purpose. These characteristics of the BER score is different from the V_{TH} points, which linearly increases as a function of V_{TH} in WBVM V_{TH} score modulation for write-hot data. For write-cold data, the V_{TH} distribution is modulated so that the BER score, which is the sum of the BER points, is minimized. Therefore, proposed WBVM provides the different types of modulation of V_{TH} distribution for write-hot and cold data. Fig. 11 shows an example of calculating the BER score. As shown in Fig. 11,

minimizing the BER score leads to minimizing the BER per code length unit during the data-retention. In this example, Combo 3 is selected from the 8 possible bit flipping combinations in Fig. 6. The measured V_{TH} distribution is shown in Fig. 12. The number of 'G' state cells is much lower compared with WBVM V_{TH} score modulation for write-hot data in Fig. 7.

Fig. 13 shows the result of WBVM BER score modulation for write-cold data. The measured BER of data-retention errors decrease by 36% and 30% compared with the Conv. 1 (Randomize) and Conv. 2 (AC), respectively. Moreover, the acceptable data-retention time is extended by 2.3 and 1.9-times compared with the Conv. 1 and Conv. 2, respectively. The relation between the measured BER and the flag data overhead is shown in Fig. 14. The data overhead is decreased with the long code length. Proposed WBVM BER score modulation decreases the BER more than Conv. 1 and Conv. 2 even if the code length is 32 with 3.13% flag overhead. The proposed WBVM BER score modulation with the CL:8 and 32 decrease the measured BER by 36% and 22%, respectively, compared with Conv. 1. Fig. 15 shows the measured program-disturb errors with WBVM V_{TH}/BER score modulation. Because proposed WBVM BER score modulation is not optimized for write-hot data, program disturb errors of WBVM BER score modulation is 27% higher than WBVM V_{TH} score. Thus, WBVM V_{TH}/BER score modulation should be selectively applied to write-hot/cold data, respectively.

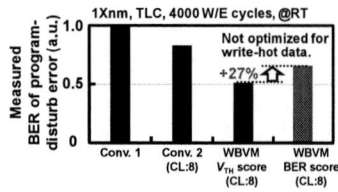

Fig. 15. Measured program-disturb errors with proposed WBVM V_{TH}/BER score modulation. If WBVM BER score modulation is applied to write-hot data, program disturb errors increase by 27% and thus is not suitable for write-hot data.

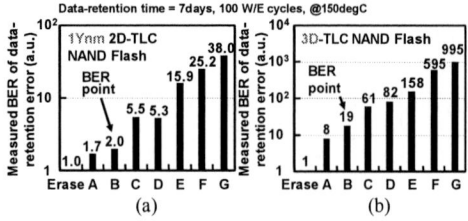

Fig. 16. Measured BER points of (a) 2D 1Ynm and (b) 3D TLC NAND Flash. The BER points increase as the V_{TH} increases in both 2D and 3D NAND Flash.

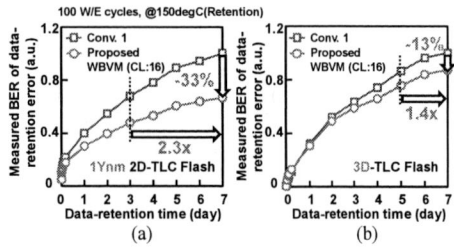

Fig. 17. Measured data-retention errors for write-cold data. In both of (a) 1Ynm 2D-TLC Flash and (b) 3D-TLC NAND Flash, the BER decreases and acceptable data-retention time is extended.

Fig. 18. Measured SSD board [2].

2D 1Ynm and vertical 3D NAND Flash memories are measured in Figs. 16-17. Fig. 16 shows the measured BER points of 2D 1Ynm and 3D NAND Flash memories. The measured BER points almost exponentially increase as the V_{TH} increases. Then, the BER of data-retention errors is measured with the proposed WBVM BER score modulation as shown in Fig. 17. The measured BERs decrease by 33% and 13% in 2D 1Ynm and 3D NAND Flash, respectively, compared with Conv. 1. Moreover, the acceptable data-retention time is extended by 2.3 and 1.4-times in 2D 1Ynm and 3D NAND Flash, respectively.

IV. CONCLUSION

Word-Line Batch V_{TH} Modulation (WBVM) is proposed to improve the reliability of both write-hot and cold data. Measured SSD board is shown in Fig. 18. The V_{TH} distribution is differently optimized for write-hot and cold data. For write-hot data, W/E stress is reduced by WBVM V_{TH} score modulation, which decreases the higher V_{TH} state cells linearly. On the other hand, for write-cold data, the data-retention error is reduced by

Table 1. Summary of proposed WBVM for write-hot data (2D 1Xnm)

Write-hot data	Conv. 1	Conv. 2 (CL:8)	Proposed WBVM (CL:8)
BER	Baseline	-18%	-49%
Acceptable W/E cycles	Baseline	1.1x	1.8x
Overhead	0	12.5%	12.5%

1Xnm, TLC, 4000 W/E cycles, @RT

Table 2. Summary of proposed WBVM for write-cold data (2D 1Xnm)

Write-cold data	Conv. 1	Conv. 2 (CL:8)	WBVM (CL:8)	WBVM (CL:16)	WBVM (CL:32)
BER	Baseline	-8.3%	-36%	-27%	-22%
Acceptable data-retention time	Baseline	1.2x	2.3x	1.6x	1.4x
Overhead	0	12.5%	12.5%	6.25%	3.13%

1Xnm, TLC, 1 W/E cycles, Data-retention time = 60days, @85degC

Table 3. Summary of proposed WBVM for write-cold data (2D 1Ynm, 3D)

Write-cold data	1Ynm NAND Flash		3D NAND Flash	
	Conv. 1	WBVM (CL:16)	Conv. 1	WBVM (CL:16)
BER	Baseline	-33%	Baseline	-13%
Acceptable data-retention time	Baseline	2.3x	Baseline	1.4x
Overhead	0	6.25%	0	6.25%

TLC, 100 W/E cycles, Data-retention time = 7days, @150degC

WBVM BER score modulation, which decreases the lower reliability state cells around exponentially. As a result, proposed WBVM V_{TH} score modulation decreases the program-disturb errors by 49% and enhances the acceptable W/E cycle by 1.8-times for write-hot data as shown in Table 1. On the other hand, proposed WBVM BER score modulation decreases the data-retention errors by 36% and extends the acceptable data-retention time by 2.4-times for write-cold data as shown in Table 2. The data overhead can be reduced by large CL. Table 3 shows the measured results of 2D 1Xnm and 3D NAND Flash. Proposed WBVM also decreases the BER and extends the acceptable data-retention time in both 2D 1Xnm and 3D NAND Flash. Thus, proposed WBVM is a flexible and comprehensive solution to best improve the reliability.

REFERENCES

[1] T. Tokutomi, M. Doi, S. Hachiya, A. Kobayashi, S. Tanakamaru, and K. Takeuchi, "Enterprise-Grade 6x Fast Read and 5x Highly Reliable SSD with TLC NAND Flash Memory for Big-Data Storage," *IEEE Int. Solid-State Circuits Conf. Dig. Tech. Papers*, pp. 140-141, Feb. 2015.

[2] Y. Deguchi, A. Kobayashi, H. Watanabe, and K. Takeuchi, "Flash Reliability Boost Huffman Coding (FRBH): Co-Optimization of Data Compressio and V_{TH} Distribution Modulation to Enhance Data-Retention Time by over 2900x," *VLSI Technology Dig. Tech. Papers*, pp. T206-T207, June 2017.

[3] S. Tanakamaru, C. Hung, and K. Takeuchi, "Highly Reliable and Low Power SSD Using Asymmetric Coding and Stripe Bitline-Pattern Elimination Programming," *IEEE J. Solid-State Circuits*, Vol. 47, No. 1, Jan. 2012.

[4] Y. Li *et al.*, "128Gb 3b/Cell NAND Flash Memory in 19nm Technology with 18MB/s Write Rate and 400Mb/s Toggle Mode," *IEEE Int. Solid-State Circuits Conf. Dig. Tech. Papers*, pp. 436-437, Feb. 2012.

[5] J.-W. Im *et al.*, "A 128Gb 3b/cell V-NAND Flash Memory with 1Gb/s I/O Rate," *IEEE Int. Solid-State Circuits Conf. Dig. Tech. Papers*, pp. 130-131, Feb. 2015.

[6] A. Kobayashi, T. Tokutomi, and K. Takeuchi, "Versatile TLC NAND Flash Memory Control to Reduce Read Disturb Errors by 85% and Extend Read Cycles by 6.7-times of Read-Hot and Cold Data for Cloud Data Centers," *VLSI Circuits Dig. Tech. Papers*, pp. C126-C127, June 2016.

[7] Y. Deguchi, T. Nakamura, A. Kobayashi, and K. Takeuchi, "12x Bit-Error Acceptable, 300x Extended Data-Retention Time, Value-Aware SSD with Vertical 3D-TLC NAND Flash Memories for Image Recognition," *IEEE Custom Integrated Circuits Conf. Dig. Tech. Papers*, May 2017.

[8] M. Lenzlinger and E. Snow, "Fowler-Nordheim hole tunneling in p-SiCOSiO2 structures," *J. Appl. Phys.*, Vol. 40, p. 278, 1969.

A 16kb Column-based Split Cell-VSS, Data-Aware Write-Assisted 9T Ultra-Low Voltage SRAM with Enhanced Read Sensing Margin in 28nm FDSOI

M. Sultan M. Siddiqui, Zhao Chuan Lee, and Tony Tae-Hyoung Kim

VIRTUS, School of Electrical and Electronic Engineering, Nanyang Technological University, Singapore

Email: thkim@ntu.edu.sg

Abstract— We propose a column-based split cell-VSS (CS-CVSS) data-aware write-assisted (DAWA) 9T ultra-low voltage SRAM with enhanced read sensing margin in 28nm FDSOI technology. The proposed write-assist technique (CS-CVSS and DAWA) improve both half-select SNM and write margin. The proposed 3T low leakage read port enhances read sensing margin by minimizing bitline leakage through negative gate to source voltage. A 16kb 9T SRAM test chip demonstrates $VDD_{MIN-Write}$ improvement of 0.39 V and $VDD_{MIN-Read}$ of 0.25 V with 1.57 μs read access time. The energy of 6.72 pJ is achieved at 0.5 V.

I. INTRODUCTION

Energy efficiency is one of the topmost design constraints in various emerging applications such as power- or energy-constrained wearable devices, Internet of Things (IoT) and health care SoCs [1-2]. Since batteries are the main power supply sources of these applications, minimizing energy is critical in maximizing the system lifetime. This can be achieved by aggressively lowering supply voltage close to the minimum energy point. SRAMs expend significant amount of power, energy, and circuit area. Therefore, careful design of ultra-low voltage SRAMs has been increasingly demanded [3]. Conventional 6T SRAMs fail to operate in this voltage scaled environment due to degradation in various design parameters such as cell stability, poor read and write margins, and exponentially increased sensitivity to PVT fluctuation. Various ultra-low voltage SRAM design techniques have been reported to tackle the limitations of the conventional 6T SRAMs [3-12].

The most significant technique in the ultra-low voltage SRAM design is the employment of 8T SRAM cell structure with a separated read port for decoupling read operation and write operations [3]. Even though SRAMs with decoupled read ports have been verified to operate in the near- or sub-threshold region [7], they still suffers from half-selected (HS) static cell stability (HS-SNM) and poor read bitline (RBL) sensing caused by small I_{ON}-to-I_{OFF} ratios. Various assist circuits such as collapsed power supply [2], boosted WL [4], hierarchical RBL structures [5], sense amplifier redundancy [6], wordline under-drive [8-9], and cell-VSS control [10-12] have been proposed at the cost of sacrificing other key parameters. This work proposes a novel 9T SRAM cell with column-based split cell-VSS (CS-CVSS) to improve (i) HS-SNM and write margin (WM) through data-aware write assist (DAWA), (ii) HS-dynamic cell stability (HS-DNM) by leveraging write through virtual ground with reduced load, and (iii) RBL sensing margin by a 3T read port for leakage minimization.

II. PROPOSED 9T SRAM

A. Proposed 9T SRAM cell and architecture

Fig. 1 depicts the proposed 9T cell. It consists of 6T for write (WR) port and data storage (PG1,2; PD1,2; PU1,2) and 3T for read (RD) port: (MR1,2,3). The pass gates (PG1, PG2) are controlled by write bit lines, WBL and WBLB. The source nodes of PG1 and PG2 are shorted to form write virtual ground (WR_VGND). The source nodes of the pull-downs (PD1, PD2) are biased through the column-based split cell-VSS (CS-CVSS) lines, CVSS1 and CVSS2. The levels of CVSS1 and CVSS2 will be controlled to change the switching thresholds (V_{ST1}, V_{ST2}) of the cross-coupled inverters (INV1, INV2). The gate node of MR3 is controlled by MR1 and MR2 for leakage suppression and RBL sensing margin enhancement. L- and inverted L-shaped cell layouts (Fig. 1(b)) are employed for high area efficiency.

The overall architecture of the proposed SRAM is depicted in Fig. 2. The proposed CS-CVSS DAWA consists of the CVSS-UP voltage generator and the data-aware CVSS-UP switching circuit. The detailed operation of them are also discussed in the following section.

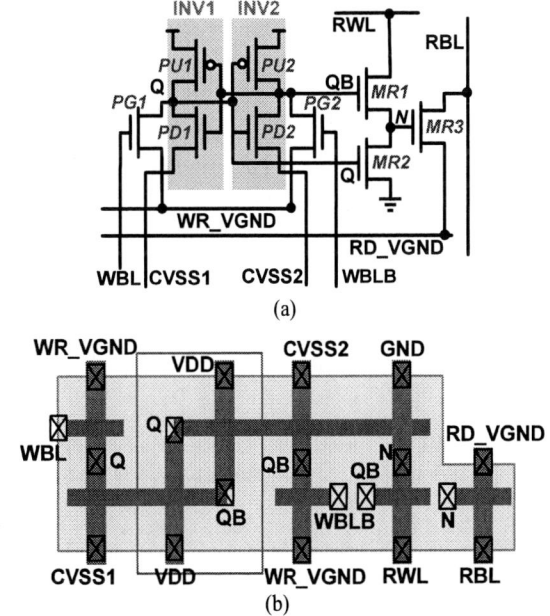

Fig. 1. (a) Proposed 9T SRAM cell and (b) L-shaped layout.

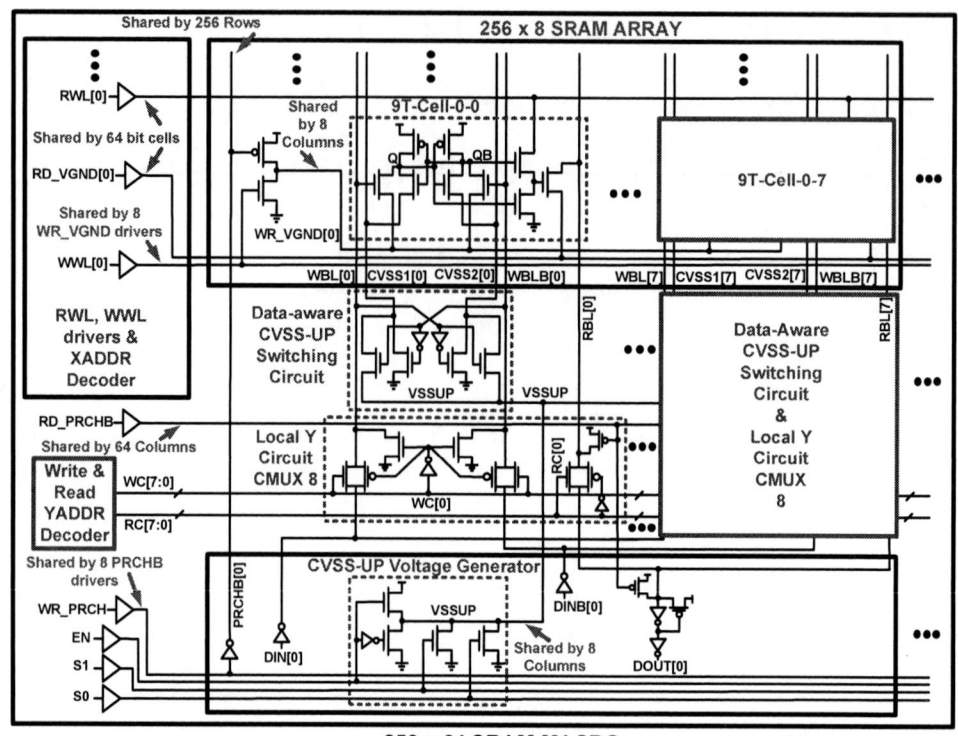

Fig. 2. Proposed 256×64 SRAM macro structure and the proposed 9T cell.

B. Column-based split cell-VSS (CS-CVSS) and data-aware write-assist (DAWA)

In our proposed SRAM macro (Fig. 2), a data-aware write assist (DAWA) is implemented using a column-based split cell-VSS (CS-CVSS): CVSS1 and CVSS2. As per the data to be written in SRAM, Data-aware CVSS-UP switching circuit clamps either CVSS1 or CVSS2 to VSSUP (>0) and other to 0. VSSUP clamp voltage is generated using CVSS-UP voltage generator. Fig. 3 (top left) depicts different level of VSSUP clamp voltages generated by CVSS-UP voltage generator depending on EN, S1 and S0 configurations.

Fig. 3 delineates the write '0' operation in our proposed SRAM macro. WBL and WBLB are loaded with data (WBL=VDD; WBLB=0V) for selected column address and WR_VGND is driven to '0' for selected row address. Data-aware CVSS-UP switching circuit switches (CVSS1, CVSS2) to (0, VSSUP). CVSS2=VSSUP weakens PU1 and elevates V_{ST1} of INV1 in selected SRAM cell (Fig. 1) thereby improving the write margin. For Half-selected cells, WR_VGNDs precharged to VDD are kept floating. Half-selected cells with Q='0' (Cell-Q0) has a voltage bump at Q but Cell-Q0 still has better Half-selected Static Noise Margin (HS-SNM-Q0) due to reduced strength of PD2 and elevated V_{ST2} of INV2 with CVSS2=VSSUP. Half-selected cells with Q='1' (Cell-Q1) are free from Half-selected disturbance as Q=WR_VGND=VDD with PG2 off. Only the Hold Static Noise Margin (HD-SNM-Q1) degrades due to CVSS2=VSSUP. However, the reduced HD-SNM-Q1 is far better than the HS-SNM-Q0. The proposed technique improves HS-SNM-0 by sacrificing HS-SNM-Q1. In this work, we have

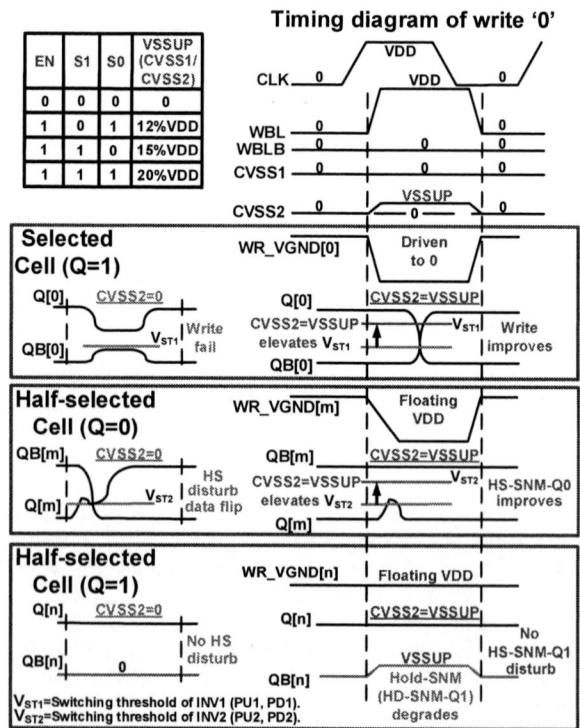

EN	S1	S0	VSSUP (CVSS1/CVSS2)
0	0	0	0
1	0	1	12%VDD
1	1	0	15%VDD
1	1	1	20%VDD

Fig. 3. Proposed column-based split cell-VSS (CS-CVSS) scheme with data-aware write assist (DAWA).

S12-4 (2065)

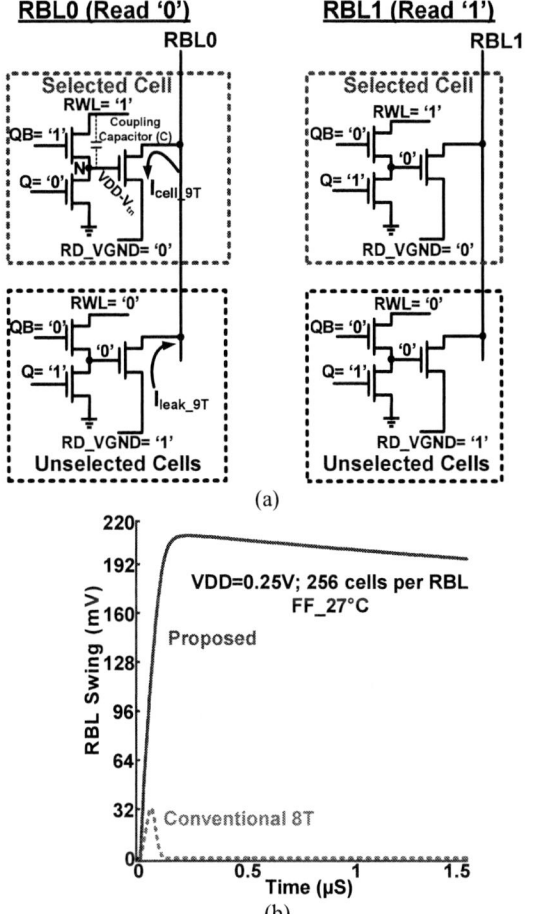

(a)

(b)

Fig. 4. (a) Principle of the proposed sensing margin technique and (b) simulated RBL swing.

selected VSSUP of 20% VDD as a design target, providing the best SNM.

C. Proposed 3T read port for enhanced RBL sensing margin

RD sensing margin is highly limited by I_{ON}-to-I_{OFF} ratio. This work proposes a novel 3T RD port for substantial improvement in I_{ON}-to-I_{OFF} ratio. In the proposed RD port, MR1 and MR2 controls MR3 gate as shown in Fig. 1. RWL is connected to MR1 and, MR3 drain and source are connected to RBL and RD virtual ground (RD_VGND). During RD (RWL='1'; RD_VGND='0'), the precharged RBL is conditionally discharged if QB='1'. In unselected cells of the selected column (RWL='0'; RD_VGND='1'), MR3 gate to source voltage is -VDD regardless of Q and QB, reducing the RD port leakage drastically. Fig. 4(a) depicts the worst case sensing margin scenarios of the 8T and the proposed 9T SRAM. Simulation results of RBL swing (RBL1 (Read '1') - RBL0 (Read '0')) in Fig. 4(b) clearly demonstrates that the proposed RD port generates larger RBL swing and wider sensing timing window than that of the 8T cell. This also allows SRAMs to employ a large number of cells per RBL because of the significant RD port leakage reduction. However, the cell read '0' current for discharging RBL0 is deteriorated

by the under-driven MR3 compared to 8T cell's fully-driven footer. To mitigate the discharging speed degradation, we increased the coupling capacitance (C) between RWL and the gate of MR3 (Node 'N' in Fig. 4(a)). Complete compensation is not required since the target applications of the proposed SRAM has low operation speed.

III. TEST CHIP MEASUREMENT RESULTS

We fabricated a 16kb (256×64) SRAM test chip in 65nm CMOS technology as shown in Fig. 5. Fig. 6 illustrates the measured Schmoo plots for write operation. The proposed 9T SRAM achieves VDD$_{MIN-Write}$ of 470 mV with CVSS1/CVSS2 voltage of 20% VDD while the conventional 8T SRAM shows VDD$_{MIN-Write}$ of 860 mV. The proposed read port shows VDD$_{MIN-Read}$ of 250 mV with read access time of 1.58 μs and maximum operating frequency of 290kHz (Fig. 7(a)). Fig. 7(b) depicts measured write power and read power as a function of VDD. Note that the write power is not shown below VDD=0.5 V as VDD$_{MIN-Write}$=470 mV due to the limited strength of CVSS1/CVSS2 drivers in this design. We can further lower VDD$_{MIN-Write}$ by using stronger CVSS1/CVSS2 drivers. Finally, Fig. 7(c) shows the measured energy of the proposed SRAM. It consumes 6.72pJ at 0.5 V. Finally, Table I summarizes the performance of the proposed 9T SRAM compared with other similar works. For fair comparison, a 8T SRAM [3] and a DSC6T SRAM [12] are also designed in the same technology.

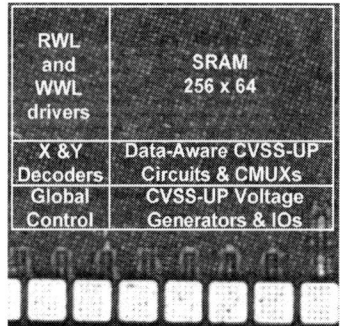

Fig. 5. Test chip micrograph.

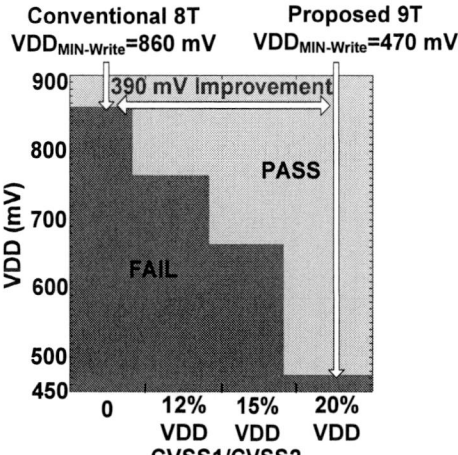

Fig. 6. Schmoo plot demonstrating the benefit of the proposed CS-CVSS DAWA.

978-1-5386-3179-9/17 $31.00 © 2017 IEEE

TABLE I: TEST CHIP MEASUREMENT RESULTS AND COMPARISON.

	8T [3]	DSC6T [12]	9T (This work)
Cell Size	1.25x	1x	1.45x
Write Assist	No	RS-CVSS + NBL-light	**CS-CVSS**
HS-DNM	Worst (256 BCs load)	Worst (256 BCs load)	Improved (8 BCs load)
Margin Trade-off	N/A	HS-SNM vs. HD-SNM	HS-SNM vs. HD-SNM
Read Delay	0.7x @VDD=0.5V	1x @VDD=0.5V	1.1x @VDD=0.5V
	Fail @VDD=0.25V	Fail @VDD=0.25V	1x @VDD=0.25V
Read Energy[*1]	1.15x @VDD=0.5V	1.23x @VDD=0.5V	1x @VDD=0.5V
	Fail @VDD=0.25V	Fail @VDD=0.25V	1x @VDD=0.25V
Write Energy	Fail @VDD=0.5V	1.1x @VDD=0.5V	1x @VDD=0.5V
Write Operation in	Only CLK ON period	CLK ON & OFF period	Only CLK ON period

256x64 SRAM, 27°C;[*1]Data: 50% 1-cells, 50% 0-cells; Simulation results: STM 28nm FDSOI.

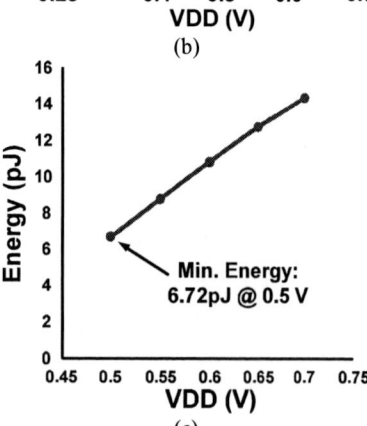

Fig. 7. Test chip measurement results: (a) read access time, (b) power, and (c) energy.

IV. CONCLUSION

We proposed the column-based split cell-VSS data-aware write-assisted 9T ultra-low voltage SRAM with enhanced read sensing margin. A 16Kb (256×64) 28nm FDSOI 9T SRAM test chip demonstrates that the proposed CS-CVSS scheme with DAWA achieve 390 mV lower $VDD_{MIN-Write}$ than 8T SRAM without write assist. The proposed 3T RD port also achieves $VDD_{MIN-Read}$ of 250 mV due to suppressed leakage.

REFERENCES

[1] A. Wang and A. Chandrakasan, "A 180-mV subthreshold FFT processor using a minimum energy design methodology," *IEEE Journal of Solid-State Circuits,* vol. 40, pp. 310-319, 2005.

[2] S. M. A. Zeinolabedin et al., "A 128-Channel Spike Sorting Processor Featuring 0.175 µW and 0.0033 mm2 per Channel in 65-nm CMOS," *IEEE Symposium on VLSI Circuits,* pp. 32-33, 2016.

[3] L. Chang et al., "An 8T-SRAM for Variability Tolerance and Low-Voltage Operation in High-Performance Caches," *IEEE Journal of Solid-State Circuits,* vol. 43, pp. 956-963, 2008.

[4] B. H. Calhoun and A. P. Chandrakasan, "A 256-kb 65-nm Sub-threshold SRAM Design for Ultra-Low-Voltage Operation," *IEEE Journal of Solid-State Circuits,* vol. 42, pp. 680-688, 2007.

[5] B.-D. Yang and L.-S. Kim, "A low-power SRAM using hierarchical bit line and local sense amplifiers," *IEEE Journal of Solid-State Circuits,* vol. 40, pp. 1366-1376, 2005.

[6] N. Verma and A. P. Chandrakasan, "A 65nm 8T Sub-Vt SRAM Employing Sense-Amplifier Redundancy," *IEEE International Solid-State Circuits Conference,* pp. 328-606, 2007

[7] J. Keane et al., "5.6Mb/mm² 1R1W 8T SRAM Arrays Operating down to 560mV Utilizing Small-Signal Sensing with Charge-Shared Bitline and Asymmetric Sense Amplifier in 14nm FinFET CMOS Technology," *IEEE International Solid-State Circuits Conference,* pp. 308-309, 2016.

[8] J. Chang et al., "A 20nm 112Mb SRAM in High-k Metal-Gate with Assist Circuitry for Low-Leakage and Low-Vmin Applications," *IEEE International Solid-State Circuits Conference,* pp. 316-317, 2013.

[9] T. Song et al., "A 14nm FinFET 128Mb 6T SRAM with Vmin-Enhancement Techniques for Low-Power Applications," *IEEE International Solid-State Circuits Conference,* pp. 232-233, 2014.

[10] C-F Chen et al., "A 210mV 7.3MHz 8T SRAM with Dual Data-Aware Write-Assists and Negative Read Wordline for High Cell-Stability, Speed and Area-Efficiency," *IEEE Symposium on VLSI Circuits,* pp. C130-C131, 2013.

[11] D. Jeon et al., "A 23mW Face Recognition Accelerator in 40nm CMOS with Mostly-Read 5T Memory," *IEEE Symposium on VLSI Circuits,* pp. C48-C49, 2015.

[12] M-F Chang et al., "A 28nm 256kb 6T-SRAM with 280mV Improvement in V_{MIN} Using a Dual-Split-Control Assist Scheme," *IEEE International Solid-State Circuits Conference,* pp. 314-315, 2015.

An Energy-optimized (37840, 34320) Symmetric BC-BCH Decoder for Healthy Mobile Storages

Seokha Hwang[1], Jaehwan Jung[2], Daesung Kim[3], Jeongseok Ha[2], In-Cheol Park[2] and Youngjoo Lee[4]

Email: youngjoo.lee@postech.ac.kr

[1]Department of Electronic Engineering, Kwangwoon University, Seoul, Korea

[2]Department of Electrical Engineering, Korea Advanced Institute of Science and Technology (KAIST), Daejeon, Korea

[3]NAND Solution Development Division, SK Hynix, Icheon, Korea

[4]Department of Electrical Engineering, Pohang University of Science and Technology (POSTECH), Pohang, Korea

Abstract— This paper presents an energy-efficient symmetric block-wise concatenated-BCH (SBC-BCH) decoder architecture for energy-starving mobile storages. The proposed 4KB SBC-BCH code remarkably enhances the hard-decision-based error-correcting performance to defer the energy-consuming memory-sensing operations for generating the soft-decision values, which are necessary to prolong the lifetime of flash memories. Thanks to the proposed powerful hard-decision code, the energy-efficient healthy memory condition is extended by 1.4 times. In addition, we introduce design methods for realizing the energy-efficiency SBC-BCH decoder. For example, the syndrome tracking allows the early start of the next decoding process, and the reordered Chien search minimizes the wasted waiting periods. Based on the proposed schemes, targeting the recent NAND flash memories, a prototype (37840, 34320) SBC-BCH decoder is realized in 65nm CMOS process. The prototype decoder achieves a throughput of 12.5 Gb/s while providing an energy efficiency of 3.56pJ/b, which is superior to the state-of-the-art ECC solutions for storages.

Keywords— *error-correction codes; low-power designs; NAND flash memoreis; storage controllers; VLSI*

I. INTRODUCTION

To dominate the storage market by replacing the traditional disk-type devices, manufacturers of NAND flash memories have continuously developed aggressive schemes, which can reduce the cost-per-bit by scarifying the integrity [1]. Based on numerous observations, it is widely reported that the channel condition of the recent NAND flash memory is gradually and rapidly degraded according to the number of program/erase (P/E) cycles [2]. To recover erroneous data, multi-step error-correction codes (ECCs) are normally applied, i.e., weak hard-decision (HD) ECCs are first used for the clean storages, and at some point strong soft-decision (SD) ECCs are utilized for extending the lifetime as much as possible [2]. Unlike to HD ECCs with energy-efficient decoding process, in general, SD ECCs requires power-hungry decoding procedures. Moreover, generating SD values from flash memory uses multiple sensing operations, consuming at least 10 times more energy [3]. In this paper, we use the terminology *healthy* to represent the clean memory condition, which is successfully recovered by only HD ECCs. Similarly, the terminology *sick* is used to denote the erroneous memory condition, requiring strong SD ECCs.

In the recent mobile storages for energy-limited devices like sensory systems, IoT platforms, and implantable systems, it is suggested to use energy-efficient HD ECCs as long as possible before moving to the SD ECC mode. Due to the poor error-correcting performance of contemporary HD ECCs [4], [5], however, the lifetime of healthy mobile storage is quite short compared to the full lifetime of storages, prolonged by the latest SD ECCs [6]–[8]. Therefore, it is urgent to develop the new HD ECC solution for mobile storages, which can narrow the current performance gap between HD and SD ECC solutions, allowing more healthy periods with low energy.

Based on the previous block-wise concatenated-BCH (BC-BCH) code [9], we present in this paper a new concept of iterative HD ECC and its energy-efficient decoder structure. The proposed symmetric BC-BCH (SBC-BCH) code consists of multiple component BCH codes to protect a data frame of 4KB, so that it can effectively support the latest advanced-format file systems [10]. Based on the tightly-coupled code structure, at the raw bit-error-rate (RBER) of 4.5×10^{-3}, the uncorrectable-BER (UBER) of the proposed SBC-BCH code successfully approaches to the level of 10^{-12}. Compared to the previous BC-BCH work, as a result, the proposed ECC extends the healthy P/E cycles of NAND flash memory by more than 40%, which is suitable to the energy-starving mobile storages.

For energy-efficient decoding operations, in addition, we present several optimization techniques. To solve the long latency problem at the straight-forward decoder architecture, we adopt the concept of syndrome tracking, which immediately updates all the syndromes when erroneous bits are detected. The proposed reordered Chien search further enhances the resource utilization by managing the critical block as early as possible. To validate the proposed schemes, a prototype SBC-BCH decoder is fabricated in 65nm CMOS process. By minimizing the unwanted waiting periods, at the RBER of 4.5×10^{-3}, the decoding throughput is greatly enhanced up to 12.5Gb/s while consuming only 3.56pJ/b.

II. PROPOSED 4KB SBC-BCH ECC

In order to construct the proposed SBC-BCH code, a 4KB user-data is zero-extended to 34320 bits for making the regular concatenation of component codes (CCs). Fig. 1 shows the

This work was supported by the National Research Foundation (NRF) grant funded by the Korea government (MSIP) (2016R1C1B1007593) and by the IC Design Education Center (IDEC)

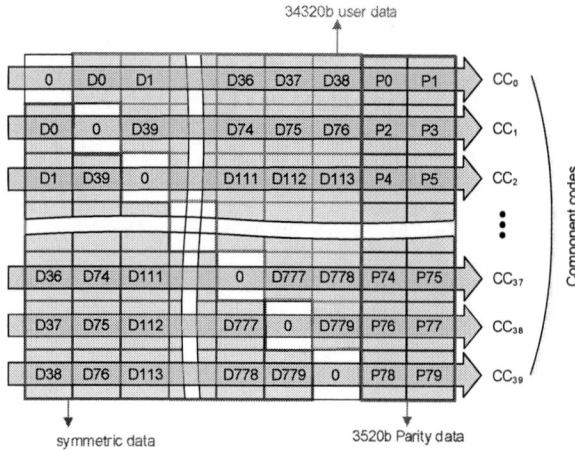

Fig. 1. The proposed structure of (37840, 34320) SBC-BCH code.

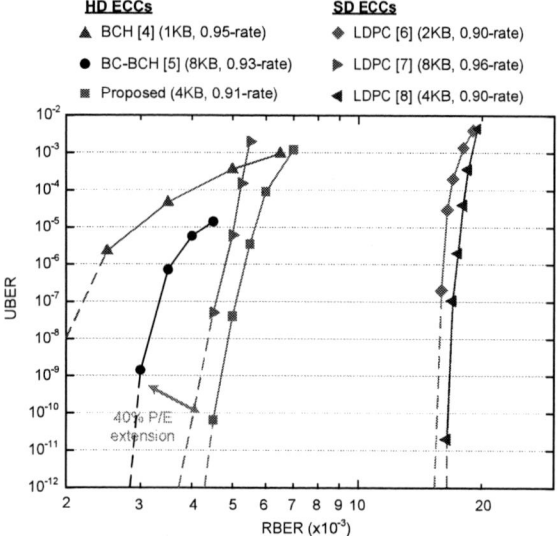

HD ECCs
▲ BCH [4] (1KB, 0.95-rate)
● BC-BCH [5] (8KB, 0.93-rate)
■ Proposed (4KB, 0.91-rate)

SD ECCs
◆ LDPC [6] (2KB, 0.90-rate)
▶ LDPC [7] (8KB, 0.96-rate)
◀ LDPC [8] (4KB, 0.90-rate)

Fig. 2. The error-correcting performances of different ECC solutions.

detailed code structure of the proposed (37840, 34320) SBC-BCH code. In contrast to the previous BC-BCH code having two-dimensionally-placed regular-sized data blocks [9], the proposed ECC first arranges the input 34320 bits into the upper-triangle of 2D plane by using 780×44b data blocks (Dx), and then the lower-triangular part is filled with reflected data blocks. In other words, the data blocks in the proposed SBC-BCH code are symmetric with respect to the diagonal blocks, i.e., all-zero blocks as depicted in Fig. 1. By allowing 2×44b parity blocks per each CC, total 40 (1848, 1760, 8) BCH codes are used as CCs. Although each CC is weak, i.e., only 8 random errors are recovered among 1848 bits, the iterative decoding manner improves the correcting performance of a whole 4KB page. More precisely, after decoding all the CCs serially, we start again the decoding process from CC$_0$ based on the updated values until reaching the maximum number of iteration, I_{MAX}. As all the CCs are related to each other, the iterative decoding gradually recovers the remaining errors,

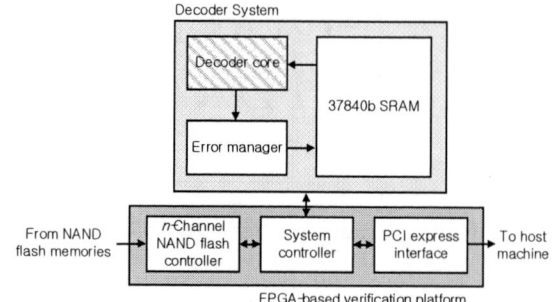

Fig. 3. The proposed decoder system with FPGA-based verification platform.

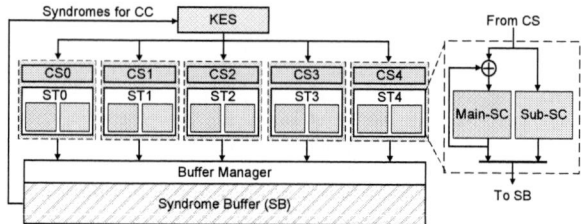

Fig. 4. The detailed architecture of the proposed SBC-BCH decoder core.

converging to the error-free result. Note that the convergence speed of the proposed SBC-BCH code is much faster than that of the previous BC-BCH code because some CCs in the previous work are independent from each other [5]. For the same I_{MAX}, therefore, it is obvious that the proposed SBC-BCH achieves much better UBER performances.

Including the proposed 4KB SBC-BCH code with $I_{MAX} = 4$, Fig. 2 shows error-correcting performances of various ECC solutions for storages. To provide UBER of 10^{-12}, which is sufficient for mobile systems, the proposed work requires at least the RBER of 0.0042, which is the best option among HD ECC solutions. Note that this performance can expand the healthy lifetime of mobile storages by around 40% compared to the previous HD ECC works as shown in Fig. 2 The recent LDPC work in [8] may outperform the proposed SBC-BCH code, however, this code uses 4b-quatized inputs and 8b internal computing resolution, requiring a huge amount of decoding energy. Therefore, our HD ECC is the attractive candidate for energy-starving mobile storages in terms of the error-correcting capability as well as the low sensing energy.

III. ENERGY-OPTIMIZED DECODER ARCHITECTURE

In this section, we present novel optimization schemes for enhancing the energy efficiency of SBC-BCH decoder itself. Connected to the FPGA-based verification platform, as shown in Fig. 3, our decoder system conceptually consists of three parts; a 37840b SRAM, an error manager, and a decoder core. The SRAM contains the received SBC-BCH code, and it is accesses by the error manager to correct erroneous bits. Similar to [5], the error manager receives only the error locations from the decoder core, which is the main part of this work.

In the straight-forward core design, which is the baseline of this work, the decoder core is nothing but the traditional BCH

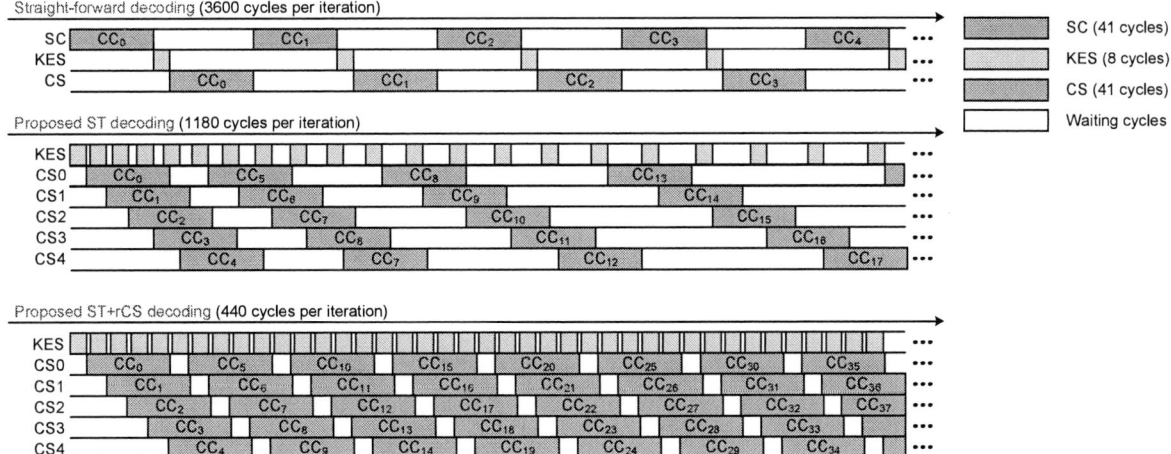

Fig. 5. The processing steps of each decoding scenario.

decoder having three well-kwon stages; syndrome calculator (SC), key-equation solver (KES), and Chien search (SC) [4]. Considering the SBC-BCH code in Fig. 1, it is natural to utilize 44-parallel SC and CS to process each data block in one cycle. As all the CCs are serially related, unlike to the previous BC-BCH case, it is impossible to use multiple BCH decoders for computing different CCs in parallel. As a result, all the decoding steps in the straight-forward architecture are fully serialized, lowering the resource utilization.

Fig. 4 depicts the proposed decoder core including five syndrome trackers (STs), one register-based syndrome buffer (SB), one KES module, and five 44-parallel CS modules. Note that each ST contains a pair of 44-parallel SC modules, i.e., main-SC and sub-SC modules in Fig. 4. Similar to the work in [5], in every CC decoding, all the syndromes are continuously tracked whenever errors are revealed by CS modules. If the CCx decoding is running at a certain CS module, as detailed in Fig. 4, the connected main-SC module collects the error patterns of CCx and accumulates the syndrome differences with the feedback path. After performing the whole CS operations, i.e., finishing the CCx decoding, the main-SC updates the total syndrome differences of CCx by accessing the SB once. On the other hand, as shown in Fig. 4, the sub-SC module computes the partial syndrome differences of each CC. Whenever an erroneous block is observed at CS operations, the sub-SC module directly updates the syndromes belonging to the related CC. Therefore, the sub-SC only accesses the SB while CS module is working. As the SB accesses are controlled by different SC modules in time-interleaved manners, each ST has one accessing port to SB, reducing the overall complexity.

Fig. 5 shows the utilization diagram in processing time for performing the internal CC decoding steps. Note that the straight-forward decoder suffers from serialized decoding processes, taking 3600 cycles per iteration as depicted in Fig. 5. Note that there are a lot of empty slots, which lowers the overall throughput as well as the energy efficiency. By replacing the SC in the straight-forward decoder to the ST of the proposed architecture, as shown in the figure, the decoding latency is shortened as ST operations can be executed in

parallel with the CS operation. Therefore, in this figure, we ignore the processing schedules of ST works. In addition, in our architecture, multiple CCs can be decoded at the same time by using the proposed ST-based architecture as the syndromes for the next CC are updated by reflecting the current error blocks, not waiting the completion of CS process. As a result, the proposed ST-based decoding takes 1180 cycles per iteration, reducing the processing latency by 67%.

Conventionally, each CS process always starts from the left-most data block of the corresponding CC in Fig. 1. As shown in Fig. 5, therefore, the wasted cycles for waiting the starting point of the current CC decoding are gradually increased, still lowering the hardware utilization. To minimize this unwanted waiting, we propose the reordered CS (rCS) as shown in the figure. In the rCS scheme, the critical block, which is shared by the next adjacent CC decoding, becomes the starting point of the current decoding. Therefore, the following CC decoding always starts just after tracking its syndromes, which takes only two fixed cycles, maximizing the hardware utilization. If the proposed ST and rCS schemes are applied, as a result, the processing cycles per iteration is reduced by 88% and 63% compared to the straight-forward decoding and the ST-only decoding, respectively.

IV. IMPLEMENATION RESULT

Based on the proposed optimization methods, the prototype 4KB SBC-BCH decoder is designed in 65nm CMOS process at the operating frequency of 300MHz. Fig. 6 shows the die-photo of the prototype decoder, which takes 375K equivalent gates while occupying $0.73mm^2$. As shown in Fig. 7(a), the decoding throughput is enhanced by applying rCS algorithm to shorten the long waiting latency. For the given RBER of 4.5×10^{-3}, the proposed fully-optimized architecture, i.e., ST+rCS in Fig. 7(a), achieves over 12Gb/s, which is 2.5 times faster than the ST-only decoding. Consequently, the energy-efficiency is gradually improved by applying the proposed optimization technique as depicted in Fig. 7(b). For the tested RBER of 4.5×10^{-3}, for example, the proposed SBC-BCH decoding process achieves an energy efficiency of 3.56pJ/b.

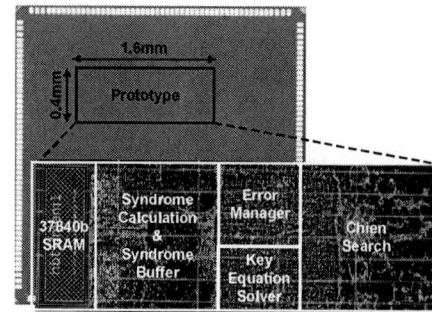

Fig. 6. The die-photo of the prototype SBC-BCH decoder.

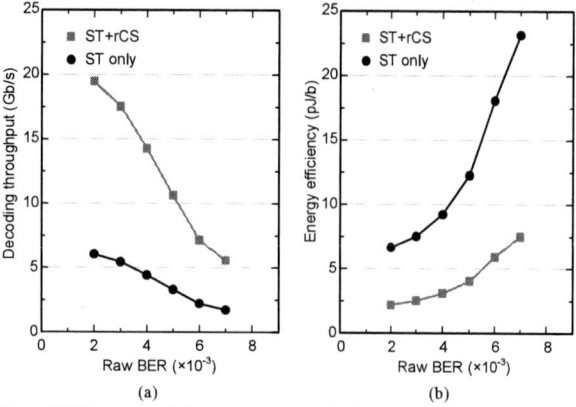

Fig. 7. RBER versus (a) the throughput and (b) the energy efficiency.

Table I summarizes implementation results of different types of ECC solutions targeting the storage applications. Note that the proposed 4KB SBC-BCH decoder definitely beats the previous ECC decoders in terms of energy efficiency. The work from [4] may present a similar level of efficiency, however, it cannot be accepted by the recent practical domains due to the poor error-correction performance. SD ECCs from [6]–[8] may be attractive in terms of error-correcting capability. However, SD ECCs could be the last option for only extending the lifetime of mobile storages due to their unacceptable energy consumption from the computing parts as well as the memory sensing operations. Considering the improved error-correcting performance with the lowest level of energy efficiency, as a result, the proposed SBC-BCH ECC can provide a long and even healthy lifetime of mobiles storages.

V. CONCLUSION

In this paper, we have presented a new HD ECC solution that can prolong the healthy status of mobile storages. The proposed SBC-BCH code contains the fully-connected CCs, enhancing the HD error-correcting performance. Hence, the strong but energy-consuming SD ECCs can be activated later, realizing the energy-efficient mobile system for a long time. In addition, several design techniques have been discussed to shorten the unnecessary cycles, leading to the energy-efficient decoder architecture. Implementation results show that the proposed decoder achieves a decoding throughput of 12.5Gb/s while consuming only 3.56pJ/b, truly attractive to the energy-starving mobile storages even at the healthy condition.

TABLE I. IMPLMENTATION RESULTS OF VARIOUS ECC DECODERS

	Proposed	[4]	[5]	[6]	[7]
ECC Type	SBC-BCH (Hard)	BCH (Hard)	BC-BCH (Hard)	LDPC (Soft)	LDPC (soft)
Code length (bits)	37840	8640	70528	18900	68254
Code rate	0.90	0.95	0.93	0.9	0.96
RBER for achieving 10^{-12} UBER[a]	4.2×10^{-3}	1.2×10^{-3}	2.6×10^{-3}	1.5×10^{-2}	1.6×10^{-2}
Technology (nm)	65	130	65	90	130
Voltage (V)	1.2	1.2	1.2	1.2	1.2
Operating speed (MHz)	300	200	250	200	N. A.
Gate count	375K	110K	335K	520K	N.A.
Chip area (mm^2)	0.73	0.85	0.65	2.56	63.1
Throughput[b] (Gb/s)	12.5	6.4	3.8	1.58	5.4
Energy efficiency[b] (pJ/b)	3.56[c]	4.92	7.5[c]	186	387
Normalized efficiency[b, d] (pJ/b)	3.56[c]	2.46	7.5[c]	134	134

[a] Extrapolated values from the UBER results in Fig. 2.
[b] Targeting the RBER of 4.5×10^{-3}
[c] Including on-chip memory-accessing energy
[d] Normalized efficiency = Energy Efficiency $\times$ (65nm/Technology) $\times$ (1.2/Voltage)^2

REFERENCES

[1] C. Kim *et al.*, "A 512Gb 3b/cell 64-stacked WL 3D V-NAND flash memory," in *Proc. IEEE Int. Solid-State Circuits Conf. (ISSCC)*, 2017, pp. 202–203.

[2] Y. Cai, Y. Luo, E. F. Haratsch, K. Mai, and O. Mutlu, "Data retention in MLC NAND flash memory: Characterization, optimization and recovery," in *Proc. IEEE Int. Symp. High Performance Comput. Archit. (HPCA)*, 2015, pp. 551–563.

[3] G. Dong, N. Xie, and T. Zhang, "Enabling NAND flash memory using soft-decision error correction codes at minimal read latency overheads," *IEEE Trans. Circuits Syst. I, Reg. Papers*, vol. 60, no. 9, pp. 2412–2421, Sep. 2013.

[4] Y. Lee, H. Yoo, I. Yoo, and I.-C. Park, "6.4 Gb/s multi-threaded BCH encoder and decoder for multi-channel SSD controllers," in *Proc. IEEE Int. Solid-State Circuits Conf. (ISSCC)*, 2012, pp. 426–427.

[5] Y. Lee, H. Yoo. J. Jung, J. Jo, and I.-C. Park, "A 2.74-pJ/bit, 17.7-Gb/s iterative concatenated-BCH decoder in 65-nm CMOS for NAND flash memory," *IEEE J. Solid-State Circuits*, vol. 48, no. 10, pp. 2531–2540, Oct. 2013.

[6] K.-C. Ho, C.-L. Chen, and H.-C. Chang, "A 520 k (18900, 17010) array dispersion LDPC decoder architectures for NAND flash memory," *IEEE Trans. Very Large Scale Integr. (VLSI) Syst.*, vol. 24, no. 4, pp. 1293–1304, Apr. 2016.

[7] J. Kim and W. Sung, "Rate-0.96 LDPC decoding VLSI for soft-decision error correction of NAND flash memory," *IEEE Trans. Very Large Scale Integr. (VLSI) Syst.*, vol. 22, no. 5, pp. 1004–1015, May 2014.

[8] Y. Lee, J. Jung, and I.-C. Park, "Energy-scalable 4KB LDPC decoding architecture for NAND-flash-based storage systems," *IEICE Trans. Electron.*, vol. E99-C, no. 2, pp. 293 – 301, Feb. 2016.

[9] D. Kim and J. Ha, "Quasi-primitive block-wise concatenated BCH codes with collaborative decoding for NAND flash memories," *IEEE Trans. Commun.*, vol. 63, no. 10, pp. 3482–3496, Oct. 2015.

[10] Y.-M. Lin, H.-T. Li, M.-H. Chung, and A.-Y. Wu, "Byte-reconfigurable LDPC codec design with application to high-performance ECC of NAND flash memory systems," *IEEE Trnas. Circuits Syst. I, Reg. Papers*, vol. 62, no. 7, pp. 1794–1804, June 2015.

S12-6 (2178)

A 130nm 1Mb HfO$_X$ Embedded RRAM Macro Using Self-Adaptive Peripheral Circuit System Techniques for 1.6X Work Temperature Range

Feng Zhang[1], Dongyu Fan[1,2], Yuan Duan[1], Jin Li[1], Cong Fang[1], Yun Li[1], Xiaowei Han[3], Lan Dai[2], Chengying Chen[1],
Jinshun Bi[1*], Ming Liu[1*], Meng-Fan Chang[4]
Email: bijinshun@ime.ac.cn   liuming@ime.ac.cn

[1]Key Laboratory of Microelectronics Devices & Integrated Technology, University of Chinese Academy of Science, Institute of Microelectronics Chinese Academy of Sciences, Beijing, 100029, P. R. China
[2]School of Electronic Information and Engineering, North China University of Technology, Beijing, 100144, P.R. China
[3]Xi'an UniIC Semiconductors Co., Ltd., Xi'an, Shaanxi 710075, P. R. China
[4]National Tsing Hua University, Hsinchu, Taiwan

Abstract—This paper designed a 1-Mb HfO$_X$-based embedded Resistive Random Access Memory (RRAM) device with a one-transistor-one-resistor (1T1R) structure, and systematically investigated its working temperature range. It noted that this embedded RRAM macro has a 1.6X working temperature range than previous design for some extreme environment. Using the peripheral-assisted technique, it can enable the error rate of the RRAM macro under 0.5% which can reduce the complexity of ECC function. Experimental results show that, the RRAM macro achieves a wider work temperature range (between -55℃ and 150℃), which improves the reliability of the entire embedded RRAM macro and has a high robustness as well.

Keywords—*RRAM macro; Reliability; Self-adaptive Write Circuit; Self-adaptive Sensing; Work Temperature Range*

I. INTRODUCTION

Embedded nonvolatile memory applications have been high-profile in recent years. RRAM, one of them, has captured much attention as an emerging candidate for the next generation of embedded nonvolatile memory because of its fast access speed, high density, good scalability, and Complementary Metal Oxide Semiconductor (CMOS) compatibility. A large number of oxide material structures have been researched. Despite the advantages of RRAM, it is a significant challenge to improve the reliability of RRAM applications. After multiple operations at extremely high or low operating temperatures, HRS/LRS will degenerate, and the operation voltages would change. Thus, it is obvious that the reliability issues of the RRAM macro must be solved first. Some of these embedded RRAM applications can only operate ambient temperature between -40℃ and 85℃ [1]. For the special applications, such as the aerospace environment, it is not applicable any more.

From an economical viewpoint to improve reliability, if a peripheral circuit can make the corresponding changes to adapt to the new situation, it can also improve the reliability [2]. Apart from the material aspects of this approach, there also has the literature [3] to improve the reliability of the RRAM macro using the circuit. However, this scheme achieves its goal at the expense of the complexity of the circuit, and there is no improvement from the read side.

Especially under the condition of the special working temperature, although the read current may change with the resistance of RRAM cell degrading [4], the circuit of the literature [3] can not make a corresponding adjustment.

This paper aims to design a complete 1-Mb HfO$_X$ embedded RRAM macro which is provided with a wide operation temperature ranges, which achieves 1.6X than previous design [1]. It meets the extreme environment's requirement. In particular, we propose an effective device and circuit techniques to improve the reliability by suppressing both write and read errors. The complete macro is fabricated using a 0.13-μm CMOS compatible process. Experimental results demonstrate that the macro can accomplish full writing and reading functions in a specified temperature range.

II. CHIP FABRICATION AND ARCHITECTURE

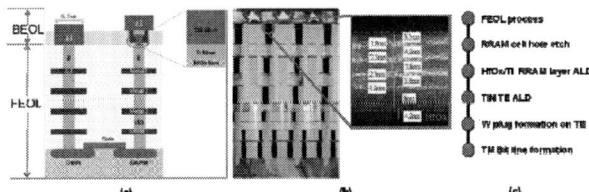

Fig. 1. (a) Schematic; (b) Cross section of 1T1R RRAM cell with structure of TiN top electrode(20nm thick)//Ti(10nm thick)/HfOx(5nm thick)//W bottom electrode; (c) Process flow based on 0.13um CMOS.

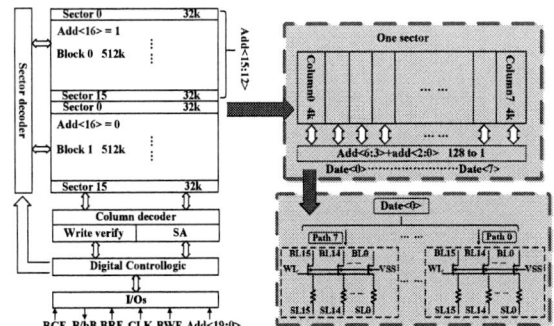

Fig. 2. Overall architecture of the 0.13μm 1-Mb RRAM macro.

978-1-5386-3179-9/17 $31.00 © 2017 IEEE

This RRAM macro is fabricated based on a 0.13-μm CMOS process. Fig.1 (a) and (b) show the schematic and cross section of the 1T1R RRAM cell. The storage structure has a TiN top electrode (20nm thick)//Ti (10nm thick)//HfOx (5nm thick)//W bottom electrode. Fig.1 (c) shows the process flow, which is fully compatible with the standard CMOS process.

Fig.2 shows the overall architecture of the 1-Mb RRAM, which is composed of a 1-Mb 1T1R array (excluding redundancy and ECC cells), control logic, analog system and I/O pins. The analog module supplies the wordline/sourceline/bitline voltage and reference current for SET/RESET, forming and read operations. The RRAM memory array is divided into two blocks. A block is composed of 16 sectors of eight columns. Each column has 4K cells (4K bytes). The word line of eight 4K cell modules is parallel, and each module has a 128-bit BLs and SLs.

III. PROPOSED RELIABILITY IMPROVEMENT TECHNIQUES

A. Material Characteristics

The variation of RRAM memory window is measured between LRS and HRS at 85℃, 100℃, and 150℃. As shown in Fig.3, the LRS and HRS values of HfOx-based RRAM maintain stable and exhibit sufficient reliability within 105 seconds. However, it is obvious that the memory window decrease while the operation temperature increases [5]. In the worst case, the memory window is less than 10X at 150℃. Consequently, if the macro still uses the conventional reading-writing circuits under high temperature, it will lead to a mismatch. This is one of the main reasons why traditional RRAM macro loses its reliability under high temperature.

In addition, the LRS will increase at low temperature (such as -55℃) due to the reducing of the activity energy of oxygen ions and vacancies. This will slow down the formation of conductive filaments and therefore disable the traditional RRAM macro.

The next involves designing a peripheral circuit system that improves the reliability of the chip. To reach this goal, this paper proposes some circuit methods.

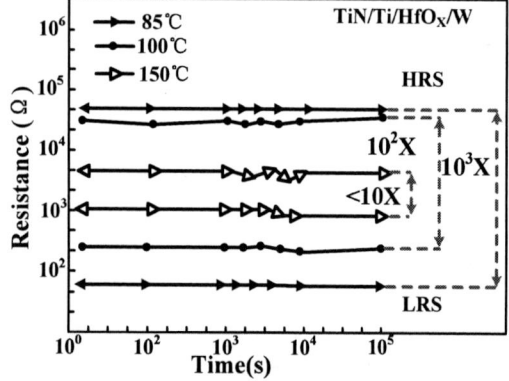

Fig. 3.   The measurement of variations of RRAM cell window.

B. Pulse Width and Amplitude with Adaptive Function Ramped Voltage Write

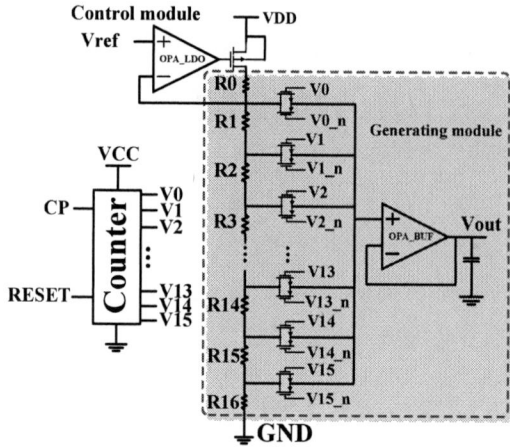

Fig. 4.   Circuit of proposed write technology and the V_{out} of SET operation.

Due to the variation of the write voltage in extreme temperature, such as 150℃ and -55℃, RRAM macro needs a write circuit with adaptive function to improving the reliability. Fig.4 shows the circuit for write operation (for both SET and RESET). To improvement structure of literature [6], the proposed write circuit is composed of three parts: a ramp-pulse-voltage generating module, a ramp pulse voltage output control module, and a counter module. This scheme can provide a variable SET/RESET voltage to accommodate different temperature conditions.

The internal write flow diagram is shown in Fig.5 (a). On the basis of the RRAM cell characteristics, the write operation is split into three steps: read and comparison, SET/RESET. In order to avoid over writing, the operation does not set the cell that is already in LRS or reset the cell that is already in HRS.

First, the data in the writing location is read out and compared with the writing data to determine if they are identical. If yes, the operation is complete. If not, the mismatch bit in the writing byte would be set or reset. The next SET operation is begun, and the read operation is executed after each set operation. If this succeed, the diagram indicates that the next operation should be started. If it failed, the operation increased the write pulse amplitude and width, and executed the SET operation again.

The SET operation continued until the RRAM cells are successfully within 16 cycles (the cycle number depended on the new test results could be changed) or not beyond the maximum cycle. If it ends because of the former, the diagram moves into the next operation, or the cell fails to be set and is considered a bad cell. The write operation finishes with a fail flag, STATE, which is set to high. After being set as successful, the RESET operation begins and is same as the SET operation. In the end, whether the cell is reset successfully or failed, the write operation is completed, and then the diagram executes a write operation on the next address. Fig. 5 (b) is the RESET operation, and (c) is the SET operation.

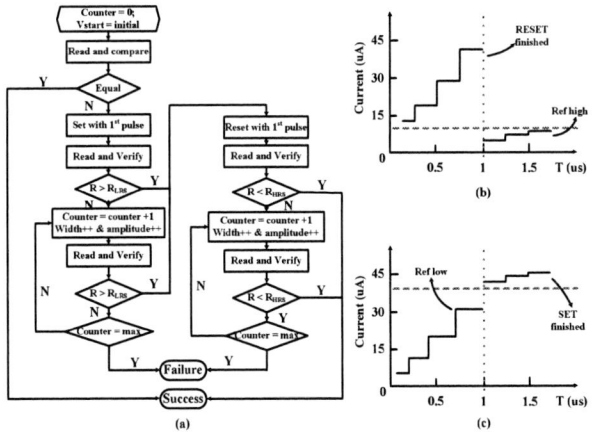

Fig. 5. The flow diagram for the self-adaptive write operations.

The write errors have been effectively suppressed by this write scheme for SET and RESET operations. As results, this RRAM macro can be self-adaptive and increased its working temperature range owing to the variation of operation voltages.

C. Self-Adaptive Sensing

To consult the experience of Ref [7], this paper proposes a self-adaptive current sense amplifier (CSA), which can fit different temperature conditions by changing the reference current. It can improve the reliability of the RRAM macro by solving the read errors. Current-mode sensing amplifiers are commonly used for RRAM with long bitlines and small cell read currents that can achieve a fast read speed with immunity to the noise.

The read circuit consists of a feedback module and a self-adaptive CSA. These are shown in Fig.6, the feedback module has a comparison circuit, which sends a feedback signal to the self-adaptive CSA when the read current of the RRAM cell changes during resistance degradation. The self-adaptive CSA changes the reference current according to the feedback signal. In a special temperature environment (such as higher than 150℃ or below -55℃), the read current of the RRAM macro will be changed after repetitive operation. The reference current of the conventional CSA cannot be changed immediately, which means only has one reference current, and the output data will be incorrect. Thus, conventional approach does not improve the reliability of the RRAM macros.

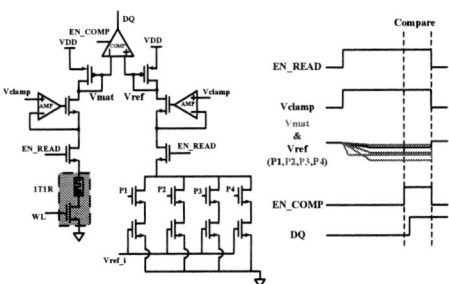

Fig. 6. Sensing circuit of the self-adaptive CSA circuit.

However, the CSA of this paper adds a self-adaptive module, which can apply different reference currents under different temperature conditions. The premise is that we have measured the internal macro read current under diverse conditions. As Fig.6 shows, using different sizes of transistors P1, P2, P3, and P4 adjusts the current magnification by opening the corresponding transistor for different environment. This can ensure the accuracy of results without affecting the read circuit delay. This scheme increases the accuracy of the CSA, thereby the reliability of the RRAM macro is increased owing to expanding the detection range of the CSA.

When the RRAM macro operates at 150℃, the read current will increase due to the degradation of HRS. That will reduce the current window which determines the high and low impedance states. The conventional CSA uses the fixed reference current, which makes the read current erroneous when the degradation starts. The self-adaptive sensing has dynamic reference current and can maintain its value at half of the current window even when the window has shrunk down. This methodology ensures RRAM macro's reliability and the correctness of the read current. The same argument for the subzero condition, by this scheme, the working temperature range is further improved.

IV. EXPERIMENTAL RESULTS

For testing, we used an Aglient N6705 Power Supply to drive the test macro and FPGA for inputting the signal. We adopted a Temptronic TP04310 as a high-low temperature tester as Fig.7. For different temperature testing, we used different chips during testing to make the chips out of accidental damage while temperature testing, because the extremely environment may destroy the function of the chip.

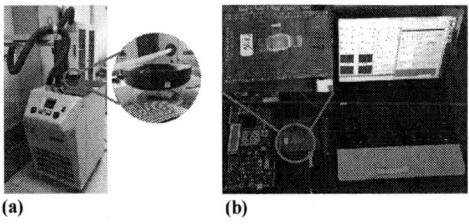

Fig. 7. (a) Test apparatus; (b) test platform.

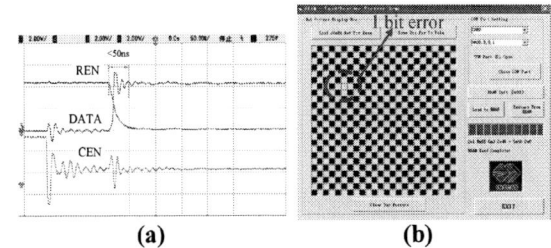

Fig. 8. The measurement of RRAM macro function test: (a) the read time; (b) the RRAM write and read program with temperature variations.

In the 1Mb RRAM measurements, we write the number "55" and "AA" to it in the continuous time as is shown in Fig. 8. The results shows that the reading time is less than 50ns, and the C++ program of "black and white dots" displayed the error rate of the writing and reading operations. In the experiment, we tested the cells of the RRAM from -55℃ and 150℃. Consequently, the error rate of this RRAM macro is under 0.5% and it is easy to utilize ECC encode to cover the error bits and it will not increase the redundancy much.

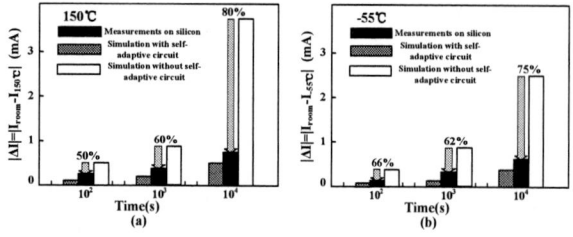

Fig. 9. The measurement and simulation of the reduction in RRAM macro current variation across temperature (a) at 150℃, the drift rate of △I reduces 80%; (b) at -55℃, △I reduces 75%.

Based on the previous test data shown in [4], the activity energy increases with heating, which makes the HRS degeneration because more oxygen vacancies are generated in HfO_X. On the other hand, at low temperature, most of oxygen ions and vacancies are frozen which reduces the activity energy lead to decreasing the readout current. Fig.9 (a) shows the RRAM macro current variations of the measured and simulation results with and without self-adaptive circuits at different degrees Celsius. As expected, the simulation without self-adaptive circuits read failure occurs after 10^3 seconds. However, the macro that used the circuit can maintain its reliability after 10^3 seconds which is consistent with its simulation results at 150℃. Fig.9 (b) shows the low temperature experimental results. Therefore, the results show that the current variation is much without the self-adaptive circuits and the circuit systems presented in this paper can change the reference current to suppress failure trends.

Fig. 10 shows the chip micro-graph and the histogram of temperature windows of different RRAM macro designs. From Table I illustrates that the feature comparison between the proposed RRAM macro and other reference designs [3], [8]. The compared results indicate that the proposed chip achieves a wider work temperature range with HfO_X technology.

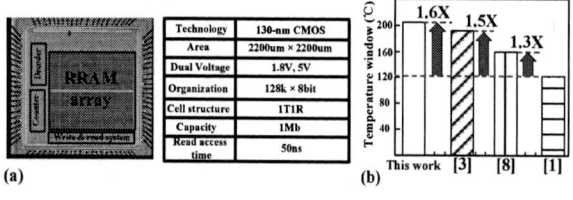

Fig. 10. (a) Chip micro-graph and feature summary of 1-Mb RRAM and (b) magnification of this work.

TABLE I.   COMPARISON WITH REFERENCE RRAM DESIGNS

	This work	[1]	[3]	[8]
Material	HfO_X	TaO_X	AlO_X /WO_X	Cu_XSi_YO
Process	130nm	180nm	180nm	40nm
Cell structure	1T1R	/	1T1R	1T1R
Chip capacity	1Mb	64Kb	128Kb	1Mb
Work temperature	-55℃–150℃	-40℃–85℃	-40℃–150℃	-40℃–125℃

V. CONCLUSIONS

This paper presents a 1-Mb HfO_X embedded RRAM in a 0.13-μm CMOS logic compatible process developed to improve the reliability of RRAM macros. The RRAM macro exploits self-adaptive circuit system techniques to 1.6X the working temperature range than previous work, thereby improving the reliability of the complete RRAM macro. This makes embedded RRAM macro possible in the applications of the extreme space, such as aerospace sectors. Experimental results show that the designed RRAM macro has a wide woke temperature range between -55℃ and 150℃.

Acknowledgment

This work was supported by grants from National Natural Science Foundation of China (NSFC), no. 61474134, and no. 61634008. The authors would like to thank Zhigang Ji from Liverpool John Moores University for the help he provided in paper reviewing and revision.

References

[1] Fujitsu Semiconductor. (2016). "Memory ReRAM 4M (512K × 8) Bit SPI MB85AS4MT," Japan.

[2] Xue X. Y., *et al*. "A 0.13μm 8Mb logic based Cu_XSi_YO resistive memory with self-adaptive yield enhancement and operation power reduction." In VLSI Circuits IEEE, Jun. 2012, pp. 42-43.

[3] Y. L. Song *et al.*, "Reliability significant improvement of resistive switching memory by dynamic self-adaptive write method," in VLSI Tech., Jun. 2013, pp. 102–103.

[4] Walczyk Christian, *et al.* "Impact of Temperature on the Resistive Switching Behavior of Embedded HfO2-Based RRAM Devices." IEEE Transactions on Electron Devices, vol. 58 no. 9 Jul. 2011, pp.3124-3131.

[5] Liu S, Wang W, Li Q J, *et al.* "Highly improved resistive switching performances of the self-doped Pt/HfO2: Cu/Cu devices by atomic layer deposition" Science China Physics Mechanics & Astronomy, vol.59 no.12, 2016: 127311.

[6] Meng-fan Chang, Juijen Wu, Yenchen Liu, *et al*. "19.4 Enbedded 1Mb ReRAM in 28nm CMOS with 0.27-to-1V Read Using Swing-Sample-and-Couple Sense Amplifier and Self-Boost-Write-Termination Scheme." ISSCC, IEEE, Feb. 2014, pp.332-333.

[7] Meng-fan Chang, Chewei Wu, Chiacheng Kuo, *et al*. "A 0.5V 4Mb Logic-Process Compatible Embedded Restive RAM (ReRAM) in 65nm CMOS Using Low-Voltage Current-Mode Sensing Scheme with 45ns Random Read Time." ISSCC, IEEE, Feb. 2012, pp.434-436.

[8] Wang Yanliang, *et al.* "Logic-Based Mega-Bit CuxSiyO emRRAM with Excellent Scalability Down to 22nm Node for post-emFLASH SOC Era."Memory Workshop IEEE, May. 2011, pp. 1-2.

S13-1 (2076)

A Reconfigurable Dual-Band WiFi/BT Combo Transceiver with Integrated 2G/BT SP3T, LNA/PA Achieving Concurrent Receiving and Wide Dynamic Range Transmitting in 40nm CMOS

Meng-Hsiung Hung, Yi-Shing Shih, Chin-Fu Li, Wei-Kai Hong, Ming-Yeh Hsu,
Chih-Hao Chen, Yu-Lun Chen, Chun-Wei Lin, Yuan-Hung Chung.

MediaTek Inc.
Hsinchu, Taiwan

Abstract — This paper describes a WiFi/BT combo SoC with integrated dual-band PAs, LNAs and T/R switches, supporting concurrent receiving for improved throughput by 30% in dense networking environment of 2.4GHz ISM band and wide-range transmitting capability (>20dB) while keeping good output power accuracy and PA power efficiency. The measured 2.4GHz/5GHz WiFi 54Mbps RX sensitivity is -78.2/-78.1dBm and Pout is 21.3/19.5dBm. BT GFSK RX sensitivity is -95.3dBm and Pout is 13dBm. The chip occupies 20mm² in 40nm 1P6M CMOS technology while RF area occupies 3.5mm².

Keywords — WiFi, BT, concurrent receiving, LNA, PA, SP3T, RF SoC

I. INTRODUCTION

Over the past few years, an explosion in smart phones and other mobile devices helps people get connected to the internet easily with wireless connectivity. Recently, the Internet of Things (IoT) is changing the way we live, work, and interact with the physical world even more than the traditional mobile devices. Moreover, all physical things embedded with smart electronic devices get connected via the internet to communicate and interact with external environment through a variety of applications. WiFi and Bluetooth (BT) have become the most ubiquitous wireless technologies for long-range and short-range communication, respectively. Therefore, IoT devices can easily access cloud services via the internet with embedded WiFi module for high throughput and simultaneously interact with each other with built-in BT connectivity. This brings new challenges to the concurrent operation of WiFi and BT, both operating in the crowded 2.4GHz ISM band. Conventionally, WiFi and BT devices share a single antenna, operating at Time Division Duplex (TDD) mode with time scheduling algorithm to avoid packet collisions. In such operation, when one of them takes up the medium time, the other suffers from acute starvation in accessing wireless medium, leading to the delay or loss in getting its own packets. In a WiFi network, this results in either very low WiFi throughput or eventual WiFi link loss for the client device. This paper presents a WiFi/BT concurrent receiving topology to

avoid these problems. It also achieves a wider operating range when receiving unbalanced WiFi/BT signal power.

A WiFi/BT combo SoC is designed with integrated dual-band PAs, LNAs and T/R switches shown in the Fig. 1, where the WiFi 2.4GHz/5GHz band transceivers use the similar T/RX architecture. 2.4GHz TX path is shared with BT TX while 2.4GHz LNA and BT LNA are put in parallel for concurrent receiving requirement. 2.4GHz SP3T architecture will be described in next section.

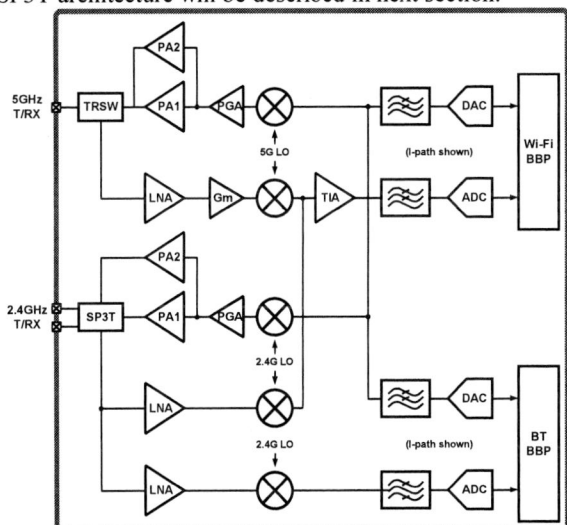

Fig. 1 Block diagram of WiFi/BT transceiver

II. INTEGRATED VIRTUAL SP3T SWITCH

To replace the traditional external Single-Pole-Two/Three-Throw (SP2T/SP3T) switch to allow WiFi and BT RF ports to share an ISM antenna, a virtual SP3T switch topology is proposed to reduce form factor. As depicted in Fig. 2, different impedance and signal attenuation requirement can be achieved with high/low power transmit and receive configurations. PA1 and PA2 are designed for high and low power operation respectively. When in high-power TX mode, switch S1 is connected to ground, then capacitor C1 and C3 become part of PA1 matching network, meanwhile C1/C3

978-1-5386-3179-9/17 $31.00 © 2017 IEEE   177

capacitance ratio is designed to provide enough attenuation to protect LNA from large PA1 voltage swing and provide isolation for PA2 to prevent coupling feedback induced oscillation. When in low-power TX mode, switch S1 is opened and matching network C1, L1 is reconfigured as a high impedance loading to PA2. Here PA2 is for low power operation and also plays a role as BT PA. While the load-line change concept for better power efficiency is similar to [1], the inductor is now chosen to reduce insertion loss since receiver is included in this SP3T design. Since PA2 loading impedance is higher, there is also a capacitor divider C2 and C4 to protect LNAs from PA2 voltage swing. In WiFi/BT RX mode, switches S1 and S2 are opened, input matching is composed of series capacitor C1 and inductor L1 to step-up the impedance to 220 ohms to provide higher matching gain for lower noise figure.

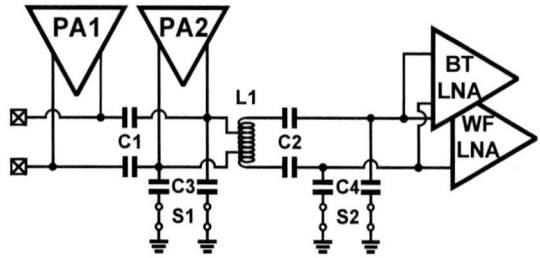

Fig. 2 Virtual SP3T switch topology

III. WiFi/BT Concurrent Receiving

Operating in the crowded 2.4GHz ISM band with a shared antenna, TDD is widely used to support multiple users simultaneously for devices integrating both WiFi and BT on a single IC. To avoid packet loss at WiFi/BT transmitting/receiving collision, a packet traffic arbiter (PTA) is required with time scheduling algorithm in SoC for the access control of WiFi/BT systems to the shared antenna. As shown in Fig 3(a), to monitor on-the-air BT traffic while accessing WiFi, the device periodically disables WiFi transceiver and activates BT receiver to do inquiry scan every $T_{Inquiry}$. Once inquiry is received and checked with procedure, BT starts to do page scan, build synchronized connection, and access traffic data at t_0. Consequently, WiFi suffers traffic loss and throughput drop during BT operation. If WiFi is granted by PTA with priority higher than BT and BT inquiry scan is skipped at t_1, BT traffic will be lost. Therefore, WiFi/BT traffic access time is slotted with time-scheduled operation, yielding the trade-off between WiFi throughput and BT link quality. With the proposed WiFi/BT concurrent receiving topology, the device can flexibly activates BT receiver for inquiry scan with WiFi traffic downlink in the meantime, yielding higher WiFi throughput, as shown in Fig. 3(b).

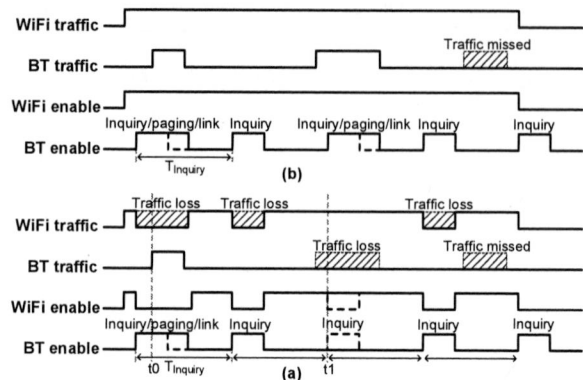

Fig. 3 Timing diagrams of WiFi/BT receiver traffic (a) time-scheduled, (b) concurrent receiving

Fig. 4 shows the concurrent receiving WiFi/BT RX LNA architecture. Immediately after the SP3T switch, two LNAs are designed in parallel for dedicated signal receiving in WiFi and BT receivers. Gain of both RX LNA can be controlled and optimized independently for better WiFi and BT performances. Compared to the shared LNA case in [2], independent WiFi/BT gain control can achieve better WiFi/BT co-existence performance because the dynamic receiving range can be wider without mutual constraint. However, due to independent gain control, impedance variation from parasitic capacitance of one LNA may be affected by the operation of the other LNA. For example, when WiFi is receiving packet and BT is under automatic gain control adjustment, WiFi RX chain may have gain and phase changes. The received WiFi signal quality will be degraded by the instant change of gain and phase. To meet the PER<0.1% requirement, the gain and phase variation must be kept below 0.6dB and 6° respectively. Similarly, this problem also appears during BT reception mode. From circuit analysis, the gain and phase change are mainly contributed from (1) parasitic capacitors variation at LNA input during LNA on/off operation and (2) the magnetic mutual coupling between the two inductor loads at BT LNA and WiFi LNA outputs. To mitigate the parasitic variation during non-synchronous WiFi and BT receiving operation, the shared impedance transformation network of the SP3T should be properly designed for the two LNAs to minimize the input loading effect between each other. The WiFi LNA is a current-reuse topology by M_{1P} and M_{1N} to achieve the better g_m and NF. The cascode M_{2N} routes the signal current to output inductor load for the higher output impedance and wider voltage swing range to achieve better linearity. The BT LNA shares the same topology with reduced size and both WiFi and BT gain adjustments are realized by device slicing. In this design, the load inductors are in quadrupole layout and placed orthogonally as "T" shape. By aligning the center of each inductor to each other, magnetic flux from another inductor can be cancelled, eliminating the coupling effect. Both receiver down-conversions are using current-mode passive mixers to convert the RF signal to IF frequency into the next filter stages to tolerate possible strong signal blockers by each other during concurrent receiving. The WiFi adopts zero-IF architecture with 3rd order low pass filter. The BT is low-IF architecture with 2nd order complex band pass filter.

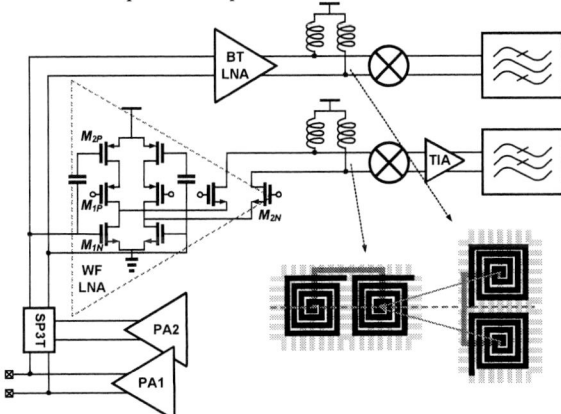

Fig. 4 Concurrent receiving WiFi/BT receiver architecture

IV. WIDE DYNAMIC RANGE TRANSMITTER

For an IoT device to operate over high/low output power (for long/short range), both load-line adjustment technique for maintaining PA efficiency and a wide power-control range over 20dB are required. The load-line adjustment method has been introduced and implemented in the SP3T architecture. For the TX power control range, traditional method can only support 10 to 15dB range [3] limited by the power detector SNR requirement. In order to have accurate power delivery for the two PAs, power detector and related control loop are designed to be configurable to cover wide output power range operation. During PA mode switching, power detector gain and tracking target are adjusted accordingly to compensate gain variation for optimum SNR and correct the target offset for accurate power level. The phase variation induced by load-line change of PA mode is calibrated by the phase compensated power detector in advance.

In this work, except for IF and LO circuits, a wide dynamic range PA is adopted to achieve a shared WiFi/BT transmitter. Since concurrent receiving LNA requires two LOs for WiFi/BT separately, an LOMUX at IQ-modulator is used to select from the two LOs. The shared WiFi/BT transmitter architecture can be extended to the upcoming BT5.0 long distance application.

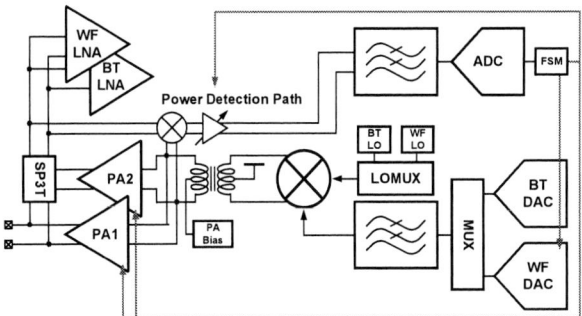

Fig. 5 Shared WiFi/BT RF transmitter architecture

V. MEASUREMENT RESULTS

Fig. 6(a) demonstrates the WiFi RX packet-error-rate (PER) under concurrent receiving, where the received BT power is ranging from -40dBm to -75dBm. WiFi channel is set at 2472MHz while BT channel is at 2402MHz. HT20 MCS7 sensitivity level is measured as -71.5dBm and -70.5dBm when BT input power up to -50dBm and -40dBm, respectively. At these testing conditions, WiFi LNA is set to high gain, while BT LNA is set to middle gain. Because the WiFi and BT gain mode can be set independently, the relative incoming receiving power of WiFi and BT can be larger to improve user experience. The performance degradation of WiFi sensitivity at concurrent receiving condition is less than 1dB. This is much smaller when compared to shared LNA case, where 10dB degradation can be expected under BT -40dBm blocker in WiFi sensitivity test.

Fig. 6(b) shows the WiFi throughput under concurrent receiving. With concurrent receiving feature, up to 30% WiFi throughput improvement (from 37.8Mbs to 49.3Mbs) can be achieved. This result is close to the pure WiFi performance under BT scan condition.

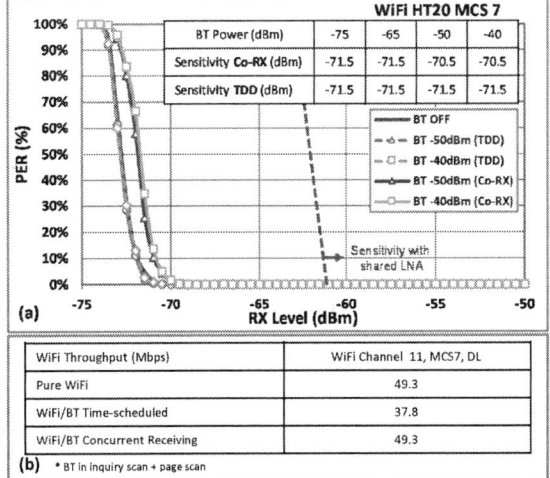

Fig. 6(a) WiFi RX PER/Sensitivity, (b) WiFi Throughput under concurrent receiving

Fig. 7 shows the measured transmit EVM and PA current vs. output power. The power control range is from 3 to 25dBm with accuracy of +/-0.6dB over temperature and channel. For 11n MCS7 compliant EVM of -28dB, the PA can deliver an output power of 21dBm. For short/middle range transmit power, 72%/33% PA current can be reduced with configurable load-line change PA.

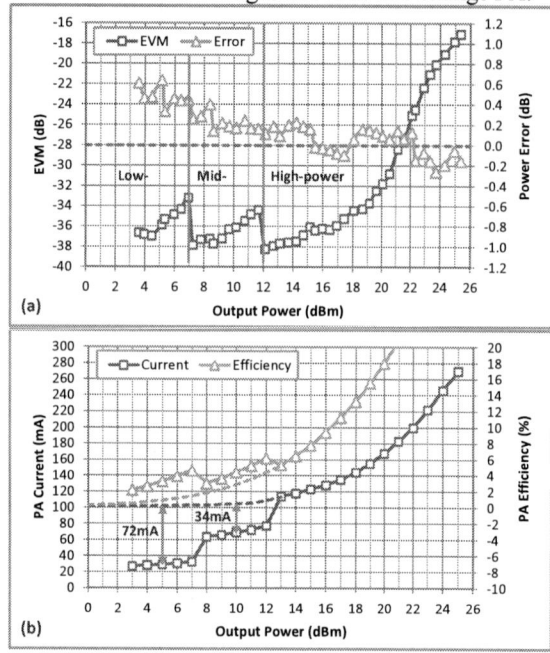

(a)

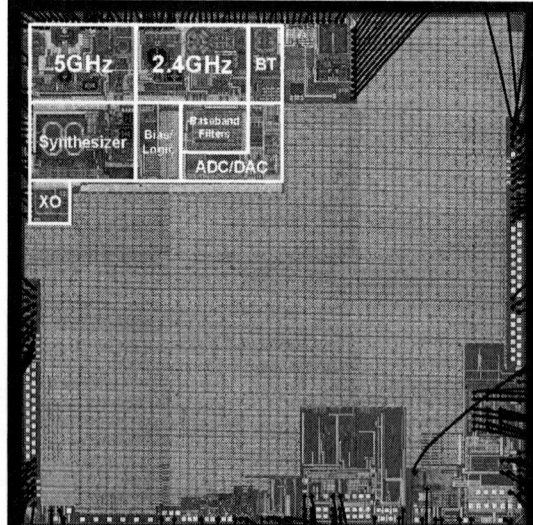

(b)

Fig. 7(a) TX EVM and power error, (b) PA current/efficiency curve over operation range

This work is implemented in 40nm CMOS technology and its die micrograph is shown in Fig. 8. The dual-band WiFi and BT RF transceiver occupies 3.5mm². The key WiFi/BT performance is summarized in Table I.

Fig. 8 Die photo

VI. Conclusion

This paper presents a 40nm WiFi/BT connectivity SoC with integrated virtual SP3T switch and shared WiFi/BT transceiver architecture fulfills the IoT application requirement with small form factor. This work demonstrates the receiver coexistence performance under concurrent receiving operation. It also presents the configurable transmitter with wider output power range.

TABLE I

PERFORMANCE SUMMARY

		This Work	ISSCC 2016 [4]	RFIC 2014 [3]
Support standards	WLAN	1x1, 11abgn/ac	2x2, 11abgn	2x2, 11abgn/ac
	BT	BT 4.2/BLE	BT 4.2/BLE	BT
Integration	2.4GHz	SP3T	SP3T	TRSW
	5GHz	TRSW	TRSW	TRSW
TX Psat (dBm)	2.4GHz PA1	28	27.8	27
	2.4GHz PA2	18.5		
	5GHz PA1	26	27	26
	5GHz PA2	16.8		
TX Pout (dBm)*	BT	13	10.2	N/A
	2.4GHz, LG54M	21.3	21.3	20.8
	5GHz, LG54M	19.5	20.5	18.4
	5GHz, 11ac VHT80MCS9	17.5	N/A	17.5
Dynamic Range(dB)	Power detector	20	N/A	10
RX Chip-in Sensitivity (dB)	BT	-95.3 (NF 5.7)	N/A (NF 3.2)	N/A
	2.4GHz, LG54M	-78.2 (NF 4.9)	N/A (NF 3)	-77.5
	5GHz, LG54M	-78.1 (NF 5.0)	N/A (NF 4)	-77
	5GHz, 11ac VHT80MCS9	-63.5	N/A	-62.5
Technology		40nm	28nm	55nm
WiFi/BT RF Die Area (mm²)		3.5 (1x1)	7.2 (2x2)	7.7 (2x2)

* TX Pout@EVM=-28dB/-32dB for OFDM54M/VHT80 MCS9, respectively

ACKNOWLEDGEMENT

The authors would like to thank Mediatek RF/DE, SE, and WCN/CD, CSD, SA, SE teams for the great support.

REFERENCES

[1] Y.-H. Chung, et al., "Dual-Band Integrated WiFi PAs with Load-Line Adjustment and Phase Compensated Power Detector", *IEEE RFIC Symposium*, pp. 223-226, June, 2015.

[2] C. Lee, et al., "A MultiStandard, Multiband SoC with Integrated BT, FM, WLAN Radios and Integrated Power Amplifier", *ISSCC Dig. Tech. Papers*, pp. 454-455, Feb. 2010.

[3] T.-M. Chen, et al., "A 2x2 MIMO 802.11 abgn/ac WLAN SoC with integrated T/R switch and on-chip PA delivering VHT80 256QAM 17.5dBm in 55nm CMOS", *IEEE RFIC Symposium*, pp. 225-228, June, 2014.

[4] R. Winoto, et al., "A 2x2 WLAN and Bluetooth Combo SoC in 28nm CMOS with On-Chip WLAN Digital Power Amplifier, Integrated 2G/BT SP3T Switch and BT Pulling Cancelation", *ISSCC Dig. Tech. Papers*, pp. 170-171, Feb. 2016.

S13-2 (2195)

A High-Speed DDFS MMIC with Frequency, Phase and Amplitude Modulations in 65nm CMOS

Abdel Martinez Alonso, Masaya Miyahara, and Akira Matsuzawa

Department of Electrical and Electronic Engineering, Tokyo Institute of Technology

2-12-1-S3-27, Ookayama, Meguro-ku, Tokyo 152-8552, Japan

Email: abdel@ssc.pe.titech.ac.jp

Abstract—**This paper describes a digital-mapping DDFS with a frequency tuning and amplitude resolutions of 24-bits and 10-bits respectively. This Si-CMOS-MMIC is the first solution supporting a sampling rate of 7GS/s and frequency, phase and amplitude modulations in the digital domain. It includes a 14-bits pipelined ripple-carry adder and a 10-bits high-speed multiplier for phase and amplitude modulations respectively. The worst case wideband/narrowband SFDR is 32dBc/42dBc. This system consumes 85.9mW/(GS/s) from a 1.2V power supply when the amplitude/phase modulations are enabled, resulting in a FoM of $469.6GS/s \cdot 2^{(SFDR/6)}$/W. A proof-of-concept chip with an active area of 0.23mm^2 was characterized in LQFP packages.**

Keywords— High-speed DDFS, CMOS, FM, PM, AM.

I. INTRODUCTION

Three main groups can be highlighted among the most widespread direct digital frequency synthesizer (DDFS) applications (state of the art representation in Table I). The first one is characterized by low-speed and low-power devices with a high spurious free dynamic range (SFDR). These features are increasingly demanded by portable and handled applications such as baseband signal processing in Bluetooth transceivers. In the second category, the most commonly used architectures are medium resolution (9 to 12 bits) digital-mapping (DM-DDFS) and nonlinear DAC based DDFSs (NLD-DDFS). Fine frequency tuning resolution and high integration density can be achieved when using these topologies. These advantages are fully exploited in applications like acousto-optic radio frequency drivers and scientific instrumentation. However, with clock frequencies up to about 2GHz, these synthesizers can barely cover the very-high frequency (VHF) and ultra-high frequency (UHF) bands. The high-speed solutions have been exclusively dominated by designs implemented in compound semiconductor technologies [1-4]. Low-resolution (5 to 8 bits) analog-mapping based DDFS (AM-DDFS) [3-4] or NLD-DDFS [1-2] architectures are commonly employed in this last group. In both cases, the phase to amplitude conversion is implemented in the analog domain. Consequently, amplitude modulation (AM) can't be performed in the digital domain when using these configurations. These devices can be employed in applications requiring high output frequency, fast settling time and only frequency and/or phase modulations (FM, PM) capabilities such as radar systems operating in the S, C and X bands [1]. The applicability of these solutions is also limited by the higher fabrication costs and more difficult access to compound semiconductor foundry services. As a

consequence, these approaches has been scarcely used in other areas. This work introduces the first DDFS operating at 7GS/s while featuring frequency, phase and amplitude modulation capabilities in the digital domain. With a maximum Nyquist-range output frequency of 3.5GHz, this solution not only can be employed in frequency exploration but also in architectures demanding combined modulation schemes. These can include polar modulators for broadcasting systems and digitally-intensive software-defined radio (SDR) front-ends operating from the high frequency (HF) band up to the S-band.

II. SYSTEM ARCHITECTURE

This design features an extended version of the 6.8GS/s DM-DDFS core proposed by these authors in [5]. Besides including a 24-bits pipelined phase accumulator (PA) and a sum of product (SoP) terms based phase to amplitude converter (PAC), it also features a high-speed 14-bits adder and a 10-bits multiplier employed to perform the phase and amplitude modulation respectively. The complementary dual-phase latch-based sequencing method described in [5] was also applied in order to speed up the performance of all the digital structures. A 10-bits two-times interleaved resistive digital to analog converter (RDAC) with 1.2V$_{pp\text{-diff}}$ output swing (600mV$_{pp\text{-diff}}$ with a 50Ω load) was integrated as well. The system also comprises a serial peripheral interface (SPI) and a 15-bits CMOS parallel interface (PI[14:0]) as the input ports for the frequency control word (FCW), the phase control word (PCW) and the amplitude control word (ACW). A register bank composed by 16 reconfigurable FCW profiles was also embedded in the system. The active FCW can be selected among 16 pre-stored values by means of a 4-bits CMOS control interface (CI[3:0]). This structure allows switching the output signal frequency without incurring in any extra delay due to the loading sequence of the SPI interface. The system architecture and its component blocks are represented in Fig. 1 a). The DDFS output signal is described by (1). The variables A, θ and F_{out} are determined according to (2)-(4). The output signal can be totally modulated in the digital domain by dynamically adjusting the FCW, PCW and ACW values.

A. High-speed amplitude-multiplier and phase-adder

In this work, a 10-bits complementary dual phase latch-based parallel amplitude-multiplier has been introduced. In order to equal the maximum throughput of the DDFS channel in [5], the structure is divided into mirrored layers having an

978-1-5386-3179-9/17 $31.00 © 2017 IEEE

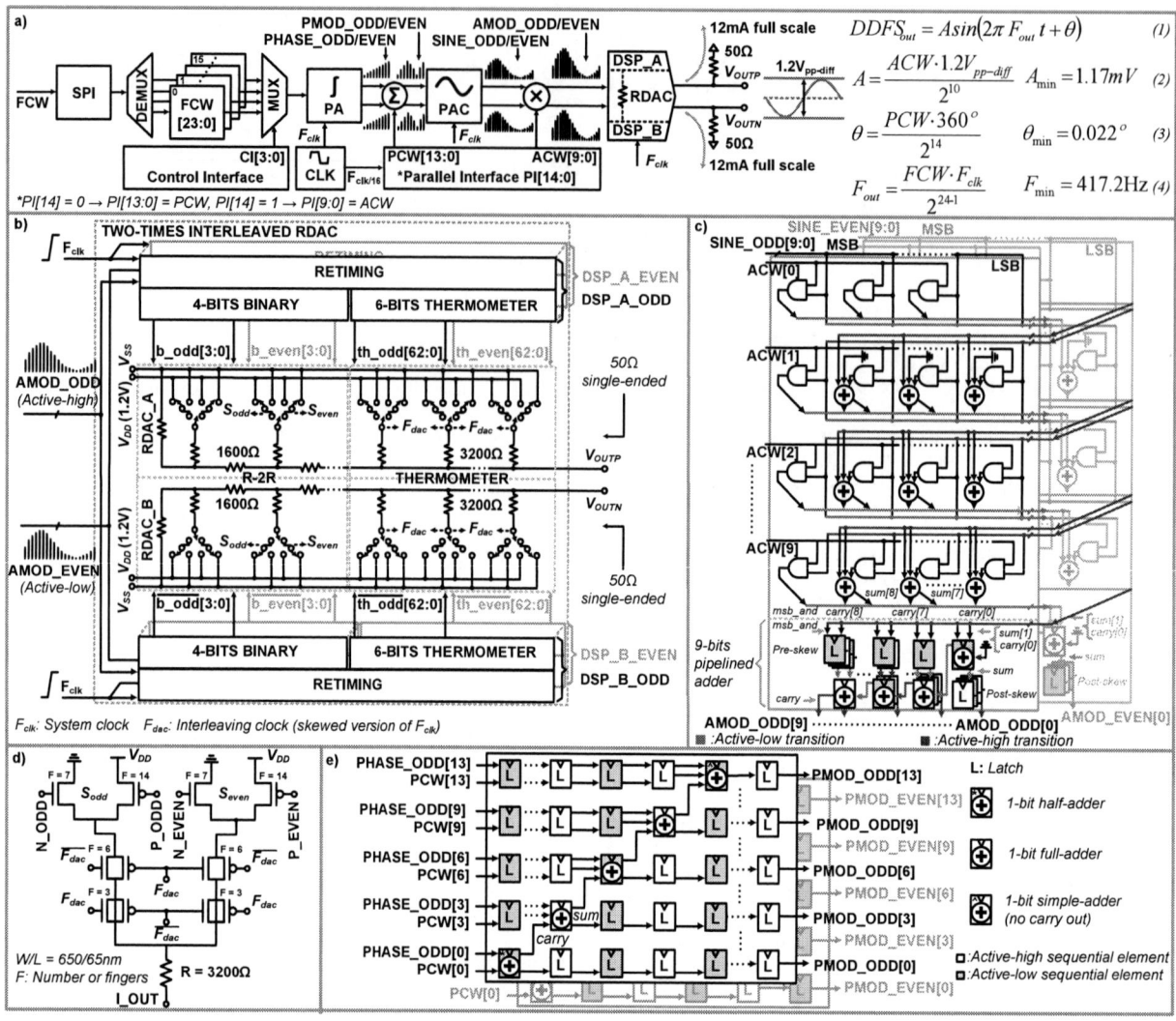

Fig. 1. Conceptual system architecture and component blocks.

180° phase offset among equivalent data transitions (Fig. 1 c)). The operands are the ACW (10 bits) and the SINE_ODD/EVEN (10-bits) outputs from the PAC. This feedforward architecture only requires static CMOS 1-bit full-adders and AND gates to perform the multiplication operations. Pre-skewing and post-skewing latches are also employed in order to pipeline the logic trough the multiplication channel at the required sampling rate (SR). It should be noticed that the intermediate arithmetic calculation results (carry and sum terms) are interleaved among the complementary layers. Hence, a new operation is executed during both high and low levels of the clock (different colored arrows in Fig. 1 c)). The final stage of this multiplier unit is implemented as a 9-bits pipelined ripple-carry adder. This system exhibits a 9½ clock periods latency resulting from 19 interleaved operations during different levels of the clock. The phase modulation can be implemented in the digital domain by

inserting an adder before the PAC block (Fig. 1 a)). This arithmetic unit accepts the PCW and PHASE_ODD/EVEN signals (14-bits) from the PA and generates the corresponding phase shifted digital ramps (PMOD_ODD/EVEN). The 14-bits adder in Fig. 1 e) consists of 1-bit adder cells embedded in an array of complementary dual phase latch-based pre-skewing and post-skewing pipeline stages. The required latency to generate a continuous sequence of phase-shifted values is equal to 7½ clock periods (15 interleaving operations).

B. Two-times interleaved RDAC

This converter comprises two digital signal processing sections (DSP_A/DSP_B) also divided into complementary mirrored layers (ODD/EVEN). In this way, it can be directly driven by the AMOD_ODD/EVEN outputs from the DM-DDFS core. Each DSP engine includes an array of RETIMING flip-flops before segmenting the data into a 4-bits LSB binary section and a 6-bits MSB unary section (Fig. 1 b)). A code-dependent switching stage (S_{odd}/S_{even}) connects the differential resistive array (RDAC_A-RDAC_B) either to V_{DD} (1.2V) or V_{SS}. The binary section is implemented as a voltage-mode R-2R ladder network. The N_ODD/P_ODD/N_EVEN/P_EVEN signals connected to the pass-gate (PG) based switches in Fig. 1 d) are driven by the thermometer encoder and the binary section outputs. This RDAC cell also includes a two-times interleaved switching stage controlled by the F_{dac} clock and implemented by using transmission-gates (TG). In this way, the distortion resulting from the code-dependent glitches can be reduced [6]. A pair of dummy TGs provides additional charge injection and clock feedthrough suppression. The R-2R and THERMOMETER sections were implemented by employing unsilicided P^+ polysilicon resistors with a $3.3 \times 12.43 \mu m^2$ area. The output currents are converted to a differential voltage developed across two off-chip 50Ω resistors.

III. EXPERIMENTAL RESULTS

A proof-of-concept chip was fabricated in a 1P9M 65nm CMOS process (Fig. 2). All the digital logic was designed by using standard-cell technology and a fully synthesizable flow. The total die area is $0.98 \times 1.46 mm^2$. The prototypes were encapsulated in a low-profile quad flat package (LQFP) and assembled in an evaluation board made of glass-reinforced epoxy laminate substrate (FR-4). The reference clock signal was driven by an Agilent E8257D signal generator. An MS2830A signal analyzer from Anritsu and an Agilent DSO90804A high performance oscilloscope were also employed. A KCU105 FPGA board from Xilinx was used in order to implement the system controller and the digital baseband processing. The narrowband-SFDR (N-SFDR) (bandwidth ≈ 1.4% of the SR) and time domain measurements of a 926.4MHz synthesized carrier are reported in Fig. 3.

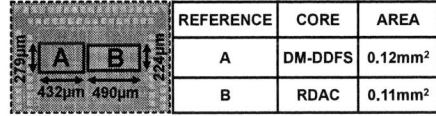

REFERENCE	CORE	AREA
A	DM-DDFS	0.12mm²
B	RDAC	0.11mm²

Fig. 2. Die micrograph.

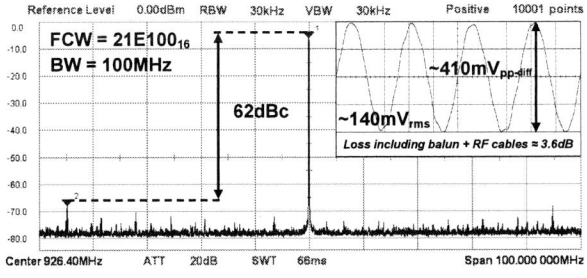

Fig. 3. N-SFDR at F_{out} = 926.4MHz. SR = 7GS/s.

A. Linear frequency modulation

The linear frequency modulation (LFM) method has been employed in surveillance radar operating in the frequency range of 2.7GHz-2.9GHz with a typical necessary bandwidth (B_N) of 6MHz [7]. Out-of-band (OoB) masks have been defined in order to evaluate the spectrum emissions [7]. Fig. 4 graphically represents the measured spectrum of a synthesized LFM pulse signal with a 2.6MHz frequency deviation (B_c) [7]. It can be observed that the achieved performance satisfies the mask requirement for a -40dB bandwidth (B_{-40dB}) of 14MHz (calculated from [7]). Fig. 5 represents a wide-band frequency sweep among 7 pre-stored FCW registers generated by driving the 4-bit control interface with a digital ramp. The signal level reduction results from uncompensated losses including the interconnection and cables attenuation. This system can switch between different output frequencies with a latency of only 12.86ns (45 clock periods at 7GS/s).

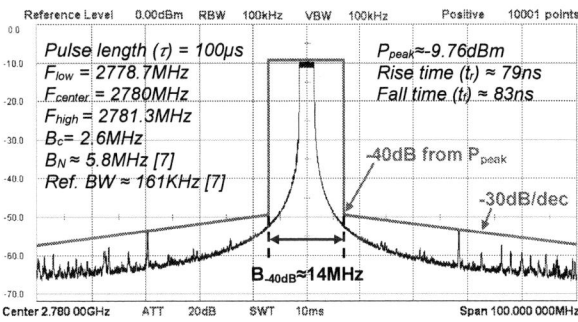

Fig. 4. LFM spectrum and OoB mask. SR = 7GS/s.

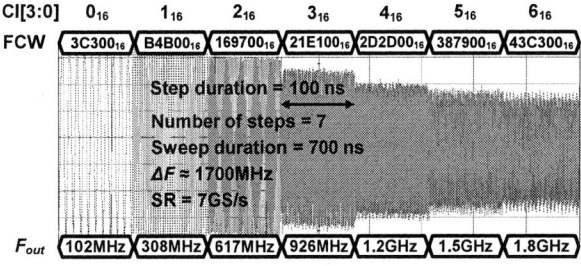

Fig. 5. Wide-band frequency sweep among 7 pre-stored FCW registers.

B. Binary phase shift keying and on-off keying

Binary phase shift keying (BPSK) and on-off keying (OOK) baseband signals have been employed to modulate the DDFS output carrier and demonstrate the PM and AM capabilities. The information source is a 16-bits pseudorandom binary sequence (PRBS) generator. It should be noted that the throughput bottleneck of this system is the general purpose interface (PMOD-GPIO) used as the output from the FPGA board (20Mbps). The maximum bitrate is limited to 1Mbps when setting an oversampling ratio (OSR) of 20 in the baseband processing unit. Higher bitrates could be possible if implementing a high speed serial interface (SERDES) in both the off-chip baseband unit and the DDFS input interface. Fig. 6 and Fig. 7 represent the measured time domain and frequency spectrum of an OOK and BPSK modulated carrier. Gaussian pulse-shaped spectrums are also reported for better completeness. In Fig. 6, rectangular and Gaussian pulse-shaped

OOK modulated carriers with equivalent binary sequences can be observed. The measured transition from 0° to 180° of a 926.4MHz BPSK modulated carrier is also shown in Fig. 7.

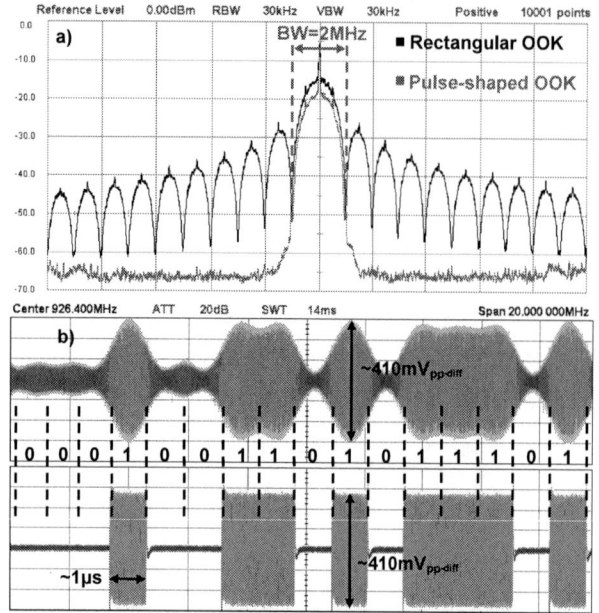

Fig. 6. 1Mbps OOK modulated signal. Spectra (a) and waveforms (b) with and without Gaussian pulse-shaping. $F_{carrier}$ = 926.4MHz. SR = 7GS/s.

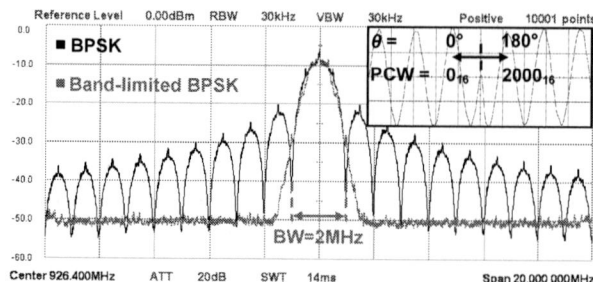

Fig. 7. 1Mbps BPSK modulated signal. Spectra and time domain waveform. $F_{carrier}$ = 926.4MHz. SR = 7GS/s.

IV. COMPARISON AND CONCLUSIONS

This work demonstrates for the first time the frequency, phase and amplitude digital modulation capabilities of a CMOS-based DDFS when running at 7GS/s. To the best of our knowledge, the only references supporting at least frequency and phase modulation and operating in excess of 2GS/s are [1] and [2]. Compared to these previous works, the proposed architecture not only achieves a higher sampling rate while equaling the best frequency and amplitude resolutions [1], but also attains superior area, power efficiency (PE) and FoM metrics as reported in Table I. Moreover, this design employs a CMOS technology instead of the more costly SiGe alternative, paving the way for new DDFS applications including but not limited to frequency exploration, polar modulation, and SDR.

TABLE I. PERFORMANCE COMPARISON

DDFS Metrics	This Work	JSSC'10 [1]	IMS'09 [2]	JSSC'11 [3]	T-MTT'10 [4]
Architecture	DM-DDFS	NLD-DDFS	NLD-DDFS	AM-DDFS	AM-DDFS
Technology	65nm CMOS	0.13μm SiGe	0.13μm SiGe	0.35μm SiGe	0.35μm SiGe
Freq./Amp. Resol. [bits]	24/10	24/10	9/7	9/8	8/6
W-SFDR [dBc] [(1)]	32[(2)]	38[(2) (3)]	-	45.7[(4)]	20[(4)]
N-SFDR [dBc]	42[(2)]	45[(2)]	35[(2)]	-	-
Area [mm²]	0.23/1.43[(5)]	7.5/11.1[(5)]	3.4/7.5[(5)]	2.1[(5)]	1.15[(5)]
SR [GS/s]	7	5	2.9	5	16.8
Power [mW]	601	4700	2000	460	486
FoM [(6)]	469.6	85.8	-	2133.3	348.4
PE [mW/(GS/s)]	85.9	940	689.6	92	28.9
FM/PM/AM	Yes/Yes/Yes	Yes/Yes/No	Yes/Yes/No	Yes/No/No	Yes/No/No

[(1)] Wideband SFDR. [(2)] Packaged/test board. [(3)] Best case.
[(4)] Die-on-board. [(6)]
[(5)] Including pads. $FoM = \dfrac{2^{W-SFDR\,[dBc]/6} \cdot SR\,[GS/s]}{Power\,[W]}$ [3]

ACKNOWLEDGMENT

This work was partially supported by HUAWEI and VDEC in collaboration with Cadence Design Systems, Inc., Synopsys, Inc., and Mentor Graphics, Inc. Special thanks to Mini-Circuits for their support through the EZ-Sample program.

REFERENCES

[1] X. Geng, F. F. Dai, J. D. Irwin, and R. C. Jaeger, "24-bit 5.0 GHz direct digital synthesizer RFIC with direct digital modulations in 0.13 μm SiGe BiCMOS technology," IEEE Journal of Solid-State Circuits (JSSC), vol. 45, no. 5, pp. 944-954, May 2010.

[2] X. Geng, F. F. Dai, J. D. Irwin and R. C. Jaeger, "A 9-bit 2.9 GHz direct digital synthesizer MMIC with direct digital frequency and phase modulations," IEEE MTT-S International Microwave Symposium (IMS) Digest, Boston, pp. 1125-1128, June 2009.

[3] C.-Y. Yang, J.-H. Weng and H.-Y. Chang, "A 5-GHz direct digital frequency synthesizer using an analog-sine-mapping technique in 0.35-μm SiGe BiCMOS," IEEE Journal of Solid-State Circuits (JSSC), vol. 46, no. 9, pp. 2064-2072, September 2011.

[4] B. Laemmle, C. Wagner, H. Knapp, H. Jaeger, L. Maurer and R. Weigel, "A differential pair-based direct digital synthesizer MMIC with 16.8-GHz clock and 488-mW power consumption," IEEE Transactions on Microwave Theory and Techniques (T-MTT), vol. 58, no. 5, pp. 1375-1383, May 2010.

[5] A. M. Alonso, M. Miyahara, and A. Matsuzawa, "A 10-bit 6.8-GS/s direct digital frequency synthesizer employing complementary dual-phase latch-based architecture," IEICE Transactions on Electronics, Vol.E99-C, No.10, pp. 1200-1210, October 2016.

[6] E. Olieman, A.-J. Annema and B. Nauta, "An interleaved full nyquist high-speed DAC technique," IEEE Journal of Solid-State Circuits (JSSC), vol. 50, no. 3, pp. 704-713, March 2015.

[7] Recommendation ITU-R SM.1541-6, Unwanted emissions in the out-of-band domain, Geneva, August 2015.

S13-3 (2160)

A −121dBm Sensitivity, 2.8μJ/bit Rx, 8.8μJ/bit Tx, Narrowband transceiver for ARIB STD and IoT

M. Kumarasamy Raja, Zhao Bin, Yan Dan Lei, Zhang Hongbao, Lim Wei Yi, Chemmanda John Leo

Institute of Microelectronics, A*STAR (Agency for Science, Technology and Research), Singapore 138634.

Abstract—A narrowband low-power low-sensitivity, IoT TxRx compliant to ARIB STD-T-67 & T-30, is presented. It employs (1) an injection locked IQ-divider without power-hungry high speed logic gates and flip flops to generate 25% duty cycle LO that drives the mixer following a simultaneously noise and impedance matched g_m-boosted LNA, and (2) A dedicated pilot-less direct automatic frequency correction of the PLL forcing the IF to be free of offset. The second feature (a) enables lowest noise bandwidth in the detector, (b) positions the signal spectrum at the center of the pass band of the high-Q IF chain away from the phase distortions in the 3-dB transition region, and (c) simplifies the base band design. Measured sensitivity is -121dBm (@4.8kbps), Rx energy efficiency is 2.81μJ/bit and Tx efficiency is 47.5%, all of which advances the state of the art, to the authors knowledge.

Keywords— Energy efficiency, Narrowband, IoT transceivers

I. INTRODUCTION

Sub-1GHz band exhibit lower path loss and have the ability to work with adverse channel conditions like in underground, multipath fading, concrete absorption etc. Advantages of lower sensitivity (P_{MINRX}) offered by narrowband (NB), include (i) lower Tx power (P_{TX}) to save dc power (P_{DCTX}), (ii) larger link budget ($P_{TX} - P_{MINRX}$) and hence larger area coverage per cell, (iii) lower latency, and (iv) larger number of channels. Thus a sub-1GHz, NB radio simplify system design and lower cost of Internet of Things (IoT) nodes and deployment [1].

Sub-1GHz NB IoT Rx, carry out the channel filtering in the digital baseband (DBB) after the complex band pass filter (CBPF) and large dynamic range, band-pass $\Delta\Sigma$ ADC (BP-$\Delta\Sigma$-ADC) that passes the wanted and adjacent channels interferer (ACI) in a low-IF topology as shown in Fig. 1 [2-6]. Digital ACI filtering avoids the noise and phase distortion inherent in analog high-Q filters. Yet, the best reported sensitivity is -120dBm, with a power consumption (P_{DC}) of 78mW [3]. The bottleneck for lower sensitivity, is the noise figure (NF) of RF/IF chain and the SNR (after ADC), which are challenged at low P_{DC}. This paper reports mainly two novel techniques among others: (i) An injection locked IQ-divider generating 25% duty cycle LO without power-hungry high speed logic gates or flip flops. This LO drives passive mixer following a simultaneously noise and impedance matched g_m-boosted LNA (NF = 1.8dB) (ii) A zero-overhead automatic frequency correction (AFC) of PLL that lowers the IF spread and improve SNR of DBB.

II. ARCHITECTURE & BASEBAND

Precise AFC correction at PLL and the reduced signal processing BW results in four advantages: (i) reduced noise BW (B_N); (ii) Signal after mixer is exactly centered at the IF frequency, away from the phase distortions at the 3-dB transition regions of the following blocks like continuous time BP-$\Sigma\Delta$-ADC (BW = 22.5KHz) as shown in Fig 1(b) and CIC; (iii) Simplified base band and (iv) Low AFC resolution (6.4Hz) leveraging on the $\Sigma\Delta$ divider in PLL minimizing the FM error.

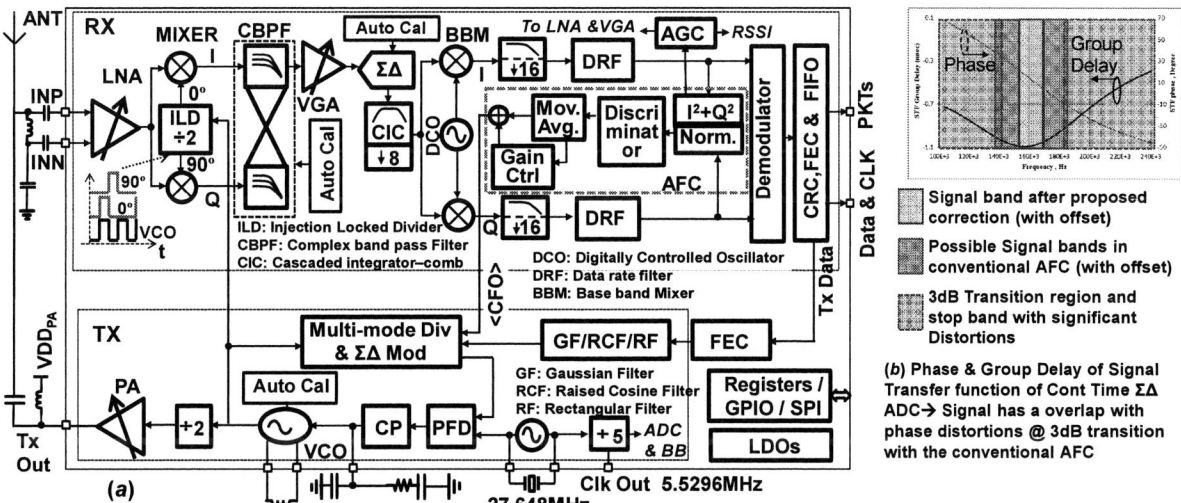

Fig. 1. (a) Proposed TxRx with Injection locked divider and pilot-less, timing-free, moving average based-AFC (highlighted in red); (b) Phase response of Signal transfer function (STF) of the ADC, showing the uncorrected IF spectrum overlapping the region with significant distortions in conventional AFC

978-1-5386-3179-9/17 $31.00 © 2017 IEEE   185

Data rate filters (DRF) rejects the ACI since the preceding CBPF (BW = 75KHz) and CT-BP-$\Sigma\Delta$-ADC rejects only the non-ACI. Thus RSSI, and carrier sense (CS), are precise, eliminating false triggers and facilitating simpler AGC. In $\Sigma\Delta$-ADC, samples per cycle (f_s/f_{IF}) is 32, and 4 after CIC. Thus I/Q output of DCO per cycle is {+1 -1 -1+1} & {+1 +1 -1 -1} respectively. Hence, digitally controlled oscillator (DCO) and baseband mixer (BBM) are implemented together by merely adding samples with sign reversed after every two samples.

A. Pilot-Free Automatic Frequency Correction (AFC)

AFC input is first normalized to RSSI (I^2+Q^2) by simple bit shift or truncation, widening the dynamic range. Thus the carrier frequency offset (CFO=<f_{offset}>) is robust to amplitude errors due to AGC switching, multi path fading, inter channel interference (ICI) caused by the LO skirts beyond signal BW. Sample by sample moving average of FSK signal with a deviation of $\pm\Delta f$, estimates <f_{offset}>, as <$\pm\Delta f/2$> tends to null under 0/1 balance. Unlike existing algorithms which demodulate a pilot to estimate <f_{offset}> [6], proposed approach is independent of data rate (f_D), bit timing and Δf even with $\Delta f \ll$ CFO. In addition, traditional CFO correction merged with data demodulation [2-3] increases the processing BW which is the noise BW (BW_N). Apart from the BW_N reduction by $\pm$4ppm (3.76KHz), the proposed pilot-free approach increases the throughput and spectral efficiency. Measured ACR and SNR are independent of f_{offset}, and is unaffected even for $\Delta f \ll f_D/4$. Gain of the feed-forward loop is adaptive, trading off speed for accuracy initially and fine tunes the CFO, for accuracy, before passing to PLL for correction.

III. CIRCUIT DESCRIPTION OF KEY BLOCKS

A. Injection Locked Frequency Divider (ILFD)

Conventionally the divider (DIV2) consists of D-FFs whose outputs are combined through AND gates, to generate 25% duty cycle, both consuming larger P_{DC} as they operate at $2\times f_{LO}$ speed. Frequency division (DIV2) through injection of $2\times f_{LO}$ signal (VCO$_{OUT}$) into a MOSFET switch that shorts the differential outputs of a ring oscillator, does not need the D-FFs, yet they need power-hungry high speed gates to generate 25% duty cycle IQ [5,7]. Moreover, four diff-pairs form the four stages of the ring oscillator. A dual-coupled differential ring oscillator topology, as in Fig. 2, exploits, the diff-pair as two inverters, effectively reducing the diff-pairs to two but generates sinusoidal IQ output. With a larger injection amplitude and bias voltage of injection transistor M_{SW1}, it could be operated like a switch, and is turned on for the positive half cycle, extending the LOW period of Q+/Q- and I+/I- to 75% as detailed below.

M_{SW1} and M_{SW2} biased in saturation ($V_{GS} > V_{TH}$) shorts the two differential outputs (Q+/Q- and I+/I-) to the common mode voltage V_{CMOUT} (LOW), during the positive half cycle of VCO$_{OUT}$. This half cycle constitutes quarter cycle of DIV2, as highlighted in Fig 2(b). During the negative half cycle of VCO$_{OUT}$, M_{SW} is off, allowing the natural output of the ring oscillator, at the output. In this period the ring oscillator output is at the mid of its positive or negative half cycle (peak or valley). Thus the 'High' and 'Low' periods of the ring oscillator respectively are unaffected for 25% of the cycle (=50% of injected VCO signal) as shown by the bottom two curves (green

and red) in Fig 2(b). When M_{SW} shorts the two outputs (Q+/Q- and I+/I-), the voltage goes below 0.7V immediately and settle to V_{CMOUT}. The cross over point of the inverter is designed to higher ($2/3^{rd}$ of V_{DD}) by sizing the pMOS to be 3 times the nMOS, so that under all PVT conditions, input level below 0.7V is treated as logic Low. V_{CMOUT} is designed to be slightly lower than $V_{DD}/2$ which enables the robust operation. This divider consumes only 1mA$\times$1.2V, against 6.5mA$\times$1.2V in [7].

B. Receiver Chain

Z_{in} of conventional source degenerated common source LNA (SDCS-LNA), neglecting C_{gd1} is [8],

$$Z_{in} = \frac{-1}{j\omega_0 C_{gs1}} + j\omega_0(L_g + L_s) + \omega_{T1}L_s\left(\frac{r_{01}}{r_{01} + j\omega_0 L_s + Z_{inCG}}\right) \quad (1)$$

The first two terms cancel each other by series resonance and the third term ($\omega_{T1} \times L_s$) is matched to 50Ω as the term inside the bracket is close to 1. Z_{inCG} (normally $1/g_{m2}$), is reduced by the factor ($1+A_C$), due to C_{CC} as below.

$$Z_{inCG} \approx \frac{1}{(1 + A_C)g_{m2}}; \quad \text{where,} \quad A_C = \frac{C_{CC}}{C_{CC} + C_{gs2}} \quad (2)$$

Lower Z_{inCG}, reduces L_S (degeneration) required for 50Ω match by increasing the third term in (1), which improves NF gain, reduces M_2's noise and slightly moves up frequency band for which Z_{in} =50-Ω, exactly aligning with best NF. Lower Z_{inCG} in (2) improves isolation, and boosts overall G_m of LNA.

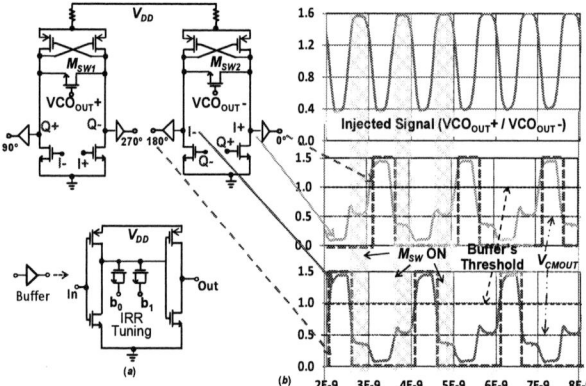

Fig. 2. Schematics of proposed low-power Injection locked divider without high speed logic and flip flops & (b) 25%-duty cycle generation with buffer's threshold about $2\times V_{DD}/3$ (Only I+ & I- shown)

	SDCS+C_{CC}	SDCS
Freq(MHz)	466	438
Gain (dB)	15.3	15.9
NF (dB)	1.41	1.73
S_{12} (dB)	-50.6	-31.9
S_{11} (dB)	-20.3	-12.6
Z_{in} (Ohm)	56-8j	33.2+10j

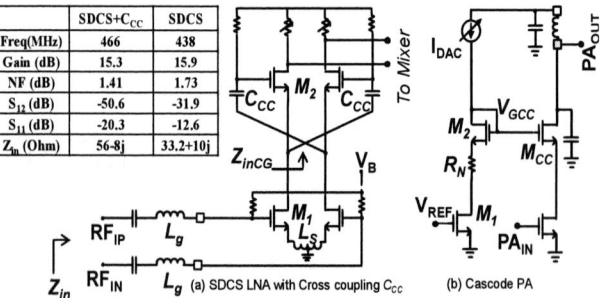

Fig. 3. Schematics of (a) LNA with C_{CC} aligning Noise and 50-Ω match & (b) I_{DAC}-gain programming in PA to alleviate RF switching noise

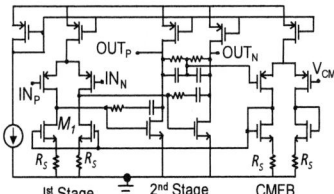

SD (Proposed) Vs. conventional 2-stage		
Parameter	SD Op-Amp	2stage Op-Amp
IDD	110μA	110μA
GBW	16MHz	5MHz
V_{nIN}	9.5nv/sqrt(Hz)	14nv/sqrt(Hz)

Improved noise and speed with proposed source degeneration (SD)

Fig. 4. Schematics of Low-Noise, source degenenerated (SD) Op-Amp & used in Complex band pass filter (CBPF) & its performance summary

Fourth order butterworth synthesized in a doubly terminated ladder, realized in op-amp RC topology forms the CBPF with 0 to 32dB gain. The op-amp employs source degeneration resistor R_S in the 1st-stage load transistors (M_l) to improve noise and speed as shown in Fig. 4. The first stage output resistance (r_{o1}) is thus increased by $g_{m1}*r_{ds1}*R_S$ (about 50K with R_S = 5K & $g_{m1}*r_{ds1}$ = 10). Hence, M_l could be down-sized, without compromising r_{o1}, reducing its parasitic capacitance pushing the dominant pole and GBW. Adding a cascode MOSFET with equivalent r_{ds}, to achieve this r_{o1} results in far higher noise, as the flicker noise of such a MOSFET is amplified by M_l and also the thermal noise contribution is larger than resistor.

C. Power Amplifier

The I_{DAC} in Fig. 3(b), is controlled by a state machine that either programs the P_{TX} or performs PA-ramping by gradually increasing the I_{DAC}. Since the resistance of M_{CC} is decreased as P_{TX} is increased, it provides smoother ramping without switching transients and does not need high speed gates and advanced nodes unlike digital PAs which switches the RF signal ON/OFF [2,5]. Although P_{TX} does not vary with temperature when high (11-13dBm), since the PA is saturated, it becomes sensitive to temperature at lower P_{TX} levels as follows. V_{TH1} reduces with temperature, forcing V_{DS1} and V_{GCC} also to reduce. Hence, r_{dsCC} increases, reducing P_{TX}. In order to circumvent this, a small series N-well resistor (R_N) with positive temperature coefficient keeps V_{GCC} constant, even when V_{TH1} (& V_{DS1}) falls at high temperature. The aspect ratio of R_N is kept high to reduce the effect of process variations. Monte carlo simulated variance (σ) of power P_{TX} is only 0.14dB and ±3σ is ±0.4 dB. At 10dBm setting, measured P_{TX} variation is +1.2dB/-1 dB (at -20°C/ 60°C), whereas without R_N, it was ±3dB.

IV. MEASURED RESULTS

Chip micrograph of the TxRx in 0.13-μm CMOS, is shown in Fig 5. Bonded in QFN24 (5mm×5mm) package it was tested on a FR4 PCB. Measurements as summarised in table I were carried out at a data rate of 4.8kbps, GFSK with Δf=2.4KHz in the desired channel, with PA and LNA individually matched to 50-Ω, unless stated otherwise. Blockers and AFC were tested with desired signal level 3dB above sensitivity.

A. Receiver Measurements

Measured sensitivity is -121dBm, maximum input power (P_{inMAX}) with AGC is -10dBm (Fig 6(a)) at a BER of 1E-2. P_{inMAX} is +5dBm when all the gain steps are exploited.

TABLE I. MEASUREMENT SUMMARY

Parameter	Spec	Results
Supply Voltage (V_{DD}) , V	1.5 / 1.2	
Frequency Range, MHz	420 – 470	
Channel Spacing, KHz	12.5 /25	
Data Rate (NRZ), kbps	1.2/2.4/4.8/7.2 (NRZ) & 0.6/1.2/2.4 (Man. encoding)	
Modulation & Filtering	(G/RC)FSK, BT=0.5, SCMSK	
Frequency Deviation Δf & step, KHz	0.1 to 3.5 & step 0.1	
Frequency offset	±4 ppm	
Sensitivity (4.8kbps,1E-2 BER), dBm	-101	-121
Rx Current, mA	-	9.5
Rx max. input dBm (with AGC)	-10	-10
without AGC		5
Intermodulation, CW jammers at ±n×Ch spacing (ARIB T-30), dB	40	42-44
Adjacent Channel Rejection (ACR)	30	44
Rejection at 1 MHz offset (non-ACR)	70	70
LO leakage at antenna port, dBm (Direct-Tie & individual matching)		< -80(direct-tie) <-100 (RF SW)
Carrier Sense, Sec	500μ	380μ
RSSI Range, dB ADC / Antenna, RSSI accuracy, dB ADC / Antenna		60 /110 (step 1) ±3 / ±6
LO Phase Noise, dBc/Hz (in band) @ 1 MHz offset (non-ACI)	-95 (ACI) -120	-95 -120
PLL settling Time, μsec	89	10
Occupied Band Width (OBW), KHz	8.5 / 16	<8.5 / 12
Transmitter Power (P_{TX}) Range, dBm Coarse Step 0.5dB; Fine step 0.2dB	10@ antenna	3 to 13
Transmitter Efficiency @ 13 dBm		47.5%
Adj Channel Power ratio ACPR, dB	>40	>46
Spurious Emission , dBm	< -26	<-42
Harmonic s 2nd , 3rd & Others, dBC	-13	-20, -30 &-35
STDBY current (SPI registers ON)	-	1μA
Shut down mode current	-	<100nA

ACR and IRR are tested by injecting a stronger GFSK modulated blockers and the results are plotted in Fig. 6(b). ACR is 44dB and the IRR after tuning is 32dB at high gain (HG) mode. Beyond CBPF's BW of 75KHz, it also aids the rejection which is better than 70dB above 1MHz offset. The IM is tested by applying two CW interferes at the adjacent two channels either above or below the desired channel (12.5/25/50KHz away) [6]. It tolerates CW interferers which are 42-44 dB above the signal level. In the HG mode, the IF chain saturates when the blocker power is about 45 dB higher, limiting the ACR, IM and even IRR (all of which are within CBPF BW). Hence ACR, IM and IRR improve at lower gain due to better linearity. IRR further improves due to better IQ matching, benefitting from larger open to closed loop gain ratio of op-amp, which was also observed in CBPF test chip. AFC is functional until ±5 KHz as shown in shown in Fig. 7(a). Beyond about ±5.5KHz, which could be drifted adjacent channels, the AFC rejects effectively due to DRF. The CS is about 0.38msec and 0.55msec, when averaged for 18 and 9 samples respectively. NF of LNA+Mixer is 2.5dB and 3.5dB when cascaded with IF (CBPF+VGA).

B. PLL and Transmitter Measurements

Fig. 7(b) plots the Tx power, Tx and PA efficiency versus the 5 bit control word to I_{DAC}. Smooth PA ramping is obtained, with selectable ramping clock as shown in Fig. 8(b). Measured OBW are summarized in Fig. 8(a). In the direct-tie mode, sensitivity and Tx power degrades by 0.6dB and 0.2dB respectively owing to the absorption of matching circuit mutually. Sensitivity, P_{DC}

performance compares favorably with state of the art, as shown in table II. Tx efficiency 47.5% is better than the closest in [4], even though [4] trades off BW for P_{TX} and hence efficiency.

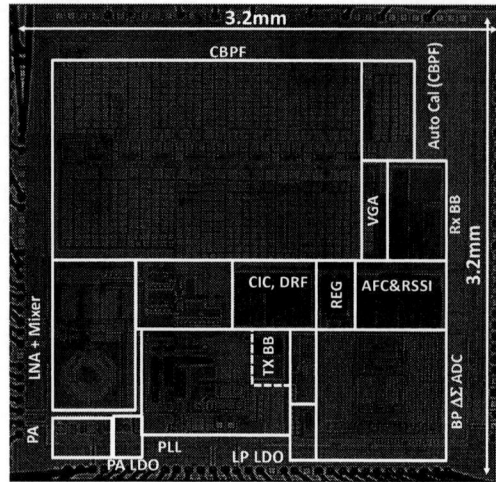

Fig. 5.   Chip Micrograph with functional block details

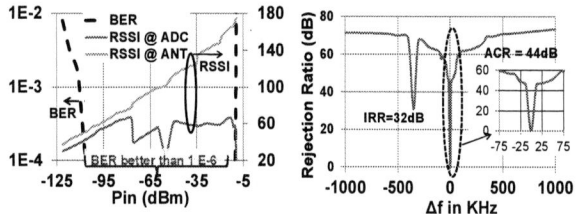

Fig. 6.   (a) BER, RSSI Vs Rx input power. Valleys at ADC input (red) is due to Gain switching & (b) Selectivity with desired signal at -118dBm, shows 44dB rejection of ACR for blockers ±12.5/25 KHz away.

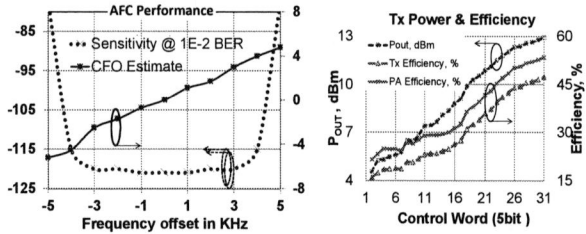

Fig. 7.   (a) Rx AFC with desired signal of -118dBm. No BER degradation for offsets of ±3.76KHz and functional for ±5KHz   (b) Tx Power and Efficiencey. Maximum global Tx efficiency is 47.5%. Step is 0.5dB for <10dBm and finer (0.25dB) from 10-13dBm.

Mod	Data Rate Kb/s	OBW, KHz
FSK	1.2	2.73
	2.4	4.87
	4.8	10
RCFSK	1.2	2.72
	2.4	3.75
	4.8	7.37
GFSK	1.2	2.23
	2.4	3.9
	4.8	7.73

Fig. 8. Tx output (a) OBW summary of modulated spectrum and (b) PA ramping shows no switching noise (ramping speeds programmable).

V.   Conclusions

This paper demonstrated a narrow band radio for IoT applications compliant to ARIB STD T-67 and T-30 with a link budget of 131dB. Low-power low-noise circuit techniques including (i) an injection locked divider free of high speed logic and FFs and (ii) a moving average based dedicated AFC that corrects the PLL precisely, were demonstrated. This results in lowest Rx sensitivity of -121 dBm, best Rx energy efficiency of 2.81μJ/bit, and highest Tx efficiency of 47.5%, over state of the art. Other features include (i) Test modes for bulk production and bit streaming mode for real time detection with clock recovery, and (iii) smooth ramping with least energy to various modes like Shut down, Standby, Tx, Rx and Idle (only PLL on).

Acknowledgment

Authors acknowledge the contributions from Jae Hong, Simon Ng, Jacky Wang Yu-shun and Pyoungwon Park. The authors also thank Atsushi Tamura, Annamalai Arasu and Minkyu Je for the technical discussions and suggestions.

TABLE II.   Performance Comparison with similar TxRx

Parameter	[2]	[3]	[4]	[5]	[6]	This
Sensitivity (dBm) &	-116^/ GFSK	-120 / FSK	- / $ BPSK	-115/ FSK	-117/ (G) FSK	-121/ (G)FSK
ACR (dB)	42	70	-	45	54	44
P_{DCRX}(mW)	14.5	78	32.4	46.2	39.6	13.5
Rx Energy Effcy(μJ/bit)	3.02	16.3	6.75	9.6	8.25	2.81
P_{DCTX}(mW) @P_{Tx}(dBm)	66@ 10	132@ 13	175@ 18	79.2 @10	97@ 14	42.1@ 13
Tx Power P_{TX} (dBm)	< 14.7	-20 to 17	18$	0 to 10	-11 to 14	4 to 13
Tx Effcy @ P_{TXmax}(%)	16.25	16.25	36$	12.6	25.9 / 14.3#	47.5 / 35#
CMOS Tech node	65nm	180nm	130nm	90nm	-	130nm

&:@4.8kb/s and 0.01BER; ^ Estimated from 600bps; # Tx Efficiency at 10dBm; $ spread spectrum BPSK

References

[1]   https://www.lora-alliance.org/What-Is-LoRa/Technology

[2]   D. Lachartre et al., "7.5 A TCXO-less 100Hz-minimum-bandwidth transceiver for ultra-narrow-band sub-GHz IoT cellular networks," 2017 IEEE International Solid-State Circuits Conference (ISSCC),pp. 134-135.

[3]   N. Kearney et al., "26.4 A 160-to-960MHz ETSI class-1-compliant IoE transceiver with 100dB blocker rejection, 70dB ACR and 800pA standby current," 2016 IEEE International Solid-State Circuits Conference (ISSCC), San Francisco, CA, 2016, pp. 442-443.

[4]   J. Van Sinderen, et al., "Wideband UHF ISM-Band Transceiver Supporting Multichannel Reception and DSSS Modulation," IEEE ISSCC Dig. Tech. Papers, Feb. 2013.

[5]   T. Tokairin1, et al., " A 3.5mm², Inductor-less Digital-intensive Radio SoC for 300-to-950MHz ISM-band applications supporting 1-to-240kbps Multi-data-rates," Symp. VLSI Circuits Dig. of Tech. Papers, 2011.

[6]   CC1120 datasheet http://www.ti.com/lit/ds/symlink/cc1120.pdf

[7]   A. Mirzaei et al., "Analysis and Optimization of Direct-Conversion Receivers with 25% Duty-Cycle Current-Driven Passive Mixers", IEEE Trans.Circuits and Syst. I, Reg. Papers, pp. 2353-2366, Sept. 2010.

[8]   Thomas H Lee, "The Design of Radio-Frequency CMOS Integrated circuits," Cambridge University Press, Second edition.

[9]   H. Lim et al., "Bandpass Continuous-Time Delta-Sigma Modulator for Wireless Receiver IC," Int. J. of Info and Electro Engg. pp. 4-7, 2013.

Detection of 3.0 THz wave with a detector in 65 nm standard CMOS process

Tong Fang[†], Zhao-yang Liu[†], Li-yuan Liu[*†], Yuan-yuan Li[‡], Jun-qi Liu[‡], Jian Liu[†] and Nan-jian Wu[†]

Email: liuly@semi.ac.cn

[†]State Key Laboratory of Superlattices and Microstructures, Institute of Semiconductors, Chinese Academy of Sciences
University of Chinese Academy of Sciences, Beijing, China

[‡]Key Laboratory of Semiconductor Materials Science, Institute of Semiconductors, Chinese Academy of Sciences
University of Chinese Academy of Sciences, Beijing, China

Abstract—This paper reports a 3.0 THz detector which can detect the terahertz wave radiated by a quantum cascade laser (QCL) working at pulse mode. The detector was implemented in a 65 nm silicon CMOS process. The chip size is $1.5 \times 1.5 \ mm^2$ including bonding pads. THz detector using FET is based on plasma wave theory proposed by Dyakonov and Shur which allows detection of THz radiations far beyond the FET devices characteristic frequency (f_T). The detector was measured by means of lock-in techniques and can achieve a room-temperature responsivity of 526 V/W and a noise equivalent power (NEP) of 73 pW/Hz$^{1/2}$. We acquired a high resolution image of a tooth pick by raster-scanning technique, and Ultra-fine brushes can be clearly distinguished. Our results demonstrate a promising application in low-cost CMOS THz camera for high resolution imaging.

Keywords—CMOS; THz; terahertz imaging; quantum cascade lasers; on-chip antenna; terahertz detector

I. INTRODUCTION

THz radiation lies in frequency range 0.1 THz-10 THz (1 THz= 10^{12} Hz) which has many unique properties: non-ionizing penetration through many dry, non-metallic and non-polar materials, high sensitivity to water, higher spatial resolution in comparison to millimeter waves, and specific spectral fingerprints for many substances. These unique properties can turned into a lot of promising commercial applications such as security screening [1], bio-sensing [2], and drug and explosives analysis [3]. Terahertz detector using FET is based on plasma wave theory proposed by Dyakonov and Shur [4] which allows detection of terahertz radiations far beyond the FET devices characteristic frequency (f_T). The detector outputs a dc voltage signal that is proportional to the detected terahertz radiation power [5]. The first silicon FET detector was demonstrated by Knap et al [6]. Then many CMOS integrated terahertz detectors have been presented [7], [8], some of them can realize active terahertz video recording at room-temperature [8]. Unfortunately, most of those detectors can only detect terahertz wave below 1 THz.

With the development of THz quantum cascade lasers (QCLs), terahertz source with strong power radiation at several THz is available. Dues to higher radiation frequency and hence shorter wave length, imaging at these frequency

range can achieve a much finer spatial resolution. Taking 3.0 THz radiation as an example, the wave length is 100 μm and theoretical spatial resolution can be as fine as $\lambda/2$ which is only 50 μm.

However compact and low cost detectors at these frequencies are still rare. Although detectors, such as bolometers, pyroelectric detectors, and Golay-cells could be used, they suffer from problems including slow response, scarcely sensitive, or requiring cryogenic cooling [9]. These drawbacks lead to a much complex imaging systems and make a compact low cost THz imager impossible. CMOS based detectors, on the other hand, can work at room-temperature with a fast response and acceptable responsivity compared to the above detectors. Furthermore, by employing CMOS technology, many other advantages are expected such as low cost, high yield and easy integration with massive digital signal processing circuits.

In this paper, we propose a CMOS THz detector in 65 nm standard CMOS process. It can detect 3.0 terahertz wave radiated by a QCL working at pulse mode. We proposed a source driven patch antenna structure with high radiation efficiency. The detector achieves a room-temperature responsivity of 526 V/W and a noise equivalent power (NEP) of 73 pW/Hz$^{1/2}$. Finally, we acquired a high resolution image of a tooth pick by raster-scanning techniques and ultra-fine brushes can be clearly distinguished.

II. DETECTOR DESIGN

The block diagram of the chip is shown in Fig. 1. The proposed detector consists of an NMOS field effect transistor with non-biased channel as rectifying element, an integrated on-chip grounded patch antenna, a matching network (MN) and an open quarter-wavelength micro-strip transmission line (TL) [11]. We put several detectors with different physical size on the chip in order to compare during measurement as well as choosing the one with best performance.

The core of the detector is an NMOS transistor which can detect the power of terahertz radiation and gives a proportional output DC voltage. The transistor has a gate length of 60 nm and a gate width of 120 nm, which is the minimum size device available in the process. The purpose is to maximum the

This work was supported by National Natural Science Foundation of China (Grant Nos. 61474108, 61331003), National Science and Technology Major Projects of the Ministry of Science and Technology of China (Grant No 2016ZX03001002-002) and Ministry of Science and Technology, China (Grant No. 2016YFA0202202).

source impedance so as to enhance the voltage responsivity [7].

The structure of the antenna can affect the performance of the detector significantly. It is difficult to design an antenna which can work at 3.0 THz and have a high response. There are many proposed CMOS integrated antenna that can work at terahertz frequency. Like bow-tie antenna [12], ring antenna [8], folded dipole antenna [10], and patch antenna [7]. Because THz-QCL outputs terahertz laser with narrow frequency span, and there is no need to choose a broadband antenna. The efficiency of the antenna design is of the most concern. As a result, we choose patch antenna due to its high radiation efficiency compared to other counterparts.

Usually patched antenna may occupy a relatively large area, however, when working at 3.0 THz, due to short wave length together with high dielectric constant passivation layer, the antenna size wouldn't be too large.

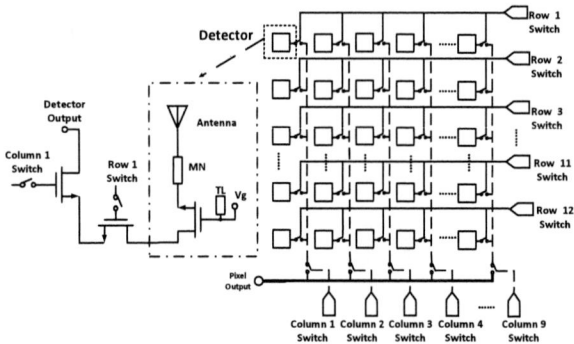

Fig. 1. Chip Diagram

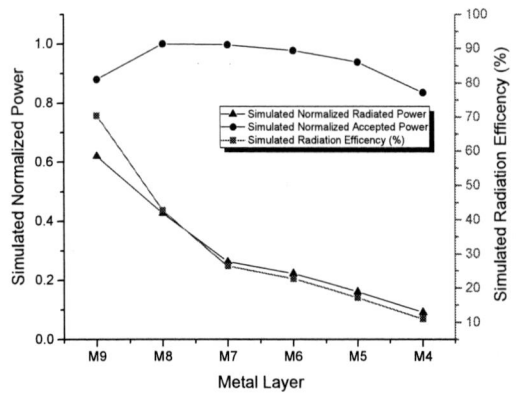

Fig. 2. Simulated normalized power and radiation efficiency versus metal layer

The 65 nm CMOS process has nine metal layers, we chose the bottom metal layer M1 to form the ground plane. As for the patch plane metal, the distance between the patch plane and the ground plane can influence the character of antenna, so we simulated the antenna whose patch plane was formed by different metal layer from M4 to M9.

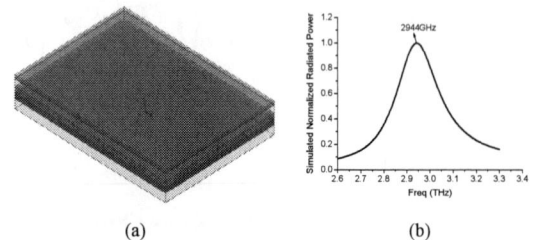

(a) (b)

Fig. 3. (a) Simulation model of the patch antenna and (b) Simulated normalized radiated power of model (a)

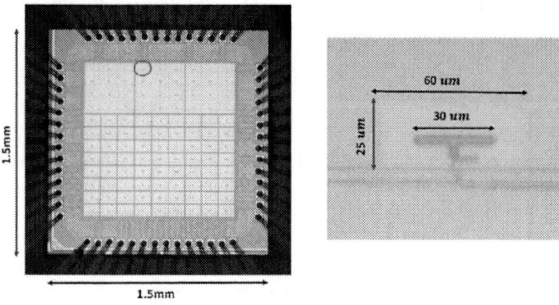

Fig. 4. Chip microphotograph including the bonding pads and a close-up photograph of 3.0 THz antenna (The marked one)

Fig. 2 is the simulated normalized power and radiation efficiency versus metal layer. We found that, when the accepted power is almost no change, the simulated radiated power and radiation efficiency decease as we chose a lower metal layer. So we choose the top metal layer M9 to form the patch plane and its simulated efficiency is 70%. The total thickness of dielectric insulator between patch and ground plane is 4.6 μm. The patch has a length of 30 μm and a width of 2 μm. A contact connecting the top metal and bottom metal at the center of the antenna is designed to provide a dc ground to the source of the NMOS. The contact does not degrade the electromagnetic performance of the antenna since the center of antenna is zero-field intensity (H-plane). Compared to Ref. [10], this structure doesn't require an extra source bias through the antenna, and thus reduces system-level complexity for future array implementation [11].

To improve power transfer efficiency, we use a micro-strip transmission line MN and vias between the metal layers to form the matching networks. The transmission line has a length of 5 μm and a width of 2 μm implemented in the top metal layer, and the vias' length between the top metal layer and the bottom metal layer is 4.6 μm. Considering the influence of pad and bonding wire, we propose an open quarter-wavelength micro-strip transmission line connecting to the gate of the NMOS [11], as shown in Fig. 3, which forms a ground point for terahertz radiation and has no impact on gate dc bias. The quarter-wavelength micro-strip transmission line has a length of 9 μm and a width of 2 μm. It is implemented in the sixth metal layer. The complete simulation model of the patch antenna and its simulated normalized radiated power shows in Fig.3. The rectifying dc voltage signal transfers through the row and column switches and then connected to the lock-in amplifier.

III. MEASUREMENT SET-UP AND RESULTS

Fig. 4 shows the die photograph of the CMOS terahertz detector with various sizes of antennas including the bondpads and a close-up photograph of an antenna whose center frequency is 3.0 THz. Fig. 5 shows the measurement set-up for scanning of the source beam using the single detector marked in Fig.4. The THz-QCL was driven by a train of current pulses from a current source working at a frequency of 5 kHz, and its duty cycle is 1%. The scattered output terahertz wave from the QCL was focused on the marked detector by two off-axis parabolic mirrors. The current source also provide a reference signal to lock-in amplifier. The computer control the stepper-motor to scan the source beam and acquire data from the lock-in amplifier simultaneously.

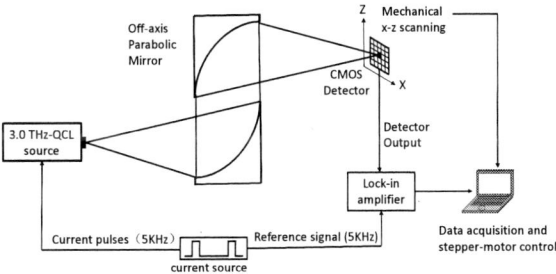

Fig. 5. Measurement set-up for scanning of the source beam

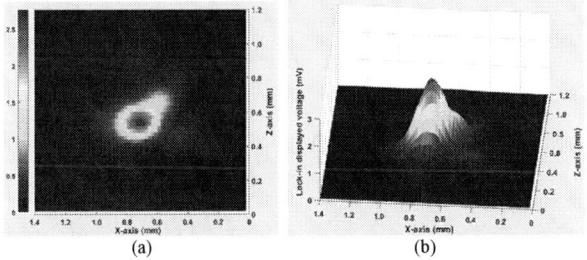

Fig. 6. (a) The raster scan image of the source beam using the single detector biased at 0.23V and (b) the 3-D vision of the picture (a). (20 μm step size)

Fig. 6 (a) Shows the raster scan image of the source beam using the single detector biased at 0.23 V and (b) the 3-D vision of the picture (a). The scan range is $1.4 \times 1.2 \ mm^2$ and the step size is 20 μm. Because this THz-QCL source hasn't been shaped, the scanned source beam is not a perfect Gaussian distribution. We replaced the detector with an absolute-power calorimeter to measure the average source power with a value of 0.165 mW. Because the duty cycle of the THz-QCL is 1%, the total source power P_{beam} is 16.5 mW (0.165 mW × 100). Since the value measured by lock-in amplifier is the effective value of the fundamental frequency amplitude of actual detection signal, so the detector's actual output voltage U_{out} should be $\frac{\sqrt{2}*V_{LOCk-IN}}{\frac{2}{\pi}\sin(\pi*0.01)}$. Then combining the source beam image and the calculated

total source power, we can obtain the voltage responsivity R_v of the detector using the following equation [12]:

$$R_v = \iint \frac{U_{out}}{P_{beam}*A_{eff}} dxdz \qquad (1)$$

A_{eff} is the effective area of the antenna which can be calculated by $A_{eff} = \frac{D\lambda^2}{4\pi}$, and D the directivity of the antenna with a simulated value of 2.75 dBi. Fig. 7 shows the measured

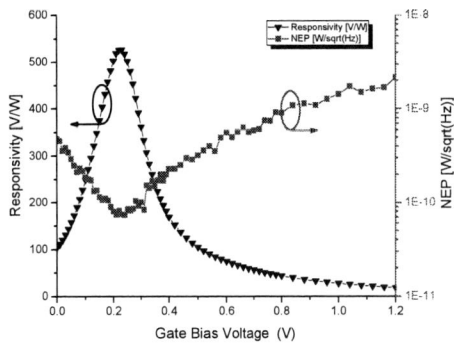

Fig. 7. Measured R_v and NEP as a function of gate bias voltage

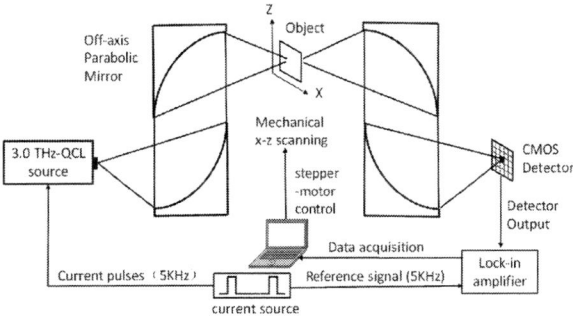

Fig. 8. Block diagram of the terahertz imaging system

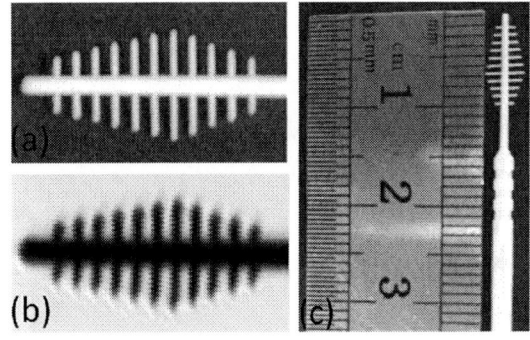

Fig. 9. Scanned image of a tooth pick (100 μm step size)

responsivity (R_v) and noise equivalent power (NEP) versus gate bias. The peak responsivity is 526 V/W when the gate bias voltage is 0.23 V. The minimum NEP is 73 pW/Hz1/2 when the gate bias voltage is 0.24 V.

Fig. 8 shows the terahertz imaging system. The QCL beam was collected and focused on the object plane, then collimated and refocused on the detector, the object was scanned on the X and Z axes by a motor-stepper controlled by the computer. Fig. 9 (a) shows the object scanned in the system which is a part from a tooth pick made by polyformaldehyde and (b) shows the raster scanned image of (a) and (c) shows the size of scanned object. The Ultra-fine brushes of the tooth pick can clearly appeared in the image. The performance comparison between this work and the published terahertz detectors is shown in Table I. Our detector achieves a relative high responsivity, and we have acquired a raster-scanned image.

IV. CONCLUSIONS

We report a 3.0 THz detector which can detect the terahertz wave radiated by a THz-QCL working at pulse mode. The detector was implemented using a 65 nm silicon CMOS process. The chip size is 1.5×1.5 mm^2 including bondpads. The detector consists of an on-chip grounded patch antenna formed by the top and bottom metal layer M9 and M1 and connected to the source terminal of the NMOS filed effect transistor. A matching network made by the micro-strip and the vias between the metal layers are designed to improve power transfer efficiency. A quarter-wavelength micro-strip transmission line is proposed to connect the gate of the NMOS to eliminate the influence of pad and bonding wire. The detector output was connected to a lock-in amplifier via the row and column switches. The detector was measured by means of lock-in techniques and can achieve a room-temperature responsivity of 526 V/W and a noise equivalent power (NEP) of 73 pW/Hz1/2. Finally, we acquired a high resolution image of a tooth pick by raster-scanning technique, the Ultra-fine brushes can be clearly distinguished in the photograph. Our results demonstrate a promising application in low-cost CMOS THz camera for high resolution imaging.

ACKNOWLEDGMENT

This work was supported by National Natural Science Foundation of China (Grant Nos. 61474108, 61331003), National Science and Technology Major Projects of the Ministry of Science and Technology of China (Grant No. 2016ZX03001002-002) and Ministry of Science and Technology, China (Grant No. 2016YFA0202202).

REFERENCES

[1] D. L. Woolard, J. O. Jensen, R. J. Hwu, and M. S. Shur, Terahertz Science and Technology for Military and Security Applications. Singapore: World Scientific, 2007.

[2] P. de Maagt, P. H. Bolivar, and C. Mann,"Terahertz science, engineering and systems—From space to earth applications," in Encyclopedia of RF and Microwave Engineering. NewYork: Wiley, 2005, pp. 5175–5194.

[3] H. Hoshina, Y. Sasaki, A. Hayashi, C. Otani, and K. Kawase, "Noninvasive mail inspection system with terahertz radiation," Applied spectroscopy 63, 81-86, 2009.

[4] M. Dyakonov and M. Shur, "Shallow water analogy for a ballistic field effect transistor: New mechanism of plasma wave generation by dc current," Physical review letters, vol.77, no.15, Oct. 1993.

[5] W. Knap, et al., "Nonresonant detection of terahertz radiation in field effect transistors," Journal of Applied Physics, vol. 91, no.11, Jun. 2002.

[6] W. Knap, et al., "Plasma wave detection of sub-terahertz and terahertz radiation by silicon field-effect transistors," Applied Physics Letters, vol. 85, no. 4, Jul. 2004.

[7] E. Öjefors, U. Pfeiffer, A. Lisauskas, and H. Roskos, "A 0.65 THz focal-plane array in a quarter-micron CMOS process technology," IEEE J. Solid-State Circuits, vol. 44, no. 7, Jul. 2009.

[8] R. Al Hadi, et al., "A 1 k-Detector Video Camera for 0.7–1.1 Terahertz Imaging Applications in 65-nm CMOS," IEEE J. Solid-State Circuits, Vol. 47, no. 12, Dec. 2012.

[9] M. Ravaro, et al., "Detection of a 2.8 THz quantum cascade laser with a semiconductor nanowire field-effect transistor coupled to a bow-tie antenna." Applied Physics Letters, 104(8): 083116, 2014.

[10] R. Al Hadi, et al., "A broadband 0.6 to 1 THz CMOS imaging detector with an integrated lens," in Proc. Microw. Symp. Di-gest (MTT), 2011 IEEE MTT-S Int. pp. 1–4, 2011.

[11] Z. Y. Liu, et al., "A CMOS Fully Integrated 860-GHz Terahertz Sensor." IEEE Transactions on Terahertz Science and Technology, vol. 7, no. 4, Jul. 2017.

[12] F. Schuster, et al., "A broadband THz imager in a low-cost CMOS technology." IEEE International Solid-State Circuits Conference IEEE, 2011:42-43.

[13] S. Boppel, et al., "Optimized Tera-FET detector performance based on an analytical device model verified up to 9 THz." International Conference on Infrared, Millimeter, and Terahertz Waves IEEE, 2013:1-1.

[14] Z. Ahmad, et al., "9.74- THz electronic Far-Infrared detection using Schottky barrier diodes in CMOS." Electron Devices Meeting IEEE, 2014:4.4.1-4.4.4.

[15] S. Boppel, et al., "CMOS Integrated Antenna-Coupled Field-Effect Transistors for the Detection of Radiation From 0.2 to 4.3 THz." IEEE Transactions on Microwave Theory & Techniques, 60.12(2012):3834-3843.

TABLE I. PERFORMANCE COMPARISON BETWEEN THE CMOS TERAHERTZ DTECTORS DETECTING ABOVE 1.0 THZ

Technology	Freq. [THz]	Max R_v [V/W]	Min NEP [pW/Hz$^{1/2}$]	Image acquisition mode	Ref.
65 nm CMOS	3.0	526	73	Lock-in tecniques	This work
90nm CMOS	2.54, 3.13	336, 308	63, 85(opticle)	-	[13]
130 nm CMOS(SBD)	4.92	383	0.43nW	Lock-in tecniques	[14]
150nm CMOS	2.9	30	487(opticle)	-	[15]

S13-5 (2217)

A 0.6-V 200-kbps 429-MHz Ultra-low-power FSK Transceiver in 90-nm CMOS

Chun-Yuan Chiu, Zhen-Cheng Zhang, and Tsung-Hsien Lin

National Taiwan University, Taipei, Taiwan

Abstract – A 0.6-V, 200-kbps, 429-MHz ultra-low-power FSK transceiver (TRX) is presented. The receiver (RX) adopts a frequency-to-time based demodulator, which detects high or low frequency to determine data output level. The proposed RX achieves -85-dBm sensitivity at 0.1% BER and draws 0.146 mW. A 38.7% global efficiency, 429-MHz FSK transmitter (TX) is also reported in this paper. To remove the power-hungry sub-circuits, such as PA and PLL, the proposed TX is essentially constituted by an oscillator which drives the antenna via an impedance matching network for power conversion. The TX draws 0.39 mW and delivers -8.2-dBm output power. It achieves 38.7% global efficiency. The transceiver is fabricated in 90-nm CMOS process.

I. INTRODUCTION

Ultra-low power (ULP) transceivers are important subsystems for Internet-of-Things (IoT) systems. In these applications, due to restricted energy sources, a low-power and low-voltage radio is desired. Some examples of conventional wireless RX and TX, shown in Fig. 1, requires several power-hungry circuits, such as LNA, PLL, PA, and ADC [1]. The power consumption issue may be addressed in the injection-locked-based RX, as reported in [2]-[3]. However, to achieve effective injection, the swing of the injecting signal must be large enough, which consumes power.

For low-power consideration, it is desire to remove PLL and PA altogether. In this work, the proposed TX is essentially an open-loop oscillator which drives the antenna via an impedance matching network for power conversion. The oscillation frequency can be adjusted via a frequency tuning loop. The TX global efficiency, which is defined in (1), can be improved by removing PA.

$$Global\ efficiency = \frac{P_{out}}{P_{dc}} \quad (1)$$

In (1), P_{out} is the output power, and P_{dc} is DC power consumption of the TX.

This paper is organized as follow. Section II introduces the architecture of the proposed low-power TRX. Section III describes the circuit implementation. Finally, measurement results and conclusion are presented in sections IV and V, respectively.

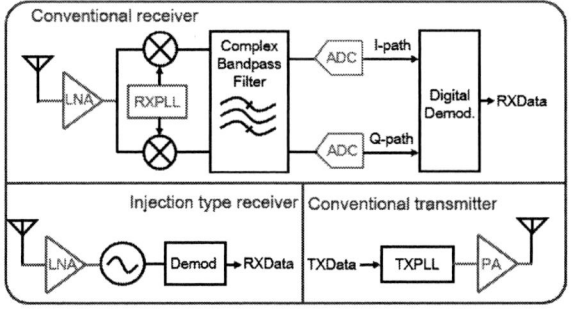

Fig. 1. Example of some existing transceiver architectures.

II. PROPOSED TRANSCEIVER ARCHITECTURE

Fig. 2 shows the block diagram of the proposed RX and TX. FSK modulation scheme is chosen in this work. The target operating frequency is at the 400-MHz MedRadio band. The FSK RX consists of a Q-enhanced LNA, a LC-VCO, a mixer, limiting amplifiers, and a frequency-to-time based demodulator. The RX design considerations are described below. First, the LNA is typically designed to provide sufficient signal amplification and suppress noise from the later stages. To facilitate low-power operation, external off-chip inductor is employed, for it offer much higher Q than that of an on-chip inductor. Second, similarly, VCO adopts off-chip inductor to take advantage of the high-Q nature to lower its power consumption. Third, considering the tradeoff between demodulation operation and power consumption of the limiting amplifiers, the intermediate frequency (IF) is chosen at 2MHz. The demodulation is based on a low-power analog implementation, which will be introduced in later section.

The TX architecture is essentially a VCO with an impedance matching network. The impedance matching network converts the power to the 50-ohm load of the antenna, which eliminates the need of a PA.

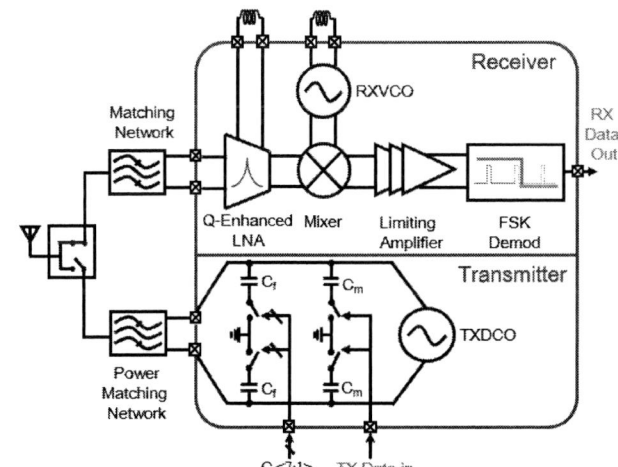

Fig. 2. Proposed transceiver architecture.

III. Circuit Implementation

A. Q-enhanced LNA

The LNA circuit is depicted in Fig. 3. It adopts a Q-enhancement technique to boost the amplifier gain. Moreover, the Gm-boosting technique is employed to further increase the LNA gain. With this technique, the transconductance, gmn, is doubled at the same bias current.

978-1-5386-3179-9/17 $31.00 © 2017 IEEE   193

In the implemented LNA, the Q-enhancement is realized with the PMOS cross-couple pair which generates a negative resistance (R_{neg}) for to cancel the loss of the LC-tank load. The R_{neg} can be expressed as

$$R_{neg} = \frac{-2}{Gmp} = \frac{-(V_{sg} - |V_{thp}|)}{I_p} \,, (I_p + IB_p = I_n) \quad (2)$$

C_p is made of a 6-bit capacitance array to control the central frequency of the resonant tank. The equivalent RLC behavior model is shown as Fig. 4. To achieve Q enhancement, the tank loss, represented by the paralleling resistor R_p, is cancelled by R_{neg} to a large extent; hence, the LNA gain is increased. The adjustable R_{neg} is realized by controlling the current, IB_p; which in term adjust the gain of the LNA to a desired value. L_p is an off-chip inductor with an inductance of 51 nH. R_p is the effective parallel resistor of the tank circuit. Its value is proportional to the square of the quality factor. The proposed LNA is designed to provide enough gain to suppress the input-referred noise (or noise figure) contributed from next stage, mixer. The power dissipation of this LNA is 63 µW.

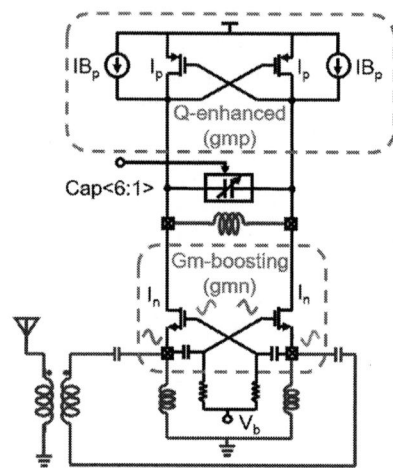

Fig. 3.   Q-enhanced LNA circuit.

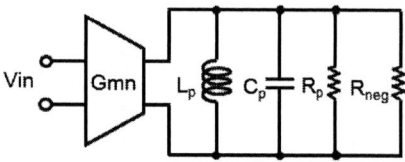

Fig. 4.   Model of the Q-enhanced LNA.

B.   Mixer, VCO, and Limiting Amplifiers

The LC-VCO, which is illustrated in Fig. 5, is implemented by off-chip inductance designed at 47 nH, with on-chip varactor. The latter component can tune the central frequency, of which operating range is 398~440MHz. The power dissipation of this oscillator is 40.2µW.

The down-conversion mixer which converts the LNA output to an IF frequency of 2 MHz.

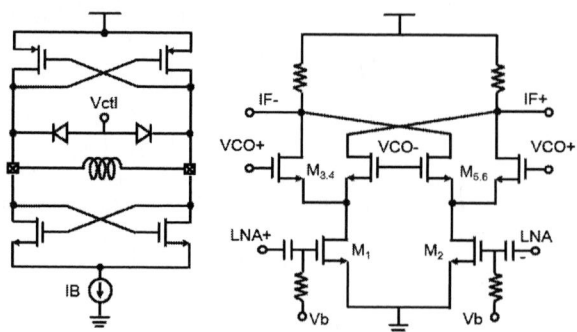

Fig. 5.   VCO and Mixer circuit

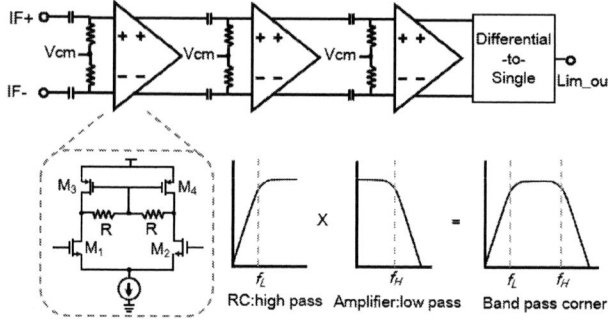

Fig. 6. Limiting amplifiers circuit

The limiting amplifier (LA), shown in Fig. 6, enlarges the signal at IF frequency to rail-to-rail waveforms. The AC coupling capacitor with the resistors defining the Vcm form a high-pass filter. Together with the amplifier's low-pass response, the LA effectively offers the band-pass characteristic, which also helps to suppress out-of-band interferes. The RC high-pass network also eliminates the DC offset from preceding stages

The LA is comprised of cascade of three amplifier stages. LA output is converted to single-ended form for the demodulation circuit. The power dissipation of the LA is only 10 µW.

C.   Frequency-to-time based FSK Demodulator

The proposed frequency-to-time based demodulator in this RX is shown in Fig. 7(a). It consists of a frequency-to-time converter and a pulse-width shaper. The demodulator operation is described as follows.

Assuming data 0 corresponds to an IF signal with a frequency of f-∆f Hz. A longer signal period is observed at the demodulator input, Lim-out (which is also the LA output). The voltage level at V_1 will be charged for a longer time and raised over V_{ref}, as illustrated in the waveforms of Fig. 7(b). Thereafter, the first comparator generates a short pulse at its output, V_{comp1}. On the other hand, for data 1, the signal frequency is at f+ ∆f. The signal has a shorter period. With a proper V_{ref1} level, V_1 will always be smaller than V_{ref}. Hence, no pulse signal is generated. As a result, with a simple charging/discharging circuit and a comparator, the FSK demodulation is accomplished with low circuit complexity.

Next, the second set of switched charging circuit (right portion of Fig 7(a); it consisted of M_2, C_{p2}, IB_{d2}, and the other comparator) is utilized to convert the pulses to non-return-to-zero (NRZ) data format. As also illustrated in the waveforms of Fig. 7(b), the pulses prevent the voltage of V_2 raises beyond reference voltage, V_{ref1}. Hence, data 0 is demodulated. Similarly, in the case of receiving data 1, voltage V_2 will be at a level close to the supply and is larger than V_{ref1}. Hence, the Data_out will be at a high level, which implies data 1 is demodulated.

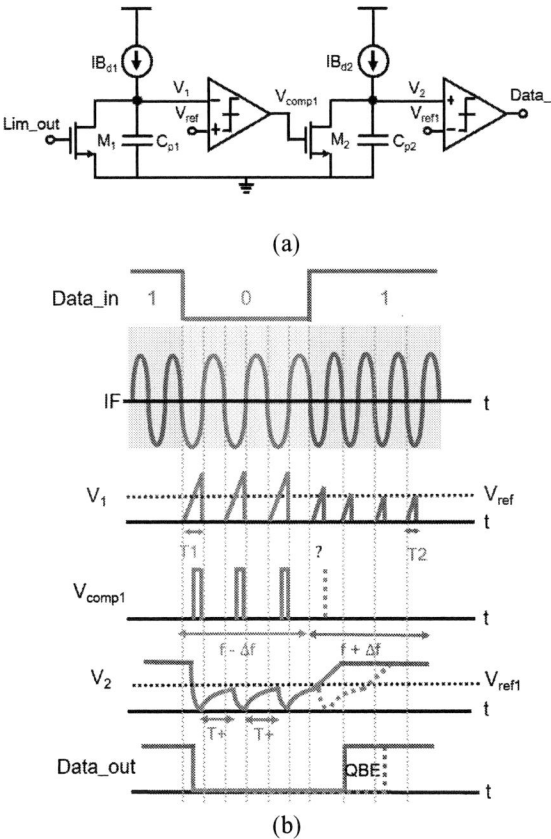

(a)

(b)

Fig. 7. (a) Proposed demodulation circuit. (b) Operation of the proposed FSK demodulation.

D. Direct Power Transfer Transmitter

The TX circuit implementation is shown in Fig. 8. M_{n1}, M_{n2}, M_{p1}, and M_{p2} constitute the cross-coupled pair that provides negative impedance. C_1, C_2, and L are composed of external components that determine the oscillation frequency and directly interfacing with the antenna. In this topology, the role of a power amplifier is merged in the VCO to save hardware and power consumption.

In this circuit, C_m define the frequency deviation of the FSK modulation. The frequency deviation is 500 kHz. The 7-bit capacitor array C_f determines the operation frequency and can be tuned over the range from 418.4 MHz to 443.6 MHz

with a 200-kHz step. The appropriate C_1 and C_2 ratio at 50-ohm antenna load achieves maximum power transfer.

Fig. 9(a) shows the LC resonance equivalent model. L_s and R_s represents the inductor series equivalent model. C_{on} and R_{on} are the capacitor and the equivalent resistance of a MOS switch in the capacitor array. Fig.9 (b) shows that L_s, R_s, C_{on} and R_{on} are converted into L_p, C_{onp} and R_p by series-parallel equivalent impedance. The ratio of R_{Lp} to R_L is approximated equal to the square of the ratio of C_2 to C_1. Obviously, a larger ratio of C_2 to C_1 results in a larger shunt resistance in the system. A larger equivalent resistance (R_{Lp}) draws less current and leads to worse global efficiency. To enhance the efficiency, it is desired to decrease this ratio. However, this is at the cost for oscillation start-up issue.

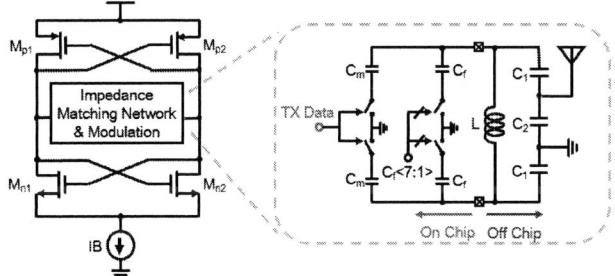

Fig. 8. Implementation of the proposed transmitter.

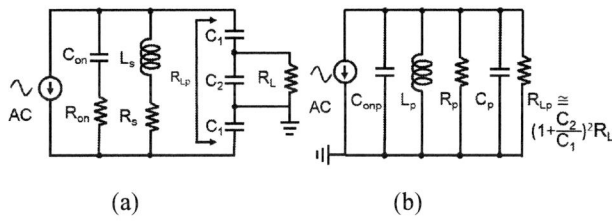

(a)   (b)

Fig. 9. (a) Series equivalent LC-tank with impedance matching. (b) Parallel equivalent transformation.

IV. MEASUREMENT RESULTS

Fig. 10 shows the measured RX waveforms. The frequency deviation is set to 500 kHz. The carrier frequency is at 429 MHz. The top waveform of Fig. 10 shows the input data at 200 kbps; the bottom one is the received output data, indicating correct demodulation. The measured BER versus input power of the RX is shown in Fig. 11. For the data rate of 200 kbps, the RX sensitivity is -85 dBm at 0.1% BER. Fig. 12 shows the RX sensitivity versus the frequency offset between the RF input frequency and the RX VCO frequency.

Fig. 13 shows the TX performance measured at a carrier frequency of 429 MHz. The output power is -8.19 dBm. The TX is operated from a 0.6-V supply voltage, and draws 0.65 mA. Fig. 14 illustrates the global efficiency versus the power consumption. The maximum efficiency is 38.7% at a power consumption of 0.39 mW. Fig. 15 shows the micrograph of the TRX chip. The active area of this RX and TX is 0.301mm^2 and 0.154mm^2, respectively.

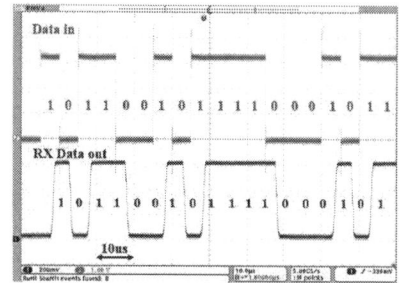

Fig. 10. Measured RX waveforms.

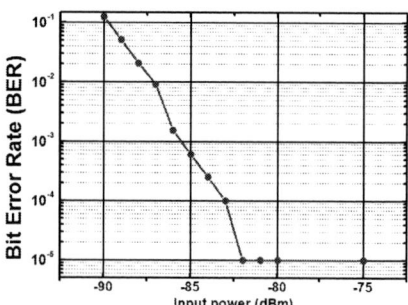

Fig. 11. Measured BER versus the input power.

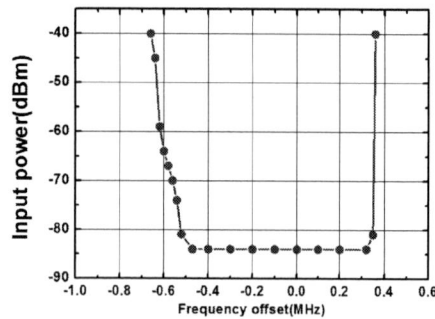

Fig. 12. RX sensitivity versus the frequency offset.

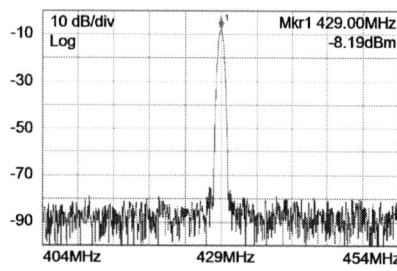

Fig. 13. TX output spectrum.

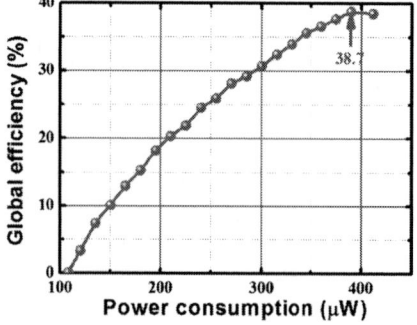

Fig. 14. TX global efficiency versus the power consumption.

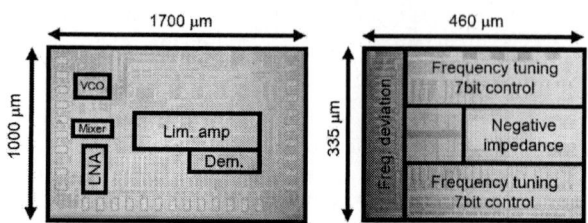

Fig. 15. Chip micrograph of the TRX.

Table I. Performance summary and comparison

	This Work	[4]	[5]	[6]	[7]	[8]
Process(nm)	90	28	180	130	180	40
Supply (V)	0.6	1	0.9	1.2	0.45	0.5
Mod.	FSK	GFSK	D-BPSK	GFSK	OOK /FSK	OOK
Freq. (MHz)	429	2400	400	2450	400	315
Data rate (MHz)	0.2	1	10	1	0.12/0.05	1
PN(dBc/Hz) @1MHz	-129.1*	-101	-	-129**	-	-
RX						
Architecture	Low-IF	PLL-Base	Injection		Low-IF	Sampling
Sen. (dBm)	-85	-95	-63	-	-55/-53.5	-55
Power (mW)	0.146	2.75	1.77	-	0.35***	0.0038
RX eff.	0.73	2.75		-	2.58/2.92	0.038
RX FoM	176.4	181.6	160.5		141/139	159
TX						
Power (mW)	0.39	3.6	0.51	4.56	-	0.052***
Output (dBm)	-8.2	0	-15	-1	-	-21
Global eff. (%)	38.7	28	6.2	17.4	-	15.3
TX energy eff. (nJ/Bit)	1.95	3.6	0.051	4.56	-	0.052
NE eff. (nJ/bit)	12.9	3.6	1.6	5.7	-	6.5

RX FoM = –Sensitivity–10 * log (Power Con. /Data Rate)
TX Energy efficiency = TX power/Data rate
(NE) Normalized Energy eff. = (TX power/data rate)/Pout
*TX LO phase noise   **phase noise @2.5MHz   ***LO off-chip

V. CONCLUSION

This paper presents an energy-efficient FSK transceiver. The proposed TX and RX do not require a PLL to set the carrier frequency. Several circuit techniques are employed I this design. In the RX, Q-enhanced LNA and low-power demodulator are proposed. In the TX, VCO is directly interfacing with the antenna to eliminate the PA. The measured RX achieves high sensitivity and low power consumption; meanwhile, the TX accomplish a good global efficiency. The proposed transceiver can be applied to energy-constrained IoT or biomedical applications.

ACKNOWLEDGEMENT

The authors thank CIC, Taiwan, for chip fabrication. This work is supported under Donation Grant FD105012.

REFERENCES

[1] C. Bachmann1 et al., *IEEE VLSI Circ. Symp*, Jun. 2015.
[2] Y. L. Tsai et al., *IEEE VLSI Circ. Symp,* Jun. 2015.
[3] R. Ni et al., *IEEE VLSI Circ. Symp,* Jun. 2014.
[4] F.W. Kuo et al., *IEEE VLSI Circ. Symp*, Jun. 2016.
[5] Y. L. Tsai et al., *IEEE RFIT*, Aug. 2016.
[6] C. Li et al., *IEEE JSSC*, Jul. 2016.
[7] J. Y. Hsieh et al., *IEEE T-CAS I*, Aug. 2016.
[8] A. Saito et al., *IEEE VLSI Circ. Symp*, Jun. 2012.

S14-1 (2028)

An 82% Energy-Saving Change-Sensing Flip-Flop in 40nm CMOS for Ultra-Low Power Applications

Van Loi Le *‡, Juhui Li‡, Alan Chang‡ and Tony T. Kim*

*VIRTUS, IC Design Centre of Excellence, School of EEE, Nanyang Technological University, Singapore
‡NXP Semiconductors, Singapore
Email: levanloi001@e.ntu.edu.sg, thkim@ntu.edu.sg

Abstract— In this paper, we propose a novel 24-transistor change-sensing flip-flop (CSFF) for ultra-low power applications. With the aid of an internal change-sensing unit, the proposed CSFF eliminates redundant transitions of internal clocked nodes when there is no change in the flip-flop content. No additional transistors are required compared to the conventional transmission-gate flip-flop (TGFF). Measurement results from a test chip fabricated in 40nm CMOS technology show that CSFF achieves the power reduction of 82% and 68% at 10% activity rate and 1.0V, and the C-Q delay improvement of 37% and 11% in the supply voltage range of 0.4V to 1.0V compared to TGFF and SSCFF. While achieving better power and energy efficiencies, CSFF still maintains robust functionality at ultra-low voltage operations. Measurement results also demonstrate that CSFF shows the minimum operating voltage of 0.19V.

Keywords— flip-flop; ultra-low voltage; ultra-low power

I. INTRODUCTION

Flip-flops are simple but critical components in modern digital integrated circuits (ICs); they typically dominate the total area and power of the overall systems. For example, the SPARC T4 processor in [1] contains 2.6 million flip-flops, consuming more than 20% of the total core power of the processor. In fact, flip-flops are typically responsible for tremendous dynamic power in systems on chips (SoCs) due to unnecessary transitions of their large number of internal clocked nodes when there are no changes in the flip-flop data.

The transmission-gate flip-flop (TGFF) (Fig. 1) is the most widely-used flip-flop since it is static, contention-free, and robust with voltage scaling; but it consumes excessive power caused by many internal clocked nodes that always toggle. These redundant transitions even occur when the input D does not change the content of the flip-flop. Several low-power flip-flops [2-6] have been proposed to resolve this major concern. However, most of them suffer from large area penalty, which is too costly since flip-flops typically occupy a significant amount of logic area in SoCs. The adaptive-coupling flip-flop (ACFF) in [2] could avoid the area overhead by only employing the adaptive-coupling element (ACE) in the master latch and removing the ACE in the slave latch as depicted in Fig. 2, but this could result in higher operating voltages (≥0.75V) due to the contention in the slave latch. The static single-phase contention-free flip-flop (SSCFF) in [7] provides robust energy-saving operations over a wide range of supply voltages with no area overhead, but SSCFF only eliminates

unnecessary transitions of its clocked nodes when input data stay at the logic '*high*' level. Indeed, SSCFF still suffers from power penalty due to redundant transitions of its clocked node *CN* when the input D does not change its state from the logic '*low*' level as illustrated in Fig. 3. TABLE I summarizes the comparison of the previous flip-flops.

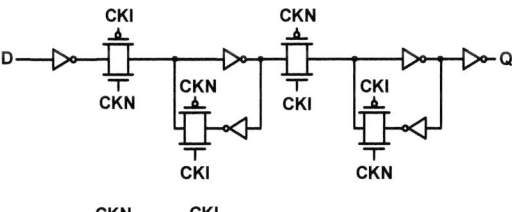

Fig. 1. Schematic of conventional TGFF

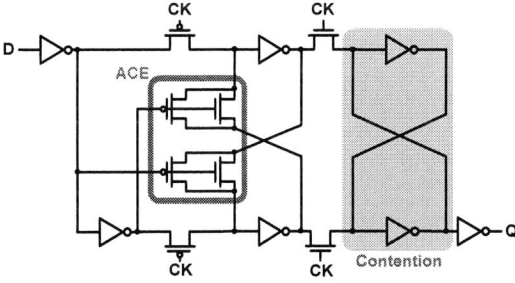

Fig. 2. Schematic of ACFF [2]

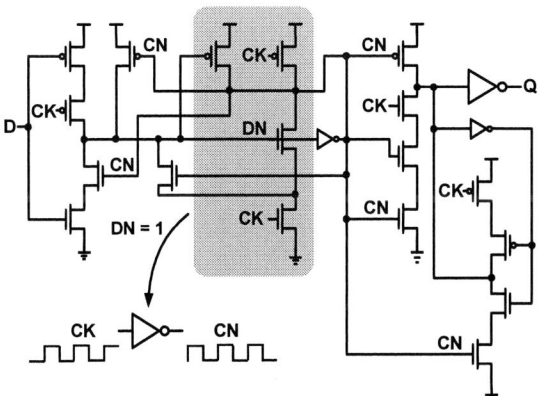

Fig. 3. Schematic of SSCFF [7]

978-1-5386-3179-9/17 $31.00 © 2017 IEEE   197

TABLE I. COMPARISON OF THE PREVIOUS FLIP-FLOPS

	TGFF	ACFF [2]	SSCFF [7]
Ultra-low Voltage Operation	YES	NO	YES
Redundant Transitions of Internal Clocked Nodes	YES	NO	YES
Transistor count	24	22*	24

*It becomes 26 if ACE is employed in the slave latch for ultra-low voltage operations.

In this paper, we present a novel D-flip-flop sensing the changes in the flip-flop data for eliminating redundant transitions of internal clocked nodes. This significantly reduces the dynamic power when there is no change in the flip-flop content. It also achieves robust functionality at ultra-low voltage operations and further improves performance compared to the conventional TGFF and SSCFF.

II. PROPOSED CHANGE-SENSING FLIP-FLOP (CSFF)

Fig. 4 illustrates the schematic of the proposed 24-transistor change-sensing flip-flop (CSFF). The master consists of T1~T3, and its latch (T7, T10~T14) while the slave is composed of T10, T15~T19, and its latch (T9, T20~T24). The change-sensing unit (T4~T9) as highlighted in Fig. 4 detects the change in the data input D. T6 and T7 detect the change from low to high while T8 and T9 detect the change from high to low. With the aid of the internal change-sensing unit, the clocked node CS only toggles when the data input D changes its state. When the current input data are continuously the same as the previous data stored in the slave latch, the internal clocked node CS is always held high by T4. Thus, CSFF completely removes the redundant transitions, which always happens in TGFF even when there are no changes in the flip-flop data, and occurs in SSCFF when the input D persistently stays at the logic '*low*' level as illustrated in Fig. 5.

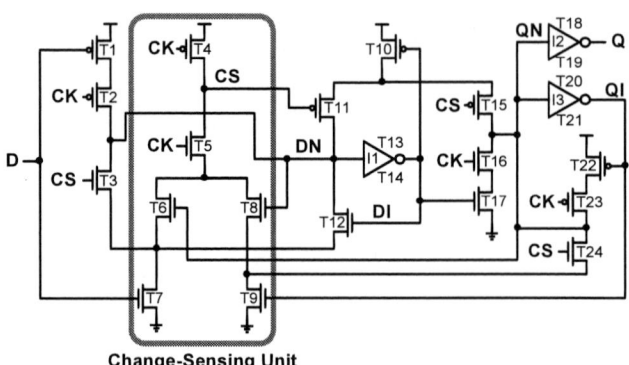

Fig. 4. Schematic of the proposed CSFF

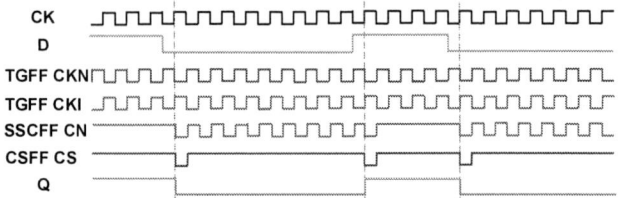

Fig. 5. Transitions of internal clocked nodes in TGFF, SSCFF, and CSFF

Fig. 6 shows the operation details of the proposed CSFF. When CK=0, the clocked node CS pre-charges through T4. The master passes the new data to the flip-flop while the slave latch holds the previous data. When CK goes high, the change-sensing unit detects the change in D and discharges the clocked node CS through the '*high-to-low-sensing*' path (T5, T8~T9) or the '*low-to-high-sensing*' path (T5~T7). Then, the master latch keeps the updated data, and the slave passes the data to the output. If the updated level is low, QN is held high by T10 and T15; thus, the output Q goes low. In the case that the updated level is high, QN is kept low by T16 and T17; then, the output Q goes high.

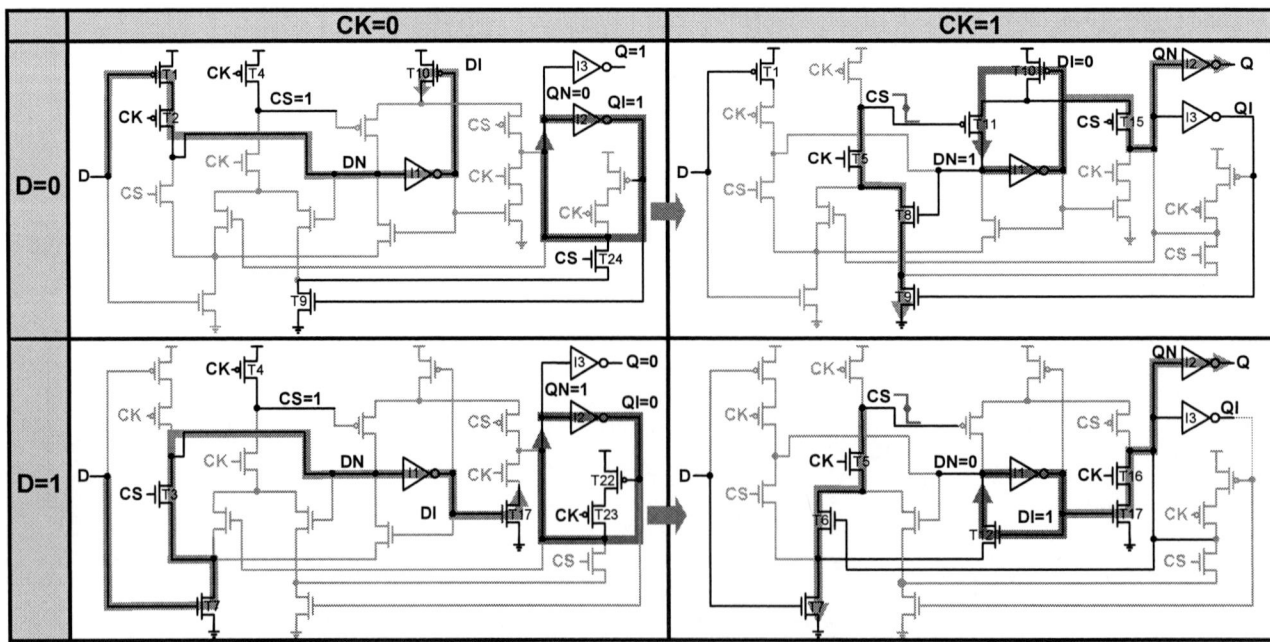

Fig. 6. Operation details of the proposed CSFF

III. Testing Circuit Implementation

A. C-Q delay/ Setup/ Hold Time Testing Circuit

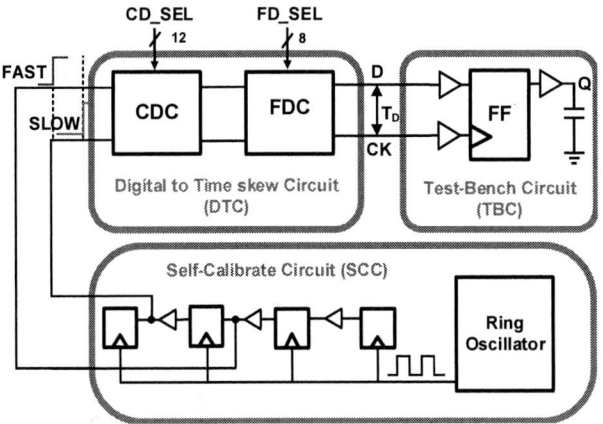

Fig. 7. C-Q delay/ setup/ hold time testing circuit

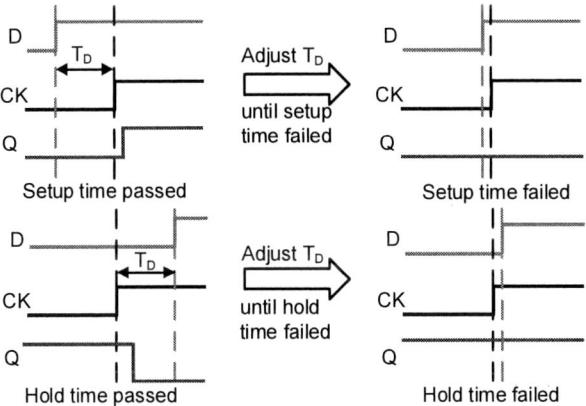

Fig. 8. Setup/hold time measuring method

The C-Q delay/ setup/ hold time testing circuit is illustrated in Fig. 7, which is based on the structure in [8]. It includes Self-Calibration Circuit (SCC), Digital to Time skew Circuit (DTC), and Test-Bench Circuit (TBC). TBC mainly consists of flip-flops and necessary buffers for measuring C-Q delay. DTC includes Coarse Delay Control (CDC) and Fine Delay Control (FDC) blocks for generating digitally-controlled delay skew T_D between the data D and the clock CK. For T_D control with a larger step T_C, CD_SEL in CDC is adjusted to control the number of the coarse delay unit in the data path and the clock path. For T_D control with a smaller step T_F, FD_SEL in FDC is used. The ring oscillator in SCC generates the internal clock for the D-flip-flop chain to create different FAST and SLOW signals for measuring the coarse delay T_C and the fine delay T_F. Once T_C and T_F are known, the setup/ hold time of the flip-flops can be accurately estimated by adjusting the time delay T_D between the clock and the data to the setup/hold time pass/fail point as illustrated in Fig. 8.

B. Power Testing Circuit

Fig. 9 illustrates the power testing circuit for measuring power consumptions of the flip-flops in different activity rates.

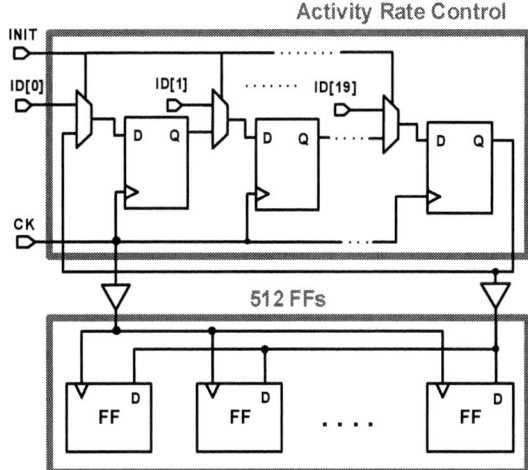

Fig. 9. Power testing circuit

When INIT=0, the activity rate is initialized by a 20-bit data input ID. When INIT goes high, the activity rate is applied, and total power consumption of the flip-flops is measured. To accurately measure power consumptions of the flip-flops (TGFF, SSCFF, and CSFF), 512 units of each flip-flop are implemented, and the average values are determined.

IV. Measurement Results

CSFF was characterized in a 40nm CMOS test chip, and TGFF and SSCFF were also implemented in the same test chip for fair comparisons. Fig. 10 (a) and (b) demonstrate the measured power consumption of the flip-flops at 1.0V and 0.4V, respectively. With the aid of the change-sensing unit, CSFF mostly consumes only leakage power at 0% activity rate, and it also provides more energy-efficient operations at low activity rates compared to TGFF and SSCFF. The average activity rate of flip-flops in SoCs is typically from 5% to 15%; CSFF saves 84% to 60% and 90% to 77% power compared to SSCFF and TGFF, respectively, in this range. From the power measurement, at 10% activity rate, CSFF has 82% and 84% lower power consumption, compared to TGFF, at 1.0V and 0.4V, respectively. When compared to SSCFF, CSFF exhibits the reduction of 68% and 69%. At 20% activity rate, CSFF shows 71% and 74% improvement in total power consumption vs TGFF, at 1.0V and 0.4V, respectively. Compared to SSCFF, CSFF has 51% and 48% improvement. In terms of energy consumption, CSFF exhibits 82% and 84% reduction, compared to TGFF, and 68% and 69% improvement, compared to SSCFF, at 1.0V and 0.4V, respectively, as shown in Fig. 10 (c). Fig. 10 (d) shows the measured leakage power of the flip-flops. CSFF has 40% and 38% lower energy consumption at 1.0V and 0.4V, respectively, compared to TGFF. CSFF also provides a slight improvement of 10% in leakage over SSCFF. Fig. 10 (e) shows the measured C-Q delay of the flip-flops. CSFF shows 37% delay reduction over TGFF in the operating voltage range of 0.4V to 1.0V. Compared to SSCFF, CSFF only provides a modest delay improvement of 11%. Fig. 10 (f) shows the die micrograph of the test chip. TABLE II includes the measured setup and hold time, and summarizes the measurement comparison of the flip-flops. With the same transistor count as TGFF and SSCFF,

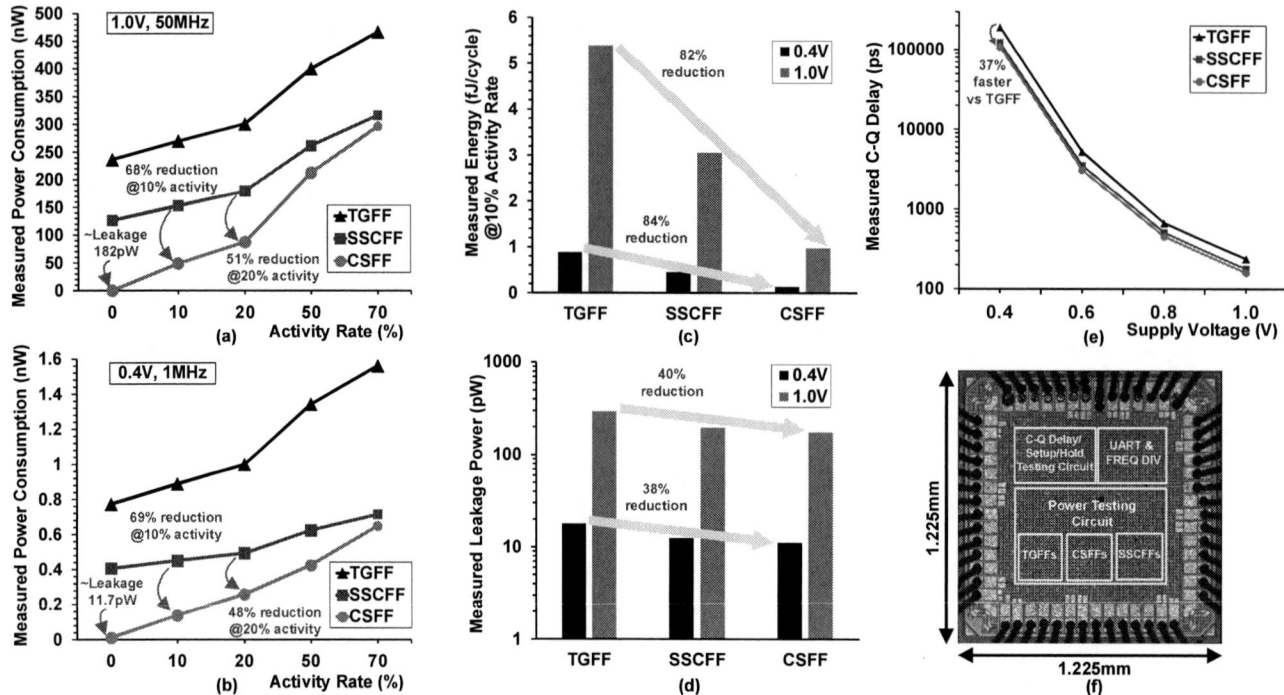

Fig. 10. (a) Measured power at 1.0V, (b) Measured power at 0.4V, (c) Measured energy, (d) Measured leakage, (e) Measured C-Q delay, (f) Die micrograph.

TABLE II. MEASUREMENT COMPARISON OF THE FLIP-FLOPS

	CSFF (This work)	SSCFF Y. Kim, ISSCC 2014	TGFF Conventional Flip-Flop
Number of Transistors	24	24	24
Normalized Layout Size	1.03	1.03	1.00
Redundant Transitions of Internal Clocked Nodes	NO	YES	YES
Measured Total Power @1.0V, 50MHz, 10% Activity	**48.8nW**	153nW	270nW
Measured Leakage @1.0V	176pW	197pW	295pW
Measured C-Q delay @1.0V	160ps	177ps	238ps
Measured Setup Time @1.0V	266.5ps	271.5ps	195ps
Measured Hold Time @1.0V	-63ps	-54ps	-18ps

and with only a 3% layout size increase vs. TGFF that corresponds to a one-minimum-poly-pitch increase in 40nm technology, CSFF eliminates redundant transitions of internal nodes, which always occurs in TGFF, and happens in SSCFF when the flip-flop data continuously stay at the logic '*low*' level. Therefore, CSFF achieves 82% and 68% power improvement over TGFF and SSCFF, respectively, at 1.0V and 10% activity rate. In terms of performance, CSFF exhibits 5% and 1.5% improvement in total setup time and C-Q delay, compared SSCFF and TGFF, respectively. CSFF also provides a better hold time compared to TGFF and SSCFF, requiring a lesser constraint for hold buffers in ICs. Measurement results also show that CSFF is functionally operational down to 0.19V.

V. CONCLUSION

In this paper, we introduced a novel change-sensing flip-flop (CSFF) for ultra-low power applications. The proposed CSFF eliminates redundant transitions of its clocked node without additional transistors when there is no change in the flip-flop content. The test chip fabricated in 40nm technology demonstrates that CSFF achieves more energy-efficient operations than TGFF and SSCFF across a wide operating voltage range ($\geq$0.19V). CSFF shows the energy reduction of 82% and 68% at 1.0V and 10% activity rate, and the C-Q delay improvement of 37% and 11% in the supply voltage range of 0.4V to 1.0V compared to TGFF and SSCFF, respectively. Therefore, CSFF can directly replace many of the flip-flops in ultra-low power applications especially when the input data have relatively low activity rates.

REFERENCES

[1] J. L. Shin, et al., "The Next Generation 64b SPARC Core in a T4 SoC Processor," *IEEE J. Solid-State Circuits*, vol. 48, no. 1, pp. 82–90, 2013.

[2] C.-K. Teh, T. Fujita, H. Hara, and M. Hamada, "A 77% Energy-Saving 22-Transistor Single-Phase-Clocking D-Flip-Flop with Adaptive-Coupling Configuration in 40nm CMOS," *ISSCC Dig. Tech.Papers* pp. 338–339, 2011.

[3] S. Nomura, et al., "A 9.7mW AAC-Decoding, 620mW H.264 720p 60fps Decoding, 8-Core Media Processor with Embedded Forward-Body-Biasing and Power-Gating Circuit in 65nm CMOS Technology," *ISSCC Dig. Tech. Papers*, pp. 262-264, 2008.

[4] Y. Ueda, et al., "6.33mW MPEG Audio Decoding on a Multimedia Processor," *ISSCC Dig. Tech. Papers*, pp. 1636-1637, 2006.

[5] M. Hamada, et al., "A Conditional Clocking Flip-Flop for Low Power H. 264/MPEG-4 Audio/Visual Codec LSI," *IEEE Custom Integrated Circuits Conference*, pp. 527-530, 2005.

[6] B.-S. Kong, S.-S. Kim, and Y.-H. Jun, "Conditional-capture flip-flop for statistical power reduction," *IEEE J. Solid-State Circuits*, vol. 36, pp. 1263–1271, 2001.

[7] Y. Kim, et al., "A Static Contention-Free Single-Phase-Clocked 24T Flip-Flop in 45nm for Low Power Applications," *ISSCC Dig. Tech. Papers*, pp. 466-467, 2014.

[8] L. Zhihong, Z. Yihao, H. Law, "Self-Calibrate Two-Step Digital Setup/Hold Time Measurement," *International Symposium on VLSI Design, Automation and Test*, pp. 232-235, 2010.

NBTI/PBTI separated BTI monitor with 4.2x Sensitivity by Standard Cell Based Unbalanced Ring Oscillator

Mitsuhiko Igarashi, Yoshio Takazawa, Yasumasa Tsukamoto, Kan Takeuchi and Koji Shibutani

Design Platform Business Dep., Renesas Electronics Corporation, Tokyo, Japan
mitsuhiko.igarashi.xv@renesas.com

Abstract— We propose an on-chip bias temperature instabilities (BTI) monitor by using standard cell based unbalanced ring-oscillator (RO). The monitor consists of NAND and NOR with extremely large difference in drive strength, which enables 4.2x sensitivity to BTI compared with normal INV based RO. This originates not only from accentuation of the degraded stage with small drive strength by the dominant delay but also from ΔVth sensitivity improvement of stacked transistors with increasing output transition time. In addition, our proposal allows to monitor either one of negative BTI (NBTI) or positive BTI (PBTI) selectively by making use of the Miller effect caused by the size combination of NOR and NAND. A test chip is implemented in a 28 nm High-k Metal-Gate (HKMG) CMOS technology. We successfully observed that measured results of each RO were well matched with simulations. Owing to this high sensitivity and successful NBTI/PBTI separation, we can detect variations and outliers of BTI at time of testing, and optimize guard band (GB) while considering NBTI/PBTI sensitivity difference of each circuit.

Keywords— *NBTI; PBTI; Standard Cell based Ring Oscillator; Unbalanced; Miller effect; Guardband; Burn-In;*

I. INTRODUCTION

In autonomous driving era, it is inevitable to pursue higher performance for automotive LSIs such as advanced driver-assistance systems (ADAS), cognitive application and so on. Figure 1 shows trend of our automotive LSI products mainly for in-vehicle infotainment (IVI) and ADAS [1]. The required higher performance should be achieved under limited power budget at high temperature condition of the car. Therefore, higher performance and power optimization are one of the key challenges for autonomous driving LSI. The dynamic voltage and frequency scaling (DVFS) technique is often used to optimize power efficiency at each operation scene and to get higher performance by voltage boost. However, voltage boost leads to aging degradation such as BTI and Hot Carrier Injection (HCI). In car usage point of view, 12K hours operating time, which corresponds to 9.1% to 15 years field lifetime, is generally supposed in AEC-Q100 [2]. However, an annual mileage of the car is expected to be increased dramatically in autonomous driving era because of paradigm shift such as car-sharing, ride-sharing and so on. Considering above paradigm shift, more operating time will be required in autonomous driving LSI, resulting in increase of aging degradation. The increase of aging degradation leads to increase GB and restricts the performance. PBTI caused by HKMG process also restricts the performance. Since BTI is a major factor of reliability in performance point of view, we focus on BTI in this paper.

Figure 2 shows voltage, temperature and operating time dependency of BTI, and its variation. Each dependency and variation are estimated from foundry process reliability models. Since BTI has large dependency on voltage and temperature, applying DVFS at high temperature environment has a large impact on BTI degradation in a field. It is known that NBTI and PBTI impact are dependent on each circuit [3-4]. The aging monitor system, which is easy to implement in products with features of high sensitivity and NBTI/PBTI separations, could be a solution to optimize actually required GB in a field [1, 13]. In addition, these features help GB optimization and BTI outlier detection at a time of testing before shipment; there are not only process variation but also aging variation between chip-chip, wafer-wafer and lot-lot. However, since a stress of Burn-In (BI) test which is applied for early defect rejection is effectively small compared to that of end of life, the normal INV based RO is not enough to detect BTI variation due to small BTI sensitivity. In this way, high sensitive and NBTI/PBTI separated BTI monitor is required. Note that there is another approach to predict lifetime in a field by environmental parameter (voltage and temperature) with and without dedicated monitor [5-7]. From functional safety point of view, combinational usage of direct monitoring of reliability and in-direct prediction could be a redundant system.

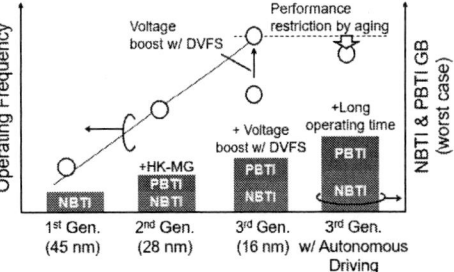

Figure 1: Schematic trend of our automotive IVI/ADAS products and BTI GB

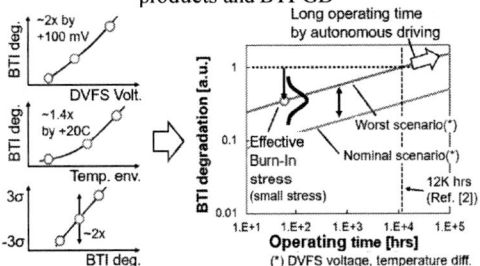

Figure 2: Fluctuation and variation factors of BTI degradation and Burn-In test stress

II. PROPOSED BTI MONITOR CIRCUIT

Figure 3 shows comparison of schematic view between proposed BTI monitor and conventional one. Several RO based BTI monitors have been proposed [8-15]. The standard configuration of RO is INV based (INV-RO) [8-11]. As shown in Figure 3 (b), the path gate (PG) is attached to the INV-RO to apply NBTI selectively [12]. The combination of INV and NOR was also proposed to enhance NBTI sensitivity [13] shown in figure 3 (c). Our proposed RO has two types; one is NBTI sensitive RO (NBTI-RO) and the other is PBTI sensitive RO (PBTI-RO.) The NBTI-RO consists of NOR with the smallest drive strength and the highest PMOS stack count whereas NAND with the largest drive strength in standard cell libraries and the all input pin of each NOR and NAND is the common node. Note that these NOR and NAND cells are normally included in a set of standard-cell libraries, which realizes easy implementation without any complicated design. The PBTI-RO is reverse configuration of the NBTI-RO. The GB optimization by defining a critical path replica has been proposed [16]. However, since there are several critical paths with different NBTI/PBTI sensitivity in an actual design, the critical path replica defined above might not be enough to track actual path's degradation.

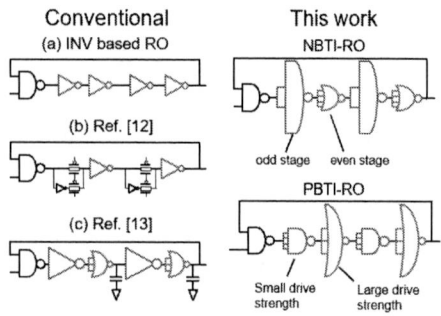

Figure 3: Topology comparison of this work with conventional BTI monitors

Figure 4 depicts delay balance and NBTI/PBTI impact of each RO topology under non-oscillation at a time of stress period (DC stress). Those are estimated by SPICE simulation for 28 nm HKMG process. tpdhl_1, tpdlh_1 mean delays at rise and fall input signal of odd stage cell respectively while tpdhl_1 and tpdlh_2 are those of odd stage cells respectively. The delay is normalized by cycle time of each RO. In the case of INV-RO, since each delay is balanced and the input pins at even and odd stage are high and low level under DC stress, around 25 % of the total delay is affected by NBTI and PBTI, respectively. In the case of ref. [12], since each input pin becomes low level by PG control, around 50 % of the total delay is affected by NBTI. This means 2x NBTI sensitivity compared to INV-RO. In ref. [13], the fall input delay of NOR becomes large because of weak drivability of stacked PMOS, resulting in around 2x NBTI sensitivity. Note that we carefully estimate this result to reproduce the result of ref. [13]. Compared with these results, the proposed RO has the largest delay ratio of tpdlh_2 in NBTI-RO and tpdhl_1 in PBTI-RO, which results in 2.8x and 2.5x sensitivity compared to INV-RO in term of delay balance.

Figure 5 demonstrates comparison of NBTI sensitivity with large output transition between INV and NOR cell. In the case

of INV, the gate-source voltage (Vgs) of the PMOS, which is the difference between the input node and power supply VDD, immediately rises to VDD when the input transition is much smaller than the output transition. On the other hand, in the case of NOR, Vgs2 of the stacked PMOS close to the output becomes small during transition of the output node as shown in Figure 5. Since small (Vgs - Vth) makes Vth degradation influential on delay (Tpd), the sensitivity of NOR to NBTI increases with increasing output transition. Here Vth is the threshold voltage of the PMOS. As a result, the NBTI-RO can achieve higher NBTI sensitivity than expected merely from the dominant delay of NBTI-suffered NOR explained in Figure 4. The similar effect can be obtained on PBTI sensitivity at rise input propagation of NAND in PBTI-RO. Note that the dull rise output of NOR suffered from NBTI further degrades the next stage NAND delay and thus the NBTI impact is further enhanced.

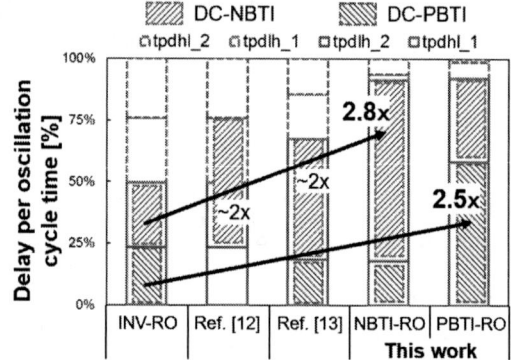

Figure 4: Delay components and NBTI/PBTI impact under DC stress

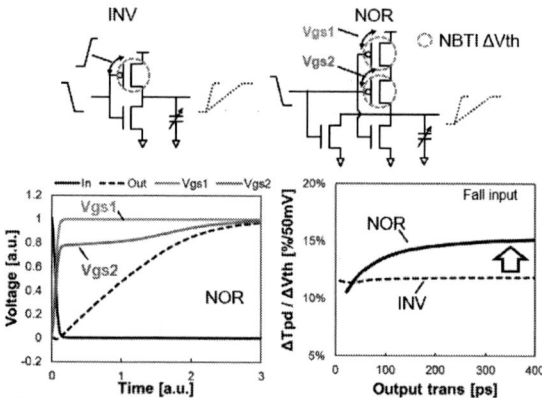

Figure 5: NBTI sensitivity improvement by small input and large output transition of NOR

Figure 6 compares fresh and aged waveform of NBTI-RO when the fall input signal propagates to NOR. The dotted line indicates fresh waveform while the solid line indicates PBTI stressed one. The output transition of NOR becomes slow because of large capacitance of NAND whereas that of NAND becomes fast because of its large drive strength. In such a case, a large Miller effect occurs during transient period of NAND. As a result, the input signal of NAND becomes substantially flat during its transition period of the NAND output. In other words, the amount of this constant Vgs of NAND NMOS is dominant factor for rise input delay of NAND. Then the Miller effect compensates the PBTI impact on the NAND delay by pulling up the flat voltage during transition. The pulling-up originates from the increased logical threshold triggering the NAND output transition. In this way, the influence of PBTI degradation is attenuated by Miller effect, enabling small PBTI sensitivity of NBTI-RO.

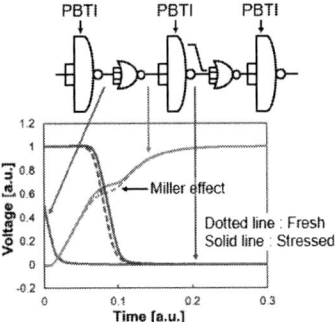

Figure 6: Attenuation of PBTI impact on NBTI-RO by Miller effect

Figure 7 shows estimation result of NBTI and PBTI sensitivity of proposed monitor and references normalized by each of INV-RO. AC stress means that RO is oscillating during stress period and thus all NMOS and PMOS suffer from PBTI and NBTI stress during ON / OFF operation, respectively. Ref. [12] and [13] have 2x NBTI sensitivity at DC stress condition, the proposed NBTI-RO and PBTI-RO have 4.2x, owing to the combinational effects described in Figs. 4 and 5. In NBTI and PBTI separation point of view, INV-RO and ref. [13] have at most 2x difference in NBTI and PBTI sensitivity while the proposed NBTI-RO and PBTI-RO realize 14x and 17x sensitivity difference. In this way, the proposed BTI monitor provides the highest BTI sensitivity and achieves NBTI and PBTI separation. Thereby, for example, observable 4.2 % degradation would occur after BI when INV-RO have 1 % degradation, which is enough to detect BTI variations and outliers before shipment. In addition, NBTI-RO and PBTI-RO also have 5x sensitivity difference under AC stress although other references generate almost no sensitivity difference. Table 1 summarizes the BTI monitor comparison. Note that proposed BTI monitor can suppress HCI influence rather than enhancing [1] and can reduce operation power compared with the case when large drive strength INV is used. This is because stacked NMOS and PMOS help to reduce voltage stress of HCI and short circuit power during transient time.

	Transistor level	INV based	Ref. [12] IRPS 2015	Ref. [13] IRPS 2013	This work
Diagram of monitor circuit					
NBTI/PBTI sensitivity enhancement	-	1x	~2x	~2x	4.2x @DC stress
NBTI/PBTI separation (DC stress)	Yes	No (Mixed)	Yes (NBTI or PBTI only stress)	No (Mixed)	Yes (14~17x resolution)
NBTI/PBTI separation (AC stress)	-	No	No	No	Yes (4~5x resolution)
Detect NBTI/PBTI variation at Burn-In	Yes	No	No	No	Yes
Implement in product	Hard	Easy (Std. cell only)	Little difficult (PG needed)	Easy (Std. cell only)	Easy (Std. cell only)
Measure after shipping	No	Yes	Yes	Yes	Yes

Table 1. Comparison of BTI monitor

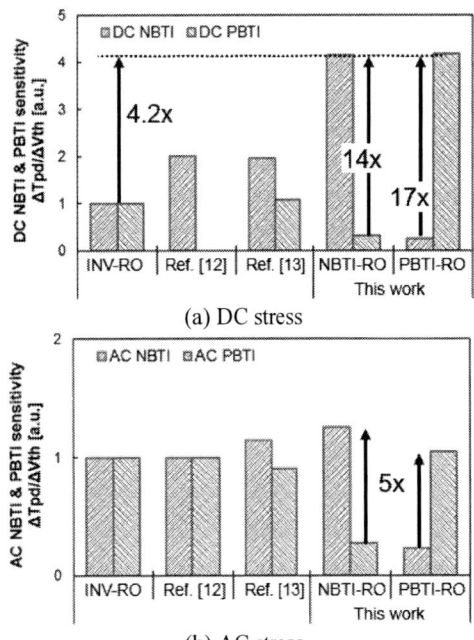

(a) DC stress

(b) AC stress

Figure 7: Estimated NBTI and PBTI sensitivity comparison; (a) DC stress, (b) AC stress

III. TEST CHIP MEASUREMENT

We have fabricated a test chip in 28nm HKMG process to confirm the characteristics of NBTI-RO and PBTI-RO. Figure 8 shows die phot and layout of test chip. There are three types of ROs, NBTI-RO, PBTI-RO and PBTI-RO with PG, with 61 stages, respectively. The PBTI-RO with PG has both "NBTI only" mode and "mixed NBTI and PBTI" mode to confirm small sensitivity of NBTI in PBTI-RO. The 2-input NAND-based RO with one of input pins fixed to high is also implemented for reference.

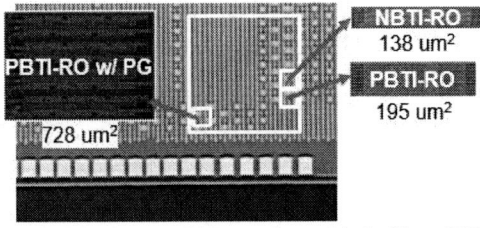

Figure 8: Die photo and layout of test chip in 28nm HKMG

Figure 9 shows measured result of NBTI-RO and PBTI-RO for 5 chips under 125C and 9 hours DC stress. Vth degradations (ΔVth) of NBTI and PBTI are estimated by the least mean square method from simulated NBTI/PBTI sensitivity ΔTpd/ΔVth and measured ΔTpd. Then, we decomposed the components of NBTI and PBTI of each measured ΔTpd by estimated ΔVth. The measured ΔTpd plots (shown by #1~#5) are well correlated with the simulation result (sum of red and blue bars) even though those ROs have significant sensitivity difference of NBTI and PBTI.

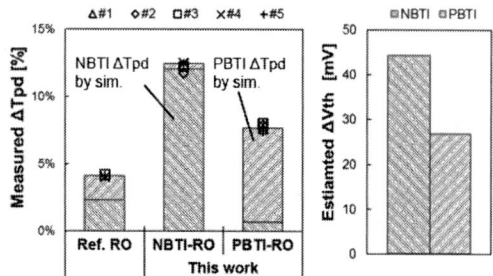

Figure 9: Measured result and estimated NBTI/PBTI impact

Figure 10 shows measured result of PBTI-RO with PG. NAND and NOR have PBTI and NBTI stress respectively in "mixed NBTI and PBTI" mode at a time of stress period, whereas both NAND and NOR have NBTI stress in "NBTI only" mode by connecting input and output node to low and high level by PG and 3-state buffer. The reference RO with PG in "NBTI only" mode is estimated for reference. Measured result shows that ΔTpd of "NBTI only" mode is much smaller than that of "mixed NBTI and PBTI" mode like the simulation result despite the degradation of NBTI is around 45 mV and larger than that of PBTI. In this way, the proposed PBTI-RO can realize the highest sensitivity of PBTI as well as separation of NBTI and PBTI.

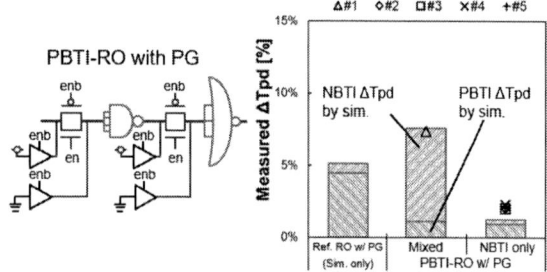

Figure 10: Measured result of PBTI-RO with PG for confirmation of small NBTI sensitivity

IV. FUTURE WORK

For feasibility study of scalability of the proposed BTI monitor, we estimate the characteristic of NBTI-RO and PBTI-RO when these are applied to scaled down to 16nm and beyond with FinFET process. We use foundry's commercial SPICE model in each process. As shown in Figure 11, our BTI monitor also has almost the same characteristics even if transistor becomes FinFET and scaled beyond 10 nm.

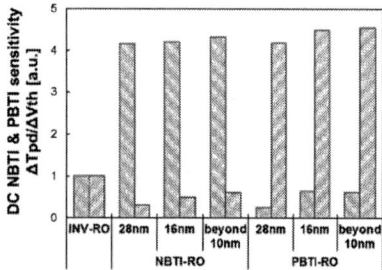

Figure 11: Estimation result of proposed BTI monitor at 16 nm and beyond 10 nm FinFET process

V. CONCLUSION

We propose an on-chip BTI monitor by standard cell based RO with unbalanced drive strength combination of NAND and NOR cells. There are two types of RO; one is NBTI sensitive RO and the other is PBTI sensitive RO. The NBTI-RO and PBTI-RO have 4.2x high sensitivity at DC stress compared with normal INV-based RO because of the unbalanced delay and ΔVth sensitivity improvement of stacked MOS by increasing output transition time. The Miller effect caused by large drive strength cell mitigates PBTI influence of NBTI-RO and vice versa, which enables separation into NBTI and PBTI. The high sensitivity and NBTI / PBTI separation help to optimize GB in design and predict GB in a field while considering NBTI and PBTI sensitivity difference of the circuit. In addition, the monitor can detect variations of BTI at a time of testing because burn-in for early defect rejection degrades the proposed BTI monitor by around 4.2 % when the INV-RO degradation is 1 %. This would allow further GB optimization and quality improvement by detecting BTI outliers. The proposed BTI monitor can be a solution to satisfy both high performance and high reliability for autonomous driving LSI.

REFERENCES

[1] M. Igarashi et al., "An On-die Digital Aging Monitor against HCI and xBTI in 16 nm Fin-FET Bulk CMOS Technology," ESSCIRC, pp. 112-115, 2015

[2] Automotive Electronics Council, "Failure Mechanism Based Stress Test Qualification for Integrated Circuits," AEC-Q100 Rev-H, 2014.

[3] H. Amrouch et al, "Reliability-aware design to suppress aging", DAC, 2016

[4] S. Kiamehr et al, "Aging-aware timing analysis considering combined effects of NBTI and PBTI," in ISQED, 2013, pp. 53-59

[5] A. Grenat et al, "Increasing the performance of a 28nm x86-64 microprocessor through system power management," ISSCC Dig. Tech. Papers, pp. 74-75, 2016.

[6] S. Mihara et.al, "Mission Profile Recorder: An Aging Monitor for Hard Events," IRPS, 4C.3, 2016.

[7] K. Takeuchi et al, "FEOL/BEOL wear-out estimator using stress-to-frequncy conversion of voltage/temperature-sensitive ring oscillators for 28nm automotive MCUs," ESSCIRC, pp. 265-268, 2016

[8] T. H. Kim et al, "Silicon Odometer: An On-Chip Reliability Monitor for Measuring Frequency Degradation of Digital Circuits," IEEE J. Solid-State Circuits, vol. 43, no. 4, pp. 874-880, March 2008

[9] S. Satapathy et al, "A Revolving Reference Odometer Circuit for BTI-Induced Frequency Fluctuation Measurements under Fast DVFS Transients," IRPS, 6A.3, 2015

[10] E. Seneyoshi et al, "A Precise-Tracking NBTI-Degradation Monitor Independent of NBTI Recovery Effect," Dig. Tech. Papers, pp. 192-193, 2010.

[11] K. K. Kim, W. Wang, K. Choi, "On-Chip Aging Sensor Circuits for Reliable Nanometer MOSFET Digital Circuits," IEEE Trans. on Circuits and Systems, vol.57, no.10, pp.798-802, 2010.

[12] P.F. Lu et al, "Long-term data for BTI degradation in 32nm IBM microprocessor using HKMG technology," IRPS, 6A.2, 2015

[13] M. Chen et al, "Aging Sensors for Workload Centric Guardbanding in Dynamic Voltage Scaling Applications," IPRS 2013.

[14] M. H. Haung et al, "The Impact and Implication of BTI/HCI Decoupling on Ring Oscillator," IRPS, 2D.6, 2014

[15] Y. C. Haung et al, "Re-investigating the Adequacy of Projecting Ring Oscillator Frequency Shift from Device Level Degradation," IRPS, 2D.6, 2014

[16] M. Cho et al, "Aging-aware Adaptive Voltage Scaling in 22nm high-K/metal-gate tri-gate CMOS," CICC, pp. 28-30, 2015.

IEEE Asian Solid-State Circuits Conference
November 6-8, 2017/Seoul, Korea

A 0.44V-1.1V 9-Transistor Transition-Detector and Half-Path Error Detection Technique for Low Power Applications

Xinchao Shang, Weiwei Shan*, Longxing Shi, Xing Wan and Jun Yang
Department of Electrical Engineering
Southeast University
Nanjing, P. R. China
Email: wwshan@seu.edu.cn

Abstract—To reduce conservative timing margin, many timing-error detection techniques by monitoring selected critical paths had been proposed. However, traditional adaptive methods incur significant area overheads and cannot prevent the error that is forming in the current clock cycle. In this paper, a low-overhead Transition-Detector (TD) with a 9-transistor current sensing circuit is proposed. TDs are inserted at the half-path point of the critical path, so that the timing error in the current clock cycle can be prevented. To solve the increasing problems of monitoring paths due to inserted point deviations, a selection method of half-path insertion point is proposed. The proposed method effectively reduces the number of inserted TDs by up to 69% compared with the number of monitors inserted at path endpoints. Test chips employing the proposed design approach are fabricated in 40nm CMOS. Silicon measurements demonstrate that the whole design has achieved up to 50.5% energy saving with 120mV additional voltage scaling compared to the conventional worst-case design at the expense of 3.1% area overhead.

Keywords—timing error detection; half-path point; selection method; transition detector; voltage scaling

I. INTRODUCTION

In traditional digital integrated circuit design, a certain timing margin should be reserved to ensure that the chip can work correctly under process, voltage and temperature (PVT) variations during design time. To decrease or eliminate worst-case safety margins, the in-situ timing monitoring techniques such as Razor II [1], Transition-Detector (TD) based error detection and correction [2][3], Razor-lite [4], and iRazor [5] have been proposed through real-time detection of chip operation. Compared to the path endpoints monitoring technique, a half-path point monitoring method has advantages in early errors prediction at half clock cycle time. Thus it can prevent the error that is forming in the current clock cycle with no need of error-recovery.

However, most of half-path monitoring approaches [6-7] focused on either super-threshold or near-threshold voltage (NTV), few mentioned the wide-voltage-range circuits from near-threshold to super-threshold. These methods may fail due to the variation of the half-path point across a wide voltage operating range.

This work is sponsored by the National Natural Science Foundation of China (61574033 and 61774038) and National High Technology Research and Development Program of China (863 Program) (2015AA016601) and the National Science and Technology Major Project (2014ZX01030-101).

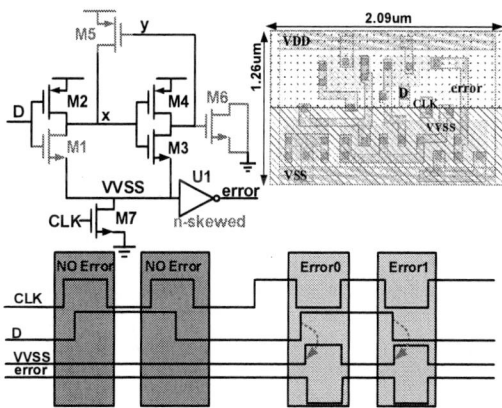

Fig .1. The schematic and the operational waveform of the proposed transition detector.

TABLE I.   THE CHARACTERISTICS OF THE TRANSITION DETECTOR COMPARED TO A CONVENTIONAL LATCH AT 0.6 V.

	Conventional latch	Proposed TD
CLK-Q Delay	0.23ns	-
D-DETECT Delay	-	0.61ns
Switching Energy	3.89fJ	5fJ (1.29x)
Leakage power	79.8nW	11.3nW (0.14x)
# of Transistor	14	9 (0.64x)
Area (um^2)	2.394	2.6334(1.1x)

Here, we propose a Transition-Detector and a half-path insertion rate reduction method for wide voltage range digital circuits. Highlights are: (1) an error detection circuit that can operate at near-threshold voltage, (2) a half-path insertion point selection method considering PVT variations. Fabricated on 40nm CMOS, test chips achieve 2.8× throughput improvement and a maximum of 50.5% higher energy efficiency at 0.56V over the margined baseline. It also pushes the minimum energy point from 0.6V to 0.49V with 38.2% energy reduction at 25°C, typical die.

II. TRANSITION DETECTOR DESIGN OVERVIEW

In order to apply a transition detector in NTV design with low area overhead, a novel wide voltage range transition

978-1-5386-3179-9/17 $31.00 © 2017 IEEE   205

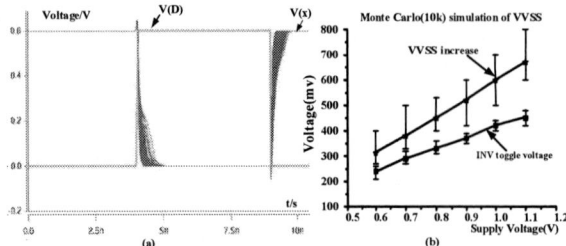

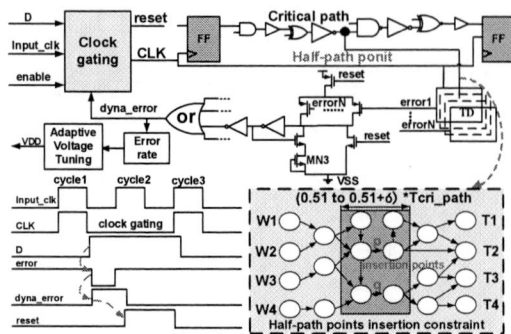

Fig. 2. (a) MC results of node x when D transition occurs and (b) results of VVSS upon error detection.

detector is proposed using a current sensing circuit with only 9 transistors for error detection. It flags any data transitions during negative clock phase as an error signal, as [5]. Our 9-transistor TD has several advantages: (1) reliable wide operation range from 0.6V to 1.1V, (2) low area and energy overheads are 10% and 29%, respectively, over a conventional latch based on extracted layouts, (3) no need of duty cycle control compared to other designs [2][3].

A. Circuit Design of the Transition Detector

The schematic and operation principle of the transition detector are illustrated in Fig. 1. In the negative phase of CLK, the node VVSS floats and will switch from 0 to 1 if an input data transition occurs. An n-skewed inverter U1 is used to reduce its switching voltage for error detection and its output flags as an error. The detection mechanism is as follows: While the clock is low, the virtual rail VVSS floats. If an input data transition occurs in this state, the node VVSS will switch from 0 to 1. When the clock goes high, M7 is always on and the node VVSS remains low. Thus, by monitoring the node VVSS, an error signal can be generated.

To extend its application to near-threshold, transistors M5 and M6 are added to increase the voltage of the floating node VVSS for error detection. In the negative clock phase, when input data changes, our added transistors M5 and M6 contribute some charging current to the floating virtual rails VVSS as follows: (1) While input data switches from 0 to 1, the node y needs a certain delay when it transits from 0 to1, which increases the voltage of node VVSS by increasing its charge time, with M1 and M5 on simultaneously. (2) While input data changes from 1 to 0, M6 is regarded as a discharging capacitance and charges the node VVSS through M3. Transistors M5 and M6 effectively boost the voltage at node VVSS. Table I shows the characteristics of the transition detector compared to a conventional 14-transistor latch at 0.6V. Note that, to ensure a fair comparison, all simulations use exactly the same environment.

B. Data Race and Leakage Considerations

TD must consider the possibility of race in the data path and leakage-induced false positive errors. To ensure the correctness of the function, the size of transistor M1 is designed twice as large as the transistor M5's. Fig. 2 (a) shows the results of Monte Carlo (MC) simulations of the voltage of node x when D transition occurs. We found no competition-induced false error detection when supply voltage scaled to

0.6V. Note that, the VVSS node is refreshed every cycle, which can relax restrictions on leakage problems. We found no leakage-induced false errors were observed down to 0.6V.

C. Low Voltage Operation

TD is designed to detect late transitions even at low voltage. Therefore, a skewed inverter U1 is used by up-sizing of the NMOS transistors. To verify that TD can operate at NTV, the voltage of VVSS is confirmed via 10k MC simulations with process variations from 0.6V to 1.1V as shown in Fig. 2 (b). MC simulation demonstrates the robustness of VVSS behavior for error detection across a wide voltage range.

III. SELECTION OF HALF-PATH POINTS

The core idea of the half-path error detection technique is that the timing error can be predicted by detecting the late arriving data transitions at a half-path point in the critical path instead of at the endpoint flip-flop, as described in Fig. 3. TDs are inserted at the half-path point of critical paths, which will generate an error signal if the detected signal transition occurs at or after the half-path point during a clock negative phase. Thus, the real timing error can be prevented through dynamic clock gating. To reduce the propagation delay of the error signal, dynamic OR gates are employed for clustering the error signals from different paths. The clock gating and supply voltage increasing techniques are used to prevent a potential timing error beforehand.

There were some studies on selecting half-path points, but the effect of PVT variations was rarely considered, especially

Fig. 3. Half-path error detection technique, with timing and error detection waveforms and insertion constraint.

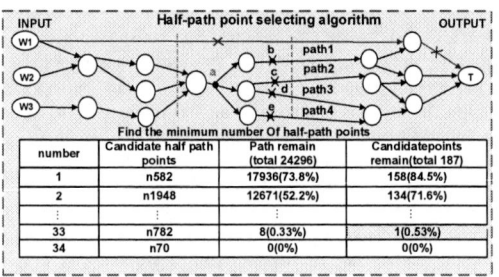

Fig. 4. The half-path point selecting algorithm.

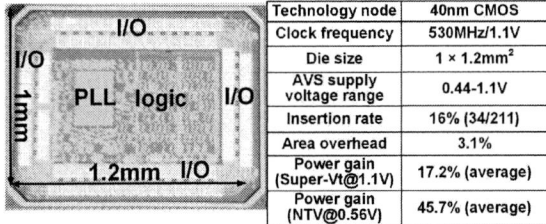

Technology node	40nm CMOS
Clock frequency	530MHz/1.1V
Die size	$1 \times 1.2mm^2$
AVS supply voltage range	0.44-1.1V
Insertion rate	16% (34/211)
Area overhead	3.1%
Power gain (Super-Vt@1.1V)	17.2% (average)
Power gain (NTV@0.56V)	45.7% (average)

Fig. 5. Die photograph and design details.

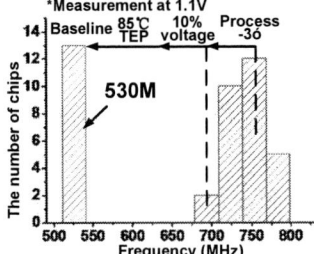

Fig. 6. Max frequencies and baseline test at 1.1V.

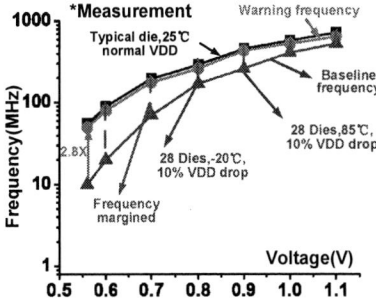

Fig. 7. Measured max, warning and baselines frequencies across 0.56-1.1V.

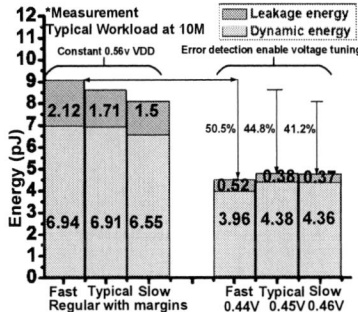

Fig. 8. Measured energy consumption at 0.56V for three typical chips (Fast, Typical and Slow dies).

for a wide range voltage applications. Many non-critical under super-Vt may become critical under NTV due to the influence of PVT variations. Therefore, we take PVT variations into consideration when selecting the insertion points.

A. Critical Path Endpoint Selection

The basic principle of critical path endpoint selection is considering the percentage of the path endpoints under different PVT conditions. The critical path endpoints are selected under the worst case (ss 0.6V 0°C).

To cover the critical path endpoints under other PVT conditions, we increase the number of path endpoints gradually until they can cover 90% of the target paths. After initial selecting, some missing endpoints which are critical path endpoints in other PVT conditions are added to meet a higher coverage. Without the selection of path endpoints, there are 110 registers which take the timing length of paths (90%-100% clock period) as a reference need to be monitored. However, only 80 endpoints are selected in our work.

B. Half-path point Selection

Some endpoints are selected by implementing the above approach. The half-path point selection method is summarized as follows: In the first step, we obtain data paths whose timing slack are less than 20% of the clock period from the selected endpoints. Next, the speculation window must account for PVTA variations that cause the deviations from the ideal half-path point. We select inserted points by setting a tiny guard band for random process variations. The speculation window is designed in the range of $(0.51, 0.51+\sigma) * T_{cp}$ (T_{cp} is the critical path delay and $\sigma=0.03$), to ensure the inserted point lags a little bit behind the exact half-path point under all PVT conditions. Finally, we apply the minimum number selecting algorithm to find ultimate half-path points. The process of half-path point selecting algorithm is illustrated in Fig. 4. The point **b** only

monitors path1 but the point **a** can monitor path1, path2, path3 and path4 simultaneously. This means that only the point **a** can monitor all the four paths from inputs to outputs. We can find the minimum number of inserted point only 34 by half-path point selecting algorithm. In contrast with conventional endpoint methods taking 10% clock cycle slack for selecting DFF endpoints, our proposed algorithm reduces the number of inserted points by up to 69%.

IV. MEASUREMENT RESULTS

To demonstrate the effectiveness of the proposed technique, we have implemented it in a 16-bit FIR circuit using 40nm CMOS process with ten layers of metal. The micrograph of the test chip with an area of 1000×1200 um^2 and the implementation details are shown in Fig. 5. Out of a total of 211 registers in the design, 34 half-path points were augmented with a TD for timing-error detection.

To ensure correct operation across all dies and operating conditions, all the chips would need to operate at the worst case voltage. Here we assume a conventional worst case PVT condition of 85°C (-20°C), 10% supply droop, and 3σ process variation at 1.1V (0.6V). Fig. 6 shows the distribution of measured performance at 1.1V from 28 dies and typical, fast, and slow dies are selected.

Based on conventional worst case design practice, we can determine the frequency of the baseline across 0.56V to 1.1V. As shown in Fig. 7, the maximum frequencies for correct operation and our detected warning signal of a typical die are

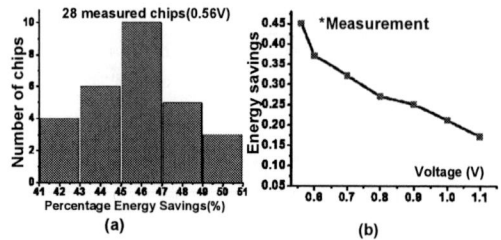

Fig. 9. (a) The distribution of energy savings at 0.56V and (b) The average energy savings of 28 dies across 0.56-1.1V.

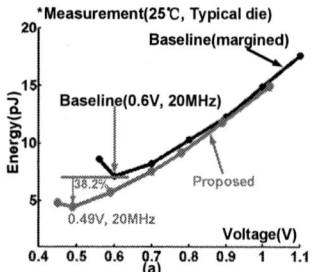

Fig. 10. Energy efficient with AVS at a typical PVT corner.

measured and compared with baseline frequencies, indicating a maximum 2.8× throughput improvement over the baseline design at 0.56V.

Fig. 8 compares the energy savings for error detection enabled operation with conventional operation at 0.56V/10 MHz under three typical chips (Fast, Typical and Slow dies). With error detection enabled voltage tuning, the design provides 50.5% and 41.2% energy savings for fast and slow dies, respectively, by scaling the supply voltage to the Point of First Error (PoFE). Energy consumption of typical die reduces from 8.62pJ at 0.56V to 4.76pJ using adaptive-tuning voltage, achieving 44.8% energy savings. With error detection enabled voltage tuning, the proposed design improves the max energy savings by 50.5% with 120mV additional voltage scaling. The distribution of the percentage energy savings for different dies at 25°C and 10 MHz is shown in Fig. 9 (a).

Due to the temperature reversal effect, the conventional worst-case PVT condition of 3σ process variation, 10% VDD drop, -20°C across 0.56V-0.8V and 3σ, 10% VDD drop, 85°C across 0.9 to 1.1V. Fig. 9 (b) shows 17.2%~45.7% average energy savings across 1.1v to 0.56V by operating at the optimal voltage point.

The baseline minimum energy consumption (Emin) per cycle of a typical die is found to be 7.11pJ when operating at CLK=20MHz and VDD=0.6V as shown in Fig. 10. The proposed design achieves Emin of 4.46pJ and CLK=20MHz at VDD=0.49V. We push the minimum energy point of a typical die from 7.11pJ to 4.46pJ, with 38.2% energy reduction.

Table II compares our method with other representative adaptive techniques. Our TD eliminates extra clock loading, with only 10% area overhead over a conventional latch. Plus, our proposed error detection technique is suitable for wide voltage design with only 3.1% total area overhead.

TABLE I.   COMPARISON CHART OF PREVIOUS WORKS

	[1]	[2]	[4]	[5]	This work
TYPE	Latch	Latch	Flip-Flop	Latch	Flip-Flop
Extra # of Transistor	31	15	8	1.46	9
Duty-cycle Control	YES	YES	NO	YES	NO
ED area Overhead	Not reported	Not reported	33%	4.3%	10%
Target Vdd	1.2V	0.6-1.0V	0.8-1.1V	0.6 - 1.1V	0.6 - 1.1V
Technology	130nm	45nm	45nm	40nm	40nm
Insertion rate	14.6%	12%	19.8%	8.6%	16%
Area overhead	Not reported	3.8%	4.42%	13.6%	3.1%
Energy saving	33%	22%	45.4% (max)	45%	50.5% (max)

V.   CONCLUSION

To minimize operating power and area overhead, a method to solve the variation of the half-path point of the critical path by analyzing the PVT variations of the paths is proposed. The proposed method can reduce the number of inserted monitors by up to 69% compared with the conventional methods. In addition, a novel transition detector technique with a 9 transistor current sensing circuit, operating reliably at NTV, is presented as well. At 0.56V, the test circuit can detect timing errors and achieve up to 50.5% higher energy efficiency over the baseline design. At the same throughput, the proposed design consumes 38.2% less energy than the baseline operating at its minimum energy point. The method proposed in this paper can effectively reduce the total area overhead.

REFERENCES

[1] S. Das et al., "Razor II: In situ error detection and correction for PVT and SER tolerance," IEEE J. Solid-State Circuits, vol. 44, no. 1, pp.32–48, Jan. 2009.

[2] K. A. Bowman et al., "Energy-efficient and metastability-immune resilient circuits for dynamic variation tolerance," IEEE J. Solid-State Circuits, vol. 44, no. 1, pp. 49–63, Jan. 2009.

[3] K. A. Bowman et al., "A 45 nm resilient microprocessor core for dynamic variation tolerance," IEEE J. Solid-State Circuits, vol. 46, no. 1, pp. 194–208, Jan. 2011.

[4] I.Kwon, S.Kim, D.Fick, M.Kim and D.Sylvester,"Razor-lite: A lightweight register for error detection by observing virtual supply rails," IEEE J. Solid-State Circuits, vol. 49, no. 9, pp.2054–2066, Sep. 2014.

[5] Yiqun Zhang, Mahmood Khayatzadeh, and D.Sylvester "iRazor: 3-Transistor Current-Based Error Detection and Correction in an ARM Cortex-R4 Processo," IEEE International Solid-State Circuits Conference (ISSCC), pp.160-162.2016.

[6] J. Zhou et al., "HEPP: A new in-situ timing-error prediction and prevention technique for variation-tolerant ultra-low-voltage designs," in Proc. IEEE A-SSCC, 129–132, Nov. 2013.

[7] L. Lai, V. Chandra, R. Aitken, and P. Gupta, "SlackProbe: A low overhead in situ on-line timing slack monitoring methodology," in Proc. IEEE DATE, pp. 282–287, Mar. 2013.

HTD: A Light-Weight Holosymmetrical Transition Detector Based In-situ Timing Monitoring Technique for Wide-Voltage-Range in 40nm CMOS

Wentao Dai, Weiwei Shan*, Xinning Liu and Jun Yang
Southeast University,
Nanjing, Jiangsu, China
wwshan@seu.edu.cn

Abstract—To eliminate the worst-case timing margins, a 13-transistor holosymmetrical transition detector (HTD) is proposed for use in timing variation resilient systems. The HTD achieves low overhead and wide-voltage-range operation via monitoring the discharge at the floating node of two-stage CMOS inverters. Using local detection and global clock stalling, the system is stalled immediately for one cycle when an error occurs, allowing the variation resilient technique to be integrated into any circuits without architectural changes. Plus, there is no need of an error recovery mechanism by keeping the system working at the point before the first failure (PBFF) and utilizing the time-borrowing characteristics of the latch. Applied on an 8th-order filter test chip in 40nm CMOS process, without changing the system architecture, chip's measurement results demonstrate that it improves energy efficiency by 45.6%/28.1% and throughput by 179.31%/28.2% in near-V_{TH} (0.474V) /super-V_{TH} (1.1V) while incurring a 4.37% area overhead compared to a baseline design.

Keywords—transition detector; wide-voltage-range; in-situ timing monitoring; resilient design; time borrowing.

I. INTRODUCTION

Wide-voltage-range operation is a key technique to achieve both high performance and energy efficiency of digital computing hardware. However, advanced nanoscale CMOS process node makes uncertainties in IC design increasingly difficult to meet timing closure. These uncertainties become worse in the near-V_{TH} regime, where the delay distributions of circuits become more dispersed across typical and worst-case process, voltage, and temperature conditions, to bring many new challenges [1].

To ensure timing closure across various operating conditions, IC designers often add some margins to the clock frequency, which result in losses in performance, energy and area cost. Several on-chip timing error detection and correction (EDAC) techniques have been proposed to reduce the excessive timing margins [2-7]. In these techniques, error-detecting registers were inserted into the endpoints of critical paths, which flagged any data transitions during their speculation window as a timing error. Then the error statistics can inform a recovery mechanism and DVFS controller to modulate the supply voltage and clock period. Thus, they can virtually eliminate all the timing margins.

However, most of the current EDAC methods may not fit for the wide-voltage-range from near-V_{TH} to super-V_{TH}. Earlier

error-detecting units [2-4] usually had a large area overhead with more than 10 extra transistors over a general flip-flop. Recently, Razor-lite [5] and iRazor [7] effectively decreased the area overhead, but they could not work reliably at near threshold voltage (NTV) due to the threshold loss of their floating node sensing. And a complex recovery mechanism was a necessity when a timing error occurred, which increased the area a lot [2-7]. Plus, the serious short-path padding problem in [2-7] further increased the area overhead.

Thus in this paper, we propose a light-weight in-situ timing monitoring strategy without need of a recovery mechanism, while at the same time the timing margin can also be almost completely eliminated. Our contributions are as follows: 1) A holosymmetrical transition detector (HTD) which can work reliably at NTV with a fast response. Combined with a latch to form a HTD-latch, it has only $1.095\times$ area compared to a flip-flop. We replace the endpoint FFs in critical paths by HTD-latches. 2) Elimination of the margin by keeping the system working steadily at the point before the first failure (PBFF). Together with the time-borrowing characteristics of the latch, it relieves the demand for error correction. 3) An adaptive duty-cycle control mechanism has been adopted that decreases the need of short-path buffering. We apply those techniques to the design of the 8-order filter circuit, which is a 3-stage, 8-bit circuit of DSP in 40nm CMOS process. The measurement results show that a 45.6%/28.1% energy-efficiency and 179.31%/28.2% throughput improvements at 0.474V/1.1V are achieved at the cost of only 4.37% area overhead.

The remainder of this paper is organized as follows. Section II demonstrates the transistor-level design of HTD and a detail analysis of its speculation window as well as its suitability. Section III describes the implementation details and discusses the measurement results. Section IV concludes this paper.

II. HTD DESIGN OVERVIEW

With the aim of eliminating the timing margin completely, it is essential to design a reliable error-detecting unit when working at wide-voltage-range and set an appropriate speculation window to avoid unnecessary short-path buffering.

A. HTD Core Circuit

The idea of HTD comes from the observation of short-circuit current in CMOS inverter. If the input of the inverter

Fig. 1. (a) The schematic of HTD. (b) (c) internal node VVDD waveforms with D transits respectively.

changes, the short-circuit current would be generated from power to ground. By adding a PMOS header on the inverter, the short-circuit current would discharge the floated inner node (VVDD) when the header closes. This switches the output inverter whose output is flagged as a warning. Based on this, the HTD is proposed to detect any switching transition on a signal line within a target speculation window.

The schematic of HTD is depicted in Fig. 1(a). The clock controlled PMOS header (twice over the ordinary size) is added between VDD and VVDD, and the other MOSFETs are only adopted by the ordinary size. Fig. 1(b) and Fig. 1(c) show the timing diagram of HTD respectively when D rises and falls. In the negative clock phase, the output of the HTD (pre_error) is 0 since VVDD is charged and initialized to 1. In the positive clock phase (error detection phase), the charge stored in VVDD will be released quickly through the virtual rail (Mn1 and Mn2) when D rises from 0 to 1. Owing to the discharging of VVDD, the node V3 cannot be charged to 1 because of the lack of charge, so VVDD will drop down to 0 through Mn1 and Mn2 at last (Fig. 1 (b)). Respectively, the node V2 will maintain 1 since the node V1 keeps 0 when D falls (Fig. 1(c)). Thus, as the time-borrowing characteristics of the latch, if data arrives after the rising edge of the clock, the HTD-latch (HTD with a latch) can both transmit the data correctly and flag the change as a timing warning to the next stage.

The comparison results between the HTD-latch and conventional FF are demonstrated in Table I. The area overhead is merely 9.5% over a DFF. Note that the switching power in non-error event is 0.79×/0.7× at super-V_{TH}/near-V_{TH} respectively, which is much less than FF. The power consumption increases in error events as the contention occurs in HTD which will be illustrated in Section II-C below. The extra leakage and switching power in error event will not lead to a large power consumption overhead due to the monitoring strategy we adopt. The impact on the critical path is negligible, since the CK-Q delay and setup time are smaller than FF circuit due to the adoption of the latch. The cell delay of HTD-

TABLE I. COMPARISON TABLE BETWEEN HTD (WITH LATCH) AND CONVENTIONAL FLIP-FLOP

	Conventional FF	HTD&Latch (Norm)
Area	5.0274 um^2	1.095×
Leakage Power	6.256nW @1.1V 0.902nW @0.6V	1.51× @1.1V 1.06× @0.6V
Switching Power	0.534uW @1.1V 0.143uW @0.6V	0.79× @1.1V 0.70× @0.6V
Error Event	N/A	1.49× @1.1V 1.78× @0.6V
CK-Q	91 ps	55 ps

latch is just 2× over a general inverter, thus the warning signal can be transmitted to the next stage with a fast response.

B. Speculation Window Design

To remove all the timing margins, the most popular methods use shadow latches to detect the setup time violation. However, they have to distinguish the data from critical paths and short paths when data arrives after the rising edge of clock (the speculation windows are often in the positive clock phase). Consequently, a large number of padding cells are needed to fix short paths, and an extra recovery mechanism has to be added when working at the point of first failure (PoFF).

To relieve the demand for error correction, we keep the system working steadily at the point before the first failure (PBFF) by increasing the supply voltage from PoFF with one minimum voltage step (Fig. 2). Thus, even if the input D is influenced by a fast variation and arrives after the rising edge of the clock, it can still be transferred correctly by time-borrowing characteristics of the latch. And then the system clock is stalled immediately for one cycle to deal with the fast variation. Thus, there is no need of complicated recovery mechanism, and the impact on performance is relatively small.

We design an adaptive duty-cycle control mechanism to decrease the need of short-path buffering. Because the positive clock phase is our speculation window, in which the short-path data arrives will lead to a timing failure. To avoid it, the focus of this work is to design an "adequate" speculation window. It can be achieved by adapting the duty cycle of the clock, as Fig. 3 shows, which is divided into four parts: ring oscillator, counter, decoder and configurable delay chain. The ring oscillator tracks the current PVT situation, and the counter transfers the detected information to the decoder, which decodes it into a control signal to configure the delay chain. Therefore, the clock duty cycle can be automatically corrected

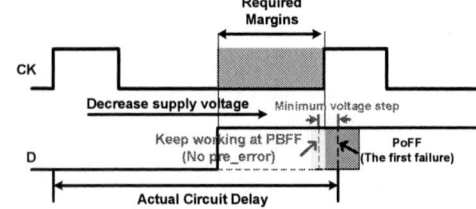

Fig. 2. The adopted strategy of PBFF.

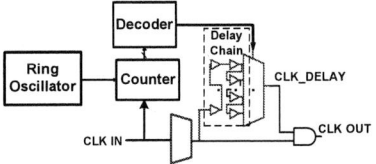

Fig. 3. Schematics of the proposed adaptive duty-cycle controller.

by configuring the size of the delay chain. By using the adaptive duty-cycle control mechanism, the padding value of the shortest path in this work decreases from 4-FO4 to 1-FO4, leading to much fewer short-paths to be fixed.

C. Virtual Rail Contention And Robustness

A potential problem of HTD is the contention. If D changes in the positive clock phase, VVDD would be discharged to 0. And then VVDD is pulled-up on the falling edge of CK while the Mn1 and Mn2 (or Mn3 and Mn4) are open, which will lead to the contention. Fortunately, as VVDD being charged, V3 will be dropped up soon, then Mn1 will be closed at once, thus the contention will disappear. The problem would lead to some power consumption in error event's scenario. Due to the final working state in this work is PBFF, the power consumption caused by contention can be neglected. Besides, the robustness of HTD was confirmed by 10^5 Monte Carlo simulations at certain clock frequencies ($\geq$1MHz), the worst temperature and each voltage (from 0.5V to 1.1V).

III. TESTCHIP IMPLEMENTATION AND MEASUREMENT

In order to verify the effectiveness of our proposed method, we apply it on an 8-order filter circuit, which is a 3-stage, 8-bit circuit of DSP. The multiplier and adder sections are the most critical paths based on post-layout timing analysis. The overall architecture is shown in Fig. 4, which is composed of a filter with its endpoint FFs in critical paths replaced by HTDs and latches, a dynamic OR gate to collect all the HTD's outputs, a duty cycle controller and a clock gating cell for adaptive control. Fabricated in 40nm CMOS process, its die photograph is shown in Fig. 5. In order to achieve a better working condition at the near-V_{TH} region, all the 4 or more fan-in units are removed, and the standard cells are re-extracted by the EDA tool Siliconsmart.

In total, 28 dies are obtained. Here baseline frequencies are defined as the max frequencies of the slowest chip among 28 dies at worst temperature and -10% VDD drop across 0.6V-1.1V, even when operating at typical PVT condition (typical die, 25°C and no VDD drop). Here the worst temperature is either -20°C for near-V_{TH} or 85°C for super-V_{TH}. The system could be halted within one cycle of a detected error, thus HTD-based mode ensures the system work correctly at the point before the first failure, so it needs no margin and its operating voltage is much lower than the baseline.

The baseline Fclk for correct operation is measured for the typical die under the margined condition at the near-V_{TH} region (10% VDD drop, the worst temperature). The original power consumption of the filter is measured to be 101.8uW when operating at Fmax = 27MHz and VDD = 0.6V. When HTD is enabled, it decreased to 55.4uW (the same typical die, the

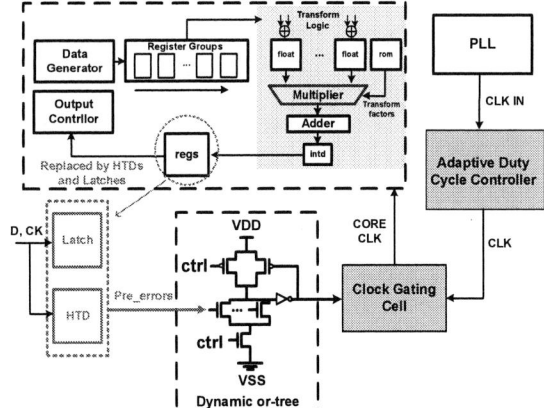

Fig. 4. Overall architecture with proposed HTD techniques.

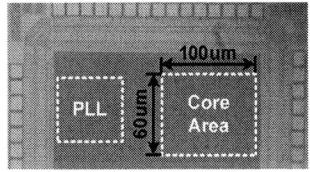

Fig. 5. Die photograph of HTD-based core.

PBFF is 0.495V, 105mV lower than the baseline, 25°C), which achieves 45.6% power gains over the baseline as shown in Fig. 6(a). Respectively, neglecting the benefits coming from aging effects and process variation, the HTD-based mode can have 1.79× higher throughputs (Fmax=101.25MHz) at the typical case (0.6V, 25°C) than the baseline at the margined case (the same typical die, 0.54V, -20°C, 36.25MHz), which is depicted in Fig. 6(b). For super-V_{TH} operation, 28.1% power gains (4.36mW over 6.06mW) and 28.2% frequency improvements (1GHz over 780MHz) are obtained at 1.1V. On the other voltage nodes, we also get more than 30% energy savings and throughputs improvements, as shown in Fig. 6.

The distributions of power and frequency among all 28 tested chips at 0.6V are summarized in Fig. 7. The fastest chip can achieve 52.79% power saving at PBFF (0.474V, 126mV lower) and 246% frequency improvement at nominal voltage (0.6V). On the other hand, the power saving of the slowest chip decreases to 40.8% compared to 45.6% of the typical chip as the wasted margins decreasing.

The testing platform and typical measured waveforms are

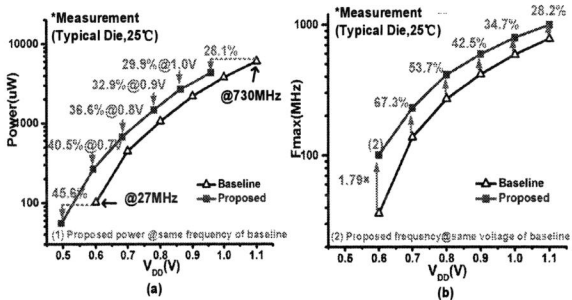

Fig. 6. Measured baseline and working frequencies and power across a wide VDD.

TABLE II.   COMPARISON CHART OF HTD AND PREVIOUS EDAC WORKS

	[3]	[4]	[5]	[6]	[7]	This work
Technology	65nm	45nm SOI	45nm SOI	65nm	40nm	40nm
Voltage	Super-V_{TH} 1.1V	Super-V_{TH} 1.0V	Super-V_{TH} 1.1V	Near-V_{TH}	0.6-1V	Wide 0.6-1.1V
Insertion rate	17%	100%	20%	13%	8.66%	8.2%
Area overhead	6.90%	8%	4.42%	8.30%	11.90%	4.37%
Power gain (Super-V_{TH})	52% (max)	54%(PoFF) 66%(max)	45.4% (max)	-	45% (PoFF)	28.1% (PBFF, Typical)
Power gain (Near-V_{TH})	-	-	-	38% (PoFF)	-	45.6% (PBFF, Typical)

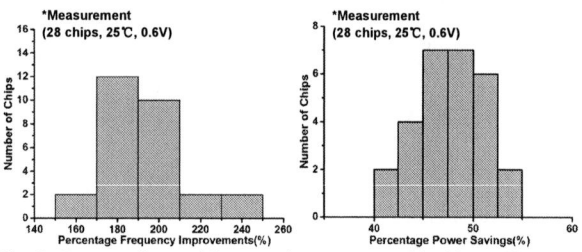

Fig. 7.   Power savings and frequency improvements of the 28 test chips at 0.6V.

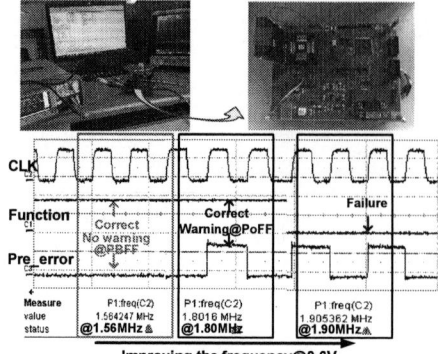

Fig. 8.   Testing platform and testing waveform before and after PBFF.

and many invasions to the circuit, while the reserved timing margins can be virtually eliminated.

ACKNOWLEDGMENT

This work is sponsored by the National Natural Science Foundation of China (61574033 and 61774038), National High Technology Research and Development Program of China (863 Program) (2015AA016601) and the National Science and Technology Major Project (2014ZX01030-101).

shown in Fig. 8. The clock in the waveform is divided by 64 and output. There is no warning when the chip operates at PBFF, the output frequency is 1.56MHz while the actual clock is 99.8MHz, and the function of system is correct. As the frequency improves, it reaches to the PoFF at 115.2MHz. Since the duty-cycle of the clock is corrected adaptively in this work, a functional failure occurs at 121.6MHz rather than faster frequency.

Table II includes a comparison table of HTD and past EDAC techniques. This work demonstrates a low-overhead sequential element and a high energy-efficiency among prior EDAC works. Compared to [3-7], the proposed HTD-based design can be well-suited for wide-voltage-range and get a substantial benefit without an error correction mechanism at the cost of 4.37% area overhead.

IV.   CONCLUSION

This paper presents a 13-transistor transition-detecting circuit called HTD that utilizes monitoring the charge release at the inner node to provide a lightweight variation resilient design. The full-discharging mechanism of HTD ensures the reliability in wide-voltage-range operation. With incurring 9.5% area overhead, the power consumption of HTD FF is merely 0.79× over a general flip-flop at the 50% activity factor. The HTD-based EDAC design is implemented within an 8th-order filter test circuit for a DSP in 40nm CMOS technology. With HTD enabled, 45.6% power savings and 1.79× throughput improvement in near-V_{TH} is obtained with 4.37% total area overhead. HTD also offers practical advantages of low design complexity without need of recovery mechanism

REFERENCES

[1] M. Seok, G. Chen, S. Hanson, M. Wieckowski, D. Blaauw and D. Sylvester, "CAS-FEST 2010: Mitigating variability in near-threshold computing", IEEE Journal on Emerging and Selected Topics in Circuits and Systems, 1(1), 42-49, 2011.

[2] D. Ernst, N. S. Kim, S. Das, S. Pant, R. Rao, T. Pham, C. Ziesler, D.Blaauw, T. Austin, K. Flautner, and T. Mudge, "Razor: A low-power pipeline based on circuit level timing speculation", Microarchitecture, 2003. MICRO-36. Proceedings. 36th Annual IEEE/ACM International Symposium on, pp. 7-18.

[3] D. Bull, S. Das, K. Shivashankar, G. S. Dasika, K. Flautner and D. Blaauw, "A power-efficient 32 bit ARM processor using timing-error detection and correction for transient-error tolerance and adaptation to PVT variation", IEEE Journal of Solid-State Circuits, 46(1), 18-31, 2011.

[4] M. Fojtik, D. Fick, Y. Kim, N. Pinckney, D. M. Harris, D. Blaauw and D. Sylvester, "Bubble razor: Eliminating timing margins in an ARM cortex-M3 processor in 45 nm CMOS using architecturally independent error detection and correction", IEEE Journal of Solid-State Circuits, 48(1), 66-81, 2013.

[5] S. Kim, I. Kwon, D. Fick, M. Kim, Y. P. Chen and D. Sylvester, "Razor-lite: A side-channel error-detection register for timing-margin recovery in 45nm SOI CMOS", In Solid-State Circuits Conference Digest of Technical Papers (ISSCC), 2013 IEEE International (pp. 264-265), 2013.

[6] S. Kim and M. Seok, "Variation-tolerant, ultra-low-voltage microprocessor with a low-overhead, within-a-cycle in-situ timing-error detection and correction technique", IEEE Journal of Solid-State Circuits, 50(6), 1478-1490, 2015.

[7] Y. Zhang, M. Khayatzadeh, K. Yang, M. Saligane, N. Pinckney, M. Alioto and D Sylvester, "irazor: 3-transistor current-based error detection and correction in an arm cortex-r4 processor", In Solid-State Circuits Conference (ISSCC), 2016 IEEE International (pp. 160-162), 2016.

A 0.5V 12-bit SAR ADC using Adaptive Time-Domain Comparator with Noise Optimization

Chen-Che Kao, Sung-En Hsieh, Chih-Cheng Hsieh

Department of Electrical Engineering
National Tsing Hua University
Hsinchu, Taiwan
jenchekao2012@gmail.com, cchsieh@ee.nthu.edu.tw

Abstract—This paper presents a low voltage and power efficient 12-bit successive-approximation register (SAR) analog-to-digital converter (ADC). The proposed adaptive time-domain (ATD) comparator automatically adjusts its input-referred noise performance according to the intermediate residual input level (ΔV_{in}) during conversion. Considering the noise requirement of 12-bit SAR ADC, the proposed implementation effectively reduces the comparator power consumption by 50% compared with the conventional approach. The prototyped ADC is fabricated in 90nm CMOS technology with a core area of 0.109mm^2. At 0.5V supply voltage and 150-to-250kS/s sampling rate with a Nyquist input, the implemented ADC achieves a SNDR of 63.8 to 66.3 dB with a corresponding ENOB of 10.3 to 10.71 bits. The resulting figure-of-merit (FoM) are 4.52 to 4.82 fJ/conversion-step.

Keywords—*SAR ADC, low noise, low power, adaptive time-domain comparator.*

I. INTRODUCTION

Biomedical sensor, wireless sensor, and internet-of-thing (IoT) applications require ADC with hundreds of kHz and resolution over 10-bit. With constrained batteries capacity, the power consumption of ADC must be minimized to sustain the system's long-term operation. For system's power efficiency consideration using energy harvester, low-voltage operating ADC with low power consumption is highly demanded. Several SAR ADC architectures [1]-[3] have been proposed to reduce the power consumption of capacitor array and comparator. Low supply voltage effectively decreases the switching energy of DAC, however, the power consumption of comparator is dramatically increased due to the strict noise requirement with decreased V_{LSB}. The most challenging design issue of low-voltage high-resolution ADC therefore becomes the noise suppression and power reduction of comparator.

Conventionally, the insertion of capacitor loading at comparator output is the most common way to decrease the noise level, which comes with the penalties of more power and slower speed. The majority voting technique [1] used multiple comparisons at critical decision to suppress the noise with a penalty of complex critical decision detection circuit. Moreover, the inherent noise of critical decision circuit may cause a fault detection when input level is small. Two step architecture [2] was proposed to reduce the power consumption of comparator at coarse conversion phase combined with a high-power and low-noise fine ADC. However, the complex calibration circuit and global control circuit are required. Moreover, in high-resolution ADC implementation, it is difficult to implement a

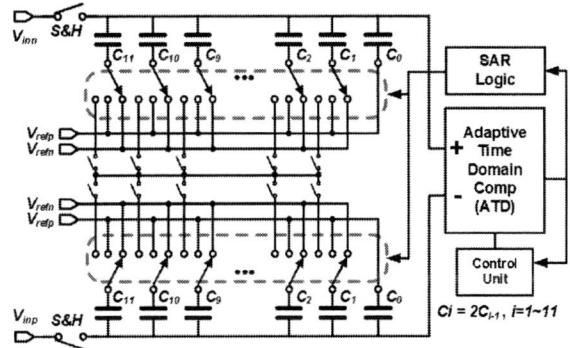

Fig. 1 Architecture of the proposed SAR ADC.

reasonable redundancy range for the noise tolerance of the coarse conversion.

By exploiting the voltage-controlled delay line (VCDL) with positive feedback and phase detector, this work proposes an adaptive time-domain (ATD) comparator with adaptive oscillation circuit (AOC), which adjusts the noise performance by enabling the oscillation loop of VCDL automatically based on the intermediate residual input level of comparator during conversion. The differential threshold window (DTW) technique is also proposed to set the optimized time difference threshold at VCDL output for AOC for noise-and-power tradeoff optimization and meta-stability prevention.

The rest of this paper is organized as follows. Section II describes the architecture of proposed design. Section III presents the detail circuit implementation. Section IV shows the measurement result and Section V provides the conclusion.

II. ADC ARCHITECTURE

Fig. 1 shows the proposed SAR ADC with adaptive time-domain (ATD) comparator, sample and hold (S&H), DAC, and SAR logic. The double-boosted S&H and local-boosted switching are implemented for leakage and linearity control in low-voltage operation. Fig. 2 shows the proposed ATD comparator which consists of voltage-controlled delay line (VCDL), differential threshold window (DTW) circuit, phase detector (PD), and oscillation control unit. The output of VCDL is fed back to enable the oscillation loop adaptively for noise reduction depends on residual input level (critical decision), which is controlled by DTW and PD operations. The details of circuit operation including oscillation enabling procedure and waveform will be discussed in the following.

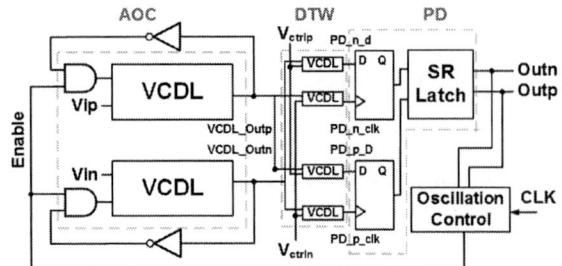

Fig. 2 Architecture of the proposed adaptive time-domain comparator.

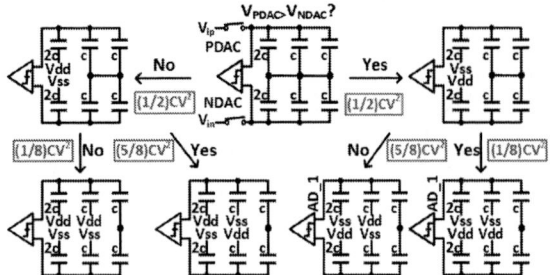

Fig.3 Proposed switching procedure.

A. DAC switching

The concept of the conventional merged capacitor switching (MCS) procedure [4] is adopted to reduce the switching energy consumption of DAC. To avoid the additional reference voltage Vcm requirement of MCS, we propose a merge operation of bottom plate at the sampling phase as shown in Fig. 3. Since the corresponding sampling capacitor arrays of V_{ip} and V_{in} are matched, the merged voltage of bottom plates will be Vdd/2 (=Vcm) at the sampling phase automatically due to the charge conservation. The common-mode voltage is then generated by charge sharing and hold on the DAC array during the conversion phase instead of being supplied by an extra voltage reference. However, the capacitor matching and leakage issues need to be well considered. The worst deviation of generated common-mode voltage is designed to be less than 2mV which induces a dynamic offset of <0.1 LSB of the target 12-bit resolution. As a result, the proposed Vcm-based switching procedure with merging of bottom plate achieves the same high energy efficiency as MCS without additional reference voltage (Vcm). With the reduced bottom plate voltage swing, the resulting switching energy is $170CV^2$ which is 133.3%, 49.9%, and 12.48% of MS [5], CAS [6], and the conventional switching procedures, respectively.

B. Operation of ATD comparator

Fig. 2 shows the architecture of the proposed adaptive time-domain (ATD) comparator composed of AOC, DTW, and PD. The AOC contains two VCDLs and feedback gating logic, which creates an adaptive time delay difference (voltage-to-time conversion gain) according to the input value. Compared to the conventional open-loop VCDL structure, the output of AOC is fed back through the control unit to enable the oscillation adaptively depends on PD output. For large input difference, the oscillation is disabled for low power consumption. In the case of small input level (critical decision)

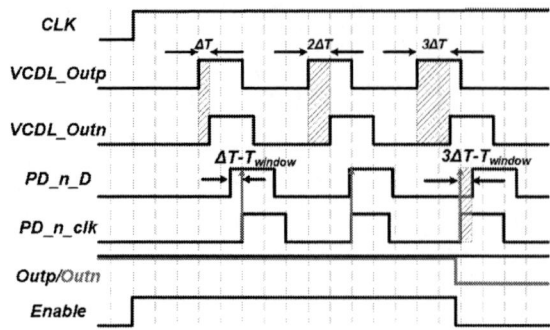

Fig. 4 Timing waveform of the proposed VTC in Oscillation mode.

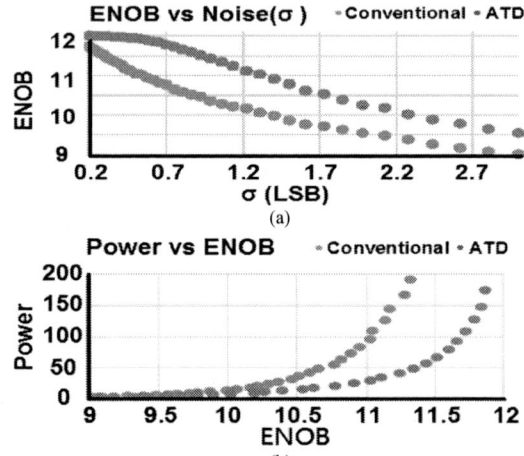

Fig. 5 MATLAB behavior model (a) ENOB vs Noise (b) Power vs ENOB

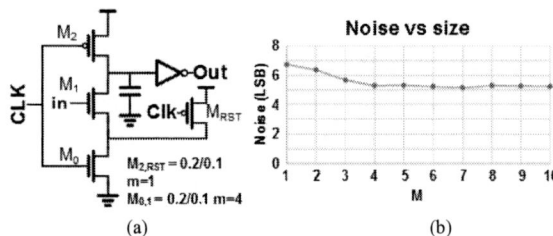

Fig. 6 (a)Voltage-controlled delay element (b)Noise simulation versus $M_{0,1}$ size

requiring a smaller noise (for example, reduced by 2), the AOC enables 4 times of oscillation to reduce the input referred noise by 2 times adaptively.

The DTW circuit is implemented to create the required time difference window T_{window} as the threshold of time-domain comparator for critical decision. Using 4 extra VCDLs with the same delay unit in AOC and controlled by the voltage difference of Vctrlp and Vctrln generated from DAC switching, the DTW generates a tunable T_{window} with a constant ratio of the delay of AOC under PVT variations. The PD is composed of two dynamic D-type flip flop (DFF) and one SR Latch. The two DFFs are triggered by the outputs from DTW to detect which signal is faster, and the result is stored in SR Latch.

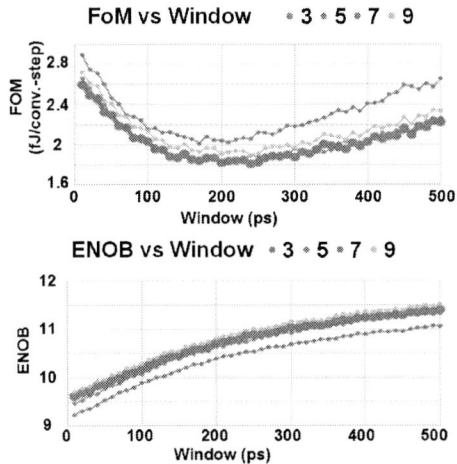

Fig. 7 MATLAB simulation of FOM and ENOB versus T_{window}

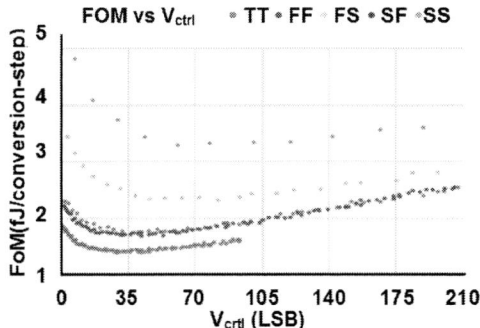

Fig. 8 MATLAB simulation of FOM versus V_{crtl}

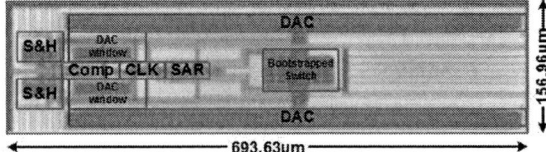

Fig. 9 Chip micrograph

Fig. 4 shows the timing waveform in oscillation mode. At the beginning, the time difference of VCDL_Outp and VCDL_Outn is smaller than T_{window} and oscillation enable =1. After three oscillation cycles of VCDL, the time difference is accumulated and larger than T_{window} to enable the SR latch for output polarity decision.

C. Performance trade-off

Figure 5 shows the simulation comparison of the conventional and proposed ATD comparators for the design tradeoff of power and ENOB. Fig. 5(a) shows the ENOB versus noise simulation of the conventional and proposed ATD comparators. The ATD comparator shows a better ENOB performance at the same input-referred noise level due to the noise suppression effect of adaptive oscillation enabling. Fig. 5(b) shows the comparison result of ENOB versus power consumption. Compared to the conventional approach, it shows the required power is decreased dramatically when ENOB is over 10-bit. By detecting the residual input level (equivalently, the time difference after VCDL) and enabling oscillation adaptively at critical decision, the proposed ATD comparator achieves an adaptive noise reduction with a better power efficiency especially for high-resolution (> 12bits) ADC implementation.

III. CIRCUIT IMPLEMENTATION

A. Voltage-controlled delay line

The voltage-controlled delay line is composed of 7 unit cells which converts the input voltage value into a time domain signal. The unit cell is implemented by a N-type voltage-controlled delay element as shown in Fig. 6(a). In order to fulfill the noise requirement of 12-bit SAR ADC, the size and cascading stage number of unit cell are optimized based on HSPICE and MATLAB simulations. The size of input (M_1) and pulling-down (M_0) devices directly affects the gain and current of the unit cell, which dominates the noise performance. From the HSPICE noise simulation as shown in Fig. 6(b), the finger number m = 4 of M_1(and M_0) is chosen to get the noise level of 5.3 LSB efficiently since the noise reduction slope is saturated even with

a larger number of "m". Once the unit cell of VCDL is decided, the numbers of cascading unit is designed based on the tradeoff of required ENOB and achieved FoM. With more cascading number of unit cell, the input referred noise of VCDL becomes smaller with less enabling of oscillation, but it also consumes more power in one cycle. To find the best combination condition, MATLAB behavior model of different numbers of unit cell is simulated as shown in Fig. 7. It shows the VCDL with over 7 unit cells achieves the saturated ENOB with a best FoM performance among all the other combinations. Therefore, the VCDL with 7 unit cells is chosen in this work.

B. Differential threshold window (DTW)

The DTW sets the minimum required time difference window of AOC output for polarity decision by phase detector. Choosing a larger window as a threshold, more cycles of oscillation is required to extend the time difference with a smaller input referred noise at a cost of larger power consumption and slower speed. To tracking the variation of process and operation condition of AOC, the DTW is implemented using the identical delay cell with a ratio control voltage from an extra DAC using the same switching method as the main DAC. Compared to the required cycles of Oscillator Collapse-Based Comparator [7] which depends on dc time and VCDL gain, the optimal cycle number of the proposed DTW is PVT independent. In order to find the best window (controlled voltage) of DTW, MATLAB behavior simulation with different corners is executed as shown in Fig. 8. It shows the achieved FoMs are optimized with a controlled voltage of around $35V_{LSB}$.

C. Phase Detector

Two TSPC DFFs are adopted to realize the function of phase detector (PD) for leading edge decision of AOC outputs. The power consumption and speed are enhanced by using the dynamic logic technique. Moreover, the setup time of PD is minimized to avoid the dead zone issue and performance degradation of DWT.

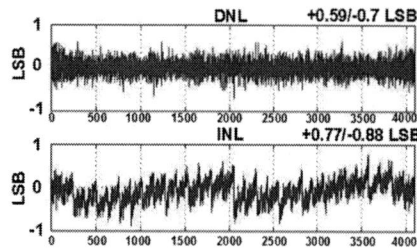

Fig. 10 Measured static performance

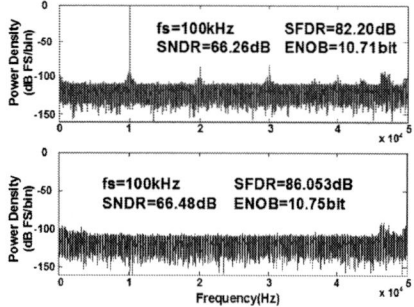

Fig. 11 Measured dynamic performance

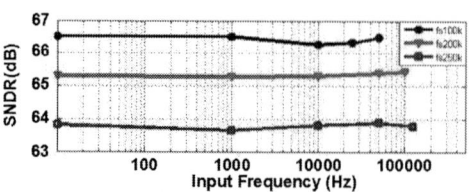

Fig. 12 Measured SNDR versus input frequency

Table. I. Comparison table

Specification	2013 ISSCC [1]	2014 ISSCC [8]	2014 ESSCIRC [9]	2016 ESSCIRC [10]	2016 VLSI [7]	This work
Tech.	65nm	65nm	65nm	65nm	40nm	**90nm**
Supply (V)	0.6	0.8	0.85	0.8	NA	**0.5**
Fs (kS/s)	40	32	1024	40	20	**100**
Resolution (bit)	12	12	13	12	15	**12**
ENOB (bit)	10.1	10.97	10.4	10.37	12.02	**10.71**
Power (µW)	0.097	0.31	45.2	0.375	1.17	**0.81**
FOM (fJ/conv.-step)	2.2	4.8	33	7.1	14.1	**4.82**

IV. MEASUREMENT RESULTS

Fig. 9 shows the micrograph of prototype chip fabricated in 1P9M 90nm CMOS with an active area of 0.109mm^2. Fig. 10 depicts the measured static performance of the proposed ADC at 0.5 V and 100kS/s. The measured peaks of DNL and INL are +0.59 / -0.7 LSB and +0.77 / -0.88 LSB, respectively.

Fig. 11 shows the measured dynamic performance of the implemented ADC with low and Nyquist input frequency at a sampling rate of 100kS/s. The resultant SNDR and corresponding ENOB at Nyquist input are 66.48dB and 10.75-bit, respectively. Fig. 12 shows the stable SNDR performance versus input frequency with different sampling rates (100kS/s~250kS/s). The total power dissipation is 0.81µW (f_s =100kS/s) with a breakdown of 70% in the proposed comparator, 18% in DAC switching energy and 12% in digital control circuits.

Table.I summarizes the performance comparison of the prototype with the state-of-the-art low power designs in 12bit resolution. With the proposed adaptive time-domain low noise comparator, this work presents a SAR ADC with FoMs of 4.82 and 4.52 fJ/conversion-step at 0.5V for 100kS/s and 250kS/s sampling rates, respectively.

V. CONCLUSION

This paper presents a 0.5V 12-bit SAR ADC using adaptive time-domain comparator with noise optimization. Using the proposed ATD comparator, the noise performance for critical decision is adaptively suppressed with an efficient power consumption. The prototyped ADC achieves a SNDR from 63.8dB to 66.3dB and a corresponding FoM of 4.52fJ/conversion-step to 4.82 fJ/conversion-step.

ACKNOWLEDGMENT

The authors thank National Chip Implementation Center (CIC) for fabrication of the test chip and support of Signal Sensing and Application Lab. (SiSAL), National Tsing Hua University, Taiwan. This research is particularly supported by Ministry of Science and Technology, Taiwan under the contract MOST 105-2218-E-007-006, 104-2221-E-007-103-MY3.

REFERENCES

[1] P. Harpe, E. Cantatore and A. v. Roermund, "A 2.2/2.7fJ/conversion-step 10/12b 40kS/s SAR ADC with Data-Driven Noise Reduction," 2013 IEEE International Solid-State Circuits Conference Digest of Technical Papers, San Francisco, CA, 2013, pp. 270-271.

[2] Y. J. Chen, K. H. Chang and C. C. Hsieh, "A 2.02–5.16 fJ/Conversion Step 10 Bit Hybrid Coarse-Fine SAR ADC With Time-Domain Quantizer in 90 nm CMOS," in IEEE Journal of Solid-State Circuits, vol. 51, no. 2, pp. 357-364, Feb. 2016.

[3] S. E. Hsieh and C. C. Hsieh, "A 0.44fJ/conversion-step 11b 600KS/s SAR ADC with semi-resting DAC," 2016 IEEE Symposium on VLSI Circuits (VLSI-Circuits), Honolulu, HI, 2016, pp. 1-2.

[4] Y. Zhu et al., "A 10-bit 100-MS/s Reference-Free SAR ADC in 90 nm CMOS," in IEEE Journal of Solid-State Circuits, vol. 45, no. 6, pp. 1111-1121, June 2010.

[5] C. C. Liu, S. J. Chang, G. Y. Huang and Y. Z. Lin, "A 10-bit 50-MS/s SAR ADC With a Monotonic Capacitor Switching Procedure," in IEEE Journal of Solid-State Circuits, vol. 45, no. 4, pp. 731-740, April 2010.

[6] C. Y. Liou and C. C. Hsieh, "A 2.4-to-5.2fJ/conversion-step 10b 0.5-to-4MS/s SAR ADC with charge-average switching DAC in 90nm CMOS," 2013 IEEE International Solid-State Circuits Conference Digest of Technical Papers, San Francisco, CA, 2013, pp. 280-281.

[7] M. Shim et al., "An oscillator collapse-based comparator with application in a 74.1dB SNDR, 20KS/s 15b SAR ADC," 2016 IEEE Symposium on VLSI Circuits (VLSI-Circuits), Honolulu, HI, 2016, pp. 1-2.

[8] P. Harpe, E. Cantatore and A. van Roermund, "11.1 An oversampled 12/14b SAR ADC with noise reduction and linearity enhancements achieving up to 79.1dB SNDR," 2014 IEEE International Solid-State Circuits Conference Digest of Technical Papers (ISSCC), San Francisco, CA, 2014, pp. 194-195.

[9] K. Yoshioka and H. Ishikuro, "A 13b SAR ADC with eye-opening VCO based comparator," ESSCIRC 2014 - 40th European Solid State Circuits Conference (ESSCIRC), Venice Lido, 2014, pp. 411-414.

[10] M. Liu, A. van Roermund and P. Harpe, "A 7.1fJ/conv.-step 88dB-SFDR 12b SAR ADC with energy-efficient swap-to-reset," ESSCIRC Conference 2016: 42nd European Solid-State Circuits Conference, Lausanne, 2016, pp. 409-412.

Range Pre-selection Sampling technique to reduce input drive energy for SAR ADCs

Harijot Singh Bindra[1], Joeri Lechevallier[1], Anne-Johan Annema[1], Simon Louwsma[2], Ed van Tuijl[1,2], Bram Nauta[1]

[1] Integrated Circuit Design, University of Twente, [2]Teledyne DALSA

Enschede, The Netherlands

Abstract—A Range Pre-selection Sampling (RPS) technique is introduced to reduce the input drive energy for SAR ADCs and is applied to a 10-bit 2MS/s SAR ADC in 65nm CMOS in this paper. Using the proposed RPS technique, the peak input sampling current and hence the input drive power requirement is reduced by a factor 2.4 as compared to conventional sampling (CS). Considering an ideal Class A operation for the buffer circuit driving the ADC, this translates into a minimum (theoretical) driver power consumption of 50µW for our RPS based ADC whereas it is 130µW for the conventional sampling, both much larger than the ADC power consumption of 3.25µW at 1MS/s operation. Our ADC occupies an area of 0.08 mm²and achieves an SFDR of 64 dB, an SNDR of 55 dB with a Walden Figure of Merit, FoM$_w$ of 6.8fJ/conversion-step at up-to 2MS/s. *Keywords— Nyquist sampling; input driver; SAR; Walden Figure-of-Merit*

I. Introduction

SAR ADCs are widely used for low power data acquisition applications e.g. in wireless sensor nodes. Most of the recent techniques found in literature [1-4] emphasize on lowering the Walden Figure of Merit, FoM$_w$ which now seems to saturate near to 1fJ per conversion-step [1,2]. In data acquisition systems targeted for low power wireless sensor nodes in IoTs, peripherals for microcontroller units (MCUs), the energy consumption of the associated signal processing and the analog front end circuitry to drive the ADC inputs can be much higher than the ADC power consumption. More importantly, for these IoT applications, the analog front end driving the ADC has to be always ON to present the signal to the ADC for conversion and further processing without significant latency or loss of critical information in case of any event detection. This calls for a greater attention to be paid to minimize the *input drive energy* of an ADC [10]. The goal of this work is to present a Range Pre-selection Sampling (RPS) based SAR ADC which helps to reduce the amount of energy required to drive the ADC inputs, so that the combined energy per conversion of driver plus the ADC is reduced.

II. Walden FoM vis-à-vis Input Drive Power

For state-of-the-art FoM$_w$ ADCs, V$_{SUPPLY}$ is below 1V, typically 0.4-0.7V [1-3,7,9]. Although this aids in lowering the power consumption of the mostly digital SAR ADC, it presents a greater challenge in driving the ADC as the supply voltage scaling demands a higher sampling capacitor, C$_S$ in order to meet its kT/C requirement. The minimum required input power to drive an ADC is estimated for state-of-the-art FoM$_w$ SAR ADCs and compared with the ADC power consumption P$_{ADC}$. An estimation for an ideal Class A driver current required for slewing and linear settling, for near Nyquist rate sampling is I$_{DR,MIN}$ =N·C$_S$·(ΔV$_{MAX}$/T$_{TRACK}$). Here ΔV$_{MAX}$ is the maximum signal change on the sampling capacitor C$_S$ and N is the number of time constants (assuming 1 for slewing and SNR/9 for linear settling) required for ½ LSB settling at the end of tracking period T$_{TRACK}$. T$_{TRACK}$ is typically 10-20% of the clock period, 1/f$_S$[8,11]. As shown in Table 1, I$_{DR,MIN}$ is typically orders of magnitude higher than the ADC supply current for the respective ADCs. For a driver operating at a supply voltage, V$_{DD}$ and considering a track period of 10% of the clock cycle, the minimum (theoretical) required input drive power for an ideal Class A driver P$_{IN,MIN}$ = V$_{DD}$·I$_{DR,MIN}$ [5,6,8] for state-of-the-art FoM$_w$ ADCs is also shown in Table 1.

It can be concluded that the actual bottleneck for low power data acquisition systems lies in driving C$_S$ which is not represented by FoM$_w$. This paper presents a 10b charge redistribution DAC (CDAC) based SAR ADC which introduces a Range Pre-selection Sampling (RPS) technique to reduce the ΔV$_{MAX}$ and thereby reducing the input driver power without affecting the Dynamic Range. Compared to conventional sampling (CS), the RPS technique results in lower peak input sampling currents thereby resulting in a lower input drive power P$_{IN}$ and consequently reduced energy consumption for the driver and ADC together.

III. Sampling technique and ADC Architecture

To demonstrate the RPS technique, we designed a SAR ADC that can be configured for either RPS or CS modes through an RPS_EN signal, Fig.1. For simplicity the single-ended architecture is shown. In actual differential implementation, V$_{DAC+}$ is compared to V$_{DAC-}$ (instead of V$_{HALF}$). The system consists of 3 CDACs, each sampling 1/3rd of single-ended input voltage range 0-V$_{PK}$. This range pre-selection sampling technique limits the maximum voltage change at each sampling capacitor to V$_{PK}$/3 while ensuring overall full-scale operation. Effectively this reduces the maximum required input drive power at high (near Nyquist-rate) frequencies, where conventionally it is the highest.

The ADC uses a split-capacitor DAC with a unit element of only 140aF. The total DAC capacitance for each CDAC is 145fF which is close to the kT/C limited value of 100fF for 10bit accuracy for 2V peak-peak differential input. To minimize glitches due to offsets between the 3 CDACs they share a common comparator. Each of the CDACs employ step-wise (dis)charging for the 3 most significant bits in the DAC array [4]. In addition, the ADC uses an event-driven control logic designed to operate at sampling rates from 10kS/s (limited by the bootstrapping S/H circuit) up-to 2MS/s at a 1V supply, maintaining almost constant FoM$_w$ for a fixed supply voltage for the wide sampling range.

978-1-5386-3179-9/17 $31.00 © 2017 IEEE

S15-2 (2203)

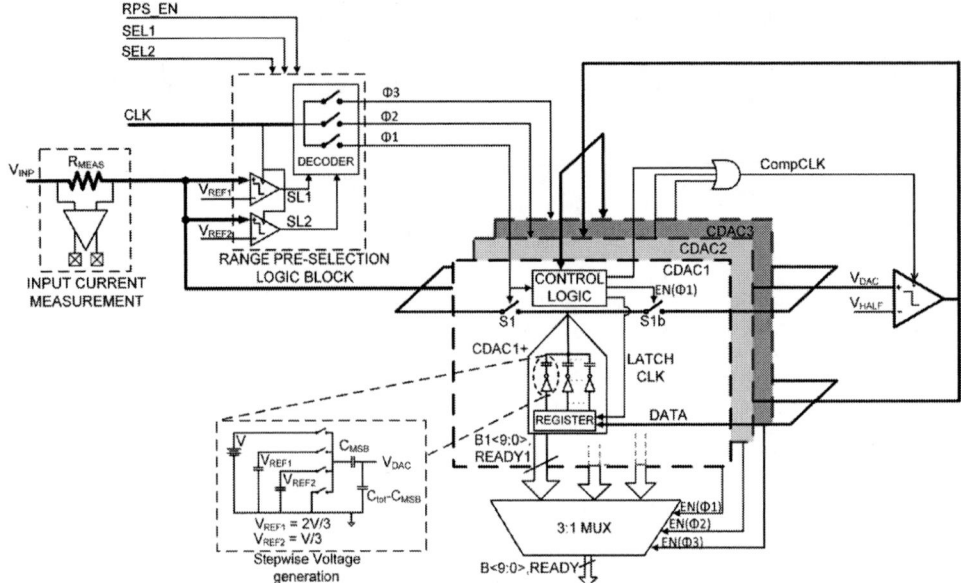

Fig. 1. Charge redistribution SAR ADC Architecture integrated with Range Pre-selection Sampling (RPS) technique

As shown in Fig.1, for the RPS mode, the range pre-selection (RPS) block determines before sampling in which range the input signal lies by comparing V_{INP} and V_{INN} to V_{REF1} (generated from the stepwise (dis)charging in the DAC array [4]). Depending on the RPS block output, CDAC1, CDAC2 and CDAC3 samples the inputs for Range 1, 2 and 3 respectively. Please note that in Fig.1 for simplicity, only V_{INP} is shown to be compared with V_{REF1} and V_{REF2} in the RPS block. However, in actual implementation due to symmetrical nature of differential inputs, both V_{INP} and V_{INN} are compared to only V_{REF1}. In the presented RPS technique, the three ranges expressed in terms of V_{INP} and V_{INN} are :

Range 1: $V_{INP} > V_{REF1}$ and $V_{INN} < V_{REF1}$,
Range 2: $V_{INP}, V_{INN} < V_{REF1}$,
Range 3: $V_{INP} < V_{REF1}$ and $V_{INN} > V_{REF1}$.

Based on the output of RPS block, either the signal Φ_1, Φ_2 or Φ_3 enable the corresponding CDAC and disables the other two. For e.g. if CDAC1 is selected, the corresponding bootstrapped S/H switch (S1) is enabled. After sampling, the corresponding enable signal (EN(Φ_1)) turns on the switch (S1b) to connect the selected CDAC (CDAC1) to the comparator input to perform the SAR conversion cycle. When the switch S1b is OFF during sampling, the main comparator inputs are pre-charged to zero to dispose of any charge from the previous SAR conversion, thereby resulting in no ISI.

Fig.2 shows the timing information together with the DAC voltages during an A/D conversion of near Nyquist rate inputs for both the RPS and CS techniques. In CS mode, RPS_EN is disabled and SEL1/SEL2 is used to select one of the CDACs to sample V_{INP}/V_{INN} on its differential DAC, CDAC+ /CDAC-. As shown, the maximum voltage change ΔV_{CS} occurs at the sampling capacitor when sampling full-scale inputs (0-V_{PK}) as V_{INP} and V_{INN} are always sampled onto the same CDAC+ and CDAC- respectively. This is in contrast for the RPS mode, wherein the RPS block selects

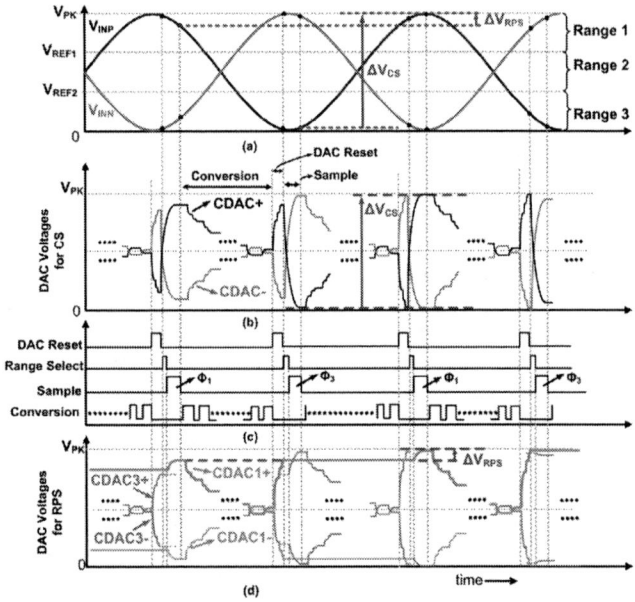

Fig. 2. (a) Input signals, V_{INP} and V_{INN} for near Nyquist rate sampling (b) DAC voltages for CS (c) Timing signals for RPS technique (d) DAC voltages for RPS based ADC highlighting the reduction in ΔV_{RPS}. For near Nyquist rate operation when V_{INP} & V_{INN} alternate between Range 1 and Range 3 at each successive sampling instant, then CDAC1 and CDAC3 are also selected alternately to respectively sample inputs in Range 1 and 3. This is highlighted by alternate selection of CDAC1 and CDAC3 in (d).

one of the CDACs to sample the inputs V_{INP} and V_{INN} depending on the range of the input signal. For example, for near Nyquist rate input frequencies when V_{INP} & V_{INN} alternate between Range 1 and Range 3, the RS block selects CDAC1 and CDAC3 alternately to sample the inputs. The maximum change (ideally) that can occur across CDAC+/CDAC- for any of the selected CDACs in RPS mode is thus $V_{PK}/3$. This happens for instance when V_{INP} changes from $V_{PK}(0)$ to V_{REF1} (V_{REF2}) for

978-1-5386-3179-9/17 $31.00 © 2017 IEEE   218

CDAC1+(CDAC3+) or from V_{REF1} to V_{REF2} for CDAC2+ at successive sampling instants. This means that the peak input sampling current required for RPS is (ideally) 3 times lower than that required for CS and the input drive power requirement can be also lowered by a factor of almost 3 for the RPS technique. For low input frequencies, when the change in input signal is less than $V_{PK}/3$ at each successive sampling instant, both the RPS mode and CS mode have similar peak input sampling current. For minimum power, the comparators in the RPS block are scaled down in size compared to the ADC main comparator. Since the information from the RPS block is only used to select the S/H switches, the accuracy of its comparators and the reference voltages (V_{REF1}, V_{REF2}) does not affect the final conversion accuracy. So even if the comparator's output in the RPS block would be incorrect, the ADC output is still *correct*. In addition to the output bits and the READY signal, the output of RPS block (SL1, SL2) is also buffered as output from the chip, indicating to which CDAC the output bits correspond to.

A measurement resistor, R_{MEAS} is placed in series with the input paths leading to the S/H switches to measure the ADC's sampling current profile. In order to settle with > 10 bit accuracy, the input impedance R_{IN} should satisfy the ½ LSB linear settling requirement at the end of tracking period, $N \cdot R_{IN,MAX} \cdot C \le 1/(10 \cdot f_s)$. R_{MEAS} is chosen as $1k\Omega$ so that together with the bootstrapped S/H switch resistance (small signal) R_{SW} of approximately $7k\Omega$ in our ADC, total input resistance $R_{IN} = (R_{MEAS} + R_{SW})$ is a factor 6 lower than the theoretical limit $R_{IN,MAX}$ (small signal) of approximately $50 k\Omega$. On-chip amplifiers measure the voltage across these resistors; their outputs are probed off-chip using a 20GS/s Keysight sampling scope.

IV. Measurements and Results

Fig. 3 shows the die micrograph fabricated in a standard 65nm CMOS process with an active area (including decaps) of 0.08 mm².

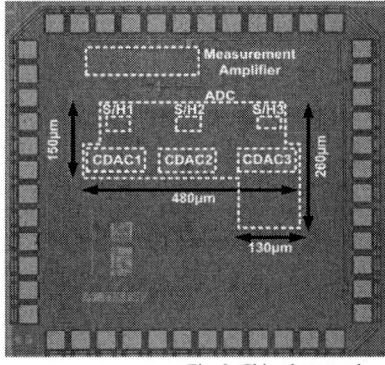

Fig. 3. Chip photograph

As evidenced by the sampling current profiles for RPS and CS in Fig.4 measured for near Nyquist rate sinusoidal input, the peak input current for RPS is reduced by a factor 2.4 for f_{IN} near to $f_s/2$. Note that the peak input current for the RPS based ADC occurs when sampling the inputs in range 2, and not in range 1 or 3 as for CS. Fig.5 shows the simulated peak input sampling current for both RPS and CS along with measured data points as function of f_{IN} for sinusoidal inputs. The overall peak input sampling current

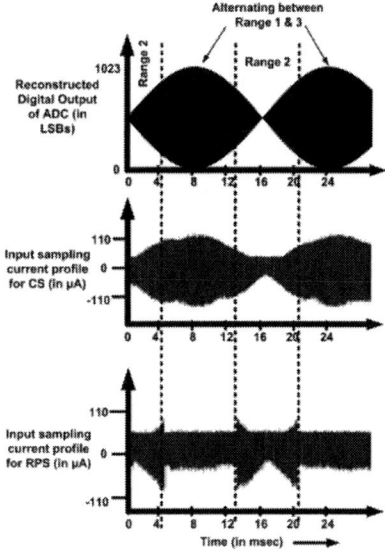

Fig. 4. Measured sampling current profile envelope at Fs–2*Fin, for both CS and RPS for Fin = 499.96875kHz and Fs = 1MHz. The high density in the plot represents the envelope of the near Nyquist input signal and the corresponding input sampling currents in CS and RPS mode, measured by a 20GS/s Digital Oscilloscope.

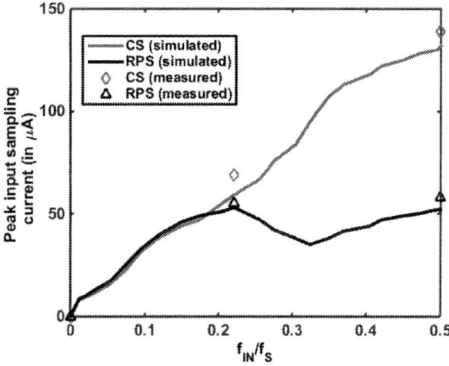

Fig. 5. Peak input sampling current of the RPS and CS based ADCs sampling current profiles as seen in Fig.4 for various input frequencies

for RPS over the entire input frequency range is 2.4 times lower in comparison to CS. Please note that the ADC drivers e.g. source followers are designed to handle maximum drive (peak sampling) currents to allow for initial slewing and linear settling. This implies that for a Class A input driver for the ADC, the input drive power P_{IN} can be decreased by at least a factor 2.4 using RPS. Using the expressions in Section II for e.g. for an ideal Class A behavior, P_{IN} is found to be reduced from $130\mu W$ for CS to $50\mu W$ for RPS, for P_{ADC} of only 3.25 μW at 1MS/s. This shows that the driver power is dominant over the ADC power consumption and hence the reduction in P_{IN} by a factor 2.4 for RPS technique is quite noteworthy. Please note that this reduction in peak input sampling current by factor 2.4 is less than the theoretical value 3 which happens in case of (ideal) impulse sampling with zero track time. Fig. 6 shows that the design achieves 64dB SFDR and 55dB SNDR with 8.9 ENOB at a 1.7V peak-peak differential input using RPS. A design error resulted in unequal interconnect parasitic between each CDAC to the common comparator thereby resulting in systematic gain mismatch. The gain mismatch of the three CDACs was measured over 7 samples. The resulting systematic gain error was calibrated one-time in the foreground for the RPS mode. As shown in Fig.7, the ADC has +0.9/-0.95 LSB DNL and +1.4/-1.1 LSB INL (after foreground calibration) for RPS mode over 7 samples. This is similar to the measured +0.8/-0.85 LSB DNL and +1.1/-1 LSB INL for the CS mode. Our RPS based ADC achieves a FoM$_W$ of 6.8fJ/conversion over a sampling rate from 10kS/s to 2MS/s, comparable to state-of-the-art SAR ADCs offering such a wide range of sampling frequency [7,9]. Please note that there is no degradation in FoM$_W$ due to the RPS technique in comparison to CS, thereby confirming that it does not degrade the SAR ADC performance. Although the RPS technique does seem to have an area penalty, however from a system perspective (including driver) the reduction in the bias current requirement of the preceding driver stage by a factor 2.4 outweighs this DAC area overhead. Also the CDACs which are disabled for a selected range act as additional decap.

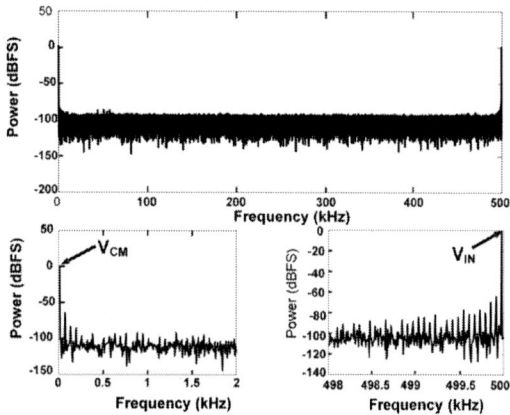

Fig. 6. FFT of the measured RPS based ADC output, normalized to the input tone Fin = 499.96875kHz and Fs = 1MHz.

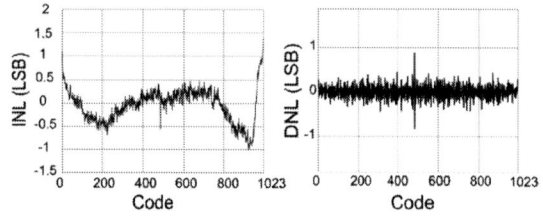

Fig. 7. INL DNL characteristics of the ADC in RPS mode. INL shown is obtained after foreground calibration done for systematic gain mismatch for 3 CDACs. DNL performance is not impacted by the gain mismatch.

Table 1 compares the performance of the proposed RPS based SAR ADC to state-of-the-art FoM$_W$ SAR ADCs. It is to be noted that the measured power for both RPS and CS in our ADC is approx. a factor 3 more than the theoretical minimum. This is because R_{IN} (large signal) is almost 3 times less than $R_{IN,MAX}$ to avoid limiting the linearity of the SAR ADC at the front end sampler for the wide sampling range up to 2MS/s and to meet ½ LSB tracking bandwidth.

Since the peak sampling current is equal to the maximum voltage step divided by this R_{IN}, hence the measured peak sampling current and consequently the input drive power is approx. a factor 3 greater than the theoretical value.

Table 1 : Comparison of RPS based ADC with state-of-art FoM$_W$ ADCs

Architecture	This Work		[2]	[7]	[9]
	SAR with RPS	SAR with CS			
Technology	65nm	65nm	90nm	65nm	90nm
Resolution[bits]	10	10	11	10	10
Supply [V]	1	1	0.3	0.7	0.7
Maximum Sampling Rate	2 MS/s	2 MS/s	600 kS/s	2 MS/s	4 MS/s
Ideal Diff. Input Swing, V_{PK-PK} [V]	2	2	1.2	1.4	1.4
P_{ADC} (in µW)	6.25	6.25	0.2	3.6	11
$I_{DR,MIN}$ (µA), Calculated input (driver) current	16	40	60	70	860
$P_{IN,MIN}$ (in µW), Calculated input power	16 @2MS/s	40 @2MS/s	36	50	600
P_{IN} (in µW), Measured input power	50	130	No data	No data	No data
ENOB [bits]	8.9	9	9.4	9.3	9
FoM$_W$ (fJ/conversion)	6.8	6.4	0.44	2.8	5.2
Area (in mm^2)	0.08 (incl.decap)	0.05	0.04	0.05	0.04

V. Conclusion

A proof-of-concept for the Range Pre-selection Sampling (RPS) technique has been demonstrated in a 10b 2MS/s SAR ADC to reduce its input drive power requirement which is very seldom addressed. The RPS based SAR ADC reduces the peak sampling current requirement by 2.4 times as compared to CS. This 2.4x reduction in peak sampling current by RPS is 1.65 times higher than the reduction through energy reduced sampling technique as reported in [10]. Considering an ideal Class A behavior, the input power can be reduced from 130µW for conventional sampling to 50µW for the case of RPS in our ADC, while the ADC dissipates 3.25µW. Since the input driver power consumption is order of magnitude greater than the ADC power, this reduction in input driver power by a factor 2.4 is significant in reducing the overall driver plus the ADC power dissipation.

References

[1] Hung-Yen Tai et.al, "A 0.85fJ/conversion-step 10b 200kS/s subranging SAR ADC in 40nm CMOS", *ISSCC 2014.*
[2] S.-E. Hsieh et al., "A 0.44fJ/conversion-step 11b 600kS/s SAR ADC with Semi-resting DAC", *VLSI Symposium 2016.*
[3] P. Harpe, et al., "A 2.2/2.7fJ/conversion-step 10/12b 40kS/s SAR ADC with Data-Driven Noise Reduction", *ISSCC 2013.*
[4] M. van Elzakker, et.al., "A 10-bit Charge-Redistribution ADC Consuming 1.9mW at 1MS/s", JSSC 2010.
[5] B. Murmann, "Energy Limits in Current A/D Converter Architectures", *ISSCC Short Course, Feb. 2012.*
[6] A.-J. Annema, et. al "Analog circuits in ultra-deep-submicron CMOS", *JSSC 2005.*
[7] P. Harpe et al., "A 0.7V 7-to-10bit 0-to-2MS/s flexible SAR ADC for ultra low-power wireless sensor nodes", *ESSCIRC 2012.*
[8] B. Murmann, "Limits on ADC Power Dissipation" in *Analog Circuit Design,* Springer, The Netherlands, 2006, pp 351-356.
[9] C.Y.Liou et. al, "A 2.4-5.2fJ/conversion-step 10b 0.5-to-4MS/s SAR ADC with charge-average switching DAC in 90nm CMOS", *ISSCC 2013.*
[10] H.S. Bindra et. al " An energy reduced sampling technique applied to a 10b 1MS/s SAR ADC", *accepted at ESSCIRC 2017.*
[11] R. Kapusta, "Advanced SAR ADCs for high throughput applications", *ISSCC Short Course, Feb. 2017.*

S15-3 (2064)

A 5-bit 2 GS/s Binary-Search ADC with Charge-Steering Comparators

U-Fat Chio, Sai-Weng Sin[1], Seng-Pan U[1, 2], Franco Maloberti[3], R. P. Martins[1, 4]

State Key Laboratory of Analog and Mixed-Signal VLSI (http://www.amsv.umac.mo)
1 - Dept. of ECE, Faculty of Science and Technology, University of Macau, Macao, China
2 – Also with Synopsys Macau Ltd.
3 – Department of Electrical, Computer and Biomedical Engineering, University of Pavia, Pavia, Italy
4 – On leave from Instituto Superior Técnico/Universidade de Lisboa, Portugal
E-mail: terryssw@umac.mo

Abstract—This paper presents a 5-bit 2GS/s binary-search ADC. The proposed architecture prevents the use of a decoder to avoid the path delay racing between control signals and clock phases; thence the bit latency reduces to 1 single comparator delay only. We also propose a dynamic charge-steering comparator to quantize each bit quickly. Besides, we present well-balanced 1-of-N-to-Binary encoders to transform the output code with low power. This ADC consumes 3.9mW at 2GS/s in 65nm CMOS. It achieves a SNDR of 28 dB at Nyquist rate resulting in a FoM of 95 fJ /conv.-step.

Keywords—binary-search ADC; asynchronous; charge-steering.

I. INTRODUCTION

Communication systems such as wireless local area networks (WLANs), optical Ethernet, electrical wireline links, etc. require high-speed and low-power ADCs. Flash ADC is often the choice due to its high conversion speed, but it suffers from relatively large power consumption and large area. Binary-search (BS) ADC structure combines the characteristics of the successive approximation register (SAR) and the flash, which is another candidate to operate at high-speed but achieving low power and small area [1]-[4].

In the primitive BS scheme [1], its energy consumption, for a N-bit resolution, happens only during input sampling, reference charging, and N times the comparator operates in the whole conversion period. Because the common feature of BS ADCs [1]-[4] is that the reference voltages are applied early at the inputs of the comparators. Subsequently, various techniques [1]-[4] were reported to utilize less number of comparators by trading with the cost of a massively increased complexity of the reference pre-charging network and control logic in each bit stage. The extra power from the decoders is inevitable in those schemes [1]-[4]. Nevertheless, the control signal path due to the decoder delay is racing with the clock path of the comparator delay. Thus, the total settling time of the reference switching must be shorter than the comparison time in each bit stage [5]. Evidently, the trade-off penalty would always be the number of comparators versus power and speed.

In this paper, we propose a BS ADC architecture constituted by 2 BS sub-ADCs to reduce the number of comparators to $2^{N-1}+1$ for an N-bit resolution, and without an additional decoder. Fig. 1(a) shows a brief comparison of different BS ADC schemes. When compared with [1], a half number of comparators is saved. This resulted in a compact die size that is advantage to achieve both high-speed and low-power. When compared with [1]-[4], our structure also avoids the delay path racing and power consumption from the decoder. Fig. 1(b) depicts the behavioral model prediction of energy per conversion versus resolution for [1]-[4] and this work,

This research work was financially supported by Research Grants of University of Macau and Macao Science & Technology Development Fund (FDCT) under the reference 055/2012/A2.

IEEE Asian Solid-State Circuits Conference
November 6-8, 2017/Seoul, Korea

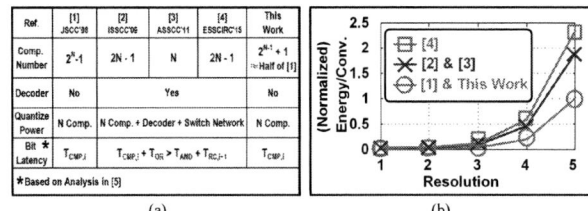

Ref	[1] JSCC'96	[2] ISSCC'06	[3] ASSCC'11	[4] ESSCIRC'15	This Work
Comp. Number	2^N - 1	2N - 1	N	2N - 1	2^{N-1} + 1 ≈ Half of [1]
Decoder	No		Yes		No
Quantize Power	N Comp.	N Comp. + Decoder + Switch Network			N Comp.
Bit ★ Latency	$T_{CMP,i}$	$T_{CMP,i}$ + T_{OR} > T_{AND} + $T_{RC,i-1}$			$T_{CMP,i}$

★ Based on Analysis in [5]

(a)

(b)

Fig. 1. Different BS ADC schemes (a) Comparison, and (b) Energy per conversion versus resolution (all normalized to [1] & this work at 5-bit).

estimated from the decoder with its logic gate count, switching network, and also including the thermal noise of the comparators [6]. It is obvious that this design achieved better energy efficiency when compared with [2]-[4] due to the reduced latency and power consumption.

Since the bit latency is only given by a single comparator delay in this work, the speed of comparator is a significant design consideration. Typically, the latch-type sense amplifier [7] and the double tail latch [8] are the popular topologies for high-speed comparator designs, due to their dynamic operation schemes. Nevertheless, the delay of latch-type sense amplifier is rather dependent on the common-mode voltage [9]. And, the tail path in the double-tail latch allows its dynamic preamplifier outputs both collapsed to ground at clock start, leading to a slow response [10]. Recently, some high-speed comparators have been proposed to employ a slightly static cascaded preamplifier to raise the response speed of the latch in SAR ADCs [11][12]. However, they are not power efficient to applied for BS scheme, due to its operation characteristic.

Here, we propose a dynamic charge-steering comparator to quantize each bit swiftly. Besides, we also utilize well-balanced 1-of-N-to-Binary encoders to transform the binary output code with low power.

II. PROPOSED ADC ARCHITECTURE

Fig. 2 shows the architecture of the proposed 5-bit BS ADC composed by 3 conversion stages, a reference ladder, 1-of-N encoders, and a clock generator. The 1st stage comprises of a passive S/H and a comparator COMP1. The 2nd stage is constituted by an upper comparator COMP2U and a lower comparator COMP2L with their own passive S/Hs. The 3rd stage comprises two 3b BS sub-ADCs CBS1 and CBS2. The proposed BS ADC operates in a fully asynchronous conversion from the 1st to the 3rd stage. The clock generator only produces 3 phases including the sampling phase Φ_S, the hold phase Φ_H, and the 1st stage comparator triggering phase Φ_1, leading to low power dissipation. During the sampling phase (Φ_S=1), the S/Hs of the three stages sample the input signal simultaneously. In the hold phase (Φ_H=1), the S/Hs of the 1st and 2nd stages

978-1-5386-3179-9/17 $31.00 © 2017 IEEE   221

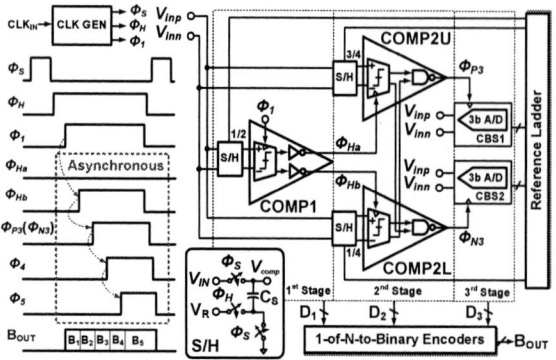

Fig. 2. Proposed ADC architecture with its timing diagram.

generate the residues by subtracting the sampled input signal with fixed references 1/2, 1/4 and 3/4 from the reference ladder. The quantization starts in the first stage at the rising edge of Φ_1. COMP1 is activated to quantize the MSB result. And then, one of the triggering phases Φ_{Ha} or Φ_{Hb} is attained to control COMP2U or COMP2L in the 2nd stage to quantize the 2nd bit output. Further, Φ_{Ha} and Φ_{Hb} are also the hold phases of CBS1 and CBS2 in the 3rd stage. Since the 1st, 2nd and 3rd stages quantize the digital outputs D_1, D_2 and D_3 in 1-of-N format, the 1-of-N-to-Binary encoders transform D_1, D_2, and D_3 to a 5-bit binary code output B_{OUT}.

In the primitive BS scheme [1], 4 comparators with reference voltages of 1/8, 3/8, 5/8 and 7/8 are ready in the 3rd bit step. Apparently, if MSB = 0, only two comparators with reference voltages of 1/8 and 3/8 are useful, and they can anticipate the selection since MSB is known before. In our proposed structure, the reference voltages of the 3rd stage no longer need to be prepared before the whole conversion starts, and they can be produced after the quantization of the 1st stage. Then, we can share 14 comparators with 28 reference voltages in the 3rd stage, consequently saving half of the comparators. Since we share the comparators with different pre-charged reference voltages, the output polarities of the comparison in the 2nd stage should be inverted while MSB = 0. Here, two NAND gates follow the outputs of COMP2U and COMP2L to match the selection principle of the binary-search algorithm. To simplify the logic, we merged two NAND gates into the last stages instead of the inverter pairs in COMP2U and COMP2L. The function of these NAND gates is to select CBS1 or CBS2 from the comparison result and, simultaneously, to provide sufficient driving capability. Here the dimension of the NAND gate is equal to the original inverter pair in the last-stage output of COMP1. Therefore, the bit latencies of the 1st and 2nd stages are almost the same as 1 comparator delay.

Fig. 3 shows the schematic of CBS1/ CBS2. There are 7 comparators connected as a 3-level binary tree. S/Hs are distributed with the comparators, which sampled the input signal during Φ_S=1. While the hold phase Φ_{Ha}/Φ_{Hb} is turned on, the corresponding reference voltages are switched to the S/Hs to be subtracted with the sampled input signal for quantization. The critical path is located at the first comparator of CBS1/CBS2. The voltage settling from the switching of references V_{RP1a}/V_{RP1b} and V_{RN1b}/V_{RN1b} must be faster than the rising edge of Φ_{P3}/Φ_{N3}. Since a single comparator delay from

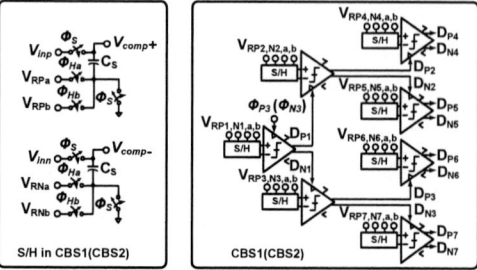

Fig. 3. Schematic of CBS1/CBS2.

COMP2U/COMP2L is occupied in this period, the time constraint can be given as,

$$T_{CMP,2} > T_{RC,1ab} \qquad (1)$$

where $T_{CMP,2}$ is the comparator delay of COMP2U/COMP2L, and the $T_{RC,1ab}$ is the RC settling time of the first S/H in CBS1/CBS2. Because of the distributed S/H, the equivalent settling capacitance is from the top plate parasitic of the sampling capacitance C_S in the S/H as well as the input capacitance of the comparator. In the design this capacitance is less than 2fF to be rapidly charged with a low power resistive ladder. Moreover, the main sampling capacitance C_S does not contribute to the reference settling time constant due to the reference level shifting operation at the high impedance node of the comparator input.

In overall the comparators count is significantly reduced, here, in 5 bit steps is 1, 2, 2, 4, 8, while it is 1, 2, 4, 8, 16 in [1], respectively. When compared to [1]-[4], we only reduce the comparator count at the 3rd-bit step, and thus avoid the extra decoder and massively increased switching network. Moreover, the path delay racing from the decoder is also prevented as well. The energy per conversion of the ith-bit stage decoder for N-bit BS ADC can be given as [6]:

$$E_{dec} = [\sum_{i=4}^{N} (2^{i-1}) \cdot (i-3)] \cdot C_{NAND} \cdot V_{DD}^2 \qquad (2)$$

where C_{NAND} is the parasitic capacitance of NAND gate. Obviously, the number of NAND gates is exponentially increasing with its resolution after the 3rd-bit quantization, leading to large power and delay.

III. CIRCUIT IMPLEMENTATION

A. Charge-steering Comparator

In the proposed BS architecture, the total conversion period after input sampling is around 5 times the comparator delay. Therefore, shortening the comparator delay can further enhance the ADC speed. Since many comparators are not triggered in the BS scheme, dynamic operation of comparator design is the key consideration. On the other hand, the fast response of

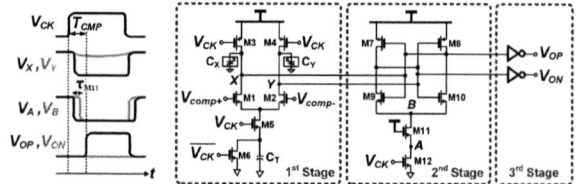

Fig. 4. Proposed charge-steering comparator with its operation waveforms.

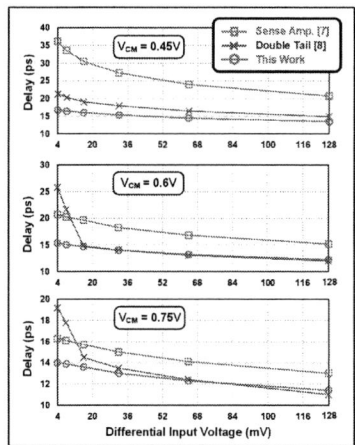

Fig. 5. Simulated comparator delays versus differential input at $V_{CM} = 0.45V$, 0.6V and 0.75V for different comparator schemes under 1.2V supply voltage.

comparator is not only related to speed, but also related to the metastability impact for the asynchronous clock phase generation.

Fig. 4 shows the proposed charge-steering comparator comprising 3 stages. The 1st stage serves as a charge-steering preamplifier, the 2nd stage works as a cross-coupled regenerative latch, and finally, the 3rd stage is a CMOS inverter pair to provide the driving capability. Differential capacitor banks C_X and C_Y are dedicated to offset voltage foreground calibration [3]. Since the charge source of C_T is equivalent to a tail current source during the amplification status [10], there are two merits for enhancing the comparator speed and reducing delay change with the input common-mode voltage variation. First, the operation current is not decreased by the reduced drain-source voltage of M5. Thus, there is no speed penalty even if the input common-mode voltage is lower. Second, the 1st stage outputs will not collapse to zero together while the clock is triggering on. Consequently, M1 and M2 can sustain in saturation region, providing a fast and robust amplification of comparator input voltage. In addition, we insert an always-on transistor M11 in the 2nd stage between node A and B, and therefore add a transistor delay τ_{M11} in the clock path of M12. This delay postpones slightly the regeneration but reserves more time for the amplification. Thus, the differential signal ΔV_{XY} between V_X and V_Y can reach the logic threshold in the regeneration nodes while the clock is toggled, reducing the metastability. Besides, the clock transition coupling of M12 from V_{CK} is isolated by M11, thus avoiding the glitch coupled to the regeneration nodes directly. In the above configuration, only dynamic power is dissipated.

Fig. 5 shows the simulated comparator delays of [7], [8], and this work versus the differential input at $V_{CM} = 0.45V$, 0.6V, and 0.75V, respectively. The comparators are operated at 3.3GHz clock frequency under 1.2V supply voltage, and their transistor dimensions are scaled to let them have the same power consumption. From the plots, the delay of [7] is sensitive to the common-mode voltage variations. And the speed of [8] dropped obviously when the input is small. Actually, while the clock turns on, the input-stage outputs drop to zero in succession with different times, depended to the input voltage difference [8]. When the input is small, the input-

Fig. 6. Proposed 1-of-N-to-Binary encoders.

stage transistors both change from saturation to triode region quickly. The decreased gain also reduces the difference of drop-to-zero voltages, and thus increasing the latch regeneration time and comparator delay. In the proposed scheme, the comparator delays are 16.8ps, 15.3ps and 14ps for a 4mV differential input with different common-mode voltages, respectively. When compared to [7] and [8], the proposed comparator attains shorten delay, less metastability, and is not sensitive to common-mode voltage variations.

B. 1-of-N-to-Binary Encoders

As mentioned before, the comparator gives rise to $V_{OP} = 1$ and $V_{ON} = 0$ if the input voltages are $V_{comp+} > V_{comp-}$, otherwise the opposite. If the comparator is not triggered, $V_{OP} = V_{ON} = 0$. Due to the characteristics of the BS conversion, there is only 1 comparator activated at each bit quantization. Subsequently, the comparator outputs appear as 1-of-N code format in each bit stage. For example, if $\Phi_{P3} = 1$, the 1st comparator in CBS1 is triggered, and the comparison result is $D_{P1,CBS1} = 1$ and $D_{N1,CBS1} = 0$. On the other hand, $\Phi_{N3} = 0$ (since only one of Φ_{P3} and Φ_{N3} can be turned into 1 during BS conversion). Thus, the 1st comparator of CBS2 is not triggered, whose output states are kept as $D_{P1,CBS2} = 0$ and $D_{N1,CBS2} = 0$. Consequently, the output code is 1000 in the 3rd bit quantization. Fig. 6 shows the schematics of 1-of-N-to-Binary encoders. In the B_3 encoder, $D_{P1,CBS1}$ ($D_{N1,CBS1}$) and $D_{P1,CBS2}$ ($D_{N1,CBS2}$) each control one NMOS switch connected to V_{DD} (V_{GND}). B_3 becomes logic '1' since only the switch controlled by $D_{P1,CBS1}$ is turned on. Actually, all comparators outputs are connected to the switches with V_{DD} or V_{GND}, depending on their polarity.

Since all asynchronous clock phases are generated by comparator outputs in the BS scheme, the balancing of comparator differential outputs loading is important. In this encoder, the comparator outputs are both connected to a NMOS switch with the same dimension, leading to an output loading well balanced. Although an NMOS switch is not adequate to pass the supply voltage, it is sufficient to process a digital signal. A strong logic '1' can be produced after two small inverters are located at the encoder output. We employ low-threshold NMOS transistors in the encoders to relieve the voltage drop of the logic '1'. Besides, we utilize minimum size transistors to reduce the loading effect to the comparator. Plus, there are only 5 switches turning on for 5-bit code transformation, leading to low-power operation.

IV. Measurement Results

Fig. 7 shows the die photograph of the prototype ADC, fabricated in a standard 1P7M 65nm CMOS, with an active area of 0.0084mm² (70µm×120µm). Fig. 8 shows the measured DNL/INL, the DNL is +0.78/-0.71 LSB and the INL is +0.95/-0.59 LSB. Fig. 9 shows the FFT spectrum measured at 2 GS/s with 1.02 GHz input. The measured SNDR and SFDR are 28 dB and 36.8 dB, respectively. Fig. 10 (a) and (b) illustrates the SNDR versus sampling frequency and input frequency,

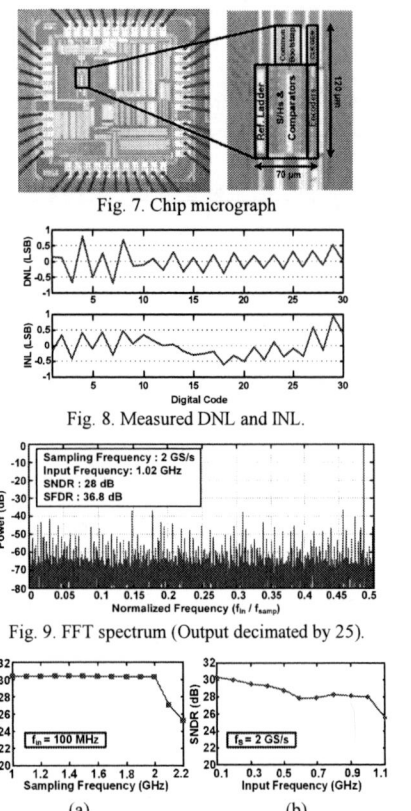

Fig. 7. Chip micrograph

Fig. 8. Measured DNL and INL.

Fig. 9. FFT spectrum (Output decimated by 25).

Sampling Frequency : 2 GS/s
Input Frequency: 1.02 GHz
SNDR : 28 dB
SFDR : 36.8 dB

(a)　(b)

Fig. 10. Measured SNDR versus (a) Sampling frequency, (b) Input frequency.

TABLE I
PERFORMANCE SUMMARY & BENCHMARK WITH STATE-OF-THE-ART

	JSSC'14 [13]	ASSCC'15 [14]	ISSCC'16 [15]	JSCC'08 [1]	ISSCC'09 [2]	ASSCC'11 [3]	ESSCIRC'15 [4]	This Work
Architecture	Subranging	Flash	SAR	BS	BS	BS	Pipelined BS	BS
Process (nm)	65	65	40	90	65	65	40	65
Resolution (bit)	6	6	6	6	5	5	6	5
Speed (GS/s)	1	3.4	1	0.25 *	0.8	0.5	1.6	2
Supply (V)	1.1	1	1	1	1	1.2	0.9	1.2
Power (mW)	9.9	12.6	1.26	0.14	1.97	1.63	3.17	3.9
SNDR (dB) @ DC	33	34.7	35.1	33.7	28.2	29.5	30.5	30.3
SNDR (dB) @ Nyquist	32.8	34.2	34.6	32.1	26.9	27.8	29.2	28
FOM	278	89	28.7	15	116	117	84.1	95
Active Area (mm²)	0.044	0.034	0.00058	0.05	0.018	0.015	0.017	0.0084
Offset Reduction	On-Chip Calibration	On-Chip Calibration	By Tolerance	Off-Chip Calibration	By Tolerance	On-Chip Calibration	On-Chip Calibration	On-Chip Calibration

* 6-bit Sub-ADC in BS Operation, Speed is 0.25 GS/s

the BS becomes a good candidate for high-speed and low-power ADC conversion due to its inherent advantages.

REFERENCES

[1] G. Van der Plas and B. Verbruggen, "A 150 MS/s 133 μ W 7 bit ADC in 90 nm Digital CMOS," in *IEEE J. Solid-State Circuits, vol. 43, no. 12, pp. 2631 - 2640, Dec. 2008.*

[2] Y. Z. Lin et al., "A 5b 800MS/s 2mW Asynchronous Binary-Search ADC in 65nm CMOS," *ISSCC Dig. Tech. Papers*, pp. 80-81, Feb. 2009.

[3] S. S. Wong et al., "A 4.8-bit ENOB 5-bit 500MS/s Binary-Search ADC with Minimized Number of Comparators," in Proc. of *2011 Asian Solid-State Circuit Conference (A-SSCC)*, pp. 73-76, Nov. 2011.

[4] K. Tanaka et al., "A 1.6 GS/s 3.17 mW 6-b Passive Pipelined Binary-Search ADC with Memory Effect Canceller and Reference Voltage Calibration," in Proc. of *IEEE European Solid-State Circuits Conference (ESSCIRC)*, pp. 327-330, Sep. 2015.

[5] Y. Z. Lin et al., "An Asynchronous Binary-Search ADC Architecture with a Reduced Comparator Count," *IEEE Transactions on Circuits and Systems I: Regular Papers*, vol. 62, Issue: 5, pp. 1829-1837, Aug. 2010.

[6] S. S. Wong et al., "A 2.3 mW 10-bit 170 MS/s Two-Step Binary-Search Assisted Time-Interleaved SAR ADC," in *IEEE J. Solid-State Circuits*, vol. 48, no. 8, pp. 1783 - 1794, Sep. 2013.

[7] T. Kobayashi et al., "A Current-Controlled Latch Sense Amplifier and a Static Power-Saving Input Buffer for Low-Power Architecture," in *IEEE J. Solid-State Circuits*, vol. 28, no. 4, pp. 523-527, Apr. 1993.

[8] D. Schinkel et al., "A Double-Tail Latch-Type Voltage Sense Amplifier with 18ps Setup+Hold Time," *ISSCC Dig. Tech. Papers*, pp. 314-315, Feb. 2007.

[9] B. Wicht et al., "Yield and Speed Optimization of a Latch-Type Voltage Sense Amplifier," in *IEEE J. Solid-State Circuits*, vol. 39, no. 7, pp. 1148-1158, Jul. 2004.

[10] Behzad Razavi, "Charge Steering: A Low-Power Design Paradigm," in Proc. of *IEEE Custom Integrated Circuits Conference(CICC)*, pp. 1-8, Sep. 2013.

[11] S. Le. Tual et al., "A 20GHz-BW 6b 10GS/s 32mW Time-Interleaved SAR ADC with Master T&H in 28nm UTBB FDSOI Technology," *ISSCC Dig. Tech. Papers*, pp. 382-383, Feb. 2014.

[12] D. G. Muratore et al., "An 8-bit 0.7-GS/s Single Channel Flash-SAR ADC in 65-nm CMOS Technology," in Proc. of *IEEE European Solid-State Circuits Conference (ESSCIRC)*, pp. 421-424, Sep. 2016.

[13] T. Danjo et al., "A 6-bit, 1-GS/s, 9.9-mW, Interpolated Subranging ADC in 65-nm CMOS," in *IEEE J. Solid-State Circuits*, vol. 49, no. 3, pp. 673 - 682, Jan. 2014.

[14] J. Liu et al., "A 89fJ-FOM 6-bit 3.4GSs Flash ADC with 4x Time-Domain Interpolation," in Proc. of *2015 IEEE Asian Solid-State Circuits Conference (A-SSCC)*, pp. 1 - 4, Nov. 2015.

[15] K. D. Choo et al., "Area-Efficient 1GS/s 6b SAR ADC with Charge-Injection-Cell-Based DAC," *ISSCC Dig. Tech. Papers*, pp. 460-461, Feb. 2016.

respectively. The drop of the SNDR from f_{in}=0.1–1 GHz is 2.3 dB, resulting in an effective resolution bandwidth (ERBW) over the Nyquist frequency. The effective number of bits (ENOB) is 4.72 bit at 100MHz and 4.36 bit at the Nyquist input. The prototype ADC consumes 3.9mW of power operating at 2 GS/s. The analog power including S/Hs, comparators and reference ladder consumes 78% of the total power, while the digital power is only 22%. This means that most of the power is dissipated in the conversion, and not in the control signal processing. Table I summarizes the performance comparison of this work with previous designs of BS ADCs as well as with the state-of-the-art single-channel ADCs at a speed of 1 to 3.5 GS/s in 5 to 6-bit resolution. The implemented 5-bit 2 GS/s BS ADC achieves a Walden FoM of 74 and 95 fJ/conversion-step for 100 MHz and Nyquist input frequency, respectively, with a very compact design of 0.0084mm² only.

V. CONCLUSIONS

This paper reported a 5-bit 2GS/s binary-search ADC. The proposed architecture exploits the advantage of the most primitive BS scheme, but reduces the comparator count. In the design, the bit latency is just only a single comparator delay. The proposed charge-steering comparator quantizes each bit rapidly further enhancing the ADC speed. Besides, the proposed 1-of-N-to-Binary encoders can well balance the asynchronous clock phase paths of the BS ADC. Obviously, by shortening the comparator delay of each bit quantization,

A 13-bit 160MS/s Pipelined Subranging-SAR ADC with Low-Offset Dynamic Comparator

Weitao Li, Fule Li*, Jia Liu, Hongyu Li and Zhihua Wang
Institute of Microelectronics
Tsinghua University, Beijing, China
*Email: lifule@tsinghua.edu.cn

Abstract—A 13-bit 160MS/s hybrid ADC in 65 nm CMOS is presented in this paper. By combining the pipelined, flash and SAR architectures, a hybrid ADC architecture is proposed to improve the power efficiency. An input offset storage technique of dynamic comparator is proposed to increase the conversion linearity. A reference voltage buffer with the charge compensation is proposed to save power and reduce the decoupling capacitor. The achieved peak SNDR, SFDR, and FoM are 64.2 dB, 80.5 dBc, and 107.3 fJ/conversion-step, respectively.

Keywords—*CMOS integrated circuits; analog-to-digital converter; hybrid ADC; dynamic comparator; reference voltage buffer; pipelined subranging-SAR ADC*

I. INTRODUCTION

High-quality imaging in medical systems and physical property measurement systems typically adopt arrays of analog-to-digital converters (ADCs) with high power efficiency. In order to achieve a combination of high resolution, high speed and low power, the challenge includes the architecture level and circuit block level. To realize high conversion performance, conventional pipelined architecture is limited by the cost of residue amplification. While the opamp-based amplification is normally power-hungry, the low-power open-loop dynamic amplifier suffers from PVT sensitivity [1]. In addition, flash ADC is limited by exponentially increasing comparators and SAR architecture is limited by serial conversion cycles.

In the circuit block level, multi-bit front end is typically adopted in the high-SNR noise-limited ADC, because it can significantly reduce the power dissipation. However, it needs to reduce the comparator offset. On the other hand, in the multi-bit front, the small size dynamic comparator is normally used to save power and area. But, large offset is introduced. Although the offset of the dynamic comparator can be calibrated by adjusting differential parameters, like the buck voltages and the loads [2], high hardware overhead is required. Conventional input offset storage can cancel the offset at a low cost [3], but it can only be used in the static preamplifier.

In addition, to provide high accuracy reference voltage, the narrow-bandwidth buffer is normally adopted because of the low noise and low power. However, it requires large decoupling capacitors. In order to save area, off-chip capacitors are usually adopted. But, extra pads are introduced and hence the bond wire effect limits the settling accuracy. On the other

Supported by the Strategic Priority Research Program of the Chinese Academy of Sciences, Grant No. XDA10010600.

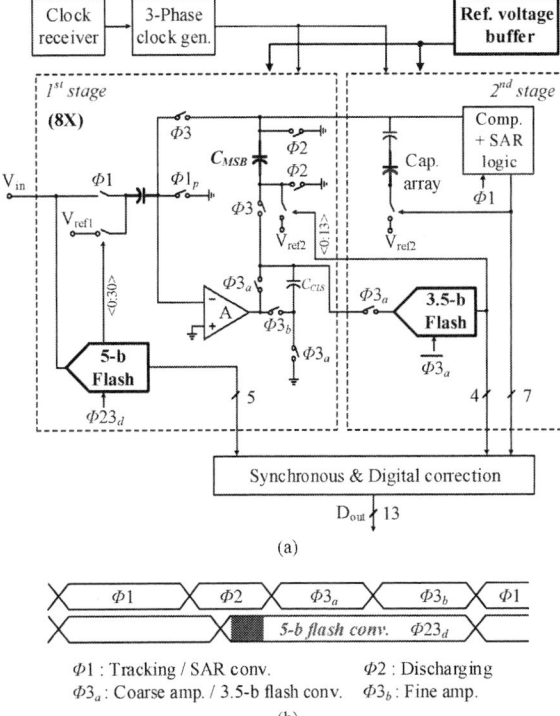

(a)

$\phi 1$: Tracking / SAR conv. $\phi 2$: Discharging
$\phi 3_a$: Coarse amp. / 3.5-b flash conv. $\phi 3_b$: Fine amp.

(b)

Fig. 1. (a) Proposed pipelined subranging-SAR architecture and (b) timing

hand, on-chip decoupling capacitors can be reduced by the calibration [4]. However, this method can only be used in the SAR ADC and one additional conversion cycle is needed.

In this paper, a pipelined subranging-SAR architecture is proposed to improve performance and save power. An offset cancellation technique of the dynamic comparator is proposed. In order to cope with large decoupling capacitors and high power dissipation, a calibration technique of the reference voltage buffer is presented. This paper is organized as follows. In section II, the hybrid ADC architecture is presented. Section III discusses the offset cancellation technique of the dynamic comparator. The detailed calibration technique of the reference voltage buffer is described in Section IV. In Section V, the measurement results are shown. Section VI concludes this work.

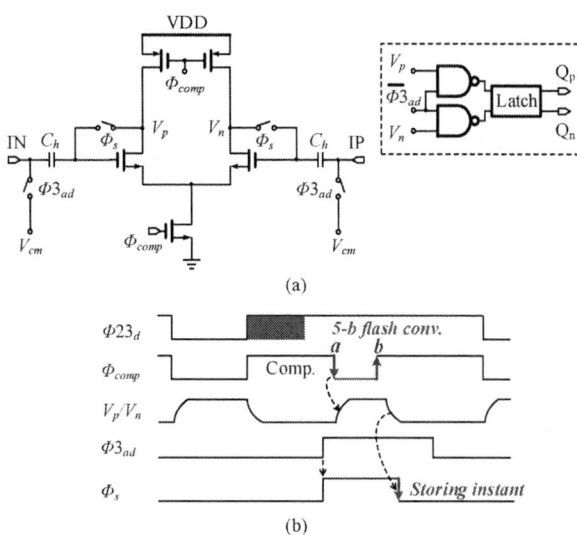

(a)

(b)

Fig. 2. (a) Proposed input offset storage in a dynamic comparator and (b) timing

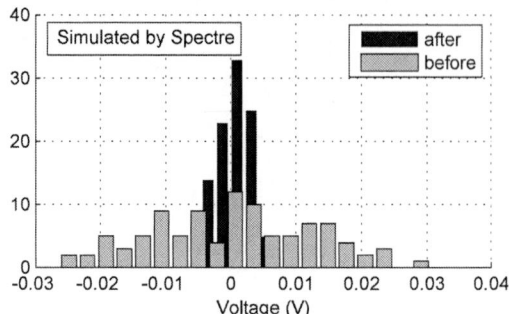

Fig. 3. Statistical simulation result on the input-referred offset

II. PIPELINED SUBRANGING-SAR ARCHITECTURE

The proposed ADC is a combination of pipelined, flash and SAR architectures, as shown in Fig. 1. It is composed of a 5-bit front end and a 10-bit subranging-SAR stage. By adopting range scaling, the 5-bit frond end which contributes 4-bit effective resolution provides an interstage gain of 8, relaxing opamp output swing requirement. With the aid of correlated level shifting (CLS) [5], the opamp gain requirements are relaxed by dividing the residue amplification into coarse ($\phi3_a$) and fine amplification ($\phi3_b$). On the other hand, since the feedback capacitor C_{MSB} of the front end is also used as the second stage MSB capacitor array, the second stage sampling operation is removed. Thus, the coarse output of the first stage can be directly converted by a 3.5-bit flash sub-ADC ($\phi3_a$) and the fine amplification result can be converted directly by a 7-bit SAR sub-ADC ($\phi1$), which accelerates the conversion of the ADC. In addition, in order to reset the shared capacitor, the discharging phase ($\phi2$) is added. Thus, $\phi2$ enables the 5-bit flash sub-ADC to complete the operation ($\phi23_d$) before the amplification phase, which copes with issues introduced by the SHA-less architecture.

The reference voltages of two pipelined stages are generated by a low-power reference voltage buffer. In addition, the clock receiver, the three-phase clock generator, and the digital correction block are also integrated in the ADC.

Benefit from the residue amplification of pipelined architecture, the parallelism of the flash architecture, the digitization of the SAR architecture, and techniques discussed above, the proposed architecture can achieve high power efficiency.

III. DYNAMIC COMPARATOR WITH OFFSET CANCELLATION TECHNIQUE

In order to save power, dynamic comparators are used in the flash sub-ADCs in Fig. 1. In the 5-bit front end, the tolerable offset is only $\pm V_{FS}/64$. In addition, the tolerance

is used to tolerate not only the comparator's offset but also other nonideal factors, including the interstage gain's error, the aperture error, the reference voltage fluctuation and so on. To eliminate the comparator offset, a novel input offset storage technique is proposed. Take the first stage comparator as an example to illustrate the approach, as shown in Fig. 2. The offset storage occurs after the 5-bit flash ADC conversion to avoid disturbing the normal operation. $\phi3_{ad}$ rising edge triggers the rise of ϕ_s. Therefore, the top plate of the storage capacitor C_h is attached to the node V_p (or V_n) through the switch ϕ_s, and the bottom plate is connected to V_{cm} via switch $\phi3_{ad}$. Thus, top plates of C_h are charged to VDD after the falling edge a and start discharging after the rising edge b. Because of the mismatch, they discharge at different rates, which is tracked by C_h. Once ϕ_s falls, the voltages at V_p and V_n are stored in the differential C_h. The stored offset is canceled in the following comparison phase. Thanks to the additional switching between the edge a and b, storage capacitors are able to store the input offset of the dynamic comparator.

The stored voltages determine the input common-mode voltage in the next comparison phase, which can be described as

$$V_{g,cm} = (V_p + V_n)/2 \qquad (1)$$

Therefore, the storing instant is triggered by V_p (or V_n) to keep the stored voltages near V_{cm}, as shown in Fig. 2. Thus, input transistors can operate in the saturation region at the beginning of the comparison phase. In addition, outputs of the preamplifier are controlled by two NAND gates so that the change in V_p (or V_n) during the storage operation does not disturb the latch.

Fig. 3 shows the statistical simulation results (by 100-point Monte-carlo simulation) on the input-referred offset. The 1-sigma(s) offset voltage is reduced from 12.7 mV to 2.4 mV at the aid of the proposed offset cancellation technique.

IV. REFERENCE VOLTAGE BUFFER WITH CHARGE COMPENSATION

The proposed reference voltage buffer is described in Fig. 4. The reference voltages of the first stage are 1.2 (V_{rp1}) and 0 (V_{rn1}) V, and those of the second stage are 0.9 (V_{rp2}) and 0.3 (V_{rn2}) V. Thus, V_{rn1} is provided directly by the input pad, VDDBUF is higher than 1.2 V and hence high threshold

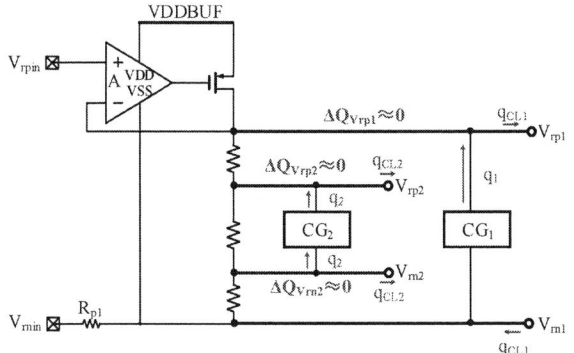

Fig. 4. Proposed reference buffer with zero-charge-variation reference voltages

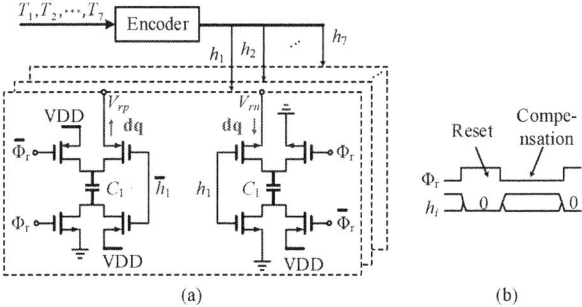

Fig. 5. (a) Proposed charge generator in the 2nd stage (CG$_2$ in Fig. 4) and (b) timing

voltage transistors are adopted. In order to compensate for the charge extracted from V_{rp1} and the charge injected into V_{rn1} and V_{rn2}, the charge generators (CGs) are proposed. In addition, the amplifier's ground node is connected to V_{rn1}, increasing the accuracy of the differential reference voltage in the first stage. R_{p1} is the parasitic resistor.

The reference voltage error caused by the switched capacitor load is discussed in [6]. However, because [6] eliminates the input-depended error by discharging the constant charge, the reference voltage range is compressed and the current increases. To respond to that, a charge generator is proposed to achieve zero charge variation of the reference voltages. For the reference buffer in Fig. 4, that can be described as

$$\Delta Q_{V_{rp1}} \approx 0, \quad \Delta Q_{V_{rp2}} \approx 0, \quad \Delta Q_{V_{rn2}} \approx 0, \quad (2)$$

Take the CG in the second stage as an example to illustrate the principle, as shown in Fig. 5. The loss of charge on V_{rp2} is

$$\frac{dQ}{V_r C_u}(n_1) = \frac{-4}{2^m - 2}[n_1^2 + n_1(2 - 2^m)] \quad (3)$$

where $V_r = V_{rp2} - V_{cm} = 0.3$, $m = 3$, and $n_1 = 0, 1, ..., 14$. And hence,

$$dq(i) = dQ(i) - dQ(i-1) = \frac{4V_r C_u}{2^m - 2}(2^m - 2i - 1) \quad (4)$$

The compensation charge for V_{rp2} by i-th CG unit in Fig. 5(a) is

$$dq = (2VDD - V_{rp})C_i \quad (5)$$

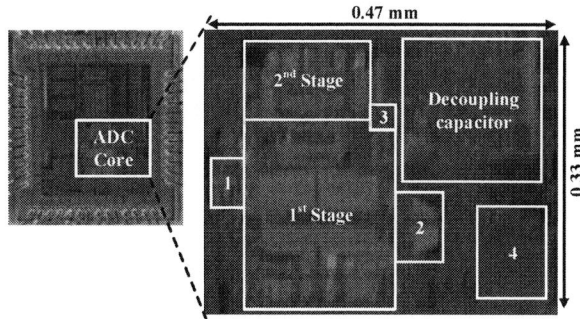

1. Clock receiver & 3-phase clock generator
2. CG1 3. CG2 4. Reference voltage buffer

Fig. 6. Chip Microphotograph

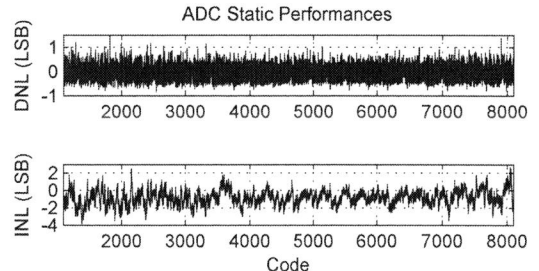

Fig. 7. DNL and INL errors

And thereby the capacitor should satisfy

$$C_i = \frac{4V_r C_u}{(2^m - 2)(2VDD - V_{rp})}(2^m - 2i - 1) \quad (6)$$

Besides, h_i is

$$h_i = T_i \overline{T_{15-i}} \quad (h_i = 1, 2, ...7) \quad (7)$$

where T_i ($i = 1, 2, ...14$) are the thermometer code of 3.5-bit flash ADC. Due to the parasitic capacitance, the charge variation on reference voltages can not be reduced to zero. In the design, the variation can be decreased significantly through parasitic optimization. In the first stage, the ration of the decoupling capacitance to the sampling capacitance is only about 46.

V. MEASUREMENT RESULTS

The chip microphotograph of prototype ADC fabricated in 65 nm CMOS is shown in Fig. 6. The ADC core occupies an area of 0.155 mm^2 (330 μm$\times$470 μm). Clocked at 160 MS/s with a 1.2 V supply, the ADC consumes 23.2 mW, without the power of off-chip capacitor mismatch calibration.

As described in Fig. 7, the DNL and INL are within +1.45/-0.82 and +2.64/-3.80 LSB. The dynamic performance of the ADC is shown in Fig. 8. For a 2.4 MHz input, the achieved SNDR and SFDR are 64.2 dB and 80.5 dBc, respectively. For the Nyquist input, the achieved SNDR and SFDR are 61.5 dB and 71.4 dBc, respectively. Besides, the measured dynamic performance versus the input frequency is summarized in Fig. 9. Table I compares this work with some

TABLE I. PERFORMANCE COMPARISON

	ASSCC 2013 [7]	ISSCC 2015 [8]	ASSCC 2016 [9]	ESSCIRC 2016 [10]	This work
Architecture	Pipelined	Pipelined	SAR	Pipelined-SAR	Pipelined Subranging-SAR
Technology (nm)	130 nm CMOS	65 nm CMOS	40 nm CMOS	65 nm CMOS	65 nm CMOS
Supply Voltage (V)	1.3	1.2	0.9	1.2	1.2
Resolution (bit)	14	12	12	12	13
Sampling Rate (MS/s)	150	250	150	180	160
SNDR @ peak (dB)	71.30	67.00	61.70	63.80	64.20
SFDR @ peak (dBc)	93.60	84.60	74.40	76.30	80.50
SNDR @ Nyq (dB)	N/A	65.70	56.20	60.92	61.50
SFDR @ Nyq (dBc)	N/A	79.00	63.50	67.17	71.40
DNL (LSB)	0.80	-0.86/0.52	-0.91/1.77	N/A	-0.82/1.45
INL (LSB)	2.60	-0.90/1.08	-2.63/2.95	N/A	-3.80/2.64
Power (mW)	68.0	N/A	1.5	6.0	22.1
Power with ref. buffer(mW)	85.0	49.7	N/A	N/A	23.2
FoM @ DC (fJ/conv. step)	188*	108.5*	10.3	26.3	107.3*
FoM @ Nqy (fJ/conv. step)	N/A	126.8*	18.9	36.7	162.6*
Core area (mm²)	4.400	0.590	0.040	0.068	0.155

* including the power dissipation of the reference voltage buffer

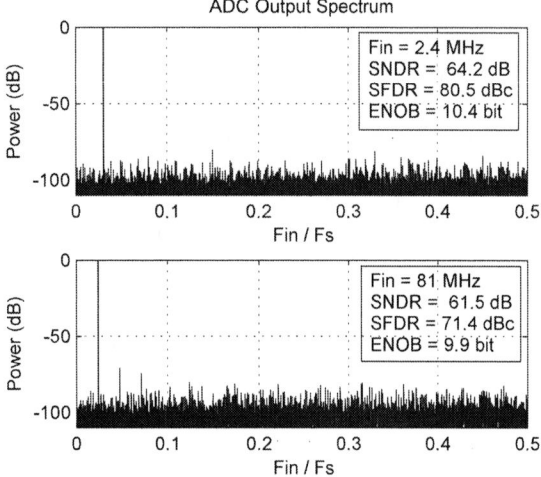

Fig. 8. ADC output spectrum at the input frequency of 2.4 MHz and 81 MHz (the output is decimated by 2X)

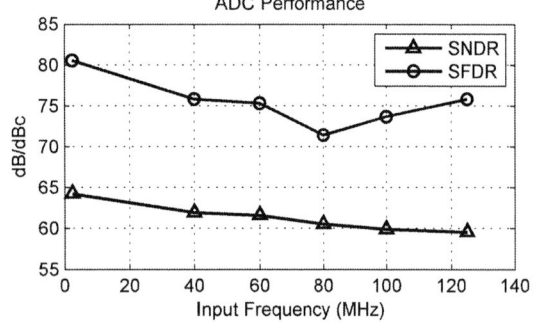

Fig. 9. Measured SFDR and SNDR versus the input frequency

state-of-the-art designs published recently. As an ADC with the input reference voltage buffer, this work shows competitive performance.

VI. CONCLUSION

This paper presents a 160MS/s 13-bit hybrid ADC. With the proposed pipelined subranging-SAR architecture, low-offset dynamic comparator, and charge compensation reference voltage buffer, the prototype achieves a competitive performance with the FoM of 107.3 fJ/conversion-step.

REFERENCES

[1] H. Huang, S. Sarkar, B. Elies, and Y. Chiu, "28.4 a 12b 330ms/s pipelined-sar adc with pvt-stabilized dynamic amplifier achieving lt;1db sndr variation," in *2017 IEEE International Solid-State Circuits Conference (ISSCC)*, pp. 472–473, Feb 2017.

[2] C. H. Chan, Y. Zhu, U. F. Chio, S. W. Sin, U. Seng-Pan, and R. P. Martins, "A reconfigurable low-noise dynamic comparator with offset calibration in 90nm cmos," in *IEEE Asian Solid-State Circuits Conference 2011*, pp. 233–236, Nov 2011.

[3] B. Razavi and B. A. Wooley, "Design techniques for high-speed, high-resolution comparators," *IEEE Journal of Solid-State Circuits*, vol. 27, pp. 1916–1926, Dec 1992.

[4] C. H. Chan, Y. Zhu, I. M. Ho, W. H. Zhang, C. L. Lio, U. Seng-Pan, and R. P. Martins, "A 0.011mm² 60db sndr 100ms/s reference error calibrated sar adc with 3pf decoupling capacitance for reference voltages," in *2016 IEEE Asian Solid-State Circuits Conference (A-SSCC)*, pp. 145–148, Nov 2016.

[5] B. R. Gregoire and U. K. Moon, "An over-60 db true rail-to-rail performance using correlated level shifting and an opamp with only 30 db loop gain," *IEEE Journal of Solid-State Circuits*, vol. 43, pp. 2620–2630, Dec 2008.

[6] Y. Wang, F. Li, C. Xue, and Z. Wang, "Charge-compensation-based reference technique for switched-capacitor adcs," in *2015 IEEE International Symposium on Circuits and Systems (ISCAS)*, pp. 2257–2260, May 2015.

[7] C. Yang, F. Li, W. Li, X. Wang, and Z. Wang, "An 85mw 14-bit 150ms/s pipelined adc with 71.3db peak sndr in 130nm cmos," in *2013 IEEE Asian Solid-State Circuits Conference (A-SSCC)*, pp. 85–88, Nov 2013.

[8] H. H. Boo, D. S. Boning, and H. S. Lee, "15.6 12b 250ms/s pipelined adc with virtual ground reference buffers," in *2015 IEEE International Solid-State Circuits Conference - (ISSCC) Digest of Technical Papers*, pp. 1–3, Feb 2015.

[9] K. H. Chang and C. C. Hsieh, "A 12 bit 150 ms/s 1.5 mw sar adc with adaptive radix dac in 40 nm cmos," in *2016 IEEE Asian Solid-State Circuits Conference (A-SSCC)*, pp. 157–160, Nov 2016.

[10] J. Zhong, Y. Zhu, C. H. Chan, S. W. Sin, S. P. U, and R. P. Martins, "A 12b 180ms/s 0.068mm² pipelined-sar adc with merged-residue dac for noise reduction," in *ESSCIRC Conference 2016: 42nd European Solid-State Circuits Conference*, pp. 169–172, Sept 2016.

S15-5 (2114)

A 1.5fJ/Conv-step 10b 100kS/s SAR ADC with Gain-Boosted Dynamic Comparator

Xiyuan Tang, Long Chen, Jeonggoo Song, and Nan Sun
University of Texas at Austin, Austin, TX
Email: xtang@cerc.utexas.edu, nansun@mail.utexas.edu

Abstract—**This paper presents a power-efficient 10-bit SAR ADC. A novel comparator topology with a dynamic common-gate stage is proposed to increase the pre-amplification gain under a low power supply voltage, thereby reducing noise and offset. Statistical estimation and loading switching techniques are synergically combined to further improve the energy efficiency. Moreover, the SAR sequencer and clock generator share only a single dynamic DFF chain to reduce the digital power. A 40nm CMOS prototype achieves a Walden FoM of 1.5fJ/conversion-step while operating at 100kS/s from a 0.5V supply.**

I. INTRODUCTION

Power-efficient ADCs are critical for energy constraint applications, such as wireless sensors and biomedical implants. In these applications, low-speed and moderate-resolution ADCs are required to digitize the sensed signals. SAR ADC is preferred due to its simplicity and scaling compatibility [1]–[5]. Since these sensors are often powered by batteries and/or energy harvesters, low power operation is critical.

The power consumption of state-of-the-art SAR ADCs is typically dominated by the comparator and digital circuits [1], [2]. A common technique to reduce power is to operate the ADC under a low-power supply voltage (e.g., 0.5V). A key problem for the comparator to operate under such a low voltage is the reduced front-end dynamic integrator gain, leading to increased input referred noise and offset. To boost the front-end gain, this work proposes a novel comparator topology with a dynamic common-gate stage inserted before the latch. This common-gate stage effectively increases the amplification gain so that the noise and offset are reduced but with little power cost.

To further reduce the comparator noise and power, this work synergically combines three other low-power techniques, including: 1) statistical estimation [1], which reduces the comparator noise by estimating the ADC conversion residue from 4 repeated LSB comparisons; 2) CMOS input pair for bidirectional integration [3], which makes use of both charging and discharging phases for amplification to halve the preamplifier power; and 3) load capacitance switching between MSB and LSB conversions [4], which saves the comparator energy during MSB operations.

Moreover, to reduce the power of the digital circuits, this work uses only one shift register array for both SAR logic control and clock generation, which is different from classic designs that typically require two register arrays. All flip-flops (DFFs) are implemented with dynamic logics. Additionally,

the strong-arm latch is used for data storage instead of DFFs to further reduce power.

The paper is organized as follows. Sec. II describes the proposed SAR ADC architecture. Sec. III presents the circuit implementation. Sec. IV shows measured results. The conclusion is drawn in Sec. V.

II. PROPOSED SAR ADC ARCHITECTURE

Fig. 1(a) shows the architecture of the proposed 10-bit SAR ADC with 5 low-power techniques. This work proposes a novel 3-stage comparator with a dynamic common-gate stage. It reduces the comparator noise and offset by extending the integration time and increasing the pre-amplification gain. This is especially important under low supply voltage where the dynamic pre-amplification gain is very limited. The details are presented in Sec. III.

The clock generator produces both the sampling clock and the data latch signal. A clock booster is used to ensure high sampling linearity under a low supply voltage. The DACs are shown in Fig. 1(b). Custom designed 0.5fF unit capacitor is adopted. A CMOS-input 3-stage dynamic comparator makes the decision. A Bayes estimation (BE) block performs statistical estimation based noise reduction.

The timing diagram is shown in Fig. 1(c). In the sampling phase, the input is top-plate sampled onto the DAC. During the first 4-b MSB conversions, the comparator operates in a low-power mode with a small load capacitor to save energy. The comparator rms noise is 0.9 LSB. The total ADC rms noise is 0.95 LSB including both comparator and quantization noise. During the next 7-b LSB conversions including a redundant bit (8C), the comparator is configured in the low-noise mode with a large load capacitor to ensure accuracy. The comparator rms noise is reduced to 0.65 LSB, which results in the ADC rms noise reducing to 0.71 LSB. This load switching can cause comparator offset variation, but its induced error is fully absorbed by the redundancy (8C), which also absorbs any error caused by the large noise during the 4 MSB comparisons.

To further reduce noise but without a large power cost, the last LSB conversion is repeated by 4 times to perform Bayes estimation [1], which exploits all the information embedded in the comparator outputs. The conversion residue voltage is estimated by examining the number of '1's out of the repeated 4 comparison results, and is then subtracted out from the ADC output. This technique reduces both the comparator noise and

978-1-5386-3179-9/17 $31.00 © 2017 IEEE

S15-5 (2114)

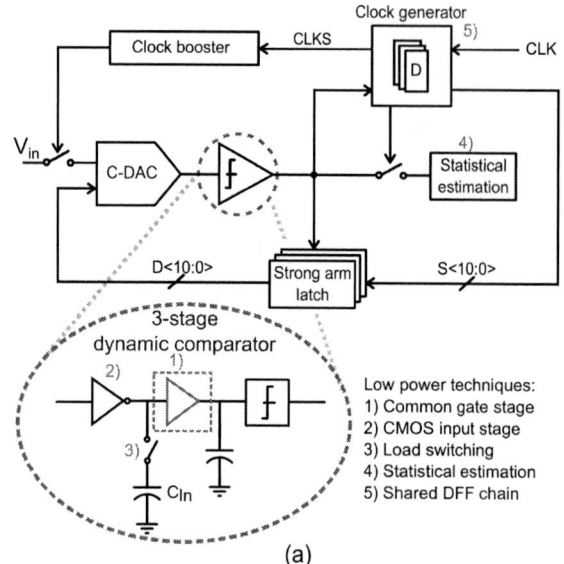

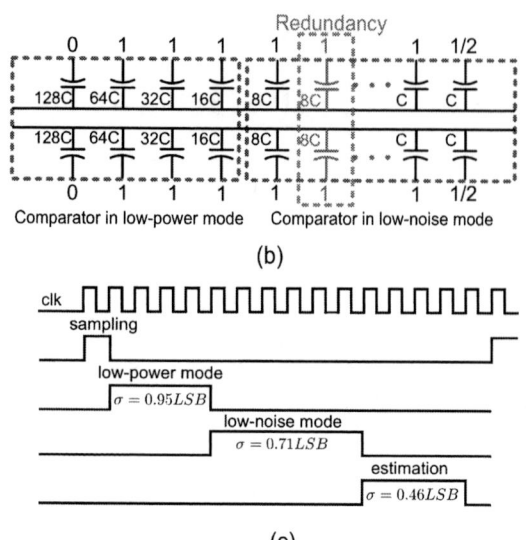

Fig. 1. (a) Proposed SAR ADC architecture, (b) DAC array and (c) system timing diagram

the quantization error. The final ADC rms noise is reduced to 0.46 LSB after Bayes estimation.

III. CIRCUIT IMPLEMENTATION

This low power SAR ADC requires judicious optimization to ensure high performance and power efficiency. Three design techniques are highlighted in this section.

A. Proposed 3-Stage Dynamic Comparator

As discussed earlier, a low power supply voltage (e.g., 0.5V) reduces the comparator front-end dynamic integrator gain, leading to increased noise and offset. To boost the front-end gain while maintaining power efficiency, a CMOS-input 3-stage dynamic comparator is proposed. As shown in Fig. 2(a), it consists of a dynamic integrator for front-end amplification and a latch to provide positive feedback for regeneration. The integration stage includes a CMOS input pair followed by a PMOS common-gate stage.

As shown in Fig. 2(b), the integration can be divided into 3 phases. In ϕ_1, the NMOS input pair performs integration by discharging V_{xp}/V_{xn} from V_{DD}. A dynamic AND gate is used to detect V_{xp}/V_{xn} level. Once they drop below $V_{DD}/2$, ϕ_2 starts. The NMOS pair is disabled, and the PMOS pair integrates the input by charging V_{xp}/V_{xn} towards V_{DD}. Once V_{xp}/V_{xn} goes above the threshold voltage of the PMOS common-gate (CG) transistors (marked in red), the CG integration phase ϕ_3 starts. The differential charges on V_{xp}/V_{xn} are transferred to V_{xp2}/V_{xn2}. Assuming large CG stage gain, the source nodes V_{xp}/V_{xn} are held still and the signal amplification continues on the new integration nodes V_{xp2}/V_{xn2}.

In the conventional strong-arm latch, the integrator output V_{cm} change is V_{DD}. This added CG stage extends the V_{cm} change by V_{th}, which effectively elongates the integration time T_{int}. The total integration time becomes:

$$T_{int} \approx \frac{C}{I_D}(V_{DD} + V_{th}) \qquad (1)$$

where I_D is the input pair current. We assume equal load C at V_{xp}/V_{xn} and V_{xp2}/V_{xn2} for simplicity. A longer T_{int} leads to a larger integration gain, which directly reduces the input referred noise and offset from the latch. The integration gain A_{int} is given by [6]:

$$A_{int} \approx \frac{g_m}{I_D}(V_{DD} + V_{th}) \qquad (2)$$

The dynamic AND gate which starts ϕ_2 has been simulated across process corners to ensure the sufficient gain.

Furthermore, the noise bandwidth of a dynamic integrator is inversely proportional to T_{int}, and thus, a longer T_{int} also reduces the noise from both the comparator input pair and the DAC. In addition to the benefits brought by the insertion of the CG stage, the use of the CMOS bidirectional integrator reduces the front-end integrator power by 2 times [3].

As discussed before, to further reduce the comparator power, load switching technique is employed. During MSB comparisons, the capacitor C_{ln} at integration node V_{xp}/V_{xn} is disconnected to save power. Only during the critical LSB comparisons, C_{ln} are connected to trade power for reduced comparator noise. C_{ln} is chosen to ensure that the comparator rms noise is 0.65 and 0.9 LSB with and without the loading of C_{ln}. The comparator power is 50% lower without the extra loading capacitor, and thus, load switching greatly reduces the MSB comparison power.

B. Statistical Estimation Based Noise Reduction

Statistical estimation is performed to reduce both the comparator noise and the quantization error [1]. In a SAR ADC, the digital output can be expressed as:

$$D_{out} = V_{in} + n_s + V_{res} \qquad (3)$$

where n_s is the sampling noise and V_{res} is the conversion residue at the comparator input. V_{res} includes the effects of both the comparator noise and the quantization error. The idea

978-1-5386-3179-9/17 $31.00 © 2017 IEEE

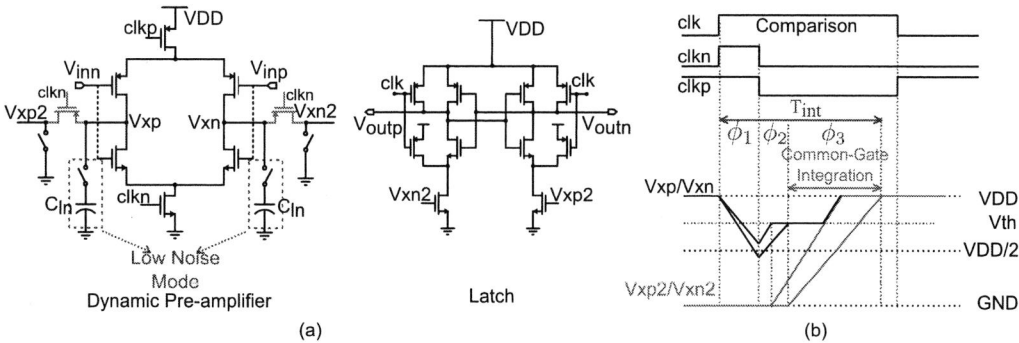

Fig. 2. (a) Comparator architecture and (b) comparator operation.

of statistical estimation based noise reduction is to form a digital estimator of V_{res}, denoted as V_{res}^*, and subtract it from D_{out} to enhance the overall ADC resolution.

The hardware cost for this statistical estimation technique is low. To obtain a target 4-dB ADC SNR improvement, we only need to repeat the ADC LSB comparisons by 4 times. We count the number of '1's from the comparator outputs and denote it as k. Depending on the value of k, we estimate the conversion residue and define V_{res}^*. Assuming the value of k is $\{0, 1, 2, 3, 4\}$, V_{res}^* is mapped to $\{-1, -0.5, 0, +0.5, +1\}$ respectively by checking the pre-stored lookup table. The extra hardware only contains a counter and the lookup table. Since only the LSB comparison is repeated by 4 times, the total comparator power is increased by only 30%. This is much more power efficient than the brute-force analog scaling, which requires 225% comparator power increase to obtain the same 4-dB ADC SNR improvement.

Comparing to the majority voting technique of [5], the merit of this statistical estimation technique is that *it makes full use of the statistical information, not just the majority information.* Thus, it can provide finer steps of ± 0.5 LSB, leading to a higher SNR enhancement. Simulations show that the overall SNR improvement is reduced by 2 dB if majority voting is used to determine the LSB. In addition, the proposed technique does not require an extra metastability detection circuit, which reduces the hardware complexity.

C. Low-Power SAR Logic Design

The digital circuit power can constitute a substantial portion of the overall SAR ADC power even in 40nm CMOS. To reduce the digital power, a 0.5V supply voltage is used. Furthermore, several design techniques are introduced. In a classic synchronous SAR logic design, a clock counter and a decoder forms a state machine triggered by the system clock clk to produce the sampling clock $clks$ and the comparator clock $clkc$. A separate shift register array is required in the SAR logic to generate the data latch signal $S < 10 : 0 >$. By contrast, as shown in Fig. 3, the proposed clock generator only contains one shift register chain. It uses both system clock clk and comparator output ready signal. In this shift register chain, the first two DFF outputs generates sampling clock $clks$ and the comparator clock $clkc$. The comparator output ready signal triggers the following DFFs in the chain,

which produces the data latch signals. The last DFF's output resets the whole chain. In this way, the shift register chain serves as both the clock counter and the sequencer.

In the classic SAR, each comparison cycle triggers 1 DFF in the clock counter, 1 DFF in the sequencer, and 1 data storage DFF. In this work, a strong-arm latch is used for data storage. The sequencer is merged with the clock generation block, as mentioned earlier. As a result, the SAR logic power is reduced by roughly 2 times: it only includes 1 DFF and 1 latch switching energy per bit. In addition, DFFs are implemented with dynamic logics to further reduce digital power.

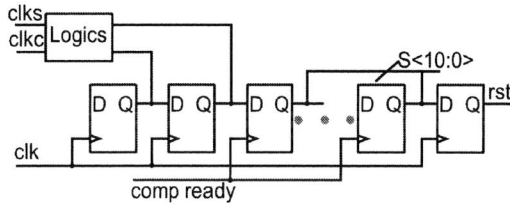

Fig. 3. Proposed clock generator.

IV. MEASUREMENT RESULTS

The prototype ADC in 40nm CMOS occupies an active area of 0.007 mm^2 as shown in Fig. 4. Fig. 5 shows the measured DNL and INL. DNL is +1.26/−1 LSB and INL is +1.76/−1.15 LSB. The DNL and INL errors are mainly caused by random mismatch among 0.5fF unit capacitors.

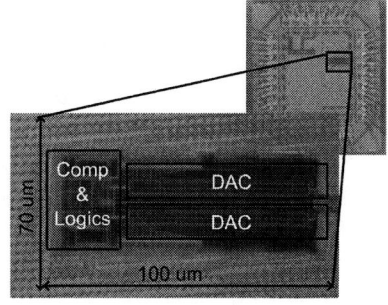

Fig. 4. Die micrograph.

The measured probability densities of D_{out} at $V_{in} = 0$ with and without Bayes estimation (BE) are shown in Fig. 6 together with fitted normal distributions. Before noise reduction, the standard deviation of D_{out} is 0.68 LSB, which is in agreement with SPICE simulation. After applying BE, the standard deviation of D_{out} is reduced to 0.45 LSB, which matches well with the theoretical prediction. Overall, the ADC input referred noise is reduced by 3.5dB after BE with only 30% increase in the total comparator power, which is much more efficient than brute-force analog scaling. If the conventional design approach is used, the comparator noise needs to be reduced to 0.35 LSB in order for the total ADC noise to be 0.45 LSB, which also includes the 0.29 LSB quantization noise. It would require 200% comparator power increase to get the same SNR improvement.

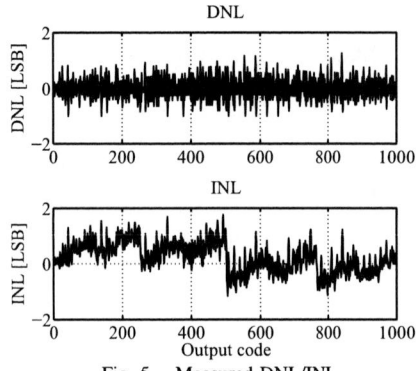

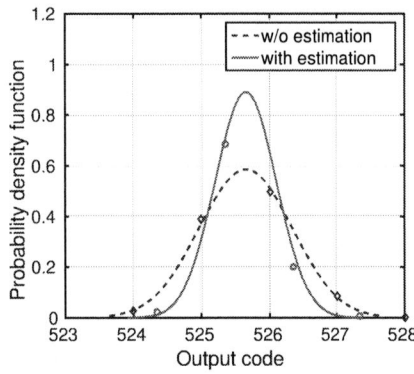

Fig. 5.   Measured DNL/INL.

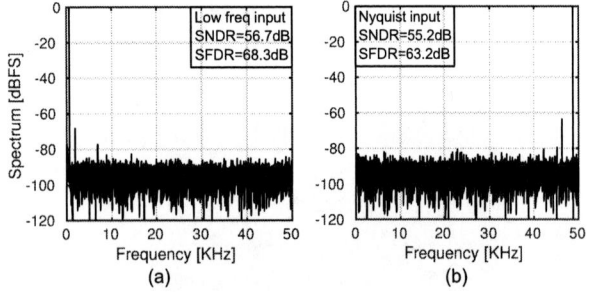

Fig. 6. D_{out} distribution with and without estimation at $V_{in} = 0$.

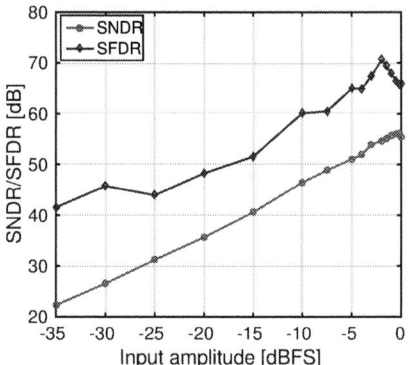

Fig. 8.   Measured SNDR versus input amplitudes.

SNDR and 63.2dB SFDR are achieved. Fig. 8 shows the SNDR with varying input amplitudes.

The ADC consumes 69nW from a 0.5V power supply. The power breakdown is as follows: 24nW for comparator, 36nW for digital circuits including extra logics for BE, and 9nW for the reference. The measured Walden figure-of-merit (FOM) is 1.5fJ/conversion-step. As shown in Table I, the proposed ADC achieves the state-of-the-art performance.

TABLE I
PERFORMANCE COMPARISON.

	[3]	[2]	[5]	This work
Process [nm]	65	40	65	**40**
Supply Voltage [V]	0.6	0.45	0.8	**0.5**
Power [nW]	97	84	106	**69**
Active area [mm²]	0.076	0.007	–	**0.007**
Sampling Rate [KS/s]	40	200	80	**100**
Resolution [bit]	12	10	10	**10**
Nyquist SNDR [dB]	62.5	55.6	56.6	**55.2**
FOM [fJ/conv-step]	2.2	0.85	2.4	**1.5**

V. CONCLUSION

This paper presented a power efficient SAR ADC. Multiple low power techniques are synergistically combined to minimize the energy consumption. It is suitable for energy constraint applications powered by battery or energy harvesting, such as autonomous wireless sensors.

Fig. 7. Measured FFT spectrum with low frequency input and Nyquist rate input with 100KHz sampling rate.

Fig. 7 shows the measured ADC output spectrum for a low frequency input and a Nyquist rate input with 100kHz sampling rate. At low frequency input, 56.7dB SNDR and 68.3dB SFDR are achieved. With Nyquist rate input, 55.2dB

REFERENCES

[1] L. Chen, X. Tang, A. Sanyal, Y. Yoon, J. Cong, and N. Sun, "A 10.5-b ENOB 645 nW 100kS/s SAR ADC with statistical estimation based noise reduction," *IEEE CICC*, pp. 1–4, Sep. 2015.

[2] H.-Y. Tai, Y.-S. Hu, H.-W. Chen, and H.-S. Chen, "A 0.85fJ/conversion-step 10b 200kS/s subranging SAR ADC in 40 nm CMOS," *IEEE ISSCC*, pp. 196–197, Feb. 2014.

[3] M. Liu, P. Harpe, and A. van Roermund, " A 0.8V 10b 80kS/S SAR ADC with Duty-Cycled Reference Generation," *IEEE ISSCC*, pp. 278–279, Feb. 2015.

[4] M. Ding, P. Harpe, Y.-H. Liu, B. Busze, K. Philips, and H. de Groot, "A 5.5fJ/conv-step 6.4MS/s 13b SAR ADC utilizing a redundancy-facilitated background errordetection-and-correction scheme," *IEEE ISSCC*, pp. 460–461, Feb. 2015.

[5] P. Harpe, E. Cantatore, and A. van Roermund, "A 2.2/2.7fJ/conversion-step 10/12b 40kS/s SAR ADC with Data-Driven Noise Reduction," *IEEE ISSCC*, pp. 270–271, Feb. 2013.

[6] B. Razavi, "The StrongARM latch: a circuit for all seasons," *IEEE SSC Mag.*, vol. 7, no. 2, pp. 12-17, Jun. 2015.

A 2.56mm^2 718GOPS Configurable Spiking Convolutional Sparse Coding Processor in 40nm CMOS

Chester Liu, Sung-Gun Cho, and Zhengya Zhang
University of Michigan, Ann Arbor, MI, USA

Abstract—A configurable neuro-inspired inference processor is designed as an array of neurons each operating in an independent clock domain. The processor implements a recurrent network using efficient sparse convolutions with zero-patch skipping for feedforward operations, and sparse spike-driven reconstruction for feedback operations. A globally asynchronous locally synchronous structure enables scalable design and load balancing to achieve 22% reduction in power. Fabricated in 40nm CMOS, the 2.56mm^2 inference processor integrates 48 neurons, a hub and an OpenRISC processor. The chip achieves 718GOPS at 380MHz, and demonstrates applications in feature extraction from images and depth extraction from stereo images.

I. INTRODUCTION

Neuro-inspired sparse coding algorithms have been applied to various types of sensory inputs, including audio, image, and video, for dictionary learning and feature extraction in a wide range of applications including compression, denoising, super-resolution, and classification tasks [1]. Sparse coding implemented as a spiking recurrent neural network can be readily mapped to hardware to achieve high performance. However, as the input dimensionality increases, the number of parameters becomes impractically large, necessitating a convolutional approach to reduce the number of parameters by exploiting translational invariance [2].

In this work, we present a configurable spiking convolutional sparse coding (sCSC) processor. Compared to the popular convolutional neural networks (CNN), sCSC offers several advantages: 1) sCSC produces sparse spikes, presenting opportunities for significant complexity and power reduction; 2) sCSC preserves structural information in dictionary-based encoding, allowing downstream processing to be done directly in the encoded, i.e., compressed, domain; and 3) sCSC uses unsupervised learning, enabling truly autonomous modules that adapt to inputs.

The configurable sCSC processor is made of an array of neurons as the compute units that perform configurable convolutions. The processor implements recurrent networks by iterative feedforward and feedback operations as depicted in Fig. 1. In a feedforward operation, each neuron convolves its input or reconstruction errors, i.e., the differences between the input and its reconstruction, with a kernel. The convolution results are accumulated, and spikes are generated when the accumulated potentials exceed a threshold. In a feedback operation, neuron spikes are convolved with kernels to reconstruct the input. Depending on application, 10 to 50 iterations are required to complete one inference. The inference output, in

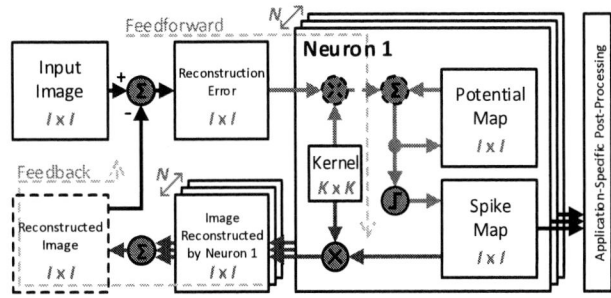

Fig. 1. Hardware mapping of spiking convolutional sparse coding (sCSC) algorithm.

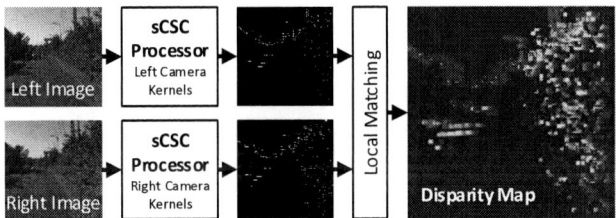

Fig. 2. sCSC processor applied to stereo images to extract depth information.

the form of neuron spikes, are passed to a downstream post-processor to complete various tasks.

We demonstrate a configurable sCSC processor chip in 40nm CMOS. The configurable convolution architecture is more versatile than fixed architectures for specialized accelerators [3]. The design optimally exploits the inherent sparsity using zero-patch skipping to make our convolution up to 40% more efficient than the state-of-the-art constant-throughput zero masking convolution [4]. A sparse spike-driven approach is adopted in feedback operations to minimize the cost of implementating recurrence by eliminating multipliers. The sCSC processor contains 48 convolutional neurons with configurable kernel size up to 15x15, which are equivalent to 10,800 non-convolutional neurons in classic implementations [5]. Each neuron operates at an independent clock and communicates using asynchronous interfaces, enabling each neuron to run at the optimal frequency to achieve load balancing. Going beyond conventional feature extraction tasks, we apply the sCSC processor to stereo images to extract depth information as illustrated in Fig. 2. Although we only demonstrate imaging applications in this work, the sCSC processor is input-agnostic and can be applied to any type of input.

978-1-5386-3179-9/17 $31.00 © 2017 IEEE

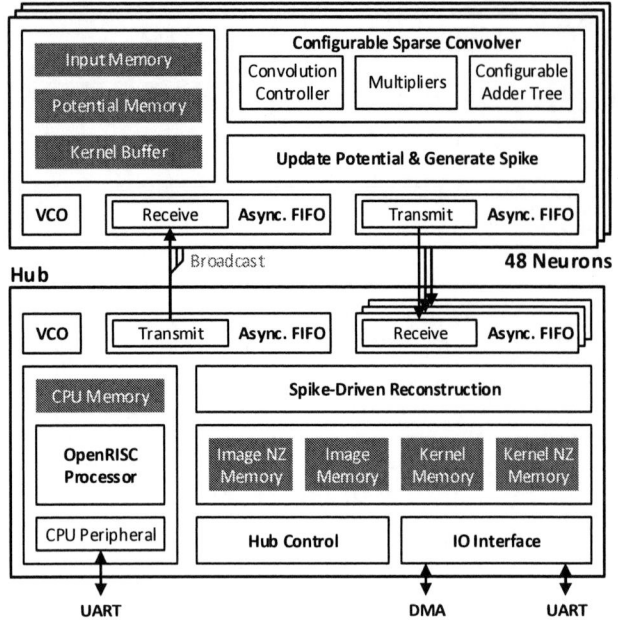

Fig. 3. sCSC processor hardware architecture.

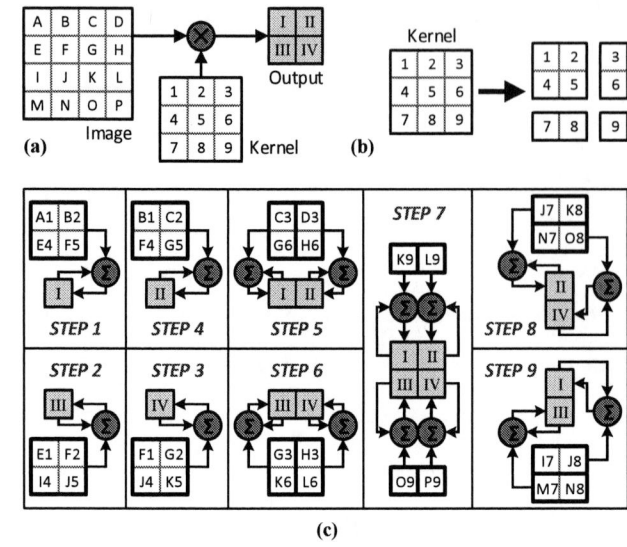

Fig. 4. (a) A 4x4 image is convolved with a 3x3 kernel to produce a 2x2 output; (b) Splitting kernel into sections that fit in the physical convolver (2x2 in this example); (c) In each step, image pixels and kernel are multiplied and accumulated in the output buffer.

II. HARDWARE ARCHITECTURE

To implement a recurrent network for sparse coding, a modular hardware architecture is designed as shown in Fig. 3, where the feedforward operations are distributed to neurons, and the neuron spikes are sent to a central hub for feedback operations. The sparse neuron spikes make it possible to deploy efficient asynchronous interfaces and share one hub for feedback operations.

In performing a feedforward operation, a neuron convolves a typically non-sparse input image (in the first iteration) or sparse reconstruction errors (in subsequent iterations) with its kernel (Fig. 1). The feedforward convolution is optimized in three ways: 1) highest throughput for sparse input by exploiting sparsity, 2) highest throughput for non-sparse input by fully utilizing the hardware, and 3) efficient support of variable kernel size. To achieve high throughput and efficiency, we design a sparse convolver supporting zero-patch skipping; and to achieve configurability, we divide variable-sized convolution into smaller fixed-sized sections and design a traverse path for the physical convolver to assemble the complete convolution result. The design of the configurable sparse convolution is detailed in Section III.

Each neuron supports a configurable kernel of size up to 15x15 using a compact latch-based kernel buffer, and variable image patch size up to 32x32. An input image larger than 32x32 is divided into 32x32 sub-images that share overlaps to minimize edge artifacts.

In a feedback operation, neuron spikes are convolved with their kernels to reconstruct the input image (Fig. 1). A direct implementation of this feedback convolution is computationally expensive and could become a performance bottleneck.

We take advantage of the binary spikes to replace all multiplications in this convolution by additions, and further make use of the high sparsity of the spikes (typically >90% sparsity) to design a sparsely activated spike-driven reconstruction to save computation and power. The design is detailed in Section IV.

The hub contains a kernel memory, and a multi-banked image memory that provides single-cycle read-accumulate-write capability. An image nonzero (NZ) memory is used to identify NZ entries in the reconstructed image to support sparse convolutions. The hub simultaneously broadcasts reconstructed image and its NZ map and receives spikes from neurons to ensure seamless feedforward and feedback operations without idling the hardware. The design of the asynchronous interfaces between the hub and neurons is detailed in Section V.

The hub uses a 16-bit bi-directional DMA interface for data I/O, and a UART interface for configuration. An OpenRISC processor is integrated on chip, and it can be tasked with on-chip learning and post-processing.

III. CONFIGURABLE SPARSE CONVOLUTION

Inside the neuron design is a 4x4 physical convolver. For simplicity, we illustrate in Fig. 4 our configurable convolution using the example of a 2x2 physical convolver computing the convolution of a 4x4 image with a 3x3 kernel (Fig. 4(a)): 1) the 3x3 kernel is systematically divided into sections suitable for the physical convolver, i.e., a 2x2 patch, a 2x1 line, a 1x2 line, and a 1x1 pixel (Fig. 4(b)); 2) the physical convolver scans the image in nine steps to cover all sections of the original convolution (Fig. 4(c)); and 3) intermediate results are accumulated in the output buffer to be the final result. To maximize throughput, multipliers need to be fully utilized, so the two 2x1 sections are processed together by the physical

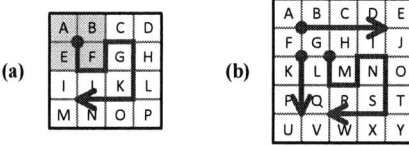

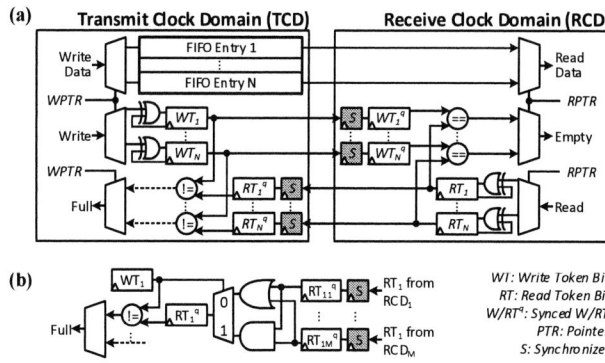

Fig. 5. Maze-walking path for (a) a 3x3 kernel on a 4x4 image and (b) a 4x4 kernel on a 5x5 image.

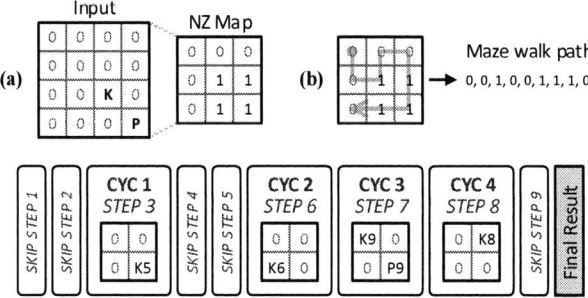

Fig. 6. (a) Entries in a NZ map indicate if at least one nonzero entry exists in a 2x2 block in the input; (b) Walking through the NZ map produces a sequence in which 0 means to skip; (c) 5 steps are skipped in calculating the convolution.

convolver in steps 5 and 6 (Fig. 4(c)). Similarly, multiple sections are processed together in steps 7 to 9 (Fig. 4(c)). The physical convolver is equipped with a configurable adder tree to handle various forms of accumulation in different steps.

To maximize locality of reference, kernel sections are fetched once and reused until done, and image sections are shifted by one row or column between steps. Such a carefully arranged sequence results in a maze-walking path, depicted in Fig. 5(a), that maximizes hardware utilization and data locality. We name the convolution "maze convolution". An optimal path exists for every kernel size; yet, to minimize storage, paths for larger kernels are created with multiple smaller paths. Fig. 5(b) shows a maze-walking path for a 4x4 kernel based on the 3x3 kernel's maze-walking path.

A sparse input enables the use of a sparse convolver to increase throughput and efficiency. In [3], a line convolver was designed to skip lines of zeros in the input. However, we observed that it is more likely to have a patch of zeros than a line of zeros in the input, so skipping zero patches is more effective. Maze convolution readily supports zero-patch skipping with the help of an input NZ map, wherein a NZ bit is 1 if at least one nonzero entry is detected in an area covered by a patch of the same size as the physical convolver. Fig. 6 shows an example in which the NZ map of an image contains two nonzero entries. Maze convolution is guided by the NZ map, skipping steps where the NZ bit is 0 to realize sparsity-proportional throughput increase. Compared to [3], maze convolution with zero-patch skipping increases the throughput by up to 40% at 90% input sparsity. The proposed maze convolution with zero patch skipping is equally applicable to deep neural networks [4], [6].

Fig. 7. (a) Token-based asynchronous FIFO; (b) FIFO full condition check for broadcast asynchronous FIFO.

IV. SPIKE-DRIVEN RECONSTRUCTION

Triggered by a neuron's spike, the hub performs reconstruction by retrieving the neuron's kernel from the kernel memory and accumulating the kernel in the image memory, with the kernel's center aligned to the spike location. Like in maze convolution, a kernel is also divided into sections to support variable kernel size in the spike-driven reconstruction. The NZ map of the reconstructed image is computed by OR'ing the NZ map of the retrieved kernels, saving both computation and latency compared to the naive way of scanning the reconstructed image. The spike-driven reconstruction eliminates the need to store spike maps, and a 16-entry FIFO is sufficient for buffering spikes, cutting the storage by 2.5x.

V. GLOBALLY ASYNCHRONOUS INTERFACES

The sCSC processor implements globally asynchronous communication between the hub and neurons to achieve scalability by breaking a single clock network with stringent timing constraints into small ones with relaxed constraints. The globally asynchronous scheme further enables load balancing by allowing the hub and individual neurons to run at the optimal clock frequencies based on workload. Following feed-forward operations, neurons send 10-bit messages to identify neuron spikes to the hub via a token-based asynchronous FIFO [7]. Following a feedback operation, the hub sends 128-bit messages that contain reconstructed image and NZ map to the neurons. To avoid routing congestion from the hub to the neurons, we design a broadcast asynchronous FIFO, which is identical to the token-based asynchronous FIFO except for the FIFO full condition check logic.

The asynchronous FIFO design is shown in Fig. 7. The token-based asynchronous FIFO is full when the transmit clock domain (TCD) write token disagrees with the synchronized receive clock domain (RCD) read token. The broadcast asynchronous FIFO has multiple RCDs and it is full when the TCD write token disagrees with any synchronized RCD read token. Synchronizer stage in all asynchronous FIFOs are configurable between 2 and 4 stages to accommodate PVT-induced delay variations.

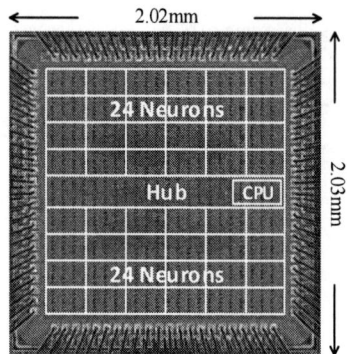

Fig. 8. Chip microphotograph.

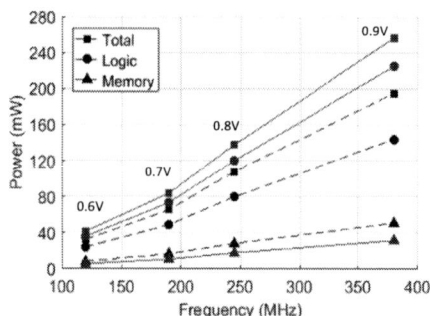

Fig. 9. Chip power measurement result of feature extraction task (dashed lines) and depth extraction task (solid lines).

VI. CHIP IMPLMENTATION AND MEASUREMENTS

A 4.1mm² test chip is implemented in 40nm CMOS, and the sCSC processor occupies 2.56mm². The chip microphotograph is shown in Fig. 8. We use a mixture of 80.5% high-V_T and 19.5% low-V_T cells to reduce the chip leakage power by 33%. Dynamic clock gating is applied to reduce the dynamic power by 24%. A balanced clock frequency setting for the hub and neurons further reduces the overall power by an average of 22%. A total of 49 VCOs are instantiated, with each VCO occupying only 250um² area. The test chip achieves 718GOPS at 380MHz with a nominal 0.9V supply at room temperature. An OP is defined as an 8-bit multiply or a 16-bit add.

We use two sample applications to demonstrate the sCSC processor: extracting sparse feature representation of images and extracting depth information from stereo images. The feature extraction task is entirely done by the hub and neurons; and the depth extraction task requires an additional local matching post-processing programmed on the on-chip Open-RISC processor. When performing feature extraction using 7x7 kernels, 10 recurrent iterations, and a target sparsity of approximately 90%, the sCSC processor achieves 24.6M pixel/s (equivalent to 375 256x256 frames per second), while consuming 195mW (shown in dashed lines in Fig. 9). In performing depth extraction using 15x15 kernels, 10 recurrent iterations, and a target sparsity of approximately 80%, the sCSC processor achieves 7.68M pixel/s (equivalent to 117

Table 1. Comparison with prior works

	This Work	JSSC 2016 [4]	ISSCC 2017 [6]
Architecture	Recurrent	Feedforward	Feedforward
Technology	40nm GP CMOS	65nm LP CMOS	28nm UTBB FD-SOI
Core Area	1.6mm x 1.6mm	3.5mm x 3.5mm	1.29mm x 1.45mm
SRAM Size	120KB	181.5KB	144KB
Voltage	0.9V	1V	1V
Frequency	380MHz	200MHz	200MHz
Performance	718GOPS[*1]	33.6GOPS[*2]	204GOPS[*1]
Throughput	256x256 @117fps	227x227 @35fps	227x227 @47fps
Power	257mW	278mW	44mW
Power Efficiency	2.79 TOPS/W	0.12 TOPS/W	4.64 TOPS/W
Area Efficiency[*3]	280 GOPS/mm²	7.24 GOPS/mm²	53.4 GOPS/mm²

[*1]8-bit operation [*2]16-bit operation [*3]Normalized to 40nm

Fig. 10. Comparison with prior works.

256x256 frames per second) while consuming 257mW (shown in solid lines in Fig. 9). Compared to the optimal baseline designs without exploiting sparsity, the throughputs of the tasks are improved by 7.7x and 9.7x, respectively. Voltage and frequency scaling measurement shows that at 0.6V supply and 120MHz clock frequency, the chip power is reduced to 53.9mW for the feature extraction task and 69.3mW for the depth extraction task.

Compared to state-of-the-art inference processors based on feedforward only networks, our sCSC processor realizes a recurrent network, supports unsupervised learning, and demonstrates expanded functionalities including depth extraction from stereo images, while still achieving competitive performance and efficiency in power and area as shown in Fig. 10.

ACKNOWLEDGMENT

This work was supported in part by SONIC, DARPA UPSIDE and Intel Corporation. The authors thank Prof. Bruno Olshausen and Dr. Garrett Kenyon for advice.

REFERENCES

[1] C. J. Rozell, D. H. Johnson, R. G. Baraniuk, and B. A. Olshausen, "Sparse coding via thresholding and local competition in neural circuits," *Neural Computation*, vol. 20, no. 10, pp. 2526–2563, Oct 2008.

[2] F. Heide, W. Heidrich, and G. Wetzstein, "Fast and flexible convolutional sparse coding," in *IEEE Conf. Computer Vision and Pattern Recognition (CVPR)*, Jun 2015, pp. 5135–5143.

[3] P. Knag, C. Liu, and Z. Zhang, "A 1.40mm² 141mW 898GOPS sparse neuromorphic processor in 40nm CMOS," in *IEEE Symp. VLSI Circuits*, Jun 2016, pp. 180–181.

[4] Y.-H. Chen, T. Krishna, J. S. Emer, and V. Sze, "Eyeriss: an energy-efficient reconfigurable accelerator for deep convolutional neural networks," *IEEE J. Solid-State Circuits*, vol. 52, no. 1, pp. 127–138, Nov 2016.

[5] J. K. Kim, P. Knag, T. Chen, and Z. Zhang, "A 640M pixel/s 3.65mW sparse event-driven neuromorphic object recognition processor with on-chip learning," in *IEEE Symp. VLSI Circuits*, Jun 2016, pp. 50–51.

[6] B. Moons, R. Uytterhoeven, W. Dehaene, and M. Verhelst, "Envision: A 0.26-to-10TOPS/W subword-parallel dynamic-voltage-accuracy-frequency-scalable convolutional neural network processor in 28nm FD-SOI," in *IEEE Int. Solid-State Circuits Conf. (ISSCC)*, Mar 2017, pp. 246–247.

[7] I. M. Panades and A. Greiner, "Bi-synchronous FIFO for synchronous circuit communication well suited for network-on-chip in GALS architectures," in *Int. Symp. Networks-on-Chip (NOCS)*, May 2007, pp. 83–94.

A 21mW Low-power Recurrent Neural Network Accelerator with Quantization Tables for Embedded Deep Learning Applications

Jinmook Lee, Dongjoo Shin, and Hoi-Jun Yoo
School of Electrical Engineering
Korea Advanced Institute of Science and Technology (KAIST)
Daejeon, Republic of Korea
jinmooklee@kaist.ac.kr

Abstract—A 21mW low-power embedded Recurrent Neural Network (RNN) accelerator is proposed to realize the image captioning applications. The low-power RNN operation is achieved by 3 key features: 1) Quantization-table-based matrix multiplication with RNN weight quantization, 2) Dynamic quantization-table allocation scheme for balanced pipelined RNN operation, and 3) Zero-skipped RNN operation using quantization-table. The Quantization table enables the 98% reduction of the multiplier operations by replacing the multiplication to the table reference. The dynamic quantization-table allocation is used to achieve high chip-utilization efficiency over 90% by balanced pipeline operation for three variations of the RNN operation. The zero-skipped RNN operation reduces the overall 27% of required external memory bandwidth and quantization-table operations without any additional hardware cost. The proposed RNN accelerator of 1.84mm^2 achieves 21mW power consumption and demonstrates its functionality on the image captioning RNN in 65nm CMOS process.

Keywords—Recurrent Neural Network; ASIC; Accelerator; Deep Learning; Embedded Systems

I. INTRODUCTION

The deep learning is playing an important role even in the area of embedded applications such as face recognition or speech recognition. Among Deep Neural Networks (DNNs), the Recurrent Neural Network (RNN), which can process the sequential data such as speech and video, is very effective in the image captioning and visual question answering [1] together with a static object classifier such as a Convolutional Neural Network (CNN). However, in order to run deep learning applications in embedded environment of limited power and performance, a dedicated low-power, high-performance deep learning accelerator is required. Even though several dedicated accelerators for CNNs have been reported [2][3], there is no RNN semiconductor chip because its heavy external memory bandwidth consumptions.

Fig. 1. shows the different computational characteristics between the RNN and the CNN. The 95.8% of CNN operations are 2D-convolutions between input feature maps and kernels [1]. In this process, "Data-reuse scheme" can be applied to reuse input feature maps for kernels or to reuse kernels for input feature maps. These data-reuse capabilities were a key feature of existing CNN accelerators to speed up the processing while

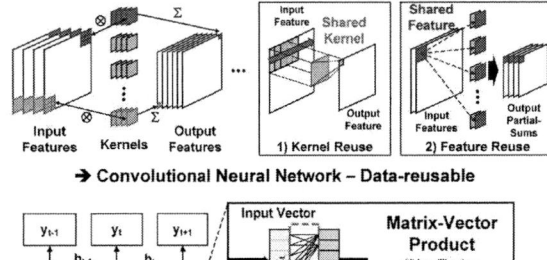

→ Convolutional Neural Network – Data-reusable

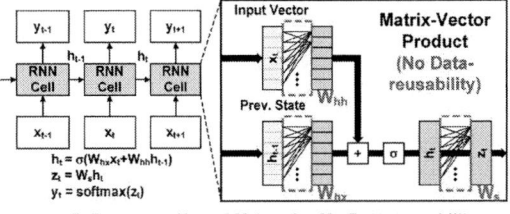

→ Recurrent Neural Network – No Data-reusability

Fig. 1. Different characteristics between CNN and RNN

reducing the hardware area. [2][3].

On the other hand, in the case of the RNN, more than the 99% of operations are matrix-vector products between the feature vectors (x_t, h_{t-1}, h_t) and the weight matrices (W_{hx}, W_{hh}, W_s) [1]. In addition, each element of the weight matrices is used only once during a RNN inference. Therefore, compared with the case of 2D-convolutions of the CNN, data-reuse with the limited on-chip memory cannot be applied for the RNN. Consequently, it is crucial to implement low-power matrix-vector product with minimum external memory accesses for running RNNs in the embedded environment of limited power and performance.

In this paper, a 21mW low-power, energy-efficient RNN accelerator is proposed for embedded deep learning applications with three key features. 1) Quantization-table (Q-table)-based matrix-vector product computation to reduce the overall 98.3% multiplication and 75% required memory bandwidth with 16-level RNN weight quantization compared to non-quantized 16-bit fixed-point precision operations. 2) Dynamic Q-table allocation scheme for energy-efficient RNN operations by pipelined RNN calculation with pipeline balancing which increase Q-table utilization to 90.5%. 3) Zero-skipping using Q-tables to reduce the required external memory bandwidth and Q-

978-1-5386-3179-9/17 $31.00 © 2017 IEEE

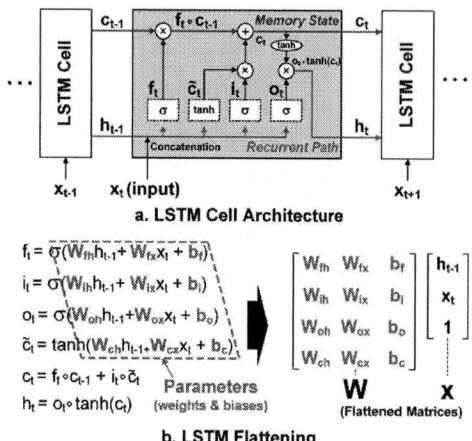

a. LSTM Cell Architecture

$$f_t = \sigma(W_{fh}h_{t-1} + W_{fx}x_t + b_f)$$
$$i_t = \sigma(W_{ih}h_{t-1} + W_{ix}x_t + b_i)$$
$$o_t = \sigma(W_{oh}h_{t-1} + W_{ox}x_t + b_o)$$
$$\tilde{c}_t = \tanh(W_{ch}h_{t-1} + W_{cx}x_t + b_c)$$
$$c_t = f_t \circ c_{t-1} + i_t \circ \tilde{c}_t$$
$$h_t = o_t \circ \tanh(c_t)$$

Parameters (weights & biases)

$$W = \begin{bmatrix} W_{fh} & W_{fx} & b_f \\ W_{ih} & W_{ix} & b_i \\ W_{oh} & W_{ox} & b_o \\ W_{ch} & W_{cx} & b_c \end{bmatrix} \quad X = \begin{bmatrix} h_{t-1} \\ x_t \\ 1 \end{bmatrix}$$

W (Flattened Matrices) **X**

b. LSTM Flattening

Fig. 2. Basics of LSTM RNN and matrix flattening

table operations by 27% without any additional hardware cost.

II. ALGORITHM ANALYSIS

Fig. 2(a) shows the architecture of the Long Short-Term Memory (LSTM) RNN which can solve gradient vanishing problem of basic RNN [4]. Recently, LSTM RNN is broadly used because it shows higher accuracy than basic RNN in many applications [1][5]. However, the additional components (f_t, i_t, o_t, c_t) of LSTM RNN increase the number and type of weight parameters so that the computation becomes more complicated than basic RNN. In this paper, matrix flattening technique is used to simplify the LSTM calculation to matrix-vector product as shown in Fig. 2(b). It concatenates the weight matrices of all components to one larger flattened matrix. With the help of this technique, the proposed hardware can be used to calculate both of the basic RNNs and the LSTMs. Furthermore, it makes more efficient to perform burst mode weight fetching by storing the RNN weight parameters to adjacent off-chip memory addresses.

III. HARDWARE ARCHITECTURE

A. Overall Architecture of Proposed RNN Accelerator

Fig. 3 describes the overall architecture of the proposed RNN accelerator. It consists of 6 main components. 1) Quantization-table (Q-table) array enables low-power RNN operation with table-based matrix-vector product calculation. 2) Q-table allocation fabric assigns the Q-tables to Q-modules 1, 2, and 3. The number of the allocated Q-tables is determined by workload of each RNN pipeline stage. 3) Q-modules control the allocated Q-tables with the help of Q-table controller to achieve energy-efficient vector-vector and matrix-vector product contained in the RNNs with the pipelined architecture. 4) Look-up Tables (LUTs) are integrated in the Q-modules for calculating activation functions of the RNNs with minimum table size while maintaining the RNN test accuracy. 5) Index buffers and feature buffers store indices of RNN weights and input feature vectors, respectively, for the RNN calculation, and 6 DMA and Network Interface transact data between on-chip memory with external memory by memory-mapped manner.

After loading the weight parameters and input feature vector stored in the external memory to the on-chip memory through

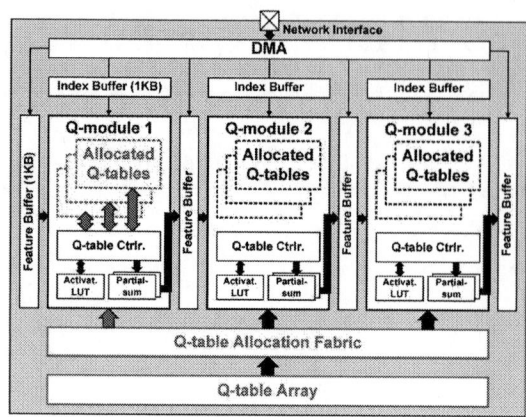

Fig. 3. Overall Architecture

the DMA and the Network Interface, the Q-table controller in the Q-modules updates the Q-tables to have all possible matrix-vector product results between the quantized RNN weights and an element of the input feature vector. The updated Q-tables generate the partial-sums of matrix-vector product between the weight matrix and the input feature vector. The calculated partial sums are stored in the partial-sum buffers of each Q-module. These partial-sums are accumulated until the end of the matrix-vector product and the accumulated partial-sums go through the activation functions with the activation LUTs and then final result of the RNN is obtained.

B. Matrix-vector Product using Quantization Table (Q-table)

1) Weight quantization of LSTM RNN

The proposed embedded RNN accelerator uses quantized weight matrices to minimize the required external memory accesses. It quantizes weight matrices to the several discrete levels by using K-means clustering algorithm. The solid in in the Fig. 4 is the histogram of the trained LSTM weight parameters. Because the trained LSTM weights follow continuous distribution, the weight value should be fetched from the off-chip memory to evaluate the LSTM. However, if the weight values are quantized like the bar graph in the Fig. 4, the value of

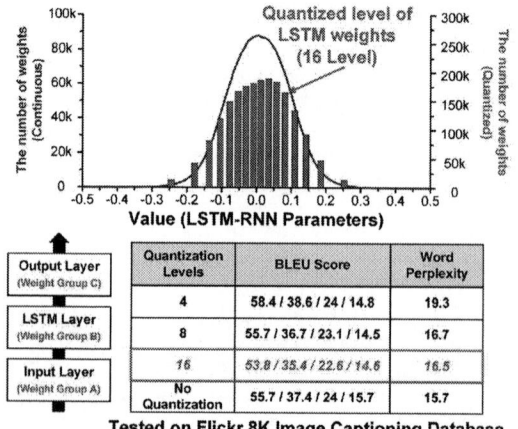

Quantization Levels	BLEU Score	Word Perplexity
4	58.4 / 38.6 / 24 / 14.8	19.3
8	55.7 / 36.7 / 23.1 / 14.5	16.7
16	53.8 / 35.4 / 22.6 / 14.6	16.5
No Quantization	55.7 / 37.4 / 24 / 15.7	15.7

Output Layer (Weight Group C)
LSTM Layer (Weight Group B)
Input Layer (Weight Group A)

Tested on Flickr 8K Image Captioning Database

Fig. 4. LSTM Weight Histogram with Quantization

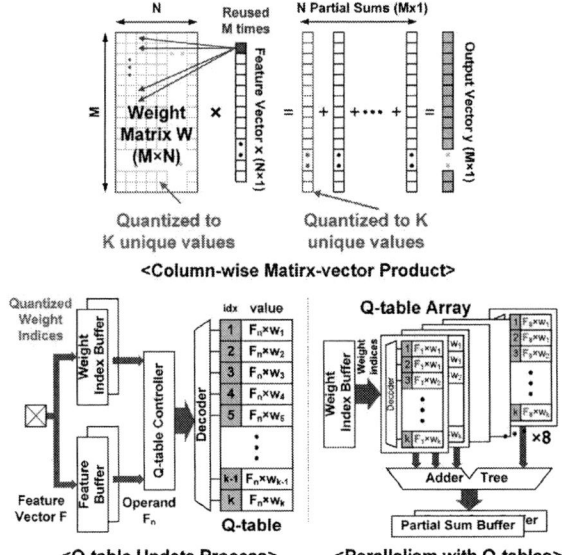

Fig. 5. Q-table-based Matrix Multiplication

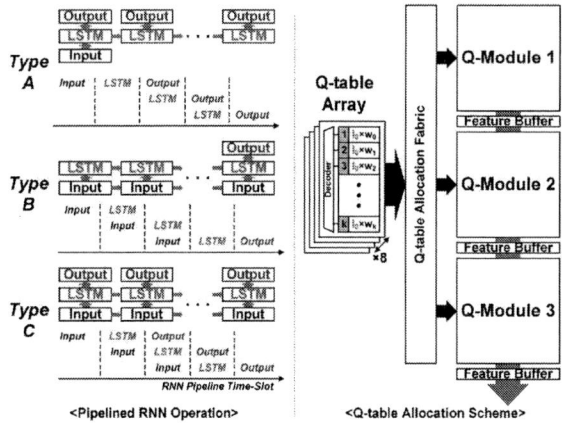

Fig. 6. RNN Pipelining and Pipeline Balancing

each weight parameter can be obtained by weight indices. Therefore, instead of fetching the weight values, only the weight indices of each weight parameter are fetched from the off-chip memory. When the weight quantization is performed with 16 levels, the weight indices can be represented with 4 bits. Therefore, the number of bits required for a weight fetching is reduced by 12 bits, compared to non-quantized weight of 16-bit precision. The table in the Fig.4 describes the performance of weight quantization to the image captioning RNN [1]. It reduces the required external memory bandwidth by 75% by performing quantization with 16 levels for each layer of LSTM to minimize degradation of word perplexity and BLEU scores which shows the performance of image captioning applications [1].

2) Matrix-vector Product using the Q-tables

Fig. 5 describes how to calculate matrix-vector product by using the proposed Q-tables. Since the weight matrix W is quantized to K levels, the partial sums, which are the product between the feature vector F and columns of W, are also quantized to K levels. Therefore, if there is a table for all possible cases where an feature vector element (F_n) are multiplied by quantized weight values ($w_1, w_2,...,w_K$), the matrix-vector product can be performed by fetching the table values with reference to the weight index. The proposed RNN accelerator fetches all of the quantized weight values from external memory before the start of matrix-vector product. Then, Q-table is updated for every column by using these weight values and a new feature vector element (F_n). After that, the product is calculated by referring Q-tables which include all the possible multiplication results ($F_n w_1,...,F_n w_k$). Accordingly, the Q-table-based RNN operation does not requires any multiplier operations except for only K multiplications for Q-table update during a partial sum calculation. In the case of image captioning RNN [1], whose largest weight matrix has dimension of 2048x1536, Q-table can reduce the overall 98% of multiplication with 16-level weight quantization.

3) Column-wise Parallelism with Q-tables

In addition, the proposed Q-table has a very efficient architecture in terms of parallel operation. Because the Q-table is updated every column of the weight matrix, it is possible to calculate multiple partial-sums simultaneously by assigning multiple Q-tables to multi-columns of the weight matrix. As shown in the right bottom in the Fig. 5, multiple Q-tables can perform the matrix-vector product with column-parallel manner by applying each partial-sum from Q-tables to an adder-tree for partial-sum accumulation. In the proposed embedded RNN accelerator total 8 Q-tables are integrated to achieve low-power high-throughput embedded RNN operation.

C. Dynamic Q-table allocation for balanced RNN pipelining

Fig. 6 describes dynamic Q-table allocation scheme for energy-efficient operation. The several types of RNN (Type-A, B, and C) can be calculated with pipelined manner to achieve high-throughput embedded RNN operation, by placing a feature buffer between them. However, not only the type of layer to be performed at given time-slot is different for each variation, but also the size of weight matrix of each layer that affects the operation time is different. In this case, the efficiency of the RNN pipelining becomes very low.

In the proposed embedded RNN accelerator, the dynamically allocated Q-table architecture is proposed to enhance the pipeline efficiency by using Q-table allocation fabric. Every Q-table is connected to Q-modules through the Q-table allocation fabric, which is constructed in a fully connected manner between each Q-table and Q-module. Therefore, it can solve the unbalanced pipeline problem and increases utilization of Q-tables by allocating Q-tables in proportional to the workload at each pipeline stage. Thus, it can increase energy-efficiency by increasing throughput per operating frequency. In the case of the image captioning RNN [1], classified as a Type-C which has weight matrix ratio of input, LSTM, and output layer is 1:3.5:2.6, the proposed Q-table allocation scheme enables overall 1.36x increase of Q-table utilization to 90.5% by assigning 1,4,3 Q-tables to the Q-module 1, 2, 3 respectively.

D. Zero-skipping of internal vectors on Q-table

Fig. 7 shows the proposed Q-table-based zero-skipping

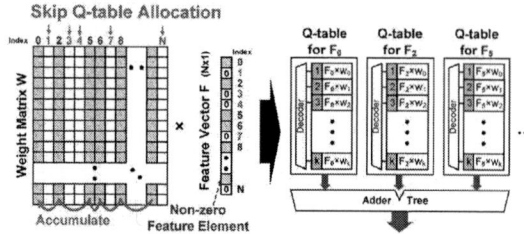

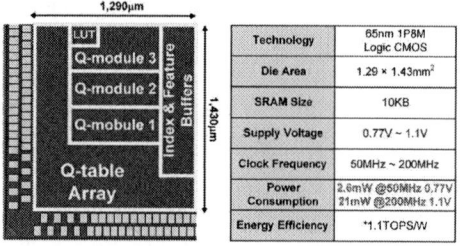

Fig. 7. Zero-skipping Scheme with Q-tables

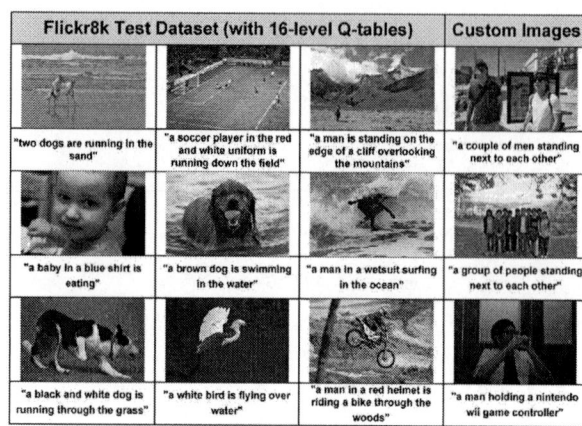

BLEU Score: 53.8/35.4/22.6/14.6, Word Perplexity: 16.5
Throughput: 23 Inferences/sec

Fig. 9. Image Captioning Measurement Results

Fig. 8. Chip Micrograph and Summary

scheme for external memory reduction and low-power embedded RNN operation. The zero-skipping can be implemented without any additional hardware cost by skipping Q-table assignment to columns of weight matrix whose values are zeros. It is possible because the feature vector element and the Q-tables are in a one-to-one relationship. The matrix-vector product is calculated by accumulation of values from each Q-table. With the help of this scheme, the 27% reduction of Q-table operations and required external memory bandwidth are achieved, respectively, in the image captioning RNN [1].

IV. IMPLEMENTATION RESULTS

Fig. 8 shows the chip summary of the proposed RNN accelerator fabricated in 65nm CMOS process as a part of a general neural network accelerator [6]. It occupies 1.84mm² with 10KB on-chip SRAM. It consumes only 21mW at 200MHz and 1.1V supply and 2.6mW at 50MHz 0.77V with the help of the Q-table based RNN operations. It also achieves 1.1TOPS/W energy efficiency due to the column-parallel Q-table operation and the pipelined RNN operation with dynamically allocated Q-tables. The measurement results on the image captioning RNN [1] is shown in the Fig. 9. The proposed Q-table-based matrix-vector product is applied to this RNN with 16-level weight quantization by using K-means clustering. It also achieves the BLEU score of 53.8/35.4/22.6/14.6 and 23infs/sec throughput at the 200MHz. The Table I describes the comparison with the previous LSTM or Fully-connected DNN hardware. Compared to the previous work, this work achieves 6.47 times higher energy-efficiency with the help of external memory bandwidth reduction and low-power matrix-vector product enabled by Q-tables. In addition, the performance of the proposed accelerator can be further increased by applying sparse LSTM as used in the [7].

V. CONCLUSION

In conclusion, a 21mW, 1.1TOPS/W energy-efficient RNN accelerator is proposed with 3 key features; Q-table based matrix

TABLE I. COMPARISON TABLE

	This Work	HOTCHIPS 2016[7]	ISCA 2016[8]
Approach	ASIC	FPGA (KU115)	ASIC
Technology	65nm	-	45nm
Frequency (MHz)	50 ~200	300MHz	800MHz
Precision (bit)	4 to 16	4 to 16	4 to 16
Power Consumption	21mW @200MHz 1.1V	45-50W	590
Energy-Efficiency	1.1TOPS/W	[a]19.2GOPS/W	0.17 TOPS/W

[a]. The case of Dense LSTM

–vector product with weight quantization, dynamic Q-table allocation for balanced pipeline, and zero-skipping of internal vectors with Q-table. The proposed Q-tables enables low-power RNN operation by removing the overall 98% multiplier operations. The proposed RNN accelerator is the first ASIC and successfully performed low-power RNN for embedded deep learning applications. Furthermore, it shows the 6.47 better energy-efficiency compared to the previous work.

REFERENCES

[1] O. Vinyals, A. Toshev, S. Bengio and D. Erhan, "Show and tell: A neural image caption generator," 2015 IEEE Conference on Computer Vision and Pattern Recognition (CVPR), Boston, MA, 2015, pp. 3156-3164.

[2] Y. H. Chen, T. Krishna, J. S. Emer and V. Sze, "Eyeriss: An Energy-Efficient Reconfigurable Accelerator for Deep Convolutional Neural Networks," in IEEE Journal of Solid-State Circuits, vol. 52, no. 1, pp. 127-138, Jan. 2017.

[3] Chen, Tianshi, et al. "Diannao: A small-footprint high-throughput accelerator for ubiquitous machine-learning." ACM Sigplan Notices. Vol. 49. No. 4. ACM, 2014.

[4] Hochreiter, Sepp, and Jürgen Schmidhuber. "Long short-term memory." Neural computation 9.8 (1997): 1735-1780.

[5] Sutskever, Ilya, Oriol Vinyals, and Quoc V. Le. "Sequence to sequence learning with neural networks." Advances in neural information processing systems. 2014.

[6] Shin, Dongjoo, et al. "14.2 DNPU: An 8.1 TOPS/W reconfigurable CNN-RNN processor for general-purpose deep neural networks." Solid-State Circuits Conference (ISSCC), 2017 IEEE International. IEEE, 2017.

[7] Kaiyuan Guo et al., "From model to FPGA: Software-hardware co-design for efficient neural network acceleration," 2016 IEEE Hot Chips 28 Symposium (HCS), Cupertino, CA, USA, 2016, pp. 1-2

[8] Han, Song, et al. "EIE: efficient inference engine on compressed deep neural network." Proceedings of the 43rd International Symposium on Computer Architecture. IEEE Press, 2016.

S16-3 (2015)

IEEE Asian Solid-State Circuits Conference
November 6-8, 2017/Seoul, Korea

EQSCALE: Energy-Quality Scalable Feature Extraction Engine for Sub-mW Real-time Video Processing with 0.55 mm² Area in 40nm CMOS

Anastacia B. Alvarez[+], Gopalakrishnan Ponnusamy, Massimo Alioto

ECE Department, National University of Singapore, Singapore
[+] also with Electrical and Electronics Engineering Institute, University of the Philippines, Philippines
Email: *massimo.alioto@nus.edu.sg*

Abstract— An energy-quality scalable (EQSCALE) feature extraction accelerator for IoT vision applications is presented. Knobs are introduced to dynamically adjust the tradeoff between energy and feature extraction quality, leveraging the intrinsic redundancy in video frames and the robustness of object recognition against missing features.

Measurements of a testchip in 40nm show 310pJ/pixel energy at nominal voltage and maximum quality, and 55.6pJ/pixel when tuned for minimum energy, while still allowing correct object recognition. The active area of the accelerator is 0.55mm². EQSCALE enables at least 5.7X energy improvement and 1.8X area reduction over state-of-the-art accelerators. To the best of our knowledge, EQSCALE is the first feature extraction accelerator operating in the sub-mW range (0.51mW) at VGA resolution and 30 fps.

Keywords—Feature extraction, ORB, energy-quality scalability, keypoint detection, keypoint description, low energy.

I. INTRODUCTION

Enabling computer vision in the Internet of Things (IoT) mandates acceleration of common tasks under very low energy and area (i.e., cost). Feature extraction is an essential task in computer vision systems, such as in object detection, classification and tracking [1], [2]. Feature extraction generates points of interest ("keypoints") and the corresponding descriptors from the incoming pixels. A typical feature extraction acceleration process is shown in Fig. 1, where pixels from a frame are processed in order to detect keypoints, which are then represented by bit vectors in the keypoint description phase.

Among existing feature extraction algorithms, SIFT is one of the most popular [1]. However, its high complexity requires massively parallel chip architectures with power consumption of 182-320mW [2], [3]. The simpler SURF algorithm [4] has

been demonstrated on a testchip with 2.7mW power consumption in [5], although at the cost of a further area increase. The ORB algorithm was introduced to further reduce the complexity, while retaining the affine invariance of SIFT and SURF [6], although no silicon demonstration has been presented to date.

In this paper, a feature extraction accelerator for IoT applications is introduced (EQSCALE). This accelerator is based on the ORB algorithm, which could consume ~100mW when implemented in a CPU. In EQSCALE, energy-quality knobs are introduced to minimize the energy for a quality target that is acceptable for the specific application at hand or visual context. EQSCALE leverages the intrinsic information redundancy in video frames, which allows correct (or gracefully degraded) object recognition even when a substantial number of keypoints are missed due to the degraded quality target.

II. EQSCALE ARCHITECTURE OF ENERGY-QUALITY SCALABLE ORB ACCELERATOR

The proposed EQSCALE architecture of the ORB algorithm has three pipelined blocks for the keypoint detection phase, as shown in Fig. 2: *Detector*, *NMS* and *Ranking*. The keypoint description comprises of two pipelined blocks: *Orientation* and *Descriptor*. To enable low-voltage operation, two latch-based memory arrays are used: *CACHE* stores a patch of the frame, and *KEYPTS* stores the detected keypoints of the current frame being processed.

A frame is progressively scanned from top to bottom and left to right as shown in Fig. 3, storing the corresponding patches into *CACHE*. To include possible keypoints at the

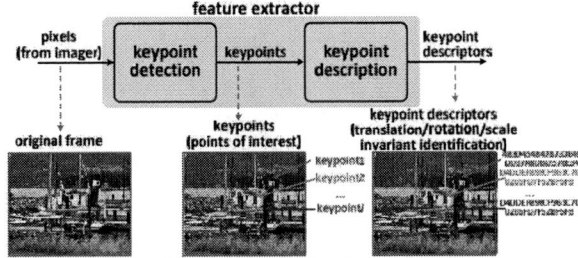

Fig. 1. Feature extraction, keypoints and descriptors.

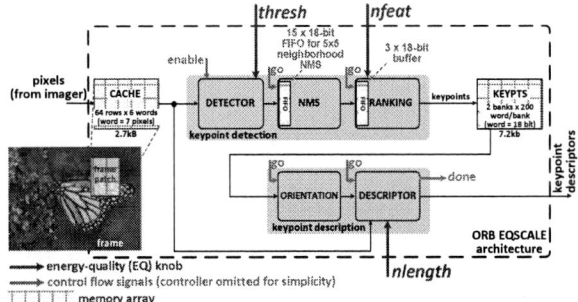

Fig. 2. EQSCALE architecture of ORB with energy-quality knobs: threshold (*thresh*), number of retained features (*nfeat*), length of description (*nlength*).

978-1-5386-3179-9/17 $31.00 © 2017 IEEE   241

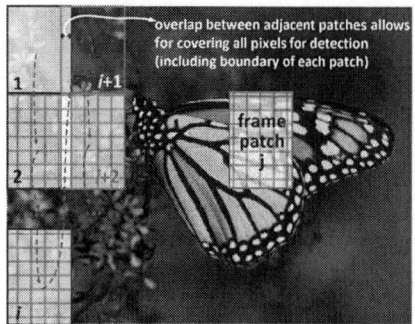

Fig. 3. Image scanning done by patch from top to bottom and left to right.

edge of the patches, overlaps are made between adjacent patches, as shown between patch 1 and *i*+1 in Fig. 3. As first processing step of the ORB algorithm, the *Detector* block acquires pixels from *CACHE* and compares the intensity of each pixel *i* with the 16-pixel circle around it [6], as illustrated in Fig. 4. Each pixel *j* in the 16-pixel circle is classified as "darker" ("lighter") if its intensity is lower (greater) than the intensity of pixel *i* by at least a pre-set threshold *thresh*. Pixel *i* is then considered a keypoint if at least 12 contiguous pixels in the16-pixel circle are either all lighter or all darker than pixel *i*. The implementation of the *Detector* is similar to the pattern matching algorithm in [7].

From Fig. 4, energy-quality (EQ) scalability is enabled by dynamically tuning the threshold *thresh*, rather than keeping it at the pre-defined value (e.g., ORB default). Higher values of *thresh* restrict pixels contributing to a candidate keypoint to those with a brighter or darker tone (i.e., fewer keypoints are generated). This lowers energy due to the reduced number of pixels considered for the successive computation of keypoints. This reduces quality due to the skipping of candidate keypoints, and thus potentially fewer keypoints.

A straightforward implementation of the latch-based *CACHE* patch buffer in Fig. 2 would dominate the overall EQSCALE area, and hence it would not be suited for IoT applications. To reduce its area, the pixel-reuse organization in Fig. 5 is introduced to enable on-the-fly/real-time computation of incoming pixels with just-enough size, instead of storing the entire frame. This reduces *CACHE* size from 300kB (i.e., the size of a VGA frame) to 2.7kB. Also, to reduce the number of *CACHE* accesses and the related energy, the same read access is reused across seven neighboring keypoints (i.e., the maximum for on-the-fly uninterrupted computation), as described in Fig. 5. This reduces the number of accesses by 7x. Overall, pixel reuse and 7-pixel parallel detection speeds up keypoint detection by 7X, enabling more aggressive voltage scaling at iso-throughput.

In the architecture in Fig. 2, *NMS* performs a non-maximal suppression within a 5x5 neighborhood, centered on the incoming keypoint, and suppressing keypoints having the lowest corner measure within the neighborhood [8]. Then, *Ranking* sorts the candidate keypoints according to corner measures [6], retaining only keypoints with higher corner measures. Image benchmarking showed that the number of keypoints can be as high as 15,000 per frame. In *Ranking*, storing the incoming keypoints in an on-chip low-voltage latch memory for further post-processing via off-line sorting methods would be impractical. Indeed, this would require an on-chip memory of approximately 270kb, whose silicon footprint of more than 0.5mm² would be comparable to the entire area of other EQSCALE blocks, thus doubling the overall area. To avoid such area cost, the Insertion Sort algorithm was adopted to enable on-the-fly sorting, which stores only the latest and most highly ranked candidate keypoints (400, as required by ORB). Accordingly, the resulting *KEYPTS* memory capacity required by Insertion Sort is reduced to 7.2kb. The drawback of the quadratic complexity of Insertion Sort was reduced by 8X by organizing the candidate keypoint memory *KEYPTS* into 10 bins. Compared to conventional Insertion Sort, this bin-based approach limits the maximum number of comparisons to 50 instead of 400 (i.e., 10 comparisons with bin extrema for bin pre-selection, 40 for intra-bin comparisons). Since *NMS* sometimes delivers keypoints faster than *Ranking* can handle, a keypoint buffer was added to store one additional row of keypoints from *NMS*.

In *Ranking*, EQ scalability is enabled by splitting KEYPTS into two 200-keypoint banks, with the second being disabled when the knob *nfeat* is set at 200, which in turn limits the number of highest-ranked keypoints to 200 rather than 400. The reduction to 200 keypoints reduces the energy due to the reduced number of keypoints considered in the subsequent processing steps (i.e., *Orientation* and *Descriptor* in Fig. 2). At the same time, this degrades the quality since the 200 lowest-ranked keypoints (i.e., least important) are skipped.

Once keypoints are available in *KEYPTS* after *Ranking*, the *Orientation* block is enabled to compute the image rotation angle as given by the equations:

$$\theta = a\tan 2(m_{01}, m_{10}) \qquad (1)$$

$$m_{pq} = \sum_{x,y} x^p y^q I(x,y) \qquad (2)$$

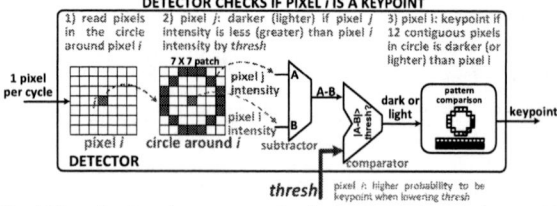

Fig. 4. Keypoint detection compares the pixel to the 16-pixel circle around it.

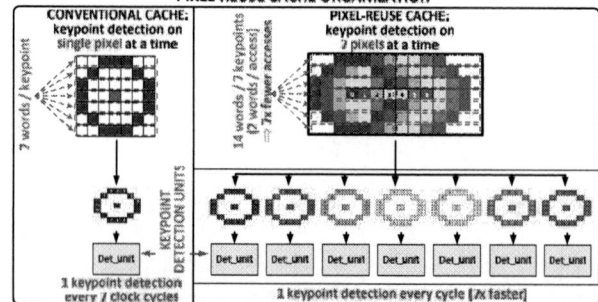

Fig. 5. Pixel reuse in CACHE read reduces access through seven parallel detections (right) for a single 14-word access, as opposed to conventional access (left), which requires 7-word access for each detection.

where m_{pq} is the pq moment of the image at (x,y) [6]. The *atan2* function is approximated using an 8-bin look-up table. *Descriptor* performs the comparisons of pixel pairs in a 31x31 patch, centered on each keypoint, generating a descriptor vector of width *nlength*, which is generally used for object matching and tracking. In *Descriptor*, energy-quality scalability is enabled by setting *nlength*=256 as ORB default value, or 128 bits to reduce the energy and the matching quality, due to the less detailed keypoint description. Changing the descriptor length from 256 to 128 simply disregards 128 pairs with higher correlation (i.e., less importance in the description of the keypoint).

III. CHIP MEASUREMENT RESULTS

The output quality was evaluated by assuming that EQSCALE is followed by a keypoint matching engine, as routinely required in systems performing object recognition and tracking [1], [2]. To this aim, the EQSCALE output keypoint descriptions were post-processed in software to perform keypoint matching through the popular 3-nearest neighbors algorithm [9] (*k*-NN, with *k*=3). Offline RANSAC algorithm [10] was then applied to identify the correct matches and measure quality. The quality Q of EQSCALE is the number of correct matches divided by the total number of matches, and for convenience is normalized to its best value (i.e., value with no degraded EQ knob). Energy E refers to the computation per frame, and is normalized to the value under nominal V_{DD} and tuning for best quality.

Fig. 6 shows the energy-quality tradeoff when EQ knobs are individually swept in a testchip in 40nm. At maximum quality Q=1, ORB turns out to have approximately the same quality as SIFT. When scaling the *thresh* energy-quality knob, ORB achieves approximately the same quality as SURF with Q~0.47 and E=0.35, which translates into 65% energy reduction compared to the case with full quality (see light blue curve in Fig. 6). At such reduced quality, successful object recognition was invariably achieved under the benchmark in [10], as confirmed by the presence of the bounding box around the recognized objects, as successfully generated by the offline RANSAC algorithm [10].

As an example, Fig. 7 shows sample matching images using the *graffiti* image from [10] for different quality targets. At maximum quality (i.e., Q=1), the image is properly matched, as indicated by the green box on the image on the right. At minimum allowable quality (i.e., Q=0.4), many keypoints are missed and object detection starts failing (smaller size of the bounding box). At lower quality (e.g., Q=0.12), keypoints are no longer properly matched and thus no object is detected (i.e., the green bounding box disappears).

At nominal V_{DD} and maximum quality, EQSCALE consumes a power of 2.9mW at VGA and 30fps, which results to an energy per pixel of 310pJ/pixel. Increasing *thresh* knob from 20 to 40 (60) reduces energy by 48% (65%) and degrades quality by 19% (53%). Analogously, when reducing *nfeat* from 400 down to 200, the energy decreases by 35% with a quality degradation of 42% (see green curve in Fig. 6). When reducing *nlength* from 256 down to 128 bits, the energy is reduced by 34% compared to the 256-bit default value, with a quality degradation of 10% (see red curve in Fig. 6). Such energy reduction is determined by the reduced number of execution cycles per frame, which also increases the throughput. Such excess throughput can be used to relax the clock cycle, enabling more aggressive voltage scaling and further energy gains, as discussed below.

When co-optimizing multiple EQ knobs and V_{DD} scaling at iso-frame rate, the comparison of the solid lines in Figs. 6 and 8 with same color shows that the addition of V_{DD} scaling (Fig. 8) allows further 50-100% energy reduction for all EQ knobs, compared to the optimization without voltage scaling in Fig. 6. The lowest energy with acceptable quality is achieved by

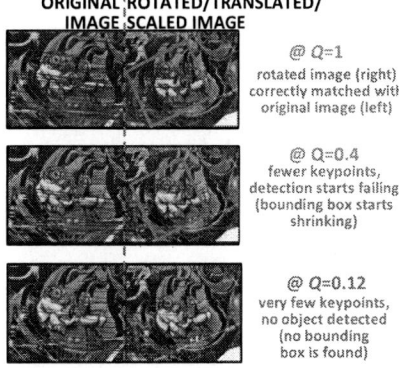

Fig. 7. Illustration of image matching at different values of Q: maximum quality Q=1 (top), minimum allowable Q=0.4 (middle), very low Q (bottom).

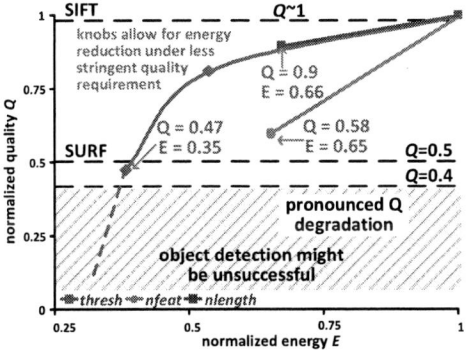

Fig. 6. Normalized energy vs. normalized quality (all quantities normalized to the value achieved at nominal V_{DD} and with knobs tuned for full quality).

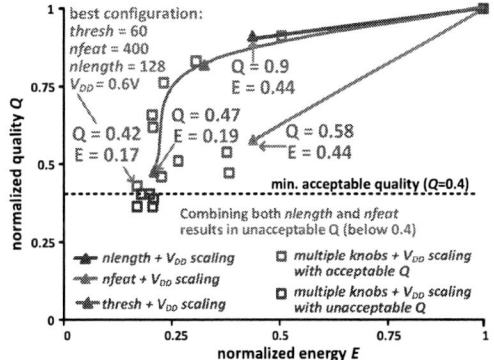

Fig. 8. Quality vs. energy with joint EQ knobs combined with voltage scaling.

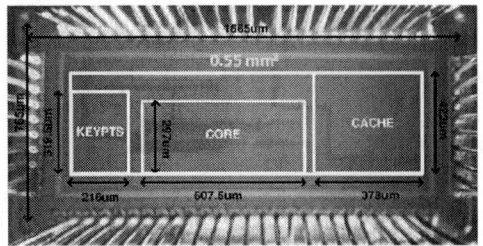

Fig. 9. Die photomicrograph of EQSCALE testchip in 40nm CMOS.

setting *thresh*=60, *nfeat*=400 and *nlength*=128 (as pointed by the grey arrow in Fig. 8). In this configuration, the total power consumption is 513uW, with an energy/pixel of 55.6pJ.

The chip microphotograph in Fig. 9 shows a total active area of 0.55mm², of which 12.7% and 29.1% are taken by *KEYPTS* and *CACHE* memories, respectively. Table I shows the comparison of EQSCALE with the state of the art. From Table I, EQSCALE is able to operate in the sub-mW power range, with a 5.3X power reduction over [5]. This is achieved thanks to the lower complexity of ORB [12] (although ORB is slightly more complex than FAST+BRIEF) and the proposed architecture with EQ scaling capability. EQSCALE has the lowest energy when compared at iso-technology, with an energy reduction of 5.7X and 7.5X over [7] and [5], respectively. The area is reduced by 8.2X over the accelerator with previously lowest power [5], and by 1.8X over the accelerator with previously lowest area [7]. Overall, EQSCALE offers at least 10.1X reduction in the area·energy product [7], thanks to simultaneous area/energy improvement.

IV. CONCLUSION

An energy-quality scalable ORB-based accelerator for real-time video feature extraction has been presented. The ability to dynamically trade off energy and quality enables sub-mW operation at 30fps VGA, with 5.7X (1.8X) energy (area) reduction compared to the state-of-the-art accelerators, and 10.1X improvement in the area-energy product. This makes EQSCALE well suited for IoT vision applications, where both low area and low energy are simultaneous design goals.

ACKNOWLEDGMENT

The authors acknowledge the kind support of TSMC for chip fabrication, the Singaporean Ministry of Education for the grant MOE2014-T2-1-161, and ERDT of DOST of the Philippines for funding Ms. A. Alvarez.

REFERENCES

[1] D. G. Lowe, "Object recognition from local scale-invariant features," in *Int. Conf. on Computer Vision*, 1999, pp. 1150–1157.

[2] J. Oh *et al.*, "A 320 mW 342 GOPS Real-Time Dynamic Object Recognition Processor for HD 720p Video Streams," *IEEE J. Solid State Circuits*, vol. 48, no. 1, pp. 33–45, 2013.

[3] G. Kim *et al.*, "A 1.22 TOPS and 1.52 mW/MHz Augmented Reality Multicore Processor with Neural Network NoC for HMD Applications," *IEEE J. Solid-State Circuits*, vol. 50, no. 1, pp. 113–124, 2015.

[4] H. Bay, T. Tuytelaars, and L. Van Gool, "SURF: Speeded Up Robust Features," in *European Conf. on Computer Vision*, 2006, pp. 404–417.

[5] D. Jeon et al., "An Energy Efficient Full-Frame Feature Extraction Accelerator With Shift-Latch FIFO in 28 nm CMOS," *IEEE J. Solid-State Circuits*, vol. 49, no. 5, pp. 1271–1284, 2014.

[6] E. Rublee, V. Rabaud, K. Konolige, G. Bradski, "ORB: An efficient alternative to SIFT or SURF," in *Int. Conf. on Computer Vision*, 2011, pp. 2564–2571.

[7] J. Park, H. Kim, L. Kim, "A 182 mW 94.3 f/s in Full HD Pattern-Matching Based Image Recognition Accelerator for an Embedded Vision System in 0.13-μm CMOS Technology," *IEEE Trans. Circuits Syst. Video Technol.*, vol. 23, no. 5, pp. 832–845, 2013.

[8] J. Park, H. Kim, H. Kim, J. Lee, and L. Kim, "A Vision Processor With a Unified Interest-Point Detection and Matching Hardware for Accelerating a Stereo-Matching Algorithm," *IEEE Trans. Circuits Syst. Video Technol.*, vol. 26, no. 12, pp. 2328–2343, 2016.

[9] F. E. Uzyıldırım, M. Özuysal, "Instance detection by keypoint matching beyond the nearest neighbor," *Signal, Image Video Process.*, vol. 10, no. 8, pp. 1527–1534, 2016.

[10] M. a Fischler, R. C. Bolles, "Random Sample Consensus: A Paradigm for Model Fitting with Applications to Image Analysis and Automated Cartography," *Commun. ACM*, vol. 24, no. 6, pp. 381–395, 1981.

[11] Katholieke Universiteit Leuven, "Affine Covariant Features." [Online]. Available: http://www.robots.ox.ac.uk/~vgg/research/affine/.

[12] D. D. Nguyen, A. El Ouardi, E. Aldea, and S. Bouaziz, "HOOFR: An enhanced bio-inspired feature extractor," *Proc. Int. Conf. Pattern Recognit.*, pp. 2977–2982, 2017.

TABLE I. COMPARISON TABLE (BEST RESULTS HIGHLIGHTED IN BOLD, POWER/ENERGY OF THIS WORK EVALUATED AT MIN. ACCEPTABLE/MAX. QUALITY)

Parameter	[2]	[7]	[8]	[5]	This work
algorithm	SIFT	FAST-BRIEF	FAST-BRIEF	SURF	ORB
technology	0.13μm	0.13μm	65nm	28nm	40nm
supply voltage	0.65 - 1.2V	1.2V	1.2V	0.47V	0.6 - 0.9V
required on-chip memory	382kB SRAM	128kB SRAM	28kB SRAM	~7kB (latch)	~4kB (latch)
frame rate (resolution)	30fps (HD)	94.3fps (FHD)	**106fps (FHD)**	30fps (VGA)	30fps (VGA)
energy-quality scalable	NO	NO	NO	NO	**YES**
power	320mW	182mW	185mW	2.7mW	**0.51 - 2.9mW**
energy/pixel at iso-technology*	3,970pJ	319pJ	589pJ	419pJ	**55.6 - 310pJ**
area at iso-technology (F^2x10^6)**	1,893	605.9	1,600	2,831.6	**343.75**
FOM = area · (energy/pixel)***	393	10.1	49.3	62	**1 - 5.6**

* normalized to 40 nm (0.7X energy decrease/generation), ** F = min. feature size of technology, *** normalized to EQSCALE at min. acceptable quality

S16-4 (2227)

A Self-Powered Always-On Vision-based Wake-up Detector for Wearable Gesture User Interfaces

Suhwan Cho, Seongrim Choi, Junsik Woo, Ara Kim, and Byeong-Gyu Nam

Department of Computer Science and Engineering
Chungnam National University
Daejeon, Korea
bgnam@cnu.ac.kr

Abstract—Hand gesture recognition is one of the secure natural user interface (NUI) mechanisms on wearable devices since it does not reveal user's intention in public domain unlike the speech recognition. However, its energy dissipation is very demanding as it requires compute-intensive machine vision processing. Recently, wake-up detectors have been proposed to improve the energy-efficiency of always-on sensing nature of the NUI systems by switching off the main functional blocks while just keeping the wake-up detector alive during idle time. However, vision-based wake-up detectors still require power-consuming vision processing so we propose a self-powered vision wake-up detector to alleviate burdens on limited battery and thus facilitate always-on wake-up detection for the wearable gesture UIs. Our work has four key features to realize the self-powered wake-up detection; 1) imaging-harvesting dual-mode CMOS image sensor (CIS) with 0.6V 3T pixels, 2) subthreshold SRAM with disturb-free 0.3V 10T bitcells, 3) hand detection engine with modified Haar-like filters invariant to skin-colors, and 4) on-die switched capacitor DC-DC converter for lightweight system design. Thanks to the features combined, this work achieves self-powered operation of always-on vision wake-up detection.

Keywords—wake-up detector; gesture recognition; energy-harvesting; dual-mode CMOS image sensor; subthreshold SRAM

I. INTRODUCTION

Hand gesture recognition is gaining attention as a natural user interface (NUI) mechanism in wearable smart devices such as smart watches and head mounted displays due to its security merits of not disclosing users' intention in public domain unlike speech recognition approaches [1]. However, always-on sensing nature of the gesture user interface demands large energy dissipation, which becomes worse due to compute-intensive vision processing for the gesture recognition functionality. Recently, wake-up solutions have been proposed to extend battery lifetime of the NUI system by turning off the main functional blocks in standby mode and keeping the wake-up detectors always alive [2-3]. There have been several studies on the wakeup detectors for speech recognition, face recognition, and surveillance applications, but they were not for the gesture recognition systems.

In this paper, we propose a self-powered always-on vision wake-up detector for the gesture user interface on wearable devices. The proposed wake-up detector accommodates four of key features to enable the self-powered operation. First, imaging-harvesting dual-mode CMOS image sensor (CIS)

based on 0.6V 3T pixels is proposed in this work. We reduce transistor counts in a pixel through the column sharing of pixel comparators. Second, we present a subthreshold SRAM with disturb-free 0.3V 10T bitcells. We eliminate all the contentions arising in a bitcell for low-voltage operations by adopting proper gating schemes into the bitcell. Third, hand detection engine is devised based on a proposed descriptor algorithm optimized for hand detections. It uses modified Haar-like filters robust to skin color variations to improve the accuracy in hand detections. Finally, switched capacitor DC-DC converter is integrated for lightweight system design and converts harvested energy from CIS to proper voltage levels in the chip. As a result, our wake-up detector achieves self-powered operation of the always-on vision-based wake-up detections.

II. SELF-POWERED VISION WAKE-UP DETECTOR

A. System Architecture

The overall system architecture of the proposed self-powered wake-up detector is shown in Fig. 1. It consists of four major functional blocks, i.e. energy-harvesting CIS, frame buffer, hand detection engine, and switched capacitor (SC) DC-DC converter. The energy-harvesting CIS works in dual modes of imaging and harvesting. In the imaging mode, the CIS captures incoming images and stores them in the frame buffer. The frame buffer is used to store the digital images from the CIS, and hand detection engine finds hand shapes from the images and generates wake-up signal for the gesture UI system when detecting hands. In the energy-harvesting mode, pixels in the CIS are used as photovoltaic cells to harvest energy from

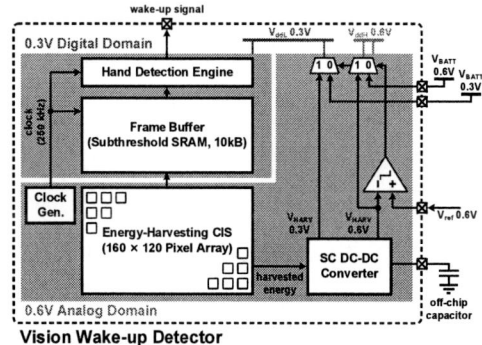

Fig. 1. Overall architecture of vision wake-up detector

978-1-5386-3179-9/17 $31.00 © 2017 IEEE   245

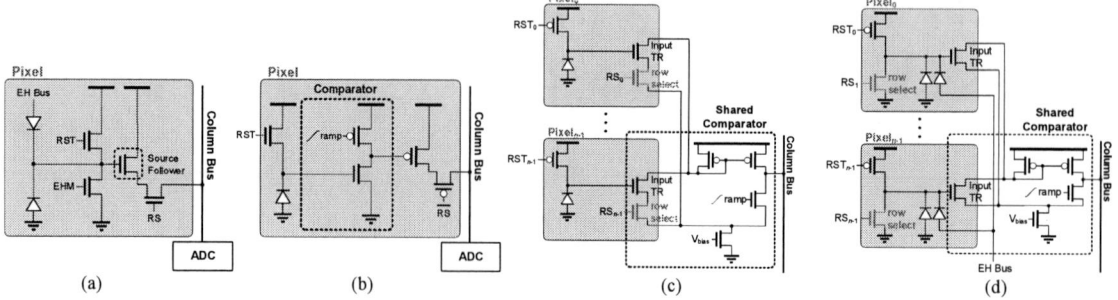

Fig. 2. Comparison of energy-harvesting pixel structures (a) dual-mode pixel (b) DPS pixel (c) DPS pixel with shared-comparator (d) low-voltage dual-mode DPS pixel

incident light. The SC DC-DC converter produces 0.6V supply voltage for the analog blocks such as CIS and SC DC-DC converter itself from the output voltage of the energy-harvesting CIS. The chip initially operates from the external battery, but it switches to the on-die supply voltage for self-powered operation when the harvested voltage level reaches a sufficient level of 0.6V.

B. Imaging-harvesting dual-mode CIS with 0.6V 3T pixels

The dual-mode CIS design with integrated imaging and harvesting operations have been studied for its area and energy-efficiency [4-5]. The dual-mode CIS operates in two phases i.e. harvesting phase and imaging phase to serve as the energy source as well as image sensor for a system. Dual-mode pixels incorporate a source follower (SF) inside them to transfer the pixel value to the comparator in the analog-to-digital converter (ADC) as illustrated in Fig. 2(a) [4], but this SF imposes limitations on the dynamic rage under low voltage operations.

In the meantime, digital pixel sensor (DPS) structure was studied for low-voltage operations of the CIS [6] and it takes out the comparator from the ADC and makes each pixel compare its own value inside the pixel to directly output the result to the following ADC [6], as described in Fig. 2(b). This approach avoids the SF from the pixel and thus improves the limited dynamic range under low voltage operations. However, the DPS structure incurs a large area overhead due to the comparator incorporated into each pixel and therefore, shared-comparator structure has been proposed for the DPS as shown in Fig. 2(c). It mitigates the area issue by just putting the input device of the comparator into each pixel to deliver pixel signal to the comparator and making the other part of the comparator shared among the pixels in the same column [7]. A row-select device is also included in the pixel to activate it when selected.

We adopt this shared-comparator DPS structure to our dual-mode pixel as in Fig. 2(d) for low-voltage operations of the dual-mode CIS. In our structure, the row-select device in Fig. 2(c) is taken out from the input stack and shunted with the diodes to reduce the number of stacked devices for low-voltage operations. It also reduces the mismatches across pixels because process variation impacts only the single input device rather than both the row-select and input devices stacked together as in Fig. 2(c). Thanks to this proposed dual-mode DPS structure, we achieve 0.6V operation and 2.37× higher fill-factor compared with the previous work [5].

C. Subthreshold SRAM with disturb-free 0.3V 10T bitcells

Conventional 6T SRAM bitcell achieves high memory-density with simple structure but its low-voltage operation is limited around 0.7V owing to several disturbance issues associated with 6T structure such as read disturb, write disturb, half select and hold stability [8]. A 12T bitcell structure have been proposed to eliminate all the contentions in a bitcell for robust subthreshold operations by adopting read buffer, cross-point write wordline, and power cutoff [9]. However, the large number of transistors in a bitcell impacts memory density so we propose a novel disturb-free 10T bitcell to address the density issue without sacrificing the robustness under subthreshold region.

The proposed 10T SRAM cell has single bitline structure while containing all the read buffer, cross-point write wordline, and power cutoff structures as the 12T bicell does. It consists of data storage (M_1-M_6) and read/write port (M_7-M_{10}) as shown in Fig. 3(a). It exploits single bitline structure by unifying the dual read/write ports exploited in the 12T bitcell. In this structure, PMOS device M_5 or M_6 is turned off during write operations to remove the write disturbance arising from the contention between pull-up device (M_1 or M_2) and write driver. The read buffer composed of M_7 and M_8 isolates the data storage from the bitline to avoid any read disturbance caused

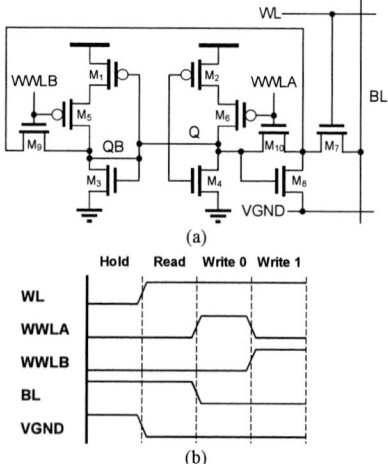

Fig. 3. Proposed 10T SRAM bitcell (a) schematic diagram (b) timing diagram

by low-impedance path between storage and bitline. Half-select is also removed as the connection between the storage node and the bitline can be made only for the full-selected bitcells, i.e. both the row-select wordline (WL) and column-select write wordline (WWLA or WWLB) should be active together for a selection. In addition, the source voltage of the read buffer is held high during the hold state to minimize bitline leakage and thereby increase hold margin. Thanks to this shared single bitline structure, our 10T bitcell reduces area and power dissipation by 25.4% and 25.6%, respectively, compared with the 12T structure.

D. Hand detection engine with skin-color invariant Haar-like filters

Adaboost is a classification algorithm to detect target objects through the votes from simple classifiers like Haar-like filters [10]. Adaboost is widely used for its low computational cost, but is weak at skin color variations since the Haar-like filter works on the pixel intensity. Therefore, we propose a modified Haar-like filter that is robust to skin color variations by utilizing the number of corners in the image rather than the pixel intensity. In this proposed Haar-like filter, each pixel stores integral number of corners within the region starting from origin to the pixel as shown in Fig. 4(a). In this way, the total number of corners inside a certain region can be obtained simply by using the values of four pixels as describes in the figure.

Fig. 4(b) shows the architecture of our hand detection engine accommodating buffer reuse scheme for integral number storage. It consists of corner detection, integral number generation, and Adaboost classifier. As described in Fig. 4(c), the wake-up detector operates in harvesting and imaging phases, and in the harvesting phase, the frame buffer is not used by the CIS. Therefore, the hand detection engine can reuse the idle frame buffer to store computed integral number of corners, thereby saving extra memory for the integral numbers. Trained Adaboost weights are stored in a lookup table to avoid accesses to external memory.

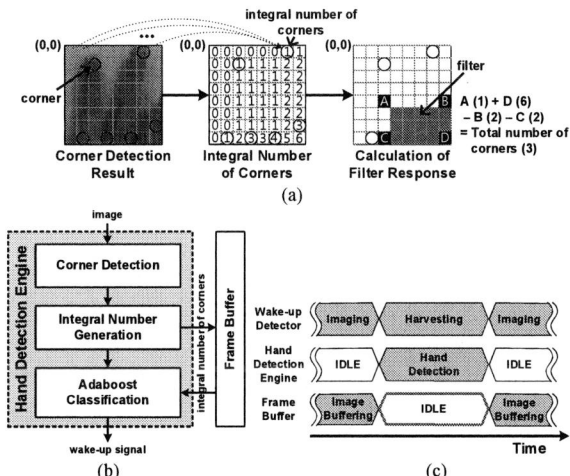

Fig. 4. Hand detection engine (a) generating integral number for corners (b) overall architecture with buffer reuse (c) operation sequence of engine

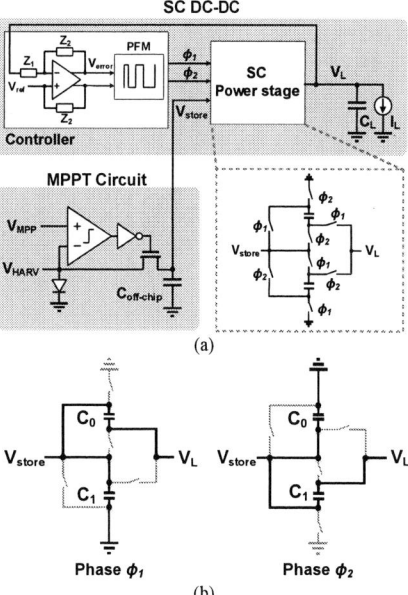

Fig. 5. The proposed SC DC-DC converter (a) overall structure (b) power stage operations.

E. Switched -Capacitor DC-DC Converter

The proposed CIS harvests 0.3V of supply voltage which needs load regulation to reduce the ripple from non-uniform occurrence of load current over time. Moreover, this 0.3V supply needs boosting up to 0.6V for proper operations in analog domain. Therefore, we exploit the switched-capacitor (SC) DC-DC converter for cost-effective on-die regulation and conversion of the energy harvested from CIS. It consists of loop controller, maximum power point tracking (MPPT) circuit and switched-capacitor power stage, as depicted in Fig. 5(a). The controller uses inverting amplifier and adopts pulse frequency modulation (PFM) for closed loop control of the power stage. The MPPT circuit monitors harvested voltage level across the photodiode and maintains the 0.3V of MPP level by controlling the charge transfer to the storage capacitor (C_{off}) for the maximum efficiency in harvesting. Fig. 5(b) illustrates how the 0.6V supply is generated in the power stage. In phase Φ_1, the 0.3V of C_{off} is transferred to the flying capacitor (C_{fly}) of the power stage through the charge sharing between the C_{off} and C_{fly} with 100:1 capacitance ratio to minimize degradation of levels from charge sharing. The power stage then boosts the 0.3V to 0.6V in phase Φ_2 through the capacitive coupling of the C_{fly}. This approach results in 82% of conversion efficiency.

III. IMPLEMENTATION RESULTS

The proposed self-powered always-on vision wake-up detector is fabricated using 65nm CMOS process. Fig. 6 shows a chip photograph and performance summary. Average power consumption is measured to be 26 µW at 250 kHz operations and the generated power is 32 µW at 60 kLux (sunny day), thereby allowing self-powered operation of the vision wake-up detection.

Fig. 7 shows the power consumption of the vision wake-up detector according to the supply voltage. The voltage for the analog domain stays at their minimum operating voltage of 0.6V, while that for the digital blocks goes further down to 0.3V thanks to the subthreshold SRAM design. We conducted 50k Monte Carlo simulations for the 0.3V operation of the conventional 12T and proposed 10T SRAM bitcells, respectively, for proper comparisons under same technology and supply voltage. Fig. 8 shows write operation results, and proposed 10T bitcell presents comparable noise margin with the 12T bitcell at a reduced area by 17%.

Table 1 summarizes the chip performance and compares the work with other state-of-the-art wake-up detectors such as human detector and speech detector as this chip is the first wake-up detector for gesture recognition purpose. As shown in the table, this work shows comparable power consumption with the others and achieves the self-powered operation of the wake-up detection for the first time.

IV. CONCLUSION

A self-powered always-on vision wake-up detector is presented for the gesture UI system on wearable devices. It incorporates imaging-harvesting dual-mode CIS with 0.6V 3T

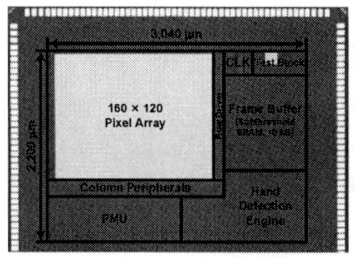

Fig. 6. Chip photograph and characteristics

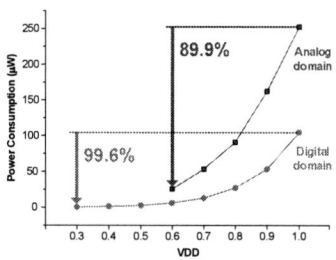

Fig. 7. Power reduction in analog and digital domains

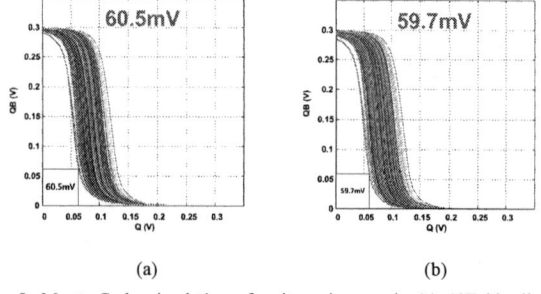

(a)   (b)

Fig. 8. Monte Carlo simulation of write noise margin (a) 12T bitcell (b) proposed 10T bitcell

TABLE I.   COMPARISON WITH RELATED WORKS

	Gesture Wake-up Detector (This Work)	Speech Wake-up Detector [2]	Human Detector [3]
Functionality	Hand Detection	Voice Detection	Feature Extraction Only
Technology (nm)	65	90	180
Supply Voltage (V)	0.3 (digital) 0.6 (analog)	N/A	0.8 (digital) 1.3 (analog)
Operating Frequency (MHz)	0.25	0.01	25
Power Consumption (μW)	26 @ 15 fps	6	51.06 @ 15 fps 3.31 @ 1 fps
Generated Power (μW)	32 @ 60 kLux	N/A	N/A

pixels for self-powered operations. Subthreshold SRAM with disturb-free 0.3V 10T bitcells is also presented for deep low-voltage operation of the chip. Hand detection engine using skin-color invariant Haar-like filter is devised to improve the robustness in hand detections. Finally, switched capacitor DC-DC converter is integrated for on-die regulation and conversion of the power harvested from the CIS. Thanks to these features combined together, the proposed vision wake-up detector chip demonstrates self-powered operation of the wake-up detection for the first time.

ACKNOWLEDGMENT

The chip fabrication and EDA tool were supported by the IC Design Education Center(IDEC), Korea.

REFERENCES

[1] S. Choi et al., "A Low-Power Real-Time Hidden Markov Model Accelerator for Gesture User Interface on Wearable Devices," *in IEEE Asian Solid-State Circuits Conf.,* pp. 261-264, Nov. 2016.

[2] K. Badami et al., "Context-Aware Hierarchical Information-Sensing in a 6μW 90nm CMOS Voice Activity Detector, " *ISSCC Dig. Tech. Papers,* pp. 430-431, Feb. 2015.

[3] J. Choi et al., "A 3.4-μW Object-Adaptive CMOS Image Sensor With Embedded Feature Extraction Algorithm for Motion-Triggered Object-of-Interest Imaging," *IEEE J. Solid-State Circuit,* vol. 49, no. 1, pp. 289-300, Jan. 2014.

[4] S. U. Ay, "A CMOS Energy Harvesting and Imaging (EHI) Active Pixel Sensor (APS) Imager for Retinal Prosthesis," *IEEE Trans. Biomed. Circuits Syst.,* vol. 5, no. 6, pp. 535-545, Dec. 2011.

[5] A. Y.-C. Chiou and C.-C. Hsieh, "A 0.4V Self-Powered CMOS Imager with 140dB Dynamic Range and Energy Harvesting," *Dig. Symp. VLSI Circuits,* pp. C86-C87, June, 2015.

[6] N. Counjot et al., "A 65 nm 0.5 V DPS CMOS Image Sensor With 17 pJ/Frame.Pixel and 42 dB Dynamic Range for Ultra-Low-Power SoCs," *IEEE J. Solid-State Circuit,* vol. 50, no. 10, pp. 2419-2430, Oct. 2015.

[7] D. Ho et al., "CMOS 3-T Digital Pixel Sensor with In-Pixel Shared Comparator," *IEEE Int. Symp. on Circuits and Systems (ISCAS),* pp. 930-933, May, 2012.

[8] B. H. Calhoun and AP. P. Chandrakasan, "A 256-kb 65-nm Sub-threshold SRAM Design for Ultra-Low-Voltage Operation," *IEEE J. Solid-State Circuits,* vol. 42, no. 3, pp. 680-688, Feb. 2007.

[9] Y.-W. Chiu et al., "40 nm bit-interleaving 12T subthreshold SRAM with data-aware write-assist," *IEEE Trans. Circuits Syst. I, Reg. Papers,* vol. 61, no. 9, pp. 2578-2585, Sept. 2014.

[10] P. Viola and M. Jones, "Rapid Object Detection using a Boosted Cascade of Simple Features," in *Proc of. CVPR,* vol. 1, pp. 511-518, Dec. 2001.

A 18-to-23 GHz -253.5dB-FoM Sub-Harmonically Injection-Locked ADPLL with ILFD Aided Adaptive Injection Timing Alignment Technique

Zhao Zhang, Jincheng Yang, Liyuan Liu[*], Peng Feng, Jian Liu, and Nanjian Wu[*]

State Key Laboratory of Superlattice and Microstructures
Institute of Semiconductors, Chinese Academy of Sciences,
University of Chinese Academy of Sciences, Beijing, China
Email: liuly@semi.ac.cn, nanjian@red.semi.ac.cn

Abstract—This paper presents an 18-to-23 GHz sub-harmonically injection-locked all-digital PLL (SIL-ADPLL). It adopts the proposed injection-locked frequency divider aided adaptive injection timing alignment technique and uses a proposed (UP-DN) block to adjust the injection timing adaptively at output frequency higher than 20 GHz with low power consumption. A new pulse generator is proposed to relax the trade-off between the phase-noise suppression and the power consumption. The SIL-ADPLL is implemented in 65 nm CMOS process. Measurement results show that the covered frequency range is from 18 GHz to 23 GHz and the rms jitter integrated from 1 kHz to 100 MHz is 57.4 fs at 20 GHz output. The power consumption is 13.7 mW and the FoM is -253.5 dB. This SIL-ADPLL also shows robustness over temperature and supply variation.

Keywords—All-digital phase-locked loop (ADPLL); sub-harmonically injection-locked ADPLL; injection-locked frequency divider (ILFD); adaptive injection timing alignment; single-ended injection

I. INTRODUCTION

The sub-harmonically injection-locked phase-locked loop (SIL-PLL) provides an effective way to achieve ultra-low jitter clock generation circuit with reasonable power consumption [1-6]. In the SIL-PLL, the adaptive injection timing adjusting technique is critical to make the SIL-PLL operate robustly. Several injection timing adjusting techniques were proposed for analog SIL-PLL [1-4]. However, the technique proposed in [1] is not suitable for the SIL-PLL with frequency higher than 20 GHz due to the limited operation speed of the timing-adjusted phase detector [1]. The approaches proposed in [2-4] can be used at frequency higher than 20 GHz. But the current-mode logic (CML) divider chain in [2] consumes large power. And the injection timing alignment technique proposed in [3] needs a quadrature voltage-controlled oscillator and four pulse generators, which are power consuming. Although the 20-GHz divider-less SIL-PLL [4] can save the power consumption of divider chain, it needs two always-on voltage-to-current converters to calibrate the current mismatch, which induces high level phase noise at lower offset frequency and thus worsen the rms jitter. Digital approaches proposed in [6] can avoid the issue of current mismatch. But the operation speed of

the bang-bang phase detector (BBPD) is limited so that it is not suitable to generate frequency higher than 20 GHz.

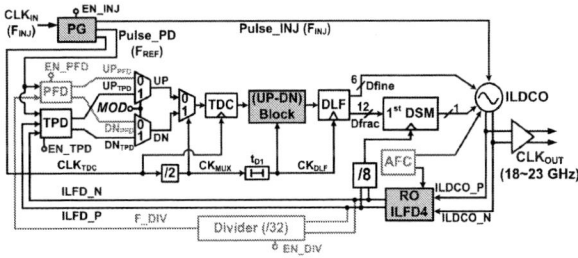

Fig. 1. Block diagram of proposed SIL-ADPLL

In this paper, we propose an 18-to-23 GHz sub-harmonically injection-locked all-digital PLL (SIL-ADPLL). The all-digital structure is adopted to avoid the issue of current mismatch. It adopts the proposed injection-locked frequency divider (ILFD) aided adaptive injection timing alignment technique and a proposed (UP-DN) block to adjust the injection timing adaptively at output frequency higher than 20 GHz with low power consumption. A new pulse generator is proposed to relax the trade-off between the phase-noise suppression and the power consumption. This paper is organized as follows. Section II introduces the architecture of proposed SIL-ADPLL. Section III gives the details of the building blocks. Section IV presents the measurement results and the conclusions are drawn in Section V.

II. ARCHITECTURE OF PROPOSED SIL-ADPLL

A. Overall Architecture

Figure 1 shows the block diagram of the proposed SIL-ADPLL with ILFD aided adaptive injection timing alignment technique. It adopts a ring oscillator based divided-by-4 ILFD (ROILFD4), an injection-locked digitally controlled oscillator (ILDCO) with single-ended injection technique, and a TPD [1] to achieve the proposed injection timing alignment technique. As our previous analysis [7] shows, if the TPD operates at frequency higher than 20 GHz, the pulse width of Pulse_PD (Fig. 1) needs to be very narrow, which is very challenging for

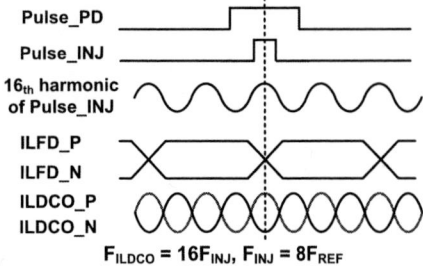

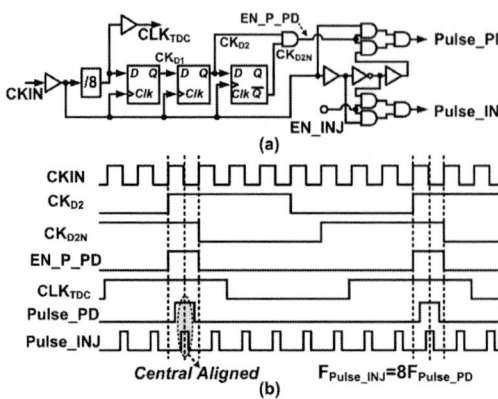

Fig. 2. Timing diagram of proposed ILFD aided injection timing alignment technique.

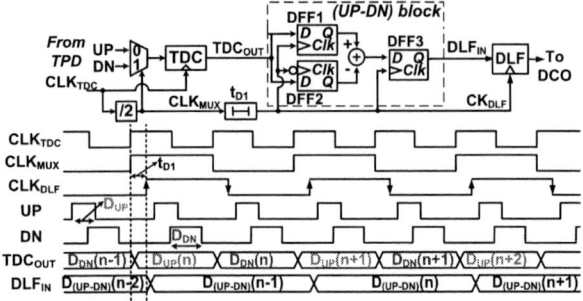

Fig. 3. Schematic and timing diagram of proposed (UP-DN) block.

Fig. 4. (a) Schematic of proposed PG, and (b) its timing diagram.

design of pulse generator (PG). So we insert the ROILFD4 between the ILDCO and the TPD not only to relax the pulse width requirement of Pulse_PD but also to significantly relax the speed requirement of TPD and thus the TPD can be used in the SIL-ADPLL with output frequency higher than 20 GHz. In addition, compared with the CML divider [2], the ROILFD4 not only consumes less power but also doesn't induce delay that degrades the injection timing. The details of the proposed injection timing alignment technique are described in Section II-B. The (UP-DN) block is proposed to quantize the pulse width difference of UP and DN with one time-to-digital converter (TDC) to save power and area. In addition, thanks to the (UP-DN) block, the digital loop filter (DLF) can operate at half of the reference frequency to further save power. The proposed PG can relax the trade-off between the phase-noise suppression and the power consumption. The details of the PG, ILDCO and ROILFD4 are shown in Section III.

The operation process of the SIL-ADPLL, which includes 4 steps, is shown as follows. At step 1 the auto-frequency calibration (AFC) block sets the optimum control words for ILDCO and ROILFD4 so that F_{ILDCO} and F_{ILFD} can be nearest to the SIL-ADPLL output target frequency F_{target} and $0.25F_{target}$, respectively. Then the PLL locks to F_{target} at step 2. At step 3 the PFD and the divide-by-32 divider are disabled to save power and the TPD starts to work to adjust the injection timing. Finally the injection pulse Pulse_INJ is enabled and the PLL starts to be injection-locked at step 4. The control signal MOD is set to be 0 at step 2 and to be 1 at step 3 and step 4.

B. ILFD Aided Injection Timing Alignment Technique

Figure 2 shows the timing diagram of the proposed injection timing alignment technique. The principle is as

follows. As the analysis in [8] shows, if the output frequency F_{ILDCO} equals to an integer M multiple of the frequency of Pulse_INJ F_{INJ}, the peak point of the ILDCO differential output can be aligned to the peak point of Mth harmonic of Pulse_INJ and the central point of the injection pulse Pulse_INJ. If the free running frequency of the ROILFD4 F_{ILFD} is near quarter of F_{ILDCO}, the crossing point of the ROILFD4 output can be aligned to the peak point of the ILDCO differential output. This timing relationship does not vary with process, voltage and temperature (PVT) variation. The TPD block can align the crossing point of the ROILFD4 output to the central point of the Pulse_PD [1]. The proposed PG can always make Pulse_PD and Pulse_INJ centrally aligned so that the central point of Pulse_INJ and the peak point of the ILDCO output align automatically. Therefore the SIL-ADPLL can achieve the injection timing alignment adaptively. In this design, M is set to be 16. The reference frequency F_{REF} of Pulse_PD is set to be 1/8 of F_{INJ} to decrease the operation speed of the PFD, TPD, TDC and digital blocks.

C. Operation Priciple of Proposed (UP-DN) Block

At step 2 and step 3 described above, the pulse width of UP and DN, D_{UP} and D_{DN}, need to be quantized by the TDC and then get the value of their difference $D_{(UP-DN)}$. However, one TDC cannot quantize $D_{(UP-DN)}$ directly at step 3 because neither the rising edge nor the falling edge of UP and DN pulse from the TPD is aligned [1]. One way to solve this problem is to use two identical TDCs to quantize D_{UP} and D_{DN}, respectively, and then get $D_{(UP-DN)}$, with the penalty of much power consumption and mismatch. In this work, a (UP-DN) block is proposed to get $D_{(UP-DN)}$ with only one TDC. The schematic and the timing diagram of the (UP-DN) block are presented in Figure 3. When CK_{MUX} is low, the TDC is set to quantize the pulse width of the nth UP, $D_{UP}(n)$, and then $D_{UP}(n)$ is send to DFF1 at the rising edge of CK_{DLF}. When CK_{MUX} is high, the TDC is set to quantize the pulse width of the nth DN, $D_{DN}(n)$, and then $D_{DN}(n)$ is send to DFF2 at the falling edge of CK_{DLF}. At next rising edge of CK_{DLF}, the difference of $D_{UP}(n)$ and $D_{UP}(n)$, $D_{(UP-DN)}(n)$, is calculated and then latched to DFF3 and $D_{UP}(n+1)$ is latched to DFF1 for getting $D_{(UP-DN)}(n+1)$. The TDC updates its output at the rising edge of CLK_{TDC}. In addition, the DLF operates at frequency of $0.5F_{REF}$, so its power consumption can be saved.

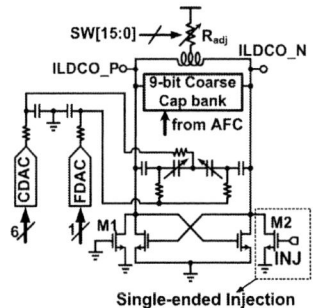

Fig. 5. Schematic of ILDCO.

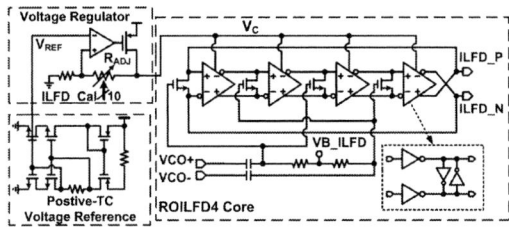

Fig. 6. Schematic of ROILFD4.

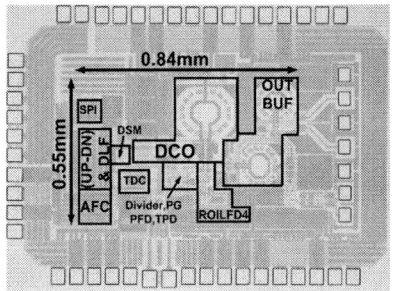

Fig. 7. Chip photograph.

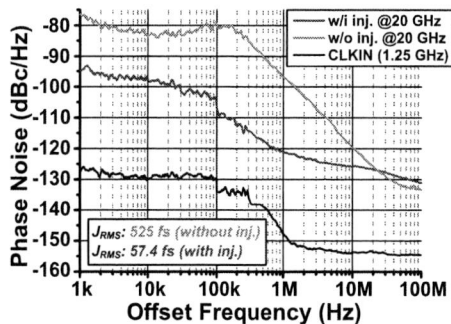

Fig. 8. Measured phase noise at 20 GHz carrier frequency without and with injecting locking, respectively.

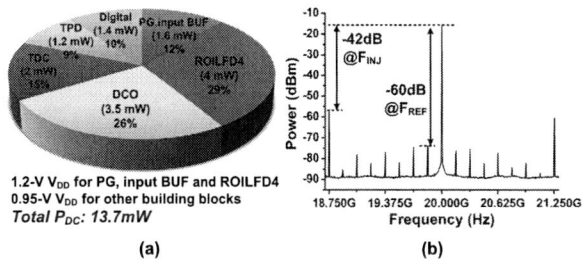

Fig. 9. (a) Power consumption breakdown. (b) Measured spectrum at 20 GHz output.

III. DESIGN OF BUILDING BLOCKS

A. Proposed PG

Figure 4(a) gives the schematic of the proposed PG. As the timing diagram shown in Fig. 4(b), the signal EN_P_PD goes to high level every 8 input clock periods to make the frequency of Pulse_INJ to be eight times of the reference clock Pulse_PD. So the injection pulse with higher frequency can be used to suppress more phase noise while do not increase the power consumption of TPD, TDC and digital blocks. The pulse width of EN_P_PD is one period of the input clock CKIN. The Pulse_PD and Pulse_INJ can be centrally aligned to meet the injection timing. The injection pulse can be enabled when the EN_INJ is high. The CLK_{TDC} is used to trig the TDC to update its output.

B. ILDCO

Figure 5 presents the schematic of the ILDCO. In order to to cover the output frequency from 18 GHz to 23 GHz over PVT variation, we use a 9-bit coarse digitally controlled capacitor array (DCCA) to cover wide frequency range. The

single-ended injection technique is achieved by injecting the injection pulse to the gate of M2. The M1 is used as the dummy to balance the load of the differential output of the ILDCO.

In order to achieve fine resolution, the varactor is controlled by 2 digital-to-analog converters (DAC) [9]. The 6-bit coarse DAC (CDAC) is from the output of the DLF and the 1-bit fine DAC (FDAC) is from the output of the 12-bit first order delta-sigma modulator (1st-DSM). The RC low-pass filter is used to suppress the quantization noise from DSM.

C. ROILFD4

Figure 6 shows the schematic ROILFD4. The ROILFD4 consists of a ROILFD4 core, a voltage regulator and a positive temperature coefficient (positive-TC) voltage reference. By adjusting the variable resistor R_{ADJ} in the regulator, V_C can be changed and thus the free running frequency of the ROILFD4 F_{ILFD} can be adjusted. R_{ADJ} is controlled by the AFC block shown in Figure 1. The positive-TC voltage reference is used to compensate F_{ILFD} over temperature variation. The cascode structure is used in the positive-TC voltage reference to reduce the influence of supply variation.

IV. IMPLEMENTATION AND MEASUREMENT RESULTS

The proposed SIL-ADPLL is implemented in a 65 nm 1P9M CMOS process. It occupies an active core area of 0.84×0.55 mm^2, of which about 0.1 mm^2 is occupied by the output buffer for testing, as shown in Fig. 7. The frequency F_{INJ} of injection pulse is 1.25 GHz when generating 20 GHz output. By varying the F_{INJ} from 1.125 GHz to 1.4375 GHz, the SIL-ADPLL can generate frequency from 18 GHz to 23 GHz.

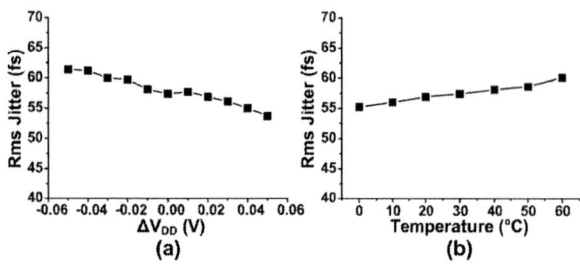

Fig. 10. Measured rms jitter at 20 GHz carrier frequency over (a) supply voltage and (b) temperature variation, respectively.

Figure 8 shows the measured phase noise at carrier frequency of 20 GHz without and with injecting locking, respectively. The phase noise without and with injecting locking is -98 dBc/Hz@1MHz and -121 dBc/Hz@1MHz, respectively. The rms jitter integrated from 1 kHz to 100 MHz without and with injection locking is 525 fs and 57.4 fs, respectively. It shows that the injection locking can considerably reduce the rms jitter. The rms jitter of the reference clock is 51.2 fs.

The supply voltage of PG, input clock buffer and ROILFD4 is 1.2 V and other building blocks operate at 0.95 V supply. The power consumption is 13.7 mW exclude the output buffer, as shown in Fig. 9(a).

Figure 9(b) shows the spectrum of 20 GHz output. It indicates that the spur level at F_{INJ} (1.25 GHz) is -42 dBc. The spur at F_{REF} (1/8 F_{INJ}, 156.25 MHz) and its integral multiple is mainly induced by the interference from reference clock Pulse_PD to the injection pulse Pulse_INJ through the supply and ground. The level is below -60 dBc, which is much less than the spur level at F_{INJ} and thus has negligible influence.

Figure 10(a) and Fig. 10(b) also show the rms jitter at 20 GHz carrier frequency over supply or temperature variation, respectively. When the supply voltage varies and the supply variation ΔV_{DD} changes from -0.05 V to 0.05 V at 30 °C, the rms jitter varies from 53.7 fs to 61.4 fs. When the temperature varies from 0 °C to 60 °C at 1.2/0.95 V supply, the rms jitter varies from 55.2 fs to 60.1 fs.

Table I summarize the performance of this work and the recently published papers. Compared with other works listed in Table I, the proposed SIL-ADPLL achieves the best figure of merit (FoM) and the widest frequency tuning range (FTR). The rms jitter and power consumption is also lower than most of the prior works listed in Table I.

V. CONCLUSION

An 18-to-23 GHz SIL-ADPLL was proposed and implemented. It adopts the proposed ILFD aided adaptive injection timing alignment technique and a proposed (UP-DN) block to adjust the injection timing adaptively at output frequency higher than 20 GHz with low power consumption. A new pulse generator was proposed to relax the trade-off between the phase-noise suppression and the power consumption. The measured results indicate that it can generate

TABLE I. PERFORMANCE SUMMARY AND COMPARISON

	This work	TMTT' 16 [2]	ISSCC' 17 [3]	TCAS-I' 15 [4]	JSSC' 09 [5] Chip A
Process	65 nm	90 nm	65 nm	65 nm	90 nm
Supply (V)	1.2/0.95	1.2	NA	1.2	1.3
Out. Freq. /range(GHz)	20 /(18~23)	25.2 /(24-26.1)	29.25 /(27.4~30.8)	20 /(19.2~23.2)	20 /NA
FTR (%)	24.4%	8.4%	11.7%	18.9%	NA
Area (mm²)	0.462	0.94	0.468	NA	0.455
P_{DC} (mW)	13.7	50.4	24.3	8	38
Inj. Freq. (MHz)	1125~1437.5	1575	2000~2400	2500	1000
J_{RMS} (fs) /int. range	57.4 (1k~100M)	56.6 (1k~40M)	86 (1k~100M)	98 (1k~100M)	110 (50k~80M)
Inj. Spur (dBc)	-42	-22	-32.62	-43	-46
FoM (dB)¹	-253.5	-248	--247.5	-251	-243.4
Architecture	Digital	Analog	Analog	Analog	Analog

¹ $FoM = 20log(J_{RMS}/1s) + 10log(P_{DC}/1mW)$, the lower the better

frequency range from 18 GHz to 23 GHz. The FoM is -253.5 dB, which is better than other SILPLLs operating at frequency higher than 20 GHz. Measurement results also show robustness over temperature and supply voltage variation.

ACKNOWLEDGMENT

This work is financially supported by Natural Science Foundations of China No. 61474108, 61331003 and 61306027, and National Science and Technology Major Projects of the Ministry Science and Technology of China Under Grant No. 2016ZX03001002-002.

REFERENCES

[1] I. T. Lee, et al., "A divider-less sub-harmonically injection-locked PLL with self-adjusted injection timing," ISSCC Dig. Tech. Papers, pp. 414-415, Feb. 2013.

[2] H.-Y. Chang, et. Al., "Design and Analysis of CMOS Low-Phase Noise Low-Jitter Subharmonically Injection-Locked VCO With FLL Self-Alignmenty Technique," IEEE Trans. on Microw. Theory and Tech.,, vol. 62, no. 5, pp. 4632-4644, Dec. 2016.

[3] S. Yoo, et al., "A PVT-Robust -39dBc 1kHz-to-100MHz Integrated-Phase-Noise 29GHz Injection-Locked Frequency Multiplier with a 600µW Frequency-Tracking Loop Using the Averages of Phase Deviations for mm-Band 5G Transceivers", ISSCC Dig. Tech. Papers, pp. 324-325, Feb. 2017.

[4] K. Huang, et al., "A 80 mW 40 Gb/s Transmitter With Automatic Serilizing Time Window Search and 2-tap Pre-Emphasis in 65 nm CMOS Technology," IEEE Trans. Circuits Syst. I, Reg. Papers, vol. 62, no. 5, pp. 1441-1450, May 2015.

[5] J. Lee and H. Wang, "Study of subharmonically injection-locked PLLs," IEEE J. Solid-State Circuits, vol. 44, no. 5, pp. 1539-1553, May. 2009.

[6] A. Elkholy, et al., "A 6.75-to-8.25GHz, 250fsrms-Integrated-Jitter 3.25mW Rapid On/Off PVT-Insensitive Fractional-N Injection-Locked Clock Multiplier in 65nm CMOS," ISSCC Dig. Tech. Papers, pp. 192-193, Feb. 2016.

[7] Z. Zhang, et al., "A 2.4-3.6-GHz Wideband Subharmonically Injection-Locked PLL With Adaptive Injection Timing Alignment Technique," IEEE Trans. Very Large Scale Integ. (VLSI) Syst., vol. 25, no. 3, pp. 929-941, Feb. 2017.

[8] A. Mirzaei, et al, "The Quadrature LC Oscillator: A Complete Portrait Based on Injection Locking," IEEE J. Solid-State Circuits, vol. 42, no. 9, pp. 1916-1932, Sep. 2007.

[9] M. Zanuso, et al, "A Wideband 3.6 GHz Digital ΔΣ Fractional-N PLL With Phase Interpolation Divider and Digital Spur Cancellation," IEEE J. Solid-State Circuits, vol. 46, no. 3, pp. 627-638, Sep. 2011.

A 1.5-GHz Sub-Sampling Fractional-N PLL for Spread-Spectrum Clock Generator in 0.18-μm CMOS

Chun-Yu Lin, Tun-Ju Wang, and Tsung-Hsien Lin

Graduate Institute of Electronics Engineering and Department of Electrical Engineering,
National Taiwan University, Taipei, Taiwan

Abstract - This paper presents a fractional-N sub-sampling phase-locked loop (SSPLL) for spread-spectrum clock generator. A digital-to-time converter (DTC) is adopted to facilitate a fractional-N SSPLL. A digital calibration scheme is employed to eliminate DTC gain error. With the calibration method enabled, the PLL is successfully locked and achieves 18.98-dB EMI reduction. This PLL was fabricated in a TSMC 0.18-μm CMOS technology. The core area is 0.467 mm^2. The chip dissipates 11.1 mW from a 1.8-V supply voltage.

I. INTRODUCTION

Electromagnetic interference (EMI) is a critical design consideration as it may degrade the performance of sensitive circuits. Several methods have been proposed to resolve the EMI issue, such as shielding or specially designed clocking scheme, etc. Among these methods, spread-spectrum clocking is most widely used. By modulating the clock frequency in a pre-determined pattern, the spectrum density of the clock signal is spread to a small frequency range. The EMI at a certain frequency can be significantly decreased.

Several methods have been proposed to implement spread-spectrum clocking based on phase-locked loops (PLL). They can be categorized into three methods. The first method is called direct voltage-controlled oscillator (VCO) modulation [1]. It directly modulates the VCO control voltage by injecting modulation signal into the loop filter. In this method, the applied modulation signal experiences the high-pass transfer function from VCO to PLL output.

The second method is the delta-sigma modulator (DSM) modulation [2]. The frequency spreading is accomplished by modulating the control code of the divider via the DSM. Modulation signal in this method travels through the low-pass transfer function from DSM to the PLL output. Since triangle modulation profile is most commonly used, the modulation signal is composed of a fundamental tone and its harmonics. Harmonics with higher frequencies are then filtered out by the low-pass transfer function, which degrades the EMI reduction. In the above two methods, tradeoff between loop bandwidth and EMI reduction is the main design consideration.

The third method, two-point modulation [3], alleviates the aforementioned tradeoff by combining the first two methods. By applying the modulation signal to both DSM and loop filter, the low-pass and high-pass transfer function is combined to achieve an all-pass response. However, this method requires gain matching between two injecting paths, which leads to higher design complexity. Furthermore, all these methods face

tradeoff between EMI reduction and loop bandwidth, which affects jitter performance.

Sub-sampling PLL (SSPLL) is first proposed in [4], in which noise gain from CP/PD to output is lowered by a factor of N^2 due to the divider-less design, where N is the ratio of the PLL output frequency F_{OUT} to the reference frequency F_{REF}. Therefore, this technique can achieve good jitter performance under high bandwidth design requirement. However, the fractional-N mode is not available in the basic structure proposed in [4]. This makes the original SSPLL not suitable for spread-spectrum applications. Digital-to-time-converter (DTC) was introduced in [5] to resolve this problem, in which digitally-controlled delay is added between the reference clock and SSPLL, as shown in Fig. 1. By properly controlling the delay, a fractional-N division can be achieved. However, the DTC control in [5] is quite complex because of the DSM adopted for fine fractional ratio. Furthermore, the DTC gain error needs to be addressed with complex calibrations.

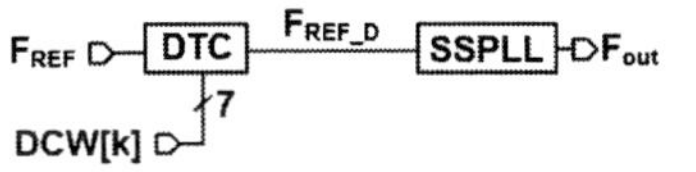

Fig. 1. Conceptual block diagram of fractional-N SSPLL.

In this paper, a 1.5-GHz fractional-N SSPLL for spread-spectrum clocking application is presented. The proposed SSPLL incorporates a simple DTC gain calibration technique to remove the gain error. Also, the DSM is not necessary for spread-spectrum clock generation in this work. Hence, the proposed PLL achieves low hardware complexity while maintaining good EMI reduction and jitter performance. This paper is organized as follows. Section II introduces the proposed spread-spectrum fractional-N SSPLL. Section III describes the circuit designs. Finally, measurement results and conclusion are presented in sections IV and V, respectively.

II. ARCHITECTURE OF THE FRACTIONAL-N SSPLL

A. Proposed Spread-Spectrum Fractional-N SSPLL

Fig. 2 shows the block diagram of the proposed spread-spectrum SSPLL. It consists of three major blocks: DTC with calibration, frequency-locked loop, and SSPLL main loop. Because of the limited frequency locking range of a SSPLL, a frequency-locked loop is employed for the SSPLL to properly acquire lock. This loop can be turned off after the SSPLL has locked to reduce power. In the main SSPLL, DTC circuit is utilized to facilitate spread-spectrum clocking generation.

Traditionally, PLL bandwidth must be carefully designed to suppress in-band noise introduced by the DSM, which modulates the DTC. However, for spread-spectrum clocking, high resolution fractional frequency step is not required [6]. This allows us to remove the DSM for DTC control, and hence, the bandwidth limitation is alleviated. Another key feature of this work is the proposed DTC gain calibration circuit for eliminating the DTC gain error, which will be described later.

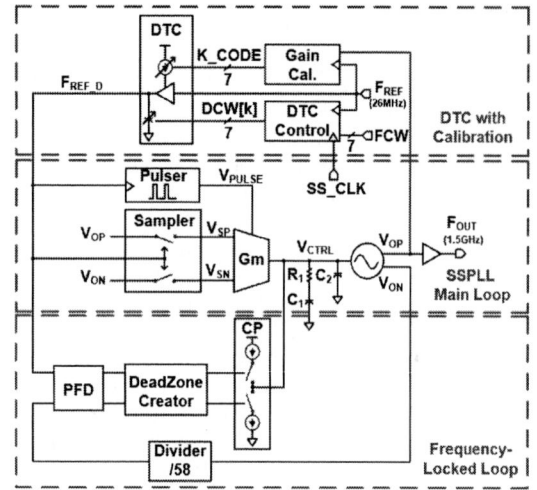

Fig. 2. Architecture of the proposed spread-spectrum SSPLL.

B. Operation of Fractional-N SSPLL

In the conceptual fractional-N SSPLL shown in Fig. 1, the reference signal F_{REF} is fed through the DTC, which is controlled by the delay-control word (DCW), and sent into the integer-N SSPLL proposed in [4]. The operation timing waveforms are illustrated in Fig. 3. Assuming that signal F_{OUT} and F_{REF} are perfectly aligned initially, the next edge of F_{REF} will deviates from F_{OUT} if a fractional divider ratio N+f is applied, where N is the integer part and f is the fractional part of divider ratio. As shown in Fig. 3(b), the proper signal edge of F_{REF_D} should be delayed from that of the F_{REF} (by adding the delay $T_d[k]$). The period of F_{REF_D} with delay, which is T_{REF_D}, can be calculated in (1).

$$
\begin{aligned}
T_{REF\_D} &= T_{REF} + T_d[k] \\
&= T_{REF} + \frac{\text{mod}\left(\sum_{i=1}^{k} DIN[i], M\right)}{M} \times T_{OUT}
\end{aligned}
\tag{1}
$$

where T_{OUT} is the period of F_{OUT}, T_{REF} is the period of F_{REF}, M is the maximum DCW code and DIN[i] is the accumulator input code.

The DTC delay $T_d[k]$ can grow to infinite if the delay continues to accumulate without bound. Fortunately, F_{OUT} is a periodic signal with a period T_{OUT}. This allows us to subtract $T_d[k]$ with integer multiple of T_{OUT} without affecting its locking operation which is the modulus operation in (1). Here, we "reset" DCW by subtract $T_d[k]$ with T_{OUT} immediately after $T_d[k]$ exceeds T_{OUT}, as shown in Fig. 3(a). This makes possible of designing the DTC with finite maximum delay of one T_{OUT}.

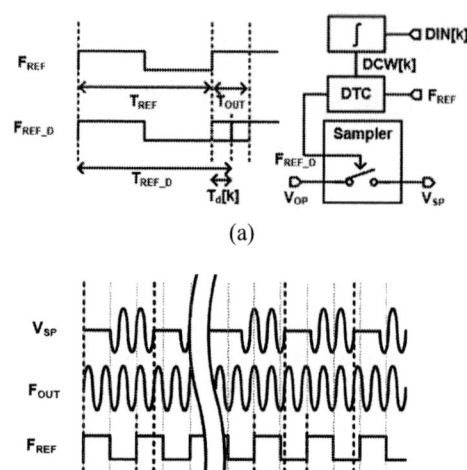

(a)

(b)

Fig. 3. (a) Operation of the DTC; (b) Timing diagram of the fractional-N SSPLL.

C. DTC Control

The DTC control circuit is depicted in Fig. 4. It mainly contains two part: Spread-Spectrum Control and Fractional-N Control. As described earlier, the delay needed is accumulated in each reference cycle and a simple accumulator can be used to achieve fractional division. This accumulator can also perform the "reset" operation without additional circuit if the DTC adjustable delay range ($T_{D,\ RANGE}$) is exactly one T_{OUT}. This is accomplished with the proposed DTC gain calibration, which will be explained in the next section.

Spectrum spreading is realized by changing DTC input word, DIN[k], with triangular modulation profile. An UP_DN control circuit is used to alter the direction of accumulation and an external clock SS_CLK is used to control modulation frequency. This control circuit detects SS_{IN} signal and toggles when SS_{IN} reaches spread boundaries. For example, when SS_{IN} reaches maximum code SS_{MAX}, UP_DN signal is inverted and the accumulator starts counting downward. When SS_{IN} reaches minimum code SS_{MIN}, UP_DN signal is inverted again and the accumulator starts counting upward. Static frequency mode without spread-spectrum is also implemented in this design by selecting SS_{IN} or frequency control word (FCW).

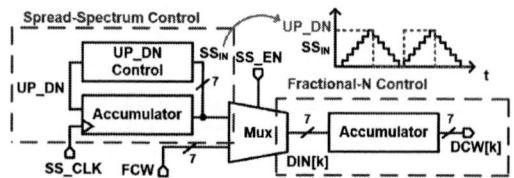

Fig. 4. Architecture of DTC control circuit.

D. DTC Gain Calibration

As mentioned previously, to ensure correct locking, the adjustable delay range $T_{D, RANGE}$ must equal to one T_{OUT}. A DTC gain calibration scheme is proposed in this design, as shown in Fig. 5(a). Signal F_{REF} is resampled twice by the clock signal F_{OUT}, generating signals V_{D1} and V_{D2}. T_{D12} is the delay between V_{D1} and V_{D2}, which is equal to one T_{OUT}. These two signals are then fed into DTC1 and DTC2 respectively, where DTC2 is a replica of DTC1. If DTC1 and DTC2 is given control word DCW1 = DCW_{MAX} (= 127, in this design) and DCW2 = DCW_{MIN} (= 0), the delay difference T_{D34}, which is the delay difference between V_{D3} and V_{D4}, should be

$$T_{D34} = T_{D12} - T_{D,RANGE} \qquad (2)$$

The delay error $T_{D, ERR} = T_{D, RANGE} - T_{OUT}$ is obtained by comparing V_{D3} and V_{D4} using an arbiter. The arbiter compares two input signals by arrival time of their rising edges. The arbiter output V_{COM} is high when V_{D3} rising edge comes before V_{D4} rising edge, and V_{COM} is low when V_{D4} rising edge comes before V_{D3} rising edge. V_{COM} is then fed into an up-down counter to adjust DTC delay by controlling K_CODE until the error $T_{D, ERR}$ is less than 1 LSB of K_CODE. When $T_{D, ERR}$ is less than 1 LSB of K_CODE, V_{COM} toggles and K_CODE is locked within two consecutive codes. Fig. 5(b) illustrates the calibration operation in the case where $T_{D, ERR} < 0$ at the beginning. At the end of calibration, the $T_{D, ERR} <$ is less than 1 LSB of the delay.

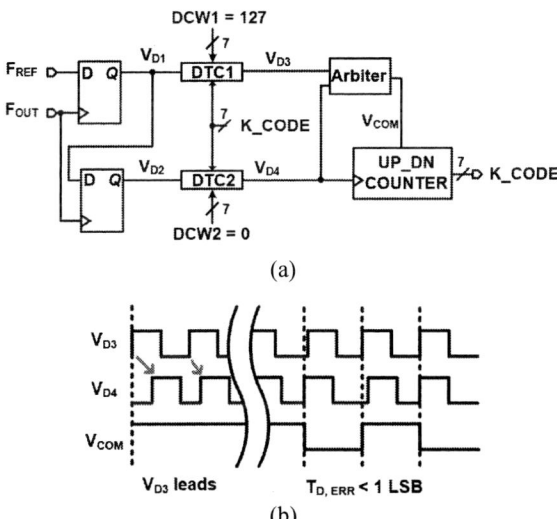

(a)

(b)

Fig. 5. (a) Block diagram of the proposed DTC gain calibration scheme. (b) Timing diagram of the proposed gain calibration.

III. Circuit Implementation

A. DTC

Fig. 6 shows the DTC design adopted in this work [5]. A 7-bit binary controlled cap bank is used for DCW control. Delay range $T_{D, RANGE}$ is designed to be 663 ps and 1 LSB delay of DCW is designed to be 5.18 ps. A 7-bit current mirror array is used for gain error calibration and 1 LSB of K_CODE is designed to be about 3 ps.

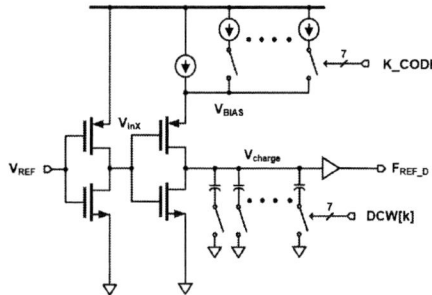

Fig. 6. DTC design

B. Sub-Sampling PLL

The main structure of the fractional-N SSPLL is shown in Fig. 2. Both the pulse width and the gm stage gain affect the overall closed-loop response. In the proposed design, the DSM noise which needs to be addressed in a conventional fractional-N PLL is no longer an issue, as the DSM is eliminated. This makes it possible to alleviate the tradeoff between DSM noise suppression and bandwidth (which affects loop filter size). The loop bandwidth is designed to be 400 kHz for optimum EMI reduction [6] with reasonable loop filter area.

C. Digital Blocks

The circuit of arbiter used in the DTC gain calibration is adopted from [7]. The accumulator used in DTC control is implemented with a 7-bit full adder with output connected to its input.

IV. Measurement Results

The proposed fractional-N SSPLL is fabricated in the TSMC 0.18-μm CMOS technology. The chip micrograph is shown in Fig. 7. The core area of the proposed design is 0.467 mm². Measured power consumption is 11.1 mW from a 1.8-V supply voltage including output buffer.

Fig. 7. Chip photo

The reference frequency is 26 MHz and is generated from the signal generator SMA-100A. Fig. 8 shows the output spectrum, with and without spectrum spreading. With spread-spectrum clocking enabled, peak spectrum power is lowered by 18.98 dB with 5000-ppm spread. As shown in Fig. 9, the triangle-modulation profile with 32-kHz modulation frequency is used. The measured phase noise with division ratio 57.692 is shown in Fig. 10, and a RMS jitter of 0.88 ps is achieved. Without performing DTC gain calibration, the SSPLL performance will degrade or even fail to lock, which can be observed also in Fig. 10.

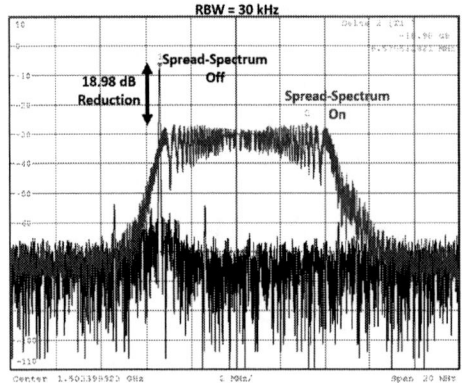

Fig. 8. Measured output spectrum with and without spread-spectrum clocking

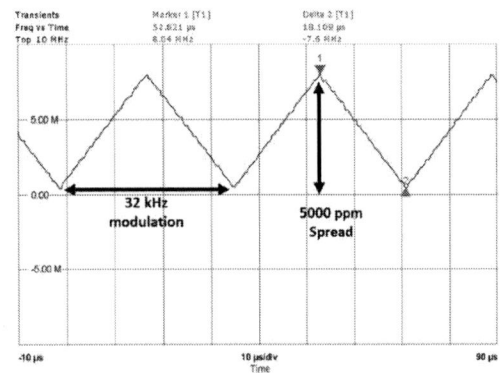

Fig. 9. Transient behavior with spread-spectrum clocking enabled.

Table I summarizes the measured chip performance and comparison with other similar works.

Table I. Performance Comparison

	[1]	[2]	[3]	[6]	This Work
Technology (nm)	65	180	180	90	180
Architecture	Direct VCO	DSM	Two-Point Mod.	Frac. N Divider	DTC SSPLL
Output Frequency (GHz)	5	1.5	1.5	6	1.5
EMI Reduction @100-kHz RBW (dB)	21*	9.8	7.52	16.12	16.37 /18.98**
RMS Jitter (ps)	1.7	3.2	5.485	0.77	0.88
Modulation Frequency (kHz)	65	30.1	31.25	32.95	32
Spread Ratio (%)	0.5	0.5	0.507	0.5	0.5
Power (mW)	7	N/A	27	27.7	11.1
Area (mm²)	0.39	1.65	0.202	0.248	0.467

*51-kHz RBW; ** 30-kHz RBW

V. CONCLUSION

This paper presents a 1.5-GHz fractional-N sub-sampling PLL with spread-spectrum clocking. With the proposed techniques, an 18.98 dB EMI reduction is achieved without complex circuit design. A DTC calibration scheme is also proposed to ensure the SSPLL can be successfully locked. Experimental results demonstrated the effectiveness of these techniques.

ACKNOWLEDGMENT

The authors thank the CIC, Taiwan, for chip fabrication; MOST, Taiwan for project support.

REFERENCES

[1] S. G. Bae; G. Kim; C. Kim, "A 5-GHz Sub-Sampling PLL based Spread-Spectrum Clock Generator by Calibrating the Frequency Deviation," in *IEEE Transactions on Circuits and Systems II: Express Briefs*, vol.PP, no.99, pp.1-1

[2] H.-R. Lee, O. Kim, G. Ahn, and D.-K. Jeong, "A low-jitter 5000 ppm spread spectrum clock generator for multi-channel SATA transceiver in 0.18 μm CMOS," in *IEEE Int. Solid-State Circuit Conf. Dig. Tech. Papers*, Feb. 2005, pp. 162–163.

[3] Y. H. Kao and Y. B. Hsieh, "A Low-Power and High-Precision Spread Spectrum Clock Generator for Serial Advanced Technology Attachment Applications Using Two-Point Modulation," in *IEEE Transactions on Electromagnetic Compatibility*, vol. 51, no. 2, pp. 245-254, May 2009.

[4] X. Gao, E. A. M. Klumperink, M. Bohsali and B. Nauta, "A Low Noise Sub-Sampling PLL in Which Divider Noise is Eliminated and PD/CP Noise is Not Multiplied by N²," in *IEEE Journal of Solid-State Circuits*, vol. 44, no. 12, pp. 3253-3263, Dec. 2009.

[5] W. S. Chang, P. C. Huang and T. C. Lee, "A Fractional-N Divider-Less Phase-Locked Loop With a Subsampling Phase Detector," in *IEEE Journal of Solid-State Circuits*, vol. 49, no. 12, pp. 2964-2975, Dec. 2014.

[6] K.-H. Cheng, C.-L. Hung, and C.-H. Chang, "A 0.77 ps RMS jitter 6-GHz spread-spectrum clock generator using a compensated phase rotating technique," in *IEEE J. Solid-State Circuits*, vol. 46, no. 5, pp. 1198–1213, May 2011.

[7] V. Gutnik and A. Chandrakasan, "On-chip picosecond time measurement," *Symposium on VLSI Circuits. Digest of Technical Papers*, Honolulu, HI, USA, 2000, pp. 52-53.

A 2.1Gbps 12-Channel Transmitter with Phase Emphasis Embedded Serializer for UHD Intra-panel Interface

Jihwan Park, Joo-Hyung Chae, Yong-Un Jeong, Jae-Whan Lee, and Suhwan Kim

Department of Electrical and Computer Engineering, Seoul National University
Seoul, Korea
E-mail : Jihwan.Park@analog.snu.ac.kr, suhwan@snu.ac.kr

Abstract— A 2.1Gbps 12-channel transmitter with phase emphasis embedded serializer for an intra-panel interface is presented. Phase emphasis is introduced into the final 2:1 stage of a 20:1 serializer to reduce the data-dependent jitter without increasing IO capacitance, by making the transition timing depend on previous data. This is combined with LVDS channel drivers which control the common-mode voltage and swing of the output signal over a wide range to match the condition of each channel. Using both phase and amplitude emphasis, the transmitter can compensate for channel losses exceeding 10dB, and eye jitter is reduced by 38%: phase emphasis is responsible for about half of this reduction. Fabricated in 28nm CMOS, the transmitter occupies 1.35mm². The proposed transmitter is verified with 55-inch UHD (3840x2160) resolution TFT-LCD intra-panel interface.

Keywords— *transmitter; phase emphasis; data-dependent-jitter; UHD resolution; intra-panel interface*

I. INTRODUCTION

As the Thin-film transistor liquid crystal displays (TFT-LCDs) have become popular and large-screen TVs using the high resolution of full-HD (FHD) have become very successful in the market, the trend of the TV market is moving toward a higher resolution called ultra-HD (UHD) resolution. Since UHD TV has to support at least 3840 x 2160 resolution, 10-bit large color depth, and 120Hz frame rate, the data interface is a bottleneck in such a high-end LCD system.

Fig. 1.(a) shows the intra-panel interface between the timing controller (TCON) and the source driver (SD) of a UHD TV. The channels consist of long PCBs and flexible flat cables (FFCs), and their lengths are not uniform. This gives the designers two challenges in achieving signal integrity: First, since signal loss can vary between the worst channel and the best channel by more than 10dB at frequency over 1.05GHz, as shown in Fig. 1.(b), the characteristics of the output signal such as the common-mode level, swing, and emphasis levels, must be independently adjustable. Second, an equalization scheme is needed to compensate for the degradation in signal integrity by the worst channel with at least a 10 dB loss and many discontinuities.

The second problem is more critical in a UHD intra-panel interface. A feed-forward equalizer (FFE) which is used as an amplitude emphasis scheme [1][2] is commonly used in a transmitter, and this is inefficient to compensate for an insertion loss of more than 10dB. Increasing the amplitude emphasis level severely causes problems such as a reduction in the signal swing and data distortion due to the fluctuation of the common-mode level at signal transition. This issue has motivated the development of ways of introducing variable tap spacing to the FFE [3] and applying phase emphasis to the quadrature-rate output driver [4] with the aim of attaining the necessary compensation despite low amplitude emphasis. These schemes reduce the amount of total jitter for which the amplitude emphasis has to compensate by decreasing the data-dependent jitter (DDJ), which is a major jitter component. However, a lot of current is consumed by the CML delay cell required to control variable tap spacing. With the latter scheme, increasing the number of input MOS to the output driver makes IO capacitance increase, which limits bandwidth. Also, the skew difference between the quadrature-clocks tends to degrade the signal quality with this scheme.

We address these problems by introducing a 12-channel transmitter with both phase emphasis and amplitude emphasis. Phase emphasis is implemented by modifying the 2:1 serializer, so there is no need to make the output driver more complicated, or increase its IO capacitance. In addition, the serializer can easily delay data through its synchronous logic to detect data transitions, so that is better able to deal with timing constraints and reduce current consumption. In addition, this transmitter is able to control the characteristics of the output signal over a sufficiently wide to maintain signal integrity across different channel environments.

II. TRANSMITTER ARCHITECTURE

As shown in Fig. 2, our 12-channel transmitter consists of 12 data channels, a SSCG_PLL, a band gap reference (BGR). The TCON link delivers 20-bit parallel data and three kinds of control signals for controlling the output signal of each channel. In order to reduce the skew in the output signal between adjacent channels, channels are configured in pairs, sharing bias circuits such as the resistor DAC (RDAC) and the bias generator.

S17-3 (2191)

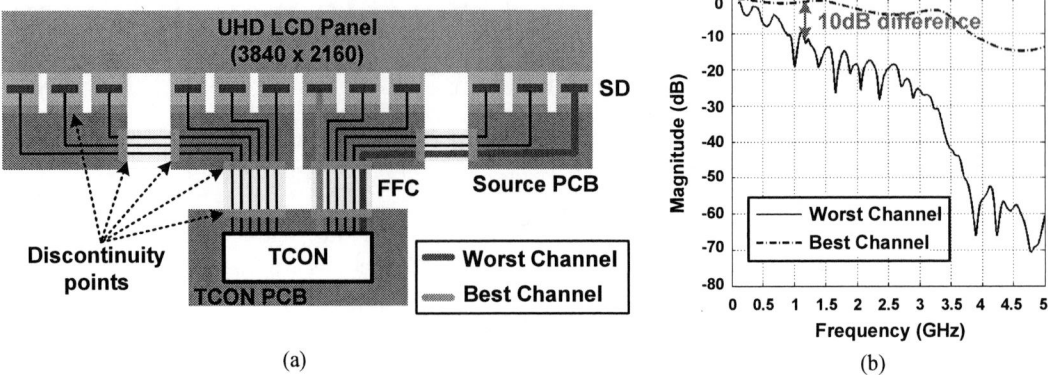

(a)

(b)

Fig. 1. (a) The intra-panel interface configuration, and (b) the insertion loss (S_{13}) of worst/best channel.

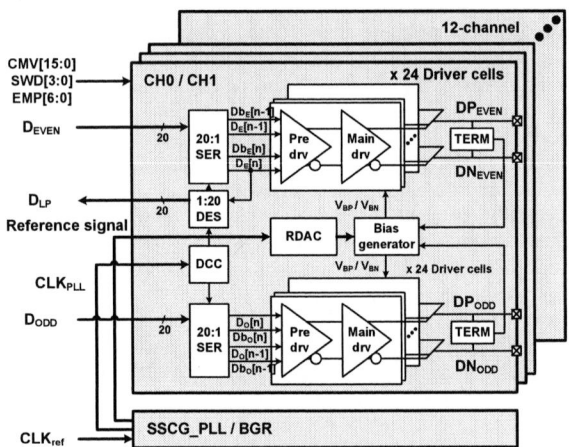

Fig. 2. Overall architecture of the proposed 12 channel transmitter.

A 20:1 serializer on each channel converts 20-bit parallel 1-bit serial data. Since the transmitter uses a low-voltage differential-signaling (LVDS) driver with 1-tap amplitude emphasis, the serializer outputs the current bit (D[n], Db[n]) and the previous bit (D[n-1], Db[n-1]). The driver is composed of 24 identical cells, each of which consists of a main driver sub-cell and pre-driver sub-cells, and each cell outputs the data by selecting either the current bit or the previous bit, depending on the emphasis command (EMP[7:0]). There is also a loopback path through the 1:20 deserializer to check the result of serialization.

III. CIRCUIT IMPLEMENTATION

A. LVDS driver with wide tuning range

The driver that controls the common-mode voltage (V_{CM}) and swing (V_{SW}) is shown in Fig. 3. The driver receives the register values (CMV[15:0], SWD[3:0]) through the TCON link. To control the common mode voltage, the differential output of driver is divided by resistors, and the central value is an input to the common-mode feedback (CMFB) control-loop. The resistor DAC generates 16 reference voltages in a range

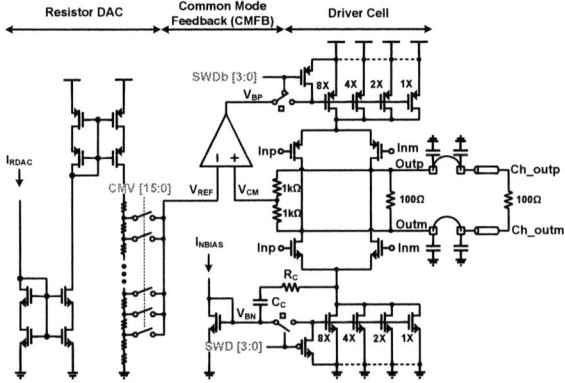

Fig. 3. LVDS driver with wide common mode and swing level controllability

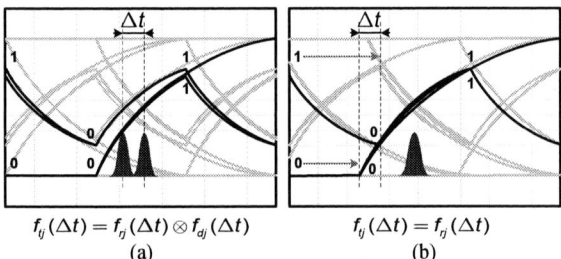

$$f_{tj}(\Delta t) = f_{rj}(\Delta t) \otimes f_{dj}(\Delta t)$$

(a)

$$f_{tj}(\Delta t) = f_{rj}(\Delta t)$$

(b)

Fig. 4. (a) Eye diagram and jitter histogram at receiver front-end with same sampling timing, and (b) those with different sampling timing for phase emphasis

from 300mV to 600mV, selects the target voltage (V_{REF}) determined by CMV[15:0], and sends it to the CMFB control-loop.

The swing control register SWD[3:0] controls the number of NMOS cells which are turned on to mirror the reference current. In order to reduce the effect of swing control on the common-mode voltage, the same number of PMOS and NMOS cells are always turned on. This allows the driver to provide an output swing range of 600mV while maintaining an acceptable common-mode voltage.

978-1-5386-3179-9/17 $31.00 © 2017 IEEE   258

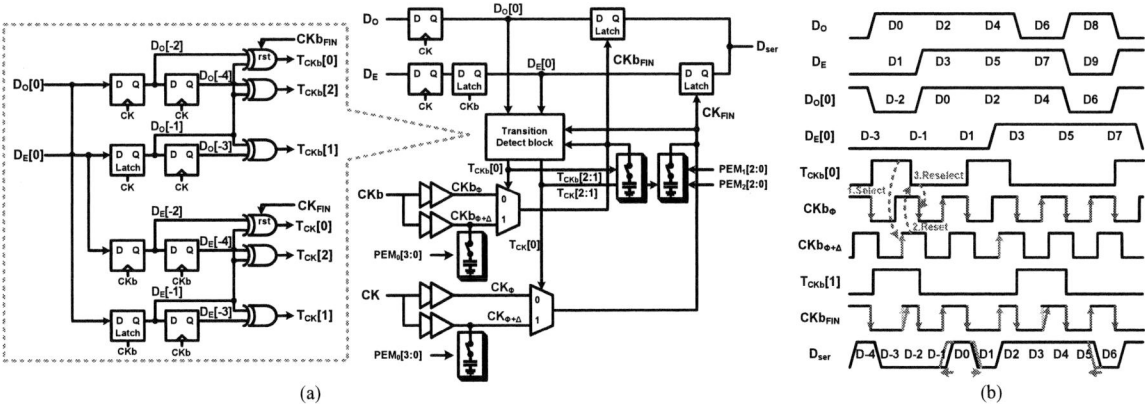

(a)   (b)

Fig. 5. (a) The proposed 2:1 serializer with phase emphasis technique and the transition detect block, and (b) the timing diagram of proposed 2:1 serializer.

B. Phase emphasis embedded 20:1 serializer

The 20:1 serializer consists of a 5:1 serializer cascaded with two 2:1 serializers. The 5:1 serializer multiplexes 5-bit input data based on a selection signal, and the 2:1 serializer has 5 parallel latches. The phase emphasis scheme is applied to the second 2:1 serializer, that sends its output to the driver.

Without phase emphasis, the eye opening of the received signal is narrowed by jitter, as shown in Fig. 4.(a). we can analyze this issue using the total jitter probability distribution function (PDF) which is the convolution of the PDF of random jitter (RJ) and deterministic jitter (DJ) [5]. Unlike RJ, which is modeled as a Gaussian distribution, the distribution of DJ can be modeled as two impulse functions [6], and then characterized by the distance between the impulses. Data-dependent jitter which is a kind of DJ can be modeled as two impulse functions. Also, since DDJ is the inter-symbol interference (ISI) applied to the next bit by the long tail of the single bit response, its value depends on previous data. Therefore, the phase emphasis scheme is designed so that the starting time of the transmission signal varies with the previous data, as shown in Fig. 4.(b). When the data is equalized by phase emphasis, the rising curves of all data pattern overlap at the decision threshold voltage of the receiver.

The second 2:1 serializer is modified, as shown in Fig. 5.(a), to provide phase emphasis by varying the phase of the sampling clock depending on whether there is a transition between the current bit (D[0]) and previous bits (D[-2], D[-3], D[-4]). The output signal of the transition detect block ($T_{CKb}[2:0]$) adjusts the delay of the sampling clock using three banks of capacitors.

As shown in Fig. 5.(b), there are two ways to adjust the delay depending on the required resolution. The first tap, which has a large phase resolution (up to 40ps), is implemented by selecting CKb_Φ or the delayed clock $CKb_{\Phi+\Delta}$; but the 2nd and 3rd taps, which produce much smaller phase resolutions (up to 16ps), are implemented by changing the value of the capacitor bank at the sampling clock node. If all three taps were implemented by switching signals, there would be a need for eight multiphase clocks, leading to an

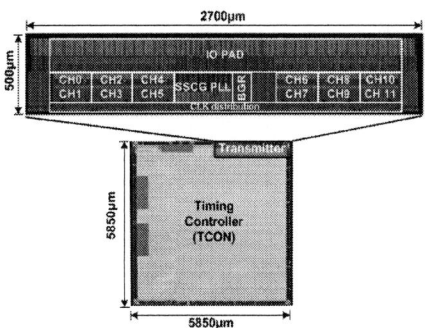

Fig. 6. The microphotograph of TCON including enlarged layout of 12-channel transmitter.

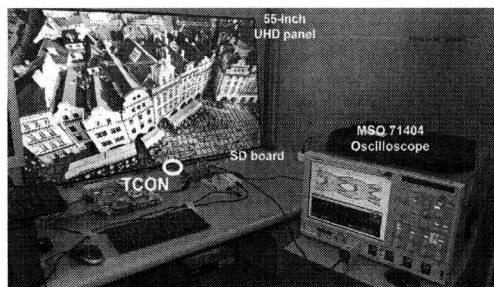

Fig. 7. 55-inch UHD TFT-LCD panel prototype to verify the proposed TCON transmitter of intra-panel interface.

overcomplicated design. Conversely, if all three taps were implemented by changing the load of sampling clock node, meta-stability in driving data might occur because of the large load on the sampling clock node. Our approach also synchronizes the falling edge of the sampling clock to the falling edge of the early-phase clock to prevent both paths from driving the output while the sampling clocks overlap.

IV. EXPERIMENTAL RESULTS

A timing controller, including our 12-channel transmitter, was implemented in a 28nm CMOS process, and its microphotograph is shown in Fig. 6. The total area of the transmitter is 1.35mm². Fig. 7 shows a 55-inch UHD

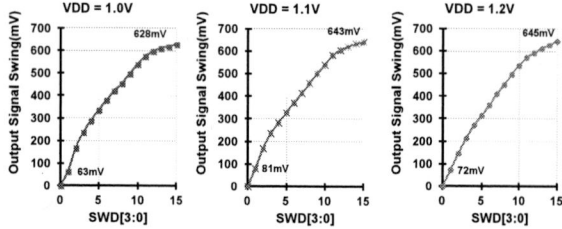

Fig. 8. Measured single-ended output swing level with supply variation.

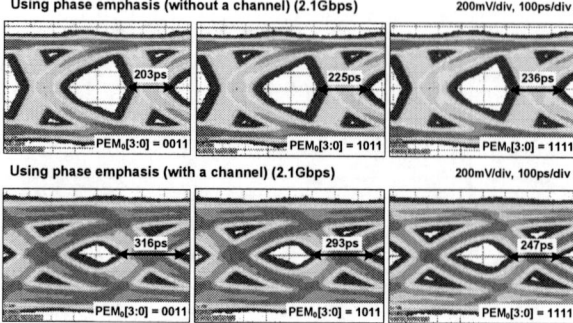

Fig. 9. Measured 2.1Gbps eye diagrams using only phase emphasis.

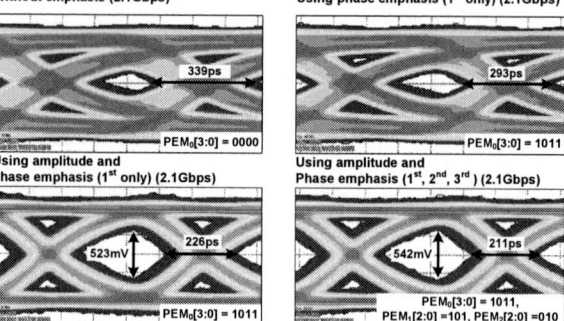

Fig. 10. Measured 2.1Gbps eye diagrams at the channel output with different emphasis condition.

resolution TFT-LCD panel prototype system with an intra-panel interface that supports a 10-bit color depth and a frame-rate of 120Hz. The measurements shown in Fig. 8 indicates that the swing of the transmitted signal is widely controlled up to 628mV regardless of the supply voltage variation.

Fig. 9 shows the measured eye diagrams with and without a connected channel while adjusting only the 1st tap coefficient of the phase emphasis. When there is no channel, the jitter increases with the coefficient, as expected; but with a channel the jitter is reduced by 22% from 316ps to 247ps. Fig. 10 shows how much the eye opening is improved by phase and amplitude emphasis. Without emphasis, the jitter is 339ps but it decreases by 38% to 211ps when equalized by emphasis.

TABLE I. PERFORMANCE SUMMARY AND COMPARISON WITH PREVIOUS DESIGNS

	This work	[1]	[2]	[4]
Technology	28nm	65nm	130nm	90nm
Supply	1.0	1.2	1.2	1.0
Data rate [Gb/s]	2.1	20	1.62	6
Cable loss [dB]	12	12.78	12.3	10
Equalization type	FFE[a] PE[b]	FFE[a]	FFE[a]	PE[b]
Swing level (V_{pk-pk}) [V]	1.2	0.3	0.2	0.6
Transmitter power [mW]	12.7	58.8	32.4	24
FoM[c] [pJ/bit·V]	5.05	9.8	100	6.67

[a]. FFE =feed forward equalizer (amplitude emphasis)

[b]. PE = phase emphasis

[c]. Figure of merit (FoM) = Transmitter Power / (Data rate x Swing level)

V. CONCLUSION

We have presented a 2.1Gbps 12-channel transmitter for a UHD intra-panel interface, fabricated in 28nm CMOS technology. Common-mode, swing, and emphasis levels can be controlled independently so as to match different channel conditions. Both phase and with amplitude emphasis are used to compensate for channel losses of 12dB. The performance of this proposed transmitter is summarized and compared with previous designs in Table I. When a 2.1Gbps signal with PRBS of 2^7-1 passes through a PCB trace with a 12dB loss, the measured eye jitter is reduced by 38% by applying both

emphasis techniques. The transmitter consumes 12.7mW at 2.1Gb/s, giving a FoM of 5.05pJ/bit·V.

ACKNOWLEDGEMENT

The authors would like to thank ABOV Semiconductor and DAVINCHIPS for technical discussions.

REFERENCES

[1] S.-Y. Kao and S.-I. Liu, "A 20-Gb/s transmitter with adaptive preemphasis in 65-nm CMOS technology," IEEE Trans. Circuits Syst. II, Exp. Briefs, vol. 57, no. 5, pp. 319–323, May 2010.

[2] S.-Y. Kao and S.-I. Liu, "A 1.62/2.7 Gbps adaptive transmitter with 2-tap pre-emphasis using a propagation-time detector," IEEE Trans. Circuits Syst. II, Exp. Briefs, vol. 57, no. 3, pp. 178–182, Mar. 2010.

[3] M. Bichan and A. C. Carusone, "A 6.5 Gb/s backplane transmitter with 6-tap FIR equalizer and variable tap spacing," 2008 IEEE Custom Integrated Circuits Conference, San Jose, CA, 2008, pp. 611-614.

[4] J. F. Buckwalter, M. Meghelli, D. J. Friedman and A. Hajimiri, "Phase and amplitude pre-emphasis techniques for low-power serial links," IEEE Journal of Solid-State Circuits, vol. 41, no. 6, pp. 1391-1399, June 2006.

[5] B. Analui, J. F. Buckwalter and A. Hajimiri, "Data-dependent jitter in serial communications," IEEE Transactions on Microwave Theory and Techniques, vol. 53, no. 11, pp. 3388-3397, Nov. 2005.

[6] Y. Cai, S. A. Werner, G. J. Zhang, M. J. Olsen, and R. D. Brink, "Jitter testing for multi-gigabit backplane Serdes-techniques to decompose and combine various types of jitter," Proc.IEEE Int.Test Conf.,Oct.2002, pp. 700–709.

A Low-Power Dual-Mode 20-Gb/s NRZ and 28-Gb/s PAM-4 Voltage-Mode Transmitter

Hae-Woong Yang[1], Ashkan Roshan-Zamir[1], Young-Hoon Song[2], and Samuel Palermo[1]

[1]Department of Electrical and Computer Engineering, Texas A&M University, College Station, TX, USA
[2]NXP Semiconductor, Chandler, AZ, USA
E-mail: {hwyang, spalermo}@tamu.edu

Abstract—A dual-mode NRZ/PAM-4 differential low-swing voltage-mode transmitter employs a quarter-rate output multiplexing architecture for low-power operation. In NRZ mode, 2-tap feedforward equalization is realized with analog replica-bias tap control that is configurable in high-performance controlled-impedance or energy-efficient impedance-modulated settings. This analog control also allows for efficient generation of the middle levels in PAM-4 operation. Fabricated in GP 65nm CMOS, the transmitter supports an output swing range of 100-400mV$_{ppd}$ with up to 12dB of equalization in NRZ mode and achieves energy efficiencies of 1.48 and 0.91pJ/b at 20-Gb/s NRZ and 28-Gb/s PAM-4 data rates. Operation in the NRZ energy-efficient impedance-modulation setting allows for power savings of up to 32% relative to the controlled-impedance setting.

Index Terms—Dual-mode transmitter, feed-forward equalizer, high-speed I/O, low-power, NRZ, PAM-4, voltage-mode driver.

I. INTRODUCTION

Emerging PAM-4 standards are under development to support serial link system requirements to increase data rates and bandwidth density. Transceiver components which can support dual-mode NRZ and PAM-4 modulation operation provide the key advantages of supporting multiple standards and lowering I/O development time [1], [2].

Low-swing voltage-mode (VM) transmitters provide the potential for very low-power operation over moderate-loss short-reach channels [3]–[6]. Efficient feed-forward equalization (FFE) tap control is necessary in the output stages, with analog control [3] generally providing better energy efficiency relative to drivers which employ digital output stage segmentation [5]. Further power savings are possible with impedance-modulated equalization [3], [7]. However, this approach may not be applicable for channels with non-smooth loss profiles. Also, dual-mode transmitters require efficient approaches to support PAM-4 modulation.

This paper presents a dual-mode NRZ/PAM-4 low-swing voltage-mode transmitter which efficiently supports both modulations and is configurable to operate in either a controlled-impedance equalization or an impedance-modulated setting. Section II describes the transmitter architecture which utilizes a scalable-supply for the serializer and pre-driver logic to save power at lower data rates. The output-multiplexing voltage-mode driver is detailed in Section III. Section IV discusses the replica-bias impedance control loops which set the FFE tap values and the PAM-4 levels. Experimental results from a GP 65nm CMOS prototype are presented in Section V. Finally, Section VI concludes the paper.

II. SYSTEM ARCHITECTURE

Fig. 1 shows the proposed dual-mode NRZ/PAM-4 two-channel transmitter architecture. The two 1/4-rate TX channels share a global voltage regulator and analog impedance control loops (AICLs) which set the NRZ equalization taps and PAM-4 levels. Depending on the channel condition, NRZ operation is configured in either a high-performance (HP) controlled-impedance equalization setting or an energy-efficient (EE) impedance-modulated setting. Due to the sensitivity to residual ISI, the transmitter always operates in the controlled-impedance HP setting for PAM-4 mode.

The serialization clocks are generated from a globally-distributed differential 1/2-rate clock that is divided by two to generate 1/4-rate quadrature clocks that are then followed by CMOS buffers with duty-cycle correction (DCC) and quadrature error correction (QEC) [3]. Sixteen bits of parallel $2^{15} - 1$ pseudo-random bit sequence (PRBS) input data are serialized by the initial 16:8 multiplexer followed by a path-selector which directs the 8b data to either the NRZ or PAM-4 data path. For NRZ operation, the 8b data passes through an additional latch-based 8:4 multiplexer stage. Four parallel slices that include retiming flip-flops, encoder logic, and level-shifting predrivers supply the 4:1 output multiplexing voltage-mode driver. The retiming flip-flops generate the necessary 1-UI delay for the 2-tap FFE in NRZ mode, while in PAM-4 mode the parallel 8b input is retimed into two 4b MSB and LSB paths. Sampling of the 4-UI encoder logic output is performed with time-interleaved pulse clocks in the level-shifting predriver stage that allows for smaller output stage transistors [8]. The entire serialization chain operates on a scalable supply that allows for reduced power consumption at lower data rates.

III. 4:1 OUTPUT-MULTIPLEXING TRANSMIT DRIVER

In order to allow the serialization and predriver logic to operate at the minimum supply voltage, the final 4:1 serialization is performed in the output multiplexing driver of Fig. 2. This driver consists of three parallel segments. Each segment has single top/bottom analog-controlled transistors, which control the output impedance and either set the de-emphasis level in NRZ mode or the middle symbol levels in PAM-4 mode, and four parallel middle digitally-switched transistors that perform the output multiplexing.

Fig. 2(a) shows the driver configured in the HP controlled-impedance setting, which is for NRZ mode with channels that

S17-4 (2099)

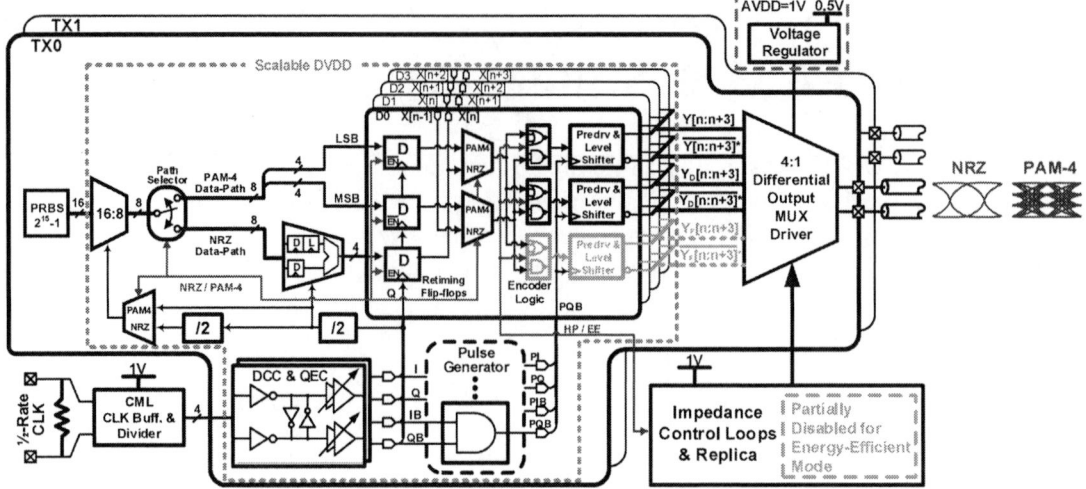

Figure 1.   Two-channel dual-mode NRZ/PAM-4 transmitter architecture.

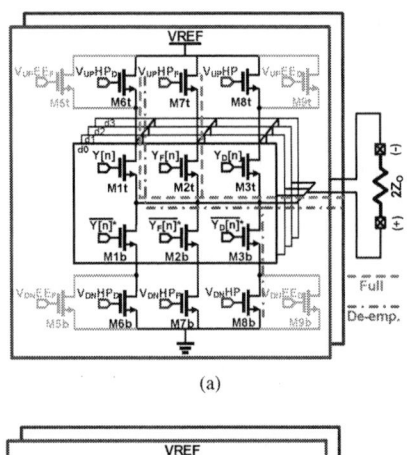

(a)

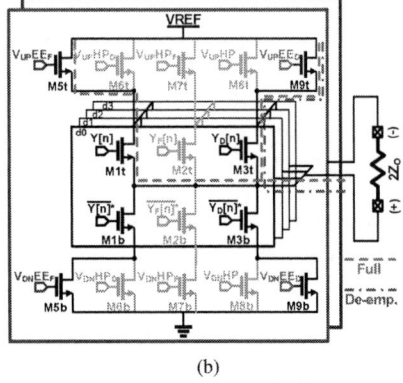

(b)

Figure 2.   Dual-mode NRZ/PAM-4 low-swing differential VM transmit driver configured in (a) controlled-impedance high-performance (HP) setting and (b) impedance-modulated energy-efficient (EE) setting for NRZ only.

have significant reflections and always for PAM-4 mode due to the sensitivity to residual ISI [3]. In HP mode, all three parallel segments are activated. For the cases of a transition NRZ bit and the maximum differential PAM-4 symbol levels, the maximum output swing is generated through the parallel activation of the M1/M6 and M2/M7 paths. The AICLs set

the $V_{XX}HP_D$ and $V_{XX}HP_F$ voltages such that

$$(R_{M1} + R_{M6})||(R_{M2} + R_{M7}) = \frac{Z_O}{1-\alpha}||\frac{Z_O}{\alpha} = Z_O, \quad (1)$$

where Z_0 and α are the characteristic channel impedance and the peaking ratio between full and de-emphasized output voltage swings, respectively. When there is an NRZ run-length greater than one, the de-emphasized level is set by de-activating the M2/M7 path and activating the opposite polarity M3/M8 path to create an output voltage division. This is also how the PAM-4 middle symbol levels are generated. The AICLs ensure a constant controlled impedance by setting the $V_{XX}HP$ voltages such that

$$(R_{M1} + R_{M6})||(R_{M3} + R_{M8}) = \frac{Z_O}{1-\alpha}||\frac{Z_O}{\alpha} = Z_O. \quad (2)$$

For NRZ mode with smooth loss-profile channels where reflections from propagating signal can be tolerated to some extent without impairing signal integrity, the EE impedance-modulated setting shown in Fig. 2(b) is used [7]. In EE mode, only two parallel segments are activated. For the case of a transition NRZ bit, the maximum output swing is generated through the parallel activation of the low-impedance M1/M5 and high-impedance M3/M9 paths. The AICLs set the $V_{XX}EE_F$ and $V_{XX}EE_D$ voltages such that

$$(R_{M1} + R_{M5})||(R_{M3} + R_{M9})$$
$$= \left(\frac{1}{2} + \frac{1}{4\alpha}\right)Z_O||\frac{1+2\alpha}{1-2\alpha}Z_O = Z_O. \quad (3)$$

When there is an NRZ run-length greater than one, the de-emphasized level is set by de-activating the low-impedance M1/M5 path, such that only the high-impedance M3/M9 path is on. In this case, the output impedance becomes

$$R_{M3} + R_{M9} = \frac{1+2\alpha}{1-2\alpha}Z_O. \quad (4)$$

During maximum differential output signal levels, the current drawn from the voltage regulator in both HP and EE modes is

$$I_{X[n+1]\neq X[n]} = \frac{V_{REF}}{4Z_O}. \quad (5)$$

978-1-5386-3179-9/17 $31.00 © 2017 IEEE   262

Table I
PRE-DRIVER LOGIC FOR THE DUAL-MODE NRZ/PAM-4 DRIVER

NRZ	$X[n]$	$X[n-1]$	$Y[n]$	$Y_D[n]$	$Y_F[n]$	$Y[n]^*$	$Y_D[n]^*$	$Y_F[n]^*$
HP	0	0	0	1	0	1	0	0
	1	1	1	0	0	0	1	0
	1	0	1	0	1	0	0	0
	0	1	1	0	0	1	0	1
EE	0	0	0	0	0	0	1	0
	1	1	0	1	0	0	0	0
	1	0	1	1	0	0	0	0
	0	1	0	0	0	1	1	0

PAM-4	LSB	MSB	$Y[n]$	$Y_D[n]$	$Y_F[n]$	$Y[n]^*$	$Y_D[n]^*$	$Y_F[n]^*$
	0	0	0	0	0	1	0	1
	0	1	0	1	0	1	0	0
	1	0	1	0	1	0	0	0
	1	1	0	0	0	1	0	0

☺ Dynamic power is greatly reduced on consecutive 0/1s

Depending on the driver configuration, the current during a de-emphasized NRZ or a middle PAM-4 level is

$$
I_{X[n+1]=X[n]} = \begin{cases} \frac{V_{REF}}{4Z_O}[1 + 4\alpha(1-\alpha)] & \text{HP setting,} \\ \frac{V_{REF}}{4Z_O}(1-2\alpha) & \text{EE setting.} \end{cases} \quad (6)
$$

Improved signaling efficiency in the EE setting is achieved by linearly scaling current with the de-emphasis level, whereas in the HP setting larger current is consumed for a higher peaking ratio to ensure proper channel termination [9]. Furthermore, dynamic power is significantly reduced in the EE mode by disabling a portion of the pre-driver logic. Table I summarizes the pre-driver logic table.

IV. IMPEDANCE CONTROL LOOPS

The replica-bias impedance control loops shown in Fig. 3 produce the analog control voltages that set the output driver impedance and either the de-emphasis level in NRZ mode or the middle symbol levels in PAM-4 mode. This background analog control compensates for process variations, tracks environmental changes, and allows for high-resolution equalization control [3].

In the controlled-impedance HP setting, all three impedance control loops are activated. For an NRZ de-emphasis or PAM-4 middle symbol level, the Fig. 3(a) loop sets the $V_{XX}HP$ voltages to control the opposite polarity segment strength and the Fig. 3(b) loop sets the $V_{XX}HP_D$ voltages to control the primary segment strength with only error amplifiers B and C activated. The Fig. 3(c) loop sets the $V_{XX}HP_F$ voltages to ensure proper impedance during the maximum differential output levels.

Only the Fig. 3(b) and Fig. 3(c) control loops are activated in the impedance-modulated EE setting. For an NRZ de-emphasis level, the Fig. 3(b) loop sets the $V_{XX}EE_D$ voltages to control the high-impedance path strength with only error amplifiers A and D activated. The Fig. 3(c) loop sets the $V_{XX}EE_F$ voltages to ensure proper impedance with both high- and low-impedance paths in parallel during the maximum differential output levels.

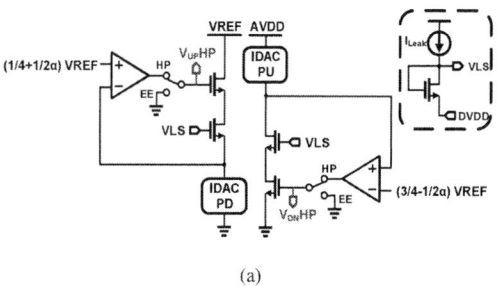

(a)

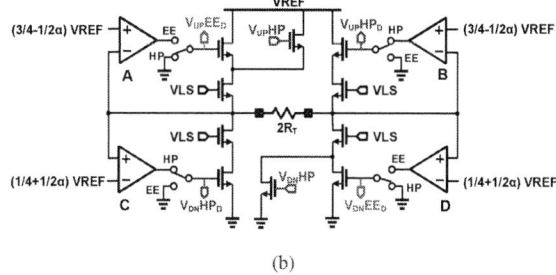

(b)

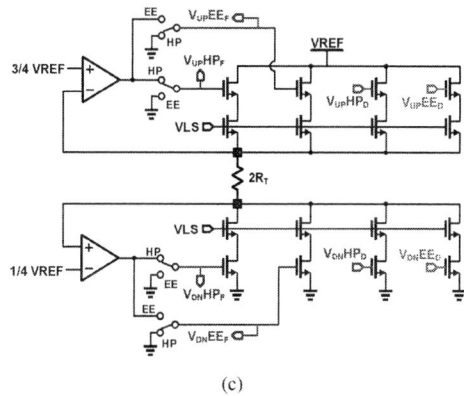

(c)

Figure 3. Global replica-bias loops for output driver impedance and level control: (a) de-emphasis and PAM-4 opposite polarity path control in HP setting (disabled in EE setting), (b) de-emphasis primary polarity path control, (c) full-swing impedance control.

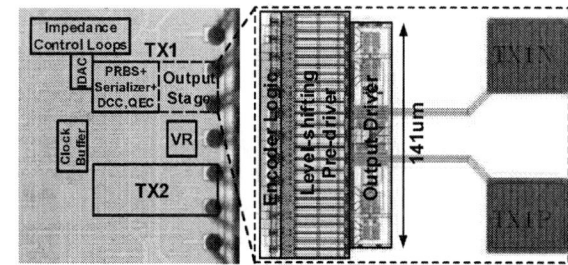

Figure 4. Chip micrograph of the 2-channel transmitter with output stage layout details.

V. EXPERIMENTAL RESULTS

Fig. 4 shows the chip micrograph of the dual-mode transmitter, which was fabricated in a GP 65nm CMOS process. The transmitter core occupies 0.029mm^2, while the global impedance control loops take 0.031mm^2.

$2^{15} - 1$ PRBS eye diagrams with the 2-tap FFE activated

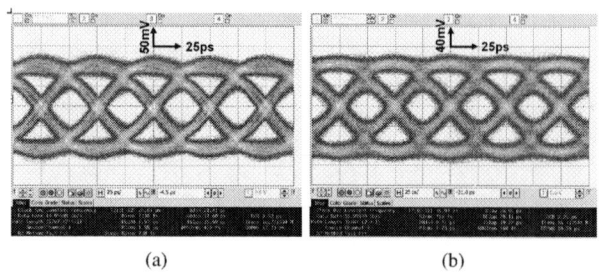

| | (a) | | (b) |

Figure 5. 20-Gb/s NRZ eye diagrams after a 16" RO4350B channel in (a) HP and (b) EE settings.

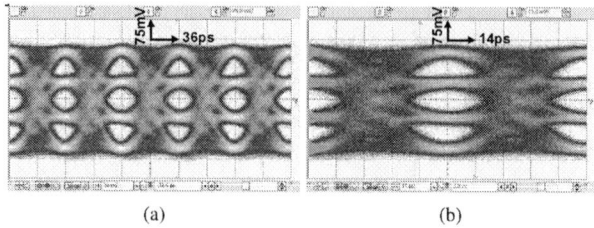

| | (a) | | (b) |

Figure 6. 28Gb/s PAM-4 eye diagrams after a 1.3" RO4350B channel. (a) Phase calibration enabling uniform eyes. (b) Histogram measurements of the PAM-4 levels.

at the maximum 20-Gb/s NRZ rate are shown in Fig. 5 for operation over a 16" RO4350B channel in both HP and EE settings. For the characterization of energy the performance constraints are given as a minimum $50mV_{dpp}$ eye height and a half UI eye width. While both settings have healthy eye openings, the controlled-impedance HP setting has larger margins. The HP setting has DJ=21.41 ps and RJ=330 fs, while the EE setting has DJ=26.06 ps and RJ=780 fs.

PAM-4 operation at the maximum 28-Gb/s rate over a 1.3" RO4350B channel is shown in Fig. 6. The DCC and QEC circuitry allows for uniform eyes out of the 1/4-rate transmitter. Vertical histogram measurements yield 94.7% level separation mismatch ratio (RLM), which validate the analog-control approach to set the PAM-4 levels.

Table II shows the power breakdown for the 20-Gb/s NRZ and 28-Gb/s PAM-4 cases, and 8-Gb/s operation over a 28" channel. The energy difference between the HP and EE settings is maximum at 8Gb/s, with a 32% savings in the EE setting. Sub-pJ/b efficiency is achieved at 28Gb/s PAM-4 operation due to the lower symbol period allowing for lower frequency clocks and a scalable supply of 0.91V. Table III compares this work with other recent low-swing VM transmitters. The proposed transmitter achieves the highest swing and compensates for the most channel loss in NRZ mode, while being the only transmitter to support PAM-4 modulation.

VI. CONCLUSION

This paper presented a low-power dual-mode NRZ/PAM-4 differential low-swing voltage-mode transmitter. The output multiplexing architecture allows for low-power operation at up to 20-Gb/s NRZ and 28-Gb/s PAM-4 data rates. Analog replica-bias control allows for both efficient NRZ equalization

Table II
TRANSMITTER POWER BREAKDOWN

	NRZ (8Gb/s, 28")		NRZ (20Gb/s, 16")		PAM-4 (28Gb/s, 1.3")
Global Clocking Buffer [mW] (amortized across 2 TX)	1.5		1.4		1.4
Serializer, Encoder, Clock [mW]	3.2 (DVDD=0.7V)		12.5 (DVDD=1.05V)		7.0 (DVDD=910mV)
Global Impedance Control, IDAC & Voltage Regulator [mW] (amortized across 2 TX)	EE 2.2	HP 5.8	EE 2.2	HP 6.0	6.0
Pre-drivers [mW]	1.9	2.7	11.1	14.6	7.9
Transmit Driver [mW]	2.2	2.9	2.3	3.1	3.2
Total Energy Efficiency [pJ/b]	1.37	2.01	1.48	1.88	0.91

Table III
TRANSMITTER PERFORMANCE COMPARISONS

Ref.	[3]	[4]	[5]	[6]	This work	
CMOS Technology	65nm GP	65nm Digital	65nm LP	65nm GP	65nm GP	
Data-rate [Gb/s]	16	12.5	10	20	20	28
V_{OUT} [mV$_{ppd}$]	300	320^2	250	250	400	400
Modulation	NRZ	NRZ	NRZ	NRZ	NRZ	PAM-4
Equalization	2-tap	2-tap	2-tap	2-tap	2-tap	NoEQ
Channel Loss @ Nyquist [dB]	-15.5	-12.1	-13	-12	-16.8	-2.2
Power [mW]	15.3^1	6.2^3	11	26.8	29.5(EE) 37.7(HP)	25.5
Total Efficiency [pJ/bit]	0.96	0.49	1.10	1.34^4	1.48(EE) 1.88(HP)	0.91

[1] Global clock distribution is excluded.
[2,3] 6.6dB pre-emphasis is taken into account.
[4] PLL is excluded. PRBS generator running at 2.5-Gb/s is not excluded.

tap settings and PAM-4 level control. The ability to support PAM-4 modulation and the flexible 2-tap NRZ FFE which is configurable in constant controlled-impedance operation or reduced-power impedance-modulated equalization allows for operation over a wide range of channel conditions.

REFERENCES

[1] J. Kim et al., "A 16-to-40Gb/s quarter-rate NRZ/PAM4 dual-mode transmitter in 14nm CMOS," in *IEEE Int. Solid-State Circuits Conf. (ISSCC) Dig. Tech. Papers*, Feb 2015, pp. 1–3.

[2] A. Roshan-Zamir et al., "A Reconfigurable 16/32 Gb/s Dual-Mode NRZ/PAM4 SerDes in 65-nm CMOS," *IEEE J. Solid-State Circuits*. [Online]. Available: http://ieeexplore.ieee.org/document/7947110/

[3] Y. H. Song et al., "An 8-16 Gb/s, 0.65-1.05 pJ/b, Voltage-Mode Transmitter With Analog Impedance Modulation Equalization and Sub-3 ns Power-State Transitioning," *IEEE J. Solid-State Circuits*, vol. 49, no. 11, pp. 2631–2643, Nov 2014.

[4] K. Fukuda et al., "A 12.3-mw 12.5-Gb/s Complete Transceiver in 65-nm CMOS Process," *IEEE Int. Solid-State Circuits Conf. (ISSCC) Dig. Tech. Papers*, vol. 45, no. 12, pp. 2838–2849, Dec 2010.

[5] Y. Lu et al., "Design and Analysis of Energy-Efficient Reconfigurable Pre-Emphasis Voltage-Mode Transmitters," *IEEE J. Solid-State Circuits*, vol. 48, no. 8, pp. 1898–1909, Aug 2013.

[6] G. S. Jeong et al., "A 20 Gb/s 0.4 pJ/b Energy-Efficient Transmitter Driver Utilizing Constant-G_m Bias," *IEEE J. Solid-State Circuits*, vol. 51, no. 10, pp. 2312–2327, Oct 2016.

[7] R. Sredojević et al., "Digital link pre-emphasis with dynamic driver impedance modulation," in *Proc. IEEE Custon Integr. Circuits Conf. (CICC)*, Sept 2010, pp. 1–4.

[8] Y. H. Song et al., "A 0.47-0.66 pJ/bit, 4.8-8 Gb/s I/O Transceiver in 65 nm CMOS," *IEEE J. Solid-State Circuits*, vol. 48, no. 5, pp. 1276–1289, May 2013.

[9] H. Hatamkhani et al., "A 10-mW 3.6-Gbps I/O Transmitter," in *Proc. IEEE Symp. VLSI Circuits*, June 2003, pp. 97–98.

S18-1 (2117)

Subthreshold Voltage Reference With Nwell/Psub Diode Leakage Compensation for Low-Power High-Temperature Systems

Inhee Lee, Dennis Sylvester, and David Blaauw

Dept. of Electrical Engineering and Computer Science, University of Michigan, Ann Arbor, MI, USA
inhee@umich.edu

Abstract— This paper proposes a voltage reference operating up to 170 °C for low-power high-temperature systems. The proposed circuit buffers body voltages to avoid degradation of temperature coefficient from nwell/psub diode leakage. For low power overhead, it measures the diode leakage and adaptively adjusts the bias current of the buffers. This enables low power consumption at low temperature, which can allow an energy harvester to recharge a battery in the target system. Prototype chips, fabricated in a 180 nm CMOS process, show a ±3σ inaccuracy of 3.4% from 0 °C to 170 °C after single trim at 80 °C and a line sensitivity of 0.088 %/V from 1.8 V to 3.6 V. It consumes 76.3 pA at room temperature and 177 nA at 170 °C from 1.8 V.

Keywords—voltage reference, body leakage, high temperature.

I. INTRODUCTION

A miniature sensing system can benefit from high-temperature applications such as oil exploration [1]. Circuits in the system need to maintain their function and performance at > 125 °C. In addition, the circuits should consume low power due to limited battery size and capacity. For example, in advanced miniature sensing system, the battery size, battery capacity, and system standby power are 2.2 mm², 2 µAhr, and 8 nW at room temperature, respectively [2]. Moreover, to recharge a battery, total system power consumption has to be less than harvested energy from an energy harvester. For instance, harvested energy using a PV cell in a miniature sensor is less than 1 µA at 10 klux [3]. Typically, an energy source for harvesting is available before a system is deployed or after retrieved at relatively low temperature or during deployment at higher temperature.

A voltage reference is a basic building block present in almost all sensor nodes. Voltage references for high-temperature applications have been proposed using an SOI process, but they consume > 2 µA [4], [5]. Designing a voltage reference that covers high temperature with low power consumption is critical for high-temperature systems. A bandgap reference (BGR) with BJT devices is a typical solution in traditional systems [6]. It shows an excellent output regulation performance, but the power consumption is typically larger than 1 µW. Nano-watt BGRs were recently reported using a switched-capacitor circuit (2 nA) [7] or a subthreshold circuit (29 nA) [8]. However, their operation is limited at < 120 °C. For low power, a CMOS voltage reference is a preferred candidate. Using subthreshold CMOS transistors, voltage references with pW power at room temperature were reported [9]–[12]. The output level is boosted using diode-connected PMOS transistors for easier use, but they

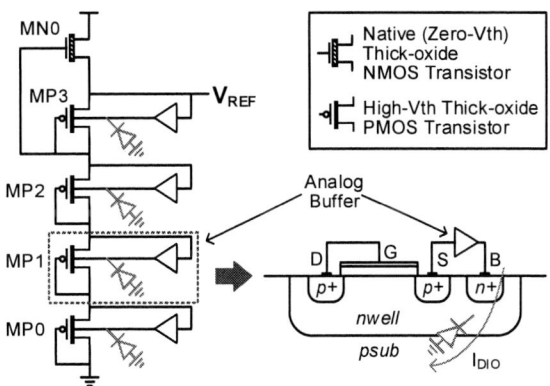

Fig. 1. Simplified diagram of the proposed voltage reference.

cannot cover > 125 °C due to significantly increased diode leakage at high temperature [10], [11].

In this paper, we propose a CMOS subthreshold voltage reference that maintains its output voltage up to 170 °C. It consumes 76.3 pA at room temperature and 177 nA at 170 °C which is well below the harvestable power in many sensor applications. This circuit isolates nwell/psub diode leakage using analog buffers. This greatly improves the temperature coefficient (TC) by avoiding voltage drop at high temperature. A key challenge is to correctly set the bias current of the buffers which needs to be high enough to compensate for the relatively high diode current at high temperature but small enough to maintain low power at room temperature. To address this, the buffers are biased by measuring the body leakage and mirroring the current to energy-efficiently isolate the body leakage. The voltage reference, fabricated in a 180 nm CMOS process, achieves a ±3σ inaccuracy of 3.4% from 0 °C to 170 °C after trimming at 80 °C.

II. PROPOSED VOLTAGE REFERENCE

Fig. 1 shows the proposed voltage reference that consists of a main voltage reference and body voltage buffers. The main voltage reference is based on the voltage reference in [11]. It includes a native NMOS transistor (MN0) and 4 high-Vth PMOS transistors (MP0-MP3). The PMOS transistors have gates diode-connected to their drains, and they are connected in series to increase the output voltage (V_REF). MN0 has a gate connected to the drain of MP3, and its source is connected to V_REF. Since Vgs of MN0 is −0.3V, it operates in the subthreshold

978-1-5386-3179-9/17 $31.00 © 2017 IEEE

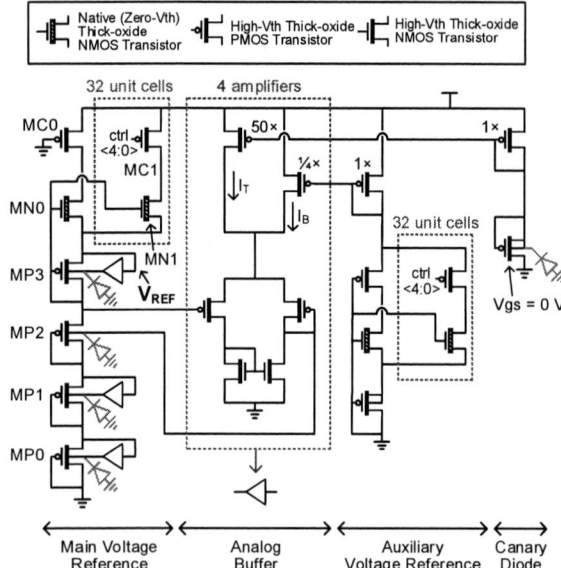

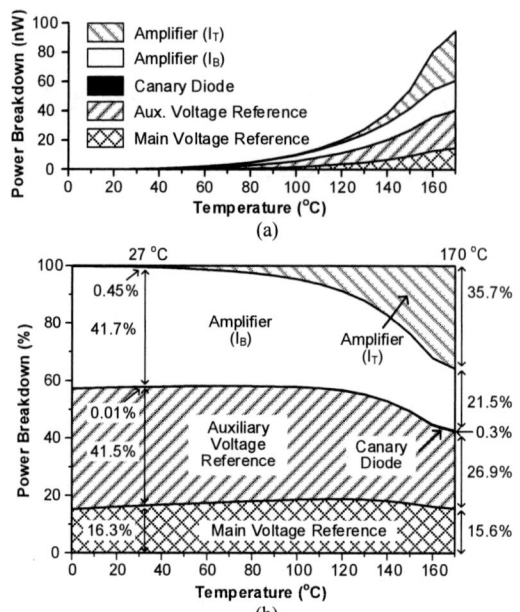

Fig. 2. Detailed circuit diagram with bias circuits.

Fig. 3. Simulated power breakdown across temperatures: (a) absolute values (b) percentage.

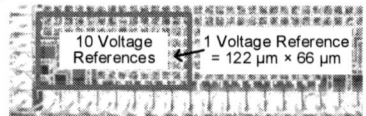

Fig. 4. Die photo.

region, and the drain current can be expressed as follows [13].

$$I_d = \mu C_{OX}\frac{W}{L}(m-1)V_T^2 e^{\left(\frac{V_{gs}-V_{th}}{mV_T}\right)}\left(1-e^{-\frac{V_{ds}}{V_T}}\right). \quad (1)$$

μ is mobility, C_{ox} is oxide capacitance, W and L are transistor size, m is subthreshold slope factor, and V_T is the thermal voltage. The factor '1-exp($-V_{ds}/V_T$)' is ignored since it is negligible for $V_{ds} > 0.2$ V (0.5% error at 170 °C). With negligible nwell/psub diode leakage (I_{DIO} in Fig. 1) at low temperature, V_{REF} can be described as [11]:

$$V_{REF} = 4\left\{\left(\frac{m_1|V_{th2}|-m_2V_{th1}}{m_1+m_2}\right) + \right.$$
$$\left.\left(\frac{m_1m_2V_T}{m_1+m_2}\right)\ln\left(\frac{\mu_1 C_{OX1}\frac{W_1}{L_1}(m_1-1)}{\mu_2 C_{OX2}\frac{W_2}{L_2}(m_2-1)}\right)\right\}. \quad (2)$$

W_1 and L_1 are the width and length of MN0. W_2 and L_2 are the MP0-MP3 size. TC can be minimized by properly sizing the transistors [9]. However, nwell/psub diode leakage (I_{DIO}) becomes noticeable at high temperature and worsen TC since current flows through each PMOS transistor Nwell to Psub and the voltage ratio among MP0-MP3 changes across temperatures. Thus, the voltage reference operates only up to 100 °C [10], [11].

Therefore, we proposed a new voltage reference using analog buffers to isolate I_{DIO}. It extends the maximum operating temperature up to 170 °C. Fig. 2 shows the circuit in detail. First, in this design, the size of nwell/psub diodes is minimized by using the minimum length of MP0-MP3 allowed in the given technology. The small device size increases V_{REF} variation from device mismatch. Thus, V_{REF} is trimmed at 80 °C using the segmented top native NMOS transistor (MN1 and MC1).

In addition, the body voltages of MP0-MP3 are buffered using amplifiers. Basic 5-transistor differential amplifiers are used as amplifiers for design simplicity and stable operation. The amplifiers need to support higher diode leakage at higher

temperature. To support the leakage, the amplifiers require high tail current at high temperature. To cover 170 °C, the amplifier needs a tail current of 7.9 nA in simulation. If we naively set the tail current of each amplifier to 7.9 nA for the entire temperature range (0 −170 °C), it will dominate the total power consumption at low temperature. For example, the main voltage reference itself consumes 23.7 pA at room temperature (simulation). Four buffers would consume 31.6 nA, and their current consumption would be three orders higher than the main voltage reference itself. This is even higher than the total power in the advanced sensor system [2].

To avoid this unnecessary power overhead, the proposed voltage reference uses an adaptive biasing scheme. It measures the leakage current using a canary PMOS transistor and adds current multiplied by 50 (I_T) to the tail current as shown in Fig. 2. This current adaptively covers increased diode leakage at high temperature and maintains TC.

By using only this bias current, the tail current is extremely small at low temperature. For instance, it is only 320 fA at room temperature. However, if the bias current is too small, the amplifier can malfunction since it can be easily affected by currents ignored and not modeled accurately in the SPICE model (e.g., gate leakage). Thus, a base current is added to the tail current. As shown in Fig. 2, the base current (I_B) is generated from a separate auxiliary voltage reference with only one bottom PMOS transistor. I_B is a quarter of the current consumption of the auxiliary one. Compared with the main voltage reference, the auxiliary reference has a similar structure and thus a similar trend of current consumption across temperatures. However, it

978-1-5386-3179-9/17 $31.00 © 2017 IEEE

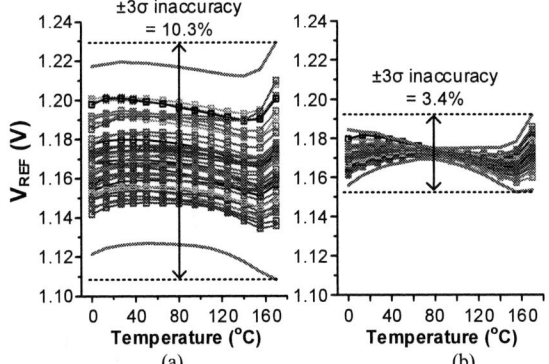

Fig. 5. Measured V_{REF} distribution across temperatures from 40 voltage references: (a) Without trimming (b) After trimming at 80 °C.

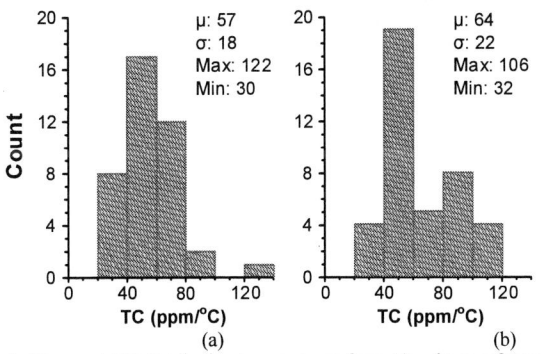

Fig. 6. Measured TC distribution temperatures from 40 voltage references: (a) Without trimming (b) After trimming at 80 °C.

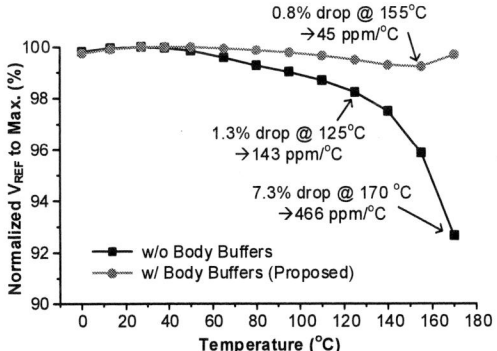

Fig. 7. Measured V_{REF} from voltage references without and with body buffers.

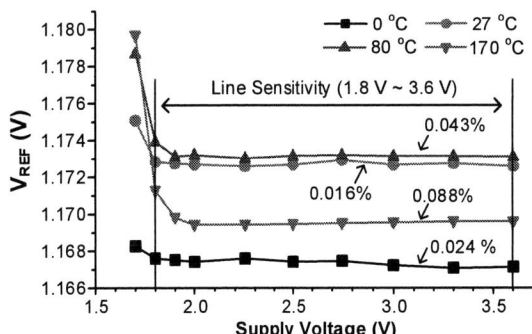

Fig. 8. Measured V_{REF} across supply voltages.

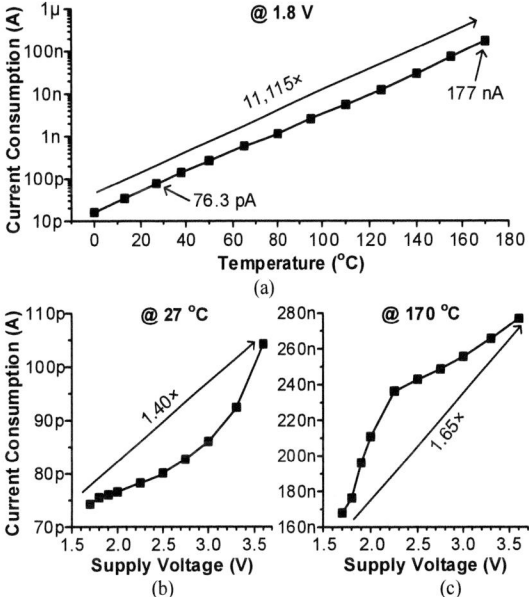

Fig. 9. Measured current consumption: (a) Across temperatures at 1.8 V. (b) Across supply voltages at room temperature. (c) Across supply voltages at 170 °C.

has more headroom due to fewer number of stacked PMOS transistors. Due to the 4:1 current copy ratio, 4 amplifiers consume as much as the auxiliary voltage reference. Fig. 3 shows simulated power breakdown of the proposed voltage reference across temperatures. I_B supplies the current consumption of the amplifier at low temperature while I_T supports additional current required at high temperature.

III. MEASUREMENT RESULTS

The proposed voltage reference is fabricated in a 180 nm CMOS process as shown in Fig. 4. The area is 122 μm × 66 μm including a 7.2 pF decoupling capacitor. 20 dies are packaged in ceramic packages. Each chip has 10 voltage references, and they share supply voltage for more accurate power measurement due

to its low power consumption. However, outputs of only two voltage references are connected to pads for V_{REF} measurement due to limited pad number. Keithley Electrometers with high input impedance are used for V_{REF} voltage and power measurement.

Figs 5 and 6 show the measured V_{REF} and TC distribution. Without trimming, the voltage references achieve a ±3σ inaccuracy of 10.3% from 0 °C to 170 °C and show an average of 57 ppm/°C. For trimming, 10 voltage references are measured at 80 °C with default control bits, and the average V_{REF} is 1.171 V. Next, all the voltage references are trimmed at 80 °C to obtain V_{REF} closest to 1.171 V. They obtain a ±3σ inaccuracy of 3.4% from 0 °C to 170 °C and show an average of 64 ppm/°C . Hence, the trimming reduces the ±3σ inaccuracy by 3.0× at the expense of a 12% TC degradation.

Fig. 7 shows both measured voltage references without and with body buffers. A typical chip is used for the voltage reference without buffers, but the average values from 40 voltage reverences are used for the one with buffers. V_{REF} of the

978-1-5386-3179-9/17 $31.00 © 2017 IEEE

TABLE I. PERFORMANCE SUMMARY AND COMPARISON WITH PREVIOUS VOLTAGE REFERENCES.

	This Work	[9]	[10]	[11]	[12]	[4]	[5]	[7]	[8]
Technology (nm)	180	130	180	180	180	130 (SOI)	1000 (SOI)	180	350
Type	CMOS	CMOS	CMOS	CMOS	CMOS	CMOS	CMOS	BJT	BJT
Supply Voltage (V)	1.8 – 3.6	0.5 – 3.0	1.2 – 2.2	1.4 – 3.6	0.15 – 1.8	2.5	5.0	1.5 – 2.5	1.4 – 3.0
V_{REF} (V)	1.17	0.18	0.98	1.25	0.018	1.5	1.8	1.19	1.18
Temp. Range (°C)	0 – 170	-20 – 80	-40 – 85	0 – 100	0 – 120	-40 – 200	-40 – 200	-20 – 100	-10 - 110
±3σ Inaccuracy @ Entire Temp. Range (%)	3.4	N/A	1.9	2.0	N/A	N/A	N/A	N/A	N/A
TC (ppm/°C)	32 – 106 μ:64	17 – 231 μ:62	48 – 124	11 – 73 μ:31	μ:1462 σ: 324	470	825	μ:25	13
Current Consumption @ Room Temp. (A)	76 p	4.4 p	95 p	24 p	174 p	20 μ	2 μ	2.0 n	29 n
Line Sensitivity (%/V)	0.09	0.03	0.38	0.31	0.31	N/A	N/A	2.03	0.20
Power Supply Rejection (dB)	-38 @ 100Hz	-53 @ 100Hz	-42 @ 100Hz	-41 @ 100Hz	-64 @ 100Hz	N/A	N/A	-67 @ 100Hz	-53 @ DC
# Samples	40	49	60	60	60	1	1	10	10
Active Area (mm²)	0.0081	0.0014	0.0049	0.0025	0.0012	N/A	N/A	0.098	0.48

reference without the body buffers decreases only by 1.3% at 125 °C (TC of 143 ppm/°C). However, it drops by 7.3 % at 170 °C (TC of 466 ppm/°C). On the other hand, V_{REF} of the proposed references with body buffers decreases only by 0.8% at 155 °C (45 ppm/°C) demonstrating the efficacy of the proposed technique isolating body leakage.

Fig. 8 displays V_{REF} across supply voltages from 1.7 V to 3.6 V at 0, 27, 80, and 170 °C. Line sensitivity is 0.088% from 1.8V to 3.6V at 170 °C (worst temperature). The minimum supply voltage can be reduced down to 1.4 V by using NMOS input amplifiers for MP3. In this design, we choose PMOS input amplifiers for all the bodies since it can save one current branch for a current mirror and thus 8.8% current draw from the supply. In addition, the target system runs the designed voltage reference under a battery that has a voltage higher than 2.5 V.

Fig. 9 (a) shows power consumption across temperatures from 0 °C to 170 °C at 1.8 V. Fig. 9 (b) and (c) show power consumption across supply voltages from 1.8 to 3.6 V at room temperature and 170 °C, respectively. Due to subthreshold operation, current consumption significantly increases at high temperature. Low power is considerably important at low temperature since it allows an energy harvester to recharge the battery with limited harvested energy. In addition, typically, there are other dominant contributors to power consumption at high temperature (e.g., 12 μW from 8 MB SRAM memory at 125 °C [14]). Thus, the power consumption of the proposed circuit at high temperature is acceptable.

Table II shows performance summary and comparison with previous voltage references. The proposed voltage reference only can operate at > 120 °C with pA current consumption at room temperature. It also achieves competitive performance in terms of ±3σ inaccuracy, TC, line sensitivity, and power supply rejection.

IV. CONCLUSION

In this paper, we proposed a voltage reference operating from 0 °C to 170 °C for high temperature applications. For low temperature coefficient, this circuit isolates nwell/psub diode leakage using analog buffers. It achieves a ±3σ inaccuracy of 3.4% from 0 °C to 170 °C after one point trimming at 80 °C. For low power, the voltage reference itself is designed in

subthreshold region, and the buffers are adaptively biased by measuring the body leakage and using its proportional current. It consumes 76.3 pA at room temperature and 177 nA at 170 °C at 1.8V. Low power consumption at low temperature allows the target miniature sensing system to recharge a battery.

REFERENCES

[1] Z. Shi et al., "Development and Field Evaluation of Distributed Microchip Downhole Measurement System," SPE Digital Energy Conference and Exhibition, Mar. 2015, SPE-173435-MS.

[2] I. Lee et al., "MBus: A Fully Synthesizable Low-power Portable Interconnect Bus for Millimeter-scale Sensor Systems," J. Semiconductor Technology and Science, vol. 16, no. 6, pp 745-753, Dec. 2016.

[3] W. Jung et al., "An ultra-low power fully integrated energy harvester based on self-oscillating switched-capacitor voltage doubler," IEEE J. Solid-State Circuits, vol. 49, no. 12, pp. 2800–2811, Dec. 2014.

[4] E. H. Boufouss et al., "Ultra-Low Power High Temperature and Radiation hard Complementary Metal-Oxide-Semiconductor (CMOS) Silicon-on-Insulator (SOI) Voltagae Referenece," Sensors, no. 12, pp. 17265–17280, Dec. 2013.

[5] M. Assaad et al., "Ultra Low Power CMOS Circuits Working in Subthreshold Regime for High Temperature and Radiation Environments," in Proceedings of the International Conference and Exhibition on High Temperature Electronics Network, July 2011.

[6] G. Ge, C. Zhang, G. Hoogzaad, and K. A. A. Makinwa, "A Single-Trim CMOS Bandgap Reference With a 3σ Inaccuracy of ±0.15% From -40°C to 125°C," IEEE J. Solid-State Circuits, vol. 46, no. 11, pp. 2693-2701, Nov. 2011.

[7] Y. P. Chen, M. Fojtik, D. Blaauw, and D. Sylvester, "A 2.98nW bandgap voltage reference using a self-tuning low leakage sample and hold," in IEEE Symp. VLSI Circuits Dig., June 2012, pp. 200-201.

[8] J. M. Lee et al., "A 29nW Bandgap Reference Circuit," in IEEE Int. Solid-State Circuits Conf. Dig. Tech. Papers, Feb. 2015, pp. 100-101.

[9] M. Seok, G. Kim, D. Blaauw, and D. Sylvester, "A portable 2-transistor picowatt temperature-compensated voltage reference operating at 0.5 V," IEEE J. Solid-State Circuits, vol. 47, no. 10, pp. 2534–2545, Oct. 2012.

[10] Q. Dong, K. Yang, D. Blaauw, and D. Sylvester, "A 114-pW PMOS-only, trim-free votlage reference with 0.26% within-wafer inaccuracy for nW systems," in IEEE Symp. VLSI Circuits Dig., June 2016.

[11] I. Lee, D. Sylvester, and D. Blaauw, "A Subthreshold Voltage Reference With Scalable Output Voltage for Low-Power IoT Systems," IEEE J. Solid-State Circuits, vol. 52, no. 5, pp. 1443-1449, May. 2017.

[12] D. Albano, F. Crupi, F. Cucchi, and G. Iannaccone, "A Sub-kT/q Voltage Reference Operating at 150 mV," IEEE Transactions on Very Large Scale Integration (VLSI) Systems, vol. 23, no. 8, pp. 1547–1551, Aug 2015.

[13] Y. Taur and T. H. Ning, Fundamentals of Modern VLSI Devices. New York, NY, USA: Cambridge Univ. Press, 2009, ch. 3, pp. 148–203.

[14] Q. Dong et al., "A 1Mb Embedded NOR Flash Memory with 39μW Program Power for mm-Scale High-Temperature Sensor Nodes," in IEEE Int. Solid-State Circuits Conf. Dig. Tech. Papers, Feb. 2017, pp. 198-199.

A Smart-Offset Analog LDO with 0.3V Minimum Input Voltage and 99.1% Current Efficiency

Saurabh Chaubey and Ramesh Harjani

University of Minnesota, Minneapolis, MN 55455, Email: harjani@umn.edu

Abstract—**In this paper we present the first fully integrated analog LDO (low dropout regulator) for sub-0.5V supply voltages. The LDO can operate from 0.3V-to-1.0V input voltage, and can sustain a load variation of 10mA-to-100mA at 1.0V input and 5mA-to-25mA at 0.3V input. It achieves a peak 99.1% current efficiency for a 100mA load at 0.9V output voltage. In order to realize the gate drive at sub-0.5V supply voltages, we introduce a negative charge pump based adaptive offset before the pass FET which provides gate-source headroom at input operation voltages normally reserved for digital LDOs. The smart-adaptive-negative offset voltage follows a $0.5\text{-}0.5\times V_{DD}$ scheme to accommodate a wide range of input voltages while providing the necessary extra gate drive for the power FET at low inputs. The 32 phase charge pump runs at a frequency of 3GHz with a ripple of $\sim 3\text{mV}$. The prototype was fabricated in TSMC's 65nm GP CMOS.**

I. Introduction

With the growing need of managing power consumption coupled with continuous technology scaling, the use of near-threshold supply circuits have become popular in VLSI subsystems, including processors, memory, and PLLs. Unfortunately, circuits which operate at near-threshold voltages, are highly sensitive to supply voltage variations, increasing the need of ripple-free power management. Low-dropout (LDO) regulators play an important role in this. Typically they provide ripple suppression and isolation from the switching regulators. In particular, for analog loads, the ripple suppression requirement becomes even more stringent.

Higher power efficiency demands lower drop out voltages, requiring LDOs to utilize PMOS pass transistors. These LDOs find it difficult to maintain the necessary gate-source headroom at lower input voltages below 0.5V. To circumvent the headroom problem, recently, digital LDOs (D-LDOs) have been developed for lower V_{DDS} [1]–[3]. D-LDOs utilize all-digital logic and/or time-to-digital conversion to replace the all-analog controller, and require no additional headroom by fully switching the pass transistor on/off. The fundamental problem with digital LDOs is that they have an inherent ripple even for a constant load due to finite resolution of switch sizes. The pass transistor characteristics roughly follow $I_L \propto \frac{W}{L} \times \Delta V^2$, where I_L, $\frac{W}{L}$ and ΔV are load current, aspect ratio and overdrive voltage receptively. In a conventional analog LDO, we use ΔV and in digital LDOs we use $\frac{W}{L}$ as the lever for load regulation. Being a square term, the ΔV is more powerful lever. So, to inherit the ΔV control lever and the simple design characteristics of analog LDOs, we propose an analog solution in the form of a smart adaptive offset LDO (SO-LDO) for handling the gate-source headroom of the PMOS pass transistor at lower supply voltages.

Fig. 1-left shows a typical LDO schematic (ignoring the voltage source offset V_{OS} for now). It contains a PMOS

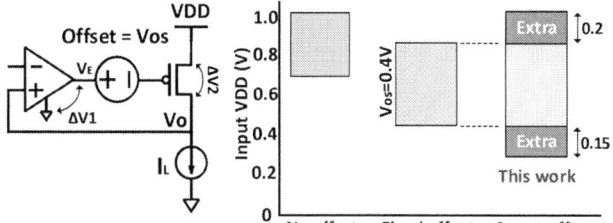

Fig. 1: Range of input supply voltages for LDO in three cases- no offset, fixed offset and smart offset (this work)

pass device and the error amplifier (EA), which forms the negative feedback loop. The EA can support an output voltage, $V_E \geq \Delta V_1$ (overdrive voltage of EA's output stage), before its gain rolls off. In a conventional analog LDO without any offset (V_{OS}), the EA voltage, V_E, for proper operation is given by $V_E = V_{DD} - V_{TP} - \Delta V_2$, where ΔV_2 is the pass transistor overdrive voltage and V_{TP} is the PMOS threshold voltage respectively. Thus for typical $V_{TP} \approx 0.3\text{V}$, VDD less than 0.7V cannot be supported (assuming $\Delta V_1 = \Delta V_2 \approx 0.2$). Let us consider what happens when we introduce an offset in the LDO. As a first step, let us consider a fixed negative offset between EA and pass transistor (as shown by voltage source V_{OS}). So for a fixed offset, we get, $V_E = V_{DD} - V_{TP} - \Delta V_2 + V_{OS}$ and for V_{OS}=0.4V, the lower limit for VDD becomes 0.45V. But the problem with this scheme is the breakdown voltage, V_{BD}, of the pass transistor, $V_{DD} \leq V_{BD} + \Delta V_1 - V_{os}$. We observe that, the lower limit of VDD does reduce but unfortunately, so does the upper limit due to the breakdown voltage concern. In our design, we resolve the upper limit problem by introducing a variable smart offset, V_{OS}=A-A×VDD (where A is a design variable), such that at lower VDDs we maintain a finite offset but at larger voltages (near 1V in 65nm), the offset approaches zero. Fig. 1-right shows the VDDs that can be supported for the three cases- LDO without offset, LDO with a fixed V_{OS} (similar to D-LDOs) and LDOs with a smart offset (this work). We observe an additional 400mV input voltage dynamic range than conventional LDO and 300mV additional range in comparison to a fixed V_{OS}.

II. Design Issues with Low Voltage LDOs

In this section we discuss the comparative present day design issues for low voltage LDO designs.

Analog LDOs: Fig. 2(a) shows the basic block diagram for a capacitor-less LDO. The loop gain of this system depends on the gain of pass transistor and the EA. Even in large output swing error amplifiers the minimum V_E is limited by the overdrive voltage of a NMOS transistor, ΔV_1 (typically $\Delta V_1 \sim 0.2\text{V}$). As V_E decreases and becomes less than or equal

978-1-5386-3179-9/17 $31.00 © 2017 IEEE

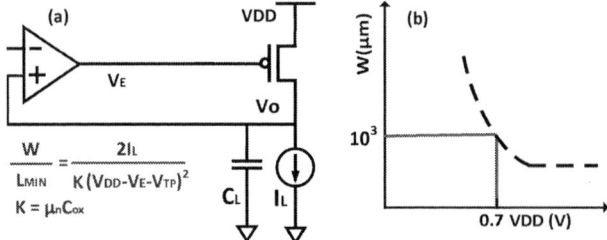

Fig. 2: (LHS) Basic capacitor-less analog LDO, (RHS) Minimum pass transistor size needed as a function of LDO input supply voltage (V_{DD})

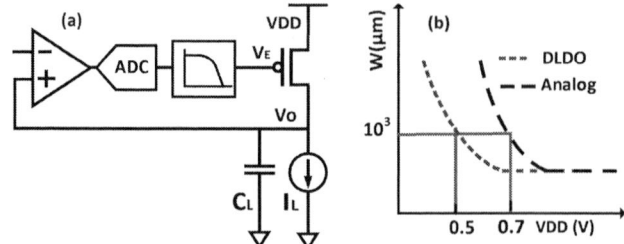

Fig. 3: (LHS) Digital LDO including ADC, comparator and switch array (or DAC). (RHS) Minimum pass transistor size vs input supply voltage for analog LDO and digital LDO. (I_L=100mA)

to ΔV_1, the EA output stage enters into the triode region thus decreasing the overall EA gain. Therefore, adequate regulation is maintained only as long as V_E is greater than equal to ΔV_1.

The maximum output voltage, $V_{O,max}$, is related to the input V_{DD} as, $V_{O,max}=V_{DD}-\Delta V_2$. Here ΔV_2 denotes the pass transistor overdrive voltage and increases with increased loads for a fixed FET size. To first order the power efficiency is given by $\eta=1-\Delta V_2/V_{DD}$. Thus it is natural to design the width of the FET corresponding to a minimum possible dropout voltage for which we maintain the pass FET in saturation and stabilize the loop gain. We choose, $\Delta V_1 \approx 0.2$V, which is a good compromise between the gm/I current efficiency and FET size. Fig. 2(b) shows the LDO pass transistor device width as a function of the input V_{DD}. We note that the device size has to increase to compensate for the lower overdrive voltage that is available at low V_{DD}s. The knee in this figure is at $V_{DD,min} \approx \Delta V_1 + |V_{TP}| + \Delta V_2$ which is around 0.7V for $|V_{TP}| = 0.3V$. After this point, for a fixed current, the transistor width increases exponentially as ΔV_1 throttles the overdrive for the pass FET. The pass transistor carries the entire load current and is normally quite large in size setting both the DC and transient response due to the large parasitics associated with this device.

Digital LDOs: Normally, trying to accommodate V_{DD}s below 0.7V using an analog LDO results in a pass transistor that is unyieldingly large. This was the primary impetus for the design of digital LDOs. Fig. 3 shows the block diagram for a digital LDO which includes a switch array, a comparator, and a digital controller. As the gate drive can be as low as zero, to first order approximation, $V_{DD,min}=0+V_{TP}+\Delta V_2 \approx 0.5$V. The analog controlled power transistor is replaced with a switch array and the number of switches that are turned-on is digitally controlled. The output voltage (V_O) is monitored by the comparator instead of the operational amplifier. Thus, the digital LDO (D-LDO) eliminates all analog circuits and can operate at input supply voltages as low as 0.5V.

Comparison between A-LDO and D-LDO Let us compare the pass FET dimensions for an A-LDO (analog-LDO) and a D-LDO operating at 0.7V with a load current of I_L=10mA. For the A-LDO the pass transistor $width$=1000μm is as shown in Fig.(3) (black). We observe that, while keeping the load fixed, as we move to lower input voltages, the width required to support the load increases rapidly. As V_{DD}s approaches ΔV_1+V_{TP} (500mV), the transistor dimensions becomes unrealizable. This is the reason why analog LDOs are not possible

for lower input V_{DD}s. However, in the case of D-LDOs, the minimum gate voltage of the pass FET can be zero so for an input V_{DD} of 0.5V at 10mA the FET size is 1000μm and the $width$-V_{DD} curve shifts to the left as shown in red in Fig. 3.

III. PROPOSED SMART OFFSET LDO:

From the previous sections, the primary bottleneck for analog LDOs is the finite minimum V_E. We propose to solve this problem via a smart negative voltage offset inserted between the error amplifier and the pass transistor as shown in Fig. 4. The EA shares the input V_{DD} with the LDO. This smart offset allows us to operate at lower voltages and also reduces the pass transistor size as discussed next. The implementation details and loop stability analysis follow after that.

Lowering of the minimum input voltage: The smart adaptive offset is designed to follow a value of $A-A \times V_{DD}$. From Fig. 4, we know $V_{OS}=V_E+V_{TP}+\Delta V_2-VDD$. Having $V_E = V_{IN}/2$ (for symmetric EA output), we get $V_{OS} \approx 0.5-0.5 \times VDD$. So A, the design parameter, is chosen to be 0.5. As the input voltage drops the smart offset increases so that we maintain a constant ΔV_2 across the pass transistor for different input voltages that range from 0.3V to 1.0V. By providing this offset, we enable the gate of the pass transistor to go to a negative voltage at lower LDO supply voltages (V_{DD}) thus avoiding the exponential bloating of the transistor size. The proposed smart offset LDO (SO-LDO) can sustain a load of 10mA and FET $width$=1000μm at a 0.3V input supply voltage (point A in Fig. 4). This clear reduction in minimum input voltage is due to the extra offset introduced.

Reduction of the pass transistor size: For the same current and a 0.5V V_{DD} we see in Fig. 4 that the FET size reduces from 1000μm to 100μm (a 10X reduction in comparison to D-LDOs) along the vertical grey line at point

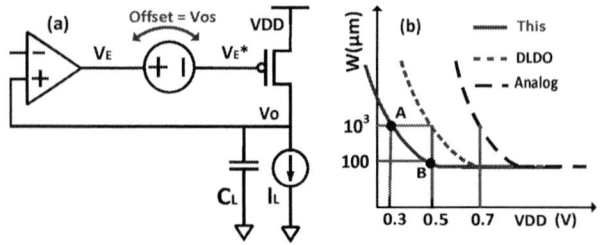

Fig. 4: Variation of the pass transistor width with the input supply voltage of the LDO. Black- analog LDO, red- digital LDO and blue- our proposed smart offset based LDO, all at IL=100mA.

978-1-5386-3179-9/17 $31.00 © 2017 IEEE

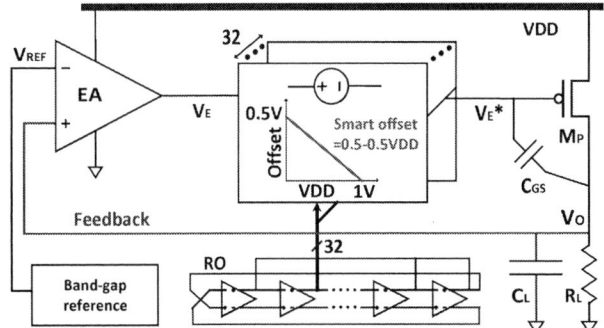

Fig. 5: Overall architectural detail of the proposed negative offset LDO

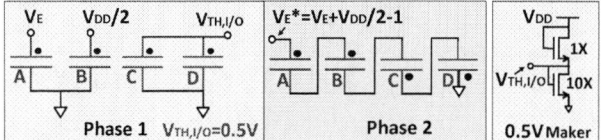

Fig. 6: Circuit details for the negative charge pump offset generator

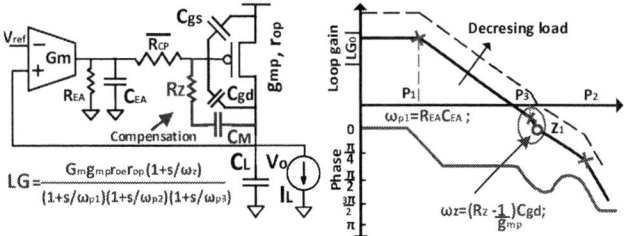

Fig. 7: Small-signal system model and frequency response

B. As shown in the Fig. 4, the V_E generated by the error amplifier is converted to $V_E^*=V_E-V_{OS}$ and thus it improves the current driving capability. Taking the case where $V_{DD}=0.3$V, $V_E=VDD/2$ and $|V_{TP}|=0.3$V we get $V_E^*=V_{DD}/2-V_{OS}=-0.2$V. Thus we see a leftward shift of the V_{DD}-vs-W curve (by 0.2V). The new design improves the drive capability $\sim$10X.

An adaptive offset helps in lowering the input V_{DD} as compared to a fixed offset. We know $V_E = V_{DD} - \Delta V_2 - V_{TP} + V_{OS}$, so the fixed offset, V_{OS} is bounded by Eqn.(1)

$$\Delta V_1 + |V_{TP}| - V_{DD} \leq V_{OS} \leq |V_{TP}| \qquad (1)$$

From Eqn.(1), we see that for a fixed offset, the minimum V_{DD} is $\Delta V_1 + \Delta V_2 + 0.1$V (design margin) ≈ 0.5V, while our proposed design allows for V_{DD}s as low as 0.3V.

Architectural details: The smart adaptive offset, $V_{OS}=$ 0.5V-0.5$\times V_{DD}$ is realized using a fast switched capacitor charge pump network. Fig. 5 shows the full circuit details. The EA output voltage V_E, feeds into the negative offset based charge pump which converts it to V_E^*. This V_E^* serves as the gate voltage for the pass FET implemented by a LVT device. The switching frequency of the pump is designed to be at least a decade higher than the loop frequency and uses 32 interleaving phases thus maintaining a low output ripple. Non overlapping phases $\Phi 1$ and $\Phi 2$ are generated by a 32 stage ring oscillator coupled to a non overlap clock generator. The charge pump consists of four capacitors operating during the two non overlapping phases. During phase $\Phi 1$ the bucket capacitor A gets charged to V_E, capacitor B gets charged to $V_{DD}/2$ and capacitors C and D get charged to ≈ 0.5V (created on-chip). In phase $\Phi 2$, all the capacitors are connected in series with the capacitors C and D having opposite polarity as shown in Fig. 5. Deep-nwell NMOSs serve as switches in order to handle negative voltages. The resultant voltage $V_E^* = V_E + 0.5\times V_{DD}-0.5$. As shown in Fig. 5-right, the 0.5V is created by V_{TN} (≈ 500mV nominal) of a 2.5V I/O NMOS device. Six-σ mismatch is around 15mV while with 100C temperature change, the variation is 80mV. The EA is a self biased folded cascode OTA (which operates in sub-threshold for V_{DD} below 0.7V and in strong inversion for V_{DD} above 0.7V).

Loop stability analysis and ripple: The loop stability can be analyzed by considering the small-signal closed model shown in Fig. 7. The charge pump is represented as an average output resistor, R_{CP}, the error amplifier is represented as a Gm

cell with a single pole roll off. This system has two poles and a left hand zero, caused by the charge pump resistor and C_{GD}, which helps with self compensation. The pole at the output of error amplifier is dominant while the pole at the LDO output is non-dominant. Thus for the frequency range that is about a decade lower than the switching frequency, the charge pump equivalent resistor helps realize a lag network. In isolation for the open loop, the gate-source capacitance acts as the filter capacitor for the charge pump thereby further reducing the ripple of the charge pump. The closed loop output impedance suppresses the charge pump ripple at the output.

IV. MEASUREMENT RESULTS

Fig. 8 shows the measured closed loop impedance. We observe that the 3GHz ripple at a load of 100mA is suppressed significantly. Fig. 9(a) shows the chip microphotograph. The total active area is $0.15\times0.4mm^2$. Fig. 9(b-c) shows that the undershoot and overshoot recovery for a 10mA-100mA transient is 34.5nS and 43.62nS @0.5V output respectively. Fig. 10 shows the output ripple and charge pump frequency vs input voltage. For V_{DD}s below 0.5V, the switches are not sufficiently turned on, requiring a lower charge pump frequency, which in turn increases the output ripple voltage.

Fig. 11 shows the measured $V_{OUT} - V_{DD}$ characteristics at I_L of 10mA at different output voltages. At V_{IN} equal to 0.4V and V_{OUT} of 0.35V, the measured line regulation is 3.1mV/V. The LDO achieves a successful load regulation of 0.65mV/mA while V_{IN} varies from 0.3V to 1.0V. The measured quiescent current does not depend on I_L, and is 11uA (minimum) at $V_{DD}=0.3$V, though the quiescent current increases with

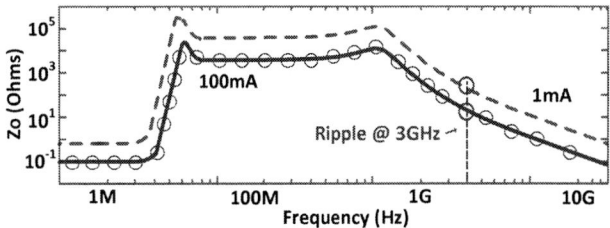

Fig. 8: Normalized closed loop output impedance for the proposed LDO at 100mA and at 10mA at $V_{DD}=0.5$V

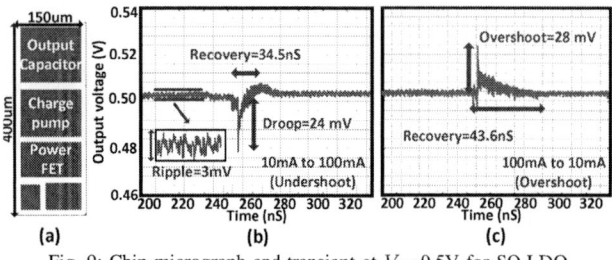
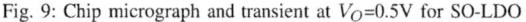
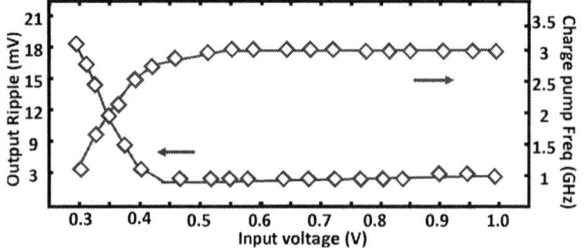

Fig. 9: Chip micrograph and transient at V_O=0.5V for SO-LDO

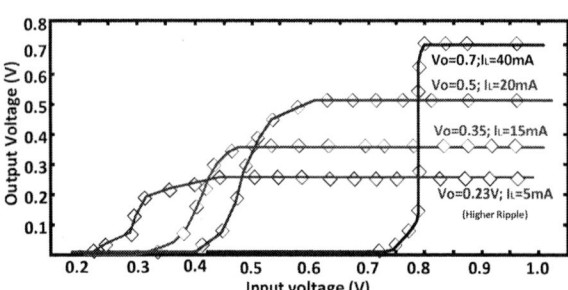

Fig. 10: Output ripple and charge-pump frequency for different input V_{DD}

I_{LOAD} in the other designs compared in TABLE I. This LDO achieves a peak current efficiency of 99.1% at 0.9-V input and 0.7-V output voltage. The input voltage, active area and the quiescent current are the lowest in comparison to any published LDOs as indicated in TABLE I. The EA and other control circuitry share the input voltage with the LDO unlike many other designs that use a higher V_{DD}. Due to the 32 phase interleaving the steady state ripple is always less than 3mV for V_{DD}s greater than 0.4V (and often as low as 1mV) which is an order of magnitude lower than other designs.

V. CONCLUSIONS

We present an analog solution for ultra low voltage LDOs using a smart offset. The offset adapts to the input voltage thus

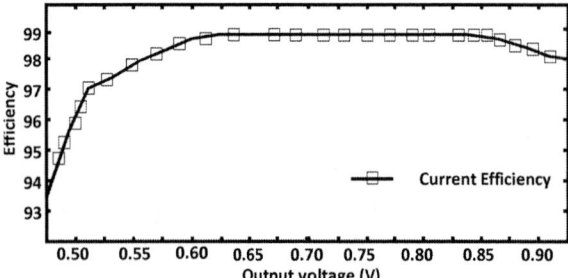

Fig. 11: O/P voltage regulation for Vo= 0.23V, 0.35V, 0.5V, 0.7V

TABLE I: Summary comparison with other related work

	[4]	[5]	[6]	[7]	This Work
Technology(nm)	65	65	45	40	65
Current Eff %	99.8	99.6	97	98.7	99.1
In Voltage (V)	1	1.2	1-1.2	1.34	0.3-1.0
Out Voltage (V)	0.6	0.7	0.8	0.9	0.23-0.9
Cout (nF)	0.3	0.86	1.67	-	0.1
Ripple (mV)	43	17	18.9	15	1-3[1]
Max I Step (mA)	120	55	13	15	90
Quiescent current (uA)	112	115	203	130	11
Undershoot time (nS)	40	50	30	70	34.5
Pwr Output (mW)	205	354	435	250	103
Area (mm2)	0.08	0.08	0.2	0.06	0.04

[1] For $V_{DD} \geq 0.4V$.

facilitating a constant minimum dropout. The charge pump generated smart offset enables an effective size reduction of the pass transistor at higher input voltages and makes the operation possible at the lowest input voltage of 0.3V. The SO-LDO has a 10X device size improvement as compared to D-LDOs. The SO-LDO can support a 0.3V-to-1.0V input with a quiescent current of 11uA. For each such input voltage, the SO-LDO can sustain a load variation of 10-100mA. It achieves a peak current efficiency of 99.1% at 100mA load and a 0.9V output voltage. The charge pump runs at a frequency of 3GHz with 32 phases so that the ripple is less than 3mV for V_{DD}s $\geq$ 0.4V which is an order of magnitude better than state of art D-LDOs. The SO-LDO is simpler to implement than D-LDOs, has lower ripple than D-LDOs and circumvents the minimum V_{DD} limit imposed by analog LDOs. The system was designed and measured in TSMC's 65nm GP.

ACKNOWLEDGEMENTS

This work was supported by SRC/TxACE task ID 2712.008. The authors thank Rakesh Palani for useful discussions.

REFERENCES

[1] Y. Okuma, K. Ishida, Y. Ryu, X. Zhang, P.-H. Chen, K. Watanabe, M. Takamiya, and T. Sakurai, "0.5-V input digital LDO with 98.7% current efficiency and 2.7-uA quiescent current in 65nm CMOS," in *IEEE Custom Integrated Circuits Conference 2010*, 2010, pp. 1–4.

[2] C. C. Chiu, P. H. Huang, M. Lin, K. H. Chen, Y. H. Lin, T. Y. Tsai, C. C. Huang, and C. C. Lee, "A 0.6V resistance-locked loop embedded digital low dropout regulator in 40nm CMOS with 77% power supply rejection improvement," in *2013 Symposium on VLSI Circuits*, 2013.

[3] C. L. Tai, A. Roth, and E. Soenen, "A digital low drop-out regulator with wide operating range in a 16nm finfet cmos process," in *2015 IEEE Asian Solid-State Circuits Conference (A-SSCC)*, 2015, pp. 1–4.

[4] F. Yang and P. K. T. Mok, "A 0.6-1V input capacitor-less asynchronous digital LDO with fast transient response achieving 9.5b over 500mA loading range in 65-nm CMOS," in *European Solid-State Circuits Conference (ESSCIRC)*, 2015, pp. 180–183.

[5] ——, "Fast-transient asynchronous digital LDO with load regulation enhancement by soft multi-step switching and adaptive timing techniques in 65-nm CMOS," in *Custom Integrated Circuits Conference (CICC) 2015 IEEE*, 2015, pp. 1–4.

[6] C. S. Wu, K. C. Lin, Y. P. Kuo, P. H. Chen, Y. H. Chu, and W. Hwang, "An all-digital power management unit with 90% power efficiency and ns-order voltage transition time for DVS operation in low power sensing SoC applications," in *2015 IEEE International Symposium on Circuits and Systems (ISCAS)*, 2015, pp. 1370–1373.

[7] M. Onouchi, K. Otsuga, Y. Igarashi, T. Ikeya, S. Morita, K. Ishibashi, and K. Yanagisawa, "A 1.39-v input fast-transient-response digital ldo composed of low-voltage mos transistors in 40-nm cmos process," in *IEEE Asian Solid-State Circuits Conference 2011*, 2011, pp. 37–40.

Fig. 12: LDO current efficiency at V_O=0.7V and V_{DD}=0.9V

S18-3 (2151)

A 762-µW 16.3-ps Resolution Digital Pulse Width Modulator Using Zooming Phase-Interpolator

Masanobu Tsuji

ROHM Co., Ltd.

Yokohama, Japan

E-mail: Masanobu.Tsuji@dsn.rohm.co.jp

Abstract—A proposed phase-interpolator (PI) based hybrid digital pulse width modulator (DPWM) effectively resolves the trade-off between resolution and power consumption. Conventional DPWM delay-line-based architectures suffer from high power consumption limited delay time per delay-tap due to process technology, while the proposed solution replaces the delay line with a PI featuring sub-gate-delay resolution. To address the issue of the increase of area when using a conventional PI, a zooming scheme is proposed. Since the proposed architecture consists only of static circuitry, zero-quiescent power, wide operating voltage range, and instantaneous start-up are provided. A prototype chip manufactured with 0.13-µm BCD technology achieves 762-µW power consumption at 16.3-ps resolution and 10-MHz switching frequency, with 1.2-V supply voltage. With 0.75-V supply voltage, 163-ps resolution and 1-MHz switching frequency is achieved with 36-µW power dissipation.

Keywords— pulse width modulation, digital pulse width modulator, phase interpolator, DC-DC converter, class-D amplifier, motor controller, LED driver.

I. INTRODUCTION

The high-efficiency switch-mode power control technique is important to the applications such as DC-DC converters, class-D audio amplifiers, motor controllers, and LED drivers. In these applications, the pulse width modulation (PWM) is employed to control the on and off time of MOSFET switches. Recently, DPWMs were proposed taking the advantages of technology scaling and digital techniques, for cost reduction by eliminating off-chip bulky passive components, flexibility, and portability to different technologies. The time resolution of DPWM generally translates to the voltage accuracy or power/area efficiency in the above applications. Higher resolution, however, typically involves with higher power consumption or larger chip area. In addition to high resolution, since recent emerging intermittent applications require "instant-on" characteristics, a DPWM with less or none settling time is also preferred.

The architecture of DPWM is mainly classified into three types as shown in Fig. 1 (a)-(c). Fig. 1 (a) shows the most basic counter-based architecture. This approach features small chip area but requires higher clock frequency for higher resolution, resulting in higher power consumption. Fig. 1 (b) shows the delay-line based architecture alleviating the clock frequency

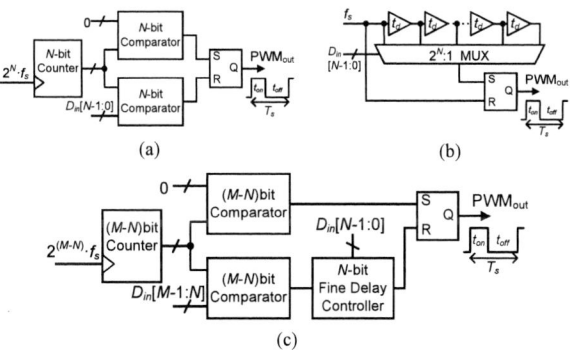

Fig. 1. Conceptual block diagrams of DPWM. (a) Counter based, (b) Tapped delay-line based and (c) Hybrid DPWM.

requirement. However, since the delay cell of 2^N stages are needed to obtain N-bit resolution, the area increases with the requirement of longer range. Moreover, a closed control loop such as delay-locked loop (DLL) or phased-locked loop (PLL) is generally needed for the accuracy of the delay time. Power consumption, area, and design complexity are increased, and waiting time until the loop settles is inevitable. The hybrid structure shown in Fig. 1 (c) combining Fig. 1 (a) and a fine delay controller (FDC) mitigates the issues of the previous two approaches. A conventional FDC is implemented with Fig. 1 (b) whose resolution is determined by the delay per tap which is limited by the process technology [2], [3-4]. To break such limitation, a PI-based solution is reported in [1]. By adopting a PI for the FDC, the time resolution shorter than a buffer delay is realized with low power. In [1], PI's time resolution is still proportionally traded off with chip area. Also, the PI is implemented with current sources, which find their limitation in low-voltage operation and instant start-up.

In this work, a proposed Zooming-PI architecture solves these issues. Using this PI, the DPWM realizes low power, high resolution, and small area.

II. PROPOSED ARCHITECTURE

A. DPWM

Fig. 2 shows the proposed DPWM block diagram. The FDC of Fig. 1 (c) is replaced by two-phase edge generator (TPEG) in Fig. 4 (a) and two-stage Zooming-PI with 8-bit

978-1-5386-3179-9/17 $31.00 © 2017 IEEE
273

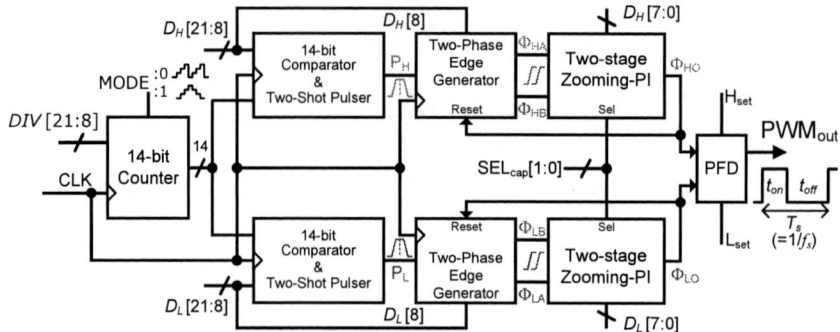

Fig. 2. Block diagram of proposed DPWM.

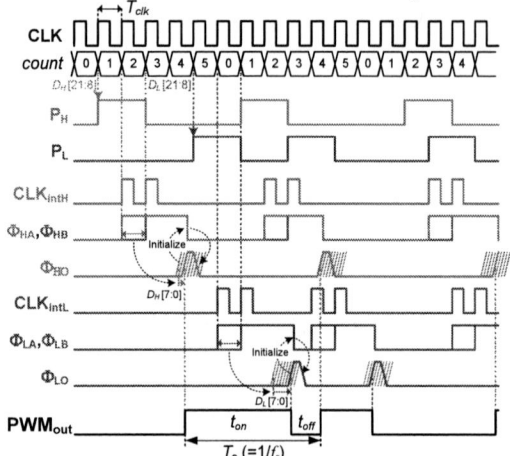

Fig. 3. Timing diagram of proposed DPWM.

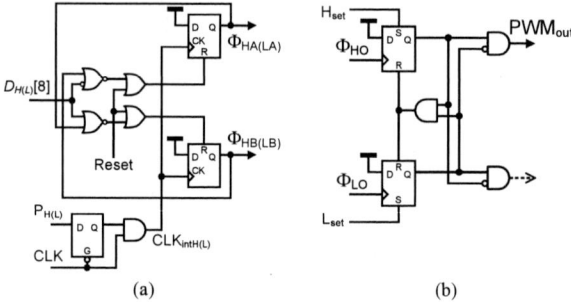

Fig. 4. Schematic of (a) Two-Phase Edge Generator and (b) Phase Frequency Detector.

resolution. Furthermore, the change to dual-edge modulation where both the rising and the falling edge of the PWM output (PWM$_{out}$) are controlled and the set-reset latch (SR-Latch) are replaced by a phase-frequency detector (PFD) of Fig. 4 (b). By MODE input in Fig. 2, the function of single-edge and dual-edge modulation can be switched. The switching frequency (f_S) of PWM$_{out}$ is generated by dividing the input clock (CLK) frequency by a 14-bit counter. Since the DPWM can accept a clock frequency ranging from 20 MHz to 240 MHz, f_S can be adjusted between 1.22 kHz and 40 MHz. The behavior of this circuit is shown as timing diagrams in Fig. 3. First, the value of D_H or D_L setting the target pulse width of PWM$_{out}$ is input to the comparator, and when the counter value exceeds it, a pulse with a width of two clock cycles is output from the two-shot pulser as P_H (or P_L). Next, in TPEG, the clock gating is canceled by P_H (or P_L). Then, two adjacent phase edge signals are generated those are shifted by T_{clk}. Thus, except for the switching timing of the PWM$_{out}$, the circuit operation behind the TPEG is stopped, reducing the power consumption. In addition, TPEG reverses the phase relationship between the two outputs of TPEG by the LSB of the data input ($D_H[8]$ or $D_L[8]$) to the comparator, to reduce the influence of the delay mismatch of the two flip-flops to improve the linearity of the pulse width control (Fig. 5b). The 8-bit two-stage Zooming-PI receives the two phases outputs a delayed pulse (Φ_{HO} or Φ_{LO}) which is controlled by the time obtained by dividing the CLK cycle (T_{clk}) by 2^8. This means that the pulse width of PWM$_{out}$ is

controlled with the time resolution from 195 ps to 16.3 ps in the frequency range of CLK. Finally, the rising edge of Φ_{HO} or Φ_{LO} is input to PFD to generate PWM$_{out}$, and at the same time, TPEG is reset and self-initialization is performed. Since the last stage PFD is edge sensitive operation, the performance with minimum and maximum pulse width is improved. Furthermore, even if the timing of Φ_{HO} and Φ_{LO} are reversed due to delay mismatch with minimum pulse setting, it is possible to prevent the error operation in which PWM$_{out}$ becomes high. The function to forcibly fix PWM$_{out}$ using H$_{set}$ or L$_{set}$ is used in interrupt sequence of digital-controlled DC-DC converters and motor controllers.

B. Zooming Phase-Interporator Architecture

Fig. 5 (a) shows the concept of Zooming-PI. At first, two adjacent phase signals are input to the two PIs of the first stage. The PI needs to have two phase inputs (Φ_{ia} and Φ_{ib}) and one variable-controlled phase output (Φ_o). Next, the PI$_{a1}$ and PI$_{b1}$ are adjusted to adjacent phases in the controllable range, and output each phase (Φ_{a1} and Φ_{b1}). By continuing the same connection at subsequent PIs, interpolation is done as if two adjacent phases of previous stages are zoomed. In this case, when controlling PI with the thermometer code, using the code conversion scheme shown in Fig. 5 (b) ensures no missing code. Thus, the total resolution N bit is the sum of N_1 to N_k. In [1], as the number of bits for PI is increased, the area increases exponentially. On the other hand, the area of proposed Zooming-PI can be suppressed by increasing the number of stages of PI. Furthermore, as the number of stages is increased, the required time-constant decreases, so the area of PI

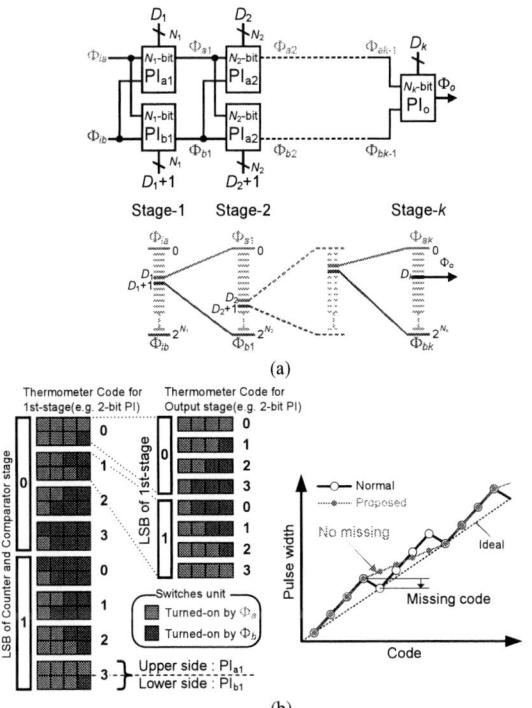

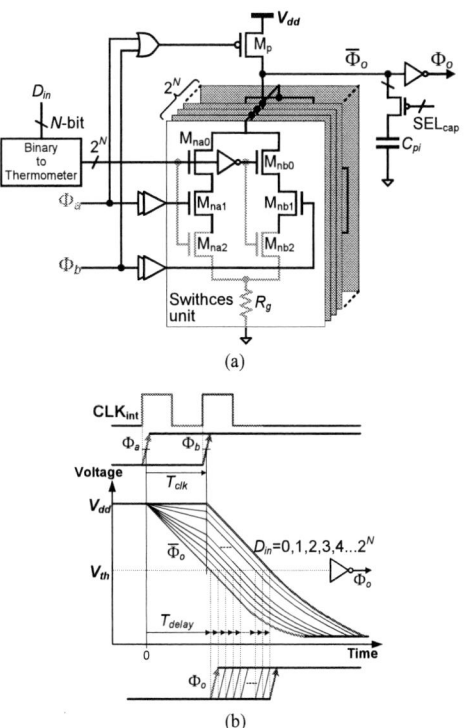

Fig. 5. (a) Conceptual diagram of proposed Zooming Phase-Interporator and (b) Code-conversion scheme in the case of two-stages Zooming PI.

Fig. 6. (a) Schematic of proposed Phase-Interporator and (b) Behavior of signal waveform.

decreases as the number of stages increases. In [5] and [6], a pipelined PI is reported, where N-stages of PIs are required for N-bit resolution. However, with the Zooming-PI scheme, the number of stages is reduced by multi-bit PIs, which reduces delay mismatching due to the difference in signal path. In this work, two stages of 4-bit PI were sufficient to implement 8-bit Zooming-PI.

C. Phase-Interporator Circuit

Fig. 6 (a) shows the circuit of PI applied to Zooming-PI. The waveform is shown in Fig. 6 (b). First, $\overline{\Phi}_o$ is pulled up to V_{dd} through M_p before two adjacent-phases Φ_a and Φ_b are input. Next, $\overline{\Phi}_o$ is disconnected from the power supply line by the rising edge of Φ_a (or Φ_b), and at the same time, the number of switches units allocated by D_{in} setting is driven by Φ_a to lower the voltage of $\overline{\Phi}_o$. Then, the remaining switches units are driven on the rising edge of Φ_b. Through these operations, Φ_o is output with delay of T_{delay} when $\overline{\Phi}_o$ crosses the threshold voltage (V_{th}) of the output inverter. T_{delay} in Fig. 6 (b) could be expressed as

$$T_{delay} = \frac{C_{pi}R_{unit}\ln(V_{dd}/V_{th}) + D_{in}T_{clk}}{2^N} \quad (1)$$

where R_{unit} is the resistance per switches unit, which is the sum of R_g and on-resistance for $M_{na0\text{-}2}$ (or $M_{nb0\text{-}2}$). N is the number of bits, T_{clk} is the input clock period. D_{in} is input data that determines the number of switches units to be assigned to Φ_a

and Φ_b, and is an integer from 0 to 2^N. Therefore, T_{delay} of Φ_o output time is controlled according to Eq. (1). Since the condition for correct phase-interpolation is that T_{delay} equals or greater than T_{clk} when D_{in} is one, T_{clk} is derived as

$$T_{clk} \leq \frac{C_{pi}R_{unit}\ln(V_{dd}/V_{th})}{2^N - 1} \quad (2)$$

Therefore, if C_{pi} is adjusted with SEL_{cap} a wide range of clock frequency can be available. To satisfy (2) with large T_{clk}, it is necessary to increase C_{pi} or R_{unit}. Although too large R_g increases the nonlinearity of PI, it helps to suppress the increase in power consumption due to the increase of C_{pi}, allowing a tradeoff between power and performance. In addition, the configuration with only static circuitry achieves zero quiescent power and allows low voltage operation, which allows for reduced power consumption and instant start-up. This is very important for applications that require intermittent operation.

III. TESTCHIP IMPLEMETAION

Fig. 7 shows a chip micrograph of two prototypes of DPWMs. Each area 0.081 mm², while DPWM1 was designed with larger PI's capacitance to improve the linearity, and R_g was reduced to optimize for the input clock frequency above 160 MHz. Fig. 8 shows the actual measurement waveform of the PWM output. DPWM0 operates over a wide input-clock-frequency range (20 MHz to over 240 MHz) targeting the applications such as digital-controlled DC-DC converters or motor controllers. Although the power consumption of

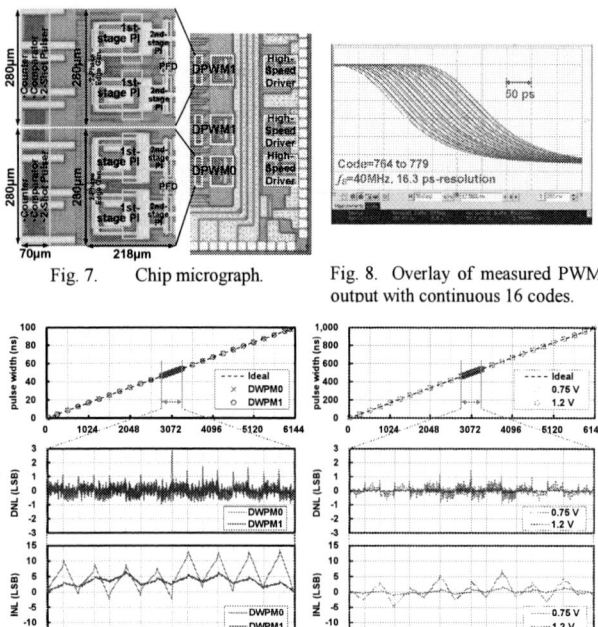

Fig. 7.   Chip micrograph.

Fig. 8.   Overlay of measured PWM output with continuous 16 codes.

Fig. 9.   Linearity comparison of (a) DPWM0 and DPWM1 at fs=10MHz, 16.3ps resolution. (b) V_{dd}=0.75V and 1.2V at DPWM0, f_S=1MHz, 163ps resolution.

DPWM1 is higher, its better linearity is suitable for class-D audio amplifiers.

IV.   EXPEIMENTAL RESULT

Fig. 9 shows a linearity comparison between DPWM0 and DPWM1 at 1.2-V, 10-MHz switching and 16.3-ps resolution. DNL of DPWM0 is +2.77/-0.99 LSB, which guarantees no missing code by proposed code conversion (Fig. 5b). INL is +13.55/-2.49 LSB. DNL and INL of DPWM1 are +0.79/-0.84 and +6.75/-0.0 LSB, indicating the expected improvement. Fig 9 (b) is a linearity comparison when the power supply voltage of DPWM0 is set to 0.75 V and 1.2 V. The condition is 1-MHz switching and 163-ps resolution. DNL and INL are +0.60/-0.23 and +1.71/-1.13 LSB at 1.2-V supply voltage. With the 0.75-V supply voltage, DNL and INL are +1.49/-0.88 LSB and +7.00/-4.95 LSB. The result shows that the proposed architecture can operate at very low voltage. Table I shows the performance with the state-of-the-arts. The power consumption of DPWM0 is 0.762 mW at 16.3-ps time resolution and 10-MHz switching frequency, with 1.2-V supply voltage. DPWM1 is 1.489 mW under the same conditions. With 0.75-V supply voltage, 163-ps resolution and 1-MHz switching frequency of DPWM0 is 36-µW power dissipation. The proposed architecture achieves the lowest power consumption and the highest time-resolution. In addition, only the proposed architecture suggests instant-on without waiting time.

V.   CONCLUSION

This paper presents a DPWM using Zooming-PI achieving 16.3-ps time resolution with 0.762-mW power consumption.

TABLE I.   SUMMARY AND COMPARISON WITH PRIOR ARTS.

	[2]	[1]	[3], [4]	This work
Technology (nm)	180	250	28	130
Swithcing Frequency	100 k ~ 10 MHz	1 k ~ 10 MHz	244 k ~ 62.5 MHz	1.4 k ~ 40 MHz
Time Resolution (ps)	390.6 ~ 39062.5	195.3 ~ 48828	62.5	16,3 ~ 195.3
Supply Voltage (V)	1.8	2.5	1	1.2 (0.75)
Power (mW) (f_S)	3.24 (10 MHz)	2.36 (10 MHz)	0.96 [a] (12.5 MHz)	0.762 (10 MHz) 0.036 (1 MHz @0.75V)
Area (mm²)	0.435 (Single-Edge)	0.15 (Single-Edge)	0.002747 [a] (Single-Edge)	0.081 (Dual-Edge)
Base Architecture	Hybrid w/ DLL	Hybrid w/ PI	Hybrid w/ PLL	Hybrid w/ Zooming-PI
Instant Start-up	No (Dep. on DLL)	No (Dep. on bias circuits)	No (Dep. on PLL)	Yes (No biasing or DLL/PLL)

a. Estimated from the sum of [3] and [4].

Prototypes of two fabricated chip in 0.13-µm BCD technology has active area of 0.081 mm². Since the proposed Zooming-PI consists of only static circuits, zero-quiescent power, operation in a wide voltage range, and instantaneous start are possible. Thus, by this foundational work, a high-resolution and low-power DPWM should be available on devices of various applications in the near future.

REFERENCES

[1]   Y. Lee, T. Kang and J. Kim, "A 9–11 bits phase-interpolating digital pulse-width modulator with 1000X frequency range," 2014 IEEE Energy Conversion Congress and Exposition (ECCE), Pittsburgh, PA, 2014, pp. 2172-2176.

[2]   Z. Lukic, C. Blake, S. C. Huerta and A. Prodic, "Universal and Fault-Tolerant Multiphase Digital PWM Controller IC for High-Frequency DC-DC Converters," APEC 07 - Twenty-Second Annual IEEE Applied Power Electronics Conference and Exposition, Anaheim, CA, USA, 2007, pp. 42-47.

[3]   S. Höppner, S. Haenzsche, S. Scholze and R. Schüffny, "An all-digital PWM generator with 62.5ps resolution in 28nm CMOS technology," 2015 IEEE International Symposium on Circuits and Systems (ISCAS), Lisbon, 2015, pp. 1738-1741.

[4]   S. Hoppner, S. Haenzsche, G. Ellguth, D. Walter, H. Eisenreich and R. Schuffny, "A Fast-Locking ADPLL With Instantaneous Restart Capability in 28-nm CMOS Technology," in IEEE Transactions on Circuits and Systems II: Express Briefs, vol. 60, no. 11, pp. 741-745, Nov. 2013.

[5]   S. Kumaki, A. H. Johari, T. Matsubara, I. Hayashi and H. Ishikuro, "A 0.5V 6-bit scalable phase interpolator," 2010 IEEE Asia Pacific Conference on Circuits and Systems, Kuala Lumpur, 2010, pp. 1019-1022.

[6]   A. T. Narayanan, M. Katsuragi, K. Kimura, S. Kondo, K. K. Tokgoz, K. Nakata, W. Deng, K. Okada and A. Matsuzawa, "A Fractional-N Sub-Sampling PLL using a Pipelined Phase-Interpolator With an FoM of -250 dB," in IEEE Journal of Solid-State Circuits, vol. 51, no. 7, pp. 1630-1640, July 2016.

Fully-Integrated AMLED Micro Display System With a Hybrid Voltage Regulator

Junmin Jiang, Liusheng Sun, Xu Zhang, Shing Hin Yuen, Xianbo Li, Wing-Hung Ki, C. Patrick Yue and Kei May Lau

Department of Electronic and Computer Engineering
The Hong Kong University of Science and Technology, Hong Kong
Email: jjiangah@connect.ust.hk; eeki@ust.hk; eekmlau@ust.hk

Abstract—**A fully-integrated active matrix light-emitting diode (AMLED) micro display system is presented. The system consists of a 36×64 pixel-drivers encompassed by a fully on-chip hybrid voltage regulator built on the same silicon chip, then integrated with the AMLED array by using the flip-chip bonding technology. As such, no external passive component is needed. The hybrid voltage regulator consists of a step-up switched-capacitor converter cascaded by a step-down linear regulator. Operating with a wide input range of the lithium-ion battery (2.7V-4.2V), the voltage regulator delivers a maximum power of 216mW with 91% peak efficiency and 78% average efficiency.**

I. Introduction

Compact micro display systems have high demand in wearable devices and virtually reality (VR) applications. In recent years, active matrix light-emitting diode (AMLED) is gaining popularity [1-4]. Compared with the liquid crystal display (LCD), a light-emitting diode (LED) based micro display is self-emissive that produces bright images without an external light source, and is thus thinner and smaller. Note that LED is semiconductor-based, with a longer lifetime, lower fabrication cost and better robustness than technologies such as organic LED (OLED) [1].

An AMLED array is usually implemented using a gallium nitride (GaN) process. The ON and OFF status of each LED pixel is controlled by a pixel-driver, usually fabricated on silicon. In some designs, the AMLED array may be bonded on the pixel-drivers [2], but the power management unit (PMU) remains external [5]. Therefore, there is a strong motivation to merge the pixel-drivers and the PMU IC on the same silicon chip to reduce both the volume and the cost of the system.

Integrating the PMU with a switching DC-DC converter is a major challenge as it is difficult to fabricate a good power inductor on-chip [6], [7]. By comparison, linear regulators only need transistors and switched-capacitor (SC) converters only need capacitors and switches that are readily available on-chip. In particular, the efficiency of an SC converter could be high when the ratio of the output voltage V_O to the input voltage V_{IN} is closed to the ideal voltage conversion ratio of the converter [8].

Another important requirement for the PMU is low output voltage ripple, especially for driving an AMLED display, as the current of a micro LED is very sensitive to its bias voltage. A large voltage ripple may result in large variation of intensity, thus degrading the uniformity of the whole display. For an SC converter, reduced voltage ripple could be achieved by multiphase interleaving [9] or modulating the turn-on resistance [10]. For a linear regulator, the ripple voltage is much reduced through the operation of the feedback circuit.

In this research, we design a fully-integrated AMLED micro display system. From the discussion above, we propose to power up the display by a hybrid voltage regulator that eliminates external components and achieves low voltage ripple. It consists of a step-up switched-capacitor converter and a step-down linear regulator and covers the whole voltage range of a Li-ion battery, that is, from 2.7V to 4.2V. The pixel-drivers and the PMU are implemented on the same silicon chip, and together they are integrated with the AMLED array by using the flip-chip technology.

This paper is organized as follows. Section II introduces the system architecture and design considerations of the proposed micro display system. Section III presents the design of the hybrid voltage regulator. Section IV discusses the digital control of the pixels, and Section V introduces the design of the AMLED and the method of system integration. Measurement results are shown in Section VI and then followed by the conclusion.

II. System Architecture and Design Considerations

Fig. 1 shows the architecture of the proposed micro display system. It consists of a hybrid voltage regulator, a pixel-driver array and an AMLED array. The AMLED array has 36×64 pixels that are built on a GaN substrate, and each pixel needs a silicon pixel-driver that is flip-bonded on the same location as the LED pixel. The pixel-driver array is powered up by a voltage regulator built on the same silicon that provides a stable internal voltage V_O.

The image data are sent from a digital controller and are received by I/O pins ($C_{CK, D, EN}$, $R_{CK, D, EN}$ and RST) in a predetermined sequence. Row and column drivers are used to transfer the data to all the pixels. For the I/O pins to communicate with any off-chip voltage level up to 5V, high to low level shifters are used to convert the input signal to the internal voltage level.

978-1-5386-3179-9/17 $31.00 © 2017 IEEE

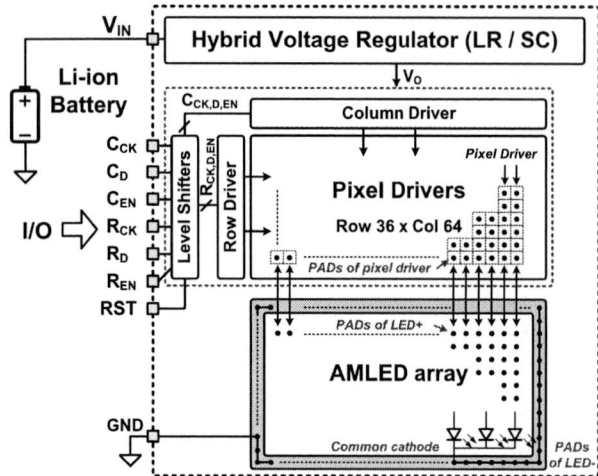

Fig. 1. System architecture of proposed micro display system.

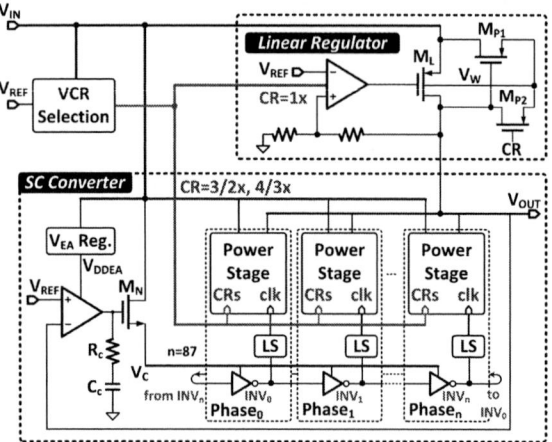

Fig. 2. Diagram of hybrid voltage regulator.

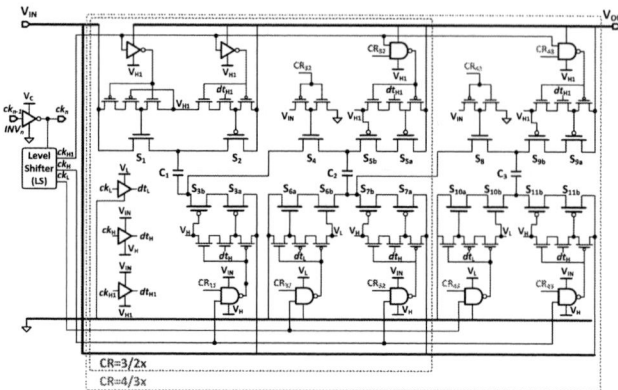

Fig. 3. Circuit implementation of power stage in SC converter.

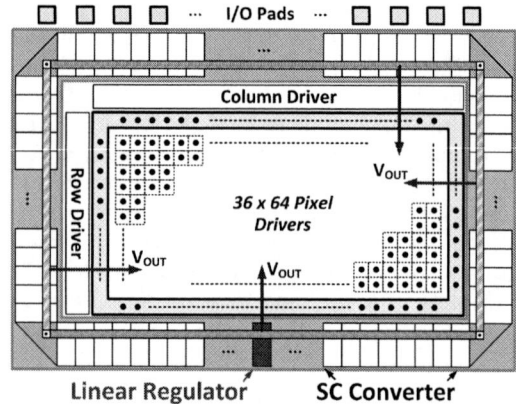

Fig 4. Floorplan of voltage regulator.

III. HYBRID VOLTAGE REGULATOR

The typical operating voltage of the LED and the pixel driver (V_O) is 3.6V. To cater for the Li-ion battery voltage that ranges from 2.7V to 4.2V, the voltage regulator should achieve both step-up and step-down voltage conversion. Fig. 2 shows the hybrid voltage regulator. It has three voltage conversion ratios (CRs). When the input voltage is lower than 3.8V, the SC converter is enabled and the CR could be adjusted to 3/2x or 4/3x. When the input voltage is higher than 3.8V, the linear regulator is enabled and the CR is 1/1x. The boundary is set higher than 3.6V to compensate for dropout voltage of the linear regulator.

Circuit implementation of the power stage of the SC converter is shown in Fig. 3. Each power cell has three flying capacitors (C_1, C_2 and C_3) that are built from both MOS and MIM capacitors to maximize the power density. Each cell has 11 power transistors (S_1 to S_{11}), some of which are implemented by stacking two low-voltage transistors together to handle high voltages. The voltage conversion ratio (VCR) is determined by the CR detection circuit that compares the input voltage V_{IN} with the reference voltage V_{REF}.

When only the linear regulator is needed (CR=1/1x), the SC converter is disabled, and all the flying capacitors are connected between V_O and ground, making them the load capacitor. This

is particularly important for the linear regulator with the dominant pole set at the output node, such that the compensation scheme is much simpler than capacitor-free designs. To avoid forward conduction of the body diodes, a body-selection circuit (M_{P1} and M_{P2}) is employed to ensure that the body of the PMOS transistor (M_L) is always tied to the highest voltage in the system, when V_O is higher than V_{IN}.

The layout floorplan is shown in Fig. 4. The AMLED array and the corresponding pixel-drivers occupy the large central area. For efficient power distribution, the power stage is divided into cascading power-cells that form a ring to encircle the pixel drivers, such that current can be delivered over the shortest path with the lowest resistance. Thick power rails are laid in between the pixel drivers to reduce the IR drop. For ripple reduction, the SC converter has a total of 87 interleaving phases. The number of phases and the aspect ratio of the power-cell are defined by the perimeter of the AMLED array. The maximum output current is 60mA.

IV. DISPLAY CONTROL

Fig. 5 shows the transistor implementation of the pixel drivers. Each pixel driver measures 40μm × 40μm and consists of three transistors (M_1, M_2 and M_3) and one capacitor (C_{ST}). The operating principle is as follows. A logic "0" enables the row R_{SEL}, turning on the corresponding M_1 and the display data is then written and stored in the capacitor C_{ST}. The voltage across C_{ST} controls the status of the driving transistor M_2. M_3 is controlled by the global enabled signal R_{EN}.

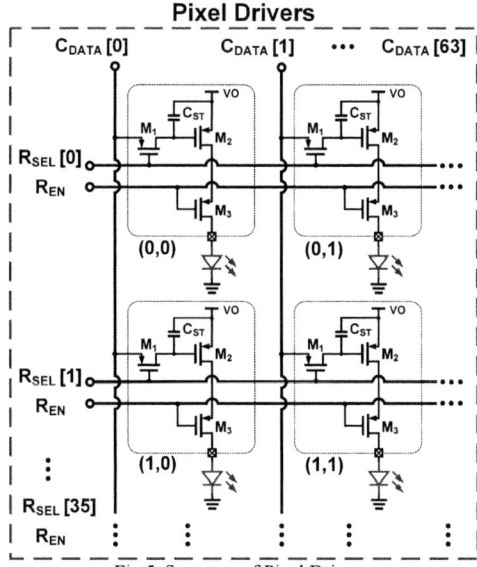

Fig 5. Structure of Pixel Drivers.

The display data is updated through row-by-row progressive scanning. First, the 64-bit display information of the first row is written into the column driver serially and also loaded into the signal lines from $C_{DATA}[0]$ to $C_{DATA}[63]$. Then, $R_{SEL}[0]$ is enabled, and each C_{ST} of the first row will store the display data from C_{DATA}. After that, $R_{SEL}[0]$ is disabled and the scanning of the first row is completed. The same procedure is then repeated for the other rows. After all pixels are loaded with the display information, R_{EN} is enabled to trigger and display the programmed image.

V. AMLED ARRAY AND SYSTEM INTEGRATION

The AMLED array consists of 36×64 pixels each measures 40µm×40µm. The aerial view of the LED array is shown in Fig. 6(a). The pixels were isolated by ICP etching down to sapphire and the trench was filled by a black mixture of three kinds of color filters, which could suppress the crosstalk between the pixels. The array adopts a common-cathode design, and all the LED pixels are uniformly distributed in the central area with p-type electrodes covered on the top, while the n-type electrodes were connected to the circumference of the LED array. The measured I-V curve of one LED pixel is shown in Fig. 6(b). The forward voltage of the pixel is 2.85V at 30 µA.

An Au-In metal bonding scheme is adopted to flip-chip bond together the AMLED array and the pixel-driver array. After indium deposition on the pixels' p-electrodes, the indium bumps can be formed through a reflow process in a furnace at 170°C for 1 second in a formic acid ambient. The scanning electron microscope (SEM) image of the indium bumps after reflow process is shown in Fig. 7(a). The diameter of the indium bumps is around 15µm and the height is about 7.5µm. The AMLED array is then flip-chip bonded on the silicon IC by using Au-In metal bonding under a pressure of 10N at 170°C for 1 minute. The two chips are aligned by reserved alignment marks.

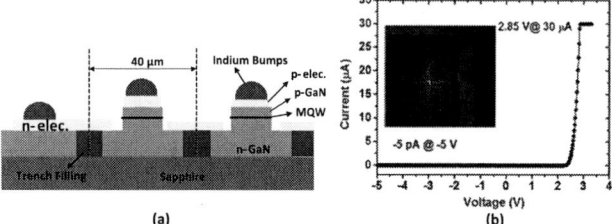

(a)   (b)

Fig. 6. (a) Cross sectional diagram of the LED µ-array with indium bumps; (b) IV characteristic of LED pixels.

(a)   (b)

Fig. 7. (a) SEM of LED µ-array with indium bumps after reflow process; (b) Alignment of AMLED array and silicon substrate.

VI. MEASUREMENT RESULTS

The silicon IC with the pixel-driver array and the PMU was fabricated in a 0.18µm bulk CMOS process, and the AMLED array was fabricated by a GaN process with a 1.6mm × 2.72mm area. Fig. 8(a) and (b) show the micrographs of the two chips. After flip-chip bonding, the power and I/O pins were connected to the PCB by wire-bonding.

Fig. 9 shows the measured steady-state waveform of V_O of the on-chip voltage regulator. The ripple voltage was less than 50mV under the full-load condition (I_O=60mA) with different V_{IN}. Fig. 10 shows the measured efficiency vs the input voltage under full and half brightness conditions. The peak efficiency was 91% when V_{IN} = 3.7V under the 1/1x mode. For the 3/2x and 4/3x modes, the efficiency over the whole range was higher than 70%. The averaged efficiency over the input range is 78%.

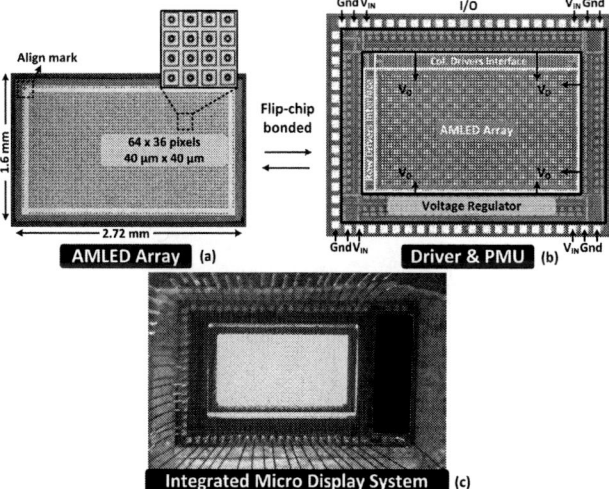

Fig.8. Chip micrographs of (a) AMLED array; (b) pixel driver and PMU and (c) integrated system.

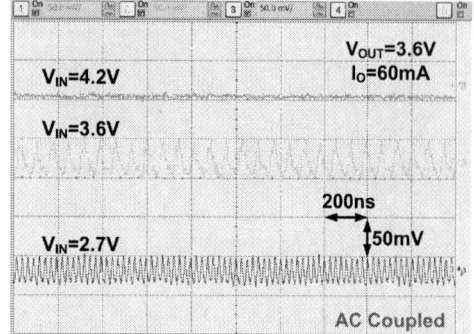

Fig.9. Measured waveforms of steady state output voltages.

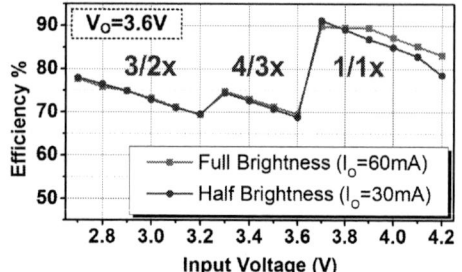

Fig.10. Measured efficiency of voltage regulator vs. input voltage.

The voltage regulator could deliver a maximum power of 216mW.

An Arduino DUE board is used to control the display images. Four examples are shown in Fig. 11. With a 4-bit grayscale control, images can be clearly rendered. Only a few dead pixels were observed, and the bonding yield rate is higher than 99.7%. It demonstrated that the proposed micro display system has potentials for portable display applications that require high performance, compact size and low cost.

CONCLUSION

A fully integrated AMLED micro display system is presented in this paper. The on-chip hybrid voltage regulator consists of a step-up switched-capacitor converter and a step-down linear regulator. It is capable of operating at a wide input range of 2.7V to 4.2V with peak efficiency of 91% and averaged efficiency of higher than 78%, and the maximum output power is 216mW. An AMLED array with 36×64 pixels was fabricated and integrated with silicon chip by using the flip-chip bonding technique. Pixel-divers are also designed to accomplish 4-bit grayscale control. By integrating the voltage regulator, the display system could be operating simply with a lithium-ion (Li-ion) battery without any external complements.

ACKNOWLEDGMENT

This work was supported in part by the Research Grants Council of the Theme-Based Research Scheme of Hong Kong under Project T23-612/12-R. The authors would like to thank Dr. L. K. Wong, Ms. Xun Liu, Mr. S. F. Luk and Dr. Jeffery C. C. Lo for their insightful discussions and technical supports.

REFERENCES

Fig.11. Source files (top) and its corresponding display images (bottom) shown in the blue micro display system.

[1] W. C. Chong, et al., "1700 pixels per inch (PPI) passive-matrix micro-LED display powered by ASIC," in *Proc. IEEE Compound Semicond. Integr. Circuit Symp.*, 2014, pp. 1–4.

[2] X. Li, et al., "Design and characterization of active matrix LED microdisplays with embedded visible light communication transmitter," *IEEE J. Lightwave Technology*, vol. 34, no. 14, pp. 3449–3457, Jul. 2016.

[3] D. Tsonev et al., "A 3-Gb/s single-LED OFDM-based wireless VLC link using a gallium nitride μLED," *IEEE Photon. Technol. Lett.*, vol. 26, no. 7, pp. 637–640, Jan. 2014.

[4] S. Zhang et al., "Directly color-tunable smart display based on a CMOS controlled micro-LED array," in *Proc. IEEE Photon. Conf.*, Sep. 2012, pp. 435–436.

[5] "TPS65632: Triple-Output AMOLED Display Power Supply," *Texas Insturments Datasheet*, Mar. 2015.

[6] X. Liu, C. Huang, P. K. T. Mok "A 50MHz 5V 3W 90% efficiency 3-level buck converter with real-time calibration and wide output range for fast-DVS in 65nm CMOS," *IEEE Symp. on VLSI Circuits*, Jun. 2016, pp. 1-2.

[7] X. Liu et al., "Analysis and design considerations of integrated 3-level buck converters," *IEEE Trans. Circuits Syst. I*, vol. 63 no. 5, pp. 671-682, May 2016.

[8] J. Jiang, et al., "A 2-/3-phase fully integrated switched-capacitor DC-DC converter in bulk CMOS for energy-efficient digital circuits with 14% efficiency improvement," in *Proc. IEEE Int. Solid-State Circuits Conf.*, Feb. 2015, pp. 366-367.

[9] Y. Lu, J. Jiang, and W.-H. Ki, "A multiphase switched-capacitor DC–DC converter ring with fast transient response and small ripple," *IEEE J. Solid-State Circuits*, vol. 52, no. 2, pp. 579–591, Feb. 2017.

[10] J. Jiang, W.-H. Ki, and Y. Lu, "Digital 2-/3-phase switched capacitor converter with ripple reduction and efficiency improvement", *IEEE J. Solid-State Circuits*, vol. 52, no. 7, pp. 1836-1848, July 2017.

S18-5 (2197)

A Low-Voltage Low-Offset Dual Strong-Arm Latch Comparator

Aikaterini Papadopoulou, Vladimir Milovanović[†] and Borivoje Nikolić
University of California at Berkeley, CA
[†]now with University of Kragujevac, Serbia
Email: katerina@eecs.berkeley.edu

Abstract—A dual strong-arm (DSA) comparator is designed targeting at low-voltage operation in deeply-scaled technologies. The addition of a second regenerative latch helps reduce both offset sensitivity and offset while maintaining comparable or better performance as a conventional double-tail latch across a wide range of voltages. A large comparator offset measurement array is fabricated in a 28nm FDSOI process. The DSA offset is measured to be 8.5mV across 6 dies, approximately 30% lower than the conventional topology at iso-area and iso-capacitance conditions. It is also shown to scale well with supply and common-mode voltage, achieving up to 65% lower offset across the voltage range.

I. INTRODUCTION

A high-speed, low-offset, and low-power comparator is a very versatile circuit, used in many applications such as analog-to-digital conversion, memory sensing circuits and data receivers. Voltage-mode comparators are particularly popular due to their zero static power consumption and rail-to-rail output.

One of the most common comparator designs is the strong-arm (SA) latch comparator [1] [2]. Although a low-power and robust circuit, the SA comparator is not necessarily the most appropriate choice for low-voltage operation due to increased mismatch in scaled technology nodes. Various techniques have been proposed, such as offset compensation or SA redundancy [3] [4], but those introduce additional design complexity and

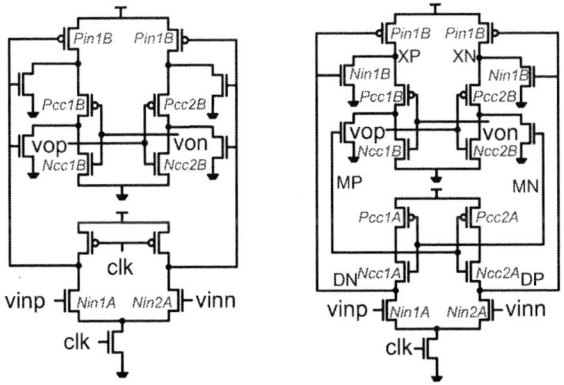

(a) Double-tail sense amp (DTSA) [5]   (b) Dual strong-arm (DSA)

Fig. 1. Conventional and proposed 2-stage comparator topologies. Precharge devices not shown.

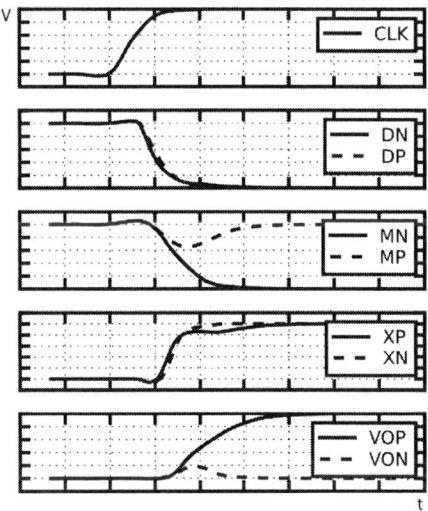

Fig. 2. Timing diagrams of the DSA comparator.

often require multiple clock phases. Two-stage comparator topologies, the state-of-the-art of which is shown in Figure 1a, generally combine an integrating first-stage and a regenerating second stage, and have been shown to be more appropriate for low-voltage operation, maintaining lower offset and larger design flexibility [5]–[7].

In this work, we present a dual strong-arm two-stage comparator topology, designed to desensitize offset from device variations, and demonstrating exceptional scaling with voltage.

II. DUAL STRONG-ARM COMPARATOR

The schematic of the DSA is shown in Figure 1b. The timing diagrams of the comparator for an input voltage difference of $\Delta V_{in} = 50$mV are shown in Figure 2. During the reset phase when CLK=0, the precharge devices charge nodes DN, DP and MN, MP to V_{DD} and therefore nodes XP, XN and VOP, VON are discharged to ground. During the amplification stage, when CLK becomes 1, the timing difference between nodes DN,DP and MN,MP enables the first stage to integrate the input while MN,MP hold the second latch in the precharged state, allowing it to see the larger signal and therefore reducing the impact of offset. When the second stage starts integrating, gain is provided through the input pair Nin1A, Nin2A as well as through the inverter pairs (Nin1B, Pin1B) and (Nin2B,

978-1-5386-3179-9/17 $31.00 © 2017 IEEE   281

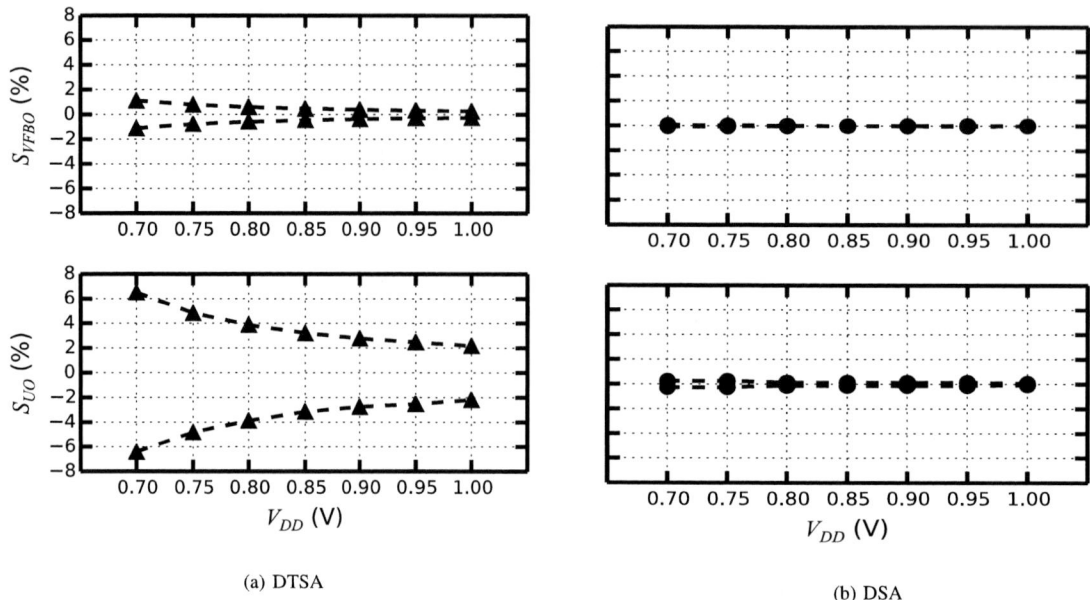

(a) DTSA   (b) DSA

Fig. 3. Comparison of simulated offset sensitivities to model parameters of device pairs (Pcc1B, Pcc2B) for two comparator topologies. The selected (Pcc1B, Pcc2B) is the most sensitive device pair of DTSA.

Pin2B) of the second stage. As soon as the voltage difference between the gate and source of Ncc1/2A or Pcc1/2B becomes larger than a V_{TH}, the corresponding latch quickly regenerates and the output is evaluated.

The comparator essentially consists of two regenerative latches, resulting in lower offset, based on the observation that the joint error probability of two half-sized sense amplifiers is lower than that of a unit-sized sense amplifier [8]. This allows for smaller individual devices to be used which desensitizes the offset from device variations, especially current variations. Furthermore, the input of the second regenerative latch is an integrated version of the comparator differential input, and it provides additional gain therefore reducing the offset contributions of device pairs in the second stage.

Offset sensitivity of a comparator is defined with respect to each comparator device as $S_{p_i} = \partial V_{offset}/\partial p_i$, where i indicates each statistical parameter of the model. The derivative is approximated using finite differences in simulation. Figure 3 shows the simulated offset sensitivity for device pair (Pcc1B, Pcc2B) for both the DTSA and the DSA comparators. All device lengths are minimum size and total active device width is $14.26\mu m$ for each latch. The comparators are sized to deliver approximately equal speed for an input voltage difference $\Delta V_{in} = 25mV$ at nominal supply. Input capacitance and load capacitance are kept the same for the two cases, in order to enable fair comparison of speed at various supplies. The tail device has been determined to significantly affect offset as it determines the current through the first stage, and unequal sizing can skew the results. Therefore, it is also kept equal in

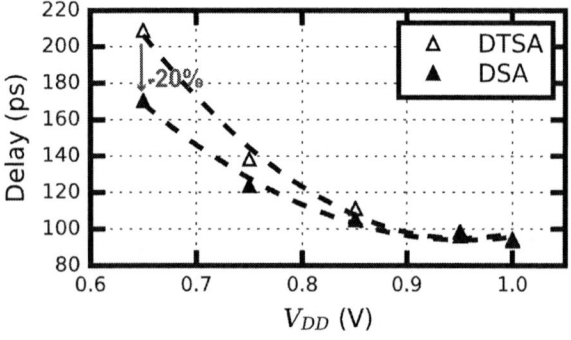

Fig. 4. Simulated delay scaling across supply voltage and least-square best-fit lines (dashed) for equal fanout DTSA and DSA.

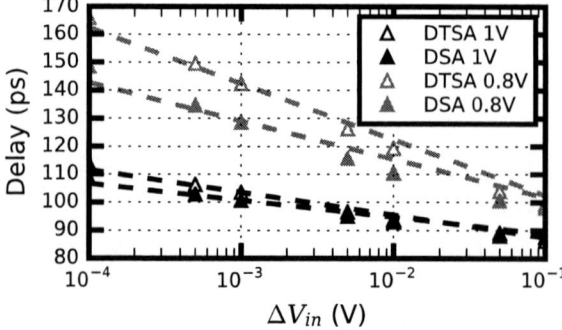

Fig. 5. Simulated delay across input voltage difference and least-square best-fit lines (dashed) for equal fanout DTSA and DSA.

Fig. 6. Die photo of the chip.

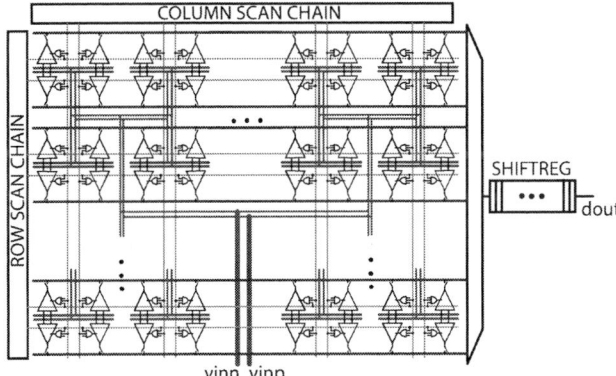

Fig. 7. Detailed block diagram of the comparator array, scan chains and output shift register.

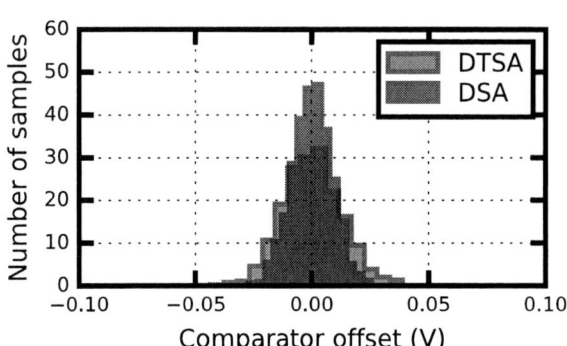

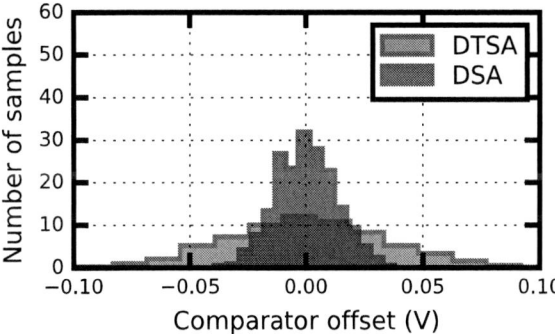

Fig. 8. Measured offset of DTSA and DSA for $V_{DD} = 1V$ (top) and $V_{DD} = 0.7V$ (bottom).

both cases. While the DTSA exhibits increased sensitivity to current variations as supply is scaling, ranging from 2.2% to 6.5 %, the DSA comparator maintains a low sensitivity ranging from 0.06% to 0.25%. Sensitivity improvement is achieved in all corresponding device pairs, but only the DTSA worst-case pair is shown in order to illustrate the improvement.

Figure 4 compares the two types of comparators in terms of speed. The DSA is shown to scale much better with supply voltage, maintaining a delay $\sim 20\%$ smaller that the DTSA at $V_{DD} = 0.65V$. Figure 5 shows a nominal simulation of the delay with respect to the input voltage difference ΔV_{in} at nominal and at scaled supply. The performance of the DSA is comparable or better than the DTSA for input differences that are smaller that 25mV, and its advantage becomes more pronounced at the scaled supply voltage.

III. SILICON MEASUREMENT RESULTS

A large array of comparators was fabricated in a 28nm FDSOI process. The die photo is shown in Figure 6 and chip details are outlined in [9]. Figure 7 shows a detailed block diagram of the chip. A total of 224 comparators per die were fabricated for each comparator design. Comparator sizing follows the rules presented in Section II to ensure fair comparison, and layout was carefully performed to avoid systematic effects that can affect the measurements. The comparators are arranged in an array addressed by scan chains, with shared

inputs. The input and buffered clock buses are delivered to each comparator through an H-tree in order to eliminate systematic offsets. The outputs are then multiplexed out to a 30kbit shift register. The design overall allows for a large sample size to be tested using a limited number of pads, while noise averaging and fast testing is possible through the use of the output shift register.

A slow ramp is applied at the input by using a programmable high-precision Keithley 2612A voltage source. The resolution of the input is set to $2\mu V$ and the swing is limited to 100mV to reduce the number of measurements. The source, all digital signals and the clock are controlled by the FPGA. The digital output of the shift register is read out and noise is averaged over the 30000 samples. A total of 1344 comparators across 6 dies were measured for each design. A measurement is assumed to fail when more than 10% of the measured samples are outside the input range.

Figure 8 shows the histogram distributions of the measured offsets at nominal and scaled supply, across all 6 dies. The measured standard deviation of the offset at nominal supply for the DSA is 8.5mV comparing to 12.7mV for the DTSA, a 33% improvement. The improvement is even more dramatic at $V_{DD} = 0.7V$, where the measured offsets for the DSA and DTSA are 13.1mV and 34.9mV, respectively, showing a 62.5% lower offset for the DSA. Figure 9 shows the measured offsets across supply voltage for a single die, in order to visualize the scaling advantage of the proposed topology. At $V_{DD} = 0.65V$

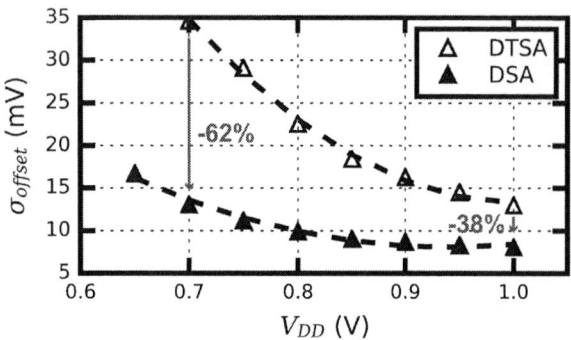

Fig. 9. Measured offset standard deviation across supply voltage for a single die.

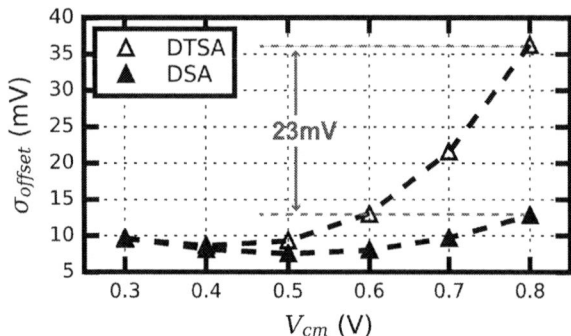

Fig. 10. Measured offset standard deviation across input common-mode for a single die.

the DTSA fails.

Figure 10 examines the operation of the two latches for various common mode voltages. Across the voltage range the offset of the DSA varies only by 5.3mV, in comparison with the DTSA offset which varies by 27.3mV. In addition, the offset of the DSA remains low at high input common-mode voltages, making it more appropriate for applications like memory. At the highest common-mode measured, the DSA has 23mV lower offset than the DTSA.

Figure 11 shows the spread of the measured standard deviation of the offset across all measured dies, showing that the DSA varies a lot less across dies. The contributors of random mismatch do not change from die to die, but systematic variation can cause, for example, all device lengths to be a little different in a different die, which in turn affects the device mismatch through the Pelgrom equation. The standard deviation of the offset for the DSA varies by < 1mV, showing exceptional tolerance to variations because of its reduced sensitivity to them.

IV. CONCLUSION

A dual strong-arm sense amplifier is presented as a variation-tolerant low-offset comparator for low-voltage ap-

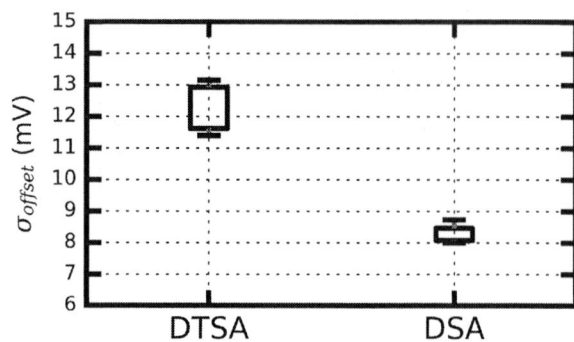

Fig. 11. Spread of the measured standard deviation of the offset across all dies for each comparator type.

plications. The comparator has reduced offset sensitivity to device variations, outperforms conventional topologies and is shown to scale well with supply voltage. Measurements in a 28nm FDSOI technology reveal a very low offset of 8.5mV at nominal supply voltage, which varies only by 7.5mV across supply and 5.3mV across common-mode voltage.

ACKNOWLEDGMENT

The authors wish to acknowledge the contributions of the students, faculty and staff of the Berkeley Wireless Research Center, wafer fabrication donation of STMicroelectronics, and funding in part by the Semiconductor Research Corporation (SRC).

REFERENCES

[1] T. Kobayashi at al., "A current-mode latch sense amplifier and a static power saving input buffer for low-power architecture," 1992 Symposium on VLSI Circuits Digest of Technical Papers, Seattle, WA, USA, 1992, pp. 28-29.

[2] J. Montanaro et al., "A 160 MHz 32 b 0.5 W CMOS RISC microprocessor," 1996 IEEE International Solid-State Circuits Conference. Digest of Technical Papers, ISSCC, 1996, pp. 214-215.

[3] B. Giridhar at al., "13.7 A reconfigurable sense amplifier with auto-zero calibration and pre-amplification in 28nm CMOS," 2014 IEEE International Solid-State Circuits Conference Digest of Technical Papers (ISSCC), 2014, pp. 242-243.

[4] M. Khayatzadeh et al., "A reconfigurable sense amplifier with 3X offset reduction in 28nm FDSOI CMOS," 2015 Symposium on VLSI Circuits (VLSI Circuits), pp. C270-C271.

[5] P. F. Chiu at al., "A double-tail sense amplifier for low-voltage SRAM in 28nm technology," 2016 IEEE Asian Solid-State Circuits Conference (A-SSCC), pp. 181-184.

[6] M. Miyahara at al., "A low-noise self-calibrating dynamic comparator for high-speed ADCs," 2008 IEEE Asian Solid-State Circuits Conference, pp. 269-272.

[7] D. Schinkel et al., "A Double-Tail Latch-Type Voltage Sense Amplifier with 18ps Setup+Hold Time," 2007 IEEE International Solid-State Circuits Conference. Digest of Technical Papers, pp. 314-605.

[8] N. Verma and A. P. Chandrakasan, "A 65nm 8T Sub-Vt SRAM Employing Sense-Amplifier Redundancy," 2007 IEEE International Solid-State Circuits Conference. Digest of Technical Papers, pp. 328-606.

[9] A. Papadopoulou and B. Nikolic, " A yield optimization methodology for mixed-signal circuits," 2017 IEEE Custom Integrated Circuits Conference, Austin, TX, 2017.

978-1-5386-3179-9/17 $31.00 © 2017 IEEE

A 5.35 mW 10 MHz Bandwidth CT Third-Order $\Delta\Sigma$ Modulator with Single Opamp Achieving 79.6/84.5 dB SNDR/DR in 65 nm CMOS

Wei Wang, Yan Zhu, Chi-Hang Chan, Seng-Pan U[1,2], Rui Paulo Martins[1,3]

State Key Laboratory of Analog and Mixed-Signal VLSI, University of Macau, Macao, China

1 – Also with Department of ECE, Faculty of Science of Technology, University of Macau, Macao, China

2 – Also with Synopsys Macau Ltd.

3 – On leave from Instituto Superior Técnico/Universidade de Lisboa, Portugal

email: yanzhu@ieee.org

Abstract - **This paper presents a power-efficient single-loop continuous-time (CT) third-order sigma delta ($\Delta\Sigma$) modulator that achieves a SNDR of 79.6 dB over a 10 MHz signal bandwidth. The modulator uses a feedforward-feedback (CIFF-FB) architecture which incorporates a single amplifier biquad (SAB) and a passive integrator to realize a third-order noise shaping. We also propose a continuous-time complementary (CTC) approach for the amplifier to improve the power efficiency. To alleviate the switch driver mismatch and jitter, we introduce an adaptive latch in the DAC driver. The modulator operates at a sampling rate of 640 MHz and consumes 5.35 mW from 1.2 V and 1.8 V power supplies. It achieves a DR of 84.5 dB and a SNDR of 79.6 dB with 10 MHz signal bandwidth, resulting in a Schreier FOM of 172.3 dB or 177.2 dB based on SNDR or DR, respectively.**

I. INTRODUCTION

The increasing call for a low power CT $\Delta\Sigma$ modulator in high-resolution and wideband applications imposes a considerable research effort in the design of optimized structures. Although a CT $\Delta\Sigma$ modulator benefits from both power efficiency and anti-alias filtering when compared with its discrete-time counterpart, it remains a challenge to target for high energy efficiency due to the power hungry integrators.

Conventional loop filters of CT $\Delta\Sigma$ modulators are often implemented with active integrators, such as active-RC and Gm-C. An N-order filter needs a number of N active integrators which not only leads to a large power and area but also increases design complexity. As shown in the power breakdown of several designs [1] [2], the amplifiers typically occupy around 40% of the total power consumption, where the 1st amplifier consumes ~20% determined by the thermal noise requirement of the modulator. Other amplifiers use the remaining 20% due to noise and driving capability requirements. Consequently, the amplifiers in the integrators become the modulator's bottleneck towards a better energy efficiency. Besides, the finite gain-bandwidth (GBW) of the amplifiers causes phase delay where

This work was financially supported by the Macao Science & Technology Development Fund (FDCT) with Ref no: 053/2014/A1, Research Grants of University of Macau with funding code number SRG2016-00081-AMSV and NSFC with funding code : F04020

the series connection of multiple amplifiers leads to a large phase shift, causing instability in the modulator [3].

To reduce the number of amplifiers in the loop filter, a noise-coupling (NC) technique for CT $\Delta\Sigma$ modulators was reported in [4], which is suitable for the modulators with an SAR ADC as the multi-bit quantizer. Noise-shaping quantizers, such as the voltage-controlled oscillator (VCO) quantizer [5], the noise-shaped integrating quantizer (NSIQ) [6], and the incorporation of both the VCO and the NSIQ [7] are also other possible alternatives. However, the bandwidth in [4], [6] and [7] is limited either by the speed of the SAR or the NSIQ quantizers, whereas the linearity of the VCO limits the input swing of the quantizer.

This paper proposes a third-order CT $\Delta\Sigma$ modulator with only one amplifier in the loop filter, achieved by incorporating a SAB and a passive integrator. The passive integrator also facilitates the excess loop delay (ELD) compensation without the active adder. We present a CTC approach and an adaptive latch to further improve the efficiency of the modulator. The prototype, fabricated in 65 nm CMOS, exhibits 79.6 dB signal-to-noise and distortion ratio (SNDR) and 84.5 dB dynamic range (DR) over a 10 MHz bandwidth, consuming only 5.35 mW. The Schreier figure-of-merit (FOM) is 172.3 dB or 177.2 dB based on SNDR or DR, respectively.

II. PROPOSED CT $\Delta\Sigma$ ADC ARCHITECTURE

Fig. 1 depicts the proposed CT CIFF-FB $\Delta\Sigma$ modulator topology with a 4-bit quantizer. We obtain the first and second integrators, as well as the feedforward path, with an improved version of SAB integrator. On the other hand, the last integrator

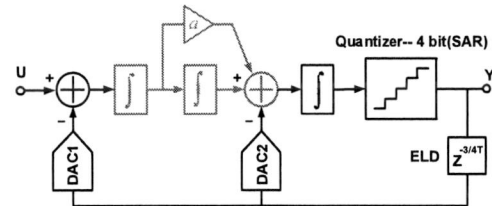

Fig.1. Proposed CT $\Delta\Sigma$ Modulator architecture.

is passive to save power. The SAR architecture is utilized in the multi-bit quantizer due to its outstanding efficiency and to avoid handling the mismatch from multiple comparators. DAC1 is the main feedback DAC and DAC2 compensates the ELD introduced by the quantizer and feedback path. This design implements a 3rd-order CT $\Delta\Sigma$ modulator with only one amplifier and 4-bit quantizer which results in an ADC with 79.6 dB SNDR and 10 MHz bandwidth. The sampling frequency is 640 MHz with an oversampling ratio (OSR) of 32 leading to a signal-to-quantization noise ratio (SQNR) of 99 dB. In this design with an 80 dB targeting SNR, we reserve a margin around 17 dB SQNR for other noise sources, such as the thermal noise, DAC mismatch and CLK jitter.

III. IMPLEMENTATION

A. Modulator

Fig. 2 shows the block diagram of the CT $\Delta\Sigma$ modulator. It achieves a 3rd-order noise-shaping by incorporating a SAB and a passive integrator, which allow single opamp implementation for low power and low design complexity. Besides, the passive integrator enables ELD compensation without the active adder. The only overheads are the DAC2 and its digital driving buffers which are relatively small as the error in the DAC2 can be attenuated by the 2nd-order noise shaping.

The main noise contributor of this design is the thermal noise of the input resistor (R1) as it cannot be shaped. Nevertheless, the current of the DAC and the capacitor load are determined by R1, implying that its value induces a tradeoff between the noise budget and power of the DAC and the opamp. We choose R1 to be 1k Ω to balance such tradeoff for the target specifications. In [2] and [3], a feedback resistor is introduced to improve the SQNR by optimizing the zero in the NTF, but its induced non-shapeable thermal noise indeed yields a diminishing return on the overall SNDR. Our SAB integrator removes the feedback resistor in the RC biquad which alleviates a non-shapeable noise source thus improving the overall SNDR of the modulator.

Even though the passive integrator does not induce significant inband noise, it imposes that the transfer function (TF) of the loop filter is non-ideal which can be shown by the follow TF of the modulator:

$$H(s) = \left(\underbrace{\frac{\frac{1}{R_1 C_1} s + \frac{1}{R_1 R_2 C_1 C_2}}{s^2}}_{1^{st} \ term} + k_{DAC2} \right) \times \frac{\frac{R_4}{R_3 + R_4} s + \frac{1}{C_3(R_3 + R_4)}}{s + \frac{1}{C_7(R_3 + R_4)}} \quad (1)$$

where k_{DAC2} is the feedback coefficient of the DAC2 used to compensate the ELD. Further, the passive filter moves the pole to p=1/C$_7$(R$_3$+R$_4$). It changes the NTF of the modulator and causes the SQNR to drop from 106 to 99 dB based on the simulations, and since we are targeting for a value around 80 dB, such influence is negligible. Nevertheless, the non-idealities of the SAB integrator greatly affect the performance of the modulator since they do not experience any noise-shaping effect. Its design considerations and non-ideal errors will be discussed next.

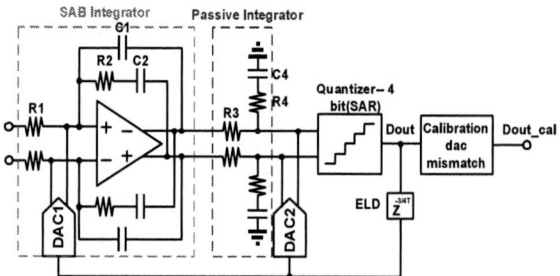

Fig.2. Block diagram of the proposed 3th-order $\Delta\Sigma$ ADC.

B. The load of SAB

In the SAB integrator, we introduce an extra RC network (R2 and C2) to realize a 2nd-order denominator with an extra zero in the 1st term of (1), which induces extra load to the amplifier when compared with a standard approach. The load of the SAB integrator can be analyzed by its equivalent impedance in the s-domain as:

$$Load = \frac{1}{Z_L} = sC_1 + \frac{1}{R_2 + \frac{1}{sC_2}} . \quad (2)$$

Obviously, the load is inversely proportional to R2 and directly proportional to C2. The extra load induced by R2 and C2 will become significant with a small R2 or a large C2. Here, the R2 is about 600 Ω, and both C1 and C2 are about 3 pF. Based on (2), the C$_2$-R$_2$ loop only increases the load by 2.7% at f_S which is a very small overhead in the modulator.

C. Finite opamp bandwidth in the loop filter

Unlike the conventional integrator, the finite GBW in the SAB integrator not only introduces a gain error and an additional pole [8], but also moves the poles from the origin to the left-half plane on the s-domain. The TF of the SAB integrator under the influence of the finite GBW can be expressed as:

$$H_{SAB}(s) = \frac{\frac{1}{CR_1} s + \frac{1}{C^2 R_1 R_2}}{s^2 + \frac{s}{GBW}\left(s^2 + \left(\frac{2}{CR_2} + \frac{1}{CR_1}\right)s + \frac{1}{C^2 R_1 R_2}\right)} . \quad (3)$$

Based on (3), Fig. 3 (a) and (b) plot the pole/zero map and the NTF, respectively, to analyze the influence of the finite GBW in the opamp. With the decrease of the GBW, the location of the zero is pushed away from DC and the poles are placed at a lower frequency which worsens the inband attenuation of the modulator. Moreover, the NTF only experiences a little deviation when the GBW is above 1.5 f_S, implying that the improvement of the inband attenuation is almost saturated. Based on the model from (3), when the GBW increases from 0.5 to 1.5 f_S, the SQNR improves by 13 dB. Plus, when the GBW is larger than 1.5 f_S, the improvements stay around only 2-3 dB. Such tradeoff between the GBW and SQNR is also similar in

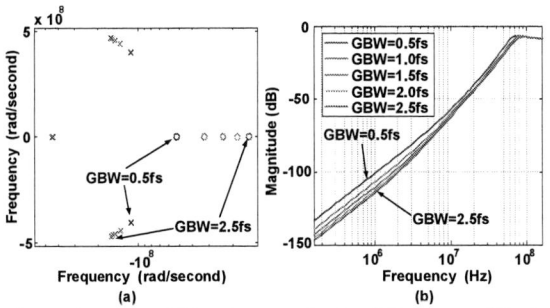

Fig.3. (a) NTF and (b) pole-zero plot for different amplifier's bandwidths.

the conventional integrator [8] which indicates that our SAB integrator does not induce extra burden on the 1st opamp. Consequently, we select a GBW of 1.5 f_S for a good energy efficiency and design margin.

D. Op-Amp

Fig. 4 shows the circuit-level schematic of the proposed CTC amplifier utilized in the SAB integrator. A two-stage topology, feed-forward path and Miller frequency compensation is adopted to improve the power efficiency. In [9], the complementary switched-capacitor technique is presented to double the g_m of the 1st stage based on a discrete-time operation. However, it is not applicable in the CT domain. As if M4a, b is supported by a DC biasing resistor, it introduces a pole-zero doublet, leading to a 6 dB loss before the zero location. In this amplifier, we introduce a CTC biasing circuit to place the zero close to the pole for cancelling the effect of the pole-zero doublet and avoiding gain loss, which only consumes about 1/35 of the power of the amplifier. Simulations show that the CTC structure can double the g_m of the 1st stage and improve the open loop gain by 6 dB. The amplifier consumes 1.6 mW at 1.2 V and achieves a 72 dB DC gain with a 1.5 f_S GBW.

E. DAC driver and feedback timing

We adopt a current steering DAC with a 1.8V power supply to obtain good noise and PSRR performances. The switch driver mismatch and jitter caused by the DAC driver directly affect the SNDR of the modulator without any attenuation. Such issues are critical in wideband designs because of their high sampling frequency. In order to alleviate these influences, the number of devices in the feedback path should be minimized. Fig. 5 shows the proposed adaptive latches in the DAC driver and the feedback timing of the DAC. Unlike other DAC drivers [1], [10], the one now proposed propagates the logic feedback without involving the latch operation. During the critical signal propagation (CLKd=0), the latch is disconnected from the feedback path and the CLK signal enables the propagation after the LSB decision from the quantizer is ready. The latch circuit resumes at CLKd=1 after the feedback is at a steady stage. The DAC control signals (Q_{nP}-Q_P and Q_{nN}-Q_N) have a high and a low cross point, respectively, which are adjusted through the sizing ratio of PMOS and NMOS in the transmission gate (TGF). They are optimized

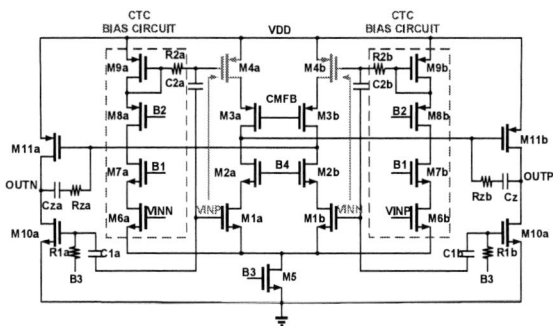

Fig.4. Two-stage CTC feed-forward amplifier.

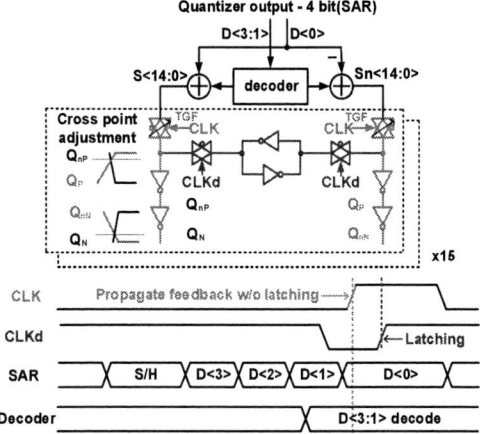

Fig.5. Adaptive latch in the DAC driver and feedback timing.

separately to avoid the DAC dynamic error induced by glitches at the rising edge of the CLK signal. With only three gates and no latch in the critical path, simulations indicate that the mismatch and jitter from the switch driver are reduced by 75%, 50% when compared with [1] or [10], respectively, with only 2/3 of power consumption in the latch.

Since we adopt the SAR quantizer in the design, it leads to a relatively long conversion time and metastability issue. In this design, we ensure a low error probability by assigning enough time for the comparator regeneration. Besides, its outputs need to be decoded from the binary to the thermometer to avoid inter-symbol-interference (ISI) in the feedback DACs. These imply that the feedback timing is very critical. In order to reduce the decoding time, we decode the SAR output D<1:3> in advance, and LSB D<0> directly feeds to the DAC without any decoding.

IV. MEASUREMENT RESULTS

The prototype, fabricated in 65 nm CMOS, has an active area of 0.033 mm² as illustrated by the die photo from Fig. 6. The modulator is clocked at 640 MHz and its signal bandwidth is 10 MHz with an OSR of 32. Fig. 7 shows the output spectrums of the modulator for a single-tone signal at a frequency of 1.4 MHz. The measured SNDR and SNR improve from 54.2 dB to 79.6

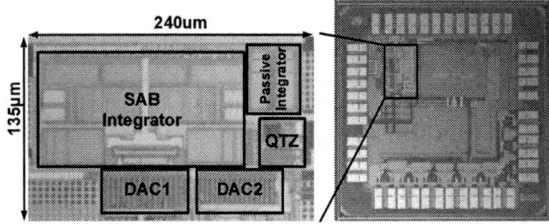

Fig.6. Die photograph.

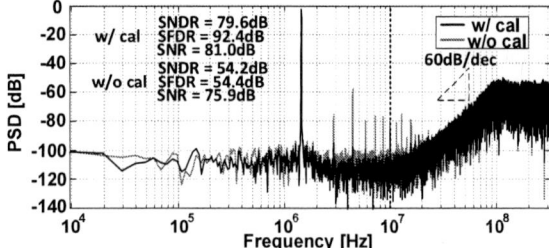

Fig.7. Measured single-ton FFT spectrum.

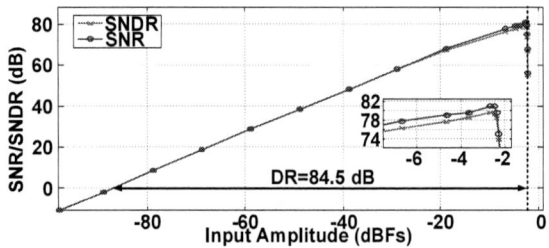

Fig.8.Measured SNDR and SNR vs. input amplitude.

dB and 75.9 dB to 81 dB with DAC mismatch calibration [11], respectively. The spurious-free dynamic range (SFDR) of 92.4 dB is dominated by the 2nd harmonic caused by the mismatch between the different current mirrors of the differential DAC. This error may also be stimulated by unsymmetrical coupling from the input, clock and output signals to the bias of the differential DAC. The 60 dB/decade spectral slope validates the 3rd-order noise-shaping by using the SAB and the passive integrator.

Fig. 8 shows the SNR and SNDR vs. input amplitude at 1.4 MHz which confirms that the proposed design achieves a DR of 84.5 dB. The total power consumption is 5.35 mW including 3.43 mW and 1.92 mW from analog and digital, respectively. The analog part includes the amplifier, DAC and comparator, and the digital part includes the logic buffer, SAR logic, DAC driver and CLK generator. TABLE I summarizes the measured performance of this work and compares it with the state-of-art CT $\Delta\Sigma$ designs with similar BW. The good power efficiency is benefited from the single amplifier modulator, the CTC approach and the adaptive latch DAC driver.

V. CONCLUSIONS

This paper presented a power-efficient CT $\Delta\Sigma$ modulator by incorporating a SAB and a passive integrator. It facilitates a

TABLE I. SUMMARY OF PERFORMANCE AND BENCHMARK WITH STATE-OF-THE-ART

	T.Kim ISSCC 2017	T.Kim VLSI 2015	B.N ISSCC 2016	Y.Shu ISSCC 2013	G.Wei VLSI 2015	This work
Area (mm²)	0.17	0.08	0.027	0.08	0.066	0.033
Technology (nm)	130	130	65	28	28	65
Supply Voltage (V)	1.2	1.2	1.0	1.2/1.6	0.9/1.8	1.2/1.8
Fs (MHz)	640	640	1000	640	432	640
Bandwidth (MHz)	15	10	10	18	5	10
Power (mW)	11.4	7.19	1.57	3.9	3.16	5.35
Peak SNDR (dB)	80.4	75.3	72.2	73.6	80.5	79.6
DR (dB)	82.9	78.5	77	78.1	83.9	84.5
FOMSch/SNDR (dB)	171.6	166.7	170.2	170.2	172.5	172.3
FOMSch/DR (dB)	174.1	169.9	172.0	174.7	175.9	177.2
FoMWa (fJ/conv.step)	44.1	75.9	23.6	27.7	36.5	36.5

small area and power efficient modulator architecture with single amplifier. The SAB realizes a 2nd-order integrator which provides 40 dB/decade slopes with little power overhead. The passive integrator implements 20 dB/decade slopes and enables ELD compensation without active adder. Moreover, to improve the power efficiency of the amplifier and address the jitter and switch driver mismatch in the DAC drivers, we proposed a CTC amplifier and an adaptive latch circuit, respectively. These techniques implemented in a CT $\Delta\Sigma$ modulator achieve a SNDR of 79.6 dB and a Schreier FOM of 172.3 dB with 10 MHz signal bandwidth.

REFERENCES

[1] C. Y. Ho, et al., "A 4.5 mW CT Self-Coupled $\Delta\Sigma$ Modulator With 2.2 MHz BW and 90.4 dB SNDR Using Residual ELD Compensation," in *IEEE JSSC*, vol. 50, no. 12, pp. 2870-2879, Dec. 2015.

[2] L. Breems *et al.*, "A 2.2 GHz Continuous-Time $\Delta\Sigma$ ADC With −102 dBc THD and 25 MHz Bandwidth," in *IEEE JSSC*, vol. 51, no. 12, pp. 2906-2916, Dec. 2016.

[3] R. Zanbaghi, et al., "An 80-dB DR, 7.2-MHz Bandwidth Single Opamp Biquad Based CT Delta Sigma Modulator Dissipating 13.7-mW," in *IEEE JSSC*, vol. 48, no. 2, pp. 487-501, Feb. 2013.

[4] B. Wu, et al., "15.1 A 24.7mW 45MHz-BW 75.3dB-SNDR SAR-assisted CT $\Delta\Sigma$ modulator with 2nd-order noise coupling in 65nm CMOS," in *IEEE ISSCC*, pp. 270-271, Feb. 2016.

[5] K. Reddy et al., "A 16-mW 78-dB SNDR 10-MHz BW CT $\Delta\Sigma$ ADC Using Residue-Cancelling VCO-Based Quantizer," in *IEEE JSSC*, vol. 47, no. 12, pp. 2916-2927, Dec. 2012.

[6] N. Maghari et al., "A Third-Order DT $\Delta\Sigma$ Modulator Using Noise-Shaped Bi-Directional Single-Slope Quantizer," in *IEEE JSSC*, vol. 46, no. 12, pp. 2882-2891, Dec. 2011.

[7] T. Kim, et al., "28.2 An 11.4mW 80.4dB-SNDR 15MHz-BW CT delta-sigma modulator using 6b double-noise-shaped quantizer," in *IEEE ISSCC*, pp. 468-469, 2017.

[8] F. Gerfers, et al., "A 1.5-V 12-bit power-efficient continuous-time third-order $\Delta\Sigma$ modulator," in *IEEE JSSC*, vol. 38, no. 8, pp. 1343-1352, Aug. 2003.

[9] J. Wu et al., "A 5.4GS/s 12b 500mW pipeline ADC in 28nm CMOS," in IEEE *VLSI*, pp. C92-C93, Jun. 2013.

[10] C. H. Lin et al., "A 12 bit 2.9 GS/s DAC With IM3 ≪ -60 dBc Beyond 1 GHz in 65 nm CMOS," in *IEEE JSSC*, vol. 44, no. 12, pp. 3285-3293, Dec. 2009.

[11] M. Bock et al., "Calibration of DAC Mismatch Errors in $\Sigma\Delta$ ADCs Based on a Sine-Wave Measurement," in *IEEE TCAS. II*, vol. 60, no. 9, pp. 567-571, Sep. 2013.

978-1-5386-3179-9/17 $31.00 © 2017 IEEE

A 72.9-dB SNDR 20-MHz BW 2-2 Discrete-Time Sturdy MASH Delta-Sigma Modulator Using Source-Follower-Based Integrators

Yong-Sik Kwak, Kang-Il Cho, Ho-Jin Kim, Seung-Hoon Lee, and Gil-Cho Ahn
Department of Electronic Engineering, Sogang University
35 Baekbeom-ro, Mapo-gu, Seoul, Korea
gcahn@sogang.ac.kr

Abstract—This paper presents a 2-2 discrete-time sturdy multi-stage noise-shaping (SMASH) delta-sigma modulator using source-follower-based open-loop integrators. The resolution of the SMASH delta-sigma modulator is enhanced by eliminating the first-stage quantization noise from the output. Using the proposed source-follower-based open-loop integrator, the operating speed of the modulator is efficiently improved. The prototype delta-sigma modulator fabricated in a 65-nm CMOS process achieves a 75.8-dB dynamic range and 72.9-dB SNDR in a 20-MHz bandwidth. The modulator occupies an active area of 0.34 mm^2, and its total power consumption is 20.4 mW from a 1.2-V supply voltage operating at a 500-MHz clock frequency.

Index Terms—analog-to-digital converter (ADC), discrete-time (DT), delta-sigma, multi-stage noise-shaping (MASH), open-loop integrator, source-follower

I. INTRODUCTION

Discrete-time (DT) delta-sigma ($\Delta\Sigma$) modulators have been widely used owing to their robust operation with accurate loop coefficients and easy pole location scaling of integrators with the operating clock frequency. However, the design of DT modulators for wideband applications poses challenges resulting from the increased clock speed. The settling time of the op-amps used for the switched-capacitor (SC) integrators limits the operating clock frequency. To increase the bandwidth (BW) of the modulator, the oversampling ratio (OSR) needs to be reduced while the order of noise-shaping and the quantizer resolution are increased. A multi-stage noise-shaping (MASH) architecture is a good candidate for this purpose, since it enables high-order noise-shaping without stability issues. High gain op-amps are required to achieve accurate gain coefficients of the integrators for minimizing the quantization noise leakage in MASH modulators. However, it is difficult to realize a fast-settling and high-gain op-amp in deep submicron CMOS technology.

A sturdy multi-stage noise-shaping (SMASH) architecture is proposed to avoid the use of a high gain op-amp in the integrator [1]. By removing digital filters used at the output of each stage, the gain accuracy requirement of the integrator for the multi-loop modulator is relaxed, while maintaining the advantage of stable high-order noise shaping of the conventional MASH architecture. However, in a conventional SMASH modulator, the quantization noise of each stage appears in the modulator output resulting in signal-to-noise ratio (SQNR) degradation unlike in a conventional MASH modulator.

In this paper, a 2-2 DT SMASH $\Delta\Sigma$ modulator architecture that enhances the resolution by eliminating the first-stage quantization noise from the output is presented. A source-follower-based open-loop integrator is proposed to efficiently increase the operating clock frequency of the SC integrators. The SMASH $\Delta\Sigma$ modulator with the feed-forward topology is used to address the design issues resulting from the non-idealities of the source-follower-based open-loop integrator [2]. The modulator is clocked at 500 MHz, and it achieves a 75.8-dB dynamic range (DR), 83.7-dB SFDR, and 72.9-dB SNDR at a 20.8-MHz BW.

II. SOURCE-FOLLOWER-BASED OPEN-LOOP INTEGRATOR

The source-follower-based open-loop integrator shown in Fig. 1 is proposed to increase the operation speed by avoiding the overhead of feedback. The integrator consists of pseudo-differential source-followers with two pairs of differential SC networks. A self-biased PMOS transistor is used for the source-follower input to minimize the non-linearity resulting from the body effect. Integration is achieved by sampling the output into each capacitor pair and connecting it in series with the input alternately. A z-domain block diagram of the proposed integrator with the assumption of unity gain source-followers is depicted in Fig. 2. Two ϕ_1 phase inputs, $V_{IN}(z)$ and $-V_{IN}(z) \cdot z^{-1}$, are integrated with the ϕ_2 phase input, $V_{IN}(z) \cdot z^{-1/2}$.

In the proposed integrator, the gain accuracy and linearity are limited because it uses open-loop topology with the source-followers. The simulated voltage gain of the source-follower is 0.95 in a typical corner. The gain of the proposed integrator with this source-follower is equivalent to that of a conventional SC integrator using a 26-dB DC gain op-amp. These integrator non-idealities are addressed by employing the proper topology, which will be explained in the following section.

III. MODULATOR ARCHITECTURE

To avoid performance degradation resulting from the inaccurate gain and non-linearity of the source-follower based integrator, feed-forward and SMASH topologies are employed [1], [2].

978-1-5386-3179-9/17 $31.00 © 2017 IEEE

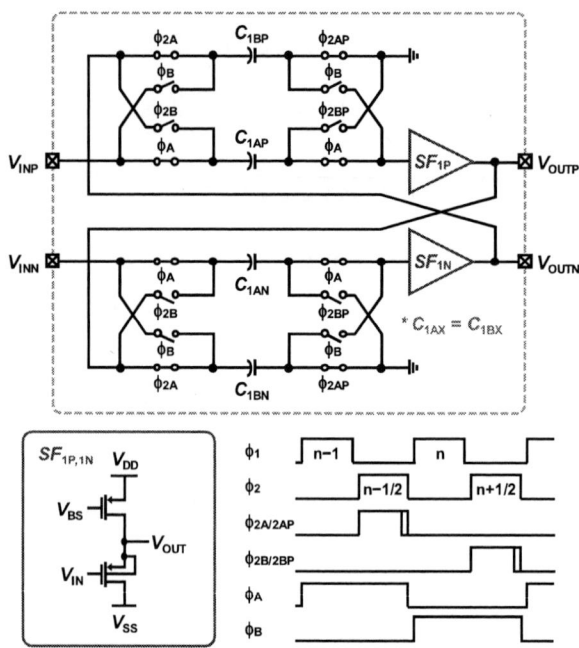

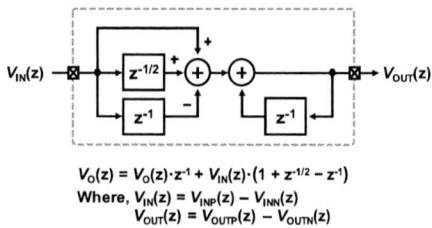

Fig. 1. Schematic of the proposed source-follower-based open-loop integrator.

$$V_O(z) = V_O(z) \cdot z^{-1} + V_{IN}(z) \cdot (1 + z^{-1/2} - z^{-1})$$
$$\text{Where, } V_{IN}(z) = V_{INP}(z) - V_{INN}(z)$$
$$V_{OUT}(z) = V_{OUTP}(z) - V_{OUTN}(z)$$

Fig. 2. Z-domain block diagram of the proposed integrator.

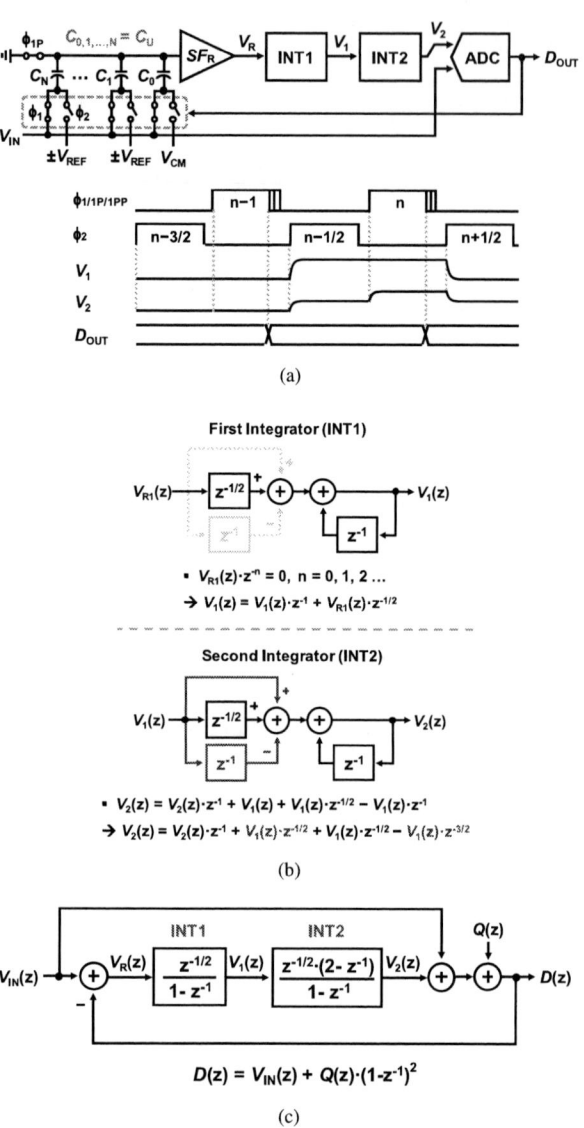

Fig. 3. (a) Schematic diagram of the second-order feed-forward $\Delta\Sigma$ modulator. (b) Transfer function of the first and second integrator. (c) Z-domain block diagram of the second-order feed-forward $\Delta\Sigma$ modulator.

A. Second-Order Feed-Forward $\Delta\Sigma$ Modulator

Fig. 3(a) shows a second-order feedforward modulator that uses the proposed source-follower-based integrators. It is composed of a SC digital-to-analog converter (DAC) with a source-follower for buffering, two cascaded integrators, and a quantizer. The transfer function of each integrator can be derived as follows. The first integrator input, V_R, is zero during the ϕ_1 phase as shown in Fig. 3(b). The output of the first integrator is given by

$$V_1(z) = V_1(z) \cdot z^{-1} + V_R(z) \cdot z^{-1/2}. \quad (1)$$

Since $V_1(n) = V_1(n-1/2)$ and $V_1(n-1) = V_1(n-3/2)$ as shown in Fig. 3(a), the output of the second integrator can be derived as

$$V_2(z) = V_2(z) \cdot z^{-1} + 2 \cdot V_1(z) \cdot z^{-1/2} - V_1(z) \cdot z^{-3/2}. \quad (2)$$

The z-domain block diagram of the second-order input feed-forward modulator using the proposed integrators and its total transfer function are depicted in Fig. 3(c). It has the same

signal and noise transfer functions as the conventional feed-forward topology [2]. Each integrator of the proposed feed-forward modulator processes quantization noise only, resulting in a relaxed gain and linearity requirement.

B. Resolution-Enhanced DT SMASH $\Delta\Sigma$ Modulator

A block diagram of the proposed 2-2 DT SMASH $\Delta\Sigma$ modulator is depicted in Fig. 4. It employs the second-order input feed-forward architecture using the source-follower-based integrators for both stages. In the conventional SMASH $\Delta\Sigma$ modulator, it is hard to cancel out the first-stage quantization noise (Q_1) due to the timing delay of signal processing in the second stage [1]. The Q_1 needs to be extracted and converted with zero delay through the second stage for the cancelation, which is not practically feasible in a DT modulator. To avoid the timing delay issue, the Q_1 is extracted half clock period

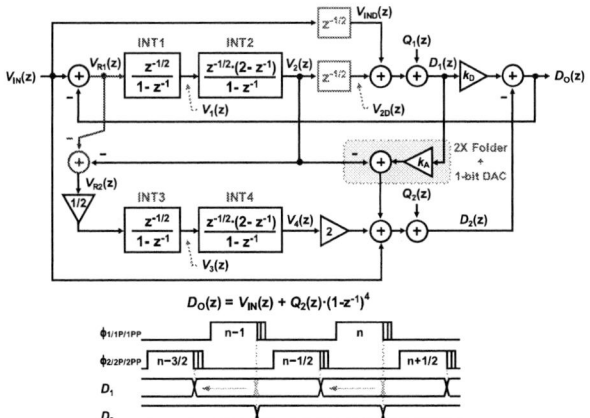

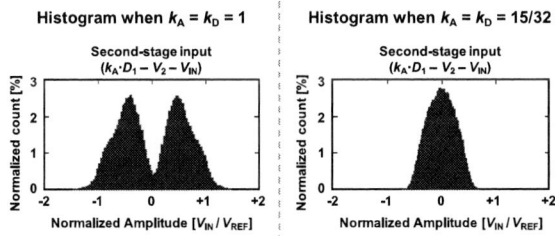

Fig. 4. Z-domain block diagram of the proposed SMASH $\Delta\Sigma$ modulator.

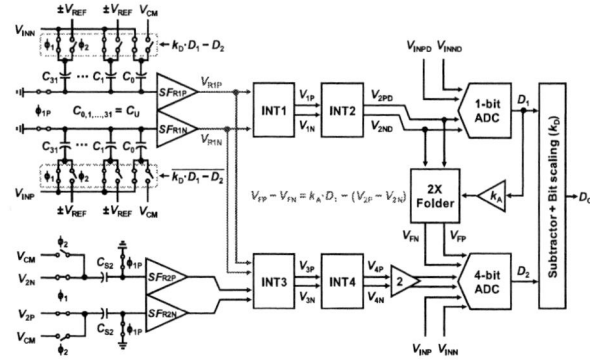

Fig. 6. Schematic diagram of the proposed SMASH $\Delta\Sigma$ modulator.

Fig. 5. Histograms of the second-stage input.

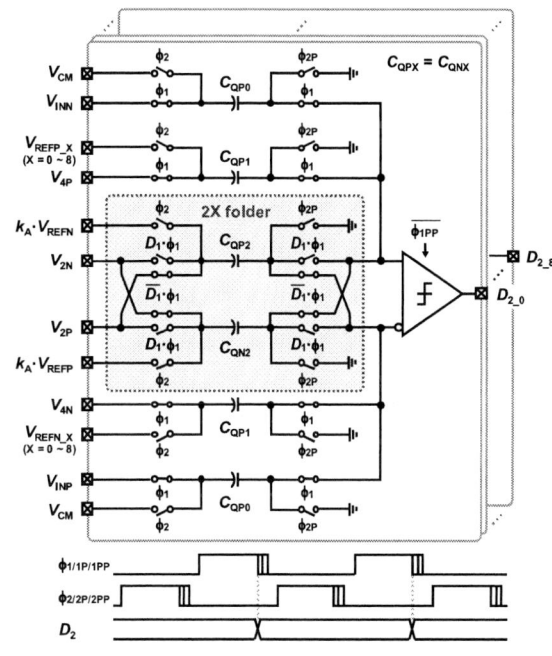

Fig. 7. Schematic diagram of the second-stage quantizer.

earlier than conventional architecture in the first stage by adding two $z^{-1/2}$ blocks in front of the quantizer. It allows half clock period timing margin for the conversion of extracted Q_1 in the second stage. The Q_1 and the errors resulting from the modified timing of the first-stage quantizer are eliminated by subtracting the second-stage output from the first-stage output.

Since the Q_1 is not included in the modulator output D_O, a low-resolution quantizer can be employed for the first stage. In this work, a 1-bit quantizer is adopted for the first stage. It relaxes the design complexity of the SC network for transferring the extracted Q_1 to the second stage. To reduce the output swing range of the integrators and achieve high resolution with a low OSR, a 4-bit quantizer is used for the second stage. A reference shuffling DWA [3] is embedded in this 4-bit quantizer to improve the linearity of the feedback DAC with high-speed operation. Gain coefficients k_D and k_A are inserted at the output of the first-stage quantizer to reduce the second-stage input signal. Fig. 5 shows histograms of the second-stage input when $k_D = k_A = 1$ (left) and $k_D = k_A = 15/32$ (right). By inserting the k_D and k_A, the required number of comparators for the second-stage 4-bit quantizer is reduced from 15 to 9. These gain coefficients are implemented with bit scaling in the digital subtractor (k_D) and scaled reference voltages for the 1-bit DAC in the second stage (k_A).

A schematic diagram of the proposed 2-2 DT SMASH $\Delta\Sigma$ modulator is shown in Fig. 6. A 5-bit feedback DAC is shared by the first and second stages, as indicated by the blue lines. The separation of the residue signal ($V_{R1P} - V_{R1N}$)

from the first integrator output using source-followers SF_{R1P} and SF_{R1N} enables this sharing. Thus, the total area of the SMASH $\Delta\Sigma$ modulator is reduced. A sample-and-hold SC network with SF_{R2P} and SF_{R2N} is employed in front of the third integrator to achieve the same timing as the first stage. A summing operation of the third-integrator input signals is implemented by using passive SC networks. Because the integrator employs a passive summation, the input signals of the third integrator are attenuated by a factor of 2. This attenuation is compensated at the output of the fourth integrator by using an open-loop amplifier, which has a gain of 2. The required gain accuracy and linearity of this amplifier are also relaxed by the feed-forward topology and noise-shaping. Fig. 7 shows a schematic of the second-stage quantizer. A 2X folding technique [4] is used to reduce the number of sampling capacitors for passive summation. It enables the passive signal subtraction by using a single SC network without gain attenuation.

IV. MEASUREMENT RESULTS

The prototype DT 2-2 SMASH $\Delta\Sigma$ modulator is implemented in a 65-nm CMOS process, and the core area is 0.34 mm^2. The modulator consumes 20.4 mW (analog 13.8 mW and digital 6.6 mW) from a 1.2-V supply at a sampling rate of 500 MHz. Fig. 8 shows the measured output spectrum with a –2.1-dBFS and 3-MHz input signal. The SNR and SNDR performances versus the input amplitude are depicted in Fig. 9. In a 20-MHz signal BW, the measured DR, peak SNR, and peak SNDR are 75.8 dB, 73.7 dB, and 72.9 dB, respectively. The prototype achieves the Schreier figure-of-merit (FOM$_S$) of 165.7 dB and Walden FOM (FOM$_W$) of 140.9 fJ/conversion-step. A die photograph of the prototype modulator is shown in Fig. 10. Table. I summarizes the performance of the prototype modulator and compares it with that of recent state-of-the-art wideband DT and continuous-time (CT) $\Delta\Sigma$ modulators (at BW $\geq$ 15 MHz).

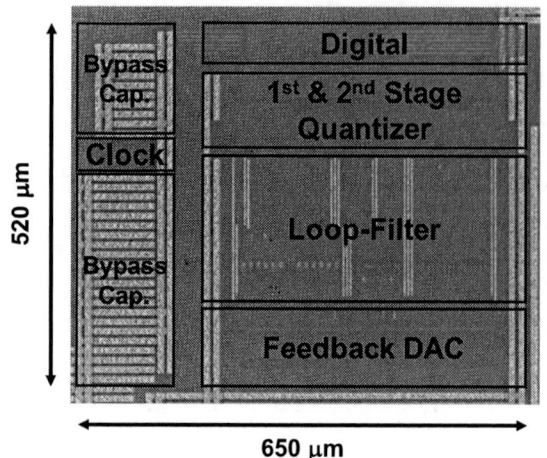

Fig. 10. Die photograph of the prototype $\Delta\Sigma$ modulator.

TABLE I
PERFORMANCE SUMMARY AND COMPARISON

	This Work	JSSC'14 [5]	VLSI'13 [6]	JSSC'13 [7]	CICC'17 [8]	ISSCC'17 [9]
Architecture	DT	DT	DT	DT	CT	CT
Process [nm]	65	65	22	65	65	130
Fs [MHz]	500	240	240	1100	1500	640
BW [MHz]	20	15	15	16.7	50	15
SNDR [dB]	72.9	67.2	66.0	74.3	73.5	80.4
DR [dB]	75.7	68.2	66.0	81.0	76.1	82.9
Power [mW]	20.4	46.0	12.7	94.0	51.8	11.4
FOM$_W$ [fJ/conv.]	140.9	838.1	259.6	663.4	134.0	44.1
FOM$_S$ [dB]	165.7	153.3	156.7	163.5	166.0	174.1

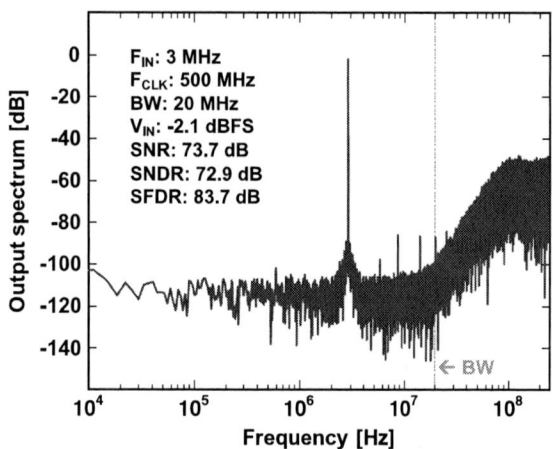

Fig. 8. Measured output spectrum.

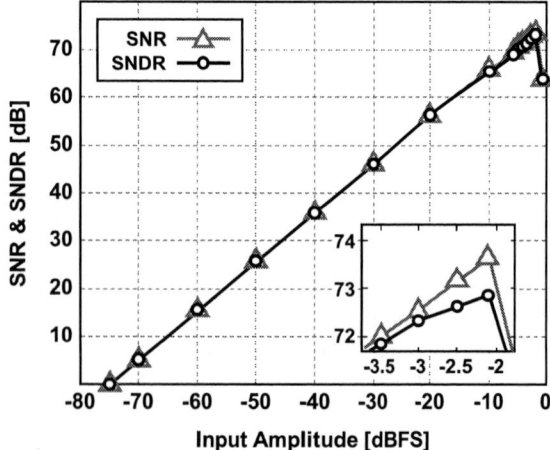

Fig. 9. SNR and SNDR versus input amplitude.

REFERENCES

[1] N. Maghari, S. Kwon, and U.-K. Moon, "74 dB SNDR Multi-Loop Sturdy-MASH Delta-Sigma Modulator Using 35 dB Open-Loop Opamp Gain," *IEEE J. Solid-State Circuits*, vol. 44, no. 8, pp. 2212-2221, Aug. 2009.

[2] J. Silva, *et al.*, "Wideband low-distortion delta-sigma ADC topology," *Electron. Lett.*, vol. 37, no. 12, pp. 737-738, Jun. 2001.

[3] W. Yang, *et al.*, "A 100mW 10MHz-BW CT $\Delta\Sigma$ Modulator with 87dB DR and 91dBc IMD," in *IEEE ISSCC Dig. Tech. Papers*, pp. 498-631, Feb. 2008.

[4] B. Verbruggen, *et al.*, "A 2.2 mW 5b 1.75GS/s Folding Flash ADC in 90nm Digital CMOS," in *IEEE ISSCC Dig. Tech. Papers*, pp. 252-611, Feb. 2008.

[5] S.-C. Lee and Y. Chiu, "A 15-MHz Bandwidth 1-0 MASH $\Delta\Sigma$ ADC With Nonlinear Memory Error Calibration Achieving 85-dBc SFDR," *IEEE J. Solid-State Circuits*, vol. 49, no. 3, pp. 695-707, Mar. 2014.

[6] C. C. Lee, *et al.*, "A 66dB SNDR 15MHz BW SAR Assisted $\Delta\Sigma$ ADC in 22nm Tri-gate CMOS," in *Proc. IEEE Symp. VLSI Circuits*, pp. 64-65, Jun. 2013.

[7] S.-H. Wu and J.-T. Wu, "A 81-dB Dynamic Range 16-MHz Bandwidth $\Delta\Sigma$ Modulator Using Background Calibration," *IEEE J. Solid-State Circuits*, vol. 48, no. 9, pp. 2170-2179, Sep. 2013.

[8] S. Dey, *et al.*, "A 50 MHz BW 73.5 dB SNDR Two-stage Continuous-time $\Delta\Sigma$ Modulator with VCO Quantizer Nonlinearity Cancellation," in *Proc. IEEE Custom Integrated Circuits Conf.*, pp. 1-4, May. 2017.

[9] T. Kim, C. Han, and N. Maghari, "An 11.4 mW 80.4 dB-SNDR 15MHz-BW CT Delta-Sigma Modulator Using 6b Double-Noise-Shaped Quantizer," in *IEEE ISSCC Dig. Tech. Papers*, pp. 468-469, Feb. 2017.

A Compact 87.1-dB DR Bandwidth-Scalable Delta-Sigma Modulator Based on Dynamic Gain-Bandwidth-Boosting Inverter for Audio Applications

Young-Ha Hwang, Jun-Eun Park, and Deog-Kyoon Jeong

Department of Electrical and Computer Engineering
Seoul National University
Seoul, Republic of Korea
yhhwang@isdl.snu.ac.kr

Abstract—**This paper presents a compact audio delta-sigma modulator that features a scalable bandwidth to also support biomedical instrumentation such as digital hearing aids and electromyography, sustaining constant FoMs. The modulator achieves a small die area and low power consumption by exploiting the proposed dynamic gain-bandwidth-boosting (GBWB) scheme in the inverter-based class-AB OTA with minimal overhead. The modulator features 56.8 dB PSRR without any external decoupling capacitor for a power supply and 66.1 dB CMRR in the audio band by utilizing a pseudo-differential structure with the dynamic GBWB scheme. For 25 kHz bandwidth, the modulator dissipates 68 μW from a 1.8 V supply and achieves a peak SNDR of 84.0 dB, a peak SNR of 85.1 dB, and a DR of 87.1 dB, maintaining the resolution higher than 13-bit against the supply voltage variation of 0.15 V. The prototype modulator is fabricated in 0.18 μm CMOS technology, occupying an active area of 0.0939 mm^2.**

Keywords—inverter-based delta-sigma modulator, dynamic gain - bandwidth-boosting, audio applications, biomedical instrumentation.

I. INTRODUCTION

The increasing demand for audio codecs in portable devices, such as smartphones, digital hearing aids, and virtual reality devices, drives the data converters toward low power consumption, small silicon area and reliability to noise and supply variation. Although the typical audible frequency range is lower than 20 kHz, an extended bandwidth is required for higher sound quality, also considering that children can perceive up to 24 kHz. A delta-sigma modulator (DSM), the best candidate for high-resolution ADCs in audio applications, is required to be compact and power-efficient with high common-mode rejection ratio (CMRR) and power supply rejection ratio (PSRR).

Recently, an inverter-based operational transconductance amplifier (OTA) has been commonly used in audio applications, not only achieving power-efficient operation with a class-C inverter but also providing sufficient performances to high-resolution DSMs [1]-[3]. However, since the class-C inverter without proper bias circuits is more dependent on the power supply than the class-AB inverter, the DC gain and the bandwidth of the class-C inverter are sensitive to the supply voltage variation [1], [2]. Furthermore, the class-C inverter has much larger device size than the class-AB inverter for an equivalent transconductance to operate in the weak inversion

region [4]. Additionally, the inverter-based integrators presented in the previous works [1]-[3] has same static current consumption even in a sampling phase, degrading power efficiency.

In this work, the compact audio DSM achieves a scalable bandwidth with constant FoMs by exploiting the class-AB inverter-based OTA. The proposed dynamic GBWB scheme boosts DC gain and the unity-gain bandwidth (UGB) of the OTA dynamically to suppress SNDR degradation in an integration phase, obtaining higher power efficiency by reducing static current in a sampling phase. Furthermore, this modulator achieves high PSRR and CMRR in the entire audio band, maintaining the resolution higher than 13-bit against supply voltage variation.

II. PROPOSED INVERTER-BASED INTEGRATOR

A. Operating Principle

Fig. 1(a) shows the proposed inverter-based integrator. The proposed OTA is a cascode inverter which has a current-saving (CS) mode for the sampling phase Φ_S and a gain-bandwidth-boosting (GBWB) mode for the integration phase Φ_I, as shown in Fig. 1(b) and (c), respectively. Since an intrinsic DC gain of a simple inverter is less than 40 dB [2], a cascode inverter is adopted to achieve a SNDR higher than 80 dB. As high gain-bandwidth product (GBW) of the OTA is unnecessary in the sampling phase, the OTA in the CS mode consumes less static current by reducing a transconductance of the OTA, as shown in Fig. 1(d). On the other hand, since both the GBW and the DC gain should be high enough in the integration phase, the GBW and the DC gain are boosted dynamically in the GBWB mode by increasing both an output impedance and the transconductance, as shown in Fig. 1(e).

B. Proposed Dynamic GBWB Scheme

In the CS mode, the OTA stores an offset in the auto-zeroing capacitor C_C by shorting the OTA's input V_G and output V_{OUT} as shown in Fig. 1(b). The input transistors M_{p1} and M_{n1} are self-biased for the integration phase by diode-connected cascode transistors M_{p2} and M_{n2}, which are in the saturation region. The OTA is self-biased by $V_{gs,p}$ and $V_{gs,n}$ for class-AB operation. The class-AB inverter has less dependency on supply voltage [1] and consists of the transistors with at least 100 times smaller aspect ratio than the class-C inverter for an equivalent transconductance

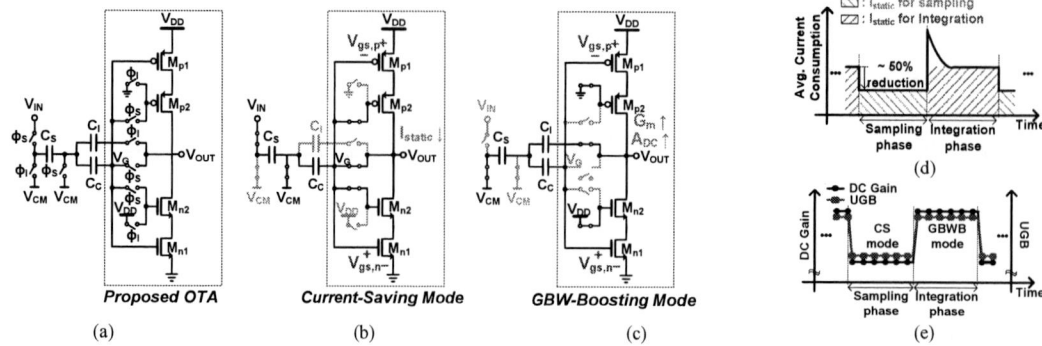

Fig. 1. (a) Proposed inverter-based switched-capacitor integrator. (b) Sampling phase with CS mode. (c) Integration phase with GBWB mode. (d) Average current consumption of proposed integrator. (e) DC gain and UGB of proposed OTA in two phases.

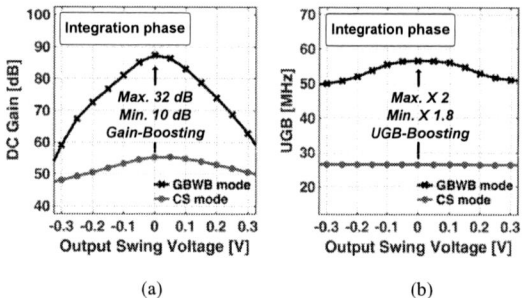

Fig. 2. (a) DC Gain and (b) UGB of proposed OTA in two modes.

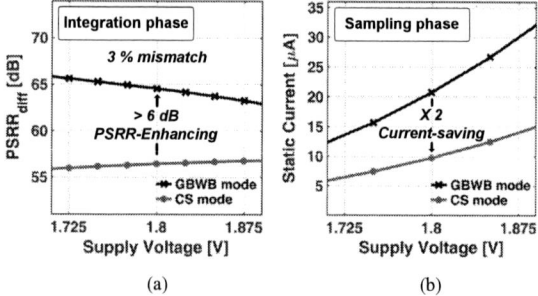

Fig. 3. (a) Effective PSRR of pseudo-differential integrator and (b) static current consumption of proposed integrator in two modes.

since the class-AB inverter operates in the strong inversion region, while the class-C inverter operates in the weak inversion region [4]. In the GBWB mode, the input transistors are kept biased for class-AB operation as shown in Fig. 1(c). The switch for the diode-connection of the cascode transistors turns off, and then the cascode transistors fully turn on to not only widen the output swing range but also boost the DC gain and the UGB dynamically. This is because the output impedance and the transconductance become larger by setting the input and the cascode transistors in the saturation region. The cascode transistors are biased quickly by the supply and the ground, which enables fast operation of the OTA.

With the dynamic GBWB scheme, the DC gain is boosted by more than 10 dB over the output swing voltage from -0.3 V to 0.3 V as shown in Fig. 2(a). Furthermore, Fig. 2(b) shows that the UGB is extended by at least 1.8 times over the output swing voltage from -0.3 V to 0.3 V. By boosting both DC gain and UGB effectively with fast bias settling, the dynamic GBWB scheme suppresses SNDR degradation due to inaccurate charge transfer and settling error. As a result, the first integrator has less than 0.4 % settling error of the output voltage for a sampling frequency of 5 MHz, resulting in SNDR degradation less than 1 dB.

The effective PSRR of a pseudo-differential integrator $PSRR_{diff}$ is enhanced in the GBWB mode during the integration phase in which the integrator output is affected by power supply noise. For a 100 mV$_{p-p}$ differential output swing with a 3 % mismatch [1] of all the transistors size in the pseudo-differential OTAs, the $PSRR_{diff}$ of the GBWB mode is enhanced by more than 6 dB than

the CS mode over the supply voltage from 1.725 V to 1.875 V, as shown in Fig. 3(a). Additionally, the in-band supply noise is suppressed by -16 dB, which is shaped to high frequencies by the noise shaping property of DSMs.

The OTA in the CS mode consumes less static current than in the GBWB mode since large transconductance is unnecessary in the sampling phase. The static current consumption in the CS mode is reduced by a factor of 2 than that in the GBWB mode as shown in Fig. 3(b), resulting in a reduction of total static current consumption by 25% than if always using the GBWB mode. The dynamic GBWB scheme improves power efficiency of the inverter based on class-AB operation by reducing static current in the sampling phase.

III. MODULATOR IMPLEMENTATION

A. Modulator

A third-order DSM with an OSR of 100 is chosen to acquire a SQNR higher than 90 dB. A feed-forward architecture with the proposed coefficients in Fig. 4 facilitates the output swing of the first integrator $S_{1,p-p}$ less than 1.2 V$_{p-p}$ for a -3 dB$_{FS}$ input signal in order to suppress SNDR degradation due to large output swing of the first integrator. Furthermore, the proposed coefficients are chosen to reduce the output swing of the second integrator $S_{2,p-p}$ less than 0.2 V$_{p-p}$ for a -3 dB$_{FS}$ input signal in order to relax a settling requirement. As shown in Fig. 5(a), a clock generator generates not only non-overlapping clock signals for the DSM but also an enable signal of the comparator Φ_{en}

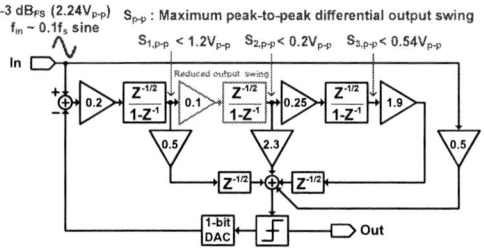

Fig. 4. Block diagram of proposed third-order switched-capacitor DSM.

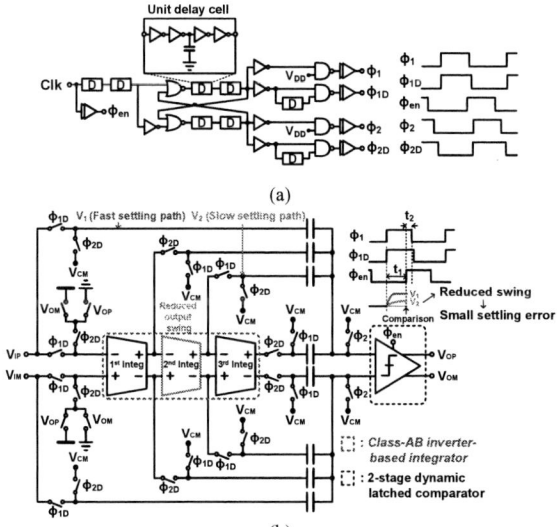

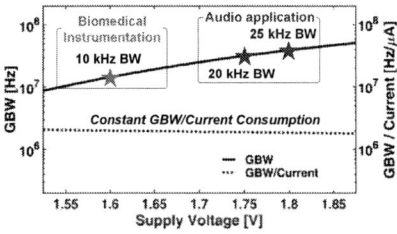

Fig. 5. (a) Schematic and timing diagram of clock generator. (b) Schematic of proposed third-order switched-capacitor DSM.

Fig. 6. Design points of proposed inverter-based OTA with constant GBW over current consumption.

which goes high before Φ_1 goes low. As shown in Fig. 5(b), the comparison starts at the end of the period in which Φ_1 is high, which is the integration phase of the second integrator, requiring faster settling of the second integrator output V_2. The feedforward input path V_1 is fast enough, which is only constrained by the on-resistance of a switch and the capacitance of a switched-capacitor adder. Owing to high DC gain and UGB of the class-AB inverter-based OTA with small output swing, reduced $S_{2,p-p}$ results in smaller settling error within shorter integration phase. Timing for Φ_{en} is carefully decided to maximize settling time t_1, considering the comparator delay t_2. A single-bit quantizer is realized with a dynamic comparator and SR latch [1].

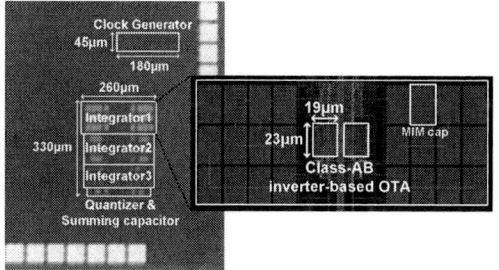

Fig. 7. Chip microphotograph and layout of first integrator.

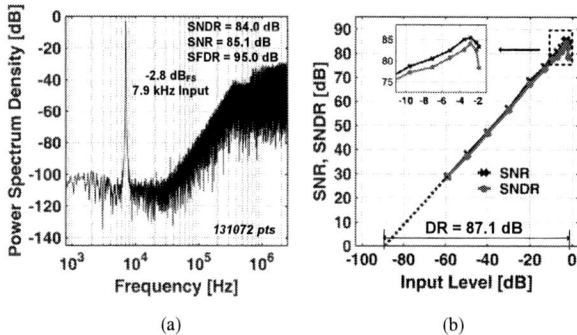

Fig. 8. (a) Measured power spectrum. (b) Measured SNR and SNDR.

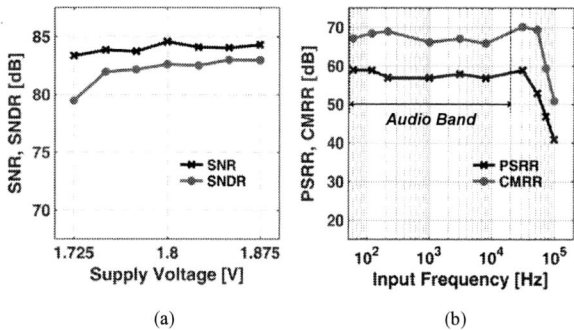

Fig. 9. (a) Measured SNR and SNDR with supply variation. (b) Measured PSRR and CMRR.

B. Inverter-Based Integrator

The pseudo-differential integrator is implemented with the proposed inverter-based OTA and the switched-capacitor passive CMFBs. The sampling capacitor is 600 fF and the auto-zeroing capacitor is 1.8 pF in the first integrator. With a 1.8 V supply, the static current consumption of the first integrator in the GBWB mode and the CS mode are 21 µA and 10 µA, respectively. The sampling capacitors for the second and the third integrator are 0.2-pF and 0.1 pF, respectively. Accordingly, the UGB and static current consumption of the OTAs in the second and the third integrator are set by a quarter of the first integrator's OTA.

C. Class-AB Inverter-Based OTA

A class-AB inverter operates in the strong inversion region,

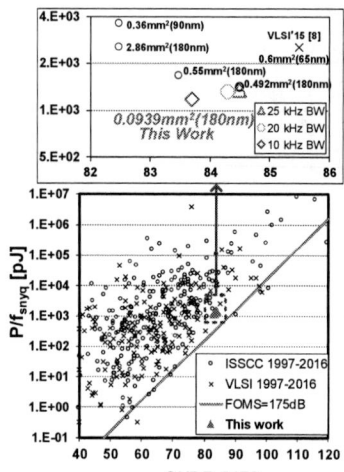

Fig. 10. Energy per Nyquist sample versus SNDR in performance comparison [5].

TABLE I.   PERFORMANCE SUMMARY AND COMPARISON

	This Work			TCAS'16 [2]	JSSC'13 [3]		VLSI'17 [6]	JSSC'16 [7]	VLSI'15 [8]
Process[nm]	180			180	65		65	65	65
Architecture	DT			DT	DT		CT	CT	CT
Integrator type	Class-AB inverter-based			Class-C inverter-based	Class-C inverter-based		Gm-C based	Gm-C based	Gm-C based
Supply[V]	1.6	1.75	1.8	1.8	0.7	0.8	1.2	1	1.1
F_S[MHz]	2	4	5	6.1	5	5	8	6.4	3.072
BW[kHz]	10	20	25	20	20	20	20	25	24
SNDR[dB]	83.2	83.8	84.0	97.7	89	91	88.5	95.2	85
SNR[dB]	83.7	84.7	85.1	98.6	92	94	N/A	100.1	N/A
DR[dB]	85.6	86.1	87.1	100.5	94	98	93.1	103	88
Power[μW]	23.4	53	68	300	152	230	55	800	121
FoM$_W$[fJ/conv]	99	104	105	119	165	198	63.1	340	173
FoM$_S$[dB]	172	172	173	178.7	175	177	178.7	177.9	171
CMRR[dB] [1]	66.1[2]			48	50		60	N/A	N/A
PSRR[dB] [1]	56.8[2]			63	55		N/A	N/A	N/A
Area[mm²]	0.0939			0.31	0.3		0.27	0.256	0.6

$$*FoM_W = \frac{Power}{2^{(SNDR-1.76)/6.02} \times 2 \times BW}, \quad FoM_S = DR_{dB} + 10\log\left(\frac{BW}{Power}\right)$$

[1] Measured in audio-band
[2] With a 1.8 V supply.

in which the GBW of the inverter increases slowly with respect to supply voltage [1]. The proposed class-AB inverter has a constant ratio of the GBW over static current consumption with respect to supply voltage, which facilitates constant FoMs of the bandwidth-scalable DSM, as shown in Fig. 6. For 20 kHz audio applications, the GBW of the first integrator's OTA is 31.8 MHz with a 4 MHz clock frequency at a 1.75 V supply voltage. For 10-kHz biomedical instrumentation, the GBW of the OTA is 14.3-MHz with a 2 MHz clock frequency at a 1.6 V supply voltage.

IV. MEASUREMENT RESULTS

The prototype inverter-based DSM is fabricated in a 0.18-μm CMOS process. The modulator has the compact area of 0.0939 mm², owing to the area-efficient class-AB inverter-based OTA as shown in Fig. 7. Fig. 8(a) shows the measured power spectrum of the prototype for a -2.8 dB$_{FS}$, 7.9 kHz sine-wave differential input with a 1.8 V supply. Fig. 8(b) shows the measured SNR and SNDR versus the input amplitude for 25-kHz bandwidth with a sampling clock frequency of 5 MHz. The measured peak SNDR, peak SNR, and DR are 84.0 dB, 85.1 dB, and 87.1 dB, respectively. Fig. 9(a) shows the measured SNR and SNDR versus the supply voltage. The modulator achieves the tolerant ENOB of 13-bit with the supply voltage variation from 1.725 V to 1.875 V. Fig. 9(b) shows the measured CMRR and PSRR versus the input frequency. The CMRR and the PSRR are measured with a 150 mV$_{p-p}$ sinusoidal signal and a 50 mV$_{p-p}$ sinusoidal signal, respectively. Note that the PSRR is measured without any off-chip decoupling capacitor for the supply. The measured CMRR and PSRR are higher than 66.1 dB and 56.8-dB in the entire audio band, respectively. The modulator consumes 68 μW with a 1.8 V supply. Fig. 10 shows this work is area- and power-efficient in terms of energy per Nyquist sample versus SNDR [5]. Table I shows that this work achieves not only better FoM$_W$ than class-C inverter-based DSM but also comparable PSRR and higher CMRR in the audio band than other works, implemented in the compact area.

V. CONCLUSION

The audio DSM achieves a compact area and scalable bandwidth to support biomedical instrumentation such as digital hearing aids and electromyography with constant FoMs by exploiting the class-AB inverter-based OTA. The proposed dynamic GBWB scheme suppress SNDR degradation due to inaccurate charge transfer and settling error, improving power efficiency of the modulator based on class-AB operation. Furthermore, this modulator achieves high PSRR and CMRR in the audio band and high linearity of more than 13-bit against the supply voltage variation of 0.15 V.

REFERENCES

[1] Y. Chae and G. Han, "Low voltage, low power, inverter-based switched-capacitor delta-sigma modulator," *IEEE J. Solid-State Circuits*, vol. 44, no. 2, pp. 458-472, Feb. 2009.

[2] S. Lee, W. Jo, S. Song, and Y. Chae, "A 300-μW audio ΔΣ modulator sith 100.5-dB DR using dynamic bias inverter," *IEEE Trans. Circuits Syst. I, Reg. Papers*, vol. 63, no. 11, pp. 1866-1875, Nov. 2016.

[3] H. Luo, Y. Han, R. C. C. Cheung, X. Liu, and T. Cao, "A 0.8-V 230μW 98-dB DR inverter-based sigma-delta modulator for audio applicati-ons," *IEEE J. Solid-State Circuits*, vol. 48, no. 10, pp. 2430-2441, Oct. 2013.

[4] S. Tedja, J. Van der Spiegel, and H. H.Williams, "Analytical and experimental studies of thermal noise in MOSFETs," *IEEE Trans. Electron Devices*, vol. 41, pp. 2069–2075, Nov. 1994.

[5] B. Murmann, "ADC Performance Survey 1997-2016," [Online]. Available: http://web.stanford.edu/~murmann/adcsurvey.html

[6] M. Jang, S. Lee, and Y. Chae, "A 55μW 93.1dB-DR 20kHz-BW single-bit CT ΔΣ Modulator with negative R-assisted integrator achieving 178.7dB FoM in 65nm CMOS," in *IEEE Symp. VLSI Circuits Dig. Tech*, Jun. 2017, pp. 40-41.

[7] Y. H. Leow, H. Tang, and Z. C. Sun, "A 1 V 103 dB 3rd-order audio continuous-time ΔΣ ADC With enhanced noise shaping in 65 nm CMOS," *IEEE J. Solid-State Circuits*, vol. 51, no. 11, pp. 2625-2638, Nov. 2016.

[8] I. Ahmed *et al.*, " A low-power Gm-C-based CT-ΔΣ audio-band ADC in 1.1V 65nm CMOS," in *IEEE Symp. VLSI Circuits Dig. Tech*, Jun. 2015, pp. 294-295.

S19-4 (2176)

A 172dB-FoM Pipelined SAR ADC Using a Regenerative Amplifier with Self-Timed Gain Control and Mixed-Signal Background Calibration

Miguel Gandara[1], Paridhi Gulati[2], and Nan Sun[1]

[1]The University of Texas at Austin, TX, USA, [2]Analog Devices, Inc., NC, USA; mfgandara@utexas.edu

Abstract—A scaling-friendly and PVT-robust pipelined SAR ADC reusing the first-stage comparator as a regenerative residue amplifier is proposed in this work. A low-power self-timed gain control block is combined with a mixed-signal background calibration to ensure a stable amplifier gain across PVT variation. A 130nm CMOS prototype achieves a peak Walden FoM of 9.2 fJ/conv-step and a Schreier FoM of 172 dB.

I. INTRODUCTION

With technology scaling, open-loop dynamic integrators have become an attractive choice for residue amplifiers of pipelined SAR ADCs [1]–[4]. A dynamic integrator has two key merits. First, it consumes only dynamic power (no static power), and thus, is much more power efficient than a traditional closed-loop amplifier. Second, it has low noise; in fact, it can be proved that an integrator achieves the lowest noise for a given power budget. Despite these merits, it brings challenges: 1) its maximum achievable gain is limited by the power supply voltage and the transistor g_m/I_D, and is typically less than 10; 2) it requires accurate offset calibration to match the offset of the integrator with the SAR comparator [1]–[3]. Offset mismatch increases the amplifier input swing and can cause significant linearity degradation for a dynamic integrator, as its linearity is much more sensitive to the input swing than a closed-loop amplifier; 3) its gain is linearly proportional to the integration time, resulting in a strict trade-off between the amplifier gain and speed. To address these issues, the authors recently proposed to reuse the strongARM latch comparator in the SAR as a dynamic amplifier, thus removing the need for any offset calibration between the SAR comparator and the residue amplifier [5]. It also naturally combines a front-end integrator, to achieve high noise efficiency, with a backend regenerative positive feedback stage, to attain high gain at high speed with negligible noise penalty as shown in Fig. 1. Furthermore, this architecture automatically realizes load switching to dynamically adjust the comparator noise and power. When it operates as the SAR comparator, the load is small inverter buffers that drive the SAR logic, and thus, is low power but high noise. When it acts as a residue amplifier, the load is a much larger second-stage SAR input capacitor, and thus, its input referred noise is reduced, which ensures an overall high ADC SNR.

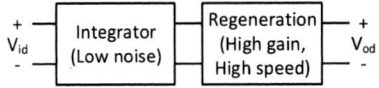

Fig. 1: Block diagram of a regenerative amplifier in [5].

Nevertheless, there are two critical challenges of dynamic amplifiers that the work of [5] has not addressed: 1) the am-

plifier gain depends on time, and thus, it requires an accurate and low jitter clock to control the amplification time, but the clock generation is nontrivial and can be power consuming; 2) the amplifier gain, relying on g_m and time, is sensitive to PVT variation. The clock requirement is especially stringent for amplifiers utilizing regeneration [5], as the gain increases exponentially with time instead of linearly. In [5], a voltage-controlled delay line (VCDL) was used to control the gain, but its high jitter caused large gain variations, resulting in ADC SNR degradation. Moreover, the VCDL delay did not track the amplifier gain across PVT variation. Digital background calibration can be used to sense the analog gain variation and adjust the digital gain to match it [1]. This technique improves ADC SNDR, but it does not address the root cause of the problem: *the amplifier gain still varies with PVT*. Amplifier gain variation causes several issues. First, extra redundancy has to be added to ensure that the largest possible gain under PVT variation does not overload the second pipelined stage. Second, when the amplifier gain drops, the noise contribution from the later pipelined stages increases, causing SNR degradation that cannot be mitigated by digital calibration. Third, the digital gain is no longer a power of 2, requiring a tunable fractional multiplier that increases power and design complexity. Besides digital background calibration, other techniques to tackle gain PVT variation use interpolation [4] or replica [6], but these techniques often require power that is comparable to the power of the dynamic amplifier.

This work introduces a low-power self-timed gain control block that provides enhanced robustness to PVT variation. The gain control block is combined with a mixed-signal background calibration loop. Fig. 2(a) shows the ADC block diagram with the classic gain calibration, where the digital gain is tuned to match a *varying* analog gain. By contrast, Fig. 2(b) shows the proposed scheme, where the analog gain is tuned to match a *constant* digital gain. This scheme takes advantage of the unique feature of dynamic amplifiers, that *the gain can be adjusted simply by changing the amplification time*. This removes the need for extra redundancy to account for increased gain across PVT, and ensures that the analog gain is always large enough to suppress the second stage non-idealities. Additionally, it does not require a fractional digital multiplier running at the ADC sampling rate, as the analog gain can be maintained to be a power of 2 and the digital scaling can be accomplished with a simple bit shift. A prototype in 130nm CMOS validates the proposed techniques and achieves a 4 times power efficiency increase compared to [5].

This paper is organized as follows. Sec. II describes the proposed gain control block. Sec. III discusses the proposed mixed-signal calibration. Sec. IV describes the pipelined SAR

978-1-5386-3179-9/17 $31.00 © 2017 IEEE

ADC architecture with self-timed gain control and mixed-signal calibration. Sec. V shows the measured results.

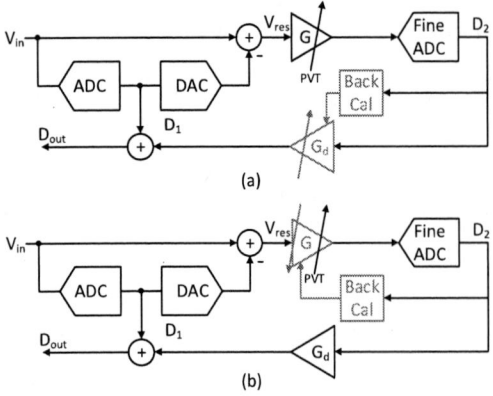

(a)

(b)

Fig. 2: Pipelined ADC with (a) digital gain calibration (b) mixed-signal gain calibration

II. PROPOSED SELF-TIMED GAIN CONTROL

Fig. 3 shows the schematic of the dynamic amplifier used in this work. When *clka* is asserted, the comparator is configured as the residue amplifier. The amplifier initially integrates the differential current from M1/M2 onto C_{s2} until the cross-coupled PMOS transistors turn on, at which point the amplifier acts in positive feedback regeneration. For low-noise operation, the integration time is generally set to be much longer than the regeneration time. Assuming that the differential current is small compared to the bias current, the time spent in the integration phase is

$$t_{int} = \frac{C_{s2}}{I_D} \cdot V_{Tp} \qquad (1)$$

where V_{Tp} is the threshold voltage of the cross-coupled PMOS transistors. Since I_D is controlled by a current mirror that is stable across PVT, the variation in integration time is mainly controlled by the PMOS threshold voltage. Since the integration time dominates the overall amplification time, a gain control block that automatically adjusts its delay based on the integration time would provide improved PVT robustness. A tunable delay can be added to this self-timed block in order to adjust the regeneration time to achieve the desired gain. The tunable delay is a small percentage of the overall delay, which reduces the overall jitter of the system.

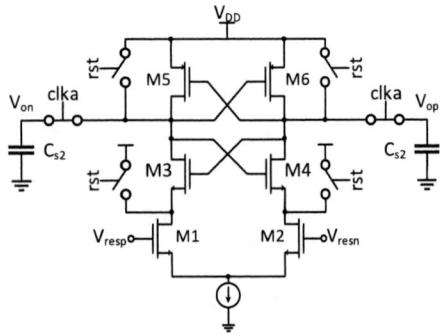

Fig. 3: Hybrid integrator and positive feedback amplifier schematic

Fig. 4 shows the block diagram of the proposed system, which combines a self-timed regeneration detection block that tracks the PMOS threshold voltage with a tunable delay that is controlled by the mixed-signal calibration described in Sec. III. Both of these functions can be realized using a single

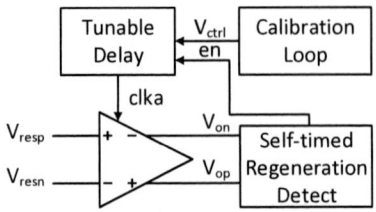

Fig. 4: Proposed gain control system

dynamic gate, as shown in Fig. 5. Two of the SAR shift register outputs, $sar\_ph\langle 1\rangle$ and $sar\_ph\langle 2\rangle$, are used to generate the reset signal, Pre, for a single SAR conversion cycle. The dynamic OR gate turns on once the output voltage crosses the PMOS threshold voltage, which directly tracks the integration time across PVT. Additionally, a control voltage, V_{ctrl}, can be applied to the backgate of the PMOS transistors to tune the gate's delay and thus control the regeneration time. Since this implementation is only a single gate, it is both low power and low jitter. The power from the proposed gain control system increases the overall amplifier power by only 10%, compared to a 400% increase in amplifier power from the VCDL in [5]. Additionally, this work is able to achieve an increase in SNDR of 4 dB over [5] due to the reduced jitter of the proposed gain control block. The gain control block is sized to ensure any mismatch in the PMOS threshold voltages does not cause significant second-order distortion in the output. The measured gain variation across temperature for the proposed gain control block and the VCDL of [5] is compared in Fig. 6, showing the significant improvement in gain tracking across temperature of the proposed gain control block.

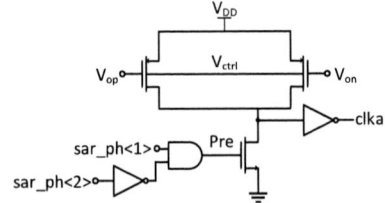

Fig. 5: Proposed self-timed gain control block

III. PROPOSED MIXED-SIGNAL CALIBRATION

The mixed-signal calibration algorithm shown in Fig. 7 is used to precisely control the gain across PVT. Similar to [7], [8], an on-chip pseudo-random number generator (PRNG) outputs a digital dither, R_n, with a value of ± 1 that is applied to the first stage digital to analog converter (DAC) and shifts the residue by one-half LSB of the first stage, $\frac{\Delta_1}{2}$. The digital output of the second stage is then multiplied by R_n and averaged. The output of the averaging block is an estimation of the gain, G_e, which is

$$G_e = \overline{\left[G\left(V_{res} + \frac{\Delta_1}{2}R_n \right) + \varepsilon_{q2} \right] R_n} \cdot \frac{2}{\Delta_1} = G \qquad (2)$$

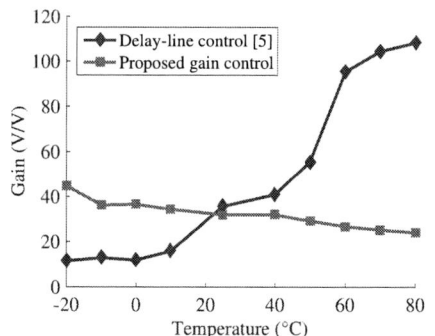

Fig. 6: Measured amplifier gain vs. temperature

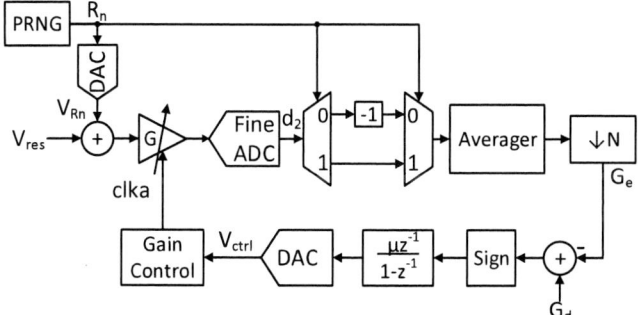

Fig. 7: Proposed mixed-signal calibration loop

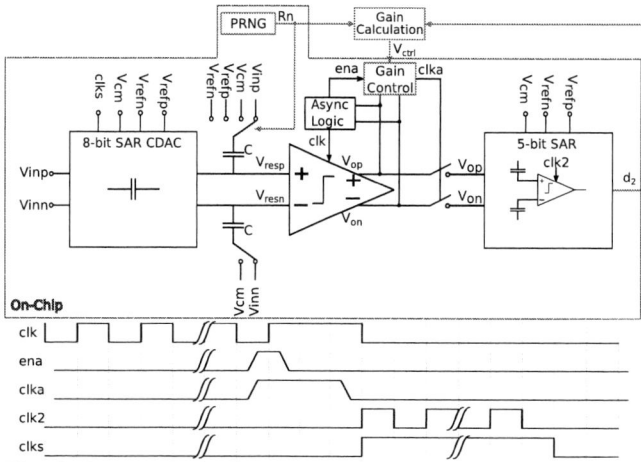

Fig. 8: Proposed pipeline ADC architecture and timing diagram.

where ε_{q2} is the second-stage quantization noise. As long as $\frac{\Delta_1}{2}$ is well controlled and the averaging block has a low enough bandwidth, G_e can be used to obtain an accurate estimate of the true amplifier gain G. In order to control the gain, the sign of the gain error is applied to a mixed-signal least-mean squares (LMS) loop that controls the backgate voltage of the gain control block. The effective LMS step size, μ, is controlled by the digital step size of the loop and the voltage step of the DAC. This allows the digital gain applied to the second stage to be implemented as a simple shift register instead of a digital fractional multiplier running at the ADC sampling rate.

The mixed-signal calibration loop is low complexity and low power. The multiplication by R_n can be implemented with a demultiplexer/multiplexer and a negation block. The averager is implemented as a first order infinite impulse response (IIR) exponential moving average filter. The output of the averager is downsampled so that the DAC in the LMS loop runs at a much lower frequency than the sampling frequency, allowing the DAC to be low power. Based on measurement and simulation results, the required DAC resolution is only 7 bits, allowing for a relatively simple DAC design. The off-chip calibration logic was synthesized and, running at low frequency, the simulated power consumption is only 3 μW in 130nm. The on-chip PRNG consumes an estimated 2.1 μW running at $f_s/8$.

IV. PROPOSED PIPELINED SAR ADC ARCHITECTURE

Fig. 8 shows the architecture of the proposed pipelined SAR ADC. The first stage resolution of 8 bits was chosen to ensure the input swing to the amplifier with the added dither

voltage would be small enough to meet the amplifier linearity requirements. The dynamic amplifier achieves a gain of 32 and an additional gain of 4 is achieved by capacitive attenuation of the reference voltage, providing one bit of redundancy for any DAC errors and the additional dither voltage. The additional loading on the amplifier from the capacitive attenuation also serves to reduce the amplifier noise to the desired level. The prototype uses asynchronous clocking [9] in order to reduce the clock generation logic complexity.

In order to minimize the increased voltage swing caused by the dither voltage, V_{R_n}, V_{cm} is used for the dither application. In traditional bidirectional single-sided switching [10], the V_{cm} application consists of stepping down the dummy capacitor from V_{refp} to V_{cm} on either side of the differential DAC. The problem with this approach is that any error in the common-mode voltage $\varepsilon_{V_{cm}}$ causes an error in the gain estimator, ε_{G_e}

$$G_e(1 + \varepsilon_{G_e}) = G\left(1 + \frac{\varepsilon_{V_{cm}}}{V_{cmi}}\right) \qquad (3)$$

where V_{cmi} is the ideal common-mode voltage. This imposes a strict requirement on the common-mode voltage to ensure an accurate gain estimation. To overcome this issue, a single dummy capacitor is instead switched from V_{cm} to either V_{refp} or V_{refn}. In this case, the common-mode voltage error gets canceled out in the estimation operation, so even large common-mode voltage errors have negligible effect on the final gain estimation.

V. MEASUREMENT RESULTS

Fig. 9 shows the prototype ADC implemented in 130nm CMOS. Fig. 10 shows the measured spectrum. Fig. 11 shows the SNDR/SFDR across input frequency and amplitude. Fig. 12 shows the measured SNDR across temperature for the ADC with only a single foreground-calibrated point at room temperature, and for the ADC with background calibration enabled. The averaging filter and DAC control discussed in Sec. III was implemented using MATLAB. With background calibration, the ADC performance stays relatively stable from -20 to 80°C. To show the convergence behavior, measurements were taken while the temperature was increased from 25°C to 65°C and then back to 25° C. Fig. 13 shows the gain across samples with and without the background calibration enabled. Due to the latency introduced by implementing the averaging and DAC control operations in MATLAB, the gain does not perfectly track the change in temperature initially. However, the number of samples required to achieve convergence is equivalent to 8 ms in the worst case. If implemented in a real-time system, this speed should allow for accurate tracking of environmental changes. The level of gain stability shown in Fig. 13 ensures the standard deviation of the SNDR is less than 0.25 dB. Table I compares the prototype ADC performance with state-of-the-art dynamic-amplifier based ADCs. The low-frequency Schreier FoM is 172 dB and the Nyquist Schreier FoM is 171.2 dB, which is comparable to that of other works fabricated in more advanced processes. Digital power comprises 64% of the total 171 μW, so fabricating this design in more advanced processes would even further improve the power efficiency.

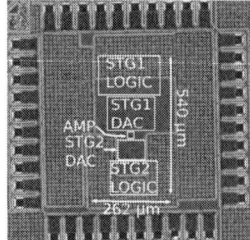

Fig. 9: ADC die photo

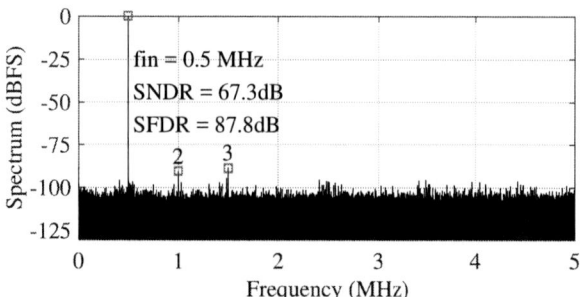

Fig. 10: Measured ADC output spectra with 32768 points.

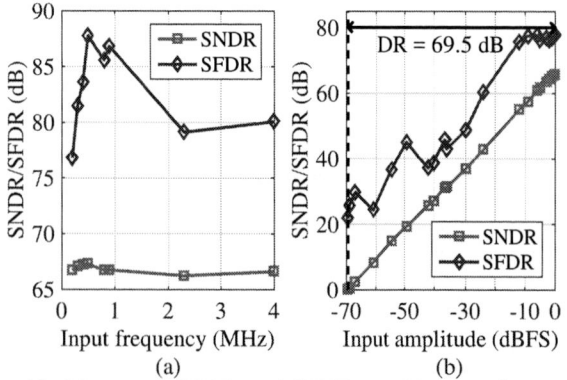

Fig. 11: Measured SNDR and SFDR vs. (a) input frequency and (b) input amplitude.

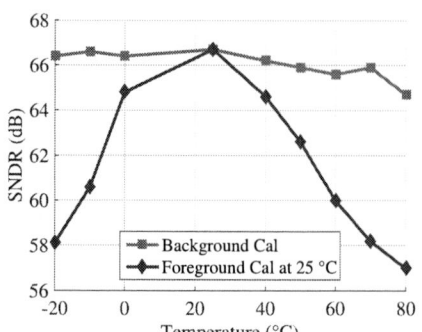

Fig. 12: Measured SNDR vs. temperature

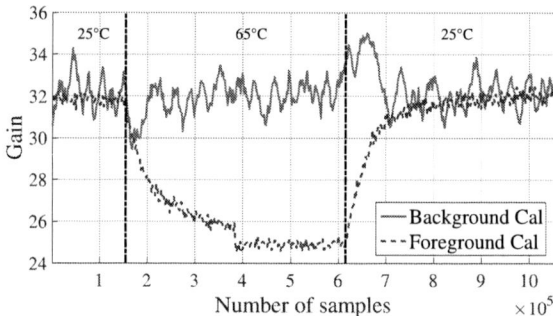

Fig. 13: SNDR over time with transient steps from 25°C to 65°C to 25° C

TABLE I: Performance Comparison

	[1]	[2]	[5]	[11]	This work
Process (nm)	28	40	130	180	130
Architecture	Pipe SAR	Pipe SAR	Pipe SAR	Pipeline	Pipe SAR
Requires OTA	No	No	No	Yes	No
Requires offline calibration	No	Yes	Yes	No	No
Requires offset calibration	Yes	Yes	No	No	No
Requires digital fractional multiplier	Yes	No	Yes	No	No
Interstage Gain	16	4	32	4	32
Supply Voltage (V)	1.0/1.8	1.1	1.2	1.3	1.2
Sampling Rate (MS/s)	80	10	10	30	10
Peak SNDR (dB)	68	62	63.2	74.2	67.3
Peak ENOB (bit)	11	10	10.2	12	10.9
SNDR (Nyq) (dB)	66	62	63.2	72	66.6
ENOB (Nyq) (bit)	10.7	10	10.2	10.7	10.8
Power (mW)	1.5	.07	0.28	6	0.17
LF Walden FoM (fJ/step)	9.1	7	23.7	48	9.2
LF Schreier FoM (dB)	172	171	166	168	172
HF Walden FoM (fJ/step)	11.5	7	23.7	61.5	9.8
HF Schreier FoM (dB)	170	171	166	166	171

REFERENCES

[1] F. van der Goes et al., in *2014 ISSCC*, Feb 2014.

[2] B. Verbruggen et al., in *2012 ISSCC*, Feb 2012.

[3] B. Malki et al., in *2014 ESSCIRC*, Sept 2014.

[4] J. Lin et al., in *2013 CICC*, Sept 2013.

[5] M. Gandara et al., in *2017 CICC, in press*, 2017.

[6] H. Huang et al., in *2017 ISSCC*, Feb 2017.

[7] A. Sanyal and N. Sun, in *2016 VLSI*, June 2016.

[8] E. Siragusa and I. Galton, *JSSC*, Dec 2004.

[9] S.-W. M. Chen and R. W. Brodersen, in *2006 ISSCC*, Feb 2006.

[10] L. Chen et al., in *2014 IEEE ESSCIRC*, Sept 2014.

[11] H. Venkatram et al., in *2014 VLSI*, June 2014.

A Fully-Synthesizable C-Element Based PUF Featuring Temperature Variation Compensation with Native 2.8% BER, 1.02fJ/b at 0.8-1.0V in 40nm

Sachin Taneja, Anastacia Alvarez, Gopalakrishnan Sadagopan, Massimo Alioto

ECE Department, National University of Singapore, Singapore

Email: *elesach@nus.edu.sg*

Abstract— **A fully-synthesizable Physically Unclonable Function (PUF) with hysteresis-enhanced stability and active compensation of temperature variations is proposed. In detail, a bitcell based on low-voltage regulated Cascode current mirror is introduced. To reduce undesired bit flips, hysteretic behavior is obtained through the insertion of a Muller C-element output stage, which mitigates the effect of noise, voltage and temperature fluctuations on the PUF output. Measurements in a 40nm test chip show that hysteresis improves the bit error rate (BER) by 14%. A feedback scheme is also introduced to compensate the effect of temperature variations at run time, improving the intra-PUF Hamming distance by 40% and BER by 6-7% under temperature variations (25-85°C). Native worst-case BER of 2.8% is measured under 0.8-1.0V and 25-85°C, with instability degradation with temperature being 0.15% per 10°C. The PUF bitcell consumes 1.02fJ/b at 0.9V, achieving an energy efficiency of 16Tbps/W and maximum throughput of 24Gbps. Also, the proposed PUF can be designed with fully automated standard cell-based flows, thus enabling substantial design effort reduction compared to prior art based on custom design styles.**

Keywords—**Physically Unclonable Functions, hardware security, regulated cascode, C-element, temperature compensation, resiliency.**

I. INTRODUCTION

The chip-unique responses of PUFs make them an essential primitive for hardware security down to the chip level, chip identification/authentication, protection of intellectual property and sensitive data, among the others [1]-[6]. Silicon implementations of PUFs suffer from various non-idealities, such as bit instability (i.e., imperfect repeatability) due to the presence of on-chip noise and process/voltage/temperature (PVT) variations. Their compensation typically requires additional methods at design, testing and boot time that increase the silicon area, the energy and the testing cost, as summarized in Fig. 1. To mitigate the related cost, PUF bits need to exhibit adequately high native stability [1], [2] (i.e., before applying the above methods).

From a design point of view, most of the existing PUFs use custom, highly constrained and regular layout styles (e.g. arrays) to assure adequate resiliency against noise and variations, which requires a uniform and very well defined surrounding layout environment [1]-[6]. The adoption of custom layout styles comes at the cost of significant design effort, and prevents the seamless integration of PUFs with other primitives (e.g., the layout of PUFs need to be separate

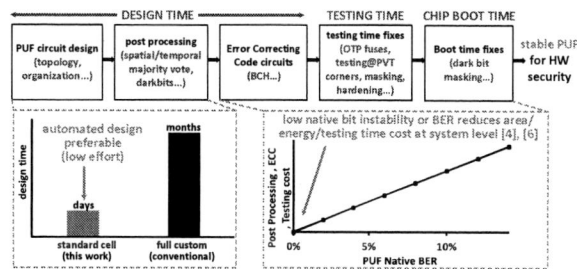

Fig. 1. Methods to improve PUF stability vs chip life phase.

from the digital circuits utilizing the PUF responses). On the other hand, design flows based on automated placement and routing (P&R) would be preferable to drastically reduce the design effort and facilitate design integration.

This work presents a PUF that can be designed with fully automated digital flows, with very limited design and integration effort. Low native bit stability is achieved by using the regulated Cascode current mirror bitcell based on the general approach in [2], with hysteresis being introduced via a C-Element output stage to enhance the PUF output stability. A body biasing feedback control loop is introduced to enable run-time mitigation of the bit instability due to temperature fluctuations. The fully automated design of the proposed PUF in a 40nm test chip achieves native BER of 2.8%, keeps the bit instability degradation within 0.15% per 10°C, and exhibits a bitcell energy/bit of 1.02fJ/b at 0.9V.

II. FULLY SYNTHESIZABLE C-ELEMENT BASED PUF BITCELL WITH FEEDBACK TEMPERATURE COMPENSATION

A. C-Element Based PUF Bitcell

The proposed PUF bitcell is based on the regulated Cascode current mirror (RCCM) as shown in Fig. 2, and belongs to the broad class of static mono-stable PUFs. Such PUFs convert the transistor current mismatch of complementary back-to-back (i.e., NMOS and PMOS) current mirrors into a nearly-full swing voltage, leveraging on their large small-signal output resistance [2]. Statistically infrequent cases with nearly-zero PMOS/NMOS threshold voltage mismatch/variations (σ_{VTH}) lead to occasional non-full swing and unstable bits due to thermal noise, voltage and temperature variations [2].

Compared to Cascode current mirrors in [2], the RCCM

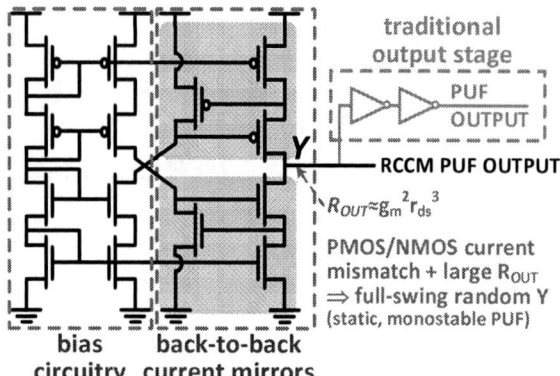

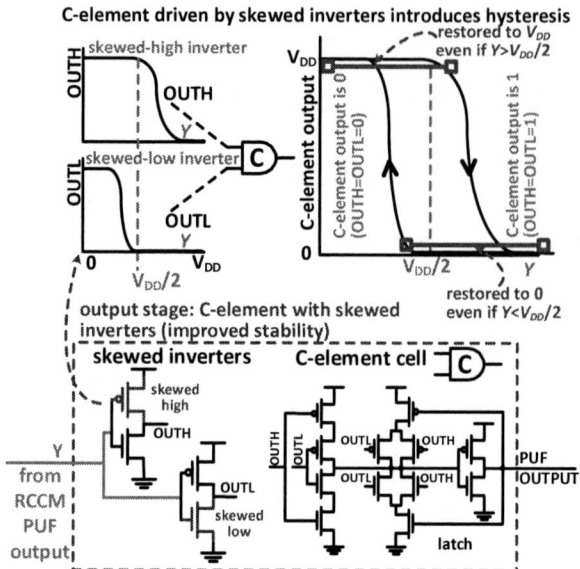

Fig. 2. PUF bitcell based on regulated cascode current mirror (RCCM) with traditional inverter output stage (in green).

circuit in Fig. 2 is based on regulated Cascode current mirrors to leverage their ability to raise the output impedance, which translates into a reduction in the bit instability due to the greater amplification in the conversion of the current mismatch into the output voltage Y. In detail, the regulated Cascode current mirror in Fig. 2 achieves the output resistance $g_m^2 r_{ds}^3$, which is higher[1] than the Cascode by $g_m r_{ds}$, thanks to its current-series sampling negative feedback (g_m and r_{ds} are the transistor transconductance and output resistance, $g_m r_{ds} \sim 10\text{-}15$ in 40nm). Compared to conventional Cascode, a 2.5X stability improvement is expected in 40nm from Monte Carlo simulations. Also, RCCM bias increases the number of transistors contributing to the PMOS/NMOS current mismatch from 4 to 8, which improves native stability and entropy of PUF outputs.

The output of the regulated Cascode current mirror is further stabilized by cascading a Muller C-element whose two inputs are driven by a low- and a high-skewed inverter (Fig. 3). Interestingly, this creates a hysteretic behavior since the Muller C-element output switches only if Y crosses the farthest logic threshold of the two skewed inverters in both rising and falling transitions. Hysteresis filters out occasional fluctuations in the voltage Y in marginally stable bitcells, as the output of the Muller-C element switches only if such fluctuations are large enough to overcome the farther of the two logic thresholds. Simulations in 40nm at 0.9V show that the range of fluctuations in Y that are filtered by the circuit in Fig. 3 is 200mV larger than the traditional inverter output stage in Fig. 2 (around both ground and supply voltage), thus improving the PVT resilience.

B. Active Compensation of Temperature Fluctuations and Fully Automated PUF Design

The impact of temperature fluctuations on bitcell stability at run-time is further mitigated through active compensation of

---

[1] As an additional advantage, regulated cascode current mirrors reduces the minimum operating voltage by the threshold voltage (V_{TH}). As the output resistance of conventional Cascode degrades severely below 1V, RCCM enables better bit stability at lower voltages till 0.6-0.7V, being $V_{TH} \sim 0.4$V.

Fig. 3. C-element with skewed inverters to introduce hysteresis and stabilize PUF output (cascaded to RCCM, replacing green output stage in Fig. 2).

the resulting deviation of the PMOS/NMOS strength ratio. This is achieved by tuning the PMOS body voltage through the feedback control circuit in Fig. 4. In this figure, the PMOS/NMOS strength ratio is sensed by the PMOS/NMOS imbalance sensor in the form of voltage V_{sense}. Fluctuations in V_{sense} due to temperature variations are then compensated by the feedback loop in Fig. 4, which sets the PMOS body voltage in the PMOS/NMOS imbalance sensor and the PUF equal to the targeted value (i.e., the value of V_{sense} measured at the nominal temperature). The single-stage subthreshold-biased gain stage in the loop was confirmed to be unconditionally stable without any compensation capacitance, as the body capacitance of the PUF is orders of magnitude larger than the one in the imbalance sensor. Interestingly, the body voltage of the PUF can also be altered to mimic PVT corners at testing time, which in turn permits to assess the stability across corners while limiting the number of actual corners (i.e., testing cost) that need to be exercised to uncover unstable bits.

In addition to temperature fluctuations, PUFs are generally sensitive to coupling, supply noise and variations, and hence the surrounding environment in which the bitcell operates. Such coupling/supply noise affects the behavior of existing PUF bitcells that rely on their transient behavior, such as ring oscillator-, delay- and metastability-based PUFs [4], [5]. For this reason, the vast majority of existing PUFs are

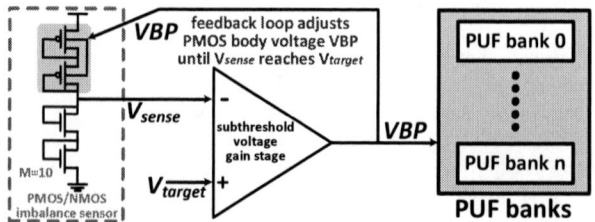

Fig. 4. Active compensation feedback loop for temperature variations.

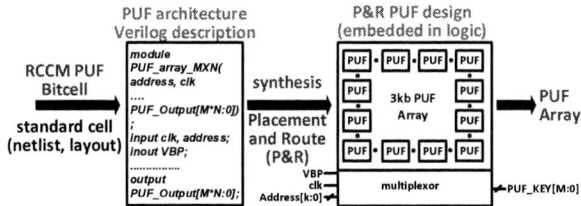

Fig. 5. Automated standard cell-based design flow for the RCCM PUF array.

designed manually or semi-manually (e.g., organized in arrays [1]-[4]) to assure noise-controlled, uniform and controlled surrounding layout environment, at the cost of significant design effort and restrictions. On the other hand, the proposed RCCM PUF bitcell is static and monostable, and is hence immune to transient noise, as opposed to PUFs relying on the randomness of their transient behavior. Thus, it is expectably rather insensitive to the surrounding environment, and well suited for automated placement and routing. The RCCM PUF bitcell with the Muller-C element in Fig. 3 is designed as a conventional standard cell, and is then used as a building block for the PUF array. The PUF organization and the readout circuitry (e.g., multiplexers, decoder) are then simply described in Verilog. Conventional synthesis and P&R are then executed to create the RCCM PUF array layout as in Fig. 5. Treating the PUF as a conventional digital design offers the additional capability to physically merge the PUF in the circuit that uses it, avoiding making the PUF an obvious target for physical attacks, as opposed to separate PUF layout designs.

III. Silicon Measurement Results

Four 3-kb RCCM PUF banks with 192-b word were designed and fabricated in 40nm CMOS, for the RCCM version with Muller C-element (Fig. 3) and the traditional output stage buffer (Fig. 2). The test chip also includes the temperature compensation in Fig. 4, and power gating circuitry to selectively shut down unused PUF banks. The die photo and chip key measurement summary are presented in Fig. 13.

Regarding the effectiveness of the Muller-C element insertion, the latter improves the bitcell stability by 15% (20%) and worst BER by 11% (14%) in nominal/golden (in-field) conditions across 500 repetitions with power-up transient in-between (see Fig. 6), compared to the traditional inverter output stage. Under such conditions, the worst-case (instantaneous) BER is 0.65%, 0.94% at 0.9V and 0.6V, and cumulative unstable bits are 2.51%, 3.22% (Fig. 7). The sensitivity of the native stability on voltage and temperature is measured as 0.94% per 100mV and 0.18% per 10°C as in Fig.

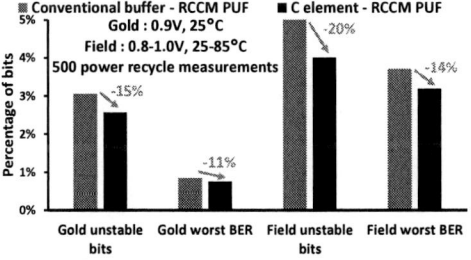

Fig. 6. Muller C-element BER and stability improvement over traditional output stage.

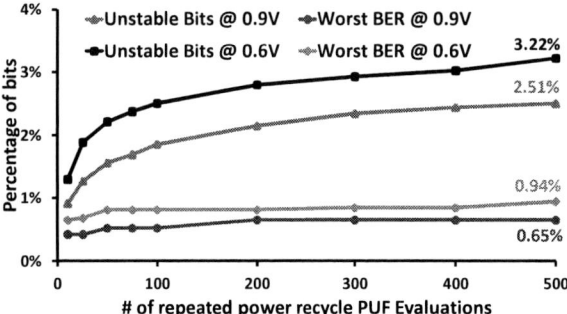

Fig. 7. BER, cumulative unstable bits vs evaluations (repeated measurements without temperature compensation circuit in Fig. 4).

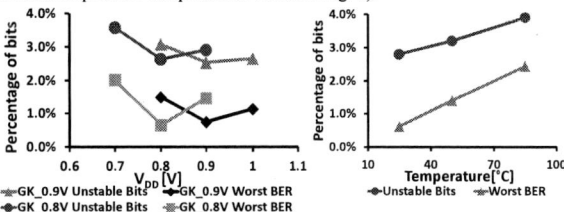

Fig. 8. Impact of V_{DD} and temperature fluctuations on instability (golden key GK at 0.9V, 25°C), without temperature compensation circuit in Fig. 4.

8, respectively.

Regarding the active temperature compensation in Fig. 4, it reduces the native stability sensitivity from 0.18% to 0.15% per 10°C and BER by 6-7% in 25-85°C (Fig. 9). It also reduces intra-PUF Hamming Distance (HD) variability (σ/μ) by 40% (Fig. 10) in 25-85°C range, in view of its ability to capture the PMOS/NMOS systematic skew from Fig. 4. The average inter-PUF HD of 49.1% is achieved with 100X separation from the intra-PUF HD at nominal conditions (Fig. 11). Such separation becomes 116X under 0.7-1.0V at 25°C, and 98X (89X) under 25-85°C with (without) compensation. The speckle diagram across banks and dice in Fig. 12 shows no spatial correlation. From Fig. 12, the autocorrelation function (ACF) of 12k PUF bits shows 95% confidence interval (CI) of σ=0.0078 (Table I). The PUF outputs pass all relevant NIST randomness tests.

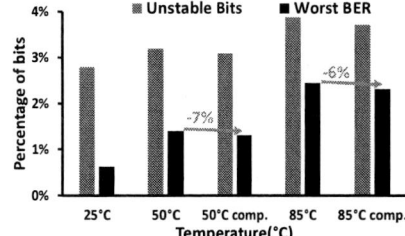

Fig. 9. Stability improvement with temperature compensation (50°C, 85°C).

Fig. 10. Intra-PUF HD distribution with/without compensation.

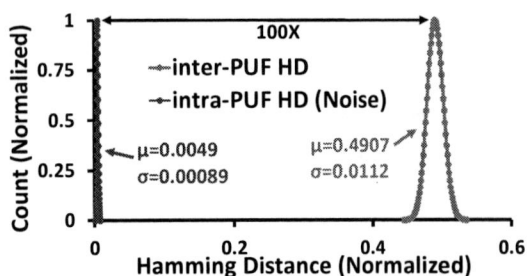

Fig. 11. Inter- and intra-PUF Hamming Distance (nominal condition).

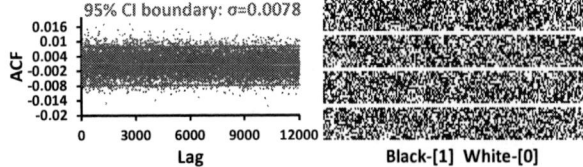

Fig. 12. Measured ACF with 95% confidence interval and speckle diagram.

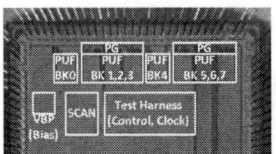

Technology	40nm CMOS
PUF Cell Area	7.23μm²
PUF Capacity	3kb/bank, 16wX192b
Maximum Throughput	24Gbps @ 0.9V,25°C
PUF Entropy	0.9972
PUF Bitcell Energy	1.02fJ/bit @ 0.9V,25°C
Readout Circuit (Decoder) Energy	55.94fJ/bit @ 0.9V,25°C
Total Power/Bank @ 125MHz	1.51mW @ 0.9V,25°C
PUF Leakage Power	27nW/b @ 0.9V,85°C

Fig. 13. Die photo and chip measurement summary.

IV. CONCLUSION

A fully synthesizable PUF based on RCCM bitcell with Muller C-element output stage has been proposed and designed using fully digital design flow with active temperature compensation to achieve BER of 2.8% across 0.8-1.0V and 25-85°C, and PUF bitcell energy of 1.02fJ/b, which is well suited for very low-energy IoT sensor nodes. Active temperature compensation further reduces the native stability sensitivity to 0.15% per 10°C, BER by 6-7% with 40% less variability in intra-PUF HD, thereby improving identifiability.

ACKNOWLEDGMENT

The authors thank TSMC for chip fabrication and acknowledge the support by the Singaporean Ministry of Education grant MOE2016-T2-1-150.

REFERENCES

[1] K. Yang, Q. Dong, D. Blaauw, D. Sylvester, "A 553F² 2-transistor amplifier-based Physically Unclonable Function (PUF) with 1.67% native instability," in *IEEE ISSCC Dig. Tech Papers*, 2017, pp. 146-148.

[2] A. Alvarez, W. Zhao, M. Alioto, "15fJ/b static physically unclonable functions for secure chip identification with <2% native bit instability and 140× Inter/Intra PUF hamming distance separation in 65nm," in *IEEE ISSCC Dig. Tech Papers*, 2015, pp. 256-258.

[3] J. Li, M. Seok, "Ultra-Compact and Robust Physically Unclonable Function Based on Voltage-Compensated Proportional-to-Absolute-Temperature Voltage Generators," in *IEEE Journal of Solid-State Circuits*, vol. 51, no. 9, pp. 2192-2202, Sept. 2016.

[4] S. Mathew *et al.*, "A 4fJ/bit delay-hardened physically unclonable function circuit with selective bit destabilization in 14nm tri-gate CMOS," in *IEEE Symp. on VLSI Circuits*, 2016, pp. 231-232.

[5] S. Mathew *et al.*, "A 0.19pJ/b PVT-variation-tolerant hybrid physically unclonable function circuit for 100% stable secure key generation in 22nm CMOS," in *IEEE ISSCC Dig. Tech Papers*, 2014, pp. 278-280.

[6] B. Karpinskyy *et al.*, "Physically unclonable function for secure key generation with a key error rate of 2E-38 in 45nm smart-card chips," in *IEEE ISSCC Dig. Tech Papers ISSCC*, 2016, pp. 158-160.

The maximum measured throughput at 0.9V is 24Gbps while consuming 1.51mW in the PUF core (including the temperature compensation loop), resulting into 1.02fJ/b energy per bit in the core. The readout energy in this specific PUF organization is 56fJ/b (24fJ/b) at 0.9V (0.6V). The power for compensation loop in Fig. 4 (sensor, gain stage) is 75nW, which is ~0.005% of the overall power in the 3-kb PUF array). Power gating improves the static power of the PUF array by at least 70X at 0.9V, 85°C for low power standby operation.

Table I summarizes the comparison with state-of-the-art PUF designs, the table shows that the energy spent in the bitcell is lower than prior art [4] by 4X, by virtue of its static behavior. The hysteretic behavior of the proposed bitcell assures a worst-case BER improvement by 1.18X compared to the previously best-in-class value [1] (i.e., from 3.13% down to 2.8%). As expected from the bitcell insensitivity to the surrounding layout, the autocorrelation function of the proposed PUF is lower by 1.28-4.65X, as an indicator of its suitability for automated placement and routing. Thanks to the hysteretic behavior and the active temperature compensation loop, the proposed PUF reduces by 1.4-3.1X the bit flips experienced when increasing the temperature by 10°C.

TABLE I. COMPARISON WITH STATE-OF-THE-ART SILICON PUF DESIGNS (BEST RESULTS IN BOLD)

	Proposed	ISSCC' 17 [1]	JSSC' 16 [3]	VLSI' 16 [4]	ISSCC' 16 [6]	ISSCC' 15 [2]	ISSCC' 14 [5]
Technology	40nm	180nm [LVT]	65nm	14nm	45nm	65nm [INV_PUF]	22nm
Normalized PUF Cell Area[a] (μm²/bit)	7.23	**1.53**	3.71	14.72	4.8	12.7	18.7
PUF Bitcell Energy (fJ/bit)	**1.02**	13.5	548	4	-	15	13
Native Unstable Bits (nominal conditions)	2.55%[b]	**1.73%**[b]	6.55%	26.80%	-	2.35%[b]	~30%
Native Worst BER (V/T variations)	**2.80%**	3.13%	-	5.76%	2.90%	-	8.50%
Supply V_{DD} (V)	0.8 - 1.0	0.8 - 1.8	0.6 - 1.2	0.55 - 0.75	-	0.6 - 1.0	0.7 - 0.9
Temperature (°C)	25 - 85	-40 - 120	0 - 80	25 - 110	-25 - 85	25 - 85	25 - 50
Bit flipping ratio per 10°C	**0.15%**	0.21%	0.44%[c]	-	0.25%	0.47%	-
Bit flipping ratio per 0.1V	0.94%	0.29%	**0.13%**[c]	-	-	1.30%	0.49%
Normalized Inter - Intra HD (identifiability)	0.491 - 0.0048	**0.498 - 0.0008**	0.500 - 0.0057	-	0.498 - N/A	0.501 - 0.0034	0.49 - 0.026
Autocorrelation function (95% CI)	**0.0078**	0.0167	0.02	-	0.017	0.0363	0.01

[a] Area normalized to 40nm for all technologies (2X shrink/generation) [b] BER/unstable bits of the proposed PUF and the PUF in [1], [2] is proportional to ($g_m r_{ds} \sigma_{VTH}$), which is invariably degraded at downscaled technology generations. The slightly better value of native unstable bits in [1] is explained by its adoption of an older technology generations and hence larger ($g_m r_{ds} \sigma_{VTH}$). In 40nm, ($g_m r_{ds} \sigma_{VTH}$) is found to be degraded by ~2.5X, when compared to 0.18μm. [c] Evaluated by using off-chip ADC and comparator manual re-calibration at each temperature (i.e., not native).

A 0.37mm^2 LTE/Wi-Fi Compatible, Memory-Based, Runtime-Reconfigurable $2^n3^m5^k$ FFT Accelerator Integrated with a RISC-V Core in 16nm FinFET

Angie Wang, Brian Richards, Palmer Dabbelt, Howard Mao, Stevo Bailey,
Jaeduk Han, Eric Chang, James Dunn, Elad Alon, Borivoje Nikolić

Department of Electrical Engineering and Computer Sciences
University of California, Berkeley, USA

Abstract—Dedicated hardware accelerators enable energy-efficient implementations of radio and imaging basebands. Multi-standard, multi-mode radio basebands require an on-the-fly reconfigurable fast Fourier transform (FFT) accelerator that implements many different FFT sizes. An instance of a runtime-reconfigurable $2^n3^m5^k$ FFT accelerator was generated by a custom hardware generator to meet the requirements of common wireless standards (Wi-Fi, LTE). The accelerator is integrated with a RISC-V processor, and the measured 16nm FinFET chip runs up to 940MHz and consumes 0.46 to 22.6mW of power when running FFT benchmarks for Wi-Fi and LTE symbol lengths.

Keywords—*hardware generator, runtime reconfigurability, fast Fourier transform, Cooley-Tukey algorithm, prime factor algorithm, Winograd's Fourier transform algorithm, Chisel, LTE, Wi-Fi, RISC-V*

I. INTRODUCTION

FFTs with a broad range of performance requirements are needed for many modern-day applications, ranging from medical imaging and machine learning to communication and radio astronomy. Even just among OFDM-based wireless applications, different standards necessitate highly configurable FFT hardware. Wi-Fi basebands must support many channel bandwidths and modulation schemes, requiring *runtime reconfigurability* across multiple data rates and different 2^n FFTs (Table I). LTE's single-carrier frequency-division multiple access scheme (SC-FDMA) must additionally perform mixed-radix FFTs (Table II).

To support software-defined radio (SDR) and radar applications, a memory-based, *runtime-reconfigurable $2^n3^m5^k$ FFT generator* has been developed in the Chisel hardware construction language [1, 2]. Generated FFT instances use 50% less data and 25% less twiddle memory than comparable instances from the Spiral FFT generator [3]. This paper describes a 16nm FinFET 0.37mm^2 FFT accelerator meeting LTE/Wi-Fi requirements, created by an improved version of our generator and integrated with a RISC-V processor to demonstrate a complete system. Measurement results show that *generator-based designs are competitive with state-of-the-art*.

II. FFT GENERATOR DESIGN

A. Memory-Based Architecture with Conflict-Free Scheduling

Memory-based architectures are more area efficient than pipelined architectures for supporting data rates of up to several hundred MHz. In such an architecture, a limited number of processing elements (PEs) implement FFT butterflies and access memory sequentially. Area efficiency is maximized by conflict-free memory access scheduling. The generator implementing this memory-based architecture uses a single butterfly that iterates through stages of the signal-flow graph, as illustrated in Fig. 1 [1].

To achieve conflict-free memory access when only one PE is active, the calculation SRAM is split into $radix_{max}$ banks. The n_i, k_i operand indices, represented by mixed-radix digits $..., m_3, m_2, m_1, m_0$, are mapped to different memory banks via:

$$bank = (m_0 + m_1 + m_2 + \cdots) \bmod radix_{max} \quad (1)$$

Conflict-free access is guaranteed because indices associated with 1 PE iteration differ by only one mixed-radix digit [1]. Fig. 2 shows that most LTE/Wi-Fi FFTs require >N compute cycles (CCs), where N is the FFT length. Thus, to support streaming IO, calculations must occur at *twice* the IO rate. When a single PE is used, some FFTs cannot complete in <2N CCs. To ensure that those FFTs can meet streaming requirements, two radix-2 butterflies are performed in parallel by reusing existing hardware. Only the radix-2 stage is targeted, because it requires the most PE iterations (N/2) and does not have twiddle multiplications, which allows for scheduling modifications without complicating twiddle address generation.

To enable conflict-free memory access with parallel PEs, butterflies are reordered. Fig. 1 shows that for N = 24, butterflies 0-5 can be performed simultaneously with butterflies 6-11, respectively. Corresponding operand indices are mapped to the same SRAM addresses, but at different banks. The schedule leverages this pattern to reduce the cycle count when a combination of radix-2/3/4 butterflies is needed (Fig. 2). Alternatively, higher radix butterflies can be used to reduce the # of CCs, as in [4].

TABLE I.   WI-FI 802.11AC FFT REQUIREMENTS

FFT Length	64	128	256	512
Bandwidth/IO Rate (MHz)*	20	40	80	160

*3.2µs symbol duration

TABLE II.   LTE FFT REQUIREMENTS

FFT Length	128	256	512	1024	1536	2048
IO Rate (MHz)*	1.92	3.84	7.68	15.36	23.04	30.72

$2^n3^m5^k$ FFT Lengths for SC-FDMA Precoding							
12	24	36	48	60	72	96	108
120	144	180	192	216	240	288	300
324	360	384	432	480	540	576	600
648	720	768	864	900	960	972	1080
1152	1200	1296					

*66.67µs symbol duration

This work was supported by DARPA CRAFT (HR0011-16-C-0052), NSF-GRFP (DGE-1106400), BWRC and ASPIRE. The authors thank David Biancolin & Ben Keller for their support.

978-1-5386-3179-9/17 $31.00 © 2017 IEEE

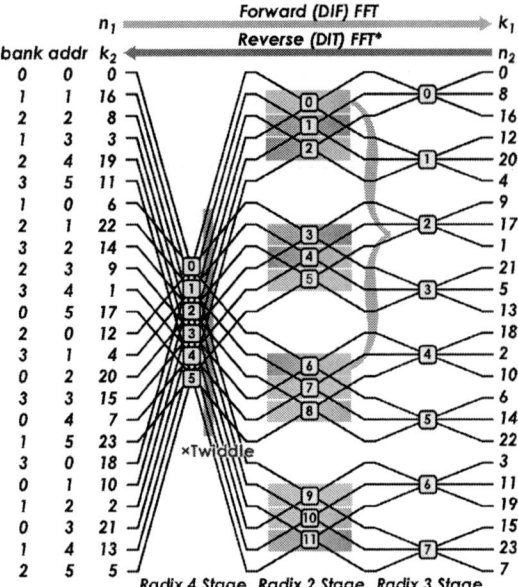

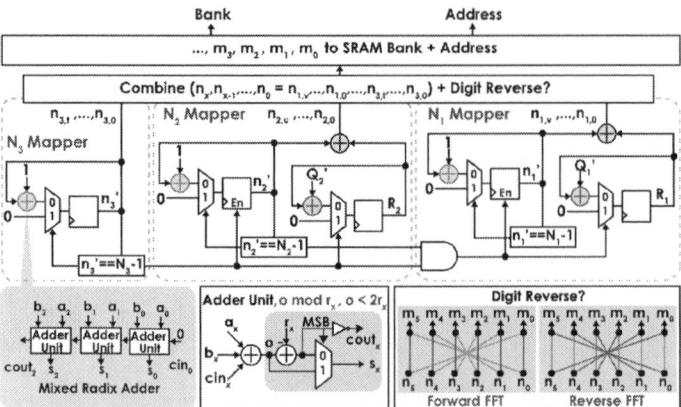

Fig. 3. IO Control logic: Index to memory bank/address mapper, consisting of n_x' and R_x *mixed-radix* counters and a digit reversal block for forward/reverse decompositions. n_2' increments (incs.) when n_3' wraps. n_1' incs. when n_3', n_2' wrap. R_x wraps when the corresponding n_x' incs. Banks are obtained via (1).

Fig. 1. Example FFT N = 24 signal flow graph. Forward and reverse decompositions mirror each other. Calculations can be performed in-place, but n_1, k_1 input/output orders are scrambled relative to each other. The ith calculation stage requires N/radix$_i$ butterfly operations. Colors in the Radix 2 stage represent different memory banks needed at each butterfly iteration. As an example, butterflies 0 & 6 use non-conflicting banks.

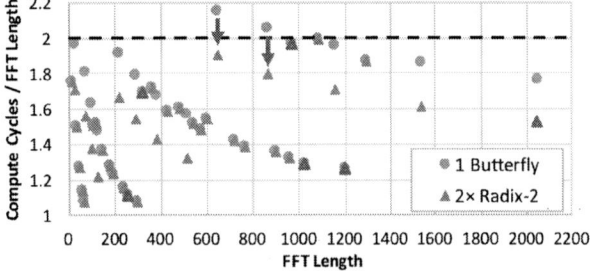

Fig. 2. Compute cycles normalized to FFT length (for FFT lengths from Tables I and II) with 1 radix-2/3/4/5 butterfly and 1 radix-$2 \times 2/3/4/5$ butterfly. Stall cycles needed for pipelining increase the # of CC's. Power measurements (Fig. 6c) are performed for the FFT lengths highlighted in red.

Finally, although memory-based architectures use more complex control logic than their pipelined counterparts, they are more easily adapted to reconfigurable, mixed-radix FFTs.

B. Reducing Twiddle LUT & SRAM Depth

To limit the number of twiddle factors needed in a *runtime-reconfigurable, mixed-radix* FFT, we use the prime factor algorithm (PFA) to factor N into 2^n, 3^m, 5^k coprime components before performing Cooley-Tukey (CTA) decomposition [1, 5]. Separate twiddle LUTs associated with $2^{n,max}$, $3^{m,max}$, $5^{k,max}$ are used, where addresses for $n < n_{max}, m < m_{max}, k < k_{max}$ are obtained via renormalization, as detailed in [1]. Thus, only $1,718 < N_{max}$ twiddles are needed for Wi-Fi + LTE.

As shown in Fig. 7, memory access dominates the FFT area & power budget. Because the input/output indices of the decimation-in-frequency (DIF) & decimation-in-time (DIT) CTA are bit-reversed——or more generally, digit-reversed——with respect to each other, 3N memory is typically used for calculation and output unscrambling [4]. However, since the forward and reverse decompositions are essentially mirrored, as

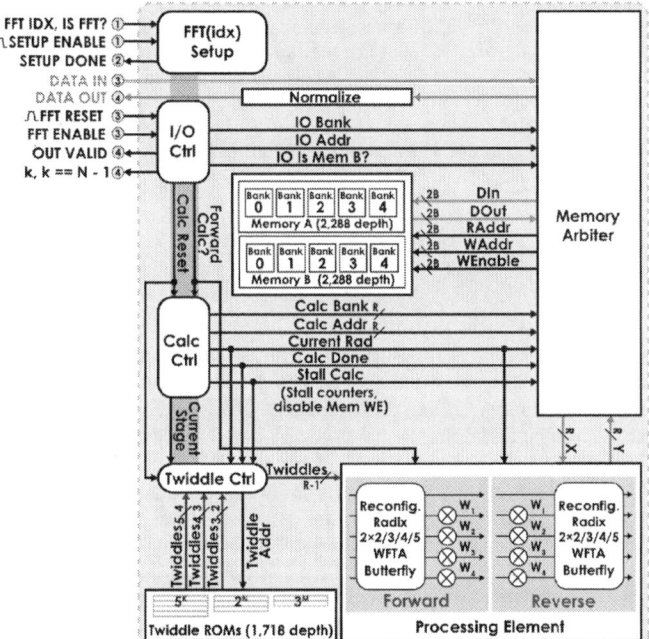

Fig. 4. LTE/Wi-Fi FFT architecture, derived from a hardware template.

shown in Fig. 1, alternating between the two decompositions every $2x^{th}$ symbol (Fig. 5b) allows for in-place IO with only 2N memory (2.23N_{max} for LTE+Wi-Fi) entries, reducing power [6].

C. Control Logic for IO Address Generation

To enable reconfigurability, rather than mapping data indices to memory addresses/banks via LUTs, a custom index vector generator is implemented [1]. Fig. 3 illustrates how this is built with mixed-radix numbers for $N_1 = 2^n$, $N_2 = 3^m$, $N_3 = 5^k$ co-primes. The IO control logic consists of a series of cascaded base-r (or mixed-radix) counters/adders that map $[0, N_y)$, $y \in [1, 3]$, with digits $n_{y,x}, \ldots, n_{y,0}$, to an index vector. The counters are built by replacing standard unsigned adders with base-r building blocks. When base-r numbers $a_2 a_1 a_0$ and $b_2 b_1 b_0$ are added, the digits of the sum and the intermediate carry outs can be computed with a simple subtraction and mux-based mod unit because $a_x + b_x + cin_x < 2r_x$ (where r_x is the radix of the xth digit).

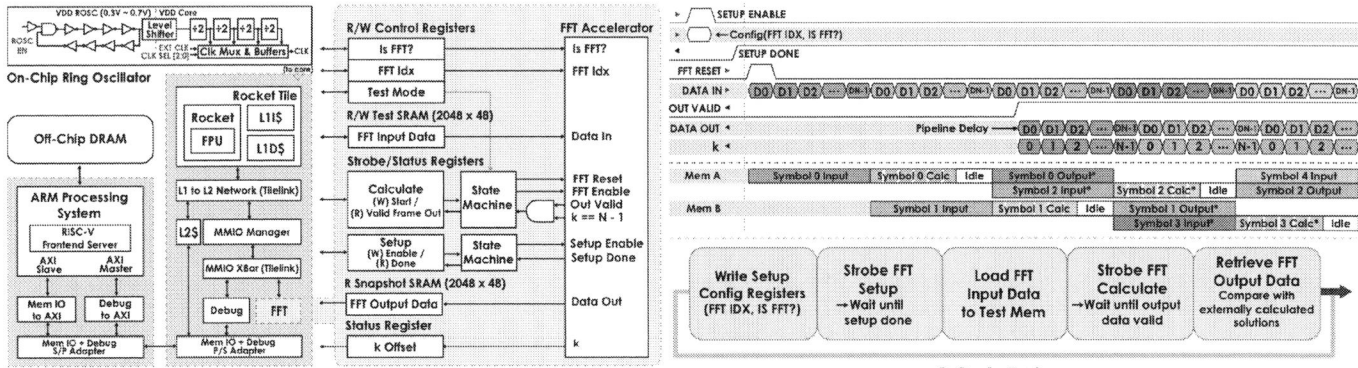

Fig. 5. (a) Rocket-Chip + FFT system, with snapshot memory. (b) FFT timing for *runtime configuration* and continuous input/output, along with the C task sequence for chip verification. Ping-pong memory timing is shown with alternating forward and reverse* decompositions. Calculation idle/stall periods are marked in white.

Bank/address mapping is implemented by the same block for forward/reverse decomposition modes. If the coprime decomposition is $N = N_1 N_2 N_3 = 2^n 3^m 5^k$ in the forward mode, it would be $N_1 N_2 N_3 = 5^k 3^m 2^n$ in the reverse mode, affecting the counters' constants. A subtle but important point is that the radix order is swapped to support in-place IO: i.e. if the counters are mixed-radix 4/2 (i.e. $6 = 3_4 0_2$, where subscripts denote radix) in the forward mode, they operate as mixed-radix 2/4 counters in the reverse mode (i.e. $6 = 1_2 2_4$). The mixed-radix digits of the N_y mapper outputs are combined into one index vector. For a reverse FFT, digit reversal is performed. Finally, bank values are determined via (1). The steps support a recursive (PFA, then CTA) decomposition reversal.

III. FFT HARDWARE INSTANCE & PROCESSOR INTEGRATION

A. Generating an FFT Instance from a Hardware Template

The FFT engine is generated from an accelerator template that includes configurable controllers to handle data flow between the IO, SRAM, PE, and an optional output normalization block (Fig. 4). The generator populates the template from a list of desired FFT lengths (N's). For LTE/Wi-Fi, the supported N's are factorized, and information about the corresponding coprimes ($2^{n,max} = 2048$, $3^{m,max} = 243$, $5^{k,max} = 25$) results in memory that is split into 2×5 SRAM banks (with depths of 4×512 and 240), a reconfigurable radix-2×2/3/4/5 butterfly, and a PE with *4* complex twiddle multipliers (supporting up to radix-5) that can be configured for forward or reverse decomposition. A total of 26 real multipliers are used by the Winograd's Fourier transform butterfly + twiddle unit for configurability across all LTE/Wi-Fi sizes.

B. Chip Design

The FFT engine has been implemented as an accelerator attached to a tethered 64-b RISC-V Rocket core (Fig. 5a) [7], all in a 16nm FinFET process. The Rocket core allows reading and writing to a series of dedicated memory-mapped I/O registers & SRAMs supporting *runtime FFT configuration* and input/output data loading/unloading. To simplify testing, a snapshot memory is included at the Rocket-Chip/FFT interface, and Rocket-Chip can pause FFT streaming to load new test vectors.

A Rocket "tile" consists of an in-order pipeline implementing the RISC-V 64-bit instruction set architecture, a floating-point unit, and L1 instruction and data caches. Main memory is realized by off-chip DRAM that is accessed via a Xilinx Zynq FPGA. A RISC-V frontend server on the FPGA's ARM processing system loads compiled C code onto the chip. The processor and accelerator are clocked via the divided output of a voltage tunable ring oscillator (Fig. 5a).

C. Generator Verification Methodology

The fixed-point FFT is characterized with complex random inputs. Its quantized results are compared with outputs from a floating-point software FFT from the Scala Breeze numerical processing library. SQNRs for 16-bit and 24-bit implementations across different FFT lengths are shown in Fig 6a. As expected, there is a ~6dB/bit SQNR improvement. Additionally, due to the accumulation of rounding errors, the SQNR (when compared to a *floating-point* implementation) degrades by ~6dB with each 2× increase in N. A 24-bit FFT has been built on this chip to evaluate the high-SQNR regime. A 16-bit implementation meets LTE/Wi-Fi requirements with ~30% lower power/area than the 24-bit design.

Bit-accurate outputs from Chisel simulations are used to automatically generate Verilog test benches for post-place & route verification. Likewise, C tests are automatically generated to simplify chip verification. The sequence of C tasks used to verify FFT functionality on the chip is illustrated in Fig. 5b.

IV. MEASUREMENT RESULTS

The FFT occupies an area of 0.37mm^2, while Rocket-Chip (and the outer memory system) occupies 0.39mm^2. The actual gate area, dominated by SRAMs, is approximately 0.24mm^2. The FFT gate area is 0.11mm^2.

Operating at a 570mV supply voltage and running C tests associated with LTE FFT requirements, the total chip power (dominated by the FFT) ranges from 0.46mW to 4.8mW, with clock frequencies up to 61.4MHz. At the same supply, the chip consumes between 2.7mW and 22.6mW when running Wi-Fi tests, which require clock frequencies up to 320MHz (Fig 6b).

Functionality has been verified up to 940MHz with a 0.9V supply. Measurements are taken by scaling the supply voltage (0.57V to 0.9V) along with frequency (40MHz to 940MHz from the ring oscillator). The total chip power ranges from 2.8mW to 170mW (Fig. 6c). Because Rocket-Chip is mostly idle while the FFT is running, the FFT accelerator and corresponding test memories draw most of the power. A power breakdown is obtained with a Primetime simulation deploying the same C tests

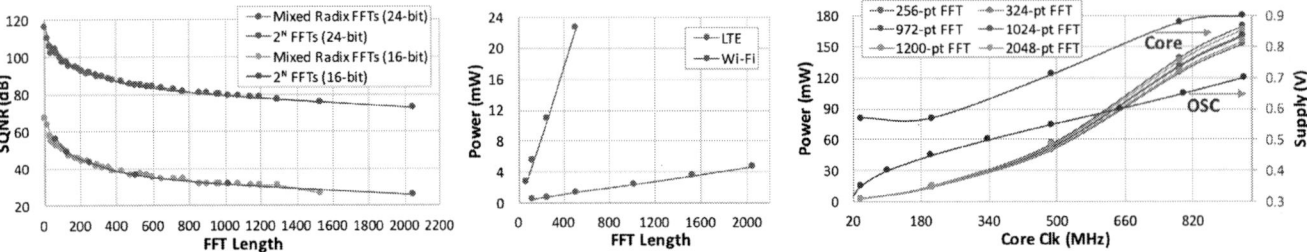

Fig. 6. (a) Fixed point SQNR (vs. floating point) for different FFT lengths & bitwidths. (b) Total (FFT + Rocket) measured power @ 570mV for LTE/Wi-Fi. (c) Total power required for various FFTs and corresponding supply voltages used at different core frequencies (ring osc. freq. / 8).

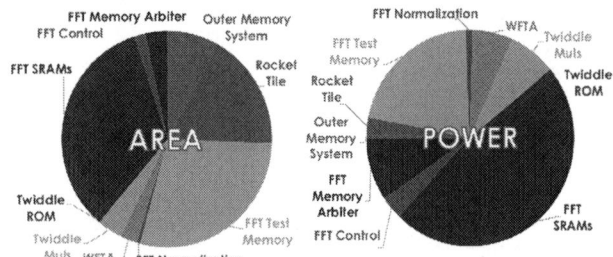

Fig. 7. (a) Gate area breakdown (0.24mm^2 total). (b) Primetime power breakdown for a 2048-pt FFT @ 520MHz and 0.72V core supply.

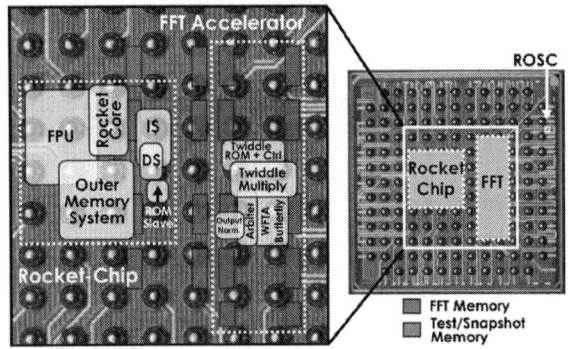

Fig. 8. Die photo. The total active area, as seen on the left, is 1.28mm^2. Rocket-Chip occupies 0.39mm^2 and the FFT occupies 0.37mm^2.

as in measurement; as expected and shown in Fig. 7, the SRAMs are the largest single contributor to both FFT power and area.

In Fig. 2, the configuration with the largest CC/N determines the ratio of calculation to IO clock rates. With a ratio of 2, all other sizes complete in a smaller fraction of the symbol period and hence exhibit lower power (Fig. 6c). For example, a 972-pt FFT that uses 1,905 compute cycles (98% of the 2N clock cycles allocated, as illustrated in Fig. 2) requires more power than all other measured FFTs. On the other hand, only 284 cycles are needed to complete a 256-pt FFT, allowing the FFT engine to sit idle (and in a lower power state) nearly half the time. Scheduling to support additional parallel PEs can be used to lower the system clock rate and power consumption.

V. CONCLUSION

Hardware generators enable rapid and reusable design of hardware instances in advanced technology nodes. A 0.37mm^2 LTE/Wi-Fi compatible $2^n3^m5^k$ FFT instance with performance and area comparable to state-of-the-art (Table III) *and integrated as an accelerator within a complete RISC-V processing system was designed and taped out within 1 month*

TABLE III. COMPARISON WITH OTHER LTE COMPATIBLE FFTS

	This Work	[8]	[5]	[4]
Architecture	Mem.	SDF	DEM	Mem.
FFT Size	64~2048, 1536, 12~1296	128~2048, 1536 + 12~1200	128~2048 + 12~1296	12~1296
Technology	16nm	28nm	0.18μm	55nm
Word Width	2×24	2×16	2×16	2×16
Mem Depth	4576 (2.23N$_{max}$)	2047 + 2213^d	2N$_{FFT}$ + 2N$_{DFT}$	3N
Gate Count	700K	170K + 511K	316K + 482K	340K
Area (mm^2)	0.37	0.31	25	1.063
Voltage (V)	0.57^{a,b}	0.61~1^a	1	1.08
Power (mW)	0.46~4.8^{a,c} 2.7~22.6^{b,c}	0.08~2.93^a	320	40.8
Clock (MHz)	3.84~61.44^a 40~320^b	1.92~30.72^a	122.88	122.88
Throughput (MS/s)	1.92~30.72^a 20~160^b	1.92~30.72^a	122.88	122.88
Incl. Processor	Yes	No	No	No

^{a} LTE, ^{b} W-Fi, ^{c} incl. Rocket + snapshot mem., ^{d} unscrambling mem. not reported

of PDK delivery. The accelerator is optimized for radix-2/3/4/5 butterfly reuse and continuous data flow with just 2.23N$_{max}$ total SRAM. It requires a twiddle LUT depth of only 1,718 (0.84N$_{max}$), despite supporting *all* LTE+Wi-Fi FFT configurations. The 0.37mm^2 encompasses all blocks that are needed to stream data in/out of the FFT accelerator in order. The chip's performance and functional correctness have been verified up to 940MHz via C tests loaded onto the Rocket core.

REFERENCES

[1] A. Wang, J. Bachrach, and B. Nikolić, "A generator of memory-based, runtime-reconfigurable $2^n3^m5^k$ FFT engines," *Proc. 41st IEEE International Conference on Acoustics, Speech and Signal Processing, ICASSP'16*, March 20-25, Shanghai, China, pp. 1016-1020.

[2] J. Bachrach, H. Vo, B. Richards, K. Asanović, and J. Wawrzynek, "Chisel: Constructing hardware in a Scala embedded language," in *Proceedings of the 49th Design Automation Conference (DAC)*, 2012.

[3] M. Püschel, *et al.*, "SPIRAL: Code generation for DSP transforms," *Proc. of IEEE*, vol. 93, no. 2, pp. 232–275, 2005.

[4] K. Xia, B. Wu, X. Zhou, and T. Xiong, "An efficient prime factor memory-based FFT processor for LTE systems," *IEEE Intl. Symposium on Circuits and Systems*, pp. 1546-1549, May 2016.

[5] J. Chen, J. Hu, S. Lee, and G. Sobelman, "Hardware Efficient Mixed Radix-25/16/9 FFT for LTE Systems," *IEEE Trans. Very Large Scale Integration (VLSI) Systems*, vol. 23, no. 2, pp. 221–229, Feb 2015.

[6] C.-F. Hsiao, Y. Chen, and C.-Y. Lee, "A generalized mixed-radix algorithm for memory-based FFT processors," *IEEE Trans. Circuits Syst. II, Exp. Briefs*, vol. 57, no. 1, pp. 26-30, 2010.

[7] K. Asanović, *et al.*, "The Rocket Chip Generator," Technical Report, EECS, University of California, Berkeley, CA, April 2016.

[8] G. Yahalom, "Analog-digital co-existence in 3D-IC," Ph.D. dissertation, EECS, MIT, Boston, MA, 2016.

A 65nm 376nA 0.4V Linear Classifier Using Time-Based Matrix-Multiplying ADC with Non-Linearity Aware Training

Anvesha A, *Student Member, IEEE* and Arijit Raychowdhury, *Senior Member, IEEE*

School of Electrical and Computer Engineering, Georgia Institute of Technology.

email: aamaravati3@gatech.edu, arijit.raychowdhury@ece.gatech.edu

Abstract—A 6-bit time-based folding matrix multiplication technique for support vector machine (SVM) classification is proposed. A voltage controlled oscillator (VCO) based analog-to-digital converter (ADC) performs in-situ matrix multiplications (MM) along with analog to digital conversion. It also offers a low supply voltage of operation (down to 0.4V) and an input range of 0.8V, drawing a total of 376nA of current. We propose a technique to extend the input range of the VCO based ADC using a folding technique. The classifier is trained considering the non-linearity of the VCO, which results in 5X lower power at iso-accuracy. An accuracy of 93% is achieved with adaboost at the digital back-end with linear SVM as the weak classifier.

Index Terms—time based ADC, dot product, classifier

I. INTRODUCTION

Embedded sensing applications require machine learning algorithms to perform in-sensor classification [1] [2] to avoid transmission bandwidth and latency to the cloud. Low power and low supply are crucial for sensing and inference. Embedded inference using neural networks, matched filters are gaining importance [1] [3]. There has been extensive work on ADC based in-sensor classifiers reported in [4] [3] [2]. All of the published results so far, demonstrate low-precision computation in voltage domain along with data-conversion. However, the minimum supply required [4] [3] is in the range of 1-1.2V. A lower supply voltage is preferred, if we need to operate the sensors using energy harvesting sources. Also, a low supply voltage is compatible with the traditional digital back-end and is scaling friendly. Due to the scalability and energy-efficiency, VCO based ADC's [5] [6] are gaining more attention. Fig. 1 (a) shows the traditional embedded sensor/classifier interface. In a traditional sensor interface, the ADC operates at a higher supply [4] and digital back end operates at low supply (0.4 - 0.6V). The traditional interface requires two different supply voltages, hence reducing power efficiency. Fig. 1 (b) shows the proposed embedded sensor/classifier interface. We use Adaboost with SVM as the weak classifier. We perform matrix multiplication in time domain and accumulation in digital domain. The accumulated value is compared to a bias value to obtain the results of the weak classifier. Then we perform majority voting and adaboost on the weak classifiers to obtain the final decision.

VCO based ADC consist of Voltage to Frequency (V to F) converter and Frequency to Digital Converter (FDC). V to F converter is achieved using a Voltage Controlled Oscillator (VCO). The FDC is a counter operating at the VCO frequency. However the VCO range is limited and the INL of such a compact ADC is poor. To calibrate the non-linearity digital post processing is typically used and the raw INL of 15-20 LSB reduces to less than a few LSBs. In this paper, we propose a technique to extend the input range and reduce the

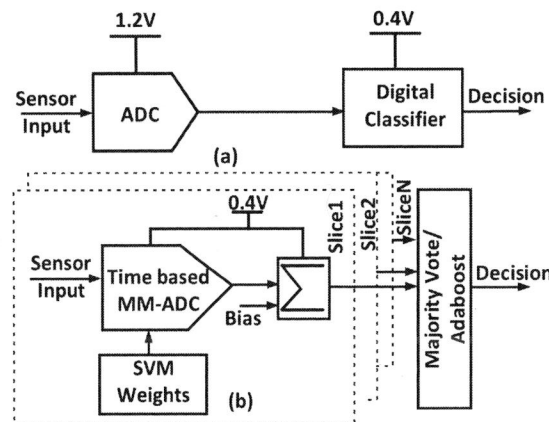

Fig. 1: (a) Traditional sensor/classifier interface (b) Proposed sensor/classifier interface

ADC non-linearity using a folding architecture. The classifier is trained in a manner that accounts for the resultant non-linearity of the ADC and no back-end digital calibration is required. This approach reduces the system power by a factor of more than 5X compared to an ADC with low INL. In the proposed architecture (Fig. 1 (b)) the sensor front-end operates fully at the digital supply.

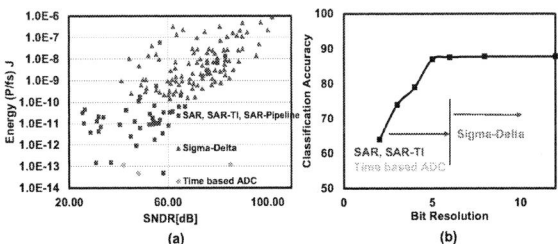

Fig. 2: Choice of ADC for in-situ matrix multiplication

Fig. 2 (a) shows the Energy vs ENOB of the state of the art ADC's. We observe that for low resolution to medium resolution successive approximation (SAR), SAR-TI (time interleaved) and VCO based ADC's have low energy/conversion. Fig. 2 (b) shows the classification accuracy of the MNIST database vs bit width. We can observe that 5-6 bits of datapath are sufficient to achieve the same accuracy as 12 bits. This makes SAR, time based ADC's most suited for embedded classification. Time based ADC's are mostly digital. Porting from one technology to another is easier and they are friendly to both voltage and technology scaling. We propose a time based mixed signal matrix multiplying ADC where : (1) the Digital to time converter produces a pulse proportional to the digital code and (2) the pulse is used to gate the VCO

978-1-5386-3179-9/17 $31.00 © 2017 IEEE

which produces a frequency proportional to the input voltage (Vin) (3) the counter produces a digital code proportional to the frequency (4) the digital code is compared to a bias value using a compactor to obtain the final decision. We have also analyzed the effect of INL on the classification accuracy of handwritten images from the MNIST database. Section II briefly describes the mathematical background of Adaboost based Support Vector Machine (SVM) as a weak classifier. Section III describes circuit implementation details. Section IV describes experimental results and Section V concludes the paper.

II. ADABOOST WITH SVM AS THE WEAK CLASSIFIER

SVM is a linear classifier that performs classification based on the hyper-plane or a set of hyper-planes. Support vectors represent data points close to the hyper-plane [7]. The MNIST database has 60,000 training data and 10,000 testing data. The samples are represented by $(X_{i1}, ... X_{iK})$ where K=60,000. Each image can be represented by a sequence of samples as: $(x_{i1}, ... x_{iN})$ where N=784. For M number of weak classifiers, the labels are represented by: $y_1, ..., y_M$. For SVM and the Adaboost classifier, we use:

$$y_t = sgn(w * X_i - b) \quad (1)$$

$$y_f = max(\alpha_1 * y_1 + \alpha_2 * y_2 + ... + \alpha_M * y_M) \quad (2)$$

Here the sgn operation indicates the sign of the operand, w is the array of support vectors, b is the bias value, $\alpha_i = 1/M$ (where, M=28) and y_f is the final classified label. The hardware implementation of the proposed technique is shown in the Fig. 1 (b). The sensor input is directly fed to the time-based MM-ADC. It is a true mixed-signal matrix multiplier, where the sensor input is an analog signal and the weights are digital values.

III. HARDWARE IMPLEMENTATION

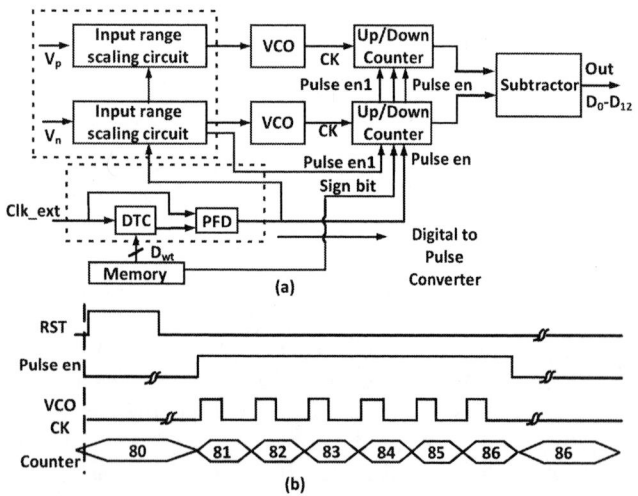

Fig. 3: (a) Proposed pseudo-differential matrix multiplying time-based ADC (b) The corresponding timing diagram

Fig. 3(a) shows the full schematic of the MM-TADC. The frequency of the VCO is directly proportional to the input voltage in the linear range of the VCO. The frequency of the VCO is:

$$F_t = K_{VCO} * V_{in} \quad (3)$$

where, K_{VCO} is the linearity co-efficient of the VCO and $V_{in} = V_p - V_n$ is the differential voltage. From the digital weight stored in the on-chip memory, we convert each digital value into a pulse using a digital-to-time converter (DTC) and a phase-frequency detector (PFD). The DTC receives an external clock reference. The DTC output and the reference clock are fed to the PFD. The PFD produces pulses whose widths are proportional to the digital codes (D_{wt}). The DTC along with the PFD forms a digital-to-pulse converter (DPC). The pulse width of the PFD is:

$$P_t = K_{DTC} * D_{wt} \quad (4)$$

The counter gets its clock from the VCO and the enable signal from the PFD output. Since the VCO input is gated with the DPC output, the counter value represents the gated VCO clock. The counter output is:

$$Out1 = K_{VCO} * Vin * K_{DTC} * D_{wt} \quad (5)$$

Eq. 5 shows that the counter output is proportional to the input voltage and the weights (D_{wt}). The sign-bit is directly fed to the counter. Since the counter is implemented as an up-down counter, we can naturally add or subtract the input multiplied by the magnitude of the weight, depending on the sign of the weight. The counter inherently is an accumulator. It is initially reset to 80 (hex). This is a mixed signal multiplier in the time-domain where the digital weights are multiplied with the absolute difference value of the analog sensor input.

VCO design and input range extension circuits: Fig. 4 shows the proposed VCO and the counter architecture. We convert the input voltage to a current using a V to I converter. The V to I converter acts as a current source to the VCO. Two phases of the VCO are used to the feed two counters. $Counter_1$ gets its enable directly from the PFD output (*Pulse-en*). $Counter_2$ gets its enable from the gated PFD output (*Pulse-en1*). Fig. 5 shows the proposed input range extension circuit. The VCO has a linear range of 250 to 300mV. We propose a technique to extend the range of the VCO. When the input is more than 600mV, we first scale down the input by 2× and perform analog-to-digital conversion. This preserves the linear operating range of the VCO. Finally, the output is scaled up by 2× to obtain the correct digital code. The proposed circuit uses a clocked comparator whose output is high if the input voltage is above 600mV; otherwise it stays at 0. The VCO receives the input voltage or a zero value from the multiplexer, depending on the *Pulse en* signal from the DPC. Output scaling is achieved as: (a) If the input is less than 600mV, one counter is used to count the VCO frequency (*Pulse-en* always enabled, *Pulse-en1* disabled). (b) Otherwise, two counters are used to count the VCO frequency (*Pulse-en1* is now enabled).

Digital-to-Pulse converter: The Digital-to-pulse converter (DPC) consists of a DTC followed by a PFD. Fig. 6 (b) shows the proposed DTC. It consists of a chain of four delay cells. Each delay cell is a cascade of two inverter cells (Fig. 6 (a)), where one inverter has a programmable capacitance as the load. An unit capacitance, C_0 of 10fF is implemented. The DTC receives an external clock. The output of the DTC is represented by CLKD. The delayed clock and the original clock are fed

978-1-5386-3179-9/17 $31.00 © 2017 IEEE

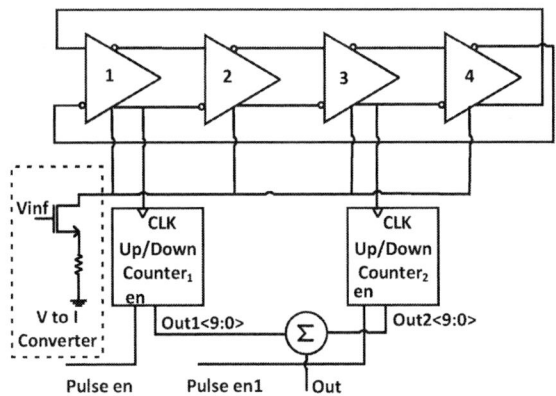

Fig. 4: A single ended VCO along with the output counter

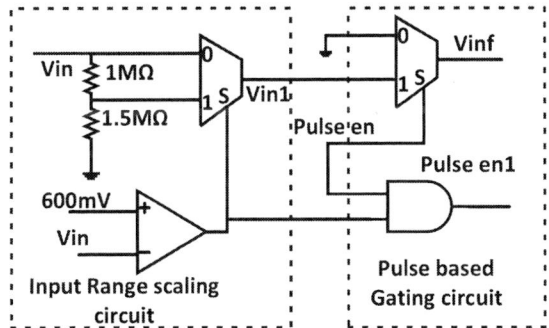

Fig. 5: Input dependent scaling of the output code achieves input range extension

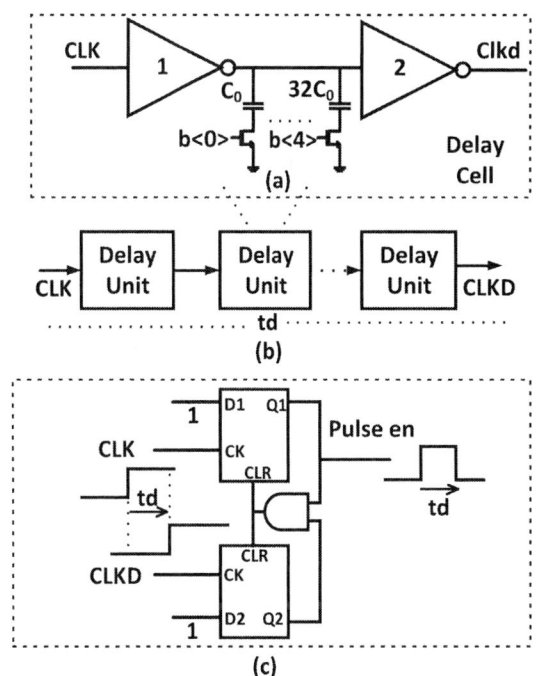

Fig. 6: Digital to pulse converter (DPC) for pulse generation

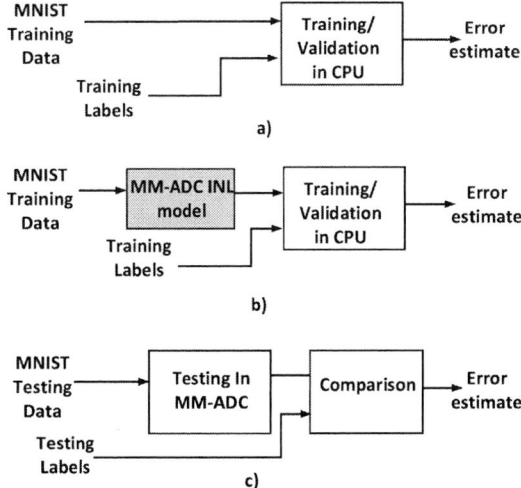

Process	GF 65nm CMOS 8M
Active Area	0.03mm²
VCO Area	0.0042mm²
Packaging	LCC
No. of Pads	32

Fig. 7: Die shot and chip characteristics

to a PFD. The PFD produces a pulse which is proportional to the delay between CLK and CLKD (Fig. 6 (c)). The test-chip is implemented in 65nm CMOS. We have implemented a bit-slice of the design and input for characterization is serially streamed in for system level analysis. The die photo and the chip characteristics are shown in Fig. 7.

IV. EXPERIMENTAL RESULTS

Fig. 8 (a) shows the traditional training procedure for classification of the MNIST database. The VCO based ADCs have low-power but poor INL performance. Here we show that the training can be aware of such non-ideality. Fig. 8 (b) shows the proposed training procedure for the MNIST classification. The proposed technique comprehends the non-linearity (INL) error included in the matrix multiplication. This error-aware model is used for training and testing (Fig. 8 (c)) the MNIST images. This error-aware technique helps to improve the accuracy of the classification. Fig. 9 (a) shows the measured single-ended VCO transfer characteristics. The single ended VCO has a linear range of 250mV. Fig. 9 (b) shows the VCO transfer characteristics with the range extension technique. The proposed extended-range differential-VCO has an input range of 800mV.

Fig. 10 (a) shows the DNL of the proposed MM-TADC. The DNL is less than 1.5LSB. Fig. 10 (b) shows the INL of the proposed MM-TADC. The INL is less than 7LSB. Fig. 11 (b) shows the classification accuracy with and without the

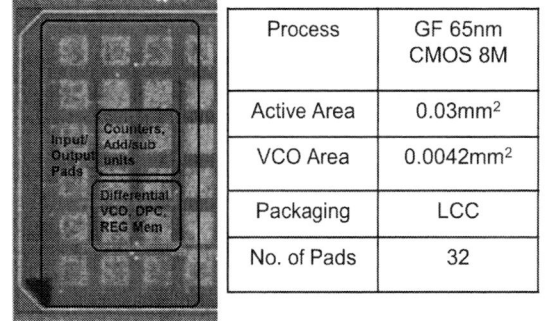

Fig. 8: (a) Traditional training procedure (b) Proposed INL aware training and (c) Testing for accuracy of inference

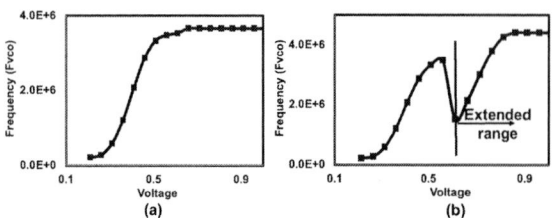

Fig. 9: Measured: (a) Single ended VCO transfer characteristic (b) Folded VCO transfer characteristic with high input range

INL error-aware training. We can observe that without error-aware training, the classification error goes up with increasing INL. However, with INL error-aware training, the classification error remains within a ±2-3% margin. The ADC energy per step scales as 2^{2*ENOB}. Here, ENOB of 5.3 can achieve the same accuracy as an 8 bits with a large INL. With the proposed circuit we achieve more than 5X reduction in power using the TD ADC. Fig. 11 (a) shows the digital code vs the differential input voltage for the proposed circuit. We can observe that after 600mV, the range gets extended. This is achieved by the proposed folding architecture before the VCO. Fig. 12 (a) and (b) show the captured VCO output and the DPC output with a pulse width of 400ns.

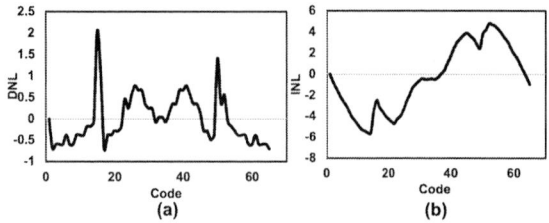

Fig. 10: (a) Measured DNL of the proposed front-end (b) Measured INL of the proposed front-end

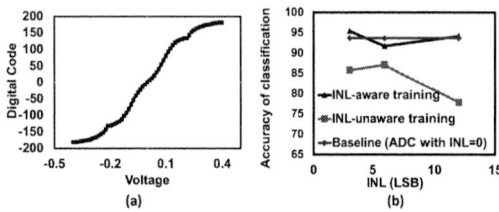

Fig. 11: (a) Digital code vs differential input voltage (b) Classification accuracy for a baseline design and TD ADC based design with and without error aware training

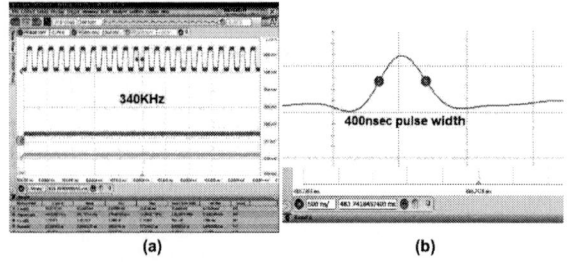

Fig. 12: (a) Scope capture of the VCO frequency at 360mV (b) Zoomed-in scope-capture of the DPC output

Component	Power
VCO	360nA
Counter	40nA
DPC	80nA
Adder/Sub-tractor	16nA
Total	376nA

TABLE I: Power drawn by various components

Table I shows the power drawn by various blocks of the proposed MM TADC. For a 700ns pulse width of the DPC, energy/MAC is 1.47pJ. For $28*28$ image size with 60 support vectors the estimated total energy is 313nJ. For Adaboost with 26 weak classifiers, the energy/classification is $8.1\mu J$. Table II shows the comparison with previously reported results [4] [8] [3]. [4] [8] perform matrix multiplication in the voltage domain. The amplifier used in [8] requires a 1-1.2V supply. The measured energy of our work is comparable to reported results and it uses a digitally scalable architecture enabled by time-based processing.

	ISSCC 2015 [4]	ISSCC 2016 [3]	ISSCC 2017 [8]	ASSCC 2016[1]	This work
System	Object/ECG detection	Object detection	Spatial Filtering	Image Classification	Image Classification
Technology	130nm	40nm	65nm	28nm	65nm
A/D Rate	20kS/s	39MS/s	NA	2.4MHz	500kHz
ENOB	7Bits	5-7bits	14bits	8 bits	5.4 bits
D/A Rate	N/A	1GS/s	1.5MHz		-
Multiplication domain	Voltage mode	Voltage mode	Voltage mode	Voltage mode	Time based
Analog/Digital Supply	1.2V	1.1V	1V	1V	0.4V
Input range	0.6V	0.5V	1V	1V	0.8V
Power(μW/MHz)	66.3	0.228	260	3	0.46
Energy (pJ/MAC)	16	0.12	2	3.2**	1.47
Classification accuracy	87%	-	89%	NA	93%

Table II. Comparison with table reported works **Off-chip Accumulator

V. CONCLUSION

This paper presents a time-based matrix multiplying technique. At 0.4V, the energy per classification for the SVM is 313nJ/frame. The proposed training technique which is aware of the INL of the ADC and achieves high classification accuracy.

VI. ACKNOWLEDGEMENTS

This work was sponsored by a gift from Qualcomm Inc.

REFERENCES

[1] D. Bankman, "An 8-bit, 16 input, 3.2 pj switched-capacitor dot product circuit in 28-nm cmos," *ASSCC*, no. 5, pp. 21–24, 2016.
[2] A. Anvesha, "A 130nm 165nj/frame c-d smashed-filter based mixed-signal class. for smart cameras," *TCAS-II*, April 2017.
[3] E. Lee, "A 2.5GHz 7.7TOPS/W switched-capacitor matrix multiplier with co-designed local memory," *IEEE ISSCC*, 2016.
[4] J. Zhang, "A matrix-multiplying ADC implementing a machine-learning classifier with data conversion," *IEEE ISSCC*, 2015.
[5] T. Gerry, "A mostly-digital variable-rate continuous-time delta-sigma modulator adc," *JSSC*, 2010.
[6] K. Reddy, "A 16-mw 78-db sndr 10-mhz bw ct adc using residue-cancelling vco-based quantizer," *JSSC*, December 2012.
[7] C. Cortes, ""support-vector networks," *Machine Learning*, vol. 43, no. 1, pp. 273–279, 1995.
[8] S. Joshi, "2pj/mac 14b 88 linear transform mixed-signal spatial filter with 84db interference suppression," *IEEE ISSCC*, 2017.

A 1GHz Fault Tolerant Processor with Dynamic Lockstep and Self-recovering Cache for ADAS SoC Complying with ISO26262 in Automotive Electronics

Jinho Han[1,2], Youngsu Kwon[1], Yong Cheol Peter Cho[1] and, Hoi-Jun Yoo[2]

[1] Processor Research Group, ETRI, Daejeon, Republic of Korea
[2] Department of Electrical Engineering, KAIST, Daejeon, Republic of Korea
soc@etri.re.kr

Abstract—We present a processing platform that implements DMR with separate clock and power sources to prevent dependent failures working with a reconfigurable cache that includes BIST with self-recovering function to detect transient faults and error prediction to prevent permanent faults. The fault tolerant processor is analyzed to be complying with the ISO26262 SOTIF for ADAS SoC which is fabricated with 28nm CMOS Technology. The single point faults metric is 99.64% with a safety mechanism.

Keywords—fault tolerant processor; ADAS; fault tolerant cache; ISO26262; dynamic lockstep

I. INTRODUCTION

Advanced driver-assistance systems (ADAS) such as Forward Collision Warning, Pedestrian Collision Warning, Lane Departure Warning, and Traffic Sign Recognition with Speed Limit Indication recognize objects and calculates the distance and the rational speed by processing images from a camera. ADAS vision processing algorithms are inherently compute intensive, and therefore, require powerful processors [2] Moreover, processors for automobiles should include the fault tolerant feature because of its harsh operation conditions and safety requirements [3].

Semiconductor chips for automotive applications must comply with ISO26262 standard to achieve functional safety [4]. The SOTIF working group in ISO26262 addresses functional safety standard for the intelligent driving system including ADAS and has proposed strict requirements to avoid or mitigate the possibility of the system failure. Previously, ADAS applications executed on DMR (Dual Modular Redundancy)-based CMPs with redundant execution at the other core for error detection and recovery [5]. Also, recent multi-thread based safety processors require complicated threading with complicated cache architectures.

In this paper, we present a processing platform that implements DMR with separate clock and power sources to prevent dependent failures working with a reconfigurable cache that includes BIST with self-recovering function to detect transient faults and error prediction to prevent permanent faults. The processing platform overcomes the low fault

coverage of the self-test and the performance overhead by the self-test time in [5] [3].The proposed fault tolerant processor is analyzed to be complying with the ISO26262 SOTIF for ADAS SoC. The processor works in tandem with hardware IPs that includes H.265 video codec and object recognition accelerator.

II. THE PROPOSED PROCESSOR ARCHITECTURE

The proposed fault tolerant processor contains three key features: 1) dynamic lockstep (DLS) with separate clock and power sources to reduce dependent failures and have the high performance, 2) the cache with self-recovering function to reduce transient faults, 3) reconfigurable function to reduce permanent faults.

Figure 1 shows a block diagram of the fault tolerant processor. It consists of 2 processor cores, 4 caches, 4 fault managers (FTM) in cache, and 1 external FTM (EFTM). For operating at 1GHz frequency on 28nm node. For the critical path is below 1ns, one processor has 13 pipeline stages with 2-issue superscalar architecture, fetch scheduler fetching 8 instructions maximally, a branch predictor with branch target buffer (BTB) using the GSHARE method with branch history registers, Load and Store Unit with 2 pipeline stages, 32KB I/D cache with 3 pipeline stages, and 32-entry I/D table look-aside buffer (TLB).

The design operates in one of two modes: Dynamic Lockstep (DLS) mode and non-DLS mode. The mode of operation is determined by programming at the software shown as Figure 2. The processors operate in DLS mode when DLS register of the both processors is enabled by a software and, operating in non-DLS mode when DLS register of the both processors is disabled by a software.

In DLS mode, one processor operates as the leading core while the other operates as a trailing core and the operating frequency of the leading core is bigger than the one of the trailing core and the difference of the operating frequency makes the effect of the temporary redundancy. both processors run the same task by controlling the core id of the both processors to the same core id by EFTM and for the result of the leading core is compared with that of the trailing one to detect faults, the data cache in each processor has stopped and

This work was supported by the ICT R&D program of MSIP/IITP. [2017-0-00261, Intelligent Many-Core Processor and SW based on Low-Power Hypervisor]

S20-4 (2167)

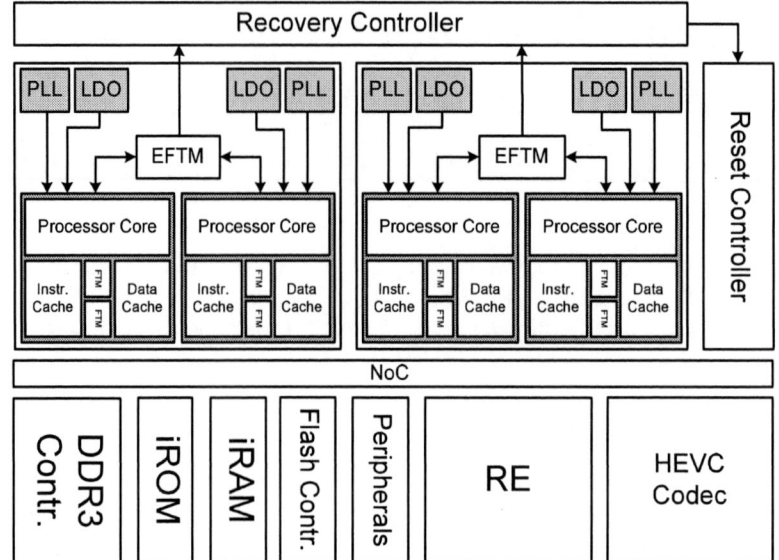

Fig. 1. A fault tolerant processor in ADAS SoC

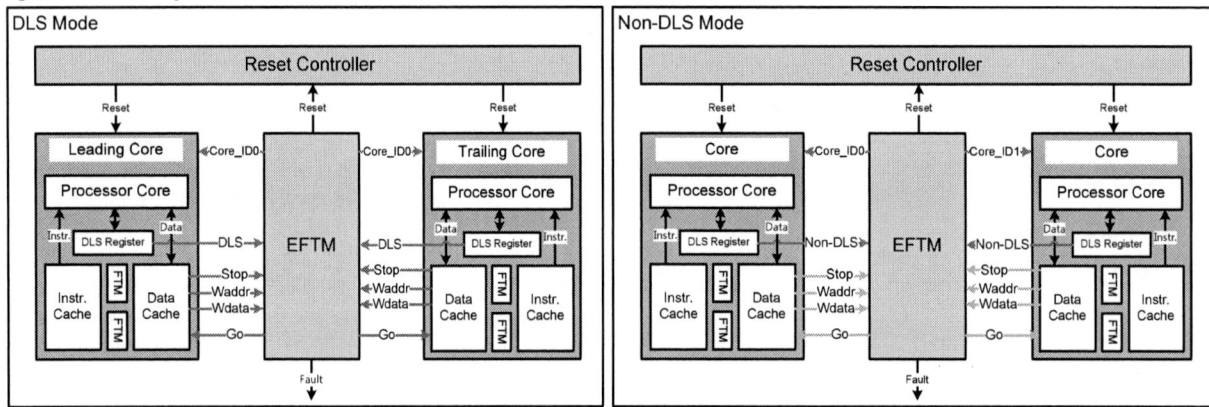

Fig. 2. Dynamic lockstep with separate clock and power sources.

send the write data and address to EFTM when the processor core request the write to the data cache. The frequency of the check point using the cache write is sufficient to identify the fault of the processor.

EFTM compares the cache write of the leading core with the cache write of the trailing core and EFTM starts the data caches of the leading and trailing core if the cache write of the trailing core is resembled from one of the leading core. But, EFTM generates the fault signal if the cache write of the trailing core is not resembled from one of the leading core. The data in SDRAM is not changed by the trailing core because the data is written in the data cache, but the dirty bit of the data cache in the trailing core is not enabled. In the other words, the data is changed by the leading core, and the trailing can use the data which is changed by the leading core.

In non-DLS mode, the two processors operate as a traditional dual-core processor would and thereby increasing throughput by running different tasks on each processor which are not critical for the functional safety of the system. EFTM

identifies the DLS register of the both processors is disabled and controls the core id of the each processor to the different core id.

A cache system is composed of two architectures: 1) the error correction in the memory by using a cache characteristics and error correction code, 2) the reconfiguration by using error predictor as Figure 3. Safety mechanisms in the cache architecture are vital as cache is intrinsically vulnerable to transient faults. In our implementation, the error correction code (ECC) named single error correction and double error detection (SECDED), corrects single errors and detects double errors in data chunks.

FTM monitors double errors and recovers the data using the characteristics of the cache which copies the data of the SDRAM in SRAM of the cache. The data can be recovered by depending on the error state of the valid-bit, tag, dirty-bit, and data memory in the cache when error correction code detect but can't correct errors. If there are no error on valid-bit, tag, dirty-bit memory and the value of dirty-bit of the erroneous data is 0,

978-1-5386-3179-9/17 $31.00 © 2017 IEEE   314

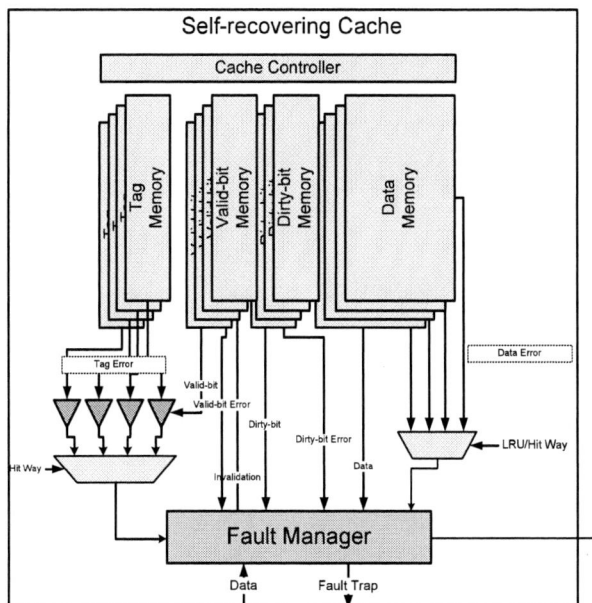

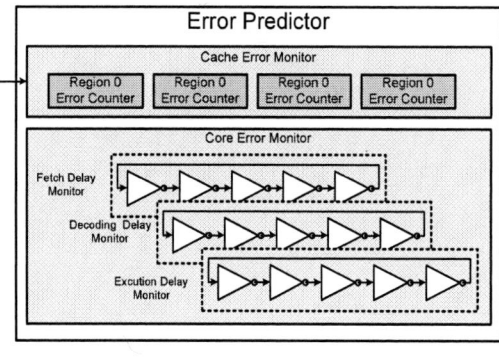

Fig. 3. A fault tolerant cache with self-recovering, and error predictor.

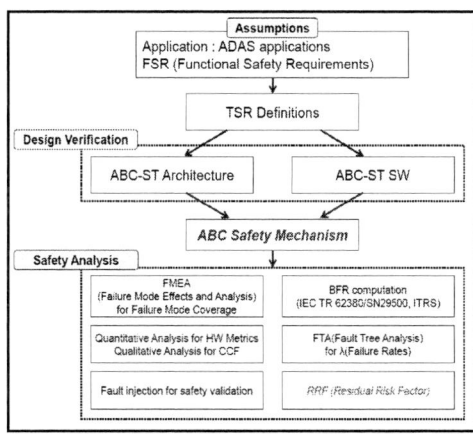

Fig. 4. SEooC flow and quantitative analysis of ISO26262.

the error is recovered by reading the data from SDRAM the error case which is composed the error state of valid-bit, tag, dirty-bit, and data memory and can be recovered and invalidated is descripted in [6].

Error predictor has cache error monitor and core error monitor as shown in Figure 3. In cache error monitor, the silent fault is counted when the error is corrected by ECC and the data of the cache is recovered by the fault manager by a region, which is divided by 8KB. The each region of the cache is powered off based on the fault frequency, or when the silent fault count of the region is bigger than the number of the pre-defined silent faults. The processors are switched in DLS mode when one of delay monitors for the pipeline stages has bigger delay than the period considering the guard band in core error monitor. Not only the transient fault but also the permanent fault can be prevented by the error predictor.

III. IMPLEMENTATION RESULTS AND FAULT ANALYSIS

The functional safety of the processor in ADAS SoC is analyzed to Safety element out of context (SEooC) method of ISO26262 as shown Figure. 4. With the implementation process for the Semiconductor with ISO26262 Compliance, the semiconductor has the fault-tolerant design, which is the wearout prevention for permanent faults, the multicore lockstep and part-wise checker for transient faults, dependent failures, and the designs for the transient, permanent, and dependent failures are analyzed and verified by FMEA, FTA, qualitative analysis, and fault injection.

The proposed fault tolerant design has self-recovering BIST and DLS for transient faults and error monitor for permanent faults. The dependent failures of two or more circuits are resulting from a single specific event or root cause. So, the

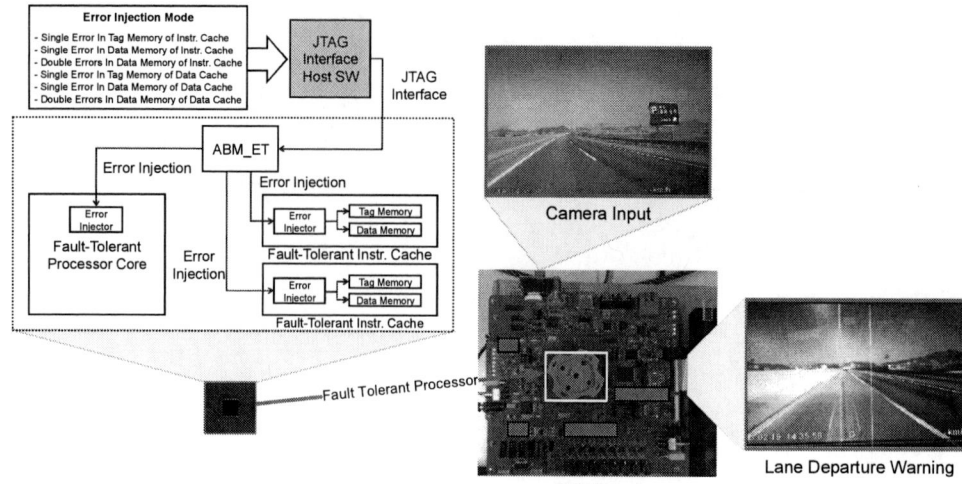

Fig. 5. Fault injection experiment.

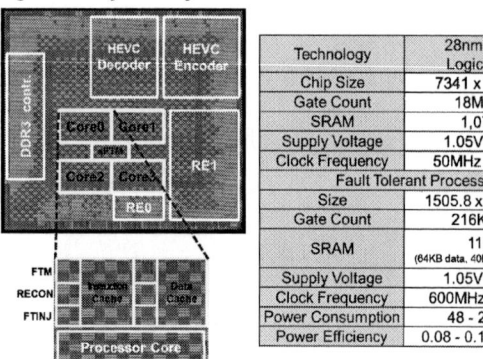

Technology	28nm 1P12M Logic CMOS
Chip Size	7341 x 7541 μm^2
Gate Count	18M gates
SRAM	1,077KB
Supply Voltage	1.05V - 1.15V
Clock Frequency	50MHz - 1.0GHz
Fault Tolerant Processor	
Size	1505.8 x 1246.4μm^2
Gate Count	216K gates
SRAM	110KB (64KB data, 40KB ECC, 4KB tag)
Supply Voltage	1.05V - 1.15V
Clock Frequency	600MHz -1.0 GHz
Power Consumption	48 - 216mW
Power Efficiency	0.08 - 0.18mW/MHz

Fig. 6. Die photo and summary.

TABLE I. THE FAULT TOLERANCE PERFORMANCE COMPARISON.

Parameter	No FT	FT Cache	DCC[5]	Takahashi [3]	This Work
Fault Detection Coverage	N/A	Medium	High	Medium	Medium ~ High
ECC	X	SECDED	N/A	N/A	SECDED
Fault Detection	X	ECC	DMR	Self-BIST	DLS with Separate Cock and Power, Self-recovering Cache
Fault Detection Time	X	1 cycle	137 cycles	<2ms	1 cycle
Fault Prediction	X	X	Thread Control	Droop Monitor Adaptive Clock Control	Error Predictor Reset/Cache Size Control
Fault Injection	X	X	X	X	O
Fault Traps	100 %	83 %	N/A	10-7 RHF/Hour	28 %
Single Point Fault	X	N/A	N/A	N/A	99.64 %
Latent Fault	X	N/A	N/A	N/A	93.23 %

dependent failures is prevented because the root cause of power sources and the single specific event of the clock sources does not exist by using the separated clock and power. With FMEA, transient faults are analyzed by fault injection at ADAS in automotive electronic system as shown Figure. 5. Permanent faults and dependent failures are analyzed by quantitative analysis with safety mechanism of DLS and self-recovering cache.

IV. CONCLUSION

We implemented ADAS SoC with the proposed fault tolerant processor. The fault tolerant processor is fabricated with 28nm CMOS Technology, having the footprint with the specification as Figure 6. The fault tolerant processor has two dual-core with DLS and the instruction cache and the data cache with a self-recovering function. The each cache can store 32KB of data with 20KB of ECC. A self-recovering cache and DLS achieves 80% reduced error traps as TABLE I when compared with the previous works. As a result, the single point faults metric about the permanent faults is 99.64% with a safety mechanism, and the latent faults metric about the permanent faults is 93.23% with a safety mechanism.

REFERENCES

[1] K. J. Lee, et al., "A 502GOPS and 0.984mW Dual-Mode ADAS SoC with RNN-FIS Engine for Intention Prediction in Automotive Black-Box System," ISSCC Dig. Tech. Papers, pp.256-257, 2016.

[2] J. Tanabe, et al., "A 1.9TOPS and 564GOPS/W Heterogeneous Multicore SoC with Color-Based Object Classification Accelerator for Image-Recognition Applications," ISSCC Dig. Tech. Papers, pp.328-329, 2015.

[3] C. Takahashi, et al., "A 16nm FinFET Heterogeneous Nona-Core SoC Complying with ISO26262 ASIL-B: Achieving 10-7 Random Hardware Failures per Hour Reliability," ISSCC Dig. Tech. Papers, pp.80-81, 2016.

[4] Road vehicles – Functional Safety, ISO26262, 2012.

[5] C. Lafrieda, et al., "Utilizing Dynamically Coupled Cores to Form a Resilient Chip Multiprocessor," IEEE Intl. Conf. on Dependable Systems and Networks Tech., pp. 317-326, 2007.

[6] J. H. Han, et al., "A Fault Tolerant Cache System of Automotive Vision Processor Complying with ISO26262,", IEEE Trans. Circuits Syst. II Exp. Briefs, vol. 63, Issue. 12, pp. 1146-1150, Dec. 2016.

S21-1 (2194)

IEEE Asian Solid-State Circuits Conference
November 6-8, 2017/Seoul, Korea

A 77-GHz Mixed-Mode FMCW Signal Generator Based on Bang-Bang Phase Detector

Jianfu Lin[1], Zheng Song[1], Nan Qi[2], Woogeun Rhee[1], and Baoyong Chi[1]

[1]Institute of Microelectronics, Tsinghua University, Beijing, China
[2]Institute of Semiconductors, Chinese Academy of Sciences, Beijing, China
chibylxc@tsinghua.edu.cn

Abstract—A 77-GHz mixed-mode frequency-modulated continuous-wave (FMCW) signal generator is proposed based on the bang-bang phase detector (BBPD). Instead of employing a linear digital-to-time converter (DTC), a 1-bit 3rd-order single-loop ΔΣ modulator (SLDSM3) and a hybrid finite-impulse response (FIR) filter are utilized to suppress the quantization noise induced by the BBPD. Two-stage infinite-impulse response (IIR) filters are inserted into the digital loop filter (DLF) to reduce the instant variation at output, smoothing the chirp waveform during its generation. To improve the linearity around the turning-around points (TAPs) of the chirp, a type-III slope estimator with switchable polarity is employed. The prototype is implemented in 65-nm CMOS technology, with the total power consumption of 43.1 mW. Measurement results show a 77-GHz carrier with -81.7-dBc/Hz phase noise at 1-MHz offset, as well as a generated triangle chirp that features 1-ms repetition period, 1.827-GHz bandwidth and 336-kHz root-mean-square (RMS) frequency error.

I. INTRODUCTION

The automotive radar system has been witnessed a continuous development for its key role in the realization of autonomous driving. The automotive radar reports the information about the distance and speed of nearby vehicles, which assists for drivers and improves the driving safety. The automotive radar based on the millimeter-wave frequency-modulated continuous-time (FMCW) transceivers has been widely used due to its robustness against various environments and its low peak power characteristic [1]-[2].

As the key block in the transceiver, FMCW signal generator has attracted more attention in recent researches [1]-[4]. The charge-pump based phase-locked loops (PLLs) are typically employed for the FMCW chirp generation [1]-[2]. But it suffers from the limitation on the chirp slope and the large deviation occurred around the turning-around points (TAPs) due to its finite loop bandwidth. An all-digital PLL with two-point modulation (TPM) has been proposed to break the loop bandwidth constraint [3]. However, the wide-tuning-range digitally-controlled oscillator (DCO) needs a complex SRAM-based calibration due to its frequency overlap between two adjacent tuning curves. Replacing the DCO with a current digital-to-analog converter (DAC) and a voltage-controlled oscillator (VCO), the mixed-mode PLL [4] overcomes the overlapping issue. However, the linearity around the TAPs of the chirp is still limited by the loop bandwidth. Both the all-digital PLL and the mixed-mode PLL need a linear time-to-digital converter (TDC), which increases the design complexity.

This paper presents a 77-GHz mixed-mode FMCW signal

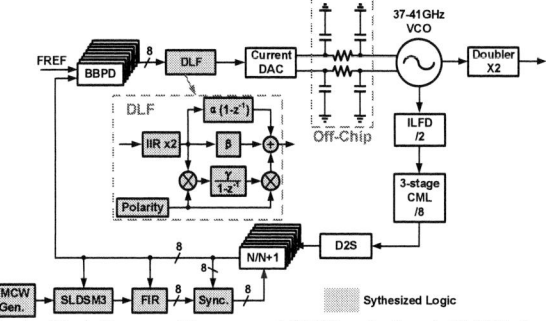

Fig. 1. Block diagram of the proposed 77-GHz mixed-mode FMCW signal generator. The blocks in the grey area are implemented with the synthesized digital logic.

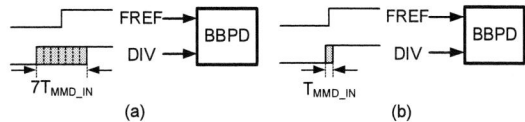

Fig. 2. Dithering range of the DIV signal with (a) 1-1-1 MASH ΔΣ modulator and (b) 1-bit SLDSM3.

generator. The bang-band phase detector (BBPD) is employed with a 1-bit 3rd-order single-loop ΔΣ modulator (SLDSM3) and a hybrid finite-impulse response (FIR) filter [5], avoiding the complicated linear TDC or digital-to-time converter (DTC). Two-stage infinite-impulse response (IIR) filters are exploited to smooth the chirp waveform. A type-III chirp slope estimator with switchable polarity [6] is utilized to improve the chirp linearity around the TAPs.

II. PROPOSED 77-GHz MIXED-MODE FMCW SIGNAL GENERATOR

A. Overall Architecture

Fig. 1 shows the block diagram of the proposed 77-GHz mixed-mode FMCW signal generator, in which the frequency doubling architecture [1] is utilized, with a VCO running around 38.5 GHz. The lowered center frequency relaxes the design of the VCO as well as the frequency divider. The output of the VCO is divided by 16 with an injection-locked frequency divider (ILFD) and 3-stage current-mode logic (CML) frequency dividers. The followed multi-modulus divider (MMD) and BBPD array enables a hybrid FIR filtering, which mitigates the phase folding problem caused by the BBPD [5]. A 1-bit SLDSM3 is employed, instead of 1-1-1 MASH ΔΣ modulator, to reduce the quantization noise of the BBPD [5]. The 4th-order digital loop filter (DLF) (Fig. 1) consists of

978-1-5386-3179-9/17 $31.00 © 2017 IEEE   317

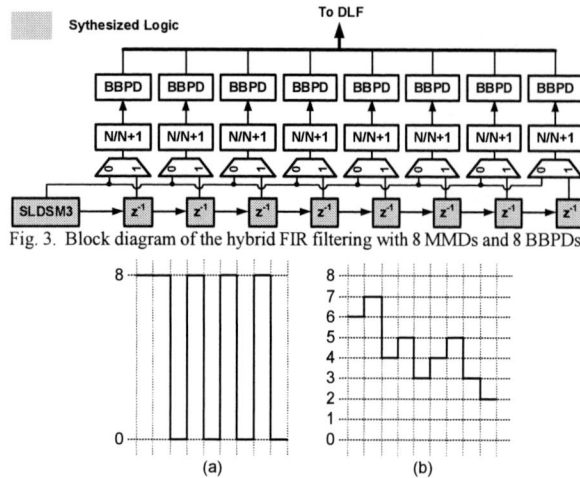

Fig. 3. Block diagram of the hybrid FIR filtering with 8 MMDs and 8 BBPDs.

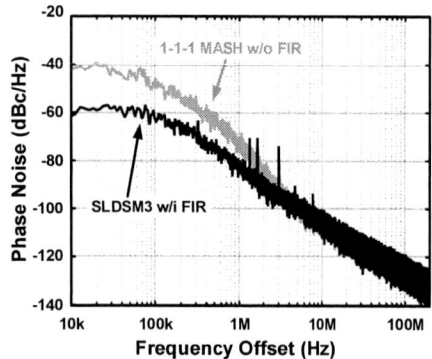

Fig. 4. Output of the 8 BBPDs with the hybrid FIR (a) turned off and (b) turned on.

Fig. 5. Simulated phase noise at 77.531 GHz.

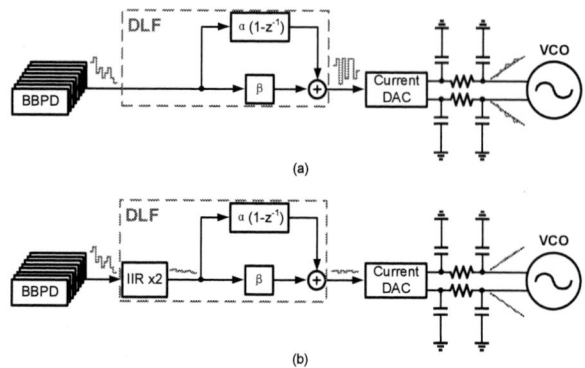

Fig. 6. Chirp generation (a) without IIR filtering and (b) with IIR filtering.

Generally, a 1-1-1 MASH $\Delta\Sigma$ modulator is used to realize a fractional-N frequency synthesizing. Nevertheless, its large dithering output range degrades the noise performance of the BBPD. Fig. 2 compares the dithering range of the DIV signal with 1-1-1 MASH and 1-bit SLDSM3. The 1-bit SLDSM3 provides a smaller RMS input jitter for the BBPD, which in turn suppresses the quantization noise. Furthermore, the 3rd-order loop in the 1-bit SLDSM3 guarantees a good spur performance [5].

Fig. 3 shows the block diagram of the 8-tap hybrid FIR filtering, which is composed of 8 flip-flops, 8 MMDs and 8 BBPDs. To show the effect of the hybrid FIR clearly, the BBPDs' output is evaluated when the hybrid FIR is turned on and off, respectively (Fig. 4). To perform a fair comparison, there are still 8 MMDs and 8 BBPDs but working with the same division ratio when the hybrid FIR is turned off. It is shown that, the output of the 8 BBPDs is switching between 0 and 8 with the hybrid FIR off. If the hybrid FIR is turned on, the cancellation of the deterministic phase error will be performed between the 8 paths. As a result, the output can be an arbitrary integer from 0 to 8, similar to a multi-bit TDC. Then the output dithering amplitude is decreased, which means an effectively enhanced resolution of the phase detecting.

To verify the improvement achieved by the 1-bit SLDSM3 and the hybrid FIR, a behavioral simulation based on MATLAB and Simulink is performed. The Fig. 5 compares the simulated phase noise between the 1-1-1 MASH without FIR and the 1-bit SLDSM3 with FIR. It turns out that the 1-bit SLDSM3 and hybrid FIR help to achieve 20-dB better in-band phase noise, and the loop bandwidth is about 3 times larger than that with the 1-1-1 MASH and FIR turned off. Simulation results coincide with the gain expression and the conclusions about the quantization noise of the BBPD.

C. Chirp Waveform Smoothing with IIR

In the typical DLF design, only a proportional path and an integral path are included for low complexity [5]. However, it is far from enough for the FMCW chirp generation. During the frequency chirp generation, the input tuning voltage of the VCO should vary linearly if we ignore the nonlinearity of the VCO. With the current DAC and analog integrator placed in front of the VCO, the DLF's output represents the chirp slope. Fig. 6(a) shows the situation that a typical DLF is employed in the FMCW generator. The output of the DLF exhibits a large instant

2 stages of IIR filters and a type-III 2nd-order loop filter, including a proportional path, an integral path and a chirp slope estimator [6]. In order to reduce the bit number of the current DAC, one of the accumulator in the DLF is moved towards the output of the differential DAC, turning into two capacitors as the analog integrator [4]. A differential RC filter between the analog integrator and the VCO provides anti-aliasing filtering for the current DAC.

B. Quantization Noise Reduction for BBPD

The larger the root-mean-square (RMS) jitter referred to the BBPD's input is, the larger the quantization noise of the BBPD would be [7]. According to the linearized gain expression of the BBPD [7]:

$$K_{BBPD} = \sqrt{\frac{2}{\pi}} \cdot \frac{1}{\sigma_{\Delta t}}, \qquad (1)$$

a larger RMS input jitter corresponds to a smaller linearized gain for the BBPD. Typically, a phase interpolator (PI) or a DTC is needed to reduce the input jitter of the BBPD [8]-[9]. However, the demand of a linear PI or DTC would increase the design complexity dramatically. Alternatively, the 1-bit SLDSM3 and hybrid FIR filtering, are employed in this work to reduce the quantization noise from the BBPD [5].

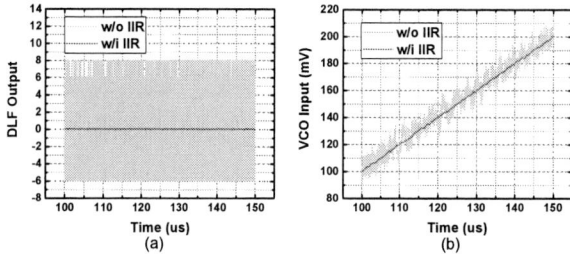

Fig. 7. Simulated transient waveforms of (a) the DLF's output and (b) the VCO's input.

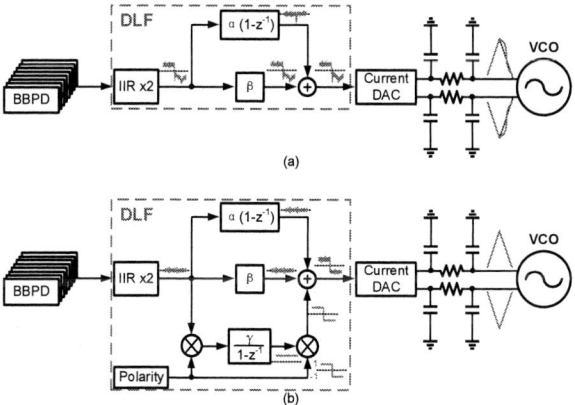

Fig. 8. Chirp generation around the TAPs (a) without the chirp slope estimator and (b) with the chirp slope estimator.

variation with the time-averaged value equal to the chirp slope. The chirp linearity will be heavily degraded with the large slope variation. In the proposed architecture, 2-stage IIR filters are inserted prior to the typical DLF in order to suppress the slope variation, as shown in Fig. 6(b). With the help of 2-stage IIR filtering, the output variation of the DLF is greatly reduced, achieving improved chirp linearity.

To verify the improvement brought by the IIR filtering, the behavioral simulation is performed with $\alpha=1$ and $\beta=2^{-7}$. The poles of the two IIR filters are both set around 1 MHz to balance between the phase margin and the filtering strength. Behavioral simulation results exhibit a significant reduction in the instant frequency error of the chirp with the IIR filtering (Fig. 7).

D. Linearity Enhancement Around TAPs

The polarity of the chirp slope needs to be switched at the TAPs. In conventional FMCW generators, the instant switching of the polarity is accomplished by the loop dynamics [1]-[2], [4]. However, the switching speed is limited by the loop bandwidth. Furthermore, in the BBPD-based architecture, the loop dynamics is slowed down due to the limited output level (only "0" or "1") of the BBPD. As a result, the frequency overshooting occurs at the TAPs (Fig. 8(a)), degrading the chirp linearity. In the proposed architecture, a type-III chirp slope estimator is employed with polarity alternating capability [6], as shown in Fig 8(b), providing an adaptively-estimated chirp slope against the nonlinearity and enabling a fast polarity switching at the TAPs.

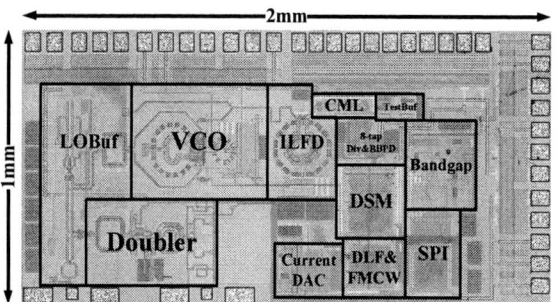

Fig. 9. Chip microphotograph of the FMCW signal generator.

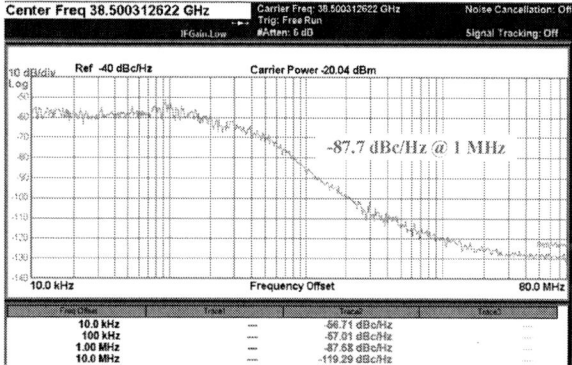

Fig. 10. Phase noise measured at the output of the VCO.

E. VCO and Frequency Doubler

A differential cross-coupled LC oscillator is employed in the proposed architecture to generate oscillating signal around 38.5 GHz. The 2.5-V differentially-tuned varactors enable 1.7-GHz tuning range within one single tuning curve. The total tuning range provided by the 2-bit capacitor array covers 37-41 GHz, with an enough overlap between the adjacent tuning curves to guarantee the continuous frequency tuning across the whole range. Tail inductor filtering and RC filtering are utilized to prevent noise injection from the tail bias.

An injection-locked frequency doubler [1] is utilized to generate 77-GHz oscillation. The 2nd-order harmonic current generated by the push-push pair is injected into the top of a coupled resonator. The free-running frequency of the coupled resonator can be adjusted from 74 GHz to 82 GHz with 5-bit digitally-controlled artificial dielectric (DiCAD) transmission lines. The input bias voltage of the push-push pair is optimized to generate the strongest 2nd-order harmonic current.

III. MEASUREMENT RESULTS

The proposed 77-GHz mixed-mode FMCW signal generator is implemented in 65-nm CMOS technology. The chip microphotograph is shown in Fig. 9, occupying an area of 2 mm² with the testing pads. A 2.5-V supply voltage is applied to the current DAC and bandgap with the remain circuits working under a 1-V power supply. The total power consumption is 43.1 mW, excluding the testing buffer. The VCO and the frequency doubler draw 10.5 mA and 14.9 mA from a 1-V power supply, respectively.

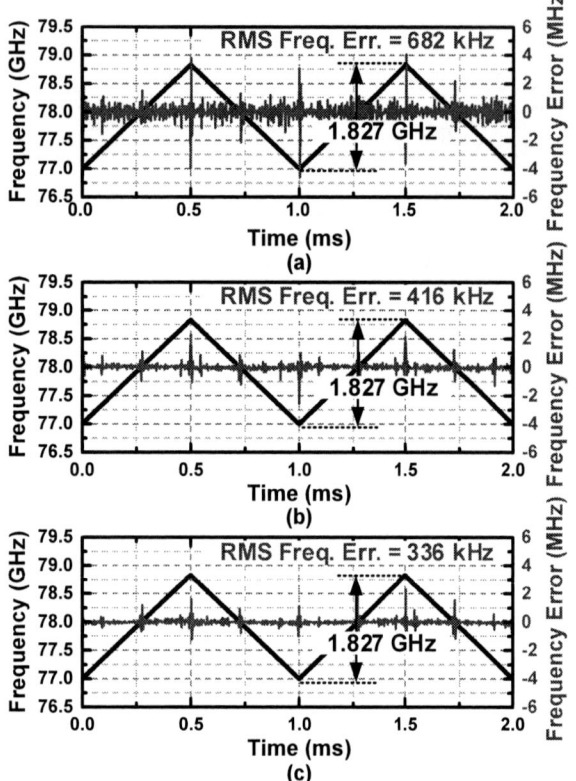

Fig. 11. Measured FMCW chirp signal and the frequency error with the period of 1 ms and the bandwidth of 1.827 GHz under three different modes: (a) with the 1-1-1 MASH, (b) with the 1-bit SLDSM3 and hybrid FIR, (c) with the 1-bit SLDSM3, hybrid FIR and type-III chirp slope estimator.

Fig. 10 shows the phase noise measured at the VCO's output with Keysight N9030A, revealing a phase noise of -87.7 dBc/Hz at 1-MHz offset. The FMCW chirp signal is measured at the 1/16 frequency prescaled output of the VCO, the frequency of which is around 2.4 GHz. The RF output is sampled by Keysight MSOV334A and demodulated with Keysight 89601B, which is multiplied by 32 to get the FMCW chirp signal and the frequency error. Fig. 11 shows the measurement results of three different modes when the period is set as 1 ms with the bandwidth of 1.827 GHz. The RMS frequency error (including the TAPs) is 682 kHz when the 1-1-1 MASH is utilized, which is improved to 416 kHz by replacing the 1-1-1 MASH with the 1-bit SLDSM3 and turning on the hybrid FIR. It is further reduced to 336 kHz with the type-III chirp slope estimator.

Table I summarizes the performance of the proposed FMCW signal generator and makes a comparison with the previously published works.

IV. CONCLUSION

A 77-GHz mixed-mode FMCW signal generator based on BBPD is proposed in this paper. 1-bit SLDSM3 and hybrid FIR filter are utilized to reduce the in-band phase noise, which is dominant by the quantization noise of the BBPD. 2-stage IIR filters are inserted in the DLF to smooth the chirp waveform.

TABLE I. PERFORMANCE SUMMARY AND COMPARISON WITH STATE-OF-THE-ART FMCW SIGNAL GENERATOR

	This Work	[1]	[2]	[3]
Function	FMCW Sig. Gen.	TRX	TX	TX
FMCW Freq. (GHz)	77-78.827	76.92-78.85	77-77.31	61.6-62.6
BW/T_{mod} (GHz/ms)	1.827/1	1.93/2	0.312/2	1/0.42
RMS Freq. Err. (kHz)	336 *	674	961 *	384
PN @ 1 MHz (dBc/Hz) **	-81.7	-81	-83.4	-90
Power (mW)	43.1	343	320	89
Area (mm²)	2	4.64	2.74	2.2
Technology (nm)	65	65	65	65

* Including the turning-around points

** Estimated from lower frequency

A type-III chirp slope estimator with switchable polarity is employed to reduce the frequency error around the TAPs. The proposed FMCW signal generator is implemented in 65-nm CMOS technology with the total power consumption of 43.1 mW. The measured RMS frequency error is 336 kHz when the chirp's period is 1 ms and the bandwidth is 1.827 GHz.

REFERENCES

[1] Haikun Jia, Lixue Kuang, Wei Zhu, Zhiping Wang, Feng Ma, Zhihua Wang, and Baoyong Chi, "A 77 GHz Frequency Doubling Two-Path Phased-Array FMCW Transceiver for Automotive Radar," *IEEE J. Solid-State Circuits*, vol. 51, no. 10, pp. 2299-2311, Oct. 2016.

[2] Joonhong Park, Hyuk Ryu, Keum-Won Ha, Jeong-Geun Kim, and Donghyun Baek, "76–81-GHz CMOS Transmitter With a Phase-Locked-Loop-Based Multichirp Modulator for Automotive Radar," *IEEE Trans. Microw. Theory Techn.*, vol. 63, no. 4, pp. 1399-1408, Apr. 2015.

[3] Wanghua Wu, Member, Robert Bogdan Staszewski, and John R. Long, "A 56.4-to-63.4 GHz Multi-Rate All-Digital Fractional-N PLL for FMCW Radar Applications in 65-nm CMOS," *IEEE J. Solid-State Circuits*, vol. 49, no. 5, pp. 1081-1096, May 2014.

[4] Hiroki Sakurai, Yuka Kobayashi, Toshiya Mitomo, Osamu Watanabe, and Shoji Otaka, "A 1.5GHz-Modulation-Range 10ms-Modulation-Period 180kHzrms-Frequency-Error 26MHz-Reference Mixed-Mode FMCW Synthesizer for mm-Wave Radar Application," in *IEEE Int. Solid-State Circuits Conf. (ISSCC) Dig. Tech. Papers*, Feb. 2011, pp. 292–293.

[5] Ni Xu, Yiyu Shen, Sitao Lv, Woogeun Rhee, and Zhihua Wang, "A Spread-Spectrum Clock Generator with FIR-Embedded Binary Phase Detection and 1-Bit High-Order ΔΣ Modulation," in *IEEE Asian Solid-State Circuit Conf. (A-SSCC) Dig. Tech. Papers*, Nov. 2015, pp. 1-4.

[6] Hwanseok Yeo, Sigang Ryu, Yoontaek Lee, Seuk Son, and Jaeha Kim, "A 940MHz-Bandwidth 28.8µs-Period 8.9GHz Chirp Frequency Synthesizer PLL in 65nm CMOS for X-Band FMCW Radar Applications," in *IEEE Int. Solid-State Circuits Conf. (ISSCC) Dig. Tech. Papers*, Feb. 2016, pp. 238–239.

[7] Marco Zanuso, Davide Tasca, Salvatore Levantino, Andrea Donadel, Carlo Samori, and Andrea L. Lacaita, "Noise Analysis and Minimization in Bang-Bang Digital PLLs," *IEEE Trans. Circuits Syst. II, Exp. Briefs*, vol. 56, no. 11, pp. 835-839, Nov. 2009.

[8] Roberto Nonis, Werner Grollitsch, Thomas Santa, Dmytro Cherniak, and Nicola Da Dalt, "digPLL-Lite: A Low-Complexity, Low-Jitter Fractional-N Digital PLL Architecture," *IEEE J. Solid-State Circuits*, vol. 48, no. 12, pp. 3134-3145, Dec. 2013.

[9] Davide Tasca, Marco Zanuso, Giovanni Marzin, Salvatore Levantino, Carlo Samori, and Andrea L. Lacaita, "A 2.9–4.0-GHz Fractional-N Digital PLL With Bang-Bang Phase Detector and 560-fs$_{rms}$ Integrated Jitter at 4.5-mW Power," *IEEE J. Solid-State Circuits*, vol. 46, no. 12, pp. 2745-2758, Dec. 2011.

S21-2 (2125)

A 7GHz-Bandwidth 31.5 GHz FMCW-PLL with Novel Twin-VCOs Structure in 65nm CMOS

Shunli Ma,Jili Sheng, Ning Li, and Junyan Ren

State Key Laboratory of ASIC and System, Fudan University, Shanghai, 200433, China

Abstract- This paper describes a novel fully integrated 35GHz frequency modulated continuous wave (FMCW) PLL with twins-VCO to 7GHz sweeping bandwidth. The proposed FMCW PLL is composed of twins-VCOs, SDM modulator and waveform generator. A novel frequency sweeping extension (FSE) technique is proposed to make the twin-VCOs take turns to sweeping to realize wide sweeping bandwidth without sacrificing the sweeping time. The RMS frequency error of the proposed PLL is 0.9MHz and phase noise is -90dBc/Hz @1MHz. The total power consumption is around 43mW. The chip was fabricated in a 65nm LP CMOS technology. The PLL consumes an active area of 0.91mm^2 and power of 43mW while operating at a 1.2V single supply.

Index term : CMOS FMCW PLL,SDM, FSE, twin-VCOs

I. INTRODUCTION

FMCW radar is widely used for distance and velocity measurement. The resolution of FMCW radar system is c/2B, where c is the speed of light and B is the chirp bandwidth of PLL [1] as shown in Fig.1. Resolution with several centimeters is required for many systems such as gesture detection sensor due to fine movement of gesture. In order to achieve a target resolution of 2cm, 7GHz frequency sweeping modulation bandwidth (BW) is required. FMCW-PLL structure widely utilizes a multi-modulus frequency divider (MMD) and delta-sigma modulation (DSM) to realize frequency sweeping modulation. Due to the frequency sweeping requirement, previously PLLs structures can only utilize one tuning curve of VCO to realize frequency sweeping. It is difficult to automatic switch the tuning curves to extend the frequency sweeping BW with good linearity. Thus, previously PLLs structures have limited chirp BW which reduces the resolution of radar. Though enlarging KVCO can extend the sweeping BW,

large KVCO sacrifices the phase noise and induces large

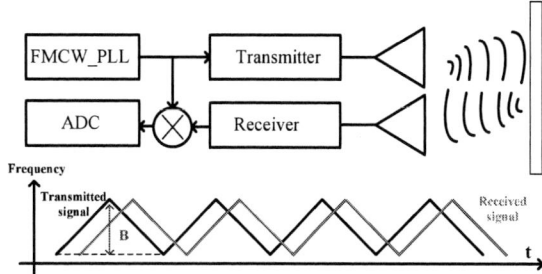

Fig.1 (a) shows the FMCW system

variation of loop bandwidth of PLL.

To address these challenges and realize large chirp BW without sacrificing phase noise and stability, this work presents a twin-VCOs PLL structure with frequency sweeping extension (FSE) technique that can realize that two VCOs take turn frequency sweeping with good linearity. A prototype chip fabricated in a 65nm CMOS demonstrates that the PLL can generate a triangular chirp profile with 7GHz bandwidth centered at 31.5GHz. The period can be tuned from 250us to 20ms with maximum 10MHzrms frequency error when period is 250us. The proposed PLL structure can realize 4.67× times BW compared to state-of-art designs [1]-[5].

In order to realize wideband sweeping frequency range, this paper proposed a novel FSE algorithm with twin-VCOs structure. The proposed FMCW PLL can realize 7GHz sweeping range. Moreover, this paper is organized as follows. Section II introduces the algorithm of the proposed

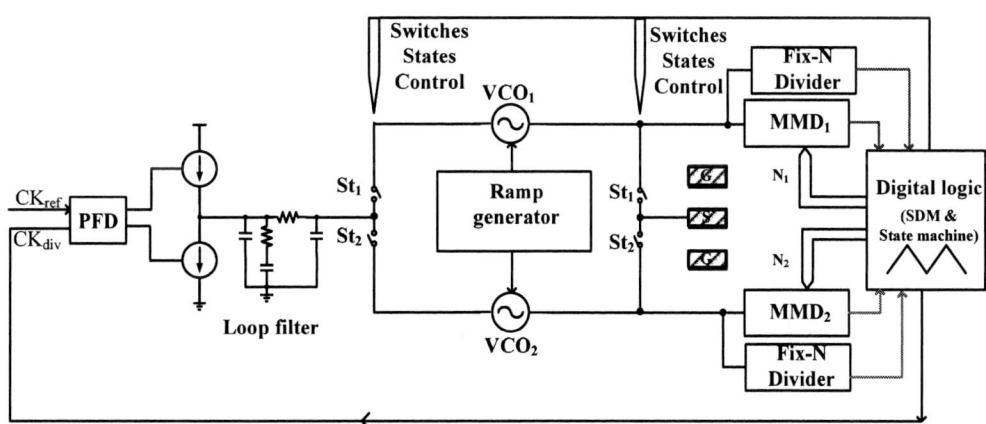

Fig. 2: The system structure of the proposed twin-VCOs FMCW PLL. It consists of the PFD, CP, LF, two VCOs, divider chains and ramp generator for FSE technique

978-1-5386-3179-9/17 $31.00 © 2017 IEEE   321

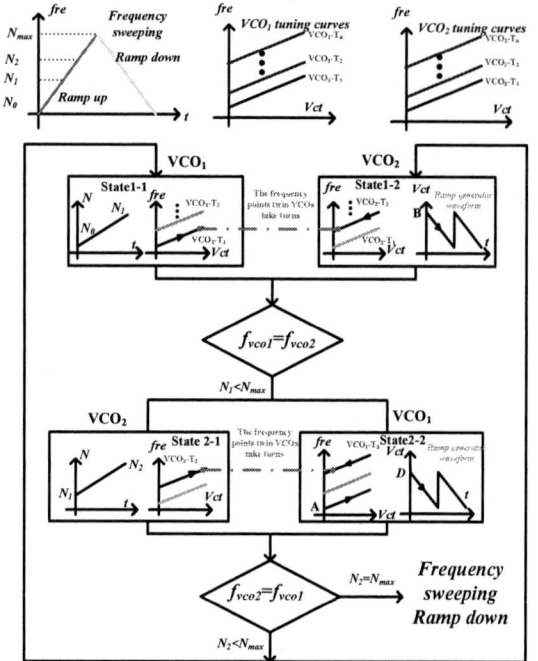

Fig. 3: The triangle frequency sweeping, VCO1 and VCO2 tuning curves (top). The working states of the two VCOs with FSE technique.

PLL system. Section III describes the detail circuit implementation. Finally, measurement results and conclusion are presented in sections IV and V, respectively.

II. THE PROPOSED FMCW-PLL STRUCTURE WITH FSE ALGORITHM ANALYSIS

Fig.2 shows the overall architecture of the proposed twin-VCOs FMCW PLL. It consists of a phase and frequency detector (PFD), charge pump (CP), three-order loop filter (LF), two VCOs, divider chains including MMD and Fix-N divider, ramp generator and digital logic. The digital logic realizes digital SDM for frequency sweeping and FSE technique by controlling the state of the ramp generator and VCO frequency tuning curves.

A. Frequency sweeping extension (FSE) Algorithm

Due to the frequency sweeping function, the most difficult thing comes from the different tuning curves changing with same frequency when two VCOs take turn. In order to address this problem, twin-VCOs PLL structure and FSE techniques are proposed. Because the frequency ramping up and down is under the same working principle, this paper takes frequency ramping up function as an example to explain its working principle as shown in Fig.2. There are four states during this process. State1-1: VCO1 is in PLL loop and its control voltage (VCT1) is controlled by the CP output. Due to the divide ratio sweeping from N0 to N1, the VCT1 ramps up in tuning curve VCO1-T1. State1-2: the control voltage (VCT2) of VCO2 is connected to ramp generator with saw waveform controlling tuning curve VCO2-T2. State1-1 and State1-2 are working at the same time. The outputs of the VCOs are connected to divider chains with fix-N divider ratio N_{fix}, respectively. When the frequency of VCO1 equals to VCO2 by checking the divider output and N1 is smaller than N_{max}, VCO2 will replaces VCO1 in PLL loop. State2-1: the VCO2 is connected to CP

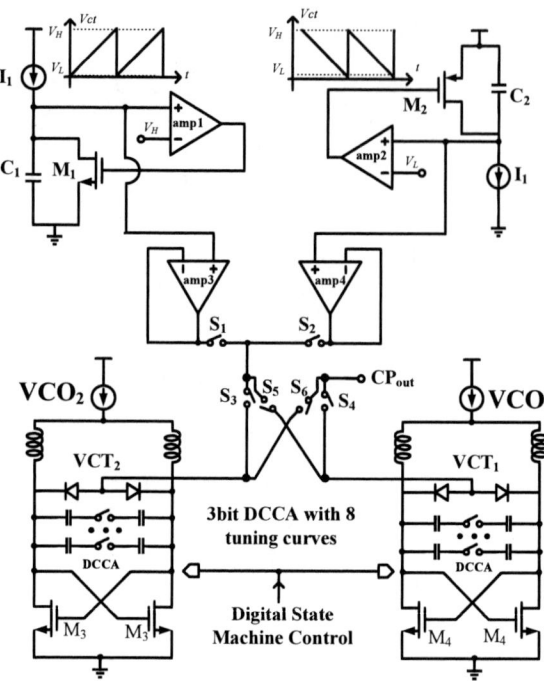

Fig. 4: The circuit implement of the proposed FSE technique.

output and the divide ratio sweeps from N1 to N2. Its control voltage ramps up in tuning curve VCO2-T2. State2-2: the VCT1 is connected to ramp generator with saw waveform D controlling tuning curve VCO1-T3. When the frequencies are the same and N2 is smaller than N_{max}, the frequency ramping up process keeps going. Frequency ramping up process stops when N equals to N_{max}. Then the frequency ramping down process will begin. During ramping down process, the saw waveform will be changed to ramp up.

B. Circuit Implementation of Frequency sweeping extension (FSE) Algorithm

Fig.4 shows the circuit schematic of twin-VCOs. The ramp generator circuits consists of two saw waveforms generator. Ramping down saw waveform generator is used for PLL ramping up FSE technique, and the other is used for PLL ramping down FSE technique. Taking frequency ramping up process as an example, the switches states are shown in Fig.4.State1-1: S4 is closed and VCO1 is connected to CP output, its oscillation frequency begins to sweep up in VCO1-T1 curve. State1-2: S2 and S3 is closed, the saw waveform controlled VCO2 in VCO2-T2 curves. The outputs of the two VCOs are connected to divide chain to get low frequency output which can be processed by digital circuits.

When the frequencies of the two VCOs are equal and divider ratio is smaller than N_{max} (Condition1) .State2-1: S6 is connected to VCO2 which is in PLL loop. State2-2: S2 and S5 are both closed; the saw waveform controlled VCO1 in VCO1-T3 curves. When Condition1 is met, the State1-1 and State1-2 will be done again until the frequency ramping up process is finished.

III. CIRCUIT IMPLEMENTATION

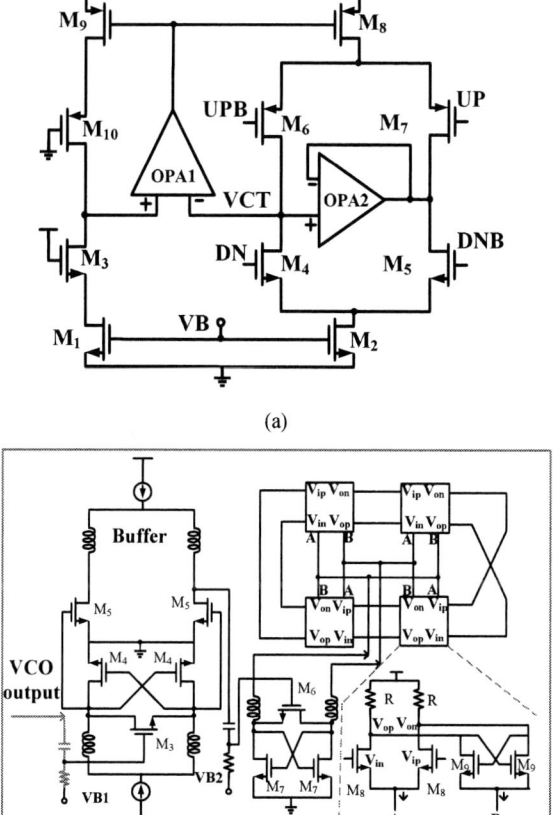

(a)

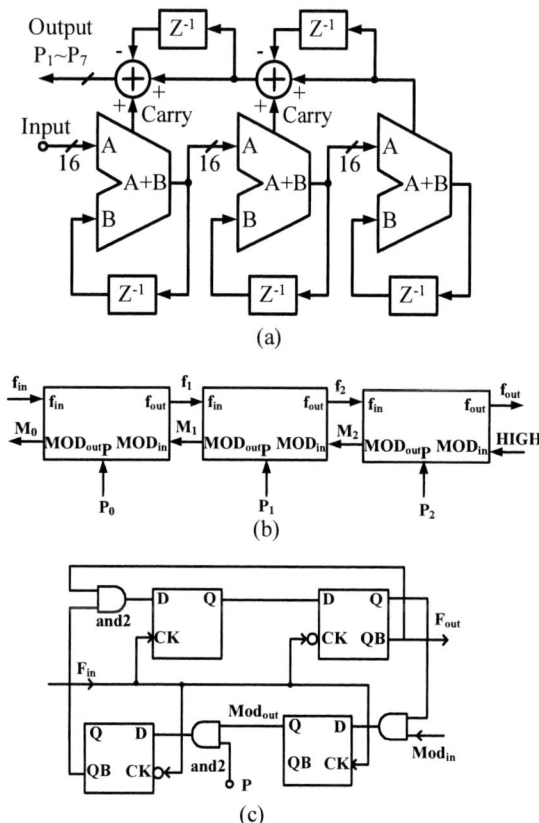

(a)

(b)

(c)

Fig.6 (a) MASH 1-1-1 design (b) MMD design (c) each cell circuit design

Fig.5 (a) charge pump design (b) High frequency divider chain design.

A tri-state PFD is utilized in the PLL. The inputs of the PFD are the VCO_L and output of the divider chain. The outputs of the PFD are two pairs (UP, Down) fully differential signals. Symmetry layout and a transparent gate are used to compensate the time mismatch between UP signals and Down signals. The output of the CP is a current pulse whose width is proportional to the phase differences of the input signals. The sink and source currents source mismatch is decreased by utilized a rail-to-rail operational amplifier. The other rail-to-rail amplifier is used to reduce the charge sharing effect. As a result, mismatch of the charge pump is less than 1% and the output voltage range cover of 90% the power supply as shown in Fig.5(a).

A. High Speed Divider Circuit implementation

Fig.5 (b) depicts high frequency divider design. The VCO is implemented as a standard LC tank structure with two-bit digital controlled capacitor banks to extend its tuning range. The 1st divider is a direct injection-locked topology with the digital controlled capacitor banks and ac-coupled to the gate of the transistor M3. The bias voltage VB1 and VB2 are optimized to 0.6V to get an optimum lock range of 10.5GHz. The 2nd divider is a LC injection-locked stage that is coupled to divider-by-4 divider implemented of CML flip-flops by current reusing. The harmonic energy is injected into the ring structure divider to realize total divider-by-16 ratio. The other dividers are programmable divider with TSPC structure.

B. Digital Circuit implementation

The MASH 1-1-1 modulator is utilized in a FMCW- PLL structure. The MASH modulator is inherently stable and has low hardware complexity as shown in Fig.6 (a). Moreover, its dynamic range can covers all input codes. Its output are connected to the MMD control bits. The MASH modulator uses a dither signal to mitigate spur issues. The dither signal is applied at different stage of a MASH modulator to suppress the spur. Several multi-mode divider are utilized to realize fractional-N function as shown in Fig.6(b)-(c).

IV. EXPERIMENT RESULTS

Fig.7 shows the phase noise measurement. The measured phase noise of the PLL signal without a chirp is -90dBc/Hz at 1MHz offset. With a 50MHz reference clock, the PLL can generate an FMCW signal in the range of 28GHz to 35GHz with a maximum chirp bandwidth of 7GHz. The power spectrum of the PLL clock including the chirp with T=250μs shows a uniform spread of the spectral power from 28 GHz to 35GHz.

Fig.8 shows a precise triangular chirp with 7GHz bandwidth and the chirp period of 250μs. The bottom shows that the RMS frequency error excluding the tuning curves jumping is less than 4.8MHzrms. During the two VCOs taking turn, the RMS frequency error is around 1 MHz. The good news is that these nonlinearity frequency points can be well predicted and easily eliminated during system implementation.

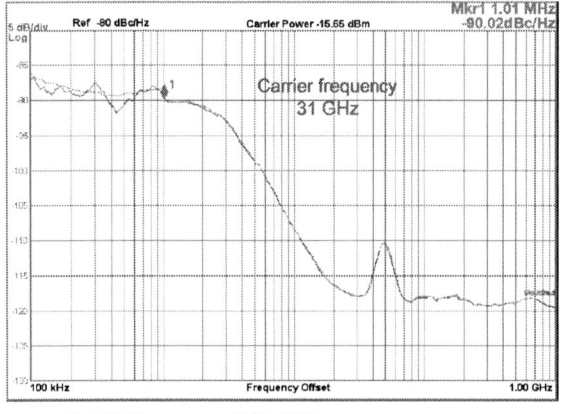

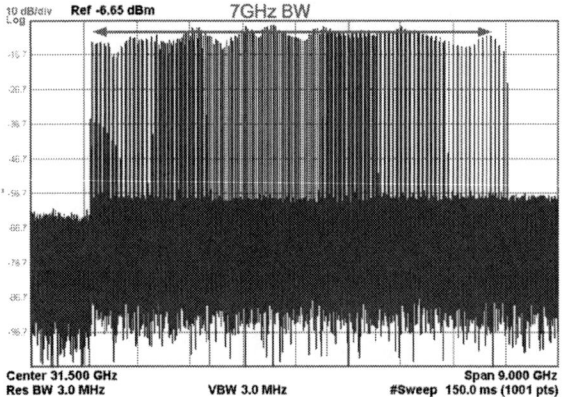

Fig. 7: The PLL phase noise without chirp (top) and power spectrum with chirp (bottom)

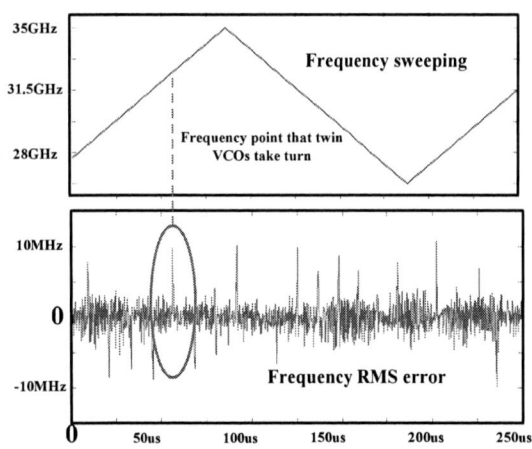

Fig. 8: The measured FMCW chirp profile with 7GHz BW (top) and RMS frequency error across entire operating range (bottom)

	This work	ISSCC10	ISSCC11	JSSC14	ISSCC16	ISSCC16
FMCW generation	Fran.-N PLL	Fran.-N PLL	Mixed-Mode	ADPLL+TPM	ADPLL+TPM	ADPLL
Frequency Range (GHz)	28GHz-35GHz	75.6-76.3	82.1-83.8	56.4-73.4	8.4-9.4	NA
Modulation type	Triangle	Triangle	Triangle	Triangle	Triangle	Saw tooth
Chirp Bandwidth (BW) [GHz]	7	0.7	1.5	1.22	0.956	1.48
Range resolution [cm]	2.1	21	10	24.5	15.6	10.1
Peak Freq. Error	4.8MHz	2MHz	4MHz	5.8MHz	6.7MHz	185KHz
RMS Frequency Error	1MHz	300KHz	179KHz	180KHz	1.9MHz	159KHz
Power consumption	43mW	73mW	152mW	48mW	14.8mW	41mW
Phase noise @1MHz offset [dBc/Hz]	-90	-85	-84	-90	-105	N/A
Technology CMOS	65nm	65nm	65nm	65nm	65nm	65nm

Table-I: Performance comparison of the proposed twin-VCO FMCW PLL with prior art.

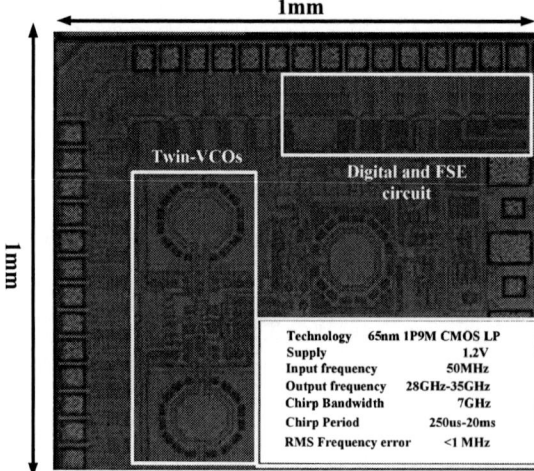

Fig. 9: Die photo and performance summary of the proposed twin-VCO FMCW PLL

The die photo of the proposed twin-VCOs FMCW PLL and performance summary are shown in Fig. 9. The prototype chip is fabricated in a 65nm LP CMOS technology. The PLL consumes an active area of $0.91mm^2$ and power of 43mW while operating at a 1.2V single supply.

V. CONCLUSIONS

In this paper, we propose a novel fully integrated FMCW PLL with FSE technique. With proposed FSE technique, the system can realize 7GHz sweeping bandwidth without sacrificing the sweeping time. The RMS frequency error of the proposed PLL is 1MHz and phase noise is -90dBc/Hz @1MHz. The total power consumption is around 43mW.

Table-I compares key performance metrics of the proposed frequency synthesizer PLL to other FMCW frequency synthesizers. The proposed structure shows its significantly improved frequency sweeping range without compromising phase noise and linearity. It can realize 7GHz frequency sweeping bandwidth which is equivalent to a 2-cm spatial resolution of the FMCW radar. What is more, this proposed structure break down the tradeoff between frequency sweeping range and tuning curve range. It can realize wider tuning by using multiple tuning curves.

Reference

[1] Y. Li et al., "A Fully Integrated 77GHz FMCW Radar System in 65nm CMOS," *ISSCC Dig. Tech Papers*, pp. 216-217, Feb. 2010.
[2] H. Sakurai et al., "A 1.5GHz-Modulation-Range 10ms Modulation Period 180kHzrms-Frequency-Error 26 MHz-Reference Mixed-Mode FMCW Synthesizers," *ISSCC Dig. Tech Papers*, pp. 292-293, Feb. 2011.
[3] W. Wu et al., "A 56.4-to-63.4GHz Multi-Rate All-Digital Fractional-N PLL for FMCW Radar Applications in 65nm CMOS," *IEEE J. Solid-State Circuits*, pp. 1081-1096, May 2014.
[4] Y. Wang et al., "A Ku-Band 260mW FMCW Synthetic Aperture Radar TRX with 1.48GHz BW in 65nm CMOS for Micro-UAVs," *ISSCC Dig. Tech Papers*, pp. 240-241, Feb. 2016.
[5] H.Yeo et al., "A 940MHz-Bandwidth 28.8μs-Period 8.9GHz Chirp Frequency Synthesizer PLL in 65nm CMOS for X-Band FMCW Radar Applications," *ISSCC Dig. Tech Papers*, pp. 238-239, Feb. 2016.

IEEE Asian Solid-State Circuits Conference
November 6-8, 2017/Seoul, Korea

A $-245\,\mathrm{dB}$ FOM $48\,\mathrm{fs}$ rms jitter semi-digital PLL with intrinsic temperature compensation in $130\,\mathrm{nm}$ CMOS

J. Anders[1], S. Bader[1], M. Dietl[2], P. Sareen[2], G. Rombach[2], S. Tambouris[2] and M. Ortmanns[1]

Email: jens.anders@uni-ulm.de

[1]Institute of Microelectronics, University of Ulm, D-89081 Ulm, Germany

[2]Texas Instruments Germany, D-85356 Freising, Germany

Abstract—In this paper, we present a temperature compensated semi-digital integer-N PLL realized in a standard $130\,\mathrm{nm}$ CMOS technology, which achieves $48\,\mathrm{fs}$ rms phase jitter and a FOM of $-245\,\mathrm{dB}$. The presented design improves the phase noise performance of previously presented semi-digital PLLs by replacing the ring oscillator VCO by a semi-digitally tuned LC-tank VCO. In contrast to mostly digital PLLs, the presented architecture uses simple linear analog circuitry, removing the need for non-linear bang-bang PFDs or high speed oversampled $\Sigma\Delta$-modulators.

I. INTRODUCTION

Most state-of-the-art high-performance electronic systems require low phase noise, low-cost and PVT stable time references. Classically, analog charge-pump PLLs (CP-PLLs) have been used as low-noise synthesizers due to their ease of implementation. More recently, mostly two approaches have been followed to further improve PLL performance: In the first approach, the classical CP-PLL is extended by digital circuitry [1], [2] to arrive at hybrid PLLs which achieve the good phase noise performance of CP-PLLs with smaller loop-filter capacitors [1] or more aggressive noise shaping thanks to a digital stabilization [2]. The second approach uses digital PLLs [3], [4]. This approach has become increasingly attractive for realizations in deep submicron CMOS due to the advantageous scaling of digital PLLs with technology in terms of power, area and programmability. Despite their many advantages, classical bang-bang PLLs (BB-PLLs) suffer from a compromised locking behavior and noise event response due to the hard-nonlinearity of the bang-bang-PFD (BB-PFDs). As a remedy, several different solutions have been proposed including the use of TDCs [5] and the introduction of hybrid PLLs, which are mostly digital and use some additional analog circuitry to achieve an improved performance [6].

As an alternative solution for ultra low phase noise frequency synthesis, as it is e.g. required by next generation mobile applications with their advanced modulation schemes, in this paper, we propose the use of the semi-digital PLL architecture [7] in combination with LC-tank VCOs. The proposed approach combines the advantages of classical CP-PLLs (low phase noise and simple linear analog circuitry) with the technology scaling benefits of digital PLLs (digital blocks as well as storage cells (SCs) in the integral (I-) control

path scale advantageously with technology), rendering it an attractive solution over a very wide range of technologies. The semi-digital PLL architecture proposed in this paper builds upon a design which was recently presented by our group in [7] but extends it for very low phase noise performance by incorporating an LC-tank VCO. Since a conventional LC-tank VCO cannot be directly incorporated into the semi-digital PLL using standard purely analog and/or digital varactor control, we have developed the concept of a semi-digitally tuned VCO, which uses both an analog tuning voltage on a standard analog varactor and a semi-digital tuning word on a newly proposed semi-digital veractor bank. Thanks to this new tuning concept, the proposed PLL architecture achieves an extremely low phase noise performance using simple linear analog circuitry, i.e. without the use of non-linear BB-PFDs or high speed oversampled $\Sigma\Delta$-modulators, as often found in digital-intensive PLLs.

Additionally, the nature of the I- control path in the proposed semi-digital LC-VCO control provides an intrinsic temperature compensation. This allows for a small VCO gain in the proportional (P-) control path enabling a further optimization of the phase noise performance without compromising the lock-range in the presence of process and temperature variations.

II. ARCHITECTURE

The semi-digital PLL architecture is shown in Fig. 1, cf. [7]. With its split P- and I- control paths, this PLL shares many of the advantages of the hybrid PLL presented in [6]. However, the major difference compared to [6] is that both the P- and the I-path receive their control signals from the same linear PFD, thereby completely removing the need for a very nonlinear

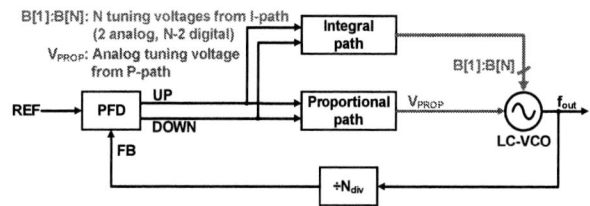

Fig. 1: Architecture of the proposed semi-digital PLL.

978-1-5386-3179-9/17 $31.00 © 2017 IEEE

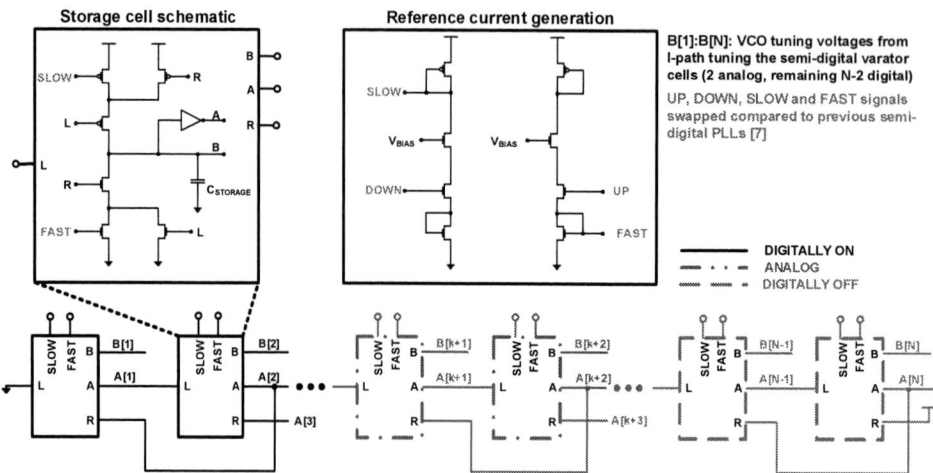

Fig. 2: Details of the I-control path of Fig. 1 with schematic views of the storage cells and the reference current generation.

BB-PFD. According to Fig. 2, the I-path is composed of N identical SCs, which provide N control signals $B[1:N]$ for the VCO. At every point in time, all SCs up to a certain position k are in digital on state ($B[i \leq k] = \text{VDD}$), those after position $k+2$ are in digital off state ($B[i > k+2] = 0$) and the two intermediate SCs at $k+1$ and $k+2$ are in analog tuning mode ($0 \leq B[k+1, k+2] \leq \text{VDD}$). Compared to previous semi-digital PLLs [7], the UP/DOWN and SLOW/FAST signals in the SC reference current generation cell and the SCs themselves are swapped, causing the state where all SCs are digitally on to correspond to the lowest possible output frequency setting of the I-path, cf. Fig. 2.

This modification allows for a semi-digital tuning of the LC-VCO by a combination of NMOS switches and biasing resistors according to Fig. 3. The outputs $B[i]$ provide a thermometer code like representation of the current PLL state with a seamless analog interpolation at the one-zero-transition thanks to the two analog SCs, cf. [7]. According to Fig. 3 the SC outputs $B[i]$ tune a bank of N semi-digital varactor cells. The unit cells of the semi-digital varactor bank essentially extend a classical digital MIM-cap tuning cell by an additional weak switch M_2 and biasing resistors R_{BIAS} to provide the semi-digital tuning capability. More specifically, in all varactor cells with $B[i] = 0$ all switches are open, resulting in approximately no effective differential tank capacitance C_{diff} produced by the cell. For all varactor cells with $B[i+1] = \text{VDD}$, the strong switch is closed, resulting in $C_{\text{diff}} = C_{\text{unit}}$. Then, there are two varactor cells at positions $i = k$ and $i = k+1$ with $0 < B[i+1] < \text{VDD}$, where the modulation of the channel resistance of the strong switch M_1 by the SC outputs $B[i+1]$ provides an analog interpolation of C_{diff} between 0 and C_{unit}. Finally, in the varactor cell at position $i = k+2$, the strong switch M_1 is fully open and the weak switch M_2 is controlled by the analog signal $B[k+2]$, which further smoothes the C_{diff} vs. SC characteristic at the transitions between different SCs. In this way, the semi-digital nature of the control signals $B[i]$ of the I-path can be used to realize a semi-digital, i.e. mostly digital with a

seamless analog interpolation between the discrete MIM cap steps, varactor tuning. In contrast to previous solutions for an interpolation between discrete MIM capacitor steps, e.g. in the form of a digital $\Sigma\Delta$-modulator, cf. [6], this is achieved without the need for additional (oversampled) circuitry. The analog control voltage V_{prop} of the P-path, which controls the VCO output frequency through a standard analog MOS varactor, is produced by a charge-pump identical to the one presented in [7], cf. Fig. 3. Here, thanks to the switching scheme introduced in [7], the charge-pump output remains quiet in the locked state, producing a very small reference spur. At this point it is important to stress that due to the semi-digital VCO tuning by the I-path, it is possible to set the P-path VCO gain to a small value without compromising the PLL lock range vs. process and temperature. This improved response vs. temperature changes is further explained in Fig. 4a and b where the temperature behavior of the proposed semi-digital PLL is contrasted against that of a conventional CP-PLL: In the conventional CP-PLL a change in temperature would require a different tuning voltage to provide a fixed output frequency, which potentially leads to locking problems if the required voltage is close to or even exceeds the charge-pump output range. In contrast, in the proposed architecture, the semi-digital VCO tuning by the I-path ensures that V_{prop} returns to its optimum mid-position by changing the position of the two analog SCs. The behavior of the SCs and the varactor bank is further illustrated in Fig. 4c assuming an increase in temperature: As more and more SCs turn off, thereby producing a smaller tuning capacitance C_{diff}, V_{prop} is gradually brought back to its optimum mid-value, where the charge-pump produces a quiet output. Therefore, the proposed architecture effectively provides an online background temperature calibration as it is required by continuously operating systems, which ensures that the VCO is always operated in a low-phase noise mode.

III. PROTOTYPE REALIZATION AND MEASUREMENTS

Fig. 5 shows a micrograph of a fabricated prototype chip realized in a 130 nm CMOS technology, where the presented

978-1-5386-3179-9/17 $31.00 © 2017 IEEE

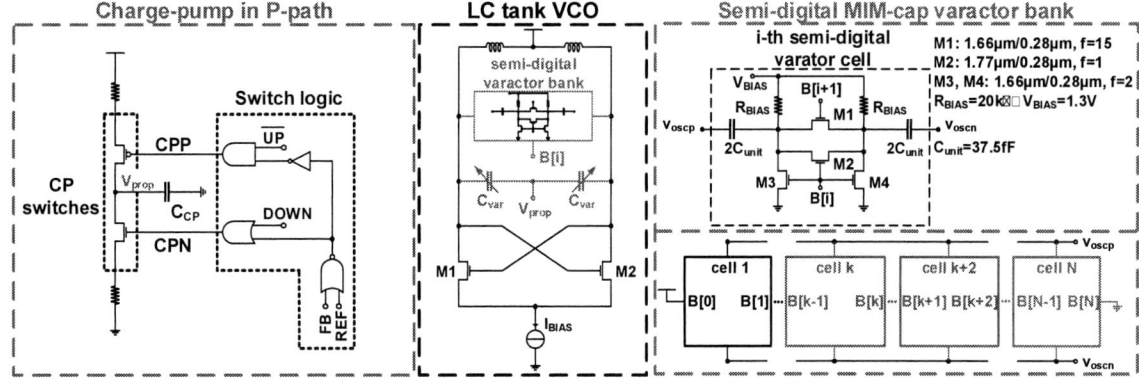

Fig. 3: Schematics of the charge-pump [7], the semi-digital LC-tank VCO and the semi-digital varactor bank with its MIM-cap based unit cell.

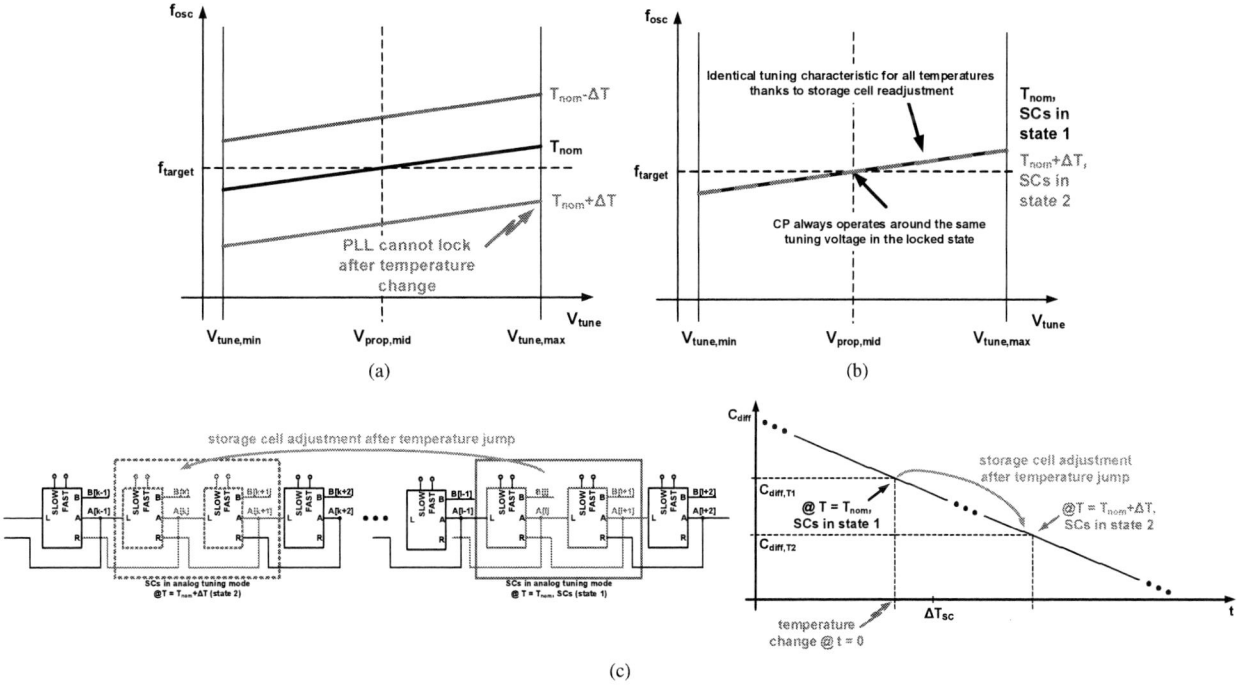

Fig. 4: (a) Comparison between the effect of a temperature change on the VCO tuning curve in a conventional CP-PLL and in the proposed architecture and (b) illustration how the I-path of the semi-digital PLL automatically restores the optimum operating point of the P-path VCO tuning curve after a temperature change. C_{diff} is the differential tank capacitance produced by the semi-digital varactor cells.

design is encircled in red. The PLL consumes a total power of 131.3 mW, where most power (130 mW) is consumed in the VCO to produce a very low open loop VCO phase noise and merely 1.3 mW go to the rest of the PLL. The measured VCO tuning range vs. PVT variations is from 3.8 GHz to 4.2 GHz.

Fig. 6a then shows the measured phase noise spectrum at a temperature of $T = 25\,°C$. The integrated rms phase noise is 48 fs and the spot phase noise PSD at 1 MHz offset is -142 dBc/Hz, both values measured at an output frequency of 983.04 MHz, a common output frequency for baseband stations in wireless infrastructure such as CDMA and UMTS.

For this measurement, a clean reference from a generator at 122.8 MHz, also a common reference frequency in baseband stations, was used. The programmable PLL divider was set to $N_{\text{div}} = 8$ and the output was measured at the output of the CML-based on-chip divide-by-4 prescaler. Thanks to the programmable divider, the PLL can accommodate reference frequencies between 10 MHz and 250 MHz and the PLL bandwidth is 500 kHz. Here, it should be pointed out that the bandwidth tracking scheme proposed in [7] was not implemented in this design because the tuning range of the utilized LC-VCO is much smaller than that of the ring

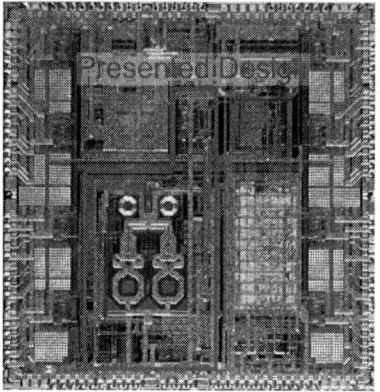

Fig. 5: Chip micrograph.

oscillator used in [7] and, moreover, the VCO gain of an LC-tank oscillator varies much less over its tuning range than that of a ring-oscillaotr VCO.

The phase noise performance vs. temperature over a range from $T = -40\,°C$ to $T = 105\,°C$, measured using the setup described above, is shown in Fig. 6. Here, the performance degradation is better than $30\,\%$. Finally, the measured reference spur level is $< -80\,dBc$ showing that the charge pump output in the P-path stays indeed very quiet in the locked state.

Table I compares the presented design against the state-of-the-art in low-noise LC-tank VCO-based PLL synthesizers, clearly highlighting the excellent noise performance of the presented prototype. Despite the elevated power consumption mandated by the ultra-low phase noise requirements and the relatively old 130 nm technology node, the achieved FOM of $-245\,dB$ compares very well against the state-of-the-art.

TABLE I: Table comparing the presented work with the state-of-the-art.
$^1@f_{\text{out}} = 983.04\,MHz$, ^{2}VCO alone, ^{3}rest of the PLL, ^{4}SOI
$$FOM = 20 \cdot \lg\left(\sigma_{\phi,\text{rms}}/s\right) + 10 \cdot \lg\left(P_{\text{PLL}}/mW\right)$$

	This work	[8]	[6]	[3]	[4]
f_{VCO} [GHz]	3.9	2.21	28	5.8	5.8
$\sigma_{\phi,\text{rms}}$ [fs]	57^1	150	200	174	160
P_{PLL} [mW]	130^2 1.3^3	7.6	31	9.5	8.2
FOM [dB]	-245	-248	-239	-246	-247
Process	130 nm	180 nm	32 nm^4	28 nm	28 nm

IV. CONCLUSION

In summary, we have presented a semi-digital ultra-low-noise PLL synthesizer, whose excellent phase noise performance was enabled by modifying a standard digital varactor tuning scheme to allow for a semi-digital control of a low-noise LC-tank VCO by the I-path of a semi-digital PLL. Moreover, measured results over a temperature range from $T = -40\,°C$ to $T = 105\,°C$ prove the intrinsic online background temperature calibration provided by the semi-digital VCO control.

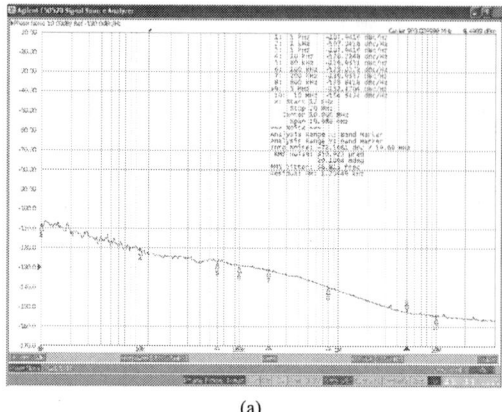

(a)

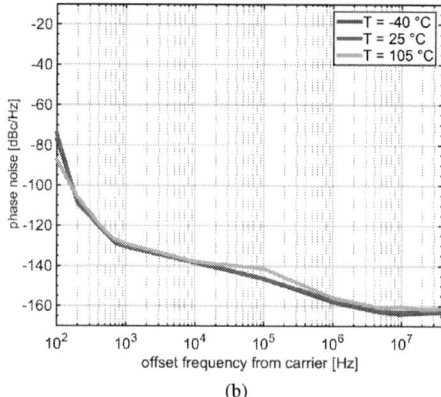

(b)

Fig. 6: (a) Room temperature measured phase noise and rms jitter at a divided down output frequency of $983.04\,MHz$, a common frequency for wireless infrastructure such as CDMA and UMTS and (b) Measured PLL phase noise for different operating temperatures between $-40\,°C$ and $105\,°C$.

REFERENCES

[1] A. Sai, T. Yamaji and T. Itakura, A 570fs rms integrated-jitter ring-VCO-based 1.21GHz PLL with hybrid loop, *2011 ISSCC Dig. Tech. Papers, 98-100*, 2011.

[2] A. Sai, Y. Kobayashi, S. Saigusa, O. Watanabe and T. Itakura, A digitally stabilized type-III PLL using ring VCO with 1.01ps rms integrated jitter in 65nm CMOS. *2012 ISSCC Dig. of Tech. Papers 248-250*, 2012.

[3] X. Gao, L. Tee, W. Wu and K.S. Lee, "A 28 nm CMOS digital fractional-N PLL with -245.5 dB FOM and a frequency tripler for 802.11 abgn/ac radio," *2015 ISSCC Digest of Technical Papers*, 1-3, 2015.

[4] X. Gao et al. , "A 2.7-to-4.3 GHz, 0.16 ps rms-jitter, -246.8 dB-FOM, digital fractional-N sampling PLL in 28 nm CMOS," *2016 ISSCC Digest of Technical Papers*, 174-175, 2016.

[5] W. Yin, R. Inti, A. Elshazly, B. Young and P. Hanumolu, A 0.7-to-3.5 GHz 0.6-to-2.8 mW Highly Digital Phase-Locked Loop With Bandwidth Tracking. *IEEE Journal of Solid-State Circuits 46*, 1870-1880, 2011.

[6] M. Ferriss, A. Rylyakov, J.A. Tierno, H. Ainspan and D.J. Friedman, "A 28 GHz Hybrid PLL in 32 nm SOI CMOS," *IEEE Journal of Solid-State Circuits, 49*, 1027–1035, 2015.

[7] S. Fahmy, M. Dietl, P. Sareen, M. Ortmanns and J. Anders, "A BW-tracking semi-digital PLL with near-optimal VCO phase noise shaping in low-cost 0.4 μm CMOS achieving 700 fs rms phase jitter," *NORCAS 2015*, 1-4, 2015.

[8] X. Gao, E.A.M. Klumperink, M. Bohsali and B. Nauta, "A 2.2 GHz 7.6 mW sub-sampling PLL with -126 dBc/Hz in-band phase noise and 0.15 ps rms jitter in 0.18 μm CMOS," *2009 ISSCC Digest of Techical Papers*, 392-393,393a, 2009.

S21-4 (2080)

An Ultra-Low Phase Noise All-Digital Multi-Frequency Generator Using Injection-Locked DCOs and Time-Interleaved Calibration

Suneui Park, Heein Yoon, and Jaehyouk Choi
Department of Electronic Engineering
Ulsan national Institute of Science and Technology (UNIST), Ulsan, Korea
Email: jaehyouk@unist.ac.kr

Abstract— This work presents an ultra-low phase noise and all-digital frequency generator, providing multiple output-frequencies. In a time-interleaved fashion, the proposed calibrator can continue to correct the multiple output-frequencies of injection-locked DCOs, which can change independently between 0.9 and 1.2GHz. Due to the time-interleaved calibrator, operating continuously in the background, each injection-locked DCO can maintain the excellent noise performance; the 1-MHz phase noise and the RMS-jitter at 930MHz were −132.2dBc/Hz and 310fs, and its variation across temperatures and voltages was less than 9%.

Keywords—Multi-freuqency generator, injection-locking, time-interleaved, calibration, DCO, low phase noise

I. INTRODUCTION

The use of multiple frequencies inside silicon can bring enormous benefits to modern ICs. As shown in Fig. 1, the carrier-aggregation (CA) technique [1], [2] for extending the effective data bandwidth of advanced wireless communications necessitate the use of multiple carrier-frequencies. Ready-made frequencies also can help expedite channel-switching or frequency-hopping [3]. For SoCs containing the various types of computing blocks, the separation of frequency domains can improve the computing performance and the energy efficiency [4]. Thus, now, how to generate multiple output frequencies efficiently is an intriguing question for IC designers. As opposed to a brute-force approach using many PLLs, reference [4] presented a PLL followed by DSM-based frequency dividers. However, since the frequencies must be decreased by passing the dividers, the operating frequency of the PLL must be even higher, demanding large expenditures of power. In addition, to remove quantization noise, a noise canceler must be present after each DSM. Alternatively, we can think a digital PLL, including pairs of a digital loop-filter and a DCO that share a TDC to tune the frequencies of DCOs sequentially. However, as the calibration period for each DCO extends M-times, where M is the number of DCOs, both the effective reference frequency and the loop bandwidth must decrease correspondingly, which would result in the significant degradation of phase noise.

This paper presents an ultra-low phase noise and all-digital injection-locked frequency generator, concurrently providing multiple output-frequencies. In this work, the proposed time-interleaved calibrator (TIC) continuously corrects the frequencies

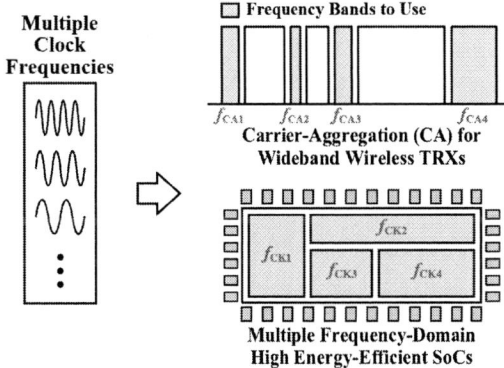

Fig.1. Modern ICs demanding the use of multiple frequencies

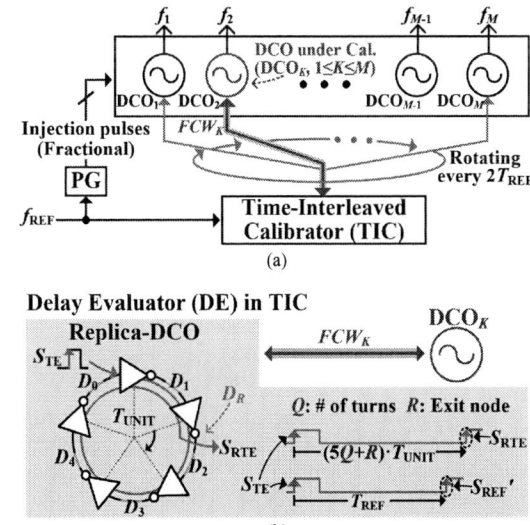

Fig. 2. Concept of (a) the multi-output freuqency generator with a time-interleaved calibrator (TIC); (b) the delay evaluator using a replica-DCO in TIC.

of the multiple DCOs in the DCO-bank. Despite the inherently limited bandwidth of the TIC, the output signals can maintain the excellent noise performance over PVT-variations, since they are firmly injection-locked [5]–[14].

978-1-5386-3179-9/17 $31.00 © 2017 IEEE   329

S21-4 (2080)

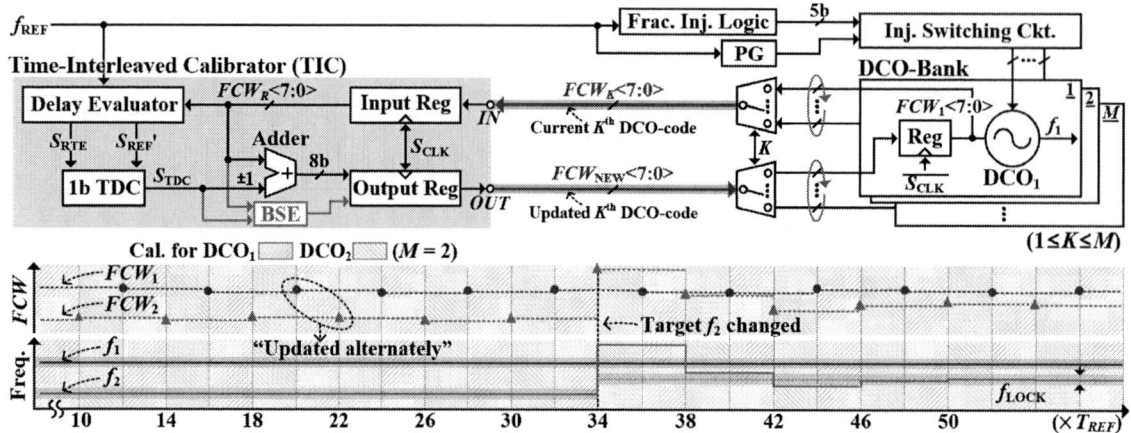

Fig. 3. Schematics and timing diagram of the proposed multi-freuqency generator

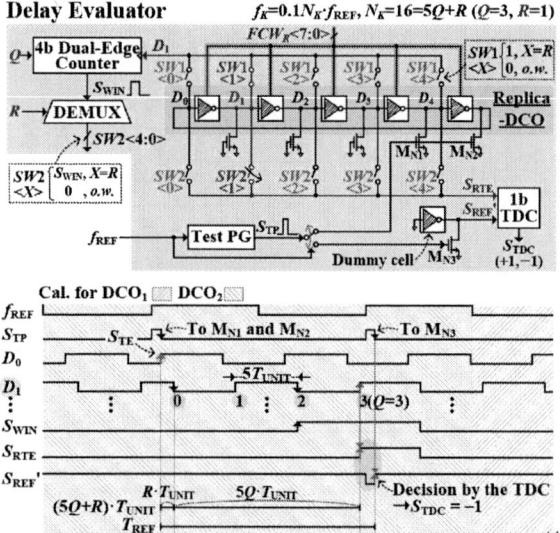

Fig. 4. Schematics and timing diagram of the delay evaluator (N_K=16)

II. CONCEPT OF THE FREQUENCY GENERATOR WITH MULTIPLE OUTPUT FREQUENCIES USING TIME-INTERLEAVED CALIBRATION

Fig. 2(a) shows the conceptual diagram of the proposed frequency generator that generates multiple output-frequencies, concurrently, using M-identical DCOs and the TIC. Having its own frequency control word (FCW), each DCO can generate a different frequency, independently. The TIC rotationally switches the target DCO to be calibrated, DCO_K, one by one in every $2T_{REF}$, where T_{REF} is the reference period. Even while the TIC calibrates another DCO, every DCO still can generate ultra-low noise output signal with its own target frequency, since it is locked by the injection pulses from the reference clock. Using the fractional-injection technique [8], the output-frequencies can change with a step of $0.1f_{REF}$, where f_{REF} is the reference frequency. Fig. 2(b) shows a delay evaluator (DE)

that is the key engine of the TIC. For the calibration of DCO_K, the DE operates as follows. First, with sharing the FCW of DCO_K, FCW_K, the replica-DCO of the DE acquires the delay information of the unit delay cell, T_{UNIT}, of DCO_K. Second, a test edge, S_{TE}, from the reference clock enters the starting node of the replica-DCO, D_0, and it travels along the five inverter-based delay cells. When it makes Q-turns, the returning test edge, S_{RTE}, is supposed to come out at D_R ($0 \leq R \leq 4$). The values of Q and R are determined, according to the target frequency of DCO_K. Since the number of the stages of the DCOs is five ($N_{DCO} = 5$), the total traveling time of the test edge is $(5Q+R) \cdot T_{UNIT}$. By comparing this duration with T_{REF}, the amount of T_{UNIT} can be evaluated, and FCW_K can be updated based on the result of this evaluation. When the calibration of DCO_K is done, the intrinsic frequency of DCO_K, f_K, can be represented as:

$$f_K \approx \frac{1}{2N_{DCO} \cdot T_{UNIT}} = \frac{(5Q+R)f_{REF}}{10} \quad (1)$$

III. IMPLEMENTATION OF ULTRA-LOW PHASE NOISE AND ALL-DGITAL FREQUENCY GENERATOR WITH MULTIPLE OUTPUT-FREQUENCIES

Fig. 3 shows the implementation of the proposed frequency generator, having multiple output-frequencies. The DCO-bank includes M-DCO-register pairs. The register for the DCO_K stores the current FCW, $FCW_K<7:0>$, and receives the updated FCW, $FCW_{NEW}<7:0>$. The TIC consists of input and output registers, the DE, a 1-b TDC, the binary-search engine (BSE), and an 8-b adder. When the turn of the calibration by the TIC goes to DCO_K, the input register of the TIC receives $FCW_K<7:0>$ and initializes the FCW of the replica-DCO in the DE, $FCW_R<7:0>$. According to the operation described in Fig. 2(b), the DE generates S_{RTE} and S_{REF}' and passes them to the following TDC. Then, based on the TDC's output, S_{TDC}, $FCW_{NEW}<7:0>$ of the output register is updated by the BSE in the coarse-tuning or the 8-b adder in the fine-tuning. The timing diagram in Fig. 3 describes how the proposed TIC calibrates two DCOs ($M = 2$). Assume that before $10T_{REF}$ the coarse-tuning was already done and the injection was enabled

978-1-5386-3179-9/17 $31.00 © 2017 IEEE   330

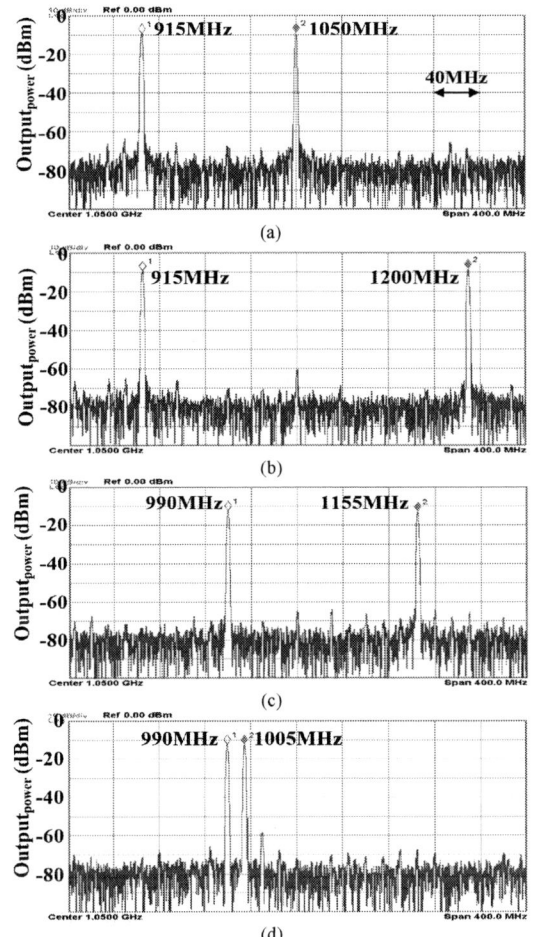

Fig. 5. Measured spectrums of the two independent output signals from the proposed multi-frequency generator. (a) 915 and 1050MHz (b) 915 and 1200MHz (c) 990 and 1155MHz (d) 990 and 1005MHz

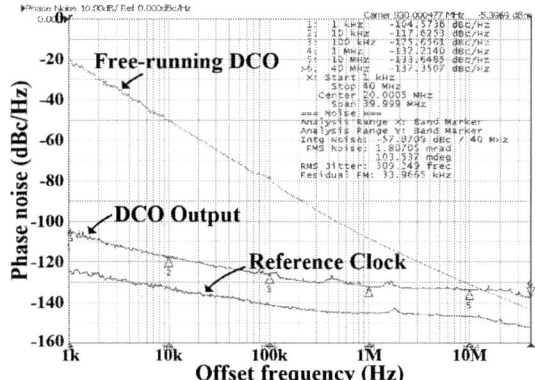

Fig. 6. Measured phase noise of one output signal with a frequency of 930MHz and. (f_{REF} = 150MHz, N_1 = 62)

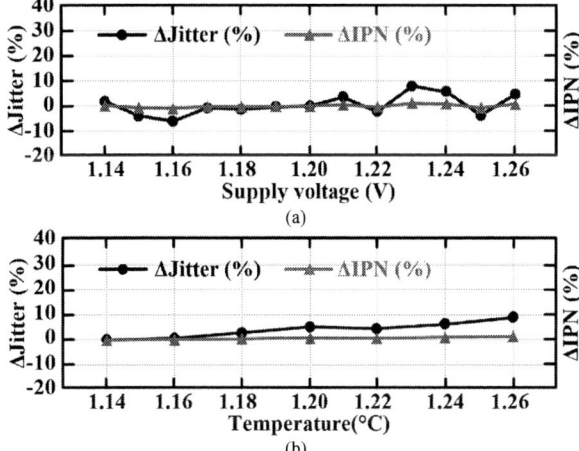

Fig. 7. Measured RMS-jitters over (a) supply voltages and (b) temperatures by the proposed calibrator. (f_{REF} = 150MHz, N_1 = 62)

for each DCO. Thus, between $10T_{REF}$ and $34T_{REF}$, the two DCOs generate their own target frequencies precisely, as being injection-locked. f_{LOCK} is the lock range of the injection-locking. As the TIC continues to update FCW_1 and FCW_2 alternately, the two injection-locked DCOs can continue to generate ultra-low phase noise signals, overcoming real-time frequency drifts due to PVT variations. When the target frequency of the DCO$_2$ changes at $34T_{REF}$, the coarse-tuning with the BSE operates again for adjusting FCW_2 swiftly. Despite the changes of f_2 and the status of the calibration for FCW_2, DCO$_1$ still can generate f_1 tightly. Fig. 4 shows how the DE was implemented. The test PG generates a stream of pulses, S_{TP}, from the rising edges of the reference clock. During the unit calibration period, $2T_{REF}$, one pulse of S_{TP} is applied to M_{N1} and M_{N2}, while the other pulse is applied to M_{N3}. When a pulse reaches the gates of M_{N1} and M_{N2}, D_0 and D_4 are reset to the ground. As the level of S_{TP} becomes low, S_{TE} starts traveling through the DCO from D_0. S_{REF}' is generated from the next pulse of S_{TP} at the drain of M_{N3} and compared with S_{RTE} by the TDC. As shown in the timing diagram, for example, when N_k is 16, the number of turns is 3

(Q = 3) and the exit node is D_1 (R = 1). Since the DCOs consists of odd-number inverters (i.e., N_{DCO} = 5), the dual-edge counter has to count both rising and falling edges. When the counter's number reaches Q–1, a window signal of S_{WIN} is generated to turn on $SW2<1>$. During this window, the next rising edge of D_1 (Q = 3) transfers to the TDC as S_{RTE}. When S_{RTE} leads or lags S_{REF}', S_{TDC} becomes −1 or +1, respectively.

IV. MEASUREMENT RESULTS

The proposed frequency generator was fabricated in 65nm CMOS and consumed 7.7mW to generate two different output-frequencies. Using an external power-combiner, the spectrums of the two output signals were measured at the same time. As shown in Fig. 5, the changes of the output-frequencies were totally independent each other between 0.9 and 1.2GHz with a 15MHz step. As shown in Fig. 6, each signal achieved excellent noise performance; 1MHz phase noise and RMS-jitter at 930MHz were −132dBc/Hz and 310fs (integrated from 1kHz to 40MHz), respectively, since the DCO's phase noise was greatly suppressed by the injection-locking, in which the bandwidth was as much as 50MHz. Fig. 7 shows the TIC was able to regulate the variations of jitter and IPN to less than 9% and 1.4%, respectively, across voltages and temperatures. From Monte

TABLE I. PERFORMANCE SUMMARY AND COMPARISON WITH STATE-OF-THE-ART RING-VCO-BASED FEQUENCY GENERATORS

	Process	No. of Freqs. (M)	Output Freq. (GHz)	Ref. Freq. (MHz)	Freq. Resol. (MHz)	1MHz PN (dBc/Hz) @f_0 (GHz)	Integ. Jitter (σ_t) (Integ. Range)	Ref. Spur (dBc)	Power Cons. (P_{DC}) (mW)	Active Area (mm²)	FOM$_{MC}$ (dB)*** (FOM$_{JIT}$ (dB))
This work	65nm CMOS	2	0.9 – 1.2	150	f_{REF}/10	−132.2 @0.93	309fs (1k–40 MHz)	−45	7.7 (Two freqs.)	0.05	−244.3 (−242.8****)
ISSCC'14 [4]	65nm CMOS	2	0.02 – 1.0	100	$3 \cdot f_{REF}$/5/10⁴	N/A	3ps*	NA	6.4 + P_{PLL}** (Two freqs.)	0.2	NA
VLSI-S'16 [6]	65nm CMOS	1	1.44	180	f_{REF}	−122.5 @1.44	450fs (1k–40 MHz)	−59	2.8	0.061	−242.5 (−242.5)
JSSC'16 [8]	65nm CMOS	1	1.2 – 2.0	400	f_{REF}/10	−122.9 @1.6	440fs (10k–40 MHz)	−39	3.6	0.032	−241.6 (−241.6)
JSSC'16 [9]	65nm CMOS	1	0.96 – 1.44	120	f_{REF}	−134.4 @1.2	185fs (10k–40MHz)	−53	9.5	0.06	−244.9 (−244.9)
ISSCC'16 [12]	28nm CMOS	1	2.4	75	f_{REF}	−115.0 @2.4	700fs (1k–40MHz)	−58	1.5	0.024	−241.3 (−241.3)

*By oscilloscope **PLL power was not reported. ***FOM$_{JIT}$=10log ($\sigma_t^2 \cdot P_{DC}$), power of DCO2 was excluded. ****FOM$_{MC}$=10log ($\sigma_t^2 \cdot P_{DC}$ /M)

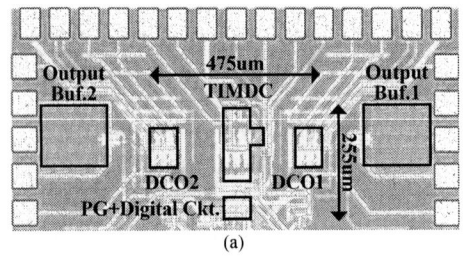

(a)

Power Breakdown

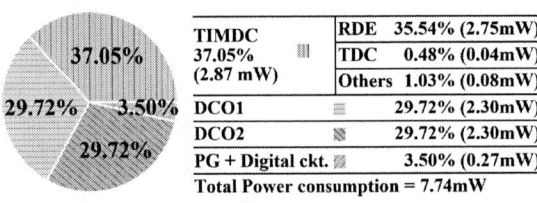

TIMDC 37.05% (2.87 mW)	RDE	35.54% (2.75mW)
	TDC	0.48% (0.04mW)
	Others	1.03% (0.08mW)
DCO1		29.72% (2.30mW)
DCO2		29.72% (2.30mW)
PG + Digital ckt.		3.50% (0.27mW)
Total Power consumption = 7.74mW		

(b)

Fig. 8. (a) Die photograph and (b) power-breakdown table.

Carlo simulation, the worst-case 3-sigma value of the frequency-difference between the replica-DCO and the two DCOs was 2.6MHz. This corresponds to 5.2% f_{LOCK}, and the impact on phase noise is negligible. In Table I, compared to [1], this work generated two output signals with far lower RMS-jitter, while spending smaller power and area. Compared to state-of-the-art injection-locked frequency generators (ILFGs) [6], [8], [9], [12], this work achieved still competitive FOM$_{JIT}$ and FOM$_{MC}$, while it was capable of generating two output frequencies at the same time. (In FOM$_{MC}$, the power, P_{DC}, was normalized by M.) The FOM$_{MC}$ and the area-efficiency of the proposed frequency generator are supposed to be improved, as M increases.

REFERENCES

[1] S. Hwu and B. Razavi, "An RF receiver for intra-band carrier aggregation," *IEEE J. Solid-State Circuits*, vol. 50, no. 4, pp. 946–961, Apr. 2015.

[2] H. Yoon, *et al.*, "A 0.56–2.92 GHz Wideband and Low Phase Noise Quadrature LO-Generator Using a Single LC-VCO for 2G–4G

Multistandard Cellular Transceivers," *IEEE J. Solid-State Circuits*, vol. 51, no. 3, pp. 614-625, Mar. 2016.

[3] S. Geng, *et al.*, "A 13.3 mW 500 Mb/s IR-UWB Transceiver With Link Margin Enhancement Technique for Meter-Range Communications," *IEEE J. Solid-State Circuits*, vol. 50, no. 3, pp. 669-678, Mar. 2015.

[4] A. Elkholy, *et al.*, "A 20-to-1000MHz ±14ps Peak-to-Peak Jitter Reconfigurable Multi-Output All-Digital Clock Generator using Open-Loop Fractional Dividers in 65nm CMOS," *ISSCC Dig. Tech. Papers*, pp. 272–273, Feb. 2014.

[5] B. Razavi, "A study of injection locking and pulling in oscillators," *IEEE J. Solid-State Circuits*, vol. 39, no. 9, pp. 1415–1424, Sep. 2004.

[6] Y Lee, *et al.*, "A PVT-robust −59-dBc reference spur and 450-fsRMS jitter injection-locked clock multiplier using a voltage-domain period-calibrating loop," in *Proc. Symp. VLSI Circuits Dig. Tech. Papers*, Jun. 2016, pp. 1–2.

[7] S. Yoo, *et al.*, "A PVT-robust −39dBc 1kHz-to-100MHz integrated-phase-noise 29GHz injection-locked frequency multiplier with a 600μW frequency-tracking loop using the averages of phase deviations for mm-band 5G transceivers," *ISSCC Dig. Tech. Papers*, pp. 324-325, Feb. 2017.

[8] M. Kim, *et al.*, "A Low-Jitter and Fractional-Resolution Injection-Locked Clock Multiplier Using a DLL-Based Real-Time PVT Calibrator With Replica-Delay Cells," *IEEE J. Solid-State Circuits*, vol. 51, pp. 401–411, Feb. 2016.

[9] S. Choi, *et al.*, "A PVT-Robust and Low-Jitter Ring-VCO-Based Injection-Locked Clock Multiplier With a Continuous Frequency-Tracking Loop Using a Replica-Delay Cell and a Dual-Edge Phase Detector," *IEEE J. Solid-State Circuits*, vol. 51, no. 8, pp. 1878-1889, Aug. 2016.

[10] W. Deng, *et al.*, "A Compact and Low-Power Fractionally Injection-Locked Quadrature Frequency Synthesizer Using a Self-Synchronized Gating Injection Technique for Software-Defined Radios," *IEEE J. Solid-State Circuits*, vol. 49, no. 9, pp. 1984-1994, Sep. 2014.

[11] A. Musa, *et al.*, "A Compact, Low-Power and Low-Jitter Dual-Loop Injection Locked PLL Using All-Digital PVT Calibration," *IEEE J. Solid-State Circuits*, vol. 49, no. 1, pp. 50-60, Jan. 2014.

[12] H. Kim, *et al.*, "A 2.4GHz 1.5mW Digital MDLL Using Pulse-Width Comparator and Double Injection Technique in 28nm CMOS," *ISSCC Dig. Tech. Papers*, pp. 328–329, Feb. 2016.

[13] A. Elkholy, *et al.*, "A 6.75-to-8.25GHz, 250fsrms-Integrated-Jitter 3.25mW Rapid On/Off PVT-Insensitive Fractional-N Injection-Locked Clock Multiplier in 65nm CMOS," *ISSCC Dig. Tech. Papers*, pp. 192–193, Feb. 2016.

[14] W. Deng *et al.*, "A 0.048mm² 3mW Synthesizable Fractional-N PLL With a Soft Injection-Locking Technique," *ISSCC Dig. Tech. Papers*, pp. 1–3, Feb. 2015.

IEEE
445 Hoes Lane
Piscataway, NJ 08854-4141

ISBN 978-1-5386-3179-9